A CENTURY OF RELATIVITY PHYSICS

To learn more about the AIP Conference Proceedings, including the
Conference Proceedings Series, please visit the webpage
http://proceedings.aip.org/proceedings

A CENTURY OF RELATIVITY PHYSICS

ERE 2005

XXVIII Spanish Relativity Meeting

Oviedo (Asturias), Spain   6 – 10 September 2005

Universidad de Oviedo
Oviedo University

Sociedad Española de
Gravitación y Relatividad
*Spanish Society for
Relativity and Gravitation*

World Year of Physics

EDITORS
Lysiane Mornas
Joaquín Diaz Alonso
*Oviedo University, Spain
and
Paris-Meudon Observatory, France*

Melville, New York, 2006
AIP CONFERENCE PROCEEDINGS ■ VOLUME 841

Editors:

Lysiane Mornas
Joaquín Diaz Alonso

Physics Department
Oviedo University
Avda Calvo Sotelo 18
33007 Oviedo (Asturias)
Spain

and

Paris-Meudon Observatory
LUTH-CNRS UMR8102
5, pl. J. Janssen
95192 Meudon cedex
France

E-mail: ere05@fisi24.ciencias.uniovi.es

Authorization to photocopy items for internal or personal use, beyond the free copying permitted under the 1978 U.S. Copyright Law (see statement below), is granted by the American Institute of Physics for users registered with the Copyright Clearance Center (CCC) Transactional Reporting Service, provided that the base fee of $23.00 per copy is paid directly to CCC, 222 Rosewood Drive, Danvers, MA 01923, USA. For those organizations that have been granted a photocopy license by CCC, a separate system of payment has been arranged. The fee code for users of the Transactional Reporting Services is: 0-7354-0333-3/06/$23.00

© 2006 American Institute of Physics

Permission is granted to quote from the AIP Conference Proceedings with the customary acknowledgment of the source. Republication of an article or portions thereof (e.g., extensive excerpts, figures, tables, etc.) in original form or in translation, as well as other types of reuse (e.g., in course packs) require formal permission from AIP and may be subject to fees. As a courtesy, the author of the original proceedings article should be informed of any request for republication/reuse. Permission may be obtained online using Rightslink. Locate the article online at http://proceedings.aip.org, then simply click on the Rightslink icon/"Permission for Reuse" link found in the article abstract. You may also address requests to: AIP Office of Rights and Permissions, Suite 1NO1, 2 Huntington Quadrangle, Melville, NY 11747-4502, USA; Fax: 516-576-2450; Tel.: 516-576-2268; E-mail: rights@aip.org.

L.C. Catalog Card No. 2006926354
ISBN 0-7354-0333-3
ISSN 0094-243X

Printed in the United States of America

CONTENTS

Preface xi
Conference Organization xiii
Sponsors xv
Other Supporting Institutions xvii

PART I: INVITED LECTURES

BRST Analysis of Unimodular Theories 3
    E. Alvarez and J. J. López-Villarejo
AdS/CFT, Black Holes and Matrix Models (Abstract) 9
    L. Álvarez-Gaumé
Gravitational Radiation from Two-Body Systems 10
    L. Blanchet
Half Century of Black-Hole Theory: From Physicists' Purgatory to
Mathematicians' Paradise 29
    B. Carter
100 Years of Relativity: Was Einstein 100% Right? 51
    T. Damour
Relativistic Statistical Mechanics, a Brief Overview 63
    R. Hakim and H. D. Sivak
Current Issues in Numerical Relativistic (Magneto-) Hydrodynamics .. 100
    J. M. Ibáñez
The Shape of Space from Einstein to WMAP Data 115
    J.-P. Luminet
Finding and Using Exact Solutions of the Einstein Equations 129
    M. A. H. MacCallum
Recent Developments in the Study of the Cosmic Microwave
Background Anisotropies 144
    E. Martínez-González
Supersymmetry and the Supergravity Landscape 162
    T. Ortín
Observing Dark Matter 182
    J. Silk
Albert Einstein: A Man for the Millenium? 195
    J. Stachel
Brane-World Cosmology 228
    D. Wands

*Italicized names indicate the authors who presented the papers.

PART II: CONTRIBUTED PAPERS

Quantum Gravity and the Fate of Lorentz Invariance in the Standard Model 247
    J. Alfaro

Ten-Dimensional Supergravity Revisited 255
    E. Bergshoeff, M. de Roo, S. Kerstan, and F. Riccioni

The Double Pulsar System J0737−3039A/B as Testbed for Relativistic Gravity 263
    M. Burgay, A. Possenti, M. Kramer, R. N. Manchester, N. D'Amico, A. G. Lyne, M. A. McLaughlin, D. R. Lorimer, F. Camilo, I. H. Stairs, P. C. C. Freire, and B. C. Joshi

The Principle of General Covariance Has Physical Content 271
    A. Chamorro

Relativistic Positioning Systems 277
    B. Coll

Geometry and Experience: Einstein's 1921 Paper and Hilbert's Axiomatic System 285
    F. De Gandt

A Weighted de Rham Operator Leading to Local Potentials for Riemann and Weyl Tensors 291
    S. B. Edgar and J. M. M. Senovilla

Kac-Moody Algebras in Gravity and M-Theories 298
    L. Houart

Hyperboloidal Data and Evolution 306
    S. Husa, C. Schneemann, T. Vogel, and A. Zenginoğlu

Stochastic Inflation with Coloured Noise 314
    H.-P. Breuer and *K. E. Kunze*

On Fixed Points of Quantum Gravity 322
    D. Litim

The Objectivity of Spacetime: Dirac Observables and Gauge Variables for the Gravitational Field 330
    L. Lusanna

Deformed Einstein Gravity 340
    F. Meyer

Do Evanescent Modes Violate Relativistic Causality? 348
    G. Nimtz

Holographic Dark Energy and Present Cosmic Acceleration 356
    D. Pavón and W. Zimdahl

Quantum Gravity as Theory of "Superfluidity" 362
    B. M. Barbashov, *V. N. Pervushin*, A. F. Zakharov, and V. A. Zinchuk

Second-Order Symmetric Lorentzian Manifolds 370
    J. M. M. Senovilla

Searches for Continuous Gravitational Wave Sources with LIGO and GEO 378
    A. M. Sintes (On behalf of the LIGO Scientific Collaboration)

*Italicized names indicate the authors who presented the papers.

PART III: SHORT COMMUNICATIONS

Black Hole Entanglement Entropy 385
    J. M. Tejeiro-Sarmiento and *J. R. Arenas-Salazar*

Is there Any Evidence for Integrated Sachs-Wolfe Signal in WMAP First Year Data? 389
    C. Hernández-Monteagudo, R. Génova-Santos, and F. Atrio-Barandela

Static Axisymmetric Spacetimes with Prescribed Multipole Moments 393
    T. Bäckdahl

Bigravity and Massive Gravity 397
    D. Blas

Nuclear Matter Equation of State in Relativistic Nonlinear Models: Results and Applications 402
    J. Pastor, J. C. Caillon, and J. Labarsouque

Neutrino Mixing and Dark Energy 406
    M. Blasone, *A. Capolupo*, S. Capozziello, and G. Vitiello

On The Rindler Horizon Energy 410
    H. Culetu

Accurate Simulations of the Barmode Instability in General Relativity 416
    R. De Pietri, L. Baiotti, G. M. Manca, and L. Rezzolla

Geodesic Completeness around Sudden Singularities 420
    L. Fernández-Jambrina and R. Lazkoz

Coll Positioning Systems: A Two-Dimensional Approach 424
    J. J. Ferrando

Non-Abelian Yang-Mills in Kundt Spacetimes 429
    A. Fuster

Stability of Self-Similar Spherical Accretion 433
    J. Gaite

Coupling Einstein-Rosen Waves to Matter: The Massless Scalar Field Case 437
    J. F. Barbero G., *I. Garay*, and E. J. S. Villaseñor

Non Spherical Collapse of Scalar Field Dark Matter 441
    A. Bernal and *F. S. Guzmán*

May We Use the LLR as a Redshift Indicator for the Gamma-Ray Bursts? 445
    M. Hafizi and R. Mochkovitch

Explicit Multipole Moments of Axisymmetric Stationary Spacetimes 449
    M. Herberthson

ISA—An Accelerometer to Detect the Disturbing Accelerations Acting on the Mercury Planetary Orbiter of the BepiColombo ESA Cornerstone Mission to Mercury: on Ground Calibration 453
    V. Iafolla, D. M. Lucchesi, S. Nozzoli, F. Santoli, M. Fois, and M. Persichini

Relic Gravitational Waves and Cosmic Accelerated Expansion 458
    G. Izquierdo

---

*Italicized names indicate the authors who presented the papers.

Doubly Special Relativity: A New Relativity or Not? 462
    N. Jafari and A. Shariati
3+1 Decomposition of Quasi-Equilibrium Black Hole Boundary
Conditions 466
    J. L. Jaramillo
From General Gravity to Einstein's. 471
    K. Just and W. Stoeger
Primordial Scalar Field: A Way Out of the Lithium Over-Production. 475
    J. Larena, J.-M. Alimi, and A. Serna
Relativistic (Covariant) Kinetic Theory of Linear Plasma Waves and
Instabilities. 479
    M. Lazar and R. Schlickeiser
Zero-Norm States and Stringy Symmetries 484
    C.-T. Chan, P.-M. Ho, J.-C. Lee, S. Teraguchi, and Y. Yang
Data and Diagnostics in LISA PathFinder. 489
    A. Lobo, M. Nofrarias, J. Ramos, J. Sanjuan, A. Conchillo, J. A. Ortega,
    X. Xirgu, H. Araujo, C. Boatella, M. Chmeissani, C. Grimani,
    C. Puigdengoles, P. Wass, S. Anza, M. Díaz Michelena, E. García-Berro,
    and R. Pérez del Real
The Gravitational Wave Radiation of Pulsating White Dwarfs. 493
    P. Lorén-Aguilar, J. Isern, L. G. Althaus, A. H. Córsico, J. A. Lobo, and
    E. García-Berro
A Combinatorial Approach to Discrete Geometry 497
    L. Bombelli and M. Lorente
Detector Configurations for Equivalence Principle Tests with Strong
Separation of Signal from Noise. 502
    E. C. Lorenzini, I. I. Shapiro, J. Ashenberg, C. Bombardelli,
    P. N. Cheimets, V. Iafolla, D. M. Lucchesi, S. Nozzoli, F. Santoli, and
    S. Glashow
Obtaining a Class of Conformally Flat Pure Radiation Metrics with
Cosmological Constant Using Invariant Operators 507
    S. B. Edgar and M. P. Machado Ramos
Gravitational Lensing of Stars Orbiting Sgr A*. 511
    V. Bozza and L. Mancini
The Pseudospin Symmetry in Atomic Nuclei. Analysis of Some of the
Explanations Proposed. 515
    S. Marcos, M. López-Quelle, R. Niembro, and L. N. Savushkin
First Order Perturbations of the Einstein-Straus Model 519
    M. Mars, F. C. Mena, and R. Vera
Spin Foam Models from the Tetrad Integration 523
    A. Miković
Brownian Motion on the Relativistic Velocity Space. 528
    E. Minguzzi
Quantization of Minimal Strings: A Mechanical Analog 532
    C. Gómez, S. Montañez, and P. Resco

*Italicized names indicate the authors who presented the papers.

Coordinates and Frames from the Causal Point of View 537
    J. A. Morales Lladosa
Perturbation Theory and Stability Analysis for String-Corrected
Black Holes in Arbitrary Dimensions 542
    F. Moura
Gravitational Collapse in Higher Curvature Theory....................... 546
    M. Nozawa and H. Maeda
Constraining Dark Energy Interacting Models with WMAP................. 550
    G. Olivares, F. Atrio-Barandela, and D. Pavón
Non-Adiabatic Radiating Collapse in de Sitter Spacetime 554
    N. Özdemir
Solving the Nosé-Hoover Thermostat for Nuclear Pasta.................... 558
    M. Á. Pérez García
Quantum Collapse in Quark Stars?....................................... 562
    A. Pérez Martínez, H. Pérez Rojas, and H. J. Mosquera Cuesta
Quantum Degrees of Freedom of a Region of Spacetime 566
    F. Piazza
Would Closed Timelike Curves Help to Do Quantum Cloning? 570
    A. R. Plastino and C. Zander
Gravity from Lorentz Symmetry Violation............................... 574
    R. Potting
Numerical Simulations of the Gravitational Wave Background
Produced by Binaries.. 578
    N. Puchades and D. Sáez
Determination of Stellar Shape via Microlensing........................... 582
    N. J. Rattenbury, P. Yock, and I. Bond
Vacuum Self-Magnetization?.. 586
    H. Pérez Rojas and E. Rodríguez Querts
Differential Rotation of R-Modes 590
    P. M. Sá and B. Tomé
Comparing Rees-Sciama and Integrated Sachs Wolfe Effects 594
    D. Sáez, N. Puchades, M. J. Fullana, and J. V. Arnau
Supersymmetric Quantization in Midisuperspace 599
    A. Macías, H. Quevedo, and A. Sánchez
Light Scattering Test Regarding the Relativistic Nature of Heat 603
    A. Sandoval-Villalbazo and L. S. García-Colín
The Initial Conditions of the Universe and Holography..................... 607
    P. Díaz, M. A. Per, and A. Seguí
Rigid Motions in Relativity: Applications 611
    D. Soler
Asymptotic Flatness and Algebraically Special Metrics 615
    W. Natorf and J. Tafel
On Nonanticommutative Sigma Models 619
    L. Álvarez-Gaumé and M. A. Vázquez-Mozo
Axially Symmetric Equilibrium Regions in FLRW Universes 623
    B. C. Nolan and R. Vera

*Italicized names indicate the authors who presented the papers.

Separation of Variables and Exact Solution of the Dirac Equation in
Some Cosmological Space-Times 627
    V. M. Villalba

PART IV: POSTERS

On the Concept of Relative Velocity 633
    V. J. Bolós

Mathematical Properties of the Elasticity Difference Tensor 635
    E. G. L. R. Vaz and *I. Brito*

Scalar Field Coupled to Gravity in Spherically Symmetric
Background in (2 + 1) Dimensions 637
    D. Daghan and A. H. Bilge

Non-Equivalence of Newtonian and Relativistic Second Order
Perturbations 639
    F. C. Mena

Emission Coordinates and the Central Observer 641
    J. M. Pozo

Perturbations of Slowly Rotating Relativistic Stars 643
    I. Rica-Méndez and A. Stavridis

Tomographic Approach to Quantum Cosmology 645
    C. Stornaiolo

Anti-Newtonian Universes Do Not Exist 647
    L. Wylleman

Conference Program 649
List of Participants 661
Author Index 677

---

*Italicized names indicate the authors who presented the papers.

PREFACE

The XXVIIIth Spanish Relativity Meeting (ERE05) was held in Oviedo (Asturias, Spain) from September 6 to September 10, 2005. It was organized by the Physics Department of the University of Oviedo under the auspices of the Spanish Society for Gravitation and Relativity (SEGRE).

The present volume gathers most of the plenary lectures, long and short communications and posters presented at this Meeting. We are delighted to have received so many contributions. We thank the participants for attending the meeting and the authors for their effort in preparing contributions to these proceedings.

On the occasion of the World Year of Physics commemorating the centenary of the publication by Einstein of several of his major papers and the birth of the Theory of Relativity, this edition of the EREs endeavoured to cover as many aspects as possible of the foundations, applications and future vistas of this branch of modern physics.

The topics treated included: Foundations of special and general relativity, observational tests, relativistic astrophysics, gravitational waves, numerical relativity, cosmology, early universe, dark matter, exact solutions of Einstein equations, black holes, quantum gravity, string theory, discrete space-time, non commutative geometry, relativistic statistical physics and nuclear physics, compact stars, historical and philosophical perspective, ...

Most of these topics could be by themselves the object of a conference. For this reason we have requested the plenary speakers to plan their lectures so that they be addressed to an audience of theoretical physicists who are not necesarily specialists of their particular topic. We are also thankful to the authors who strived to give a short historical introduction and thus situate the concepts treated in the more general context of the evolution of fundamental physics during the XXth century.

We would like to acknowledge the financial support of our sponsors which made it possible to organize this conference. It is moreover a pleasure to express our thanks to the various scientific institutions who have backed us. The full list is too long to be reproduced in this short introduction and can be found on a specific page.

Lysiane Mornas
Joaquin Diaz Alonso

CONFERENCE ORGANIZATION

Scientific Committee

- M. Alcubierre (ICN-UNAM, Mexico, Mexico)
- J.M. Alimi (LUTH, Obs. Paris, France)
- E. Álvarez (UAM, Madrid, Spain)
- L. Álvarez-Gaumé (CERN, Geneva, Switzerland)
- R. Beig (ITP, Vienna, Austria)
- B. Carter (LUTH, Obs. Paris, France)
- B. Coll (SYRTE, Obs. Paris, France)
- C. Cutler (AEI, Potsdam, Germany)
- T. Damour (IHES, Paris, France)
- R. Domínguez-Tenreiro (UAM, Madrid, Spain)
- J.Mª Ibáñez (DAA, Valencia, Spain)
- R. Maartens (ICG, Portsmouth, UK)
- J. Martín (Salamanca Univ., Spain)
- A.R. Plastino (U. Pretoria, South Africa)
- F. Quevedo (DAMTP, Cambridge, UK)
- J.M. Sánchez-Ron (UAM, Madrid, Spain)
- C. Vilain (LUTH, Obs. Paris, France)

Local Organizing Committee

- J. Díaz-Alonso
- M. Lorente
- Y. Lozano
- L. Mornas
- A. Nieto
- M.A.R. Osorio
- L. Toffolatti

Financial Management

Fundación Universidad de Oviedo

SPONSORS

  Auditorio Principe Felipe
Prince Felipe Auditorium graciously made available by the
Excmo. Ayuntamiento de Oviedo
Municipality of Oviedo

 CajAstur - Obra Social y Cultural
Saving Bank of Asturias

 Embajada de España en Cuba
Spanish Embassy in Cuba with funding from the
Agencia Española de Cooperación Internacional
Spanish Agency for International Cooperation

 East West Task Force program of the European Physical Society

   Fundación para el Fomento en Asturias de la Investigación Científica Aplicada y la Tecnología
Fundation for the Promotion of Applied Scientific Research and Technology in Asturias
(Plan de Investigación, Desarrollo Tecnológico e Innovación 2001-2004)
Plan 2001-2004 for Research, Technological Development and Innovation

 Sociedad Española de Gravitación y Relatividad
Spanish Society for Relativity and Gravitation with funding from the
Ministerio de Educación y Ciencia
Spanish Ministry of Education and Science

Universidad de Oviedo
Oviedo University
Vicerrectorado de Investigación & Departamento de Física

OTHER PARTICIPATING INSTITUTIONS

 World Year of Physics

 American Physical Society (Topical Group in Gravitation : GGR)

 Asociación Física Argentina
Argentine Physical Association

 CERN (Theoretical Physics Division)

 European Physical Society

 Instituto de Física de la Plata, Argentina
La Plata Institute of Physics, Argentina

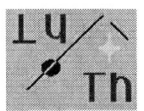 Laboratoire de L'Univers et de ses Théories (LUTH)
Observatoire de Paris-Meudon, France
Universe and Theory Laboratory – Paris-Meudon Observatory, France

 Sociedad Boliviana de Física
Bolivian Physical Society

 Sociedade Brasileira de Física
Brasilian Physical Society

 Sociedad Chilena de Física
Chilian Physical Society

 Sociedad Colombiana de Física
Colombian Physical Society

 Sociedad Cubana de Física
Cuban Physical Society

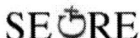 Sociedad Española de Gravitación y Relatividad
Spanish Society for Relativity and Gravitation

 Società Italiana di Relatività e Fisica della Gravitazione
Italian Society of Relativity and Physics of Gravitation

 Sociedad Mexicana de Física
(División de Gravitación y Física Matemática)
Mexican Physical Society
(Gravitation and Mathematical Physics Division)

 Sociedade Portuguesa de Física
Portuguese Physical Society

 Sociedade Venezolana de Física
Venezuelian Physical Society

PART I: INVITED LECTURES

BRST Analysis of Unimodular Theories

Enrique Alvarez* and J.J. López-Villarejo*

*IFTE UAM/CSIC, módulo CXVI, Universidad Autónoma
Canto Blanco, 28049 Madrid, Spain

Abstract. 1. Introduction. 2. Abelian gauge invariance: transverse Fierz-Pauli symmetry. 3. The non-Abelian cse. 4. Gauge fixing.

1. INTRODUCTION.

There are interesting theories such as Einstein's 1919 traceless one, (when derived from a variational principle) in which the full group of invariance is not the set of all diffeomorphisms, but only those that are area preserving; that is, that keep constant the determinant of the metric (cf.[2] for a general discussion with references). The field content usually includes a scalar particle, akin to the usual dilaton. TFP has been studied in some detail in [3]. The aim of the present note is to discuss gauge fixing in them, both in the (somewhat trivial) linear setting and in the much more complicated nonlinear regime.

2. ABELIAN GAUGE INVARIANCE: TRANSVERSE FIERZ-PAULI SYMMETRY.

A sometimes confusing issue is the following. The Fierz-Pauli (FP) symmetry is not exactly the linearization of full diffeomorphism invariance, which would have been

$$\delta h_{\alpha\beta} = \partial_\alpha \xi_\beta + \partial_\beta \xi_\alpha + \xi^\rho \partial_\rho h_{\alpha\beta} \tag{1}$$

insofar as the last term is absent. This issue is clearly explained in page 80 of Ortín's book [9].

Indeed, gauge fixing with the full FP symmetry is trivial, and e.g. harmonic gauge can be imposed:

$$\omega_\mu \equiv \partial^\lambda h_{\lambda\mu} - \frac{1}{2}\partial_\mu h = 0 \tag{2}$$

through a gauge fixing

$$L_{gf} = B^\mu \omega_\mu + \frac{\alpha}{2} B^\mu B_\mu \tag{3}$$

The ghost lagrangian is,

$$L_{gh} = b^\rho \Box c_\rho \tag{4}$$

and the BRST transformations can be taken simply as:

$$sh_{\alpha\beta} = \partial_\alpha c_\beta + \partial_\beta c_\alpha$$
$$sB_\mu = 0$$
$$sb_\mu = -B_\mu$$
$$sc_\mu = 0 \qquad (5)$$

Were we to implement the transverse part of the symmetry (TFP) only, the parameters are not arbitrary but rather

$$\partial_\alpha \xi^\alpha = 0 \qquad (6)$$

This complicates matters in several ways. First of all, we cannot reach the full harmonic gauge. The best we can do is to impose, for example, the spatial piece, i.e.

$$\omega_i = 0 \qquad (7)$$

or even better, [1] the three independent conditions:

$$\partial_\alpha \omega^{\alpha\beta} = 0 \qquad (10)$$

where

$$\omega_{\mu\nu} \equiv \partial^\lambda \left(\partial_\mu h_{\lambda\nu} - \partial_\nu h_{\lambda\mu} \right) \qquad (11)$$

The second thing is that ghosts now must obey

$$c^\mu \equiv \partial_\rho c_1^{\rho\mu} \qquad (12)$$

where we have indicated as a subscript the ghost number. The antighosts will be treated momentarily

Those objects are transverse:

$$\partial_\rho c^\rho = \partial_\rho b^\rho = 0 \qquad (13)$$

owing to the fact that the two-index ghosts are assumed to be completely antisymmetric (ghostly forms).

There is the apparent complication that the ghostly forms are only defined modulo total differentials:

$$\varepsilon^{\mu\nu\alpha\beta} \partial_\alpha c_\beta^1 \qquad (14)$$

---

[1] Another possibility would be to impose as a gauge condition the self-dual part of $d\omega$, i.e..

$$\omega^+_{\alpha\beta} \equiv P^{+\;\mu\nu}_{\alpha\beta} \omega_{\mu\nu} = 0 \qquad (8)$$

where the projector on the space of self-dual forms, is given by

$$P^{+\;\mu\nu}_{\alpha\beta} \equiv \frac{1}{2} \left(\delta^{\mu\nu}_{\alpha\beta} - i\varepsilon_{\alpha\beta}{}^{\mu\nu} \right) \qquad (9)$$

This cuts in half the number of independent components, so that it amounts to three independent conditions only.

(this is indeed the correct counting: $6 - (4 - 1) = 3$).

The gauge fixing is then

$$L_{gf} \equiv B^\alpha \partial^\rho \omega_{\alpha\rho} + \frac{\alpha}{2} B_\alpha^2 \tag{15}$$

The corresponding ghost lagrangian is

$$L_{gh} = -b_\alpha \Box^2 c^\alpha \tag{16}$$

which has got the drawback of being of fourth order in derivatives, which is irrelevant nevertheless, because it is independent of the gauge fields. [2] The corresponding BRST transformations are:

$$\begin{aligned} sh_{\alpha\beta} &= \partial_\alpha \partial^\mu c^1_{\mu\beta} + \partial_\beta \partial^\mu c^1_{\mu\alpha} \\ sB_\mu &= 0 \\ sb_\mu &= -B_\mu \\ sc^1_{\rho\mu} &= 0 \end{aligned} \tag{20}$$

3. THE NON-ABELIAN CASE

Let us now use the convenient language of differential forms, indicating sometimes its degree by a subscript (this is trivially related to the ghost number):

$$c \equiv c_1 \equiv c_\mu dx^\mu \tag{21}$$

with the constraint

$$\delta c_1 = 0 \tag{22}$$

---

[2] Were we to impose the self-dual form as a geuge condition, then the gauge fixing piece of the lagrangian can be taken as

$$L_{gf} = B^{\alpha\beta} \omega^+_{\alpha\beta} + \frac{\alpha}{2} B^2_{\alpha\beta} \tag{17}$$

where the fields $B_{\alpha\beta}$ represent the three components of a selfdual form; and the ghost lagrangian reads

$$L_{gh} = b^+_{\alpha\beta} P^{\alpha\beta\mu\nu}_+ \Box \left(\partial_\mu c_\nu - \partial_\nu c_\mu \right) \tag{18}$$

It is more or less unavoidable also here that this piece of the lagrangian is of third order in derivatives. The BRST transformations are then

$$\begin{aligned} sh_{\alpha\beta} &= \partial_\alpha \partial^\mu c^1_{\mu\beta} + \partial_\beta \partial^\mu c^1_{\mu\alpha} \\ sB_{\mu\nu} &= 0 \\ sb^+_{\mu\nu} &= -B_{\mu\nu} \\ sc^1_{\rho\mu} &= 0 \end{aligned} \tag{19}$$

so that
$$c_1 = \delta c_2 \tag{23}$$
This means that
$$\delta s c_1 = 0 \tag{24}$$
Indeed, given that acting on the metric
$$s g_{\alpha\beta} = \pounds(c) g_{\alpha\beta} = \nabla_\alpha \nabla^\lambda c^\lambda{}_\beta + \nabla_\beta \nabla_\lambda c^\lambda{}_\alpha \tag{25}$$
nilpotency needs
$$s c_1 = \frac{1}{2}\delta(c \wedge c) \tag{26}$$
Please note that owing to the odd Grassmann parity of the ghosts,
$$c \wedge c \neq 0 \tag{27}$$
so that
$$s c_2 = \frac{1}{2} c \wedge c - \delta c_3 \tag{28}$$
This three form c_3 cannot be trivial, because using nilpotency again, this time on the ghost itself,
$$s^2 c_1 = 0 = s(\frac{1}{2} c \wedge c) - s\delta c_3 \tag{29}$$
conveying the fact that
$$s c_3 = \frac{1}{3!} c \wedge c \wedge c - \delta c_4 \tag{30}$$
Once more, using nilpotency,
$$s^2 c_3 = 0 = \frac{1}{4!}\delta(c \wedge c \wedge c \wedge c) - s\delta c_4 \tag{31}$$
and, finally
$$s c_4 = \frac{1}{4!} c \wedge c \wedge c \wedge c \tag{32}$$
and $s^2 = 0$ because there are no forms of dgree five in four dimensions.

So we need altogether 11 independent ghosts: 6 grassmann odd, ghost number one c_2, plus 4 Grasmann even, ghost number two c_3 plus one Grassmann odd, ghost number three, c_4.

For the antighosts the story is even simpler. We define the corresponding forms:
$$b_1 = \delta b_2 \tag{33}$$
and
$$\begin{aligned} s b_2 &= B_2 \\ s B_2 &= 0 \end{aligned} \tag{34}$$

The antighosts b are Grassmann odd, and enjoy ghost number -1, whereas B is Grassmann even and has vanishing ghost number.

This analysis coincides basically with the one performed earlier by Dragon and Kreuzer [5][8], which however employ a non covariant, less convenient language.

4. GAUGE FIXING

The gauge fixing fermion has got to be a Lorentz scalar of ghost number -1. We can define the most general operator composed out of fields with zero ghost number:

$$H_{\alpha_1\alpha_2} = A_{\alpha_1\alpha_2\alpha_3\alpha_4}h^{\alpha_3\alpha_4} + B_{\alpha_1\alpha_2\alpha_3\alpha_4}B^{\alpha_3\alpha_4} + C_{\alpha_1\alpha_2\alpha_3\alpha_4\alpha_5\alpha_6}h^{\alpha_3\alpha_4}h^{\alpha_5\alpha_6} + \ldots \quad (35)$$

That is, the most general polynomial in the fields B and h. The most general composite operator with ghost number -1 is of the form:

$$G_{\alpha_1\alpha_2} \equiv K_{\alpha_1\alpha_2\alpha_3\alpha_4}b^{\alpha_3\alpha_4} + K_{\alpha_1\alpha_2\alpha_3\alpha_4\alpha_5\alpha_6}b^{\alpha_3\alpha_4}b^{\alpha_3\alpha_4}c^{\alpha_5\alpha_6} + \ldots \quad (36)$$

so that the gauge fixing fermion is given by

$$\Psi \equiv G_{\mu\nu}H^{\mu\nu} \quad (37)$$

where the indices are raised and lowered with Minkowski's metric. The contribution to the lagrangian is

$$s\Psi = (K_{\alpha_1\alpha_2\alpha_3\alpha_4}B^{\alpha_3\alpha_4} + \ldots)H^{\alpha_1\alpha_2} - G_{\alpha_1\alpha_2}\left(A_{\alpha_1\alpha_2\alpha_3\alpha_4}\left(\nabla^{\alpha_3}\nabla_\lambda c^{\alpha_4\lambda} + \nabla^{\alpha_4}\nabla_\lambda c^{\alpha_3\lambda}\right) + \ldots\right) \quad (38)$$

ACKNOWLEDGMENTS

This work was begun while one of us (E.A.) was a guest at the Universita di Padova. He would like to thank Mario Tonin and the other members of the particle physics group for their wonderful hospitality. Parts of it heve been presented at ERES-05. We are grateful to Joaquín Díaz-Alonso, Lisianne Mornas and the rest of the organizers for the invitation and warm support. This work has been partially supported by the European Commission (HPRN-CT-200-00148) and by FPA2003-04597 (DGI del MCyT, Spain).

REFERENCES

1. E. Alvarez, "Quantum Gravity: A Pedagogical Introduction To Some Recent Results," Rev. Mod. Phys. **61** (1989) 561.
2. E. Alvarez, "Can one tell Einstein's unimodular theory from Einstein's general relativity?," arXiv:hep-th/0501146.
3. E. Alvarez, D. Blas, J. Garriga and E. Verdaguer, "Transverse Fierz-Pauli symmetry"
4. N. Arkani-Hamed, H. Georgi and M. D. Schwartz, "Effective field theory for massive gravitons and gravity in theory space," Annals Phys. **305** (2003) 96 [arXiv:hep-th/0210184].
5. N. Dragon and M. Kreuzer, "Quantization Of Restricted Gravity," Z. Phys. C **41** (1988) 485.

6. M. Fierz and W. Pauli, "On Relativistic Wave Equations For Particles Of Arbitrary Spin In An Electromagnetic Field," Proc. Roy. Soc. Lond. A **173** (1939) 211.
7. M. Henneaux and C. Teitelboim, "The Cosmological Constant And General Covariance," Phys. Lett. B **222**, 195 (1989).
8. M. Kreuzer, "Gauge Theory Of Volume Preserving Diffeomorphisms," Class. Quant. Grav. **7** (1990) 1303.
9. T. Ortin, "Gravity and strings,"
10. H. van Dam and M. J. G. Veltman, "Massive And Massless Yang-Mills And Gravitational Fields," Nucl. Phys. B **22**, 397 (1970).
    V.I. Zakharov, JETP lett 12, 312 (1970).
    A. I. Vainshtein, "To The Problem Of Nonvanishing Gravitation Mass," Phys. Lett. B **39** (1972) 393.
11. J. J. van der Bij, H. van Dam and Y. J. Ng, "Theory Of Gravity And The Cosmological Term: The Little Group Viewpoint," Physica **116A**, 307 (1982).
12. P. Van Nieuwenhuizen, "On Ghost-Free Tensor Lagrangians And Linearized Gravitation," Nucl. Phys. B **60**, 478 (1973).
13. M. J. G. Veltman, "Quantum Theory Of Gravitation,"

AdS/CFT, Black Holes and Matrix Models (abstract)

Luis Álvarez-Gaumé

Theory Group, Physics Department, CERN, CH-1211 Geneva 23, Switzerland

This talk is based on papers [1, 2], where the reader is sent for details and references. The main theme is associated with the fact that many properties of Black Holes (at least in AdS space-times) can be cast in terms of properties of $N = 4$ supersymmetric Yang-Mills Theory according to the Maldacena conjecture. In particular thermal properties of black holes can be translated into thermal properties of supersymmetric field theories.

In [5] it is shown how one can reformulate properties of black holes in terms of certain types of unitary matrix models, and the simples examples are considered in detail. In [1] we generalize the models presented in [5] and show that under certain reasonable conditions one can reproduce the properties of the Hawking-Page transition, and the Hagedorn temperature in term of some specific matrix models, making some universality assumptions. The string interpretation close to the singularities associated to the various critical points are analyzed and studied in detail using Matrix Model technology.

In [2] we reconsider the problems treated in [1, 5] and we analyze it in the most general possible thermal effective action of the gauge theory on a three sphere. With some mild assumptions we understand the behavior of the small, unstable black holes in AdS in terms of the Horowitz-Polchinski transitions from Black Holes to and AdS background with a gas of highly excited thermal strings.

REFERENCES

1. L. Alvarez-Gaume, C. Gomez, H. Liu and S. Wadia, *"Finite temperature effective action, AdS(5) black holes, and 1/N expansion,"* Phys. Rev. D **71** (2005) 124023 [arXiv:hep-th/0502227].
2. L. Alvarez-Gaume, M. Marino, P. Basu and S. Wadia, *"Blackhole/String transition, AdS/CFT Correspondence and Critical Unitary Matrix Models"*, to appear.
3. J. M. Maldacena, *"The large N limit of superconformal field theories and supergravity,"* Adv. Theor. Math. Phys. **2** (1998) 231 [Int. J. Theor. Phys. **38** (1999) 1113] [arXiv:hep-th/9711200].
4. Many details and references can be found in the following review: O. Aharony, S. S. Gubser, J. M. Maldacena, H. Ooguri and Y. Oz, *"Large N field theories, string theory and gravity,"* Phys. Rept. **323** (2000) 183 [arXiv:hep-th/9905111].
5. O. Aharony, J. Marsano, S. Minwalla, K. Papadodimas and M. Van Raamsdonk, *"The Hagedorn / deconfinement phase transition in weakly coupled large N gauge theories,"* Adv. Theor. Math. Phys. **8** (2004) 603 [arXiv:hep-th/0310285].

Gravitational Radiation from Two-Body Systems

Luc Blanchet

*Gravitation et Cosmologie ($\mathscr{G}\mathbb{R}\varepsilon\mathbb{C}\mathscr{O}$), Institut d'Astrophysique de Paris,
98bis boulevard Arago, 75014 Paris, France*

Abstract. Thanks to the new generation of gravitational wave detectors LIGO and VIRGO, the theory of general relativity will face new and important confrontations to observational data with unprecedented precision. Indeed the detection and analysis of the gravitational waves from compact binary star systems requires beforehand a very precise solution of the two-body problem within general relativity. The approximation currently used to solve this problem is the post-Newtonian one, and must be pushed to high order in order to describe with sufficient accuracy (given the sensitivity of the detectors) the inspiral phase of compact bodies, which immediately precedes their final merger. The resulting post-Newtonian "templates" are currently known to 3.5PN order, and are used for searching and deciphering the gravitational wave signals in VIRGO and LIGO.

Keywords: Gravitational waves, Compact binary systems, Post-Newtonian approximation
PACS: 04.30.-w, 04.25.Nx

1. INTRODUCTION

A compelling motivation for accurate computations of the gravitational radiation field generated by compact binary systems (*i.e.*, made of neutron stars and/or black holes) is the need for accurate *templates* to be used in the data analysis of the current and future generations of laser interferometric gravitational wave detectors. It is indeed recognized that the *inspiral* phase of the coalescence of two compact objects represents an extremely important source for the ground-based detectors such as LIGO and VIRGO, provided that their total mass does not exceed say 10 or 20 $M_\odot$ (this includes the very interesting case of double neutron-star systems), and for space-based detectors like LISA, in the case of the coalescence of two galactic black holes, if the masses are within the range between say 10^5 and $10^8 M_\odot$.

For these sources the *post-Newtonian* (PN) approximation scheme has proved to be the appropriate theoretical tool in order to construct the necessary templates. A program started long ago with the goal of obtaining these templates with 3PN and even 3.5PN accuracy.[1] Several studies, *e.g.* [1, 2], have shown that such a high PN precision is probably sufficient, not only for detecting the signals in LIGO/VIRGO, but also for analyzing them and accurately measuring the parameters of the binary (such high-accuracy templates will also be of great value for detecting massive black-hole mergers in LISA). The templates have been first completed through 2PN order [3]. The 3.5PN accuracy (in the case where the compact objects have negligible intrinsic spins) has been

---

[1] Following the standard custom we use the qualifier nPN for a term in the wave form or (for instance) the energy flux which is of the order of $1/c^{2n}$ relatively to the lowest-order Newtonian quadrupolar radiation.

achieved more recently [4, 5].

The calculation of the 3PN order turned out to be very intricate and quite subtle. The first step has been to compute all the terms, in both the 3PN equations of motion [6, 7, 8, 9, 10] and 3.5PN gravitational radiation field [11, 12, 13], by means of the Hadamard self-field regularization [14, 15]. A regularization is needed in this problem in order to remove the infinite self-field of point masses. However, a few terms were left undetermined by Hadamard's regularization, which correspond to some incompleteness of this regularization occurring at the 3PN order. These terms could be parametrized by some unknown numerical coefficients called *ambiguity parameters*. The second step has been to use the more powerful *dimensional regularization* [16], which is technically based on analytic continuation in the dimension of space, which finally enabled to fix the values of all the ambiguity parameters [17, 18, 5, 19].

In Section 2 of this article we review and comment on the striking appearance of Hadamard self-field regularization parameters at 3PN order, and on their computation using dimensional regularization. Section 3 is devoted to the notion of the multipole moments of an isolated post-Newtonian extended source, at the basis of the construction of gravitational-wave post-Newtonian templates. In Section 4 we present two checks of the values of the latter ambiguity parameters, coming from the comparison between the binary's dipole moment and its center-of-mass vector on the one hand, and based on an argument from classical field-theory diagrams on the other hand. Finally, in Section 5, we consider the limiting case where one of the masses is exactly zero, and the remaining one moves with uniform velocity, and show that such "boosted Schwarzschild solution" limit yields the determination of the third ambiguity parameter in the radiation field. These tests, altogether, provide a verification, independent of dimensional regularization, for all the ambiguity parameters in the 3PN gravitational radiation field.

2. HADAMARD REGULARIZATION PARAMETERS

The standard Hadamard regularization yields some ambiguous results for the computation of certain integrals at the 3PN order, as Jaranowski and Schäfer [6, 7] first noticed in their computation of the equations of motion of point particles within the ADM-Hamiltonian formulation of general relativity. Hadamard's regularization is based on the notion of *partie finie* of a singular function, given by the angular integral of the finite part coefficient in the singular expansion of that function near a singular point, and the related notion of *partie finie* of a divergent integral. It was shown [6, 7] that there are *two and only two* types of ambiguous terms in the 3PN Hamiltonian, which were then parametrized by two unknown numerical coefficients ω_{static} and ω_{kinetic}.

Motivated by the previous result, Blanchet and Faye introduced an extended version of Hadamard's regularization [20, 21], which is mathematically well-defined and free of ambiguities; in particular it yields unique results for the computation of any of the integrals occuring in the 3PN equations of motion. Unfortunately, the extended Hadamard regularization turned out to be in a sense incomplete, because it was found [8, 9] that the 3PN equations of motion involve *one and only one* unknown numerical constant, called λ, which cannot be determined within the method. The comparison with the work [6, 7], on the basis of the computation of the invariant energy of compact binaries moving on circular orbits, revealed [8] that

$$\omega_{kinetic} = \frac{41}{24}, \qquad (1)$$

$$\omega_{static} = -\frac{11}{3}\lambda - \frac{1987}{840}. \qquad (2)$$

Therefore, the ambiguity $\omega_{kinetic}$ is fixed, while λ is equivalent to the other ambiguity ω_{static}. Notice that the value (1) for the kinetic ambiguity parameter $\omega_{kinetic}$, which is in factor of some velocity dependent terms, is the only one for which the 3PN equations of motion are Poincaré invariant. Fixing up this value was possible because the extended Hadamard regularization [20, 21] was defined in such a way that it keeps the Poincaré invariance.

The appearance of one and only one physical unknown coefficient λ in the equations of motion constitutes a quite striking fact, that is related specifically with the use of some Hadamard-type regularization. Technically speaking, the presence of the parameter λ is associated with the so-called "non-distributivity" of Hadamard's regularization. [2] Mathematically speaking, λ is probably related to the fact that it is impossible to construct a distributional derivative operator satisfying the Leibniz rule for the derivation of the product. The Einstein field equations can be written into many different forms, by shifting the derivatives and operating some terms by parts with the help of the Leibniz rule. All these forms are equivalent in the case of regular sources, but since the distributional derivative operator violates the Leibniz rule they become inequivalent for point particles. Finally, physically speaking, we can argue that λ has its root in the fact that, in a complete computation of the equations of motion valid for two regular *extended* weakly self-gravitating bodies, many non-linear integrals, when taken *individually*, start depending, from the 3PN order, on the internal structure of the bodies, even in the "compact-body" limit where the radii tend to zero. However, when considering the full equations of motion, we expect that all the terms depending on the internal structure can be removed, in the compact-body limit, by a coordinate transformation (or by some appropriate shifts of the central world lines of the bodies), and that finally λ is given by a pure number, for instance a rational fraction, independent of the details of the internal structure of the compact bodies. From this argument (which could be justified by invoking the effacing principle in general relativity [22]) the value of λ is necessarily the one we shall obtain below, Eq. (4), and will be valid for any compact objects, for instance black holes.

The ambiguity parameter ω_{static}, which is in factor of some static, velocity-independent term, was computed by Damour, Jaranowski and Schäfer [17] by means of *dimensional regularization*, instead of some Hadamard-type one, within the ADM-Hamiltonian formalism. Their result is

$$\omega_{static} = 0. \qquad (3)$$

---

[2] By non-distributivity we mean that the Hadamard regularization of a product of functions differs in general from the product of regularizations.

As Damour *et al.* [17] argue, clearing up the static ambiguity is made possible by the fact that dimensional regularization, contrary to Hadamard's regularization, respects all the basic properties of the algebraic and differential calculus of ordinary functions: associativity, commutativity and distributivity of point-wise addition and multiplication, Leibniz's rule, and the Schwarz lemma. In this respect, dimensional regularization is certainly better than Hadamard's one, which does not respect the distributivity of the product and unavoidably violates at some stage the Leibniz rule for the differentiation of a product.

The ambiguity parameter λ is fixed from the result (3) and the necessary link (2) provided by the equivalence between the harmonic-coordinates and ADM-Hamiltonian formalisms. However, λ was also computed directly by Blanchet, Damour and Esposito-Farèse [18] applying dimensional regularization to the 3PN equations of motion in harmonic coordinates (in the line of Refs. [8, 9]). The end result,

$$\lambda = -\frac{1987}{3080}, \qquad (4)$$

is in full agreement with Eq. (3). Besides the independent confirmation of the value of ω_{static} or λ, the work [18] provides also a confirmation of the *consistency* of dimensional regularization, because the explicit calculations are entirely different from the ones of Ref. [17]: harmonic coordinates are used instead of ADM-type ones, the work is at the level of the equations of motion instead of the Hamiltonian, a different form of Einstein's field equations is solved by a different iteration scheme.

Let us comment here that the use of a self-field regularization, be it dimensional or based on Hadamard's partie finie, signals a somewhat unsatisfactory situation on the physical point of view, because, ideally, we would like to perform a complete calculation valid for extended bodies, taking into account the details of the internal structure of the bodies (energy density, pressure, internal velocity field). By considering the limit where the radii of the objects tend to zero, one should recover the same result as obtained by means of the point-mass regularization. This would demonstrate the suitability of the regularization. This program was undertaken at the 2PN order by Grishchuk and Kopeikin [23, 24] who derived the equations of motion of two extended fluid balls, and obtained equations of motion depending only on the two masses m_1 and m_2 of the compact bodies. At the 3PN order we expect that the extended-body program should give the value of the regularization parameter λ (maybe after some gauge transformation to remove the terms depending on the internal structure). Ideally, its value should be confirmed by independent and more physical methods. One such method is the one of Itoh and Futamase [25, 26], who derived the 3PN equations of motion in harmonic coordinates by means of a particular variant of the famous "surface-integral" method introduced long ago by Einstein, Infeld and Hoffmann [27]. This approach is interesting because it is based on the physical notion of extended compact bodies in general relativity, and is free of the problems of ambiguities due to Hadamard's self-field regularization. The end result of Refs. [25, 26] is in agreement with the complete 3PN equations of motion in harmonic coordinates [8, 9] and, moreover, is unambiguous, as it does determine the ambiguity parameter λ to exactly the value (4).

We next consider the problem of the binary's radiation field, where the same phenomenon occurs, with the appearance of some Hadamard regularization ambiguity parameters at 3PN order. More precisely, Blanchet, Iyer and Joguet [12], in their computation of the 3PN compact binary's *mass quadrupole moment* I_{ij}, found it necessary to introduce *three* Hadamard regularization constants ξ, κ and ζ, which are additional to (and independent of) the equation-of-motion related constant λ. The total gravitational-wave flux at 3PN order, in the case of circular orbits, was found to depend on a single combination of the latter constants, $\theta = \xi + 2\kappa + \zeta$, and the binary's orbital phase, for circular orbits, involves only the linear combination of θ and λ given by $\hat{\theta} = \theta - 7\lambda/3$, as shown in [4].

Dimensional regularization (instead of Hadamard's) was applied in Refs. [5, 19] to the computation of the 3PN radiation field of compact binaries, finally leading to the following unique values for the ambiguity parameters

$$\xi = -\frac{9871}{9240}, \qquad (5)$$
$$\kappa = 0, \qquad (6)$$
$$\zeta = -\frac{7}{33}. \qquad (7)$$

These values represent the end result of dimensional regularization. However, we shall review in the present Article some alternative calculations which provide some checks, independent of dimensional regularization, for all the parameters (5)–(7).

The result (5)–(7) completes the problem of the general relativistic prediction for the templates of inspiralling compact binaries up to 3PN order (and actually up to 3.5PN order as the corresponding tail terms have already been determined [11]). The relevant combination of the parameters entering the 3PN energy flux in the case of circular orbits is now fixed to be

$$\theta \equiv \xi + 2\kappa + \zeta = -\frac{11831}{9240}. \qquad (8)$$

The orbital phase of compact binaries, in the adiabatic inspiral regime (*i.e.*, evolving by radiation reaction), involves at 3PN order a combination of parameters which is determined as

$$\hat{\theta} \equiv \theta - \frac{7}{3}\lambda = \frac{1039}{4620}. \qquad (9)$$

The fact that the numerical value of this parameter is quite small, $\hat{\theta} \simeq 0.22489$, indicates that the 3PN (or, even better, 3.5PN) order should provide an excellent approximation for both the on-line search and the subsequent off-line analysis of gravitational wave signals from inspiralling compact binaries in the LIGO and VIRGO detectors.

3. THE MULTIPOLAR POST-NEWTONIAN FORMALISM

3.1. Multipole moments of a post-Newtonian extended source

The multipole moments of a post-Newtonian (PN) source, by which we mean a source which is at once slowly moving, weakly stressed and weakly self-gravitating, are crucial for the present gravitational wave generation formalism. They are obtained in Ref. [28]

as functionals of the PN expansion of the pseudo-stress energy tensor $\tau^{\mu\nu}$ of the matter and gravitational fields in the *harmonic coordinate* system. The pseudo-tensor $\tau^{\mu\nu}$ has a non-compact support because of the contribution of the gravitational field which extends up to infinity from the source. Let us denote the formal PN expansion of the pseudo tensor by means of an overbar, so that $\overline{\tau}^{\mu\nu} = \text{PN}[\tau^{\mu\nu}]$. The two types of multipole moments of the gravitating source, mass-type I_L moments and current-type ones J_L, are then given by [3]

$$\text{I}_L(t) = \frac{1}{c^2} \mathop{\text{FP}}_{B=0} \int d^3\mathbf{x}\, r^B \left\{ \hat{x}_L \left(\overline{\tau}^{00}_{[\ell]} + \overline{\tau}^{ii}_{[\ell]} \right) \right.$$
$$- \frac{4(2\ell+1)}{c(\ell+1)(2\ell+3)} \hat{x}_{iL} \dot{\overline{\tau}}^{i0}_{[\ell+1]}$$
$$\left. + \frac{2(2\ell+1)}{c^2(\ell+1)(\ell+2)(2\ell+5)} \hat{x}_{ijL} \ddot{\overline{\tau}}^{ij}_{[\ell+2]} \right\}, \tag{10}$$

$$\text{J}_L(t) = \frac{1}{c} \mathop{\text{FP}}_{B=0} \varepsilon_{ab\langle i_\ell} \int d^3\mathbf{x}\, r^B \left\{ \hat{x}_{L-1\rangle a} \overline{\tau}^{b0}_{[\ell]} \right.$$
$$\left. - \frac{2\ell+1}{c(\ell+2)(2\ell+3)} \hat{x}_{L-1\rangle ac} \dot{\overline{\tau}}^{bc}_{[\ell+1]} \right\}. \tag{11}$$

Since Eqs. (10)–(11) are valid only in the sense of PN expansions, the operational meaning of the underscript $[\ell]$ in (10)–(11) is actually that of an infinite PN series, which is given by

$$\overline{\tau}^{\mu\nu}_{[\ell]}(\mathbf{x},t) = \sum_{k=0}^{+\infty} \alpha_{k,\ell} \left(\frac{r}{c} \frac{\partial}{\partial t} \right)^{2k} \overline{\tau}^{\mu\nu}(\mathbf{x},t), \tag{12}$$

$$\alpha_{k,\ell} = \frac{(2\ell+1)!!}{(2k)!!(2\ell+2k+1)!!}. \tag{13}$$

A basic feature of the expressions of the moments is that the integral formally extends over the whole support of the PN expansion of the stress-energy pseudo-tensor, $\overline{\tau}^{\mu\nu}$, *i.e.* from $r \equiv |\mathbf{x}| = 0$ up to infinity. Recall that a formal PN series such as $\overline{\tau}^{\mu\nu}$ is physically meaningful only within the near-zone. Therefore the integrals (10)–(11) physically refer to a result obtained from near-zone quantities only (in the formal limit where $c \to +\infty$). However, it was found extremely useful in Ref. [28] to mathematically extend the integrals up to $r \to +\infty$. This was made possible by the use of the prefactor r^B, together with a process of analytic continuation in the complex B plane. [4] This shows up in Eqs. (10)–(11) as the crucial Finite Part (FP) operation, when $B \to 0$, which technically

---

[3] Our notation is: $L \equiv i_1 \cdots i_\ell$ for a multi-index composed of ℓ multipolar indices $i_1, \cdots, i_\ell$; $x_L \equiv x_{i_1} \cdots x_{i_\ell}$ for the product of ℓ spatial vectors $x^i \equiv x_i$; and $\hat{x}_L \equiv \text{STF}(x_{i_1} \cdots x_{i_\ell})$ for the symmetric-trace-free (STF) part of that product, also denoted by carets surrounding the indices, $x_{\langle L \rangle} \equiv \hat{x}_L$.
[4] The prefactor r^B should in principle be adimensionalized as $(r/r_0)^B$ where r_0 is a constant arbitrary scale, but here we set $r_0 = 1$.

allows one to uniquely define integrals which would otherwise be divergent at their upper boundary, $r = |\mathbf{x}| \to +\infty$. See Ref. [28] for the proof and details.

3.2. Surface-integral expressions of the multipole moments

Let us next review the recent derivation [29] of an alternative form of the PN source moments (10)–(11) in terms of two-dimensional surface integrals. Such a possibility of expressing the moments, for general ℓ and at any PN order, as some surface integrals is quite useful for practical purposes, as we shall show in the application we consider in Section 5. In keeping with the fact that the "volume integrals" Eqs. (10)–(11) physically involve only near-zone quantities, the "surface integrals" into which we shall transform the moments I_L and J_L physically refer to an operation which extracts some coefficients in the "far near-zone" expansion of the gravitational field, *i.e.* in the expansion in increasing powers of $1/r$ of the PN-expanded near-zone metric. Technically, as our starting point (10)–(11) is made of integrals extended up to $r \to +\infty$, our mathematical manipulations below will involve "surface terms" on arbitrary large spheres $r = \mathscr{R}$. All these manipulations will be mathematically well-defined because of the properties of complex analytic continuation in B.

The basic idea is to go from the "source term", $\overline{\tau}^{\mu\nu}$, to the corresponding "solution" $\overline{h}^{\mu\nu}$, via integrating by parts the Laplace operator present in the Einstein field equation in harmonic coordinates, namely $\overline{\tau}^{\mu\nu} = \frac{c^4}{16\pi G} \Box \overline{h}^{\mu\nu}$, where $\overline{h}^{\mu\nu}$ is the (PN expansion of the) basic gravitational field variable, satisfying the harmonic-coordinate condition $\partial_\nu \overline{h}^{\mu\nu} = 0$. From Eq. (12) we have

$$\int d^3\mathbf{x}\, r^B\, \hat{x}_L\, \overline{\tau}^{\mu\nu}_{[\ell]} = \frac{c^4}{16\pi G} \sum_{k=0}^{+\infty} \alpha_{k,\ell} \left(\frac{d}{c\,dt}\right)^{2k} \int d^3\mathbf{x}\, r^{B+2k}\, \hat{x}_L\, \Box \overline{h}^{\mu\nu}, \tag{14}$$

in the right-hand-side of which we insert $\Box = \Delta - \left(\frac{\partial}{c\,\partial t}\right)^2$, and operate the Laplacian by parts using $\Delta(r^{B+2k}\, \hat{x}_L) = (B+2k)(B+2\ell+2k+1) r^{B+2k-2}\, \hat{x}_L$. In the process we can ignore the all-integrated surface terms because they are identically zero by complex analytic continuation, from the case where the real part of B is chosen to be a large enough *negative* number. Using the expression of the coefficients (13), we are next led to the alternative expression

$$\int d^3\mathbf{x}\, r^B\, \hat{x}_L\, \overline{\tau}^{\mu\nu}_{[\ell]} = \frac{c^4}{16\pi G} \sum_{k=0}^{+\infty} B(B+2\ell+4k+1)\, \alpha_{k,\ell} \left(\frac{d}{c\,dt}\right)^{2k} \int d^3\mathbf{x}\, r^{B+2k-2}\, \hat{x}_L\, \overline{h}^{\mu\nu}. \tag{15}$$

A remarkable feature of this result, which is the basis of our new expressions, is the presence of an *explicit factor B* in front of the integral. The factor means that the result depends only on the occurrence of *poles*, $\propto 1/B^p$, in the boundary of the integral at infinity: $r \to +\infty$ with $t = $ const.

Thanks to the factor B we can replace the integration domain of Eq. (15) by some outer domain of the type $r > \mathscr{R}$, where $\mathscr{R}$ denotes some large arbitrary constant radius.

The integral over the inner domain $r < \mathcal{R}$ is always zero in the limit $B \to 0$ because the integrand is constructed from $\overline{\tau}^{\mu\nu}$, and we are considering extended regular PN sources, without singularities. Now, in the outer (but still near-zone) domain we can replace the PN metric coefficients $\overline{h}^{\mu\nu}$ by the expansion in increasing powers of $1/r$ of the PN-expanded metric, which is identical to the multipolar expansion of the PN-expanded metric, that we shall denote by $\mathcal{M}(\overline{h}^{\mu\nu})$. Hence we have

$$\int d^3\mathbf{x}\, r^B \hat{x}_{L}\, \overline{\tau}^{\mu\nu} = \frac{c^4}{16\pi G} \sum_{k=0}^{+\infty} B(B+2\ell+4k+1)\, \alpha_{k,\ell} \left(\frac{d}{cdt}\right)^{2k} \int_{r>\mathcal{R}} d^3\mathbf{x}\, r^{B+2k-2} \hat{x}_L \mathcal{M}(\overline{h}^{\mu\nu}). \tag{16}$$

We want now to make use of a more explicit form of the far near-zone expansion $\mathcal{M}(\overline{h}^{\mu\nu})$, whose general structure is known. It consists of terms proportional to arbitrary powers of $1/r$, and multiplied by powers of the *logarithm* of r; more precisely,

$$\mathcal{M}(\overline{h}^{\mu\nu})(\mathbf{x},t) = \sum_{a,b} \frac{(\ln r)^b}{r^a} \varphi_{a,b}^{\mu\nu}(\mathbf{n},t), \tag{17}$$

where a can take any positive or negative integer values, and b can be any positive integer: $a \in \mathbb{Z}$, $b \in \mathbb{N}$. The coefficients $\varphi_{a,b}^{\mu\nu}$ depend on the unit direction $\mathbf{n} \equiv \mathbf{x}/r$ and on the coordinate time t (in the harmonic coordinate system). The structure (17) for the multipolar expansion of the near-zone (PN-expanded) metric is a consequence of the so-called matching equation

$$\mathcal{M}(\overline{h}^{\mu\nu}) \equiv \overline{\mathcal{M}(h^{\mu\nu})}, \tag{18}$$

which says that the multipolar re-expansion of the PN metric $\overline{h}^{\mu\nu}$ agrees, in the sense of formal series, with the *near-zone* re-expansion (also denoted with an overbar) of the external multipolar metric $\mathcal{M}(h^{\mu\nu})$ (see [28] for details). Inserting Eq. (17) into (16), we are therefore led to the computation of the integral

$$\int_{r>\mathcal{R}} d^3\mathbf{x}\, r^{B+2k-2} \hat{x}_L \mathcal{M}(\overline{h}^{\mu\nu}) = \sum_{a,b} \int_{\mathcal{R}}^{+\infty} dr\, r^{B+2k+\ell-a} (\ln r)^b \int d\Omega\, \hat{n}_L\, \varphi_{a,b}^{\mu\nu}(\mathbf{n},t), \tag{19}$$

where $d\Omega$ is the solid angle element associated with the unit direction $\mathbf{n}$ (and $\hat{n}_L \equiv \hat{x}_L/r^\ell$). The radial integral can be trivially integrated by analytic continuation in B, with result

$$\int_{\mathcal{R}}^{+\infty} dr\, r^{B+2k+\ell-a} (\ln r)^b = -\left(\frac{d}{dB}\right)^b \left[\frac{\mathcal{R}^{B+2k+\ell-a+1}}{B+2k+\ell-a+1}\right]. \tag{20}$$

Remember that we are ultimately interested only in the analytic continuation of such integrals down to $B=0$. And as an integral such as (20) is multiplied by a coefficient which is proportional to B, we must control the poles of Eq. (20) at $B=0$. Those poles are in general multiple because of the presence of powers of $\ln r$ in the expansion, and the consecutive multiple differentiation with respect to B shown in Eq. (20). The poles at $B=0$ clearly come from a single value of a, namely $a = 2k+\ell+1$. For that value, the "multiplicity" of the pole takes the value $b+1$. Here a useful simplification comes from the fact that the factor in front of the integrals in (16) is of the form $\sim B(B+K)$. In other words, this factor contains only the first and second powers of B. Therefore, only the simple and double poles $1/B$ and $1/B^2$ in (20) can contribute to the final result. Hence, we conclude that it is enough to consider the values $b = 0, 1$ for the exponent b of $\ln r$ in the expansion (17).

To express the result in the most convenient manner let us introduce a special notation for some relevant combination of coefficients $\varphi_{a,b}^{\mu\nu}(\mathbf{n},t)$, which as we just said correspond exclusively to the values $a = \ell + 2k + 1$ and $b = 0$ or 1. Namely,

$$\Psi_{k,\ell}^{\mu\nu}(\mathbf{n},t) \equiv \alpha_{k,\ell}\left[-(2\ell+4k+1)\varphi_{2k+\ell+1,0}^{\mu\nu}(\mathbf{n},t) + \varphi_{2k+\ell+1,1}^{\mu\nu}(\mathbf{n},t)\right], \quad (21)$$

in which we have absorbed the numerical coefficient $\alpha_{k,\ell}$ defined by (13). With this notation we then obtain

$$\operatorname*{FP}_{B=0} B(B+2\ell+4k+1)\alpha_{k,\ell}\int_{r>\mathscr{R}} d^3\mathbf{x}\, r^{B+2k-2}\, \hat{x}_L\, \mathscr{M}\left(\overline{h}^{\mu\nu}\right) = 4\pi\left\langle \hat{n}_L \Psi_{k,\ell}^{\mu\nu}\right\rangle, \quad (22)$$

where the brackets refer to the spherical or angular average (at coordinate time t), i.e.

$$\left\langle \hat{n}_L \Psi_{k,\ell}^{\mu\nu}\right\rangle(t) \equiv \int \frac{d\Omega}{4\pi} \hat{n}_L \Psi_{k,\ell}^{\mu\nu}(\mathbf{n},t). \quad (23)$$

The quantities (23) are integrals over a unit sphere, and can rightly be referred to as *surface integrals*. These surface integrals are the basic blocks entering our alternative expressions for the multipole moments. If we wish to physically think of them as integrals over some two-surface surrounding the source, we can roughly consider that this two-surface is located at a radius $\mathscr{R}$, with $a \ll \mathscr{R} \ll cT$. Anyway, the important point is that, as we can see from Eq. (23), the surface integrals, and therefore the multipole moments, are strictly independent of the choice of the intermediate scale $\mathscr{R}$ which entered our reasoning.

Finally, we are in a position to write down the following final results for an alternative form of the source multipole moments (10)–(11), expressed solely in terms of the surface integrals of the type (23),

$$I_L = \frac{c^2}{4G} \sum_{k=0}^{+\infty}\left\{\left(\frac{d}{cdt}\right)^{2k}\left\langle \hat{n}_L\left(\Psi_{k,\ell}^{00} + \Psi_{k,\ell}^{ii}\right)\right\rangle\right.$$
$$-\frac{4(2\ell+1)}{(\ell+1)(2\ell+3)}\left(\frac{d}{cdt}\right)^{2k+1}\left\langle \hat{n}_{iL}\Psi_{k,\ell+1}^{i0}\right\rangle$$
$$\left.+\frac{2(2\ell+1)}{(\ell+1)(\ell+2)(2\ell+5)}\left(\frac{d}{cdt}\right)^{2k+2}\left\langle \hat{n}_{ijL}\Psi_{k,\ell+2}^{ij}\right\rangle\right\}, \quad (24)$$

$$J_L = \frac{c^3}{4G}\varepsilon_{ab\langle i_\ell} \sum_{k=0}^{+\infty}\left\{\left(\frac{d}{cdt}\right)^{2k}\left\langle \hat{n}_{L-1\rangle a}\Psi_{k,\ell}^{b0}\right\rangle\right.$$
$$\left.-\frac{2\ell+1}{(\ell+2)(2\ell+3)}\left(\frac{d}{cdt}\right)^{2k+1}\left\langle \hat{n}_{L-1\rangle ac}\Psi_{k,\ell+1}^{bc}\right\rangle\right\}. \quad (25)$$

4. MULTIPOLE MOMENTS OF TWO-BODY SYSTEMS

4.1. Quadrupole and dipole moments, and the center-of-mass vector

Let us show how a particular combination of ambiguity parameters can be determined within Hadamard's regularization and confirm the result of dimensional regularization. For this purpose we use the computations in Ref. [13] of the mass-type quadrupole I_{ij} and dipole I_i moments of point particle binaries at the 3PN order. These were derived by applying the expression (10) [with $\ell = 1,2$] to a binary systems of point masses, following the rules of the Hadamard regularization, in the so-called "pure Hadamard-Schwartz" (pHS) variant of it. Following the definition of Ref. [18], the pHS regularization is a specific, minimal Hadamard-type regularization of integrals, based on the usual Hadamard partie finie of a divergent integral, together with a minimal treatment (supposed to be "distributive") of compact-support terms. The pHS regularization also assumes the use of standard Schwartz distributional derivatives [15].

We shall denote by I_{ij}^{pHS} the result of such pHS calculation of the mass-type quadrupole moment. Now it was argued in Ref. [12] that the Hadamard regularization of the 3PN quadrupole moment is incomplete, in the sense that the pHS calculation I_{ij}^{pHS} must be augmented, in order to be correct, by some unknown, ambiguous, contributions. The first source of ambiguity is the "kinetic" one, linked to the inability of the Hadamard regularization to ensure the global Poincaré invariance of the formalism. As discussed in Ref. [12] (see also Section 2) we must account for the kinetic ambiguity by adding "by hands" a specific ambiguity term, depending on a single ambiguity parameter called ζ. The second source of ambiguity is "static". It comes from the *a priori* unknown relation between some Hadamard regularization length scales, s_1 and s_2 (one for each particles), and the ones, called r'_1 and r'_2, parametrizing the final 3PN equations of motion in harmonic coordinates [8, 9]. The static ambiguity is accounted for by two other ambiguity parameters ξ and κ (see Section 2).

The Hadamard-regularized 3PN quadrupole moment reads

$$I_{ij}[\xi, \kappa, \zeta] = I_{ij}^{\text{pHS}} \qquad (26)$$
$$+ \frac{44}{3} \frac{G^2 m_1^3}{c^6} \left[\left(\xi + \frac{1}{22} + \kappa \frac{m_1 + m_2}{m_1} \right) y_1^{\langle i} a_1^{j \rangle} + \left(\zeta + \frac{9}{110} \right) v_1^{\langle ij \rangle} \right] + 1 \leftrightarrow 2,$$

where one sees in the second term the effect of adding the ambiguities, parametrized by the same parameters ξ, κ and ζ as introduced in Ref. [12]. Here, m_1 and m_2 are the masses, y_1^i, v_1^i and a_1^i denote the position, velocity and Newtonian acceleration of the first particle, and we pose $y_1^{\langle i} a_1^{j \rangle} \equiv \text{STF}(y_1^i a_1^j)$ and $v_1^{\langle ij \rangle} \equiv \text{STF}(v_1^i v_1^j)$. The symbol $1 \leftrightarrow 2$ refers to the same terms but concerning the second particles. All the terms composing the pHS part have been explicitly computed up to 3PN order for general binary orbits [13].

Let us now consider the case of the mass dipole moment I_i. Repeating the same arguments as for the quadrupole, we can write I_i as the pHS part I_i^{pHS} and augmented by an ambiguous part. However, in the dipole case we find that no ambiguity of the kinetic type occurs, and that the only ambiguity is static. We find that the expression analogous to (26) reads

$$\mathrm{I}_i[\xi + \kappa] = \mathrm{I}_i^{\mathrm{pHS}} + \frac{22}{3} \frac{G^2 m_1^3}{c^6} \left(\xi + \kappa + \frac{1}{22} \right) a_1^i + 1 \leftrightarrow 2 \,. \tag{27}$$

As we see, there is only one ambiguity parameter, in the form of the *sum* of ξ and κ, where ξ and κ are exactly the same as in the quadrupole moment (26). Let us now fix that particular sum of ambiguity parameters.

The case of the dipole moment I_i is very interesting. Indeed let us argue that I_i, which represents the distribution of positions of particles as weighted by their *gravitational masses* m_{g}, must be *identical* to the position of the center of mass G_i of the system of particles (*per* unit of total mass), because the center of mass G_i represents in fact the same quantity as the dipole I_i but corresponding to the *inertial masses* m_{i} of the particles. The equality between mass dipole I_i and center-of-mass position G_i can thus be seen as a consequence of the equivalence principle $m_{\mathrm{i}} = m_{\mathrm{g}}$, which is surely incorporated in our model of point particles. Now the center of mass G_i is already known at the 3PN order for point particle binaries, as one of the conserved integrals of the 3PN motion in harmonic coordinates.[5] The point is that G_i, given in Ref. [10], is free of ambiguities; for instance the ambiguity parameter λ in the 3PN equations of motion disappears from the expression of G_i. Let us therefore impose the equivalence between I_i and G_i, which means that we make the complete identification

$$\mathrm{I}_i[\xi + \kappa] \equiv \mathrm{G}_i \,. \tag{28}$$

Comparing I_i with the expression of G_i given by Eq. (4.5) in [10], we find that Eq. (28) is verified for all the terms *if and only if* the particular combination of ambiguity parameters $\xi + \kappa$ takes the unique value

$$\xi + \kappa = -\frac{9871}{9240} \,. \tag{29}$$

This result, obtained within Hadamard's regularization, is nicely consistent with the result of dimensional regularization, see Eqs. (5)–(6). It shows that, although as we have seen Hadamard's regularization is physically incomplete (at 3PN order), it can nevertheless be partially completed by invoking some external physical arguments — in the present case the equivalence between mass dipole and center-of-mass position. On the other hand, dimensional regularization *is* complete; it does not need to invoke any external physical argument in order to determine the value of all the ambiguity parameters. Nevertheless, it remains that the result (29), based simply on a consistency argument between the 3PN equations of motion and the 3PN radiation field, does provide a verification of the consistency and completeness of dimensional regularization itself.

---

[5] We neglect the radiation-reaction term at 2.5PN order.

4.2. Diagrammatic representation of the multipole moments

Let us describe the multipole moments in terms of classical field-theory diagrams, representing the non-linear interactions of classical general relativity (we refer to [30] for definition and use of these diagrams). We represent the basic delta-function sources entering the matter stress-energy tensor $T^{\mu\nu}$ — *i.e.*, the matter part of the pseudo-tensor $\bar{\tau}^{\mu\nu}$ of Section 3 — as two world-lines, and each (post-Minkowskian) propagator $\Box^{-1}$ as a dotted line. The various non-linear potentials entering the gravitational part of $\bar{\tau}^{\mu\nu}$ can then be represented by drawing some dotted lines which start at the matter sources, join at some intermediate vertices, corresponding to some non-linear couplings, and end at the field point x. Finally, we can represent the inclusion of the multipolar factors, such as $\hat{x}_L$, by adding a circled cross $\otimes$. It is then understood that one integrates over the crossed vertex, *i.e.*, the field point.

Using such a representation, the multipole moments are given by the sum of many diagrams. We are now looking at "dangerously" diverging diagrams, which generate poles $\propto 1/\varepsilon$ in a dimensionally continued approach, with $d = 3 + \varepsilon$ being the dimension of space. Examining the types of singular integrals corresponding to the possible diagrams, we find [19] that the only dangerously diverging diagrams are those containing (at least) three propagator lines that can simultaneously shrink to zero size, as a subset of vertices coalesce together on one of the particle world-lines. But as there are, in the present problem dealing with the 3PN order, at most three source points, this means that the dangerously divergent diagrams are only those represented in Fig. 1 (or their mirror image obtained by exchanging $1 \leftrightarrow 2$).

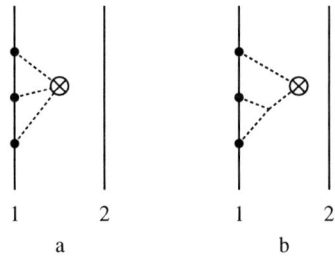

FIGURE 1. Dangerously divergent diagrams contributing to the 3PN multipole moments. The world-lines of particles 1 and 2 are represented by vertical solid lines, the propagator $\Box^{-1}$ by dotted lines, the source points by bullets, and the $\otimes$ symbol means a multiplication by a multipolar factor, such as $\hat{x}_L$, together with a spatial integration $\int d^d\mathbf{x}$.

Since the dangerous divergencies associated with the vicinity of the first world-line (say) are entirely contained in the diagrams shown in Fig. 1, they are, therefore, proportional to m_1^3 (*i.e.*, one factor m_1 *per* source point), without any explicit [6] dependence on the second mass m_2. As a consequence, we can prove [19], because the presence

---

[6] There is also an implicit dependence on m_2 via the fact that the acceleration a_1^i is proportional to m_2. But, at the level of the diagrams, a_1^i must be considered as a pure characteristic of the first world-line.

of ambiguity parameters is directly linked with the occurence of poles $\propto 1/\varepsilon$, that the structure of the ambiguous terms in the mass quadrupole moment (26) must be such that it is proportional to some factor m_1^3. Now, the definition of the parameter κ in Ref. [12] was to parametrize a conceivable *a priori* static ambiguity appearing in the renormalization of the logarithmic divergencies of the quadrupole moment, and these ambiguities were found in Eq. (26) to be of the form $(\xi + \kappa) m_1^3 + \kappa m_1^2 m_2$ (for what concerns the first particle). This shows that the parameter κ corresponds to a mixing between diagrams with three legs on the first world-line (as in Fig. 1) and diagrams having two legs on the first world-line and one on the second. Our diagrammatic study has shown that the latter diagrams have no dangerous divergencies, *i.e.*, that they do not introduce any conceivable ambiguity. Therefore we conclude, confirming Eq. (6), that

$$\kappa = 0. \tag{30}$$

5. FIELD GENERATED BY A SINGLE BODY

As another application, making use of the explicit surface-integral formula (24), and yielding another check of ambiguity parameters, we wish to compute the source-type multipole moments of a spherically symmetric extended body moving with *uniform* velocity. Remember that our formalism assumes, in principle, that we are dealing with regular, weakly self-gravitating bodies. We expect, because of the effacing properties of Einstein's theory [22], that our final physical results, especially when they are expressed as surface integrals like in (24), can be applied to more general sources, such as neutron stars or black holes. Indeed, we are going to confirm this expectation in the simplest possible case, that of an isolated spherically symmetric body which is known, by Birkhoff's theorem, to generate a universal exterior gravitational field, given by the Schwarzschild solution.

5.1. The boosted Schwarzschild solution

Following Ref. [29] we shall apply our formulas to a *boosted Schwarzschild solution* (BSS). Actually, in order to justify our use of the BSS (in standard harmonic coordinates), we must dispose of a small technicality. This technicality concerns the non-uniqueness of harmonic coordinates for the Schwarzschild solution, even under the assumption of stationarity (in the rest frame) and spherical symmetry. Indeed, under these assumptions, and starting from the usual Schwarzschild radial coordinate, say r_S, the (rest frame) radial coordinate of the most general harmonic coordinate system, say $r = k(r_S)$, must satisfy the differential equation (see, *e.g.*, Weinberg [31], page 181)

$$\frac{d}{dr_S}\left[\left(r_S^2 - \frac{2GM}{c^2}r_S\right)\frac{dk}{dr_S}\right] = 2k. \tag{31}$$

The standard solution of Eq. (31), which is considered in all textbooks such as [31], reads simply

$$r = k^{\text{standard}}(r_S) = r_S - \frac{GM}{c^2}. \tag{32}$$

In the black hole case, the solution (32) is the only one which is regular on the horizon, *i.e.* when $r_S = 2GM/c^2$. However, in the case of the external metric of an extended spherically symmetric body, regularity on the horizon is not a relevant issue. What is relevant is that the solution of the *external* problem (31) be smoothly matched to a *regular* solution of the corresponding *internal* problem. As usual, this matching determines a unique solution everywhere. In general, this unique, everywhere regular, solution will correspond, in the exterior of the body, to a particular case of the general, two-parameter solution of the second-order differential equation (31). The latter is of the form

$$r = k^{\text{general}}(r_S) = C_1 \left(r_S - \frac{GM}{c^2} \right) + C_2 k_2(r_S), \tag{33}$$

where $k_2(r_S)$ denotes the (uniquely defined) "radially decaying solution" of Eq. (31), and where C_1 and C_2 are two integration constants. Indeed, when considering the flat-space limit of Eq. (31), it is easily seen that there are two independent solutions which behave, when $r_S \to +\infty$, as r_S and r_S^{-2} respectively. An explicit expression for the decaying solution is [7]

$$k_2(r_S) = \frac{1}{r_S^2} F\left(2, 2, 4, \frac{2GM}{c^2 r_S}\right). \tag{34}$$

We can always normalize C_1 to the value $C_1 = 1$. Then, with the above definitions, C_2 has the dimension of a length cubed. By considering in more detail the matching of the general solution of the harmonically relaxed Einstein equations at the 2PN level (see, *e.g.*, the book by Fock [32], page 322), one easily finds that the second integration constant is of the order of $C_2 \sim (GM/c^2)^2 a$, where a denotes the radius of the extended body under consideration. It is also easily checked that the constant C_2 parametrizes, at the linearized order, a *gauge vector* of the form $\varphi^i \propto C_2 \partial_i(1/r)$, and can thus be referred to as a "gauge parameter".

Contrarily to the multipole moments of *stationary* sources, which are geometric invariants (and can be expressed as surface integrals on a sphere at spatial infinity), the source multipole moments defined in Ref. [28] (and re-expressed in Section 3 as surface integrals over spheres in some intermediate region, $a \ll r \ll cT$) are probably not geometric invariants. They are useful intermediate constructs, which allow one to compute physically invariant information, but their definition is linked to the choice of harmonic coordinates covering the source. There are also various *gauge multipoles* (denoted W_L, X_L, Y_L, Z_L in Ref. [28]) which will influence, at some non-linear order, the values of the two sequences of *physical multipoles*: I_L, J_L. Therefore, one should expect that, at some non-linear order, the physical multipoles I_L, J_L of a boosted general, harmonic-coordinate spherically symmetric metric will start to depend on the value of the gauge parameter C_2.

---

[7] Here $F(2,2,4,z)$ denotes a particular case of Gauss' hypergeometric function

$$F(\alpha, \beta, \gamma, z) = 1 + \frac{\alpha \beta}{\gamma} \frac{z}{1!} + \frac{\alpha(\alpha+1)\beta(\beta+1)}{\gamma(\gamma+1)} \frac{z^2}{2!} + \cdots.$$

Here, we are only interested in computing the quadrupole moment I_{ij} of a boosted general spherically symmetric metric. We shall see below that the index structure of I_{ij} will be provided by the STF tensor product of the boost velocity V^i with itself, denoted $V^{\langle i} V^{j \rangle}$ (assuming that the origin of the coordinates is at the initial position of the center of symmetry of the BSS). Therefore, any contribution to I_{ij} coming from the gauge parameter C_2 must contain, at least, the factors C_2 and $V^{\langle i} V^{j \rangle}$, and also the total mass M. Taking into account the dimensionality of $C_2 \sim (GM/c^2)^2 a$, which is that of a length cubed, it is easily seen that there is no way to generate such a contribution to I_{ij}. Therefore, we conclude that the source quadrupole moment of a boosted general, harmonic-coordinate spherically symmetric metric is strictly equal to the source quadrupole moment of a boosted *standard harmonic-coordinate* Schwarzschild solution, obtained by setting $C_2 = 0$ (and $C_1 = 1$) in (33), *i.e.* by choosing the standard harmonic radial coordinate (32).

In the following, we shall therefore consider only such a boosted Schwarzschild solution (BSS) in standard form. We shall sometimes refer to the source of this solution as a black hole (though, strictly speaking, one should always have in mind some extended spherical star). For simplicity, we shall translate the origin of the coordinate system so that it is located at the initial position of the black hole at coordinate time $t = 0$. With this choice of origin of the coordinates all the current-type moments J_L of the BSS are zero. We shall concentrate our attention on the mass-type quadrupole moment I_{ij}, that we shall compute at the 3PN order.

The BSS metric, written in terms of the Gothic metric deviation $h^{\mu\nu} = \sqrt{-g}\, g^{\mu\nu} - \eta^{\mu\nu}$, satisfying standard harmonic coordinates so $\partial_\nu h^{\mu\nu} = 0$, is best formulated in a manifestly Lorentz covariant way as

$$h^{\mu\nu} = \left(1 - \frac{\left(1 + \frac{GM}{c^2 r_\perp}\right)^3}{1 - \frac{GM}{c^2 r_\perp}}\right) u^\mu u^\nu - \frac{G^2 M^2}{c^4 r_\perp^2} n^\mu n^\nu, \qquad (35)$$

where u^μ is the time-like unit four-velocity of the center of symmetry of the BSS, where n^μ is the space-like unit vector pointing from the BSS to the field point along the direction *orthogonal* (in a Minkowskian sense) to the world line of the BSS, and where $r_\perp$ denotes the orthogonal distance to the world line (square root of the interval). However, in our explicit calculations (done with the software Mathematica) it is preferable to employ a more "coordinate-rooted" formulation of the BSS metric, which is of course completely equivalent to (though less elegant than) Eq. (35).

Le us denote by $x^\mu = (ct, \mathbf{x})$ the global reference frame, in which the black hole is moving, and by $X^\mu = (cT, \mathbf{X})$ the rest frame of the black hole — both x^μ and X^μ are assumed to be harmonic coordinates. Let $x^i(t)$ be the rectilinear and uniform trajectory of the center of symmetry of the BSS in the global coordinates x^μ, and $\mathbf{V} = (V^i)$ be the constant coordinate velocity of the BSS,

$$V^i \equiv \frac{dx^i(t)}{dt}. \qquad (36)$$

The rest frame X^μ is transformed from the global one x^μ by the Lorentz boost,

$$x^\mu = \Lambda^\mu{}_\nu(\mathbf{V}) X^\nu. \tag{37}$$

For simplicity we consider a pure Lorentz boost $\Lambda^\mu{}_\nu(\mathbf{V})$ without rotation of the spatial coordinates. As explained above, we can assume that the metric of the BSS in the rest frame X^μ takes the standard harmonic-coordinate Schwarzschild expression, which we write in terms of the Gothic metric deviation $H^{\mu\nu}$, satisfying $\partial_\nu H^{\mu\nu} = 0$. Hence,

$$H^{00} = 1 - \frac{\left(1 + \frac{GM}{c^2 R}\right)^3}{1 - \frac{GM}{c^2 R}}, \tag{38}$$

$$H^{i0} = 0, \tag{39}$$

$$H^{ij} = -\frac{G^2 M^2}{c^4 R^2} N^i N^j, \tag{40}$$

where M is the total mass, $R \equiv |\mathbf{X}|$ and $N^i \equiv X^i/R$. A well-known feature of the Schwarzschild metric in harmonic coordinates is that the spatial Gothic metric H^{ij} is made of a single quadratic-order term $\propto G^2$ as shown in Eq. (40). The Gothic metric deviation transforms like a Lorentz tensor so the metric of the BSS in the global frame x^μ reads as

$$h^{\mu\nu}(x) = \Lambda^\mu{}_\rho \Lambda^\nu{}_\sigma H^{\rho\sigma}(\Lambda^{-1} x), \tag{41}$$

in which the rest-frame coordinates have been expressed by means of the global ones, i.e. $X^\mu(x) = (\Lambda^{-1})^\mu{}_\nu x^\nu$, using the inverse Lorentz transformation. The only problem is to derive the explicit relations giving the rest-frame radial coordinate R and the unit direction N^i as functions of their global-frame counterparts r and n^i, of the global coordinate time t, and of the boost velocity V^i. For these relations we find

$$R = r\left[1 + c^2(\gamma^2 - 1)\left(\frac{t}{r}\right)^2 - 2\gamma^2 (Vn)\left(\frac{t}{r}\right) + \gamma^2 \frac{(Vn)^2}{c^2}\right]^{1/2}, \tag{42}$$

$$N^i = \frac{r}{R}\left[n^i - \gamma V^i \left(\frac{t}{r}\right) + \frac{\gamma^2}{\gamma+1} \frac{V^i}{c^2}(Vn)\right], \tag{43}$$

where $\gamma \equiv (1 - V^2/c^2)^{-1/2}$ and $(Vn) \equiv \mathbf{V} \cdot \mathbf{n} = V^j n^j$ is the usual Euclidean scalar product. The latter formulation of the BSS metric is well adapted to our calculations because we shall have to perform, for computing the source multipole moments, an integration over the *coordinate* three-dimensional spatial slice $\mathbf{x} \in \mathbb{R}^3$, with *coordinate* time $t = $ const, which is easily done using the explicit relations (42)–(43).

5.2. Quadrupole moment of a boosted Schwarzschild black hole

We compute the quadrupole moment I_{ij} of the BSS, following the prescriptions defined by Eq. (24). To this end we first expand $h^{\mu\nu}$ when $c \to +\infty$, taking into account all the c's present both in the expression of the rest frame metric $H^{\mu\nu}$ as well as those coming from the Lorentz transformation (41)–(43). In this process the boost velocity $\mathbf{V}$ is to be considered as a constant, "spectator", vector. Note in passing that, in the present problem, the characteristic size a of the source at time t is given by the displacement from the origin, $a \sim Vt$, where $V \equiv |\mathbf{V}|$, while the near-zone corresponds to $r \ll ct$. Therefore, the far near-zone, where we read off the multipole moments as some combination of expansion coefficients $\varphi_{a,b}^{\mu\nu}(\mathbf{n},t)$, is the domain $Vt \ll r \ll ct$. We have evidently to assume that $V \ll c$ for this region to exist.

We then first get the near-zone (or PN) expansion of the BSS metric, $\overline{h}^{\mu\nu}$, by expanding in inverse powers of c up to 3PN order. Next we compute the multipolar (or far) re-expansion of each of the PN coefficients when $r \to +\infty$ with $t = $ const. In this way we obtain what we have denoted by $\mathcal{M}(\overline{h}^{\mu\nu})$ in Eq. (17). In the BBS case it is evident that the far-zone expansion given by (17) involves simply some powers of $1/r$, without any logarithm of r.

With $\mathcal{M}(\overline{h}^{\mu\nu})$ in hand we have the coefficients of the various powers of $1/r$, and we obtain thereby the needed quantities $\Psi_{k,\ell}^{\mu\nu}$ defined by Eq. (21). It is then a simple matter to compute all the required angular averages present in the formula (24) and to obtain the following 3PN mass quadrupole moment of the BSS,

$$\mathrm{I}_{ij}^{\mathrm{BSS}} = Mt^2 V^{\langle i} V^{j\rangle} \left[1 + \frac{9}{14}\frac{V^2}{c^2} + \frac{83}{168}\frac{V^4}{c^4} + \frac{507}{1232}\frac{V^6}{c^6}\right]$$
$$+ \frac{4}{7}\frac{G^2 M^3}{c^6} V^{\langle i} V^{j\rangle} + \mathcal{O}\left(\frac{1}{c^8}\right). \tag{44}$$

The first term represents the standard Newtonian expression, augmented here by a bunch of relativistic corrections. (Recall that we have chosen the origin of the coordinate system at the initial location of the BSS at $t=0$.)

The last term in Eq. (44), with coefficient $\mathscr{C} = 4/7$, is the most interesting for our purpose. It is purely of 3PN order, and it contains the first occurence of the gravitational constant G, which therefore arises in the quadrupole of the BSS only at 3PN order. This term is interesting because it corresponds to one of the regularization ambiguities, due to an incompleteness of Hadamard's self-field regularization, which appears in the calculation of the mass-type quadrupole moment of point particle binaries at the 3PN order [12, 13]. As we see from Eq. (26) the associated ambiguity parameter is ζ, which represents in fact the analogue of the kinetic ambiguity parameter $\omega_{\mathrm{kinetic}}$ in the equations of motion, see Eq. (1). It is now clear that ζ can be determined from what we shall now call the *BSS limit* of a binary system, which consists of setting one of the masses of the binary to be *exactly zero*, say $m_2 = 0$.

We have obtained the BSS limit of the 3PN mass-type quadrupole moment of compact binaries computed for general binary orbits in Refs. [12, 13]. We have also inserted for the position of the first body $y_1^i = v_1^i t$ in order to conform with our choice for the origin of the coordinates. In this way we get

$$\mathrm{I}_{ij}^{\mathrm{BSS\,limit}} = m_1 t^2 v_1^{\langle i} v_1^{j\rangle} \left[1 + \frac{9}{14}\frac{v_1^2}{c^2} + \frac{83}{168}\frac{v_1^4}{c^4} + \frac{507}{1232}\frac{v_1^6}{c^6}\right]$$

$$+ \left(\frac{232}{63} + \frac{44}{3}\zeta\right) \frac{G^2 m_1^3}{c^6} v_1^{\langle i} v_1^{j \rangle} + \mathcal{O}\left(\frac{1}{c^8}\right). \tag{45}$$

The comparison of Eqs. (45) and (44) reveals a complete match between the two results if and only if we have the expected agreement between the masses, *i.e.* $M = m_1$, and the velocities, $v_1^i = V^i$ (since the velocity of the body remaining after taking the BSS limit should exactly be the boost velocity), and the ambiguity constant ζ takes the *unique* value

$$\zeta = -\frac{7}{33}. \tag{46}$$

Our conclusion, therefore, is that the ambiguity parameter ζ is uniquely determined by the BSS limit. Because of the close relation between the BSS limit with Lorentz boosts, it is clear that ζ is linked to the Lorentz-Poincaré invariance of the multipole moment formalism of Ref. [28] as applied to compact binary systems in [12, 13]. This link strongly suggests that the specific value (46) represents the only one for which the expression of the 3PN quadrupole moment is compatible with the Poincaré symmetry. In other words the present calculation indicates that the Poincaré invariance should correctly be incorporated into the laws of transformation of the source-type multipole moments for general extended PN sources as given by Eqs. (10)–(11) or (24)–(25).

Let us finally emphasize that Eq. (46) has been obtained here without using any regularization scheme for curing the divergencies associated with the self field of point particles. However, we find, very nicely, that the value for ζ is in agreement with the one derived in the problem of point particles binaries at 3PN order by means of the dimensional self-field regularization, Eq. (7). This shows in particular that dimensional regularization is able to correctly keep track of the global Poincaré invariance of the general relativistic description of isolated systems.

REFERENCES

1. C. Cutler, T.A. Apostolatos, L. Bildsten, L.S. Finn, E.E. Flanagan, D. Kennefick, D.M. Markovic, A. Ori, E. Poisson, G.J. Sussman, and K.S. Thorne. The last three minutes: Issues in gravitational-wave measurements of coalescing compact binaries. *Phys. Rev. Lett.*, 70:2984–2987, 1993.
2. C. Cutler and E.E. Flanagan. Gravitational waves from merging compact binaries: How accurately can one extract the binary's parameters from the inspiral waveform? *Phys. Rev. D*, 49:2658–2697, 1994.
3. Luc Blanchet, Thibault Damour, Bala R. Iyer, Clifford M. Will, and Alan. G. Wiseman. Gravitational radiation damping of compact binary systems to second post-newtonian order. *Phys. Rev. Lett.*, 74:3515–3518, 1995.
4. Luc Blanchet, Guillaume Faye, Bala R. Iyer, and Benoit Joguet. Gravitational-wave inspiral of compact binary systems to 7/2 post-newtonian order. *Phys. Rev. D*, 65:061501(R), 2002. Erratum Phys. Rev. D **71**, 129902(E) (2005).
5. Luc Blanchet, Thibault Damour, Gilles Esposito-Farèse, and Bala R. Iyer. Gravitational radiation from inspiralling compact binaries completed at the third post-newtonian order. *Phys. Rev. Lett.*, 93:091101, 2004.
6. P. Jaranowski and G. Schäfer. Third post-newtonian higher order adm hamilton dynamics for two-body point-mass systems. *Phys. Rev. D*, 57:7274–7291, 1998.
7. P. Jaranowski and G. Schäfer. Binary black-hole problem at the third post-newtonian approximation in the orbital motion: Static part. *Phys. Rev. D*, 60:124003–1–12403–7, 1999.

8. Luc Blanchet and Guillaume Faye. Equations of motion of point-particle binaries at the third post-newtonian order. *Phys. Lett. A*, 271:58, 2000.
9. Luc Blanchet and Guillaume Faye. General relativistic dynamics of compact binaries at the third post-newtonian order. *Phys. Rev. D*, 63:062005, 2001.
10. V.C. de Andrade, L. Blanchet, and G. Faye. Third post-newtonian dynamics of compact binaries: Noetherian conserved quantities and equivalence between the harmonic-coordinate and adm-hamiltonian formalisms. *Class. Quant. Grav.*, 18:753–778, 2001.
11. Luc Blanchet. Gravitational-wave tails of tails. *Class. Quant. Grav.*, 15:113–141, 1998. Erratum Class. Quant. Grav. **22**, 3381 (2005).
12. Luc Blanchet, Bala R. Iyer, and Benoit Joguet. Gravitational waves from inspiralling compact binaries: Energy flux to third post-newtonian order. *Phys. Rev. D*, 65:064005, 2002. Erratum Phys. Rev. D **71**, 129903(E) (2005).
13. Luc Blanchet and Bala R. Iyer. Hadamard regularization of the third post-newtonian gravitational wave generation of two point masses. *Phys. Rev. D*, 71:024004, 2004.
14. J. Hadamard. *Le problème de Cauchy et les équations aux dérivées partielles linéaires hyperboliques*. Hermann, Paris, 1932.
15. L. Schwartz. *Théorie des distributions*. Hermann, Paris, 1978.
16. G. 't Hooft and M. Veltman. *Nucl. Phys.*, B44:139, 1972.
17. T. Damour, P. Jaranowski, and G. Schäfer. Dimensional regularization of the gravitational interaction of point masses. *Phys. Lett. B*, 513:147–155, 2001.
18. Luc Blanchet, Thibault Damour, and Gilles Esposito-Farèse. Dimensional regularization of the third post-newtonian dynamics of point particles in harmonic coordinates. *Phys. Rev. D*, 69:124007, 2004.
19. Luc Blanchet, Thibault Damour, Gilles Esposito-Farèse, and Bala R. Iyer. Dimensional regularization of the third post-newtonian gravitational wave generation of two point masses. *Phys. Rev. D*, 71:124004, 2005.
20. Luc Blanchet and Guillaume Faye. Hadamard regularization. *J. Math. Phys.*, 41:7675–7714, 2000.
21. Luc Blanchet and Guillaume Faye. Lorentzian regularization and the problem of point-like particles in general relativity. *J. Math. Phys.*, 42:4391–4418, 2001.
22. T. Damour. Gravitational radiation and the motion of compact bodies. In N. Deruelle and T. Piran, editors, *Gravitational Radiation*, pages 59–144, Amsterdam, 1983. North-Holland Company.
23. S.M. Kopeikin. The equations of motion of extended bodies in general-relativity with conservative corrections and radiation damping taken into account. *Astron. Zh.*, 62:889–904, 1985.
24. L.P. Grishchuk and S.M. Kopeikin. Equations of motion for isolated bodies with relativistic corrections including the radiation reaction force. In J. Kovalevsky and V.A. Brumberg, editors, *Relativity in Celestial Mechanics and Astrometry*, pages 19–33, Dordrecht, 1986. Reidel.
25. Yousuke Itoh and Toshifumi Futamase. *Phys. Rev. D*, 68:121501(R), 2003.
26. Yousuke Itoh. *Phys. Rev. D*, 69:064018, 2004.
27. A. Einstein, L. Infeld, and B. Hoffmann. The gravitational equations and the problem of motion. *Ann. Math.*, 39:65–100, 1938.
28. Luc Blanchet. On the multipole expansion of the gravitational field. *Class. Quant. Grav.*, 15:1971–1999, 1998.
29. Luc Blanchet, Thibault Damour, and Bala R. Iyer. Surface-integral expressions for the multipole moments of post-newtonian sources and the boosted schwarzschild solution. *Class. Quant. Grav.*, 22:155, 2005.
30. Thibault Damour and Gilles Esposito-Farèse. Testing gravity to second post-newtonian order: A field theory approach. *Phys. Rev. D*, 53:5541–5578, 1996.
31. S. Weinberg. *Gravitation and Cosmology*. John Wiley, New York, 1972.
32. V.A. Fock. *Theory of space, time and gravitation*. Pergamon, London, 1959.

Half Century of Black-Hole Theory: From Physicists' Purgatory to Mathematicians' Paradise.

Brandon Carter

LuTh, Observatoire Paris-Meudon, France

Abstract. Although implicit in the discovery of the Schwarzschild solution 40 years earlier, the issues raised by the theory of what are now known as black holes were so unsettling to physicists of Einstein's generation that the subject remained in a state of semiclandestine gestation until his demise. That turning point – just half a century after Einstein's original foundation of relativity theory, and just half a century ago today – can be considered to mark the birth of black hole theory as a subject of systematic development by physicists of a new and less inhibited generation, whose enthusastic investigations have revealed structures of unforeseen mathematical beauty, even though questions about the physical significance of the concomitant singularities remain controversial.

Keywords: Black holes; Singularities
PACS: 04.70.-s; 04.20.Dw

INTRODUCTION: SCHWARZSCHILD'S UNWELCOME SOLUTION

This illustrated review is intended to provide a brief overview of the emergence, during the last half century, of the theory of ordinary (macroscopic 4-dimensional) black holes, considered as a phenomenon that (unlike the time reversed phenomenon of white holes) is manifestly of astrophysical importance in the real world. The scope of this review therefore does not cover quantum aspects such as the Bekenstein-Hawking particle creation effect, which is far too weak to be significant for the macroscopic black holes that are believed to actually exist in the observable universe. Nor does it cover the interesting mathematics of higher dimensional generalisations, a subject that is (for the time being) so far from relevance to the known physical world (in which – according to the second law of thermodynamics – the distinction between past and future actually matters) that its practitioners have formed a subculture in which the senior members seem to have forgotten (and their juniors seem never to have been aware of have been aware of) the distinction between black and white holes, as they have adopted a regretably misleading terminology whereby the adjective "black" is abusively applied to any brane system that is **hollow** – including the case of an ordinary (black or white) hole, which, to be systematic, should be classified as a (black or white) hollow zero brane of codimension 3.

The rapid general acceptance of the reality and importance of the positrons whose existence was implied by Dirac's 1928 theory of the electron is in striking contrast with the widespread resistance to recognition of the reality and importance of the black holes

FIGURE 1. Numerically simulated view of (isolated) spherical black hole illuminated only by uniform distant sky background from which light is received only for viewing angle $\alpha > \beta$ where, for an observer falling radially from rest at large distance, the angle β subtended by the hole will be given by the formula (45) obtained in the appendix.

whose existence was implied by Einstein's 1915 theory of gravity. It is symptomatic that black holes were not even named as such until more than half a century later. The sloth with which the subject has been developed over the years is illustrated by the fact that although the simplest black hole solution was already discovered (by Schwarzschild) in 1916, the simulation in Figure 1 of the present review (80 years later) provides what seems to the first serious reply to the very easy question of what it would actually look like, all by itself, with no illumination other than that from a uniform sky background.

Much of the responsibility for the delay in the investigation of the consequences of his own theory is attributable [1] to Einstein himself. Although his work had revolutionary implications, Einstein's instincts tended to be rather conservative. It was as a matter of necessity (to provide an adequate account first of electromagnetism and then of gravitation) rather than preference that Einstein introduced the radically new paradigms involved first in his theory of special relativity, just a hundred years ago, and then in the work on general relativity that came to fruition ten years later. When cherished prejudices were undermined by the consequences, Einstein was as much upset as any of his contempory colleagues. It could have been said of Albert Einstein (as it was said of his illustrious and like minded contempory, Arthur Eddington) that he was always profound, but sometimes profoundly wrong.

The most flagrant example was occasionned by Friedmann's prescient 1922 discovery of what is now known as the "big bang" solution of the general relativity equations, which Einstein refused to accept because it conflicted with his unreasonable prejudice in favor of a cosmological scenario that would be not only homogeneous (as actually sug-

FIGURE 2. John Wheeler with Robert Dicke (Princeton 1971).

gested by subsequently available data) but also static (as commonly supposed by earlier generations) despite the incompatibility (in thermal disequilibrium) of these alternative simplifications with each other and with the obvious observational consideration (known in cosmologically minded circles as the Cheseaux-Olbers paradox) that – between the stars – the night sky is dark. Einstein's incoherent attitude (reminiscent of the murder suspect who claimed to have an alibi as well as the excuse of having acted in self defense) lead him not only to tamper with his own gravitation equations by inclusion of the cosmological constant, but anyway to presume without checking that Friedmann's (actually quite valid) solution of the original version must have been mathematically erroneous.

Compared with his tendency to obstruct progress in cosmology, Einstein's conservatism was rather more excusable in the not so simple case of what are now known as black holes. It is understandable that (like Eddington) he should have been unwilling to explore the limitations on the validity of his theory that are indicated by the weird and singular – or as Thorne [1] puts it "outrageous" – features that emerge when strong field solutions of the general relativity equations are extrapolated too far into the non linear regime.

At the outset Einstein's interest in the spherical vacuum solution of his 1915 gravitational field equations was entirely restricted to the weak field regime, far outside the "horizon" at $r = 2m$ in the simple exact solution

$$ds^2 = r^2(d\theta^2 + \sin^2\theta\, d\phi^2) + dr^2/(1-2m/r) - (1-2m/r)dt^2, \qquad (1)$$

that was obtained within a year, but that was immediately orphanned by the premature death of its discoverer, Karl Schwarszschild, after which its embarrassing physical implications were hardly taken seriously by anyone – with the notable exception of

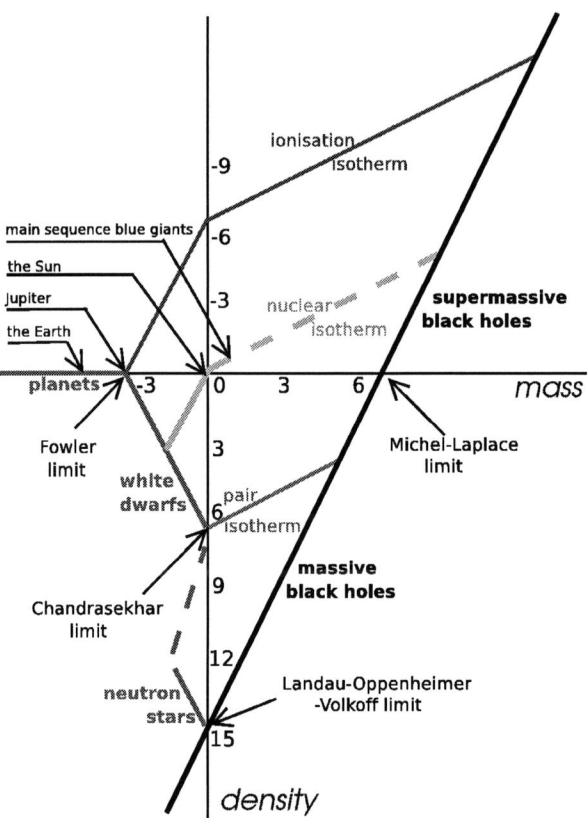

FIGURE 3. Logarithmic plot of density versus mass

Oppenheimer [2] – until the topic was taken up by a less inhibited generation subsequent to the death of Einstein himself, just half a century ago, at Princeton in 1955. It was only then (and there) that John Wheeler inaugurated the systematic development of the subject – for which he coined the name "black hole" theory – in a series of pionneering investigations that started [3] by addressing the crucial question of stability, while not long afterwards, on the other side of the "iron curtain" another nuclear arms veteran, Yacob Zel'dovich, initiated an independent approach [4] to the same problem (using the alternative name "frozen star" which in the end did not catch on).

OUTCOME OF STELLAR EVOLUTION: CHANDRA'S UNWELCOME LIMIT

The question of gravitational trapping of light had been raised in the eighteenth century by Michel and Laplace, whose critical mass $m \approx \rho^{-1/2}$ assumed the standard mass

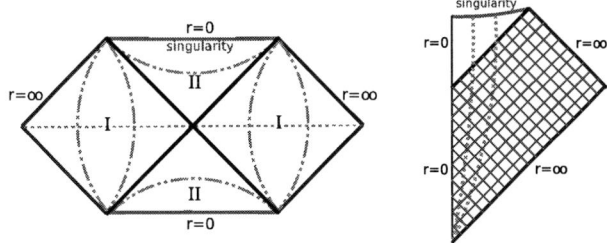

FIGURE 4. Conformal Projection diagrams showing firstly the combined black and white hole geometry obtained (belatedly [6] in 1960, by Wheeler in collaboration with Kruskal) as the artificial analytic (vacuum) extension of the Schwarzschild solution, and secondly a more astrophysically natural extension with homogeneous interior (found in 1939 [2] by Oppenheimer in collaboration with Sneyder) in which the shaded egion is the "domain of outer communications" and the unshaded region is the prototype example of a hole qualifiable as **black** in the strict sense.

density that is understood (on the basis of quantum theory as developped by 1930) to result in hadronic matter from balance between Fermi repulsion and electrostatic attraction which (in Planck units, with proton and electron masses $m_p \approx 10^{-19}$, $m_e \approx 10^{-22}$ gives $\rho \approx m_p n$ with $n \approx \lambda^{-3}$ for the Bohr radius $\lambda \approx e^2/m_e$ with $e^2 \simeq 1/137$. However most theorists refused to face the issue of gravitational collapse even after progress in quantum theory lead to Chandrasekhar's 1931 discovery of the maximum mass $m \approx m_p^{-2}$ for cold body – which is attained when relativistic gas pressure $P \approx n/\lambda \approx n^{4/3}$ provides the support required by virial condition $P \approx m_p^{2/3} \rho^{4/3}$.

For a lower mass $m \lesssim m_p^{-2}$, stellar evolution at finite temperature Θ, with gas pressure $P \approx n\Theta$ subject to $\rho \approx \Theta^3/m_p^3 m^2$, can terminate in cold equilibrium supported by non-relativistic Fermi pressure $P \approx n^{5/3}/m_e$ giving $\rho \approx m_e^3 m_p^5 m^2$ for a white dwarf, or $P \approx n^{5/3}/m_p$ giving $\rho \approx m_p^8 m^2$ for neutron star, as shown in Figure 3.

However a self gravitating mass of hot gas will be radiation dominated with $P \approx \Theta^4$, whenever its mass exceeds the Chandrasekhar limit, $m \gtrsim m_p^{-2}$, so that, as first understood by Chandra's Cambridge research director, Arthur Eddington, its condition for (thermally supported) equilibrium will be given by $\rho \approx \Theta^3/m^{1/2}$. What Chandra could never get Eddington to accept is that, for such a large mass, no cold equilibrium state will be available, so after exhaustion of fuel for thermonuclear burning (at $\Theta \approx e^4 m_p$) gravitational collapse will become inevitable.

SPHERICAL COLLAPSE PAST THE HORIZON

Eddington's example shows how, as has described in detail by Werner Israel [5] (and in striking contrast with the open mindedness of Michel and Laplace a century and a half earlier) physicists of Einstein's generation tried to convince themselves that nature would never allow compacification within a radius comparable to the Schwarzschild value. While Einstein lived, even after Chandrasekhar's discovery had shown that such

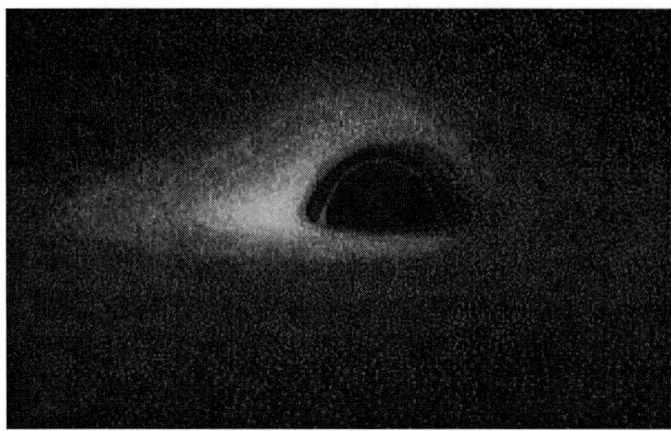

FIGURE 5. First realistic simulation of distant view of spherical black hole with thin accretion disc by Jean-Pierre Luminet, 1978.

a fate might often be difficult to avoid, the implications were taken seriously only by Oppenheimer and his colleagues, who showed [2] how, as shown in Figure 4, the solutions of Schwarzschild and Friedmann could be combined to provide a complete description of the collapse of a homogeneous spherical body through what is now called its event horizon all the way to a terminal singularity.

Despite the persuasion of such experienced physicists as Wheeler and Zel'dovich, and the mathematical progress due to younger geometers such as Robert Boyer and particularly Roger Penrose, the astrophysical relevance of the region near and within the horizon continued to be widely disbelieved until (and even after) the 1967 discovery [7] by Israel of the uniqueness of the Schwarzschild geometry as a static solution: many people (for a while including Israel himself [5]) still supposed (wrongly) that the horizon was an unstable artefact of exact spherical symmetry. It is therefore not surprising that the question of what such a black hole would actually look like was not addressed until much more recently, particularly considering that nothing would be seen at all without some source of illumination.

The realisation that many spectacular astrophysical phenomena ranging in scale from supermassive quasars in distant parts of the universe down to stellar mass X ray sources within our own galaxy may be attributed to accretion discs [8, 9, 10] round more or less massive black holes has however provided the motivation for increasingly realistic numerical simulations (Figures 5 and 6) of what would be seen from outside in the presence of an illuminating source of this kind [11, 12].

As the most easily calculable example, I have shown in the appendix how to work out the case shown in Figure 1 of an isolated spherical black hole for which the only source of illumination is a uniform distant sky background, viewed as a function of proper time,

$$\tau = -\frac{4m}{3}\left(\frac{r}{2m}\right)^{3/2}, \qquad (2)$$

FIGURE 6. Simulation of close up view of spherical black hole with thin accretion disc by Jean Alain Marck, 1996.

by a (doomed) observer falling towards the singularity inside the black hole, with zero energy and angular momentum.

In such a case the redshift Z determining the observed energy $\mathscr{E}/(1+Z)$ of a photon emitted from the sky background with the uniform average energy $\mathscr{E}$ say will be given by the formula

$$Z = -\frac{\cos\alpha}{\sqrt{r/2m}}, \qquad \alpha > \beta, \qquad (3)$$

where α is the apparent angle of reception, which must of course excede the apparent angle β subtended by the black hole. This means that the redshift will be positive (so that the sky will appear darker than normal) due to the Doppler effect, for photons coming in from behind the observer (with $\alpha > \pi/2$). However photons received in the range $\beta < \alpha < \pi/2$ will be blueshifted by an amount that will diverge, as shown in Figure 7, as the singularity is approached.

DISCOVERY OF HORIZON STABILITY AND OF KERR SOLUTION

Following the demise of Einstein (and the development of nuclear weapons) a new (less inhibited) generation of physicists, lead by Wheeler and Zel'dovich, came to recognise the likelihood – and need in any case for testing – of stability with respect to non-spherical perturbations of what was termed a "black hole". Work by Vishweshwara [13], Price [14], and others confirmed that "anything that can be radiated away will be radiated away" – leaving a final equilibrium state characterised only by mass and angular

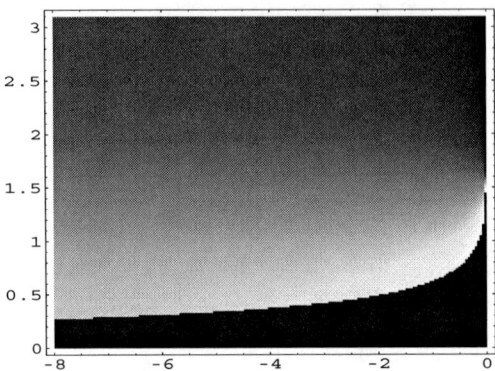

FIGURE 7. Plot of reception angle α against countdown proper time τ for arrival at singularity of radially falling zero-energy observer in units such that $2m = \sqrt{2/3}$ so that null orbit radius $r = 3m$ is crossed when $\tau = -1$. Constant brightness contours indicate intensity of light received from uniform background, which will be inversely proportional to 4th power of redshift factor $(1+Z)$, having unshifted value for rays arriving at angle $\pi/2$.

momentum. The (still open) mathematical question of the extent to which this remains true (with singularities hidden inside horizon) for very large deviations from sphericity was raised by the "cosmic censorship" conjecture formulated by Roger Penrose [15, 16] but in any case the relevance of black holes for astrophysical phenomena (notably quasars) was generally accepted in astronomical circles from 1970 onwards.

The generic form of what was afterwards recognised to be the final black hole equilibrium state state in question was discovered in 1963, when Roy Kerr announced [17, 18] that "among the solutions ... there is one which is stationary ... and also axisymmetric. Like the Schwarzschild metric, which it contains, it is type D ... m is a real constant ... The metric is

$$ds^2 = (r^2 + a^2\cos^2\theta)(d\theta^2 + \sin^2\theta \, d\phi^2) + 2(du + a\sin^2\theta \, d\phi)(dr + a\sin^2\theta \, d\phi)$$
$$- \left(1 - \frac{2mr}{r^2 + a^2\cos^2\theta}\right)(du + a\sin^2\theta \, d\phi)^2, \qquad (4)$$

where a is a real constant. This may be transformed to an asymptotically flat coordinate system ... we find that m is the Schwarzschild mass and ma the angular momentum ".

Since the black hole concept had still not been clearly formulated then, it was at first (wrongly) supposed that the physical relevance of this vacuum solution would be as the exterior to a compact self gravitating body like a neutron star, as suggested by Kerr's (off the mark) conclusion [17] that it would be "desirable to calculate an interior solution."

What actually makes the Kerr metric so important however, as can be see from Figure 9 (using C.P. diagrams, which were originally developed for this purpose) is the feature first clearly recognised [19, 20] by Bob Boyer in 1965, which is that for $a^2 \leq m^2$ the distant sky limit known as "asymptopia" is both visible and accessible only in a non-singular "domain of outer communications" bounded by past and future null (outer)

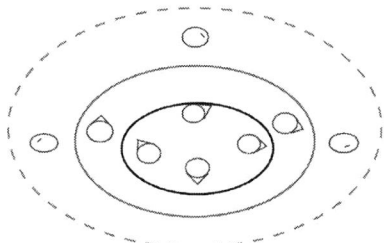

FIGURE 8. Sketch showing projections of null cones in equatorial space section of Kerr metric. The pale curve marks the "ergosurface" bounding the region where stationary motion is allowed, and the heavy black curve marks the horizon bounding the trapped "black hole" region.

horizons, on which $\Delta = 0$, where

$$r = m + c, \qquad c = \sqrt{m^2 - a^2} \tag{5}$$

The topology within the black (and white) hole regions was first elucidated [21] in terms of Conformal Projections on the symmetry axis in 1966 and then completely [22, 23, 24] by Boyer, Lindquist and myself in 1967 and 1968 – the year when the much needed term "black hole" was finally introduced by Wheeler to describe the region from which light cannot escape to "asymptopia". (A "white hole" region would be one that could not receive light from"asymptopia".) In the generic rotating case (unlike the static Schwarzschild limit) the well behaved domain outside the black hole horizon includes an "ergosphere" region where, as shown in Figure 8, the Killing vector generating the stationarity symmetry becomes spacelike, so that (globally defined) particle energies can be negative.

In contrast with the good behavior of the outer region, $r > m + c$, I found that, as well as having the irremovable ring shaped curvature singularity already noticed by Kerr where $r^2 + a^2\cos^2\theta \to 0$, , the inner parts of the rotating Kerr solutions would always be causally pathological, due to the existence near the ring singularity of a small region (see Figure 10) where the axial symmetry generating Killing vector becomes timelike[23, 25]. This feature gives rise to a causality violating "time machine region" (a feature so "outrageous" as to be unmentionable even by Thorne [1]) that would extend all the way out to "asymptopia" (meaning $r \to +\infty$) in the – presumably unphysical – case for which $a^2 > m^2$. (I would emphasize that this kind of time machine, like those recently considered by Ori [32], would survive even if one takes the covering space, unlike a time machine of the wormhole kind discussed by Thorne [1] which is merely an artefact of multiply connected space time topology).

In so far as the (physically relevant) black hole cases characterised by $a^2 \leq m^2$ are concerned, the good news [23, 25] (for believers in causality) is that the closed timelike lines are all contained within the inner region $r < m - c$. The boundary of the time machine region is constituted by the "inner horizon", where $r = m - c$. which acts as a Cauchy hypersurface from the point of view of inital data for formation of the black hole by gravitational collapse. Unlike the outer horizon $r = m+c$, whose stability throughout the allowed range $0 \leq a^2 < m^2$ has been even confirmed by Whiting [26], it was to be expected [27, 28] that a Cauchy horizon of the kind occurring at $r = m-c$ would be unstable, and it has been shown that outcome is likely to be the formation of a curvature singularity of the weak kind designated by the term "mass inflation" [29, 30, 31].

SEDUCTIVE MATHEMATICAL FEATURES OF KERR TYPE METRICS

In his original 1963 letter [17], and with Alfred Schild [18] in a sequel, Kerr obtained the useful alternative form

$$ds^2 = g_{\mu\nu} dx^\mu dx^\nu = \eta_{\mu\nu} + 2(m/U) n_\mu n_\nu \tag{6}$$

with **null** covector $n_\mu dx^\mu = du + a\sin^2\theta \, d\phi$, for $U = (r^2 + a^2\cos^2\theta)/r$, in a **flat** background. The latter was obtained in the Minkowski form,

$$\eta_{\mu\nu} dx^\mu dx^\nu = d\bar{x}^2 + d\bar{y}^2 + d\bar{z}^2 - d\bar{t}^2, \tag{7}$$

by setting $\bar{t} = u - r$, $\bar{z} = a\cos\theta$, $\bar{x} + i\bar{y} = (r - ia)e^{i\phi}\sin\theta$, which gave

$$n_\mu dx^\mu = d\bar{t} + \frac{\bar{z}d\bar{z}}{r} + \frac{(r\bar{x} - a\bar{y})d\bar{x} + (r\bar{y} + a\bar{x})d\bar{y}}{r^2 + a^2}. \tag{8}$$

(This form of pure vacuum solution was generalised to higher dimensions by Myers and Perry[33]. It is perhaps of greater current cosmological interest – in view of the evidence that the expansion of the universe is accelerating – that this form has also beeen extended to include a cosmological constant in a 4 dimensional **De Sitter** background by myself [24, 34], while further generalisations to a De Sitter background in 5 and higher dimensions [35, 36] have been obtained more recently.)

As well as time and axial symmmetry, the Kerr solution has a discrete PT symmetry that was predictable from Papapetrou's "circularity" theorem [37], and made manifest in 1967 [22] by the Boyer Lindquist transformation

$$dt = du - (r^2 + a^2)\Delta^{-1}dr, \quad d\varphi = -d\phi + a\Delta^{-1}dr, \tag{9}$$

with

$$\Delta = r^2 - 2mr + a^2. \tag{10}$$

This gives Kerr's null form as

$$n_\mu dx^\mu = dt - a\sin^2\theta \, d\varphi + \rho^2\Delta^{-1}dr, \quad \rho = \sqrt{r^2 + a^2\cos^2\theta},. \tag{11}$$

The metric itself is thereby obtained in the convenient form

$$ds^2 = \rho^2\left(\frac{dr^2}{\Delta} + d\theta^2\right) + (r^2 + a^2)\sin^2\theta \, d\varphi^2 + \frac{2mr}{\rho^2}(dt - a\sin^2\theta \, d\varphi)^2 - dt^2, \tag{12}$$

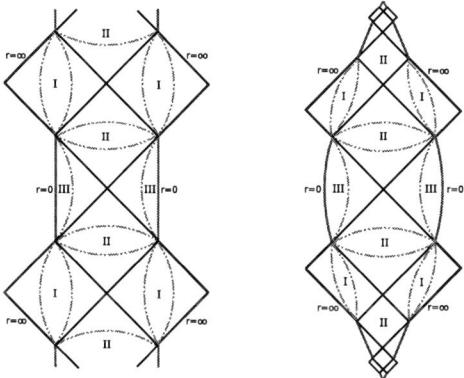

FIGURE 9. Representation of equatorial space time section of Kerr metric by Conformal Projection diagrams. (Such C.P. diagrams were originally developed for this purpose.) The second version achieves complete compactifiction by letting scale for successive universes tend to zero at extremities of chain.

in which there are cross terms involving the non-ignorable differentials, dr and $d\theta$, but – as the price for this simplification – if $a^2 \leq m^2$ there will be a removable coordinate singularity on the null "horizon" where Δ vanishes.

Whereas the possibility of making the foregoing simplification was predictable in advance, there was no reason to anticipate the discovery [23, 38] that, in addition to the ordinary "circular" symmetry generated by Killing vectors, $k^\mu \partial/\partial x^\mu = \partial/\partial t$ and $h^\mu \partial/\partial x^\mu = \partial/\partial \varphi$, the Kerr metric would turn out to have the hidden symmetry that is embodied in the **canonical** tetrad

$$g_{\mu\nu} = \sum_{i=1}^{3} \vartheta^{\hat{i}}_\mu \vartheta^{\hat{i}}_\nu - \vartheta^{\hat{0}}_\mu \vartheta^{\hat{0}}_\nu \tag{13}$$

specified by

$$\vartheta^{\hat{1}}_\mu dx^\mu = (\rho/\sqrt{\Delta})dr, \qquad \vartheta^{\hat{2}}_\mu dx^\mu = \rho\, d\theta, \tag{14}$$

$$\frac{\vartheta^{\hat{3}}_\mu dx^\mu}{\sin\theta} = \frac{(r^2+a^2)d\varphi - a\,dt}{\rho}, \qquad \frac{\vartheta^{\hat{0}}_\mu dx^\mu}{\sqrt{\Delta}} = \frac{dt - a\sin^2\theta\, d\varphi}{\rho}. \tag{15}$$

In terms of this canonical tetrad, the Kerr-Schild form of the metric is expressible as

$$g_{\mu\nu} = \eta_{\mu\nu} + 2mr(\vartheta^{\hat{0}}_\mu + \vartheta^{\hat{1}}_\mu)(\vartheta^{\hat{0}}_\nu + \vartheta^{\hat{1}}_\nu), \tag{16}$$

while the Killing-Yano 2-form brought to light by Roger Penrose and his coworkers is expressible as

$$f_{\mu\nu} = 2a\cos\theta\, \vartheta^{\hat{1}}_{[\mu} \vartheta^{\hat{0}}_{\nu]} - 2r\, \vartheta^{\hat{2}}_{[\mu} \vartheta^{\hat{3}}_{\nu]}. \tag{17}$$

The property of being a Killing-Yano 2-form means that it is such as to satisfy the very restrictive condition condition

$$\nabla_\mu f_{\nu\rho} = \nabla_{[\mu} f_{\nu\rho]}, \tag{18}$$

thus providing a symmetric solution

$$K_{\mu\nu} = f_{\mu\rho} f^\rho{}_\nu, \qquad \nabla_{(\mu} K_{\nu\rho)} = 0, \tag{19}$$

of the Eisenhart type Killing tensor equation, as well as secondary and primary solutions $\tilde{k}^\mu = K^\mu{}_\nu k^\nu = a^2 k^\mu + ah^\mu$ and $k^\mu = \frac{1}{6}\varepsilon^{\mu\nu\rho\sigma}\nabla_\nu f_{\rho\sigma}$ of the ordinary Killing vector equation $\nabla_{(\mu} k_{\nu)} = 0$.

For affine geodesic motion, $p^\nu \nabla_\nu p^\mu = 0$, one thus obtains (energy and axial angular momentum) constants $\mathscr{E} = k^\nu p_\nu$ and $\mathscr{M} = h^\nu p_\nu$, while the Killing tensor gives the constant $\mathscr{K} = K^{\mu\nu} p_\mu p_\nu = \ell_\mu \ell^\nu$, with (angular momentum) $\mathscr{J}_\mu = f_{\mu\nu} p^\mu$ obeying $p^\nu \nabla_\nu \ell_\mu = 0$.

There will also [40] be corresponding (self adjoint) operators

$$\mathscr{E} = ik^\nu \nabla_\nu, \qquad \mathscr{M} = ih^\nu \nabla_\nu, \qquad \mathscr{K} = \nabla_\mu K^{\mu\nu} \nabla_\nu, \tag{20}$$

whose action on a scalar field commute with that of the the Dalembertian $\Box = \nabla^\nu \nabla_\nu$: in other words $[\mathscr{E}, \Box] = 0$, $[\mathscr{M}, \Box] = 0$, and (consistently with the integrability condition $K^\rho{}_{[\mu} R_{\nu]\rho} = 0$) also $[\mathscr{K}, \Box] = 0$.

The ensuing integrability of the geodesic equation [23] and of the scalar wave equation is equivalent to their solubility by separation of variables[23, 38]. The possibility of extending these rather miraculous separability properties to the neutrino equation [41] and even to the massive spin 1/2 field [42, 43] as governed by the Dirac operator $\mathscr{D} = \gamma^\mu \nabla_\mu$ is attributable to corresponding spinor operator conservation laws

$$[\mathscr{E}, \mathscr{D}] = 0, \qquad [\mathscr{M}, \mathscr{D}] = 0, \qquad [\mathscr{J}, \mathscr{D}] = 0, \tag{21}$$

of energy, axial angular momentum, and (unsquared) total angular momentum, as respectively given [44] by

$$\mathscr{E} = ik^\nu \nabla_\nu + \tfrac{1}{4} i (\nabla_\mu k_\nu) \gamma^\mu \gamma^\nu, \qquad \mathscr{M} = ih^\nu \nabla_\nu + \tfrac{1}{4} i (\nabla_\mu h_\nu) \gamma^\mu \gamma^\nu, \tag{22}$$

and

$$\mathscr{J} = i\gamma^\mu (\gamma^5 f_\mu{}^\nu \nabla_\nu - k_\mu). \tag{23}$$

Such a neat commutation formulation is not (yet?) available for Teukolsky's extension [45, 46] of solubility by separation of variables to massless spin 1 and spin 2 fields representing electromagnetic and gravitation perturbations – of which the latter are particularly important for Bernard Whiting's demonstration [26] of stability. An even more difficult problem is posed by the charged generalisation [47] of the Kerr black hole metric, which retains many of its convenient properties (and is noteworthy for having the same gyromagnetic ratio as the Dirac electron [23, 48]) but which gives rise to a system of coupled electromagnetic and gravitational perturbations that has so far been found to be entirely intractable.

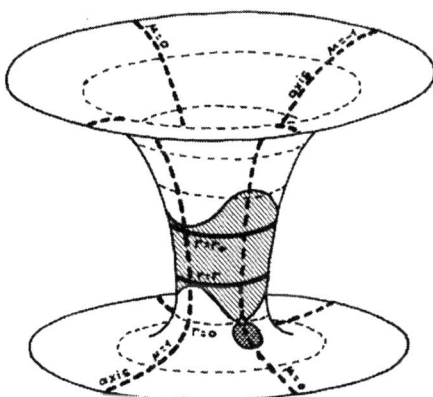

FIGURE 10. Reproduction from 1972 Les Houches notes [24]: sketch of $\{r,\theta\}$ section through ring singularity at junction between heavily shaded region responsible for causality violation where axisymmetry generator is spacelike, and lightly shaded ergo region where time summetry generator is spacelike.

NO HAIR AND UNIQUENESS THEOREMS FOR BLACK HOLE EQUILIBRIUM

The overwhelming importance of Kerr solution derives from its provision of the generic representation of the final outcome of gravitational collapse, as was made fairly clear in 1971 by the prototype **no-hair theorem** [49, 50] proving that no other vacuum black hole equilibrium state can be obtained by continuous axisymmetric variation from the spherical Schwarzschild solution that had been shown by the earlier work of Israel [7] (before the generic definition of a black hole was available) to be only static possibility.

Conceivable loopholes (such as doubts about the axisymmetry assumption) in the reasonning leading to this conclusion (which was rapidly – perhaps too uncritically – accepted in astronomical circles) were successively dealt with by the subsequent mathematical work of Stephen Hawking [51], David Robinson [52] and other more recent contributors [53, 54, 55] to what has by now become a rather complete and watertight uniqueness theorem for pure vaccum black hole solutions in 4 spacetime dimensions. It should however be remarked [56] that there are some mathematical loose ends (concerning assumptions of analyticity and causality) that still need to be tidied up.

The demonstration uses ellipsoidal coordinates for the 2-dimensional space metric $\mathrm{d}\hat{s}^2 = \mathrm{d}\lambda^2/(\lambda^2-c^2) + \mathrm{d}\mu^2/(1-\mu^2)$, in terms of which the generic stationary axisymmetric asymptotically flat vacuum metric is known from the work of Papapetrou [37] to be expressible in the form

$$\mathrm{d}s^2 = \rho^2 \mathrm{d}\hat{s}^2 + X(\mathrm{d}\varphi - \omega \mathrm{d}t)^2 - (\lambda^2-c^2)(1-\mu^2)\mathrm{d}t^2. \qquad (24)$$

for which, by the introduction of an Ernst [57] type potential given by $X^2 \partial \omega/\partial \lambda = (1-\mu^2)\partial Y/\partial \mu$, the relevant Einstein equations will be obtainable from the (positive definite) action $\int d\lambda \, d\mu (|\hat{\nabla} X|^2 + |\hat{\nabla} Y|^2)/X^2$.

The black hole equilibrium problem is thus [49, 24, 50] reduced to a non linear 2 dimensional elliptic boundary value problem for the scalars X, Y, subject to conditions of regularity on the horizon (with rigid angular velocity Ω) where $\lambda = c$ and to appropriate boundary conditions on the axis where $\mu = \pm 1$ and at large radius $\lambda \to \infty$ in terms of angular momentum ma.

The uniqueness theorem states that this 2 dimensional boundary problem has no solutions other that those given given (with $\lambda = r - m$, $\mu = \cos\theta$) by the Kerr solution having mass $m = \sqrt{c^2 + a^2}$ and horizon angular velocity $\Omega = a/2m(m+c)$. The proof is obtained from an identity equating a quantity that is a positive definite function of the relevant deviation (of some other hypothetical solution from the Kerr value) to a divergence whose surface integral can be seen to vanish by the boundary conditions.

The original no-hair theorem (applying just to the small deviation limit) was based on an infinitesimal divergence identity that I obtained by a hit and miss method [49] that was generalised by Robinson [52] to the finite difference divergence identity that was needed to complete the proof in the pure vacuum case. For the electromagnetic (Einstein Maxwell) generalisation, the analogous step from an infinitesimal no-hair theorem[58] to a fully non-linear uniquenes theorem was more difficult, and was not obtained until our hit and miss approach was superceded by the more sophisticated methods that were developed later on by Mazur [59, 60] and Bunting [61, 62].

FURTHER DEVELOPMENTS

After it had become clear that (in the framework of Einstein's theory) the Kerr solutions (with $a \leq m$) are the only vacuumm black hole equilibrium states, the next thing to be investigated was the way the black holes will evolve when the equilibrium is perturbed. A particularly noteworthy result, based on concepts (see Figure 11) developed in collaboration with Penrose [63] was the demonstration by Stephen Hawking [64, 51] that the area of a black hole horizon (which is proportional to what Christodoulou [65] had previously identified as irreducible mass) can never decrease. More particularly it was shown [66] that the area would grow, not only when the hole swallowed matter but more generally whenever the null generators of the horizon were subjected to shear. It was remarked that this effect could be described in terms of an effective viscosity and that the horizon could also be characterised [67, 68] by an effective resistivity.

Later astrophysical developments were concerned more with surrounding or infalling matter – for example in accretion discs – than with the black hole as such, at least until recently. However the prospect of detecting gravitational radiation in the foreseeable future has encouraged a resurgence of interest in purely gravitational effects, particularly those involved in binary coalescence. The climax of a coalescence is too complicated to be dealt with except by advanced methods of numerical computation, but the quasi stationary preliminary stages are more amenable [69, 70, 71, 72, 73, 74], as also are the final stages of ringdown, which can be analysed in terms of quasi normal modes (and their superpositions in power law tails of the kind first described by Price [14]) which have been the subject of considerable attention, particularly concerning the influence of rotation [75, 76, 77, 78, 79].

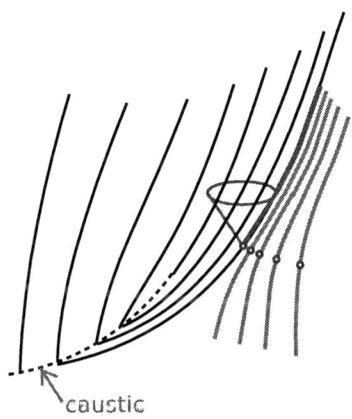

FIGURE 11. Sketch (in plot of time against space) of segment of black hole horizon showing how a null generator (obtained as limit of escaping timelike curves) can begin (on a caustic) but can never end towards the future.

APPENDIX: NULL GEODESICS IN SPHERICAL CASE

Although it would be insufficient for the complete Kruskal (black and white hole) extension, in order to cover a purely black hole (Oppenheimer Sneyder type) extension of the Schwarzschild solution, it will suffice to use an outgoing null coordinate patch of the kind introduced for the Kerr metric (4) for which, when $a = 0$, the metric will be given in terms of $x^0 = u$, $x^1 = r$, $x^2 = \theta$, $x^3 = \phi$, , simply by

$$ds^2 = -(1 - 2m/r)\,du^2 + 2\,du\,dr + r^2 d\theta^2 + \sin^2\theta\,d\phi^2. \tag{25}$$

Within such a system, an observer falling in freely from a large distance with zero energy and angular momentum will have a geodesic trajectory characterised by fixed values of the angle coordinates θ and ϕ and by a radial coordinate r that is given implicitly as a monotonically decreasing function of proper time by (2) and as a monotonically decreasing function of the ignorable coordinate u by the relation

$$\frac{u_0 - u}{2m} = \frac{2}{3}\sqrt{\frac{r}{2m}}\left(\frac{r}{2m} + 3\right) - \frac{r}{2m} - 2\ln\left\{1 + \sqrt{\frac{r}{2m}}\right\}, \tag{26}$$

in which u_0 is a constant of integration specifying the value of u for which the trajectory terminates at the singular limit $r \to 0$. For such a trajectory the (future oriented) timelike unit tangent vector will be given by

$$e^0_{(0)} = \left(1 + \sqrt{2m/r}\right)^{-1}, \qquad e^1_{(0)} = -\sqrt{2m/r}, \tag{27}$$

and the tetrad specifying a corresponding local reference frame can be completed in a natural manner by using the associated (outward oriented) orthogonal spacelike unit vector, which will be given by

$$e^0_{(1)} = \left(1+\sqrt{2m/r}\right)^{-1}, \qquad e^1_{(1)} = 1, \qquad (28)$$

together with two other horizontally oriented unit vectors whose specification will not matter for the present purpose because of the rotation symmetry of the system.

Let us consider the observation of photon that arrives with trajectory deviating by an angle α say from the outward radial direction. The first two components of its (null) momentum vector, p^μ say, will evidently be given in terms of its energy $\tilde{\mathcal{E}}$ with respect to such a frame by

$$p^i = \tilde{\mathcal{E}}(e^i_{(0)} + \cos\alpha\, e^i_{(1)}), \qquad (29)$$

for $i = 0, 1$. This locally observed energy is to be compared with the globally defined photon energy (as calibrated with respect to the asympotic rest frame at large distance) that will be given in terms of the timelike Killing vector with components $k^\mu = \delta^\mu_0$ by

$$\mathcal{E} = -k_\mu p^\mu = (1 - 2m/r)p^0 - p^1, \qquad (30)$$

the important feature of the latter being that it is conserved by the affine transport of the momentum vector along the null geodesic photon trajectory to which it is tangent. It can thus be seen that the corresponding locally observed energy $\tilde{\mathcal{E}}$ will be related to the globally defined energy constant $\mathcal{E}$ by

$$\mathcal{E} = \tilde{\mathcal{E}}(1+Z), \qquad (31)$$

with the redshift Z given by (3), and that the associated component ratio will be given by

$$\frac{p^1}{p^0} = (1+\sqrt{2m/r})\left(\frac{\cos\alpha - \sqrt{2m/r}}{\cos\alpha + 1}\right). \qquad (32)$$

As the (unsurprising) spherical limit of the (still rather mysterious) separability of Kerr's rotating generalisation, the evolution of the relevant affinely transported momentum components will be given in terms of the energy constant $\mathcal{E}$ and the associated squared angular momentum constant $\mathcal{K}$ [23] (using a dot for differentiation with respect to the affine parameter) by

$$p^0 = \dot{u} = \frac{\pm\sqrt{\mathcal{R}} + \mathcal{P}}{r^2 - 2mr}, \qquad p^1 = \dot{r} = \frac{\pm\sqrt{\mathcal{R}}}{r^2}, \qquad (33)$$

where

$$\mathcal{P} = \mathcal{E}r^2, \qquad \mathcal{R} = \mathcal{P}^2 - \mathcal{K}(r^2 - 2mr). \qquad (34)$$

We thereby obtain

$$\frac{p^0}{p^1} = \frac{du}{dr} = (1-2m/r)^{-1}\left(1 \pm \frac{\mathcal{P}}{\sqrt{\mathcal{R}}}\right), \qquad (35)$$

and hence, by comparison with (32),

$$\pm\frac{\sqrt{\mathscr{R}}}{\mathscr{P}} = \sqrt{r/2m}\left(\frac{\cos\alpha - \sqrt{2m/r}}{\sqrt{r/2m} - \cos\alpha}\right). \tag{36}$$

This expression can be used to evaluate the squared angular momentum constant as a function of the locally defined energy $\tilde{\mathscr{E}}$ and angle α in the form

$$\mathscr{K} = r^2\tilde{\mathscr{E}}^2\sin^2\alpha \tag{37}$$

in which the variable $\tilde{\mathscr{E}}$ will itself be given via (31) in terms of the globally defined energy constant $\mathscr{E}$ by the – red or blue – shift formula (3) so that for the square of the constant ratio of angular momentum to energy one obtains

$$\frac{\mathscr{K}}{\mathscr{E}^2} = \left(\frac{r\sin\alpha}{1 - \sqrt{2m/r}\cos\alpha}\right)^2. \tag{38}$$

With respect to an unconventional affine parameter orientation condition to the effect that the energy should always be non-negative, $\mathscr{E} \geq 0$, it can be seen (in view of the consideration that the squared angular momentum constant must necessarily be non-negative, $\mathscr{K} \geq 0$) that a null geodesic segment will be appropriately be describable as "incoming" or "outgoing" according to whether the upper or lower of the sign possibilities $\pm$ is applicable, i.e. according to whether the right hand side of (36) is positive or strictly negative. It is however to be remarked that this convention will be consistent with the usual requirement that the affine parameter orientation be future oriented, giving $\dot{u} \geq 0$, only outside the horizon and for "ingoing" null segments within the horizon, where $r < 2m$, but that for "outgoing" null segments within the horizon it would entail the opposite orientation convention, giving $\dot{u} \leq 0$. With respect to the usual parameter orientation condition giving $\dot{u} \geq 0$ the "outgoing" null segments within the horizon will need to be parametrised the other way round, which means that they will be characterised by negative energy $\mathscr{E} < 0$ and by the upper of the sign possibilities $\pm$.

Whichever convention is used, it can be seen that within the horizon the radius r will aways be a decreasing function of u, even for the (relatively) "outgoing" null segments, and that only an "incoming" null segment can cross the horizon at a finite value of u. It can be seen from (36) that outside the horizon (i.e. for $r > 2m$) the criterion for a null segment to be classified as "incoming" is that it should have $\cos\alpha \leq \sqrt{2m/r}$ and that the corresponding requirement within the horizon is $\cos\alpha \leq \sqrt{r/2m}$.

It can be deduced from the expression (34) that the function $\mathscr{R}$ will remain positive wherever r is positive if $\mathscr{K}/\mathscr{E}^2 \leq 27m^2$. In such a case, the null geodesic will either be permanently "incoming", proceding all the way from "infinity" (i.e. the limit $r \to \infty$) down to the internal singularity (i.e. the limit $r \to 0$), or else it will be permanently "outgoing", proceding all the way to the singularity or to infinity depending on whether it inside or outside the finite horizon radius value $r = 2m$ to which it extends in the infinite past, i.e. as $u \to -\infty$. As well the such "ingoing" and permanently "outgoing" possibilities, the critical case

$$\mathscr{K}/\mathscr{E}^2 = 27m^2 \tag{39}$$

includes also the exceptional possibility of a marginally outgoing – effectively "trapped" – null trajectory with fixed radius $r = 3m$.

When the angular momentum exceeds this critical value, i.e. if $\mathcal{K}/\mathcal{E}^2 > 27m^2$, there will be a forbidden range $r_- < r < r_+$ of values of r for which $\mathcal{R} < 0$. It can be seen from 34 that relevant limits are explicitly obtainable, as the non negative solutions of the cubic equation

$$\mathcal{E}^2 r_\pm^3 - \mathcal{K}(r_\pm - 2m) = 0, \tag{40}$$

in the form

$$r_\pm = 2\sqrt{\mathcal{K}/3\mathcal{E}^2}\cos\{\psi_\pm/3\}, \tag{41}$$

with

$$\psi_\pm = \pi \mp \arcsin\sqrt{1 - 27\mathcal{E}^2 m^2/\mathcal{K}}, \tag{42}$$

which evidently entails the conditions $2m < r_- < 3m < r_+$.

This means that for a value of $\mathcal{K}/\mathcal{E}^2$ above the critical bound (39) the possible null trajectories will be classifiable as "free" or "trapped". The "free" geodesics are initially "incoming" from "infinity" but become "outgoing" after reaching the inner bound at $r = r_+$ so as to remain in the range $r \geq r_+$. The "trapped" geodesics are either permanently "ingoing" within the horizon or else are initially "outgoing" from just outside the horizion but become "ingoing" after reaching the outer bound $r = r_-$ so as to remain within the range $r \leq r_-$.

For a position in the range $r \leq 3m$ the only kinds of "bright" geodesic, meaning those coming from the distant sky at "infinity", are of the permanently "incoming" kind characterised by $\mathcal{K}/\mathcal{E}^2 \leq 27m^2$. whereas for a position in the range $r \leq 3m$ there will also be "bright" geodesics of the "free" kind characterised by $\mathcal{K}/\mathcal{E}^2 > 27m^2$. Apart from the special case of the circular null geodesics at $r = 3m$, all the other kinds of null geodesic can be classified as "dark" since they can be seen to have emerged from near the horizon limit radius $r \to 2m$ in the distant past (the limit $u \to -\infty$) and so can be interpreted as trajectories of very highly redshifted radiation from the infalling matter that be presumed to originally formed the black hole whose static final state is under consideration here.

It can be seen that the ratio $\mathcal{K}/\mathcal{E}^2$ specified as a function of $\cos\alpha$ by (38) will be monotonically increasing in the "incoming" range, i.e. for $-1 \leq \cos\alpha < \sqrt{2m/r}$ where $r > 2m$ and for $-1 \leq \cos\alpha < \sqrt{r/2m}$ where $r < 2m$. At the upper end of this "incoming" range the ratio $\mathcal{K}/\mathcal{E}^2$ the tends to a maximum that will be finite – with value $r^3/(r-2m)$ – outside the horizon, but that will be infinite inside the black hole. The ratio $\mathcal{K}/\mathcal{E}^2$ will then decrease monotonically for the higher "outgoing" part of the range of $\cos\alpha$.

The critical value (39) will be attained for two values of $\cos\alpha$, of which the lower one, $\cos\alpha = \mathcal{X}_-$ say, will be in the "incoming" range, and the higher one, $\cos\alpha = \mathcal{X}_+$ will be in the "outgoing" range. It can be seen from (38) that these values will be obtainable as the upper and lower roots of

$$r^2(1 - \mathcal{X}_\pm^2) = 27m^2(1 - \sqrt{2m/r}\,\mathcal{X}_\pm)^2, \tag{43}$$

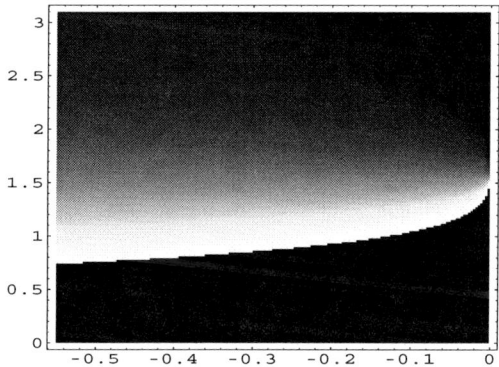

FIGURE 12. As in Figure 7 but with higher time resolution, showing just final stages after observer passes horizon at $r = 2m$ (which occurs when $\tau = -(2/3)^{3/2} \simeq 0.54$ in units used here) as the black hole's apparent angular radius β increases towards its upper limit $\pi/2$ when $r \to 0$.

which will be real and distinct except at $r = 3m$ where they will coincide. The range of angles characterising the "bright" geodesics will therefore be given by

$$-1 \leq \cos\alpha < \cos\beta \qquad (44)$$

(so that β will be interpretable as the apparent angular radius of the black hole) with a bounding value $\cos\beta$ that will be given by $\cos\beta = \mathscr{X}_-$, within the radius of the circular null trajectory, i.e. for $r < 3m$, while in the outer regions for which $r > 3m$ it will be given by $\cos\beta = \mathscr{X}_+$.

The required solutions of (43) are expressible in terms of the dimensionless variable $\bar{r} = r/2m$ by $\mathscr{X}_\pm = \left(27\sqrt{\bar{r}} \pm |2\bar{r}^2 - 3\bar{r}|\sqrt{\bar{r}^2 + 3\bar{r}}\right)/(4\bar{r}^3 + 27)$. It can thus be seen that (for the freely falling observer) the apparent angular size β of the black hole – as shown in the simulation of Figure 1, and as plotted against the proper time (2) in Figure 7 and Figure 12 – will be given as a function of the dimensionless radial variable $\bar{r} = r/2m$ by the analytic formula

$$\cos\beta = \frac{27\sqrt{\bar{r}} + (2\bar{r}^2 - 3\bar{r})\sqrt{\bar{r}^2 + 3\bar{r}}}{4\bar{r}^3 + 27}. \qquad (45)$$

REFERENCES

1. K.S. Thorne, *Black holes and time warps, Einstein's outrageous legacy*, (Norton, New York, 1994).
2. J.R. Oppenheimer, H. Snyder, "On continued gravitational contraction", *Phys. Rev.* **56** (1939) 455-459.
3. T. Regge, J.A. Wheeler, "Stability of a Schwarzschild singularity", *Phys. Rev.* **108** (1957) 1063-1069.
4. I.D. Novikov, Ya.B. Zel'dovich, "Physics of Rlativistic collapse", *Nuovo Cimento, Supp.* bf 4 (1966) 810, Add. 2.

5. W. Israel, "Dark stars: the evolution of an idea", in *300 years of gravitation*, ed S.W. Hawking, W. Israel (Cambridge U.P., 1987) 199-276.
6. M.D. Kruskal, "Minimal extension of the Schwarzschild metric" *Phys. Rev.* **119** (1960) 1743.
7. W. Israel, "Event horizons in static vacuum spacetimes", *Phys.Rev* **164** (1967) 1776-79.
8. J.M. Bardeen, "Rapidly rotating stars, disks, and black holes", in *Black Holes (Les Hoches72)* ed. B. and C. DeWitt (Gordon and Breach, New York, 1973) 241-289.
9. I.D. Novikov, K.S. Thorne,, "Astrophysics of black holes", in *Black Holes (Les Hoches72)* ed. B. and C. DeWitt (Gordon and Breach, New York, 1973) 343-450.
10. D.N. Page, K.S. Thorne, "Disk accretion onto a black hole. I Time averaged structure of accretion disk", *Astroph. J.* **191** (1974) 499-506.
11. J.P. Luminet, "Image of a spherical black hole with thin accretion disk", *Astron. Astroph.* **75** (1979) 228-235.
12. J.A. Marck, "Short cut method of solution of geodesic equations for Schwarzschild black hole", *Class. Quantum Grav.* **13** (1996) 393-402. [gr-qc/9505010]
13. C.V. Vishveshwara, "Stability of the Schwarschild metric", *Phys. Rev.* **D1** (1970) 2870-2879.
14. R.H. Price, "Nonspherical perturbations of relativistic gravitational collapse", *Phys. Rev.* **D5** (1972) 2419-2454.
15. R. Penrose, "Gravitational collapse and spacetime singularities", *Phys. Rev. Lett.* **14** (1965) 57-59.
16. R. Penrose, "Gravitational collapse: the role of general relativity", *Nuovo Cimento* **1** (1969) 252-276.
17. R. Kerr, "Gravitational field of a spinning mass as an example of algebraically special metrics", *Phys. rev. Let.* **11** (1963) 237-238.
18. R.P. Kerr and A. Schild, "Some algebraically degenerate solutions of Einstein's gravitational field equations", *Proc. Symp. Appl. Math.* **17**, (1965) 199.
19. R.H. Boyer, T.G. Price, "An interpretation of the Kerr metric in General Relativity", *Proc. Camb.Phil.Soc.* **61** (1965) 531-34.
20. R.H. Boyer, "Geodesic orbits and bifurcate Killing horizons", *Proc. Roy. Soc. Lond* **A311**, 245-52 (1969).
21. B. Carter, "Complete Analytic Extension of Symmetry Axis of Kerr's Solution of Einstein's Equations", *Phys. Rev.* **141** (1966), 1242-1247.
22. R.H. Boyer, R.W. Lindquist, "Maximal analytic extension of the Kerr metric", *J. Math. Phys.* **8** (1967) 265-81.
23. B. Carter, "Global Structure of the Kerr Family of Gravitational Fields",*Phys. Rev.* **174** (1968) 1559-71.
24. B. Carter, "Black hole equilibrium states", in *Black Holes (1972 Les Houches Lectures)*, eds. B.S. DeWitt and C. DeWitt (Gordon and Breach, New York, 1973) 57-210.
25. B. Carter, "Domains of Stationary Communications in Space-Time", B. Carter, *Gen. Rel. and Grav.* **9**, *pp* 437-450 (1978).
26. B. Whiting, "Mode stability of the Kerr blackhole", *J. Math. Phys.* **30** (1989) 1301-05.
27. M. Simpson, R. Penrose, *Int. J. Th. Phys.* **7** (173) 183.
28. S. Chandrasekha, J.B. Hartle, *Proc R. Soc. Lond.* **A284** (1982) 301.
29. E. Poisson, W. Israel, "Internal structure of black holes", *Phys. Rev.* **D41** (1990) 1796-1809.
30. A. Ori, "Structure of the singularity inside a realistic rotating black hole" *Phys. Rev. Lett* **68** (1992) 2117-2120.
31. A. Ori, E.E. Flanaghan, "How generic are null spacetime singularities?" *Phys. Rev.* **D53** (1996) 1754-1758. [gr-qc/9508066]
32. A. Ori, "A class of time-machine solutions with compact vacuum core", *Phys. Rev. Lett.* **95** (2005) 021101. [gr-qc/0503077]
33. R.C. Myers and M.J. Perry, "Black holes in higher dimensional space-times", *Ann. Phys.* **172** (1986) 304.
34. G.W. Gibbons, S.W. Hawking "Cosmological event horizons, thermodynamics, and particle creation", *Phys.Rev.* **D15** (1977) 2738-2751.
35. S.W. Hawking, C.J. Hunter and M.M. Taylor-Robinson, "Rotation and the AdS/CFT correspondence", *Phys. Rev.* **D59** (1999) 064005. [hep-th/9811056]
36. G.W. Gibbons, H. Lu, D.N. Page, C.N. Pope, "Rotating Black Holes in Higher Dimensions with a Cosmological Constant", *Phys. Rev. Lett.* **93** (2004) 171102. [hep-th/0409155]

37. A. Papapetrou, "Champs gravitationnels stationnaires à symmetrie axiale", *Ann. Inst. H. Poincaré* **4** (1986) 83-85.
38. B. Carter, "Hamilton-Jacobi and Schrodinger Separable Solutions of Einstein's Equations",*Commun. Math. Phys.* **10** (1968) 280-310.
39. R. Penrose, "Naked Singularities", *Ann. N.Y. Acad. Sci.* **224** (1973) 125-134
40. B. Carter, "Killing Tensor Quantum Numbers and Conserved Quantities in Curved Space", *Phys. Rev.* **D16** (1977) 3395-3414.
41. W. Unruh, "Separability of the neutrino equation in a Kerr background", *Phys. Rev. Lett.* **31** (1973) 1265-1267.
42. S. Chandrasekhar, "The solution of Dirac's equation in Kerr geometry" *Proc. Ror. Soc. Lond.* **A349** (1976) 571-575.
43. D. Page, "Dirac equation around a charged rotating black hole", *Phys. Rev.* **D14** (1976) 1509-1510.
44. B. Carter, R.G. McLenaghan, "Generalised Total Angular Momentum Operator for the Dirac Equation in Curved Space-Time", *Phys. Rev.* **D19** (1979) 1093-1097.
45. S.A. Teukolsky, "Perturbations of a rotating black hole, I: Fundamental equations for gravitational and electromagnetic perturbations ", *Astroph. J.* **185** (1973) 635-47.
46. W.H. Press, S.A. Teukolsky, "Perturbations of a rotating black hole, II: Dynamical stability of the Kerr metric ", *Astroph. J.* **185** (1973) 649-73.
47. E. Newman, E. Couch, K. Chinnapared, A. Exton, A. Prakash, R. Torrence, "Metric of a rotating charged mass", *J. Math. Phys.* **6** (1965) 918-919.
48. C. Reina, A. Treves, "Gyromagnetic ratio of Einstein-Maxwell fields", *Phys. Rev.* **D11** (1975) 3031-3032.
49. B. Carter, "An Axisymmetric Black Hole has only Two Degrees of Freedom", B. Carter, *Phys. Rev. Letters* **26** (1971) 331-33.
50. B. Carter, "Mechanics and equilibrium geometry of black holes, membranes, and strings", in *Black Hole Physics, (NATO ASI C364)* ed. V. de Sabbata, Z. Zhang (Kluwer, Dordrecht, 1992) 283-357. [hep-th/0411259]
51. S.W. Hawking "Black holes in General Relativity", *Commun. Math. Phys.* **25** (1972) 152-56.
52. D.C. Robinson, "Uniqueness of the Kerr black hole", *Phys, Rev. Lett.* **34** (1975) 905-06.
53. D. Sudarsky, R.M. Wald, "Mass formulas for stationary Einstein-Yang-Mills black holes and a simple proof of two staticity theorems", *Phys. Rev.* **D47** (1993) 5209-13. [gr-qc/9305023]
54. P.T Chrusciel, R.M. Wald, "Maximal hypersurfaces in stationary asymptotically flat spacetimes", *Comm. Math. Phys.* **163** (1994) 561-604. [gr-qc/9304009]
55. P.T. Chrusciel, R.M. Wald, "On the topology of stationary black holes", *Class. Quantum Grav.* **11** (1994) L147-52 . [gr-qc/9410004]
56. B. Carter, "Has the black hole equilibrium problem been solved?" in *General Relativity, Gravitation, and Relativistic Field Theories*, ed. T. Piran (World Scientific, Singapore, 1999) 136-165. [gr-qc/9712028]
57. F.J. Ernst, "New formulation of the axially symmetric gravitational field problem", *Phys. Rev.* **167** (1968) 1175-1178.
58. D.C. Robinson, "Classification of black holes with electromagnetic fields", *Phys, Rev.* **D10** (1974) 458-60.
59. P.O. Mazur, "Proof of uniqueness of the Kerr-Newman black hole solution", *J. Phys.* **A15** (1982) 3173-80.
60. P.O. Mazur, "Black hole uniqueness from a hidden symmetry of Einstein's gravity", *Gen. Rel Grav.* **16** (1984) 211-15.
61. G. Bunting, "Proof of the Uniqueness Conjecture for Black Holes,"(Ph.D. Thesis, Univ. New England, Armadale N.S.W., 1983).
62. B. Carter, "The Bunting Identity and Mazur Identity for non-linear Elliptic Systems including the Black Hole Equilibrium Problem", *Commun. Math. Phys.* **99** (1985) 563-91.
63. S.W. Hawking, R. Penrose, "The singularities of gravitational collapse and cosmology", *Proc. R. Soc. Lond.* **A314** (1969) 529-528.
64. S.W. Hawking "Gravitational radiation from colliding black holes", *Phys. Rev. Lett.* **26** (1971) 1344-1346.
65. D. Christodoulou, "Reversible and irreversible transformations in black hole physics", *Phys. Rev. Lett.* **25** (1970) 1596-1597.

66. S.W. Hawking, J.B. Hartle, "Energy and angular momentum flow into a black hole", *Commun. Math. Phys.* |bf 27 (1972) 283-290.
67. T. Damour, "Black hole eddy currents", *Phys. Rev.* **D18** (1978) 3598-3604.
68. R.L. Znajek, "Charged current loops around Kerr holes", *Mon. Not. R. Ast. Soc.* **182** (1978) 639-646.
69. A.D. Kulkarney, L.C. Shepley, J.W. York, "Initial data for black holes", *Phys. Lett.* **A96** (1983) 228-230.
70. P. Marronetti, R.A. Matzner, "Solving the initial value problem of two black holes", *Phys. Rev. Lett.* **85** (2000) 5500-5503.
71. G.B. Cook, "Initial data for binary black hole collisions", *Phys. Rev.* **D44** (1991) 2983-3000.
72. H.P. Pfeiffer, S.A. Teukolsky, G.B. Cook, "Quasi-circular Orbits for Spinning Binary Black Holes", *Phys. Rev.* **D62** (2000) 104018.
73. P. Grandclément, E. Gourgoulhon, S. Bonazzola, "Binary black holes in circular orbits." *Phys. Rev.* **D65** (2002) 044020, 044021. [gr-qc/0106016]
74. G.B. Cook, "Corotating and irrotational binary black holes in quasicircular orbits", *Phys. Rev.* **D65** (2002) 084003. [gr-qc/0108076]
75. E.W. Leaver, "An analytic representation for the quasi-normal modes of Kerr black holes", *Proc. R. Soc. Lond.* **A402** (1985) 285-298.
76. E. Seidel, S. Iyer, "Black hole normal modes: A WKB approach. IV Kerr black holes", *Phys. Rev.* **D41** (1990) 374-382.
77. K.D. Kokkotas, "Normal modes of the Kerr black hole", *Class. Quantum Grav.* **8** (1991) 2217-2224.
78. K. Onozawa, "Detailed study of quasinormal frequencies of the Kerr black hole", *Phys. Rev.* **D55** (1997) 3593-3602. [gr-qc/9610048].
79. W. Krivan, P. Laguna, P. Papadopoulos, N. Andersson, "Dynamics of perturbations of rotating black holes", *Phys. Rev. D56* (1997) 3395-3404.

100 Years of Relativity: Was Einstein 100 % Right ?

Thibault Damour

Institut des Hautes Etudes Scientifiques, 91440 Bures-sur-Yvette, France

Abstract. A review of the current experimental tests of General Relativity (including the binary-pulsar tests of the strong-field regime of gravity) is presented. Future developments in experimental gravity are briefly mentioned.

INTRODUCTION

To celebrate the centenary of Einstein's miraculous year 1905, it is interesting to review a century of experimental tests of Relativity (both Special and General). In this written version (based on [1]) of a talk given at the 28th Spanish Relativity Meeting (Oviedo, 6-10 September 2005) I will focus on the tests of General Relativity. For recent accounts of the tests of Special Relativity, I refer the reader to [2].

General Relativity is classically defined by two basic postulates. One postulate states that the Lagrangian density describing the propagation and self-interaction of the gravitational field is the Einstein-Hilbert one:

$$\mathscr{L}_{\text{Ein}}[g_{\mu\nu}] = \frac{c^4}{16\pi G_N} \sqrt{g} g^{\mu\nu} R_{\mu\nu}(g), \qquad (1)$$

where G_N denotes Newton's constant. [We use the 'mostly plus' signature.]

A second postulate states that $g_{\mu\nu}$ couples universally, and minimally, to all the fields of the Standard Model by replacing everywhere the Minkowski metric $\eta_{\mu\nu}$ ("Equivalence Principle"). Schematically (suppressing matrix indices and labels for the various gauge fields A_μ and fermions ψ, and for the Higgs doublet H), this leads to a (Standard Model) "matter" Lagrangian of the form:

$$\begin{aligned}
\mathscr{L}_{\text{SM}}[\psi, A_\mu, H, g_{\mu\nu}] &= -\tfrac{1}{4} \sum \sqrt{g} g^{\mu\alpha} g^{\nu\beta} F^a_{\mu\nu} F^a_{\alpha\beta} \\
&\quad - \sum \sqrt{g} \, \overline{\psi} \gamma^\mu D_\mu \psi \\
&\quad - \tfrac{1}{2} \sqrt{g} g^{\mu\nu} D_\mu H D_\nu H - \sqrt{g} V(H) \\
&\quad - \sum \lambda \sqrt{g} \, \overline{\psi} H \psi.
\end{aligned} \qquad (2)$$

From the total action $S_{\text{tot}}[g_{\mu\nu}, \psi, A_\mu, H] = c^{-1} \int d^4x (\mathscr{L}_{\text{Ein}} + \mathscr{L}_{\text{SM}})$ follows Einstein's field equations,

$$R_{\mu\nu} - \tfrac{1}{2} R g_{\mu\nu} = \frac{8\pi G_N}{c^4} T_{\mu\nu}. \qquad (3)$$

Here, $R = g^{\mu\nu}R_{\mu\nu}$, $T_{\mu\nu} = g_{\mu\alpha}g_{\nu\beta}T^{\alpha\beta}$, and $T^{\mu\nu} = (2/\sqrt{g})\delta\mathscr{L}_{SM}/\delta g_{\mu\nu}$ is the (symmetric) energy-momentum tensor of the Standard Model matter.

We shall only consider the classical limit of gravitation (*i.e.*, classical matter and classical gravity). Considering quantum matter in a classical gravitational background already poses interesting challenges, notably the possibility that the zero-point fluctuations of the matter fields generate a nonvanishing vacuum energy density ρ_{vac}, corresponding to a term $-\sqrt{g}\,\rho_{vac}$ in $\mathscr{L}_{SM}$ [5]. This is equivalent to adding a "cosmological constant" term $+\Lambda g_{\mu\nu}$ on the left-hand side of Einstein's equations Eq. (3), with $\Lambda = 8\pi G_N \rho_{vac}/c^4$. Recent cosmological observations suggest a positive value of Λ corresponding to $\rho_{vac} \approx (2.3 \times 10^{-3} \text{eV})^4$. Such a small value has a negligible effect on the tests discussed below.

EXPERIMENTAL TESTS OF THE COUPLING BETWEEN MATTER AND GRAVITY

The universality of the coupling between $g_{\mu\nu}$ and the Standard Model matter postulated in Eq. (2) ("Equivalence Principle") has many observable consequences. First, it predicts that the outcome of a local non-gravitational experiment, referred to local standards, does not depend on where, when, and in which locally inertial frame the experiment is performed. This means, for instance, that local experiments should neither feel the cosmological evolution of the universe (constancy of the "constants"), nor exhibit preferred directions in spacetime (isotropy of space, local Lorentz invariance). These predictions are consistent with many experiments and observations. The best limit on a possible time variation of the basic coupling constants concerns the fine-structure constant α_{em}, and has been obtained by analyzing a natural fission reactor phenomenon which took place at Oklo, Gabon, two billion years ago [6]. A conservative estimate of the (95% C.L.) Oklo limit on the variability of α_{em} is (see second reference in [6])

$$-0.9 \times 10^{-7} < \frac{\alpha_{em}^{Oklo} - \alpha_{em}^{now}}{\alpha_{em}} < 1.2 \times 10^{-7}, \quad (4)$$

which corresponds to the following limit on the average time derivative of α_{em}

$$-6.7 \times 10^{-17} \text{yr}^{-1} < \dot{\alpha}_{em}/\alpha_{em} < 5.0 \times 10^{-17} \text{yr}^{-1}. \quad (5)$$

A comparable limit ($\Delta\alpha_{em}/\alpha_{em} = (8 \pm 8) \times 10^{-7}$ over 4.6 Gyr) has been placed by using meteoritic data to constrain nuclear decay rates back to the time of solar system formation [7]. Direct laboratory limits on the time variation of α_{em} have recently reached the level $\dot{\alpha}_{em}/\alpha_{em} = (-0.9 \pm 2.9) \times 10^{-15} \text{yr}^{-1}$ by combining measurements testing the relative stability of different atomic clock frequencies over several years [8]. The announcement [9] of the detection, in quasar absorption lines, of a significant variation (between redshifts $z \simeq 0.5-3.5$ and now) of α_{em}, at the $\sim 10^{-5}$ level, has attracted a lot of interest. Several recent results [10] [11] yield only upper bounds on the cosmological variation of α_{em}, e.g. $\Delta\alpha_{em}/\alpha_{em} = (-6 \pm 6) \times 10^{-7}$ corresponding to an average time derivative $-2.5 \times 10^{-16} \text{yr}^{-1} < \Delta\alpha_{em}/\alpha_{em}\Delta t < 1.2 \times 10^{-16} \text{yr}^{-1}$ [11]. See Ref. [12] for a general review of the issue of "variable constants".

Some of the highest precision tests of the isotropy of space have been performed by looking for possible quadrupolar shifts of nuclear energy levels [13]. The (null) results can be interpreted as testing the fact that the various pieces in the matter Lagrangian Eq. (2) are indeed coupled to one and the same external metric $g_{\mu\nu}$ to the 10^{-27} level. Similarly, several recent experiments putting bounds on violation of Lorentz invariance can be viewed as tests of the Equivalence Principle (see [2] for recent reviews).

The universal coupling to $g_{\mu\nu}$ postulated in Eq. (2) implies that two (electrically neutral) test bodies dropped at the same location and with the same velocity in an external gravitational field fall in the same way, independently of their masses and compositions. The universality of the acceleration of free fall has been verified at the 10^{-12} level both for laboratory bodies [14],

$$\left(\frac{\Delta a}{a}\right)_{\text{BeCu}} = (-1.9 \pm 2.5) \times 10^{-12}, \tag{6}$$

and for the gravitational accelerations of the Moon and the Earth toward the Sun [15],

$$\left(\frac{\Delta a}{a}\right)_{\text{EarthMoon}} = (-1.0 \pm 1.4) \times 10^{-13}. \tag{7}$$

Refs. [16] [17] have also obtained the limit (where $f_e - f_m = 0.281$)

$$\frac{\Delta a}{a} = \frac{1}{f_e - f_m}\frac{\Delta a_{\text{CD}}}{a} = (3.6 \pm 5.0(\text{stat}) \pm 0.7(\text{syst})) \times 10^{-13}, \tag{8}$$

on the fractional difference in free-fall acceleration toward the Sun of earth-core-like and moon-mantle-like laboratory bodies. See also Ref. [18] for *short-range* tests of the universality of free-fall.

Finally, Eq. (2) also implies that two identically constructed clocks located at two different positions in a static external Newtonian potential $U(\boldsymbol{x}) = \sum G_N m/r$ exhibit, when intercompared by means of electromagnetic signals, the (apparent) difference in clock rate,

$$\frac{\tau_1}{\tau_2} = \frac{\nu_2}{\nu_1} = 1 + \frac{1}{c^2}[U(\boldsymbol{x}_1) - U(\boldsymbol{x}_2)] + O\left(\frac{1}{c^4}\right), \tag{9}$$

independently of their nature and constitution. This universal gravitational redshift of clock rates has been verified at the few percent level by flying atomic clocks on airplanes [19], and at the 10^{-4} level by comparing a hydrogen-maser clock, flying on a rocket up to an altitude $\sim 10,000$ km, to a similar clock on the ground [20]. The Global Positioning System would not work without carefully taking into account the gravitational and motional frequency shifts of the atomic clocks aboard the satellites [21]. For more details and references on experimental gravity see, *e.g.*, Refs. [22] [23].

TESTS OF THE DYNAMICS OF THE GRAVITATIONAL FIELD IN THE WEAK FIELD REGIME

The effect on matter of one-graviton exchange, *i.e.*, the interaction Lagrangian obtained when solving Einstein's field equations Eq. (3) written in, say, the harmonic gauge at first order in $h_{\mu\nu}$,

$$\Box h_{\mu\nu} = -\frac{16\pi G_N}{c^4}(T_{\mu\nu} - \tfrac{1}{2}T\eta_{\mu\nu}) + O(h^2) + O(hT), \tag{10}$$

reads $-(8\pi G_N/c^4)T^{\mu\nu}\Box^{-1}(T_{\mu\nu} - \tfrac{1}{2}T\eta_{\mu\nu})$. For a system of N moving point masses, with free Lagrangian $L^{(1)} = \sum_{A=1}^{N} -m_A c^2 \sqrt{1 - v_A^2/c^2}$, this interaction, expanded to order v^2/c^2, reads (with $r_{AB} \equiv |\mathbf{x}_A - \mathbf{x}_B|$, $\mathbf{n}_{AB} \equiv (\mathbf{x}_A - \mathbf{x}_B)/r_{AB}$)

$$\begin{aligned} L^{(2)} &= \tfrac{1}{2}\sum_{A\neq B} \frac{G_N m_A m_B}{r_{AB}} \left[1 + \tfrac{3}{2c^2}(v_A^2 + v_B^2) - \tfrac{7}{2c^2}(\mathbf{v}_A \cdot \mathbf{v}_B) \right. \\ &\quad \left. - \frac{1}{2c^2}(\mathbf{n}_{AB}\cdot\mathbf{v}_A)(\mathbf{n}_{AB}\cdot\mathbf{v}_B) + O\left(\frac{1}{c^4}\right) \right]. \end{aligned} \tag{11}$$

The two-body interactions Eq. (11) exhibit v^2/c^2 corrections to Newton's $1/r$ potential induced by spin-2 exchange. Consistency at the "post-Newtonian" level $v^2/c^2 \sim G_N m/rc^2$ requires that one also consider the three-body interactions induced by some of the three-graviton vertices, and other nonlinearities (terms $O(h^2)$ and $O(hT)$ in Eq. (10)),

$$L^{(3)} = -\frac{1}{2}\sum_{B\neq A\neq C} \frac{G_N^2 m_A m_B m_C}{r_{AB} r_{AC} c^2} + O\left(\frac{1}{c^4}\right). \tag{12}$$

All currently performed gravitational experiments in the solar system, including perihelion advances of planetary orbits, the bending, delay and frequency shift of electromagnetic signals passing near the Sun, and very accurate ranging data to the Moon obtained by laser echoes, are compatible with the post-Newtonian results Eqs. Eq. (10)–(12).

Similarly to what is done in discussions of precision electroweak experiments, it is useful to quantify the significance of precision gravitational experiments by parameterizing plausible deviations from General Relativity. The addition of a mass-term in Einstein's field equations leads to a score of theoretical difficulties [24], which have not yet received any consensual solution. We shall, therefore, not consider here the ill-defined "mass of the graviton" as a possible deviation parameter from General Relativity. Deviations from Einstein's pure spin-2 theory are then defined by adding new bosonic light or massless macroscopically coupled fields. The possibility of new gravitational-strength couplings leading (on small, and possibly large, scales) to deviations from Einsteinian (and Newtonian) gravity is suggested by String Theory [25], and by Brane World ideas [26]. For compilations of experimental constraints on Yukawa-type additional interactions, see Refs. [14] [27] [28]. Recent experiments have set limits on non-Newtonian forces below 0.1 mm [29].

Here, we shall focus on the parametrization of long-range deviations from relativistic gravity obtained by adding a massless scalar field φ, coupled to the trace of the energy-momentum tensor $T = g_{\mu\nu}T^{\mu\nu}$ [30]. The most general such theory contains an arbitrary function $a(\varphi)$ of the scalar field, and can be defined by the Lagrangian

$$\mathcal{L}_{\text{tot}}[g_{\mu\nu},\varphi,\psi,A_\mu,H] = \frac{c^4}{16\pi G}\sqrt{g}(R(g) - 2g^{\mu\nu}\partial_\mu\varphi\partial_\nu\varphi) \\ + \mathcal{L}_{\text{SM}}[\psi,A_\mu,H,\widetilde{g}_{\mu\nu}], \qquad (13)$$

where G is a "bare" Newton constant, and where the Standard Model matter is coupled not to the "Einstein" (pure spin-2) metric $g_{\mu\nu}$, but to the conformally related ("Jordan-Fierz") metric $\widetilde{g}_{\mu\nu} = \exp(2a(\varphi))g_{\mu\nu}$. The scalar field equation $\Box_g \varphi = -(4\pi G/c^4)\alpha(\varphi)T$ displays $\alpha(\varphi) \equiv \partial a(\varphi)/\partial \varphi$ as the basic (field-dependent) coupling between φ and matter [31]. The one-parameter (ω) Jordan-Fierz-Brans-Dicke theory [30] is the special case $a(\varphi) = \alpha_0\varphi$, leading to a field-independent coupling $\alpha(\varphi) = \alpha_0$ (with $\alpha_0^2 = 1/(2\omega+3)$).

In the weak field, slow motion limit, appropriate to describing gravitational experiments in the solar system, the addition of φ modifies Einstein's predictions only through the appearance of two "post-Einstein" dimensionless parameters: $\overline{\gamma} = -2\alpha_0^2/(1+\alpha_0^2)$ and $\overline{\beta} = +\frac{1}{2}\beta_0\alpha_0^2/(1+\alpha_0^2)^2$, where $\alpha_0 \equiv \alpha(\varphi_0)$, $\beta_0 \equiv \partial\alpha(\varphi_0)/\partial\varphi_0$, φ_0 denoting the vacuum expectation value of φ. These parameters also show up naturally (in the form $\gamma_{\text{PPN}} = 1+\overline{\gamma}$, $\beta_{\text{PPN}} = 1+\overline{\beta}$) in phenomenological discussions of possible deviations from General Relativity [32][22]. The parameter $\overline{\gamma}$ measures the admixture of spin 0 to Einstein's graviton, and contributes an extra term $+\overline{\gamma}(\mathbf{v}_A - \mathbf{v}_B)^2/c^2$ in the square brackets of the two-body Lagrangian Eq. (11). The parameter $\overline{\beta}$ modifies the three-body interaction Eq. (12) by a factor $1+2\overline{\beta}$. Moreover, the combination $\eta \equiv 4\overline{\beta} - \overline{\gamma}$ parameterizes the lowest order effect of the self-gravity of orbiting masses by modifying the Newtonian interaction energy terms in Eq. (11) into $G_{AB}m_A m_B/r_{AB}$, with a body-dependent gravitational "constant" $G_{AB} = G_N[1+\eta(E_A^{\text{grav}}/m_A c^2 + E_B^{\text{grav}}/m_B c^2) + O(1/c^4)]$, where $G_N = G\exp[2a(\varphi_0)](1+\alpha_0^2)$, and where E_A^{grav} denotes the gravitational binding energy of body A.

Some of the current limits on the post-Einstein parameters $\overline{\gamma}$ and $\overline{\beta}$ are (at the 68% confidence level):

$$\overline{\gamma} = (2.1 \pm 2.3) \times 10^{-5},$$

deduced from the additional Doppler shift experienced by radio-wave beams connecting the Earth to the Cassini spacecraft when they passed near the Sun [33],

$$\overline{\gamma} = (-1.7 \pm 4.5) \times 10^{-4},$$

from very-long-baseline interferometric measurements of the deflection of radio waves by the Sun [34], and

$$4\overline{\beta} - \overline{\gamma} = (4.4 \pm 4.5) \times 10^{-4},$$

from Lunar Laser Ranging measurements [15] of a possible polarization of the Moon toward the Sun [35]. More stringent limits on $\overline{\gamma}$ are obtained in models (e.g., string-inspired ones [36]) where scalar couplings violate the Equivalence Principle.

Among other experimental confirmations of weak-field general relativistic predictions let us mention: (i) the advance of Mercury's perihelion [37], (ii) the (geodetic) precession of the lunar orbit with respect to a barycentric frame [15], (iii) the dragging of the orbital planes of the LAGEOS satellites caused by the spin of the Earth [38].

TESTS OF THE DYNAMICS OF THE GRAVITATIONAL FIELD IN THE RADIATIVE AND/OR STRONG FIELD REGIMES

The discovery of pulsars (*i.e.*, rotating neutron stars emitting a beam of radio noise) in gravitationally bound orbits [39] [40] has opened up an entirely new testing ground for relativistic gravity, giving us an experimental handle on the regime of radiative and/or strong gravitational fields. In these systems, the finite velocity of propagation of the gravitational interaction between the pulsar and its companion generates damping-like terms at order $(v/c)^5$ in the equations of motion [41]. These damping forces are the local counterparts of the gravitational radiation emitted at infinity by the system ("gravitational radiation reaction"). They cause the binary orbit to shrink and its orbital period P_b to decrease. The remarkable stability of the pulsar clock has allowed Taylor and collaborators to measure the corresponding very small orbital period decay $\dot{P}_b \equiv dP_b/dt \sim -(v/c)^5 \sim -10^{-12}$ [40] [42] [43], thereby giving us a direct experimental confirmation of the propagation properties of the gravitational field, and, in particular, an experimental confirmation that the speed of propagation of gravity is equal to the velocity of light to better than a part in a thousand [44]. In addition, the surface gravitational potential of a neutron star $h_{00}(R) \simeq 2Gm/c^2R \simeq 0.4$ being a factor $\sim 10^8$ higher than the surface potential of the Earth, and a mere factor 2.5 below the black hole limit ($h_{00} = 1$), pulsar data are sensitive probes of the strong-gravitational-field regime.

Binary pulsar timing data record the times of arrival of successive electromagnetic pulses emitted by a pulsar orbiting around the center of mass of a binary system. After correcting for the Earth motion around the Sun, and for the dispersion due to propagation in the interstellar plasma, the time of arrival of the Nth pulse t_N can be described by a generic, parameterized "timing formula," [45] whose functional form is common to the whole class of tensor-scalar gravitation theories:

$$t_N - t_0 = F[T_N(v_p, \dot{v}_p, \ddot{v}_p); \{p^K\}; \{p^{PK}\}]. \quad (14)$$

Here, T_N is the pulsar proper time corresponding to the Nth turn given by $N/2\pi = v_p T_N + \frac{1}{2}\dot{v}_p T_N^2 + \frac{1}{6}\ddot{v}_p T_N^3$ (with $v_p \equiv 1/P_p$ the spin frequency of the pulsar, *etc.*), $\{p^K\} = \{P_b, T_0, e, \omega_0, x\}$ is the set of "Keplerian" parameters, (notably, orbital period P_b, eccentricity e and projected semi-major axis $x = a\sin i/c$), and $\{p^{PK}\} = \{k, \gamma_{\text{timing}}, \dot{P}_b, r, s, \dot{\delta}_\theta, \dot{e}, \dot{x}\}$ denotes the set of (separately measurable) "post-Keplerian" parameters. Most important among these are: the fractional periastron advance per orbit $k \equiv \dot{\omega} P_b/2\pi$, a dimensionful time-dilation parameter γ_{timing}, the orbital period derivative $\dot{P}_b$, and the "range" and "shape" parameters of the gravitational time delay caused by the companion, r and s.

Without assuming any specific theory of gravity, one can phenomenologically analyze the data from any binary pulsar by least-squares fitting the observed sequence of pulse arrival times to the timing formula Eq. (14). This fit yields the "measured" values of the parameters $\{v_p, \dot{v}_p, \ddot{v}_p\}$, $\{p^K\}$, $\{p^{PK}\}$. Now, each specific relativistic theory of gravity predicts that, for instance, k, γ_{timing}, $\dot{P}_b$, r, and s (to quote parameters that have been successfully measured from some binary pulsar data) are some theory-dependent functions of the Keplerian parameters and of the (unknown) masses m_1, m_2 of the pulsar and its companion. For instance, in General Relativity, one finds (with $M \equiv m_1 + m_2$, $n \equiv 2\pi/P_b$)

$$\begin{aligned}
k^{\text{GR}}(m_1, m_2) &= 3(1-e^2)^{-1}(G_N M n/c^3)^{2/3}, \\
\gamma_{\text{timing}}^{\text{GR}}(m_1, m_2) &= en^{-1}(G_N M n/c^3)^{2/3} m_2(m_1 + 2m_2)/M^2, \\
\dot{P}_b^{\text{GR}}(m_1, m_2) &= -(192\pi/5)(1-e^2)^{-7/2}\left(1 + \tfrac{73}{24}e^2 + \tfrac{37}{96}e^4\right) \\
&\quad \times (G_N M n/c^3)^{5/3} m_1 m_2/M^2, \\
r(m_1, m_2) &= G_N m_2/c^3, \text{ and} \\
s(m_1, m_2) &= nx(G_N M n/c^3)^{-1/3} M/m_2.
\end{aligned} \quad (15)$$

In tensor-scalar theories, each of the functions $k^{\text{theory}}(m_1, m_2)$, $\gamma_{\text{timing}}^{\text{theory}}(m_1, m_2)$, $\dot{P}_b^{\text{theory}}(m_1, m_2)$, etc. is modified by quasi-static strong field effects (associated with the self-gravities of the pulsar and its companion), while the particular function $\dot{P}_b^{\text{theory}}(m_1, m_2)$ is further modified by radiative effects (associated with the spin 0 propagator) [31] [46].

Let us summarize the current experimental situation (see Ref. [47] for a more extensive review). In the first discovered binary pulsar PSR1913 + 16 [39] [40], it has been possible to measure with accuracy the three post-Keplerian parameters k, γ_{timing} and $\dot{P}_b$. The three equations $k^{\text{measured}} = k^{\text{theory}}(m_1, m_2)$, $\gamma_{\text{timing}}^{\text{measured}} = \gamma_{\text{timing}}^{\text{theory}}(m_1, m_2)$, $\dot{P}_b^{\text{measured}} = \dot{P}_b^{\text{theory}}(m_1, m_2)$ determine, for each given theory, three curves in the two-dimensional mass plane. This yields *one* (combined radiative/strong-field) test of the specified theory, according to whether the three curves meet at one point, as they should. After subtracting from the observed orbital period derivative, $\dot{P}_b^{\text{obs}} = (-2.4184 \pm 0.0009) \times 10^{-12}$, a Newtonian perturbing effect caused by the Galaxy [48], $\dot{P}_b^{\text{Gal}} = (-0.0128 \pm 0.0050) \times 10^{-12}$ (which dominates the final error bar), one finds that General Relativity passes this $(k - \gamma_{\text{timing}} - \dot{P}_b)_{1913+16}$ test with complete success at the 10^{-3} level [40] [42] [43]

$$\left[\frac{\dot{P}_b^{\text{obs}} - \dot{P}_b^{\text{galactic}}}{\dot{P}_b^{\text{GR}}[k^{\text{obs}}, \gamma_{\text{timing}}^{\text{obs}}]}\right]_{1913+16} = 1.0013 \pm 0.0021. \quad (16)$$

Here $\dot{P}_b^{\text{GR}}[k^{\text{obs}}, \gamma_{\text{timing}}^{\text{obs}}]$ is the result of inserting in $\dot{P}_b^{\text{GR}}(m_1, m_2)$ the values of the masses predicted by the two equations $k^{\text{obs}} = k^{\text{GR}}(m_1, m_2)$, $\gamma_{\text{timing}}^{\text{obs}} = \gamma_{\text{timing}}^{\text{GR}}(m_1, m_2)$.

The discovery of the binary pulsar PSR1534 + 12 [49] has allowed one to measure the four post-Keplerian parameters k, γ_{timing}, r and s, and thereby to obtain *two* (four observables minus two masses) tests of strong field gravity, without mixing of radiative effects [50]. General Relativity passes these tests within the measurement accuracy [40] [50]. The most precise of these new, pure, strong-field tests is the one obtained by combining the measurements of k, γ, and s. Using the most recent data [51], one finds agreement at the 1% level:

$$\left[\frac{s^{\text{obs}}}{s^{\text{GR}}[k^{\text{obs}}, \gamma^{\text{obs}}_{\text{timing}}]}\right]_{1534+12} = 1.000 \pm 0.007. \quad (17)$$

It has also been possible to measure the orbital period change of PSR1534 + 12. General Relativity passes the corresponding $(k - \gamma_{\text{timing}} - \dot{P}_b)_{1534+12}$ test with success at the 15% level [51].

The discovery of the binary pulsar PSR J1141 − 6545 [52] (whose companion is probably a white dwarf) has led to the measurement of the three post-Keplerian parameters k, γ_{timing} and $\dot{P}_b$ [53]. As in the PSR 1913 + 16 system, this yields *one* combined radiative/strong-field test of relativistic gravity. One finds that General Relativity passes this $(k - \gamma_{\text{timing}} - \dot{P}_b)_{1141-6545}$ test with success at the 25% level [53]. Scintillation measurements have allowed one to measure the sine of the orbital inclination angle of this system [54]. The result is in excellent agreement with the general relativistic expectation for the parameter s calculated from the masses derived from k and γ_{timing} [53].

Recently, a remarkable "double binary pulsar" has been discovered: PSR J0737 − 3039 [55]. This highly relativistic double neutron star system (where both bodies, say A and B, are pulsars) has already established itself as a unique laboratory for testing gravity in the sense that it allows for the simultaneous measurement of an unprecedented number of independent timing parameters, namely *six*: the five post-Keplerian parameters k, γ_{timing}, $\dot{P}_b$, r and s, together with the ratio $R = x_B/x_A$ between the (Keplerian) projected semi-major axes of the two pulsars PSR J0737 − 3039B (pulse period 2.8 sec) and PSR J0737 − 3039A (pulse period 22 msec). These six measurements determine six curves in the m_A, m_B mass plane, and thereby yield four tests of relativistic gravity. Three of these tests probe the strong-field regime of gravity without mixing of radiative effects, while one test (involving $\dot{P}_b$) is a new test of the radiative aspects of gravity. General relativity passes all these tests with flying colors. Referring to [55] for details, let us only mention here that the test involving $\dot{P}_b$ confirms again the radiative structure of General Relativity at the $\sim 1\%$ level [M. Kramer, private communication, February 2006], while the most precise pure strong-field test is at the 10^{-3} level:

$$\left[\frac{s^{\text{obs}}}{s^{\text{GR}}[k^{\text{obs}}, R^{\text{obs}}]}\right]_{0737-3039} = 0.9998^{+0.0006}_{-0.0011}. \quad (18)$$

Several other binary pulsar systems, of a nonsymmetric type (nearly circular systems made of a neutron star and a white dwarf), can also be used to test relativistic gravity [56]–[59]. The constraints on tensor-scalar theories provided by binary-pulsar "experiments" have been analyzed in Ref. [46], and shown to exclude a large portion of the parameter space allowed by solar-system tests.

The prediction [60] that the general relativistic spin-orbit coupling should cause a secular change in the orientation of the pulsar beam, with respect to the line of sight ("geodetic precession"), has been confirmed by the study of the evolution of the intensity profiles of several binary pulsars: PSR1913 + 16 [61], PSR B1534+12 [62] (where the spin precession rate could be measured in a model-independent way), and PSR J1141−6545 [63].

The tests considered above have examined the gravitational interaction on scales between a fraction of a millimeter and a few astronomical units. On the other hand, the general relativistic action on light and matter of an external gravitational field on length scales between 100 kpc and 1 Mpc has been verified to $\sim 30\%$ in some gravitational lensing systems [64] [65]. Some tests on cosmological scales are also available. In particular, Big Bang Nucleosynthesis has been used to set significant constraints on the variability of the gravitational "constant" G_N [66]. For other cosmological tests of the "constancy of constants," see the review [12]. Let us also mention that the recent analysis [15] of lunar laser ranging data has set the limit $\dot{G}_N/G_N = (4 \pm 9) \times 10^{-13}$ yr^{-1}.

CONCLUSIONS

All present experimental tests are compatible with the predictions of the current "standard" theory of gravitation: Einstein's General Relativity. The universality of the coupling between matter and gravity (Equivalence Principle) has been verified at better than the 10^{-12} level. Solar system experiments have tested all the weak-field predictions of Einstein's theory at better than the 10^{-3} level (and down to the 2×10^{-5} level for the post-Einstein parameter $\bar{\gamma}$). The propagation properties of relativistic gravity, as well as several of its strong-field aspects, have been verified at the 10^{-3} level in several independent binary pulsar experiments. Recent laboratory experiments have set strong constraints on sub-millimeter modifications of Newtonian gravity.

Several important new developments in experimental gravitation are expected in the near future. The NASA Gravity Probe B mission [67] (a space gyroscope experiment; launched on April 20, 2004) is expected to directly observe the gravitational spin-orbit and spin-spin couplings, thereby measuring the parameter $\bar{\gamma}$ to better than the 10^{-5} level. The universality of free-fall acceleration should soon be tested to much better than the 10^{-12} level by some satellite experiments: the approved CNES MICROSCOPE [68] mission (10^{-15} level; to be launched in 2009), and the planned (cryogenic) NASA-ESA STEP [69] mission (10^{-18} level). The recently constructed kilometer-size laser interferometers (notably LIGO [70] in the USA, and VIRGO [71] and GEO600 [72] in Europe), should soon directly detect gravitational waves arriving on Earth. As the sources of these waves are expected to be extremely relativistic objects with strong internal gravitational fields (*e.g.*, coalescing binary black holes), their detection will allow one to experimentally probe gravity in highly dynamical circumstances. Note also that arrays of millisecond pulsars are sensitive detectors of (very low frequency) gravitational waves [73] [74]. The APOLLO project (see [17]) plans to use enhanced laser and telescope technologies to improve the lunar laser ranging experiment by about an order of magnitude.

Finally, among the many longer-time-scale projects that might be important for experimental gravity (such as LISA [75], GAIA [76], MORE [77], ...) let us mention the LATOR project which plans to measure $\bar{\gamma}$ to the 10^{-9} level [78].

REFERENCES

1. See chapter 18 (Experimental Tests of Gravitation Theory) of the Review of Particle Physics; partial 2005 update available on the web (http://pdg.lbl.gov/). For the last published version, one should refer to S. Eidelman *et al.* [Particle Data Group], "Review of particle physics," Phys. Lett. B **592**, 1 (2004).
2. T. Jacobson, S. Liberati and D. Mattingly, "Lorentz violation at high energy: Concepts, phenomena and astrophysical constraints," Annals Phys. **321**, 150 (2006) [arXiv:astro-ph/0505267];
   David Mattingly, "Modern Tests of Lorentz Invariance", Living Rev. Relativity 8, (2005), 5. URL: http://www.livingreviews.org/lrr-2005-5
3. S.N. Gupta, Phys. Rev. **96**, 1683 (1954);
   R.H. Kraichnan, Phys. Rev. **98**, 1118 (1955);
   R.P. Feynman, F.B. Morinigo, and W.G. Wagner, *Feynman Lectures on Gravitation*, edited by Brian Hatfield (Addison-Wesley, Reading, 1995);
   S. Weinberg, Phys. Rev. **138**, B988 (1965);
   V.I. Ogievetsky and I.V. Polubarinov, Ann. Phys. (NY) **35**, 167 (1965);
   W. Wyss, Helv. Phys. Acta **38**, 469 (1965);
   S. Deser, Gen. Rel. Grav. **1**, 9 (1970);
   D.G. Boulware and S. Deser, Ann. Phys. (NY) **89**, 193 (1975);
   J. Fang and C. Fronsdal, J. Math. Phys. **20**, 2264 (1979);
   R.M. Wald, Phys. Rev. **D33**, 3613 (1986);
   C. Cutler and R.M. Wald, Class. Quantum Grav. **4**, 1267 (1987);
   R.M. Wald, Class. Quantum Grav. **4**, 1279 (1987);
   N. Boulanger *et al.*, Nucl. Phys. **B597**, 127 (2001).
4. S. Weinberg, *Gravitation and Cosmology* (John Wiley, New York, 1972).
5. S. Weinberg, Rev. Mod. Phys. **61**, 1 (1989).
6. A.I. Shlyakhter, Nature **264**, 340 (1976);
   T. Damour and F. Dyson, Nucl. Phys. **B480**, 37 (1996);
   Y. Fujii *et al.*, Nucl. Phys. **B573**, 377 (2000).
7. K.A. Olive *et al.*, Phys. Rev. **D66**, 045022 (2002); and Phys. Rev. **D69**, 027701 (2004).
8. H. Marion *et al.*, Phys. Rev. Lett. **90**, 150801 (2003);
   S. Bize *et al.*, Phys. Rev. Lett. **90**, 150802 (2003);
   M. Fischer *et al.*, Phys. Rev. Lett. **92**, 230802 (2004).
9. J.K. Webb *et al.*, Phys. Rev. Lett. **87**, 091301 (2001);
   M.T. Murphy *et al.*, Mon. Not. Roy. Astron. Soc. **327**, 1208 (2001).
10. R. Quast, D. Reimers and S. A. Levshakov, Astron. Astrophys. **415**, L7 (2004).
11. R. Srianand, H. Chand, P. Petitjean and B. Aracil, Phys. Rev. Lett. **92**, 121302 (2004).
12. J.P. Uzan, Rev. Mod. Phys. **75**, 403 (2003).
13. J.D. Prestage *et al.*, Phys. Rev. Lett. **54**, 2387 (1985);
    S.K. Lamoreaux *et al.*, Phys. Rev. Lett. **57**, 3125 (1986);
    T.E. Chupp *et al.*, Phys. Rev. Lett. **63**, 1541 (1989).
14. Y. Su *et al.*, Phys. Rev. **D50**, 3614 (1994).
15. J.G. Williams, S. G. Turyshev and D. H. Boggs, Phys. Rev. Lett. **93**, 261101 (2004).
16. S. Baessler *et al.*, Phys. Rev. Lett. **83**, 3585 (1999).
17. E. G. Adelberger, Class. Quantum Grav. **18**, 2397 (2001).
18. G.L. Smith *et al.*, Phys. Rev. **D61**, 022001 (1999).
19. J. C. Hafele and R. E. Keating, Science **177**, 166-168 and 168-170 (1972); C. O. Alley, in *Proceedings of the 33rd Annual Symposium on Frequency Control* (Electronic Industries Association, Washington D. C., 1979).
20. R.F.C. Vessot and M.W. Levine, Gen. Rel. Grav. **10**, 181 (1978);
    R.F.C. Vessot *et al.*, Phys. Rev. Lett. **45**, 2081 (1980)
21. N. Ashby, Living Rev. Relativity **6** (2003) 1.
22. C.M. Will, *Theory and Experiment in Gravitational Physics* (Cambridge University Press, Cambridge, 1993); and Living Rev. Rel. **4** (2001) 4.
23. T. Damour, in *Gravitation and Quantizations*, ed. B. Julia and J. Zinn-Justin, Les Houches, Session LVII (Elsevier, Amsterdam, 1995), pp. 1–61.

24. H. van Dam and M.J. Veltman, Nucl. Phys. **B22**, 397 (1970);
    V.I. Zakharov, Sov. Phys. JETP Lett. **12**, 312 (1970);
    D.G. Boulware and S. Deser, Phys. Rev. **D6**, 3368 (1972);
    C. Aragone and J. Chela-Flores, Nuovo Cim. **10A**, 818 (1972);
    A.I. Vainshtein, Phys. Lett. **B39**, 393 (1972);
    C. Deffayet et al., Phys. Rev. **D65**, 044026 (2002);
    M. Porrati, Phys. Lett. **B534**, 209 (2002);
    N. Arkani-Hamed, H. Georgi, and M.D. Schwartz, Annals Phys. **305**, 96 (2003);
    T. Damour, I.I. Kogan, and A. Papazoglou, Phys. Rev. **D67**, 064009 (2003);
    G. Dvali, A. Gruzinov, and M. Zaldarriaga, Phys. Rev. **D68**, 024012 (2003).
25. T.R. Taylor and G. Veneziano, Phys. Lett. **B213**, 450 (1988);
    S. Dimopoulos and G. Giudice, Phys. Lett. **B379**, 105 (1996);
    I. Antoniadis, S. Dimopoulos, and G. Dvali, Nucl. Phys. **B516**, 70 (1998).
26. V.A. Rubakov, Phys. Usp **44**, 871 (2001);
    I.I. Kogan, in *Proceedings of the XXXVIth Rencontres de Moriond, Electro-Weak Interactions and Unified Theories* (March 2001); `astro-ph/0108220`;
    R. Maartens, Living Rev.Rel. **7** (2004) 7.
27. E. Fischbach and C.L. Talmadge, *The search for non-Newtonian gravity* (Springer-Verlag, New York, 1999).
28. J.C. Long, H.W. Chan, and J.C. Price, Nucl. Phys. **B539**, 23 (1999).
29. C.D. Hoyle et al., Phys. Rev. D 70, 042004 (2004);
    J. Chiaverini et al., Phys. Rev. Lett. **90**, 151101 (2003);
    J.C. Long et al., Nature **421**, 922 (2003);
    E. G. Adelberger et al., Ann. Rev. Nucl. Part. Sci. **53**, 77 (2003).
30. P. Jordan, *Schwerkraft und Weltall* (Vieweg, Braunschweig, 1955);
    M. Fierz, Helv. Phys. Acta **29**, 128 (1956);
    C. Brans and R.H. Dicke, Phys. Rev. **124**, 925 (1961).
31. T. Damour and G. Esposito-Farèse, Class. Quantum Grav. **9**, 2093 (1992).
32. A.S. Eddington, *The Mathematical Theory of Relativity* (Cambridge University Press, Cambridge, 1923);
    K. Nordtvedt, Phys. Rev. **169**, 1017 (1968);
    C.M. Will, Astrophys. J. **163**, 611 (1971).
33. B. Bertotti, L. Iess and P. Tortora, Nature, **425**, 374 (2003).
34. S. S. Shapiro et al., Phys. Rev. Lett. **92**, 121101 (2004).
35. K. Nordtvedt, Phys. Rev. **170**, 1186 (1968).
36. T.R. Taylor and G. Veneziano, Phys. Lett. **B213**, 450 (1988);
    T. Damour and A.M. Polyakov, Nucl. Phys. **B423**, 532 (1994).
37. I. I. Shapiro, in *General Relativity and Gravitation 1989*, N. Ashby et al. eds., (Cambridge University Press, Cambridge, 1990), p. 313.
38. I. Ciufolini and E. C. Pavlis, Nature **431**, 958 (2004).
39. R.A. Hulse, Rev. Mod. Phys. **66**, 699 (1994).
40. J.H. Taylor, Rev. Mod. Phys. **66**, 711 (1994).
41. T. Damour and N. Deruelle, Phys. Lett. **A87**, 81 (1981);
    T. Damour, C.R. Acad. Sci. Paris **294**, 1335 (1982).
42. J.H. Taylor, Class. Quantum Grav. **10**, S167 (Supplement 1993).
43. J. Weisberg and J.H. Taylor, in *Binary Radio Pulsars, ASP Conference Series* **328**, 25 (2005), edited by F. A. Rasio and I. H. Stairs (available on `http://www.astro.northwestern.edu/AspenW04/ program.html`).
44. By contrast to the binary pulsar case, which does involve the gauge-invariant, helicity-two propagating degrees of freedom of the gravitational field, the recent measurement of light deflection from Jupiter (E.B. Fomalont and S.M. Kopeikin, Astrophys. J. **598**, 704 (2003)) does not depend (in spite of contrary claims: S.M. Kopeikin, Astrophys. J. **556**, L1 (2001)), at the considered precision level, on the propagation speed of gravity (see C.M. Will, Astrophys. J. **590**, 683 (2003) and S. Samuel, Phys. Rev. Lett. **90**, 231101 (2003)).
45. T. Damour and J.H. Taylor, Phys. Rev. **D45**, 1840 (1992).

46. T. Damour and G. Esposito-Farèse, Phys. Rev. **D54**, 1474 (1996); and Phys. Rev. **D58**, 042001 (1998); G. Esposito-Farèse, `gr-qc/0402007`.
47. I.H. Stairs, Living Rev. Rel. **6** (2003) 5.
48. T. Damour and J.H. Taylor, Astrophys. J. **366**, 501 (1991).
49. A. Wolszczan, Nature **350**, 688 (1991).
50. J.H. Taylor *et al.*, Nature **355**, 132 (1992).
51. I.H. Stairs *et al.*, Astrophys. J. **505**, 352 (1998);
    I.H. Stairs *et al.*, Astrophys. J. **581**, 501 (2002).
52. V.M. Kaspi *et al.*, Astrophys. J. **528**, 445 (2000).
53. M. Bailes *et al.*, Astrophys. J. **595**, L49 (2003); and M. Bailes, in *Binary Radio Pulsars, ASP Conference Series* **328**, 33 (2005), edited by F. A. Rasio and I. H. Stairs (available on `http://www.astro.northwestern.edu/AspenW04/ program.html`).
54. S. M. Ord, M. Bailes and W. van Straten, The Astrophysical Journal, **574**, Issue 1, pp. L75-L78 (2002).
55. M. Burgay *et al.*, Nature **426**, 531 (2003);
    A. G. Lyne *et al.*, Science **303**, 1153 (2004);
    M. Kramer *et al.*, eConf **C041213**, 0038 (2004) [arXiv:astro-ph/0503386].
56. C.M. Will and H.W. Zaglauer, Astrophys. J. **346**, 366 (1989).
57. T. Damour and G. Schäfer, Phys. Rev. Lett. **66**, 2549 (1991).
58. N. Wex, in *Pulsar Astronomy - 2000 and Beyond, IAU Colloquium 177*, M. Kramer, N. Wex, R. Wielebinski, eds. (San Francisco: Astronomical Society of the Pacific), 113 (2000).
59. I. H. Stairs *et al.*, `astro-ph/0506188`.
60. T. Damour and R. Ruffini, C. R. Acad. Sc. Paris **279**, série A, 971 (1974);
    B.M. Barker and R.F. O'Connell, Phys. Rev. D 12, 329 (1975).
61. M. Kramer, Astrophys. J. **509**, 856 (1998);
    J.M. Weisberg and J.H. Taylor, Astrophys. J. **576**, 942 (2002).
62. I. H. Stairs, S. E. Thorsett, and Z. Arzoumanian, Phys. Rev. Lett. **93**, 141101 (2004).
63. A. W. Hotan, M. Bailes and S. M. Ord, Astrophys. J. **624**, 906 (2005).
64. A. Dar, Nucl. Phys. (Proc. Supp.) **B28A**, 321 (1992); see also `astro-ph/9407072`.
65. S. W. Allen, S. Ettori and A. C. Fabian, Mon. Not. Roy. Astron. Soc. **324**, 877 (2001).
66. J. Yang *et al.*, Astrophys. J. **227**, 697 (1979);
    T. Rothman and R. Matzner, Astrophys. J. **257**, 450 (1982);
    F.S. Accetta, L.M. Krauss, and P. Romanelli, Phys. Lett. **B248**, 146 (1990).
67. `http://einstein.stanford.edu`.
68. P. Touboul *et al.*, C.R.Acad. Sci. Paris **2** (série IV) 1271 (2001).
69. P.W. Worden, in *Proc. 7th Marcel Grossmann Meeting on General Relativity*, edited by R.J. Jantzen and G. MacKeiser, (World Scientific, Singapore, 1995), pp. 1569-1573.
70. `http://www.ligo.caltech.edu` .
71. `http://www.virgo.infn.it` .
72. `http://www.geo600.uni-hannover.de` .
73. V.M. Kaspi, J.H. Taylor and M.F. Ryba, Astrophys. J. **428**, 713 (1994).
74. A.N. Lommen and D.C. Backer, Bulletin of the American Astronomical Society **33**, 1347 (2001); and Astrophys. J. **562**, 297 (2001); A. N. Lommen, in *Bad Honnef 2002, Neutron stars, pulsars and supernova remnants*, 114-125 and `astro-ph/0208572` .
75. `http://lisa.jpl.nasa.gov` .
76. A. Vecchiato, M. G. Lattanzi, B. Bucciarelli, M. T. Crosta, F. de Felice and M. Gai, Astron. Astrophys. **399**, 337 (2003).
77. "Mercury Orbiter Relativity Experiment" to fly with the BepiColombo ESA mission to Mercury; see `http://sci.esa.int/ science-e/www/area/ index.cfm?fareaid=30` .
78. S. G. Turyshev *et al.* [LATOR Collaboration], `gr-qc/0506104` .

Relativistic Statistical Mechanics, a brief overview

Rémi Hakim, Horacio D. Sivak[+]

LUTH, Observatoire de Paris-Meudon
(France)

Abstract: 1. Introduction 2. The one-particle distribution function, 3. Relativistic statistical mechanics (classical), 4. Treatment of radiation, 5. Relativistic statistical mechanics (quantum), 6. Relativistic kinetic theory, 7. Curved space-time, 8. A brief conclusion, 9. References.

Keywords: relativity, statistical mechanics, kinetic theory, radiation, plasmas, nuclear matter, quasi-particles, strong magnetic fields, Wigner function.
PACS: 11.10.Wx, 21.65.+f, 24.10.Pa, 51.10+y, 52.27.Ny, 95.30.Sf,

1. INTRODUCTION

Relativistic statistical mechanics is nowadays an honorable subject applying from astrophysics to heavy-ion collisions without forgetting nuclear matter, etc. Since the first articles by F. Jüttner (1911) [see also W. Pauli (1921)], the subject has not attracted much attention until the beginning of the sixties, the more so since possible applications seemed to be quite speculative at that time.

In 1928, F. Jüttner generalized his 1911 ideal gas results to the case of the ideal quantum gas, soon applied to the theory of white dwarfs by S. Chandrasekhar (1930, 1934) with the now well-known consequence of the existence of a limiting mass for this kind of star, the so-called Chandrasekhar mass.

A less known work in the domain is the article by A.G. Walker (1936) where, for the first time, general relativity was introduced and kinetic theory applied to the expanding universe.

Slightly later, D. van Dantzig improved relativistic hydrodynamics and studied the ideal gas case (1939); his results were described and extended by J.L. Synge (1957). P.G. Bergmann (1951, 1962) gave various tools to be used in relativistic statistical mechanics (essentially, techniques involving differential forms, well-suited to such a case). At about the same time, A.E. Scheidegger and C.D. McKay (1951) and A.O. Barut (1958) devised techniques to perform "statistics of fields", still in the non-interacting case.

The interest raised by nuclear fusion, in the late fifties, led to various studies on relativistic plasmas [S. Titeica (1956), S.T.Beliaev and G.I. Budker (1956), P.C. Clemmow and A.J. Wilson (1957), Yu.L. Klimontovich (1960)]. While S. Titeica gave a covariant version of the Vlasov equation, S.T.Beliaev and G.I. Budker included a Landau-like collision term.

[+] In memory of my late friend and collaborator, Horacio Dario Sivak [Buenos-Aires (Argentina), 1946 — Villejuif (France), 2000].

However, Yu.L. Klimontovich achieved a decisive progress —using M. Schönberg's (1952, 1953) method of second quantization in phase space — and was able to provide a BBGKY hierarchy for the covariant one-, two-, etc -particles distribution function of an electron plasma embedded in a neutralizing uniform background. From this hierarchy, he was able to derive the relativistic Landau-collision term and hence the plasma Fokker-Planck equation; he also obtained the Balescu-Guernsey-Lennard equation whose collision term involves the influence of the plasma modes.

Although Yu.L. Klimontovich performed a great step, the general situation — discussed in details by P. Havas (1964) — was still unclear since, apart from plasmas, and no other non-quantum physical system was known. Furthermore, it was believed that only Hamiltonian equations of motion were needed in relativistic statistical mechanics. As a matter of fact, D.G. Currie, T.F. Jordan, and E.C.G. Sudarshan (1963,1984) proved a "no-interaction theorem"[1] to the effect that a Hamiltonian formalism only applies to systems constituted by non-interacting particles. Therefore, the sole remaining possibility was the simultaneous statistical treatment of both particles and field(s) through which they were supposed to be interacting.

Such an approach was already known in the non-relativistic case (see *e.g.* E.G. Harris, I. Prigogine, etc.) and could easily be extended to the relativistic case [see *e.g.* A. Mangeney (1965)] although the detailed calculations were not trivial at all. The results were not manifestly covariant and hence the proof that they actually satisfy the relativity principle had to be provided for each particular case. Accordingly, the Brussels' school (I. Prigogine and his collaborators) invented a formalism that provided the Lorentz transformation properties of their equations and also of the physical observables [see *e.g.* R. Balescu, T. Kotera and E. Pena (1967)]. However their formalism, although ingenuous and corresponding to an implicit and quite admissible philosophical position as to relativity — space and time must be kept separated —, was extremely involved and had the consequence that the Lorentz transformation acquired a curious dynamical meaning while, according to common wisdom, it is a merely kinematical transformation[2].

Meanwhile, N.A. Chernikov (1956 — 1964), G.E. Tauber and J.W. Weinberg (1961), W. Israel (1963) studied the covariant Boltzmann equation, whether in a flat space-time case or in a curved one. These studies were taken up later by numerous authors and applied to the calculation of transport coefficients (bulk and shear viscosity, heat conduction coefficient, diffusion coefficient, etc.) *via* the use of approximation methods (Chapman-Enskog, moments methods, etc.) adapted to the case of relativity.

As to quantum systems, impulsions to their active study were provided by the so-called statistical model of multiple production of particles [see the general review by R. Hagedorn (1995) and by its extension by the same author (1965) to the statistical bootstrap model. In the mid-seventies, still in view of multi-production of particles, P.A. Carruthers and F. Zachariasen (1974 —1983) first

---

[2] The dynamical interpretation of I. Prigogine and his coworkers is perfectly admissible but it does not correspond to the general trend of physicists who look for symmetries in the laws of physics.

used a covariant form of the Wigner function (1932). At about the same time F. Cooper, D.H. Sharp, M. Feigenbaum and others worked in the same direction. This latter was then generalized to fermions, or given a gauge-invariant form [E.A. Remler (1977), V.V. Klimov (1982), J. Winter (1984), U. Heinz (1983, 1985), H.-Th. Elze, M. Gyulassy and D. Vasak (1986)]. The covariant Wigner function was used in the study of relativistic quantum plasmas, embedded or not in strong magnetic fields, to the derivation of the main properties of nuclear matter through the use of the J.D. Walecka's model (1974) or other phenomenological ones.

The covariant Wigner function was used in the study of relativistic quantum plasmas, embedded or not in strong magnetic fields, to and in the derivation of the main properties of nuclear matter as described by the J.D. Walecka's model (1974) or other phenomenological ones.

It remains to explain briefly the reasons of the exponential development of this domain since roughly twenty-five years. Besides a theoretical interest, there exists essentially four factors that lead to such an expansion. The first one is the inflating status of relativistic astrophysics, itself growing under the pressure of new observational data concerning dense stars (white dwarves, neutron stars, magnetars) and/or supernovae explosions, the primeval universe. All these objects, and many other exotic ones, do require studies of relativistic dense matter and/or of relativistic plasmas. The second factor — which is not necessarily the most important one — is constituted by the fact that relativistic plasmas will be the objects of experimental studies in a close future. A third reason can be found in the systematic experiments on heavy ion collisions and the accompanying theoretical studies on relativistic hot and dense nuclear matter with its possible exotic states (meson condensation, superfluidity, etc). Finally, the active search for the quark-gluon plasma, and its signature in heavy ion collisions, is constantly stimulating this domain.

2. THE ONE-PARTICLE DISTRIBUTION FUNCTION

The first works began, as could be expected, by those notions derived from kinetic theory, such as the distribution function, the Maxwell-Boltzmann distribution function, and the kinetic equations it is supposed to obey. Accordingly, the same path is followed in this section. The covariant form of the distribution function was first given by A.G. Walker (1936), D. van Dantzig (1939), S. Titeica (1956), J.L. Synge (1957), etc. A simpler definition was apparently given by R.L. Stratonovich (1959) in an unpublished work quoted by Yu. L. Klimontovich (1960), improved by R. Hakim (1967) and also considered by N.G. van Kampen (1969). This latter path is followed in this review.

The One-particle Distribution Function

Let us first consider a classical, *i.e.* non-quantum, relativistic particle. The numerical four-current it defines in space-time is provided by the so-called Feynman current

$$J^\mu(x) = \int ds\, \delta^{(4)}[x-x(s)]\frac{d}{ds}x^\mu(s),$$

where x is the space-time point and where s is an arbitrary parameter — generally taken to be the proper time — along the space-time trajectory, $x(s)$, over which the integral is extended. Similarly, the energy-momentum tensor of the particle is given by

$$T^{\mu\nu}(x) = \int ds\, \delta^{(4)}[x-x(s)]p^\mu(s)\frac{d}{ds}x^\nu(s).$$

For a system of N particles, the four-current and the energy-momentum tensor of the particles are then provided by

$$\begin{cases} J^\mu(x) = \sum_{i=1}^{i=N} \int ds\, \delta^{(4)}[x-x_i(s)]\frac{d}{ds}x_i^\mu(s) \\ T^{\mu\nu}(x) = \sum_{i=1}^{i=N} \int ds\, \delta^{(4)}[x-x_i(s)]p_i^\mu(s)\frac{d}{ds}x_i^\nu(s) \end{cases},$$

which can be rewritten as

$$J^\mu(x) = \int d^4p \int ds \sum_{i=1}^{i=N} \delta^{(4)}[p-p_i(s)]\delta^{(4)}[x-x_i(s)]u_i^\mu(s)$$

$$= \int d^4u\, \frac{p^\mu}{m} \int ds \sum_{i=1}^{i=N} \delta^{(4)}[p-p_i(s)]\delta^{(4)}[x-x_i(s)]$$

$$\equiv \int d^4p\, \frac{p^\mu}{m} R(x,p)$$

for the four-current and as

$$T^{\mu\nu} = \int d^4p\, \frac{p^\mu p^\nu}{m} R(x,p)$$

for the energy-momentum tensor. In these last two equations we used the definition

$$R(x,p) \equiv \int ds \sum_{i=1}^{i=N} \delta^{(4)}[p-p_i(s)]\, \delta^{(4)}[x-x_i(s)].$$

$R(x, p)$ depends on the initial data chosen for the trajectories of the relativistic particles and thus is a *random function* in the context of a statistical ensemble where these data are known only in a statistical manner within the J.W. Gibbs' conventional views.

The covariant distribution function $f(x, p)$ is thus defined as

$$f(x,p) \equiv <R(x,p)>,$$

where the average value $<...>$ is taken over the initial data, whatever they might be[3] so that, by construction, it allows the calculation of any kind of average values of observables quantities whatsoever.

Therefore, it appears that the one-particle relativistic phase-space, or µ-space, is formally the 8-dimensional space subtended by $\{x, p\}$. In fact the four-momentum p is generally constrained by a mass-shell condition of the type $p^2 = m^2$ or by any other, such as

$$(p-eA)^2 = m^2$$

when dealing with a charged system embedded in an electromagnetic four-potential A.

Let $A^{...}(x,p)$ be a tensorial observable connected to the particles; its local density is given by

$$A^{\mu...}(x) = \int d^4 p \; \frac{p^\mu}{m} A^{...}(x,p) f(x,p),$$

where the global quantity of $A^{...}(x,p)$ in the system is given by

$$A^{...} = \int_\Sigma \int d^4 p \; \frac{p^\mu}{m} A^{...}(x,p) f(x,p),$$

where Σ is an arbitrary space-like three-surface; i.e. $A^{...}$ is the flux of the four-current $A^{\mu...}(x)$ through Σ. As an example, the entropy of the (non-quantum) system is given by

$$S = \int_\Sigma d\Sigma_\mu \; S^\mu(x),$$

where $S^\mu(x)$ is the entropy four-current

$$S^\mu(x) = -k_B \int d^4 p \; \frac{p^\mu}{m} f(x,p) \log f(x,p).$$

Accordingly, the normalization of the covariant distribution function reads

---

[3] In the classical relativistic context of the so-called action-at-a-distance formalism of interacting particles, the initial value problem is not yet solved and the initial data necessary to determined completely the future of the system might consist of the initial positions and velocities of the particles *and* some part of the trajectories in the past.

$$\int_\Sigma \int d\Sigma_\mu d^4p \, \frac{p^\mu}{m} f(x,p) \equiv \int_\Sigma d\Sigma_\mu J^\mu(x) = N$$

when there are N particles in the system[4]. In the above equation, $d\Sigma_\mu$ is the differential form

$$d\Sigma_\mu = \frac{1}{3!} \varepsilon_{\mu\nu\alpha\beta} dx^\nu \times dx^\alpha \times dx^\beta,$$

the surface element on Σ. Note that, owing to the mass-shell condition $p^2 = m^2$, the integration element d^4p in µ-space actually reduces to a 3-dimensional one[5]

$$d^4p \to m \frac{d^3p}{p_0}.$$

Finally, it appears that the integration extends over a 6-dimensional µ-space $\Sigma \times \{p^2 = m^2\}$, as usual. Whether this last 6-dimensional phase-space or the covariant 8-dimensional one is called "phase-space" is only a matter of definition. When one considers the six-dimensional phase-space $\Sigma \times \mu$, its invariant "volume element" is given by

$$d\Omega(x,p) = d\Sigma_\mu(x) \times d\Sigma^\mu(p),$$

where

$$d\Sigma^\mu(p) = \frac{1}{3!} \varepsilon^{\mu\nu\alpha\beta} \, dp_\nu \times dp_\alpha \times dp_\beta$$

is the differential form "element of three-surface". The above element of integration on phase-space is of course written in an obvious system of coordinates adapted to its structure as a product of two three-surfaces and its explicit calculation simply yields

$$d\Sigma_\mu(p) = p_\mu \frac{d^3p}{p_0}.$$

The volume element is sometimes taken to be truly d^4p and the constraint $p^2 = m^2$ occurs either explicitly,

$$d^4p \, 2\theta(p^0) \delta(p^2 - m^2) f(p)$$

---

[4] Instead of N the normalization is often chosen to 1, in order for f to be a probability.
[5] The use of d^4p is generally more convenient; however it can be a source of confusion if one is not cautious enough [see e.g. B. Kursunoglu (1967)].

or implicitly in the distribution function. In any case, care must always be taken when dealing with either the integration element or with the distribution function : is the mass-shell restriction included into the former or the latter ? Or is it explicit ?

For an infinite system, the normalization of $f(x, p)$ occurs *via* the four-current or the local $n(x)$ density

$$n(x) = \left[J^\mu(x) J_\mu(x) \right]^{1/2},$$

i.e. via

$$n(x) = \int d^4p \; u_\mu(x).(p^\mu/m) f(x,p),$$

with $u^\mu(x) \equiv J^\mu(x)/n(x)$, is the average four-velocity of the system.

The Jüttner-Synge Equilibrium Distribution

The relativistic Maxwell-Boltzmann distribution function, hereafter called the *Jüttner-Synge distribution*, has been first derived by F. Jüttner in 1911 and studied in detail by J.L. Synge (1957). It can be derived in numerous possible ways : by remarking that the Boltzmann factor $\exp(-\beta E)$ can be obtained from thermodynamic considerations, independently of relativity theory, and hence it is sufficient to replace E by its relativistic expression and to normalize the result; by maximizing the entropy of the system while taking account of the constraints provided by the average energy and the number of particles within the system [J.L. Synge (1957)]; by solving the covariant Boltzmann equation [W. Israel (1963)]; by using a covariant formulation for the passage of a micro-canonical ensemble to a canonical one [R. Hakim (1973)], as first shown by A.I. Khinchin (1949) in the non-relativistic domain; etc.

The simplest derivation of the Jüttner-Synge distribution consists in maximizing the free energy of the system,

$$F = U - TS,$$

while the number N of particles is kept conserved and one immediately finds

$$f_{eq}(p) = A \exp.(-\beta u^\mu p_\mu)$$

[with $p_0 \equiv \sqrt{p^2 + m^2}$], where A is directly connected to the Lagrange multiplier introduced to take account of the constraint and where $\beta \equiv (k_B T)^{-1}$; it is determined by the normalization condition as

$$A = \frac{n_{eq} \beta}{4\pi m^2 K_2(m\beta)}$$

where K_2 is a Kelvin[6] function of order 2 and where $K_n(\xi)$ are defined by

$$K_n(\xi) = \frac{2^n n!}{(2n)!} \frac{1}{\xi^{n-2}} \int_0^\infty d\chi \ sh^{2n}\chi \exp(-\xi ch\chi);$$

A is connected to the chemical potential μ through

$$A = \frac{n_{eq}\beta}{4\pi m^2 K_2(m\beta)} = \exp(\beta\mu).$$

The energy-momentum tensor can easily be calculated [cf. J.L. Synge (1957)] as

$$T^{\mu\nu} = \left[n_{eq} m \frac{K_3(m\beta)}{K_2(m\beta)} + \frac{n_{eq}}{\beta} \right] u^\mu u^\nu - \frac{n_{eq}}{\beta} \eta^{\mu\nu}$$

or, alternatively, as

$$T^{\mu\nu} = \left\{ mn_{eq} \frac{K_1(\beta m)}{K_2(\beta m)} + \frac{4n_{eq}}{\beta} \right\} u^\mu u^\nu - \frac{n_{eq}}{\beta} \eta^{\mu\nu}.$$

The Lagrange multiplier β is determined from the equation of state of the relativistic gas. A comparison of the energy-momentum tensor, which has the so-called *perfect fluid* form[7]

$$T^{\mu\nu} = (\rho + P)u^\mu u^\nu - P\eta^{\mu\nu},$$

finally yields $P\beta = n_{eq}$, which is nothing else but the perfect gas equation of state and hence this terminates the identification of β with $1/k_B T$ (k_B is the usual Boltzmann constant). In this last equation ρ is the (invariant) energy density of the system and P is its pressure.

One-Particle Liouville Theorem

Before studying kinetic equations, we first indicate briefly how the one-particle Liouville theorem occurs in μ-space. Since the number of particles in the system is assumed to be conserved,[8] the 8-current in μ-space is necessarily conserved and its "continuity equation" then reads

---

[6] See Abramovitz and Stegun (1965).
[7] This means that the energy-momentum tensor does not contain any dissipation term, which would introduce gradients of some macroscopic quantities such as the average four-velocity, the temperature, etc.
[8] See *e.g.* Ch. Marle (1969) for the case of decaying or mutually transforming particles.

$$\partial_\mu \left[\frac{dx^\mu}{d\tau} f(x,p) \right] + \nabla_\mu \left[\frac{dp^\mu}{d\tau} f(x,p) \right] = 0,$$

where τ the proper time; or equivalently, after it is remarked that the "velocity" in this µ-space is given by

$$\left(u^\mu(x,p) \equiv \frac{dx^\mu}{d\tau} \text{ (4-velocity)}, \quad F^\mu(x,p) = \frac{dp^\mu}{d\tau} \text{ (4-force)} \right)$$

it is written as

$$\left[u^\mu(x,p) \partial_\mu + \frac{\partial}{\partial p^\mu} \left(F^\mu(x,p) \right) \right] f(x,p) = 0.$$

This is *not* the Liouville equation, which reads

$$\frac{d}{d\tau} f(x,p) \equiv \left[u^\mu(x,p) \partial_\mu + F^\mu(x,p) \frac{\partial}{\partial p^\mu} \right] f(x,p) = 0.$$

The Liouville equation is obeyed only by those 4-forces that satisfy the following condition

$$\frac{\partial}{\partial p^\mu} \left(F^\mu(x,p) \right) = 0.$$

For instance, this is the case of a system composed of charged particles submitted to an external electromagnetic field $F^{\mu\upsilon}$ where $F^\mu = (e/m) p_\upsilon F^{\mu\upsilon}$. Note that when this condition is not satisfied, the (one-particle) Liouville theorem is no longer valid.

3. RELATIVISTIC STATISTICAL MECHANICS (CLASSICAL)

This is the standard font and layout for the individual paragraphs. The style is In the Newtonian context a system of N particles is described by a trajectory in a $6N$-dimensional phase-space

$$\Gamma^{(6N)} \equiv \mu^{(6)} \times \mu^{(6)} \ldots \times \mu^{(6)},$$

where $\mu^{(6)}$ is the one-particle classical phase space. In special relativity, the natural generalization of the Newtonian phase-space that contains no arbitrary object, such as space-like three-surfaces, is the $8N$-dimensional Γ-space

$$\Gamma^{8N} \equiv \mu^{(8)} \times \mu^{(8)} \ldots \times \mu^{(8)}.$$

However in such a space, a point representing the state of the system does not lie on a one-dimensional trajectory, as in the classical case, but rather on an N-dimensional manifold [R. Hakim (1967)]. This can easily be seen from the fact that such a point,

$$\{x^A\}_{A=1,2...8N} = \{[x_{(1)}(\tau_1), p_{(1)}(\tau_1)], [x_{(2)}(\tau_2), p_{(2)}(\tau_2)],..., [x_{(N)}(\tau_N), p_{(N)}(\tau_N)]\},$$

does depend on the N proper times of the particles within the system. It follows that the statistical description of a relativistic system of point particles deeply differs from the usual one although the former gives rise to the latter in appropriate conditions.

Furthermore, from a dynamical point of view, there are also deep differences between relativistic force laws (which are essentially *non-local*) and Newtonian ones. The only known non-quantum physical system consists of charged particles interacting *via* electromagnetism. In such a case either one starts from the field-particle equations (*i.e.* from Maxwell's equations) and the equations of motion or, more directly, from the so-called Fokker action principle[9]

$$\delta I = \delta \left\{ \sum_{i=1}^{i=N} \int p_i^\mu p_{i\mu} d\tau + \frac{1}{2} \sum_{i,j} e^2 \int\int p_i^\mu p_{j\mu} D_{ret}(x_i - x_j) d\tau_i d\tau_j \right\} = 0,$$

where D_{ret} is the retarded elementary solution of the wave equation

$$D_{ret}(x) = \theta(x^0)\delta(x^2).$$

This variational principle exhibits the non-local nature of the equations of motion. Both viewpoints are then equivalent as far as the motions of the particles are concerned. The equation of motions read

$$\begin{cases} \dfrac{d}{d\tau_i} p_i^\mu(\tau_i) = \dfrac{e}{m} F^{\mu\upsilon}[x(\tau_i)] p_{i\upsilon}(\tau_i) \\ F^{\mu\upsilon}(x) = \partial^\mu A^\upsilon(x) - \partial^\upsilon A^\mu(x) \end{cases} \quad i = 1,2,...,N$$

with

$$A^\mu[x_i(\tau_i)] = \frac{4\pi e}{m} \sum_{i=1}^{i=N} \int d^4 x' d\tau_i \; p_i^\mu(\tau_i) \; D[x'] \; \delta^{(4)}[x'-x_i(\tau_i)]$$

Many-particle Distribution Functions

On the relativistic $8k$-dimensional reduced phase-space the k-particle *random* densities are first defined as

---

[9] A.D. Fokker, Z. Phys. **58**, 386, (1929); see also A.O. Barut (1965); J. Rzewuski (1964).

$$R_k\left[x_{\mu_1},p_{\mu_1};x_{\mu_2},p_{\mu_2};...;x_{\mu_N},p_{\mu_N}\right]=$$

$$\int...\int d\tau_1 d\tau_2...d\tau_N \left\{ \sum_{\substack{i_1,i_2,...,i_k \\ \text{all differents}}} \prod_{j=1}^{j=k} \delta^{(4)}\left[x_j - x_{i_j}(\tau_j)\right]\delta^{(4)}\left[p_j - p_{i_j}(\tau_j)\right]\right\}$$

From this definition the usual (multitime) k-particle distribution function

$$f_k\left[x_{\mu_1},p_{\mu_1};x_{\mu_2},p_{\mu_2};\ ...\ ;x_{\mu_N},p_{\mu_N}\right]$$

is defined as

$$f_k\left[x_{\mu_1},p_{\mu_1};x_{\mu_2},p_{\mu_2};\ ...\ ;x_{\mu_N},p_{\mu_N}\right] = <R_k\left[x_{\mu_1},p_{\mu_1};x_{\mu_2},p_{\mu_2};...;x_{\mu_N},p_{\mu_N}\right]>,$$

The brackets represent an average value over the initial conditions whatever they might be : it is sufficient to assume their existence with their usual properties like linearity, etc… To be more specific, we first note that the one-particle distribution function is of the above type. Explicitly the two-particle distribution reads

$$f_2\left[x_1,p_1;x_2,p_2\right]=<\int\int d\tau_1 d\tau_2 \sum_{i\neq j}\delta^{(4)}\left[x_1-x_i(\tau_1)\right]\delta^{(4)}\left[p_1-p_i(\tau_1)\right]$$
$$\delta^{(4)}\left[x_2-x_j(\tau_2)\right]\delta^{(4)}\left[p_2-p_j(\tau_2)\right]>$$

and is normalized as

$$N(N-1)=\int_\Sigma\int_{\Sigma'}\int\int d\Sigma_\mu d\Sigma_\nu d^4p_1 d^4p_2\ p_1^\mu p_2^\nu f_2[x_1,p_1;x_2,p_2]$$

where Σ and Σ' are two arbitrary space-like three-surfaces. The normalization of the f_k's are quite similar except that they are normalized to $(k!C_N^k)$.

However, unlike the Newtonian case, other kinds of distribution functions must also be introduced in order to get a complete system of equations. Some of them are exhibited in the next section.

The Relativistic BBGKY[10] Hierarchy

In order to obtain a BBGKY hierarchy for the various densities, a generating equation for the random distribution R_1 is first derived as [Yu. L. Klimontovich (1960); R. Hakim (1967)]

---

[10] Bogoliubov, Born, Green, Kirkwood, Yvon.

$$p^\mu \partial_\mu R_1(x,p) + 4\pi e^2 \nabla_\mu \Big(p_\upsilon \int d\tau d\tau' d^4x' d^4p' \big[p'^\mu \partial^\upsilon - p'^\upsilon \partial^\mu \big] D_{ret}(x-x') \times$$
$$\times \sum_{i,j} \delta^{(4)}\big[x-x_i(\tau)\big] \delta^{(4)}\big[p-p_i(\tau)\big] \delta^{(4)}\big[x'-x_j(\tau')\big] \delta^{(4)}\big[p'-p_j(\tau')\big] \Big) = 0$$

where the elementary properties of the δ-distributions have been used together with the replacement of the various expressions of $F^{\mu\nu}$, A^μ by their explicit forms. The double sum in this last equation can be split as

$$\sum_{i,j} = \sum_{i\neq j} + \sum_{i=j} .$$

The first term of the right hand side of this equation ($i \neq j$) is simply what has been defined as R_2 after it has been integrated over τ and τ' and while the second one deserves a brief explanation. It is explicitly written as

$$W_2(x,p;x',p') \equiv \int d\tau d\tau' \sum_i \delta^{(4)}\big[x-x_i(\tau)\big] \delta^{(4)}\big[p-p_i(\tau)\big]$$
$$\times \delta^{(4)}\big[x'-x_i(\tau')\big] \delta^{(4)}\big[p'-p_i(\tau')\big],$$

and it represents essentially the probability density for a *given* particle to be in the state (x, p) and next to undergo a transition to the state (x', p'). While W_2 must vanish out of the null cone $(x-x')^2 = 0$ (x or x' being in the future of x' or x, respectively, causality is implied by the time-like character of the trajectory of the particle), this is *a priori* not the case for R_2, which refers to *different* particles. From a dynamical viewpoint, the term that involves W_2 expresses the back reaction of the particle on itself; accordingly, since one deals with the electromagnetic interaction, it is an infinite term, which is often discarded *a priori*. However, besides an infinite term, it also gives rise to a finite — albeit small — contribution that lead to the so-called Abraham-Lorentz-Dirac equations[11]. Finally, setting

$$P_2 = <W_2>,$$

and taking the average value of the generating equation of the hierarchy, it turns out that the first equation of the hierarchy reads

$$p^\mu \partial_\mu f_1(x,p) + 4\pi e^2 \nabla_\mu \Big\{ p_\upsilon \int d^4x' d^4p' \big[p'^\mu \partial^\upsilon - p'^\upsilon \partial^\mu \big]$$
$$\times D(x-x')\big[f_2(x,p;x'p') + P_2(x,p;x'p') \big] \Big\} = 0.$$

The next equation of the relativistic BBGKY hierarchy is obtained by multiplying the generating equation by R_1 and averaging; after a little algebra, one gets

---

[11] See *e.g.* A.O. Barut (1965); F. Rohrlich (1965).

$$p^\mu \partial_\mu f_2(x,p;x'p')$$
$$+4\pi e^2 \nabla_\mu \left\{ p_\nu \int d^4x'' d^4p'' \left[p''^\mu \partial^\nu - p''^\nu \partial^\mu \right] D_{ret}(x-x'') \right.$$
$$\times \left[f_3(x,p;x',p';x''p'') + F_3^2[(x'',p'') \to (x,p);(x',p')] \right.$$
$$\left. \left. + F_3^2[(x'',p'') \to (x',p');(x,p)] \right] \right\} = 0.$$

A few comments are now in order. While in the Newtonian case the first equation of the hierarchy is an equation that needs the knowledge of f_2, in order to evaluate f_1, in the relativistic case one also needs the knowledge of one more function, say P_2. A glance at the next equation of the hierarchy shows that the knowledge of one more distribution, say $F_3^2[(x,p) \to (x',p');(x'',p'')]$, is also needed. This last distribution is the distribution of one particle undergoing the transition $(x,p) \to (x',p')$ while another particle is present in the state (x'',p''). It is not useful to give the third equation of the hierarchy since in actual practice only the first two are needed. This second equation is also supplemented by a symmetric equation on the primed variables, unlike the non-relativistic case[12]. However, one also needs, at least in principle, an equation for P_2. It can easily be obtained from the generating equation of the relativistic hierarchy, or from an equation for W_2, as

$$p^\mu \partial_\mu P_2(x,p;x'p') +$$
$$4\pi e^2 \nabla_\mu \left\{ p_\nu \int d^4x'' d^4p'' \left[p''^\mu \partial^\nu - p''^\nu \partial^\mu \right] D(x-x'') \right.$$
$$\left. \left[P_3(x,p;x',p';x''p'') + F_3^2[(x,p) \to (x',p');x'',p'')] \right] \right\} = 0,$$

where P_3 is the distribution function for a given particle to undergo the following transitions

$$(x,p) \to (x',p') \to (x'',p'').$$

4. TREATMENTS OF RADIATION

Radiation can be dealt with by using the conventional methods of turbulence theory or of stochastic processes : one looks for equations for the various moments of the electromagnetic field

$$\langle F^{\mu\nu}(x) \rangle, \langle F^{\mu\nu}(x) F^{\mu'\nu'}(x') \rangle, \langle F^{\mu\nu}(x) F^{\mu'\nu'}(x') F^{\mu''\nu''}(x'') \rangle, \ldots$$

and "mixed" quantities such as

$$\langle R_1(x) F^{\mu\nu}(x') \rangle, \langle R_2(x,p;x',p') F^{\mu\nu}(x'') \rangle, \ldots, \langle R_1(x) F^{\mu\nu}(x') F^{\mu'\nu'}(x'') \rangle, \ldots$$

---

[12] Note, however, that in the classical "multi-time" hierarchy, one also gets similar distributions.

where the resulting hierarchy has to be cut with some further physical assumption. Such a treatment has been extensively studied in the literature and, in the relativistic case, by A. Mangeney (1965). One usually decomposes the electromagnetic field into the field oscillators and this system is then treated as a conventional mechanical system. Although such an approach is not manifestly covariant, it needs a delicate study of the infinities brought by the electron self-field. This is the reason why another treatment was preferred [R. Hakim (1967)] that eliminates these infinities *ab initio*.

In the preceding section, it was mentioned that the interaction term involving P_2 is connected to a *self-interaction* of the particle. This has to be elaborated a little bit further. When we go back to the original δ-terms from which it results, we can realize that this self-interaction is a consequence of the action of the retarded electromagnetic field $F_{ret}^{\mu\nu}$ emitted by a given particle and acting on the same one. When this retarded field is split as[13]

$$F_{ret}^{\mu\nu} = \frac{1}{2}\left[F_{ret}^{\mu\nu} + F_{adv}^{\mu\nu}\right] + \frac{1}{2}\left[F_{ret}^{\mu\nu} - F_{adv}^{\mu\nu}\right],$$

it can be shown that the self-interaction of a particle contains two parts : the first one is *infinite* and of the general form

$$\infty \times \frac{dx^\mu(\tau)}{d\tau},$$

while the second one is finite and takes account of the back-reaction of the radiation emitted. The infinite term can be absorbed in a formal mass renormalization. Finally, the equations of motion obeyed by a system of electrons, embedded within a uniformly charged neutralizing background, read

$$\begin{cases} \dfrac{dp_i^\mu(\tau_i)}{d\tau_i} = \dfrac{e}{m} F_{ret}^{(i)\mu\nu}(x_i) p_{i\nu}(\tau_i) + \dfrac{2}{3} e^2 \left[\dot{\gamma}_i^\mu(\tau_i) + \gamma(\tau_i).\gamma(\tau_i) u_i^\mu(\tau_i)\right] \\ \partial_\mu F_{ret}^{(i)\mu\nu*} = 0 \qquad\qquad\qquad\qquad\qquad i = (1,2,...N); j = (1,2,...,\neq i,...N) \\ \partial_\mu F_{ret}^{(i)\mu\nu}(x_i) = \dfrac{e}{m} \int \sum_{j \neq i} \delta^{(4)}\left[x_i - x_j(\tau_j)\right] p_j^\nu(\tau_j) \end{cases}$$

with $\gamma^\mu \equiv d^2 x^\mu(\tau)/d\tau^2$. These equations must be supplemented by the following asymptotic conditions

$$\lim_{\tau \to \infty} \gamma^\mu(\tau) = 0$$

---

[13] A.O. Barut, *Electrodynamics and Classical Theory of Fields and Particles*, [Mac Millan; New-York (1965)]; F. Rohrlich *Classical charged particles* [Addison-Wesley, Reading (Mass.), (1965)].

which expresses the fact that the electrons are free at infinity. This last condition allows the elimination of the so-called "runaway solutions" [see F. Rohrlich (1965)].

These equations are discussed in details in the books by A.O. Barut (1965) and F. Rohrlich (1965). The presence of the proper time derivative of the four-acceleration in the equations of motion requires a modification of the above statistical treatment of the system. In particular, phase-space needs to be enlarged so as to take account of the acceleration variables as

$$\Gamma_{new} = \Gamma \times \{\gamma\}^{(4N)},$$

which necessitates the introduction of new distribution functions on this new phase-space. Accordingly, we introduce a random distribution R_1, as in a preceding section,

$$R_1[x,p] = \int d\tau \sum_i \delta^{(4)}[x - x_i(\tau)] \, \delta^{(4)}[p - p_i(\tau)] \, \delta^{(4)}[\gamma - \gamma_i(\tau)],$$

and also its average value over the initial conditions $f_1 = <R_1>$. Note that this new f_1 is normalized through

$$\int_\Sigma d\Sigma_\mu \int d^{(4)}p \int d^{(4)}\gamma \, \delta(p.\gamma) \frac{p^\mu}{m} f_1(x,p,\gamma) = N$$

since the integration must obey the constraint $p.\gamma = 0$. The continuity equation in the new one-particle phase-space[14]

$$\partial_\mu(p^\mu R_1) + \nabla_\mu(\gamma^\mu R_1) + \frac{\partial}{\partial \gamma^\mu}(\dot{\gamma}^\mu R_1) = 0,$$

gives rise to a new generating equation for the relativistic BBGKY hierarchy with radiation effects; or, explicitly, one obtains

$$p^\mu \partial_\mu R_1(x,p) + \gamma^\mu \nabla_\mu R_1 + \frac{\partial}{\partial \gamma^\mu}\left(\left[\frac{\gamma^\mu}{m\tau_{r_0}} - \gamma.\gamma \frac{p^\mu}{m}\right]R_1\right) =$$

$$\frac{4\pi e^2}{m\tau_0} \frac{\partial}{\partial \gamma^\mu}\left\{p_\nu \int d^4x' d^4p' d^{(4)}\gamma' \left[p'^\mu \partial^\nu - p'^\nu \partial^\mu\right] D_{ret}(x-x') R_2(x,p;x'p')\right\} = 0.$$

---

[14] Although the same notation for $R_1(x,p)$ and $R_1(x,p,\gamma)$ is used, there will be no confusion owing to the context.

The term W_2 has disappeared from this generating equation : the self-interaction has been eliminated in favor of a mass renormalization and of the finite γ-terms. This equation, like the non-relativistic one, is an equation for f_1 as a function of f_2. However, in the higher order equations "mixed" distributions still appear. This equation looks quite different from the former generating equation and it should be cast into a more useful form.

As a matter of fact, there exists an alternative phase-space and hence, alternative distribution functions, more suitable for a perturbation in powers of the small parameter τ_0. This last quantity is actually much smaller[15] than any physically meaningful times in the system. This stems from the remark that the equations of motion of the particles can be cast into the following form

$$\frac{dp_i^\mu(\tau_i)}{d\tau_i} = \frac{e}{m} F_{ret}^{(i)\mu\nu}(x_i) p_{i\nu}(\tau_i) + m\tau_0 \Delta^{\mu\nu}(p_i/m)\dot{\gamma}_{i\nu},$$

which shows clearly that the acceleration γ can be expressed in terms of p and $\dot{\gamma}$. In this last equation one has set

$$\Delta^{\mu\nu}(p) \equiv \eta^{\mu\nu} - \frac{p^\mu p^\nu}{m^2},$$

the projector orthogonal to p^μ. This means that another phase-space can be used, namely

$$\Gamma_{new} = \Gamma \times \{\dot{\gamma}\}^{(4N)},$$

with of course distribution functions depending on the variables $(x, p, \dot{\gamma})$. One can then derive various kinetic equations and obtained average values for the quantities connected to radiation (spectrum, etc).

5. RELATIVISTIC STATISTICAL MECHANICS (QUANTUM)

In this section, the above results are extended to the quantum case, mainly for fermions since the boson case is quite similar. In this review, we limit ourselves to the use of the covariant Wigner function method. It is indeed out of the scope of this review to deal with other powerful methods[16] such as Green's functions or thermo-field dynamics, etc. Also equations of state of high-density relativistic matter as occurs, for instance, in neutron or strange stars or in the primeval universe, are not considered even though they constitute a natural application of relativistic quantum statistical mechanics.

---

[15] Nevertheless, a dimensional analysis of the various terms of the basic equations shows that the radiation terms are comparable to the other ones whenever $k_B T \approx mc^2$.
[16] See *e.g.* J. Kapusta (1989)]; M. Le Bellac (2000)]

It should also be mentioned that, besides the original papers by F. Jüttner (1928), relativistic quantum statistical mechanics mainly originated from the remark by S. Chandrasekhar (1930, 1934) that the electrons within a white dwarf are both degenerate and relativistic. At about the same moment, the identification of pulsars with neutron stars [T. Gold (1968)] led, during the seventies, to numerous phenomenological models of relativistic dense matter. Finally, the need for possible signals for the production of a quark-gluon plasma in heavy ion collisions generated a systematic study of this physics.

As in the non-relativistic case, the basic tool is the density operator

$$\rho = \sum_n |n\rangle \, \varpi_n \, \langle n| ,$$

$$\sum_n \varpi_n = 1 , \quad \varpi_n \geq 0$$

where the ϖ_n's are the statistical weights of the n-th state. From the density operator ρ one calculates various quantities — such as Green's functions, introduced in statistical mechanics in 1959 by E.S. Fradkin, or Wigner functions on a "semi-classical" phase-space [E.P. Wigner (1932)] — from which the physics of the system under study can be extracted. Average values are then obtained as

$$<A> \; = Tr[\rho A] = \sum_n \varpi_n \langle n|A|n\rangle,$$

where A is a given observable. These relations are still valid in special relativity although some care is needed in their manipulation.

Thermal Equilibrium

In thermal equilibrium the density operator ρ possesses the general form

$$\rho_{eq} = const \times \exp(\sum_i \alpha_i A_i),$$

where the A_i's are *additive* first integrals of the system and where the α_i's are corresponding Lagrange multipliers. Whether in Newtonian physics or in relativity, there exists only seven time-independent such additive integrals : energy, impulsion and kinetic momentum. One possibly has to add other additive quantities such as the charge, the baryon number, etc. As usual in relativity, energy and impulsion will be treated on the same footing while rotational symmetry will not be dealt with. Therefore, the basic equilibrium density operator ρ_{eq} has the following form

$$\rho_{eq} = \frac{1}{Z} \exp(-\beta_\mu P^\mu + \beta\mu N),$$

where P^μ is the total four energy-momentum of the system; where N is the particle number operator, present when the number of particles is not constant and $<N> = const.$; where $\beta_\mu = \beta u_\mu$, with $\beta = (k_B T)^{-1}$; and where μ is the chemical potential, while Z is the partition function. P^μ is obtained from the energy-momentum tensor $T^{\mu\nu}$ of the system as

$$P^\mu = \int_\Sigma d\Sigma_\mu \, T^{\mu\nu}$$

where Σ is an arbitrary space-like three-surface, which can always be chosen as being the space-like three-plane $t = const.$, since $T^{\mu\nu}$ is conservative. This energy-momentum tensor contains the contributions of all the fields present in the system and their interactions — both quantum and possibly external — within the system.

Here, only the free field case is considered, a case first considered by A.E. Scheidegger and C.D. McKay (1951) and A.O. Barut (1958) although the first study of the relativistic and Fermi-Dirac distributions is due to F. Jüttner (1928).

Choosing the arbitrary space-like three-surface, involved in the definition of P^μ, as being a three-plane $t = const.$, and in the local frame of reference in which $\beta^\mu = (\beta, \vec{0})$, the Hamiltonian $P^0 \equiv H$ reads

$$H = \sum_{\{n\}} \omega_{\{n\}} \left[a^+_{\{n\}} a_{\{n\}} + \tfrac{1}{2} \right]$$

where $\{n\}$ indicates the set of those quantum numbers that determine the state of the system whose energy is $\omega_{\{n\}}$. This Hamiltonian corresponds to a scalar real boson field and the "vacuum" factor ½ can be absorbed into Z, so that it will be omitted. Also, the operator "number of particles" is given by

$$N = \sum_{\{n\}} a^+_{\{n\}} a_{\{n\}} \, .$$

Finally, the equilibrium density operator is of the general form

$$\rho_{eq} = \frac{1}{Z} \exp\left(-\beta \sum_{\{n\}} \omega_{\{n\}} a^+_{\{n\}} a_{\{n\}} \right),$$

which is formally identical to the usual one[17]. This has the consequence that the same calculations do apply in this case and that the statistical distribution of the states $\{n\}$ is still given by a Bose-Einstein factor

$$\text{Prob}[\{n\}] = \frac{1}{\exp(\beta \omega_{\{n\}}) - 1} \, .$$

---

[17] See e.g. K. Huang (1963)].

Of course, one has similar relations for fermions. For a free particle $\{n\} \equiv \{\vec{p}\}$ and

$$\omega_{\{n\}} \equiv \sqrt{\vec{p}^2 + m^2}$$

so that, for such a free particle, the relativistic Bose-Eintein distribution reads

$$f(p) = \frac{1}{\exp[\beta u^\nu p_\nu] - 1}$$

where the four-momenta p^μ are connected through the mass-shell relation $p^2 = m^2$. It is normalized *via*

$$n \equiv u_\alpha J^\alpha = \frac{d}{(2\pi)^3} \int \frac{dp}{p_0} u_\alpha p^\alpha \frac{1}{\exp[\beta u^\nu p_\nu] - 1},$$

where J^α is the particle four-current. In this last relation, d is a degeneracy factor depending on the internal quantum numbers (spin, etc) and as usual p^0 is the relativistic energy. From the expression of the energy-momentum tensor

$$T^{\alpha\beta} = \frac{d}{(2\pi)^3} \int \frac{dp}{p_0} p^\alpha p^\beta \frac{1}{\exp[\beta u^\nu p_\nu] - 1},$$

which has necessarily the perfect fluid form, one obtain the energy density as

$$\varepsilon = T^{\alpha\beta} u_\alpha u_\beta = \frac{d}{(2\pi)^3} \int \frac{dp}{p_0} (p.u)^2 \frac{1}{\exp[\beta u^\nu p_\nu] - 1}$$

and the pressure

$$P = \tfrac{1}{3}\Delta_{\alpha\beta}(u) T^{\alpha\beta} = \frac{d}{3(2\pi)^3} \int \frac{dp}{p_0} \Delta_{\alpha\beta}(u) p^\alpha p^\beta \frac{1}{\exp[\beta u^\nu p_\nu] - 1}$$

$$= \frac{d}{3(2\pi)^3} \int \frac{dp}{p_0} \frac{\vec{p}^2}{\exp[\beta u^\nu p_\nu] - 1}.$$

An example where the Bose-Einstein distribution is found with another set of quantum numbers $\{n\}$ can be found with the case of charged bosons embedded in an external magnetic field [Ph. Adam, R. Hakim (1982)].

The Covariant Wigner Distribution

The Wigner function (1932) is the quantum analog of the usual distribution function on phase-space with this difference that it is not always positive, in particular in domains of the size of $\hbar^3$. Furthermore, because of the fact that it does not constitute a true probability density but rather a theoretical way for the computation of statistical data on quantum systems, it is not unique and many other definitions are possible[18]. In the relativistic context, the first uses of such a tool seem to be the ones by D. Biskamp (1967) and by R. Balescu (1968, 1969). However, not only are these relativistic Wigner functions not manifestly covariant but they also contain an unnecessary normal product in their definition [Ch.G. van Weert (1975, 1980)]: this normal product eliminates *ipso facto* all vacuum terms and hence some finite effects, after renormalization. Covariant, albeit not general, relativistic Wigner functions were introduced later by P.A. Carruthers and F. Zachariasen (1974, 1976), by R. Hakim and R. Dominguez-Tenreiro (1976), R. Hakim and J. Heyvaerts (1976). A covariant definition for spin ½ particles was provided by Ch.G. Van Weert and W.P.H. de Boer (1975). Finally, the full covariant definition occurred at about the same time [R. Hakim (1976, 1978)]. This latter was then given a gauge-invariant form [E.A. Remler (1977), V.V. Klimov (1982), J. Winter (1984), U. Heinz (1983, 1985), H.-Th. Elze, M. Gyulassy and D. Vasak (1986). An interesting attempt (see below) of a relativistic albeit non-manifestly covariant, gauge-covariant Wigner function has also been studied by I. Bialynicki-Birula, P. Gornicki and J. Rafelski (1991). In what follows, we limit ourselves to the case of spin ½ particles.

As in the non quantum case, the starting point of the construction of a covariant Wigner function is the data of the basic observables of spin ½ particles; namely the four-current

$$J^\mu(x) = \overline{\psi}(x)\gamma^\mu\psi(x)$$

and the energy-momentum tensor

$$T^{\mu\nu} = \frac{i}{2}\overline{\psi}(x)\gamma^\mu \overset{\leftrightarrow}{\partial}{}^\nu \psi(x),$$

where ψ is the fermion field and where the γ's are the usual Dirac matrices. More generally, the four-current of an observable $A^{\cdots}$, is given by

$$J_A^{\mu\cdots}(x) = \overline{\psi}(x)\gamma^\mu A^{\cdots}\psi(x)$$

and its quantum statistical average is given by

$$<J_A^{\mu\cdots}(x)> \;=\; Tr\{\rho J_A^{\mu\cdots}(x)\}, \; etc.$$

Let us now introduce the covariant Wigner function operator as

$$F_{op}(x,p) = \frac{1}{(2\pi)^4}\int d^4R \; \exp[-ip.R] \; \overline{\psi}\!\left(x+\tfrac{1}{2}R\right) \otimes \psi\!\left(x-\tfrac{1}{2}R\right),$$

---

[18] See *e.g.* L. Cohen (1966).

and the covariant Wigner function as its quantum statistical average value

$$F(x,p) = <F_{op}(x,p)> = Tr\{\rho F_{op}(x,p)\}$$

$$= \frac{1}{(2\pi)^4} \int d^4R \ \exp[-ip.R] \ <\bar{\psi}(x+\tfrac{1}{2}R) \otimes \psi(x-\tfrac{1}{2}R)>$$

Note that $F(x, p)$ is a matrix in the spinorial (and possible other internal) indices; explicitly, one has

$$F_{\alpha\beta}(x,p) = \frac{1}{(2\pi)^4} \int d^4R \ \exp[-ip.R] \ \bar{\psi}_\beta(x+\tfrac{1}{2}R) \ \psi_\alpha(x-\tfrac{1}{2}R);$$

notice the opposite position of the indices for the ψ's and for F.

With this definition, the above average values of observables are given by

$$<J^\mu(x)> = Sp \int d^4p \ \gamma^\mu F(x,p)$$

$$<T^{\mu\nu}> = Sp \int d^4p \ \gamma^\mu p^\nu F(x,p),$$

where Sp indicates a trace over spinorial (and possibly internal) indices. As to the four-current of a given general observable A, it has also the form

$$<J_A^{\mu\cdots}(x)> = Sp \int d^4p \ \gamma^\mu A^{\cdots}(x,p) F(x,p)$$

although in each particular case, one has to find out the specific form of the function $A^{\cdots}(x,p)$. In actual practice, only $<J^\mu(x)>$ and $<T^{\mu\nu}>$ have to be calculated and possibly their fluctuations.

The one-particle covariant Wigner function $F(x, p)$ can be expanded on the basis of the sixteen Dirac's matrices

$$\{\gamma_A\}_{A=1,2,\ldots,16} = \left(I, \ \gamma^\mu, \ \sigma_{\mu\nu} \equiv \tfrac{i}{2}[\gamma^\mu,\gamma^\nu], \ \gamma_5, \ \gamma_5\gamma^\mu\right)$$

as

$$F(x,p) = \sum_A f_A(x,p) \ \gamma^A$$

$$\equiv f(x,p)I + f^\mu(x,p)\gamma_\mu + if^{\mu\nu}(x,p)\sigma_{\mu\nu} + if_5(x,p)\gamma^5 + f_{5\mu}(x,p)\gamma^5\gamma^\mu,$$

with

$$f_A(x,p) = Tr[F(x,p)\gamma_A].$$

Besides the one-particle Wigner function, other quantum distributions are needed such as the two-body covariant Wigner function

$$F_2(x,p;x',p') = \int \frac{dR}{(2\pi)^4} \frac{dR'}{(2\pi)^4} \exp[-ip.R]\exp[-ip'.R']$$
$$\times <\overline{\psi}(x'+\tfrac{1}{2}R')\overline{\psi}(x+\tfrac{1}{2}R)\ \psi(x-\tfrac{1}{2}R)\psi(x'-\tfrac{1}{2}R')>$$

Basic Equations

Let us now examine the equations obeyed by the Wigner function $F(x, p)$ and, in order to be specific, let us consider non-interacting particles. Then the fermion field ψ satisfies the following Dirac's equations

$$\begin{cases} [i\gamma\partial - m]\psi(x) = 0 \\ \overline{\psi}(x)[i\gamma\partial + m] = 0 \end{cases}$$

which, once considered respectively at points $x - R/2$ and $x + R/2$, after multiplying the first one by $\overline{\psi}(x+\tfrac{1}{2}R)\exp[-ip.R]$ from the left and the second one by $\psi(x-\tfrac{1}{2}R)\exp[-ip.R]$ from the right, and integrating over R, lead to

$$\begin{cases} \{i\gamma\partial + 2[\gamma.p - m]\}F(x,p) = 0 \\ F(x,p)\{i\gamma\partial - 2[\gamma.p - m]\} = 0 \end{cases}$$

where the operator $\{...\}$ in the second equation acts on the left. One can get some further insights from the expansion of $F(x, p)$ on the matrices γ_A of the Dirac's algebra and taking the trace of these equations. One obtains a set of thirty-two equations, which can be analyzed in detail and show that besides describing the general flow of particles and antiparticles in phase-space, they also contain their mass-shell.

This latter statement can be seen, for instance, in the case of the stationary solutions of the basic equations satisfied by $F(p)$, which then read

$$\begin{cases} [\gamma.p - m]F(p) = 0 \\ F(p)[\gamma.p - m] = 0 \end{cases}$$

and multiplying, for example, the first equation by $[\gamma.p + m]$, one gets

$$[p^2 - m^2]F(p) = 0,$$

which shows that

$$F(p) \propto \delta(p^2 - m^2),$$

and hence that all its components f_A's are on the particle mass-shell. The general solution of the above system is then easily shown to be of the general form

$$F(p) = \frac{[\gamma.p+m]}{m} f(p)$$

and, in thermal equilibrium, written in a compact form, one finds

$$F_{eq}(p) = [\gamma.p+m] \frac{d\delta(p^2-m^2)}{(2\pi)^3} \frac{\varepsilon(p.u)}{\exp[\beta(p.u-\mu)]+1}.$$

where d is a degeneracy factor that takes account of spin (then $d = 2$) and other possible internal degrees of freedom; and where $\varepsilon(...)$ is the sign function. This compact form contains the contribution of the fermions and their anti-particles and also a vacuum term, absent when the definition of the Wigner function contains a normal product; however, after a renormalization, it leads to finite effects.

Another case of particular interest is the one in which there exists a space-like pseudo-vector S^μ. The general form for $F(x, p)$, still for a stationary solution, is then found to be

$$F(p) = \frac{(\gamma.p+m)}{2m} \frac{(1+\gamma_5\gamma_\mu S^\mu(p))}{2} f(p).$$

and represents a polarized state [R. Hakim, L. Mornas, P. Peter, H. Sivak (1992)].

The BBGKY Relativistic Quantum Hierarchy

In order to illustrate the above techniques in a more concrete way, they will be applied here to the so-called "scalar plasma" first studied by G. Kalman (1974). This toy model can also be slightly extended to yield the phenomenological model for nuclear matter proposed by J.D. Walecka (1974).

The scalar plasma consists of a baryon field $\psi(x)$ coupled to a scalar field $\phi(x)$ and is described by the following Lagrangian

$$L = \tfrac{1}{2} i\overline{\psi}\gamma^\mu \overset{\leftrightarrow}{\partial}_\mu \psi - (m - g_s\varphi) + \tfrac{1}{2}\left(\partial_\mu\varphi.\partial^\mu\varphi - m_s^2\varphi^2\right)$$

where m_S is the mass of the scalar particles and g_S their coupling constant with the baryon field.

Using the equations of motion

$$\begin{cases} i\gamma.\partial\psi(x) - (m - g_s\varphi(x))\psi(x) = 0 \\ i\partial\overline{\psi}(x).\gamma + \overline{\psi}(x)(m - g_s\varphi(x)) = 0 \\ (\Box + m_s^2)\varphi(x) = g_s\overline{\psi}(x)\psi(x) \end{cases}$$

and the definition of the one-particle Wigner operator, one is led to the following system

$$\begin{cases} \{i\gamma\partial+2[\gamma.p-m]\}F_{op}(x,p)=-2g_s \int \frac{d^4R}{(2\pi)^4} d^4\xi \, \exp[-i(p-\xi).R] \, F_{op}(x,\xi)\varphi(x-\tfrac{1}{2}R) \\ F_{op}(x,p)\{i\gamma\overleftarrow{\partial}-2[\gamma.p-m]\}=+2g_s \int \frac{d^4R}{(2\pi)^4} d^4\xi \, \exp[-i(p-\xi).R]\varphi(x+\tfrac{1}{2}R)F_{op}(x,\xi) \\ (\partial^2+m_s^2)\varphi(x)=g_s Sp \int d^4 p \, F_{op}(x,p) \end{cases}$$

hereafter referred to as the *generating equation* of the relativistic quantum BBGKY hierarchy. It can be formally rewritten as

$$\begin{cases} LF_{op} = g_s \int F_{op}\varphi \\ L^{\circ}F_{op} = g_s \int \varphi F_{op} \\ KG\varphi = g_s Sp \int F_{op} \end{cases}$$

in order to facilitate the subsequent developments. After taking the average value of this system, one finds

$$\begin{cases} L F = g_s \int <F_{op}\varphi> \\ F L^{\circ} = g_s \int <\varphi F_{op}> \\ KG<\varphi> = g_s Sp \int F \end{cases}$$

which is now briefly discussed. First, it connects F to $<F_{op}\varphi>$ and to $<\varphi F_{op}>$, showing that the determination of the one-particle Wigner function needs the knowledge of another function. Next, the first two equations of the system must be consistent with each other and thus must be such that the following equality be always satisfied

$$L^{-1}\int <F_{op}\varphi> = \int <\varphi F_{op}> L^{\circ -1} \,.$$

In order to calculate $<F_{op}\varphi>$, the first equation of the generating system is multiplied by φ from the right and then averaged; this yields

$$L <F_{op}\varphi> = g_s \int <F_{op}\varphi\varphi>$$

and the knowledge of $<F_{op}\varphi>$ demands that of $<F_{op}\varphi\varphi>$, etc. Similarly, from the third equation of the generating equation of the hierarchy, with an analogous manipulation, one obtains

$$KG < F_{op}\varphi > \; = \; g_S Sp \int < F_{op}F_{op} >,$$

from which one gets

$$< F_{op}\varphi > \; = \; g_S KG^{-1} Sp \int < F_{op}F_{op} >$$

and hence

$$LF = g_S^2 KG^{-1} Sp \int < F_{op}F_{op} >.$$

Note that $< F_{op}F_{op} >$ is not the two-particle Wigner function of the baryons but is related to the latter and can be used to calculate the fluctuations of various observables. The system is closed as usual by cutting the hierarchy with some particular *Ansatz* such as the Hartree-Vlasov one, $\langle F(x,p)\varphi(x')\rangle \approx \langle F(x,p)\rangle\langle\varphi(x')\rangle$, which is the simplest one.

Strong Magnetic Fields

V. Canuto and HY. Chiu (1968*ff*) have studied in great detail the electron gas embedded in a strong[19] magnetic field[20] while the QED plasma modes (still in a strong magnetic field) have been studied by many authors [V. Canuto (1969, 1970), P. Bakshi, R. Cover and G. Kalman (1976), H. Sivak (1985), Shabad and Perez-Rojas (1976*ff*), etc]. The equation of state and the equilibrium properties of such a plasma have been investigated by D.H. Constantinescu (1972*ff*, P. Rehak (1975), G.A. Schulman (1972*ff*), S. Visvanathan (1962), etc. The covariant Wigner function techniques have been used by several authors [R. Dominguez-Tenreiro, R. Hakim (1977), R. Hakim, H. Sivak (1982)] in this case : thermal equilibrium, excitation modes of the relativistic quantum plasma embedded in a strong magnetic field [H. Sivak (1985)], calculation of the transport coefficients, etc.

6. RELATIVISTIC KINETIC THEORY

Once the relativistic distribution function is defined and the Liouville equation derived, the next step is the obtaining of a kinetic equation, which the covariant distribution function is supposed to obey. All relativistic kinetic equations, whether classical or quantum, contain the following three basic ingredients : (i) the description of the general flow of particles in phase-space by the Liouville equation (or by the equivalent quantum analog), (ii) a collision term and (iii) the influence of possible collective effects. The latter can occur either within the Liouville equation (as for instance *via* a mean field term) and/or in the collision term itself, as in the relativistic Landau equation given by S.T. Beliaev and G.I.

---

[19] A magnetic field is said to be strong when the Compton wavelength of the charged particles is of the order of their Larmor radius; for electrons, this corresponds to B $\sim 10^{14}$ Gauss.
[20] See the excellent review by V. Canuto and J. Ventura (1977).

Budker (1956) or Yu. L. Klimontovich (1960). Since in section 2 and 4 the Liouville equation and a quantum analog were given, we now emphasize points (ii) and (iii) only.

Various Collision Terms

The best known one is a relativistic generalization of the usual Boltzmann equation. It was first given by A. Lichnérowicz and R. Marrot (1940) in a non manifestly covariant form while it was later studied (and rediscovered) in a fully relativistic way by many authors : N.A. Chernikov (1956*ff*), G. Tauber and J.W. Weinberg (1961), W. Israel (1963), Ch. Marle (1969), etc. This equation is found to have the following form [S.R. de Groot, W.A. van Leuwen, Ch. Van Weert (1980); see also J.M. Stewart (1971)]

$$\left[p^\mu \partial_\mu + F^\mu(x,p) \frac{\partial}{\partial p^\mu} \right] f(x,p) = C\{f(x,p)\}$$

where $C\{f(x,p)\}$ is the Boltzmann collision term, which we give in terms of the transition probability $W(p',p'' \to p,\bar{p})$ for the scattering of two particles within the system as

$$p.\partial f(x,p) = \frac{1}{2} \int \frac{d^3 p'}{p'_0} \frac{d^3 p''}{p''_0} \frac{d^3 \bar{p}}{\bar{p}_0} W(p',p'' \to p,\bar{p}) \delta^{(4)}(p+p''-p'-\bar{p})$$
$$\times \left[f(x,p')f(x,p'') - f(x,p)f(x,\bar{p}) \right]$$

[S.R. de Groot, W.A. van Leuwen, Ch. Van Weert (1980)]; $F^\mu(x,p)$ is an external force in which the system is possibly embedded. More generally, the collision term of any valid relativistic kinetic equation should be such that the conservation laws are obeyed; this leads, after successively multiplying both sides of the kinetic equation by 1 and p^μ and integrating over p, to

$$\begin{cases} \partial_\nu J^\nu(x) = 0 = \int d^4 p \; C\{f(x,p)\} = 0 \\ \partial_\nu T^{\mu\nu}(x) = 0 = \int d^4 p \; p^\mu C\{f(x,p)\} = 0 \end{cases}$$

The Boltzmann's equation has been studied mainly by the above-mentioned authors and, in particular, by W. Israel (1963). The non-relativistic approximation methods have been extended by various authors to the special relativity case and applied in several formal situations, such as the derivation of the various relativistic forms of hydrodynamics

Unfortunately, the relativistic Boltzmann equation seems to be mainly a conceptual tool, useful for the evaluation of those differences existing between the relativistic and classical cases that do not apply to any real physical situation[21]. In particular, it requires point-particles collisions while in relativistic dense matter one rather deals with collisions of collective (extended) modes. Therefore, it seems preferable to use a merely *phenomenological* kinetic equation that should contain the main characteristic features of the physical system under study.

The simplest one is the covariant version of the Bhatnagar-Gross-Krook equation studied in detail by J.L. Anderson and H.R. Witting (1974)

$$\left[p^\mu \partial_\mu + F^\mu(x,p) \frac{\partial}{\partial p^\mu} \right] f(x,p) = C\{f(x,p)\} \equiv -p.u \, \frac{f(x,p) - f_{eq}(p)}{\tau_0},$$

where τ_0 is a *relaxation time* to be evaluated with the help of other considerations and which might possibly be a function of x and p. τ_0 can be roughly evaluated as

$$\tau_0 \approx \frac{1}{n v_{th} \sigma},$$

where n is the numerical density of particles, v_{th} the thermal velocity and σ the total collision cross-section. Note that, in the quantum case, one has a similar equation with this difference that f_{eq} is a quantum distribution of the Fermi-Dirac or Bose-Einstein-type.

The quantum analog of the Boltzmann collision term is the Uhlenbeck-Uehling term that contains the effect of those statistics (Bose or Fermi) obeyed by the particles of the system and reads[22]

$$p.\partial f(x,p) = \frac{1}{2} \int \frac{d^3 p'}{p'_0} \frac{d^3 p''}{p''_0} \frac{d^3 \overline{p}}{\overline{p}_0} W\left(p', p'' \to p, \overline{p}\right) \delta^{(4)}\left(p + p'' - p' - \overline{p}\right)$$
$$\times \left[f(x,p') f(x,p'') [1 \pm f(x,p)] [(1 \pm f(x,\overline{p})] \right.$$
$$\left. - f(x,p) f(x,\overline{p}) [1 \pm f(x,p')] [1 \pm f(x,p'')] \right]$$

and one can verify that the Bose-Einstein (plus sign) or the Fermi-Dirac functions (minus sign) are stationary equilibrium solutions.

In the quantum case, for spin ½ particles, to be specific, one can also obtain a rich variety of collision terms. However, insofar as nuclear matter is concerned, the covariant version of the BGK equation for such particles, *i.e.*

---

[21] For instance, it has been used to explain the matter/antimatter asymmetry in the primeval universe [see *e.g.* E.W. Kolb and M. Turner (1983)]. However, the situation prevailing at the time considered (10^{-18} s) is one of high densities and/or temperatures where collective effects are predominant and not collisions.
[22] W is the transition probability per unit of time. See S.R. de Groot *et al.* for details.

$$\begin{cases} \{i\gamma.\partial + 2[\gamma.p - m]\}F(x,p) = -i\gamma.u \dfrac{F(x,p) - F_{eq}(p)}{\tau} \\ F(x,p)\{i\gamma.\partial - 2[\gamma.p - m]\} = -\dfrac{F(x,p) - F_{eq}(p)}{\tau} i\gamma.u \end{cases}$$

(u is the average four-velocity of the system) exhibits both a correct behavior for physical quantities as functions of the variables at hand (density, temperature, etc) and results of the right order of magnitude [see *e.g.* L. Mornas (1992)].

Collective Effects

The simplest case much studied in the current literature is that of the relativistic plasma, whether classical or quantum. It is embodied in the Vlasov (or Hartree-Vlasov) equation where the collective term appears in the Liouville equation (or its quantum analog) : see *e.g.* V.P. Silin (1960), Yu.L. Klimontovich (1960), R. Hakim and A. Mangeney (1968, 1971), for the classical plasma and D. Biskamp (1967), R. Hakim and J. Heyvaerts (1978,1980), H. Sivak (1985), etc. for the quantum one. The case of relativistic quasi-particles [see *e.g.* A.D. Migdal (1978), R. Hakim and H. Sivak (1993, 1995, 1999), R. Hakim, L. Mornas (1993), etc.] is typical of collective effects and, for instance, the BGK relativistic kinetic equation for spin ½ quasi-particles reads

$$\begin{cases} \{i\gamma.\partial + 2[\gamma.p - m]\}F(x,p) = -i\dfrac{\partial D(k)}{\partial k}..u \dfrac{F(x,p) - F_{eq}(p)}{\tau} \\ F(x,p)\{i\gamma.\partial - 2[\gamma.p - m]\} = -\dfrac{F(x,p) - F_{eq}(p)}{\tau} i\dfrac{\partial D(k)}{\partial k}..u \end{cases}$$

where

$$D(k)\psi(k) \equiv \{\gamma.k - \Sigma(k)\}\psi(k) = 0$$

is the field equation obeyed by the quasi-particles and where $F_{eq}(p)$ is their equilibrium distribution, which differs from the ordinary one by its mass-shell and its normalization only.

7. CURVED SPACETIME

In the presence of gravitation, *i.e.* in the case of a curved space-time, most of the above results have been extended by numerous authors. The one-particle phase-space has a particular mathematical structure since the configuration space is curved while the energy-momentum space is co-tangent to the space-time manifold

$$\mu = H^4(x) \times V^4,$$

where $H^4(x)$ is the energy-momentum space, generally characterized by

$$H^4(x) : g_{\mu\nu}(x)p^\mu p^\nu = m^2$$

This μ-space is thus endowed with a *fiber bundle* structure, whose fiber is nothing but the above *x*-dependent hyperboloid and whose invariance group is the Lorentz group. While the definition of the one-particle density is identical in the curved and flat cases, there exists some minor modifications as to the Liouville (or the kinetic) equation(s) and as to the Jüttner-Synge equilibrium distribution function.

One should also notice that, in general relativity, one is mostly interested in *local* quantitities, or four-currents of various physical quantities and hence, only the latter actually make sense such as, for instance, the energy-momentum tensor, the four-current of a given physical quantity (charge, number of particles, etc.), entropy four-current and other thermodynamical quantities. Another point to be remarked is the fact that gravitation being a long range and very weak force, its gradients are generally negligible on the distance of two *short-range* colliding particle. It follows that the flat space-time collision term is still valid in the case of general relativity.

The one-particle distribution function is now a little bit modified *via* the invariant (under coordinate changes) element of integration on the energy-momentum space $H^4(x)$

$$\frac{d^3p}{p_0} \to \sqrt{|g(x)|}\frac{d^3p}{p_0},$$

where $g(x)$ is the determinant of the metric tensor; of course, in a local frame of reference $|g(x)|=1$. Consequently, the four current of particles and the energy-momentum tensors read respectively

$$J^\mu = \int_{g_{\mu\nu}(x)p^\mu p^\nu = m^2} \sqrt{|g(x)|}\frac{d^3p}{p_0} p^\mu f(x,p)$$

$$T^{\mu\nu}(x) = \int_{g_{\mu\nu}(x)p^\mu p^\nu = m^2} \sqrt{|g(x)|}\frac{d^3p}{p_0} p^\mu p^\nu f(x,p).$$

Let us now briefly derive the Liouville equation when an *external* gravitational field is present and in the absence of other forces, since the latter can be added without any particular difficulty. The equations of motion for one particle is the usual geodesic equation

$$\frac{dp^\mu}{d\tau} + \frac{1}{m}\Gamma^\mu_{\alpha\beta}p^\alpha p^\beta = 0,$$

which can be obtained from the *formal* Hamiltonian[23]

$$H(x,p) = \frac{1}{2m} g_{\mu\nu}(x) p^\mu p^\nu.$$

Consequently, the Liouville equation follows in a straightforward way as

$$p.\partial f(x,p) - \Gamma^\mu_{\alpha\beta} p^\alpha p^\beta \frac{\partial}{\partial p^\mu} f(x,p) = 0.$$

Many other derivations of this equation have been obtained. Another form of this equation, coming directly from the above Hamiltonian, also reads

$$p.\partial f(x,p) - \frac{1}{2} \partial_\mu g_{\alpha\beta}(x) p^\alpha p^\beta \frac{\partial}{\partial p_\mu} f(x,p) = 0.$$

When the Liouville equation in curved space-time is coupled to the gravitation field *via* the Einstein's equations

$$R_{\mu\nu}(x) - \frac{1}{2} R(x) g_{\mu\nu} = 4\pi G T^{\mu\nu}(x)$$

$$= 4\pi G \int_{g_{\mu\nu}(x) p^\mu p^\nu = m^2} \sqrt{|g(x)|} \frac{d^3p}{p_0} p^\mu p^\nu,$$

one obtains the gravitational equivalent of the usual Vlasov equation for an electromagnetic plasma [Ph. Droz-Vincent, R. Hakim (1968)]. This system is often referred to as the Einstein-Liouville equations. In particular, it has been used to find the normal modes of a gravitational plasma [E. Asseo, D. Gerbal, J. Heyvaerts, M. Signore (1976)].

Let us now consider the Jüttner-Synge equilibrium distribution in presence of gravitation. It constitutes always a *local* equilibrium and not a global one as in the flat space case. The gravitation field enters the equilibrium distribution $f_{eq} = A \exp[-\beta.p]$ through the scalar product

$$\beta.p = g_{\mu\nu}(x) \beta^\mu p^\nu.$$

The equilibrium distribution must obey the equation of motion written under the form of the Liouville equation; introducing f_{eq} in the latter, one gets

$$p.\partial A + \frac{1}{2}\left[\nabla_\mu \beta_\nu + \nabla_\nu \beta_\mu\right] A p^\mu p^\nu = 0,$$

---

[23] This means only that the geodesic equation can be recovered as a Hamiltonian equation but *not* that H is the energy of the particle.

which must be satisfied whatever p. This implies that, in general, the four-vector β^μ has to satisfy the Killing conditions

$$[\nabla_\mu \beta_\nu + \nabla_\nu \beta_\mu] = 0,$$

and $A = const$. This latter condition yields $n = const$ and $T = const$. There is however another case, the one in which β^μ is Killing conformal

$$[\nabla_\mu \beta_\nu + \nabla_\nu \beta_\mu] = \phi(x) g^{\mu\nu}(x),$$

where $\phi(x)$ is an arbitrary function, and $p^2 = 0$.

A first conclusion is that the one particle equilibrium does not always exist; for massive particles, the four-vector $\beta^\mu u^\mu$ has to be a Killing field while for zero mass particles it must be Killing conformal.

Let us finally add that statistical mechanics of gravitationally interacting particles have been attempted by A.H. Tauber (1974) and in a series of paper by H.E. Kandrup see (1986). Also, the Wigner function in curved spaces have been defined and used by J. Winter (1985), E. Calzetta and B.L. Hu (1986,1988,1989), F. Antonsen (1987), H.E. Kandrup (1988), O.A. Fonarev (1993), etc.

8. A BRIEF CONCLUSION

From this brief overview of relativistic statistical mechanics, one might have the feeling that this (small) domain of relativity has reached more or less the same status as its Newtonian counterpart. Such an impression is largely correct despite their basic differences : *non-local* versus local interactions, *non-Hamiltonian* equations of motion in the relativistic case, *fields* versus particles in the quantum case, etc. Of course, mathematical problems are not all solved and, in particular in the classical domain, those dealing with the nature of initial data. However, one should notice that this problem does not prevent the building of statistical mechanics since it is sufficient to assume the existence of average values over the initial data, whatever their specific nature. In the quantum domain, there exists a number of specific difficulties which don't exist either in the Newtonian case or in a perturbative treatment; for instance, how to express a cluster decomposition for the *N*-body Wigner function or to separate out exchange terms ? More generally, the case of gauge thoeries, as required in the case of the quark-gluon plasma, has not been considered in this article : it constitutes indeed a full subject in itself, in constant development and with its own problems both of physical and mathematical nature.

Let us finally insist that the various methods presented above are of common use in the study of nuclear matter and heavy ion collisions and relativistic plasmas, two domains of particular importance in astrophysics.

ACKNOWLEDGMENTS

I would like to thank the organizers of this conference for their warm hospitality in Oviedo and, in particular, Prof. J. Diaz-Alonso and Dr. L. Mornas. I am indebted to this latter for many suggestions and for the correction of many misprints in a first version of the manuscript of this talk.

REFERENCES

Abramovitz M. and I.A. Stegun, *Handbook of mathematical functions* [Dover; New York (1965)].
Adam Ph., R. Hakim, *Covariant Wigner Function Approach to the Relativistic charged Gas in a Strong Magnetic Field II. Spin Zero in Thermal Equilibrium,* Physica, **113A**, 491 (1982)
Anderson J.L., H.R. Witting, *A relativistic relaxation-time model for the Bolzmann equation,* Physica, **74**, 466 (1974)
Anderson J.L., H.R. Witting, *Relativistic quantum transport coefficients for Fermi-Dirac and Bose-Einstein gases,* Physica **74**, 489 (1974)
Asseo E., D. Gerbal, J. Heyvaerts, M. Signore, *General-relativistic kinetic theory of waves in a massive particle medium,* Phys. Rev., **D13**, 2724 (1976)
Balescu R., *On the Statistical Mechanics of a Relativistic Quantum Plasma,* Physica, **31**, 1599, (1965)
Balescu R., T. Kotera, *On the covariant formulation of classical relativistic statistical mechanics,* Physica, **33**, 558(1967)
Balescu R., T. Kotera, E. Pena, *Lorentz Transformations in Phase Space and in Physical Space,* Physica **33**, 581 (1967)
Balescu R., R. Sergysels, *Evolution Equations for the Coherence Function in a Relativistic Plasma,* Nuovo Cimento, **50**, 378, (1967)
Balescu R., M. Baus, A. Pytte, *On some Mathematical Aspects of Classical Relativistic Statistical Mechanics,* Bull. Cl. Sciences, Acad. Roy. Belg., **53**, 1043, (1967)
Balescu R., *A covariant formulation of quantum relativistic statistical mechanics, I. Phase space description of a relativistic quantum plasma,* Acta Phys. Austriaca, **28**, 309 (1968)
Balescu R., A. Pytte, *Geometrical and Tensorial Lorentz Transformations,* Nuovo Cimento, **55 B**, 51, (1968)
Balescu R., *Kinetic Equations and Lorentz Transformations,* Physica, **38**, 119, (1968)
Balescu R., *Relativistic Statistical Thermodynamics,* Physica, **40**, 309, (1968)
Balescu R., *A covariant formulation of quantum relativistic statistical mechanics, I. The Liouville equations for a relativistic quantum plasma,* Acta Phys. Austriaca, **29**, 313 (1969)
Balescu R., *A covariant formulation of quantum relativistic statistical mechanics, II. The Liouville equations for a relativistic quantum plasma,* Acta Phys. Austriaca, **29**, 313 (1969)
Balescu R., L. Brenig, *Relativistic Covariance of Nonequilibrium Statistical Mechanics,* Physica, **54**, 504, (1971)
Barut A.O., *"Electrodynamics and Classical Theory of Fields and Particles",* p. 122 ff. [Mac Millan; New-York (1965)]
Barut A.O., *Covariant quantum statistics of fields,* Phys. Rev., **109**, 1376 (1958)
Beliaev S.T., G.I. Budker, *The relativistic kinetic equation,* Sov. Phys. Dokl., **1**, 218 (1956)
Bergmann P.G., *Generalized statistical mechanics,* Phys. Rev., **84**, 1026 (1951)
Bergmann P.G., *The special theory of relativity* in : S. Flügge (Ed.), Handbuch der Physik (Encyclopedia of Physics) 4, Prinzipien der Elektrodynamik und Relativitätstheorie, 159 (Springer-Verlag, Berlin, 1962).
Bialynicki-Birula I., E.D. Davis, J. Rafelski, *Evolution modes of the vacuum Wigner function in strong-field QED,* Phys. Lett., B311, 329 (1993)
Calzetta E., S. Habib, B.L. Hu, *Quantum kinetic field theory in curved spacetime: Covariant Wigner function and Liouville-Vlasov equations,* Phys. Rev. **D37**, 2001 (1988)
Calzetta E., B.L. Hu, *Wigner distribution function and phase-space formulation of quantum cosmology,* Phys. Rev. **D40**, 380 (1989)
Calzetta E., *Wigner functions in quantum gravity and quantum cosmology,* Proceedings of the fourth Lake Louise Winter Institute; A. Astbury, B.A. Campbell, W. Israel, A.N. Kamal, F.C. Khanna eds. [Frontiers in physics, Singapore (1989)]
Canuto V., H.-Y. Chiu, L. Fassio-Canuto, *Thermodynamic approach to the equation of state of a magnetized Fermi gas,* Astrophys. Space Sci., **3**, 258 (1969)
Canuto V., H.-Y. Chiu, *Magnetic moment of a magnetized Fermi gas,* Phys. Rev., **173**, 1229 (1968)

Canuto V., H.-Y. Chiu, *Properties of high-density matter in intense magnetic fields*, Phys. Rev. Lett., **21**, 110 (1968)
Canuto V., H.-Y. Chiu, *Quantum theory of an electron gas in intense magnetic fields*, Phys. Rev., **173**, 1210 (1968)
Canuto V., H.-Y. Chiu, *Thermodynamic properties of a magnetized Fermi gas*, Phys. Rev., **173**, 1220 (1968)
Canuto V., J. Ventura, *Quantum theory of the dielectric constant of a magnetized plasma and astrophydical applications I. Theory*, Space Sci. Rev., **18**, 104 (1972)
Canuto V., J. Ventura, *Quantizing magnetic fields in astrophysics*, Fund. Cosmic Phys., **2**, 203 (1977)
Carruthers P.A., F. Zachariasen, *Transport equation approach to multiparticle production* [American Physical Society meeting; Williamsburg (1974)]
Carruthers P.A., F. Zachariasen, *Relativistic quantum transport theory approach to multiparticle production*, Phys. Rev., **D13**, 950 (1976)
Carruthers P.A., F. Zachariasen, *Quantum collision theory with phase-space distribution functions*, Rev. Mod. Phys., **55**, 245 (1983)
Chandrasekhar S., *Introduction to the Study of Stellar Structure*, [Dover; New York]
Chernikov N.A., *The generalized stochastic problem of particle motion*, Sov. Phys. Dokl., **1**, 103 (1956) VV
Chernikov N.A., *Reduction of the relativistic collision integral to the Boltzmann form*, Sov. Phys. Dokl., **5**, 764 (1960)
Chernikov N.A., *Relativistic kinetic equation and the equilibrium state of a gas in a static, spherically symmetric gravitational field*, Sov. Phys. Dokl., **5**, 786 (1960)
Chernikov N.A., *Kinetic equation for a relativistic gas in an arbitrary gravitational field*, Sov. Phys. Dokl., **7**, 397 (1962)
Chernikov N.A., *Relativistic Maxwell-Boltzmann distribution and integral form of the conservation laws*, Sov. Phys. Dokl., **7**, 428 (1962)
Chernikov N.A., *Flux vector and mass tensor of a relativistic ideal gas*, Sov. Phys. Dokl., **7**, 414 (1962)
Chernikov N.A., *Derivation of the equations of relativistic hydrodynamics from the relativistic transport equation*, Phys. Lett. **5**, 115 (1963)
Chernikov N.A., *Equilibrium distribution of the relativistic gas*, Acta Physica Polonica, **26**, 1069 (1964)
Chernikov N.A., *Microscopic foundation of relativistic hydrodynamics*, Acta Physica Polonica, **27**, 465 (1964)
Chernikov N.A., *The relativistic collision integral*, Sov. Phys. Dokl., **2**, 248 (1957)
Chernikov N.A., *The relativistic gas in the gravitational field*, Acta Physica Polonica, **23**, 629 (1963)
Cohen L., *Generalized phase-space distribution functions*, J. Math. Phys., **7**, 781 (1966)
Constantinescu D.H., P. Rehàk, *Condensed matter in a very strong magnetic field at high pressure and zero temperature*, Meeting on the *Role of Magnetic Fields in Physics and Astrophysics*, Nordita, Copenhagen (1974)
Constantinescu D.H., P. Rehàk, *Equation of state for condensed matter in a very strong magnetic field*, Ann. New York Acad. Sci., **257**, 85 (1975)
Constantinescu D.H., *Condensed matter in a very strong magnetic field, at high pressure and zero temperature*, Nuovo Cim., **B32**, 177 (1976)
Constantinescu D.H., *Thermodynamics of the statistical model of atoms in very strong magnetic fields: an approach to the properties of condensed matter in the outer crust of pulsars*, Phys. Rev., **D18**, 1820 (1978)
Cooper F., D.H. Sharp, *Pion production from a classical source : transport and hydrodynamical properties*, Phys. Rev **D8**, 194 (1975)
Cooper F., D.H. Sharp, *Transport and hydrodynamic aspects of multiparticle production*, Phys. Rev **D8**, 194 (1975)
Cooper F., *The hydrodynamical model of multiparticle production* in: Ettore Majorana Conference Proceedings *"Is there an ultimate temperature in hadron physics ?"*, [Bielefeld] (1975)
Cooper F., G.S. Guralnik, S.H. Kasdan, *Collective phenomena in $\lambda\phi^4$ field theory treated in the random-phase approximation*, Phys. Rev **D14**, 1607 (1976)
Cooper F., M. Feigenbaum, *Transport approach to multiparticle production : collective phenomena and renormalization*, Phys. Rev **D14**, 583 (1976)
Currie, D.G. T.F. Jordan and E.C.G. Sudarshan, *Relativistic Invariance and Hamiltonian Theory of Interacting Particles*, Rev. Mod. Phys., **35**, 350 (1963) ; see also G. Marmo et al.
de Groot S.R., W.A. van Leeuwen, C.G. van Weert, *Relativistic kinetic theory* [North Holland; Amsterdam (1980)]
Dominguez Tenreiro R., R. Hakim, *Transport coefficients of the relativistic degenerate electron gas in a strong magnetic field*, J. Phys., **A10**, 1525 (1977)
Dominguez Tenreiro R., R. Hakim, *Transport properties of the relativistic degenerate electron gas in a strong magnetic field: covariant relaxation-time model*, Phys. Rev. **D15**, 1435 (1977)

Dominguez Tenreiro R., R. Hakim, *Covariant Wigner Function Approach to the Relativistic charged Gas in a Strong Magnetic Field, I. Electron Gas*, Physica, **113A**, 477 (1982)

Droz-Vincent Ph., R. Hakim, *Collective motions of the relativistic gravitating gas*, Ann. Inst. H. Poincaré, **9**, 17 (1967)

Elze H.-T., M. Gyulassy, D. Vasak, *Transport equations for the QCD quark Wigner operator*, Nuclear Phys., **B276**, 706 (1986)

Elze H.-T., M. Gyulassy, D. Vasak, *Transport equations for the QCD gluon Wigner operator*, Phys. Lett, **B177**, 402 (1986)

Elze H.-T., U. Heinz, *Quark-gluon transport theory*, Phys. Rev., **A33**, 1879 (1986)

Elze H.-T., D.E. Miller, K. Redlich, *Gauge theories at finite temperatures and chemical potential*, Phys. Rev., **D35**, 748 (1987)

Elze H.-T., M. Gyulassy, D. Vasak, H. Heinz, H. Stöcker, W. Greiner, *Towards a relativistic selfconsistent quantum transport theory of hadronic matter*, Mod. Phys. Lett., **A2**, 451 (1987)

Elze H.-T., *Gluon transport equations, plasma oscillations, and screening*, Z. Phys., **C38**, 211 (1988)

Elze H.-T., K. Kajantie, T. Toimela, *Chromomagnetic screening at high temperature*, Z. Phys., **C37**, 601 (1988)

Elze H.-T., U. Heinz, K. Kajantie, T. Toimela, *High temperature gluon matter in the background gauge*, Z. Phys., **C37**, 305 (1988)

H-Th. Elze, U. Heinz, *Quark gluon transport theory*, Phys. Rep., **183**, 81 (1989)

Fokker A.D., *Ein invarianter Variationssatz für die Bewegung mehrerer elektrischer Massenteilchen*, Z. Phys. **58**, 386, (1929)

Fradkin E.S., *The Green's function method in quantum statistics*, Nuclear Phys., **12**, 465 (1959)

Fradkin E.S., *Some general relations in statistical quantum electrodynamics*, Soviet Phys., JETP, **11**,114 (1960)

Gold T., *Rotating Neutron Stars as the Origin of the Pulsating Radio Sources*, Nature, **218**, 731 (1968)

Hagedorn R., *Statistical thermodynamics of strong interactions at high energies*, Suppl. Nuovo Cim., **3**, 147(1965)

Hagedorn R., *The long way to the statistical model* in : "*Hot hadronic matter, theory and experiment*" [J. Rafelski and J. Letessier eds.; Plenum Press, New York (1995)]

Hakim R., *Remarks on relativistic statistical mechanics I. Reduced densities*, J. Math. Phys., **8**, 1315 (1967)

Hakim R., *Remarks on relativistic statistical mechanics II. Hierarchies for the reduced densities*, J. Math. Phys., **8**, 1379 (1967)

Hakim R., A. Mangeney, *Kinetic equations including radiation effects I. Vlasov approximation*, J. Math. Phys., **9**, 116 (1968)

Hakim R., A. Mangeney, *Collective Oscillations of a Relativistic Radiating Electron Plasma*, Phys. Fluids, **14**, 2751 (1971)

Hakim R., *Equilibrium statistical mechanics of relativistic particles with variable masses I. Nonquantal theory*, J. Math. Phys., **15**, 1310 (1974)

Hakim R., J. Heyvaerts, *Covariant Wigner Function Approach to Relativistic Quantum Plasmas*, Phys. Rev., **A18**, 1250 (1978)

Hakim R., *Statistical Mechanics of Relativistic Dense Matter*, Riv. Nuovo Cim., **1**, 1 (1978)

Hakim R., J. Heyvaerts, *Excitation Spectrum of the Relativistic Quantum Plasma*, J. Phys., **A13**, 2001 (1980)

Hakim R., *Fonction de Wigner d'Equilibre du Plasma Relativiste Plongé dans un Champ Magnétique Intense* [Colloque Soc. Fr. Phys. "*Plasmas Denses et Champs Magnétiques Intenses*"; Meudon (1979)]

Hakim R., *Plasma Physics Techniques for QED and QCD systems [International Symposium on the Statistical Mechanics of Quarks and Gluons*; Bielefeld, August 1980] in "*Statistical Mechanics of Quarks and Hadrons* "[H. Satz ed., North Holland, Amsterdam (1981)]

Hakim R., H.D. Sivak, *Covariant Wigner Function Approach to the Relativistic Quantum Electron Gas in a Strong Magnetic Field*, Ann. Phys.(NY), **139**, 230 (1982)

Hakim R., L. Mornas, P. Peter, H.D. Sivak, *Relaxation Time Approximation for Relativistic Dense Matter*, Phys. Rev. **D46**, 4603 (1992)

Hakim R., L. Mornas, *Transport Coefficients of Relativistic Dense Matter - Collective Effects*, Phys. Rev. **C47**, 2846 (1993)

Hakim R., H.D. Sivak, *Relativistic Quasi-Particles and their Statistics*, Class. Quant. Grav. **10** Sup., 223 (1993)

Hakim R., H.D. Sivak, *Relativistic Quasiparticles, a Covariant Approach.*; [Proceedings of the Spanish Relativity Conference 1998. Editors: A. Molina, J. Martin, E. Ruiz and F. Atrio.World Scientific, 1999].

Hakim R., H.D. Sivak, *Transport Properties of Relativistic QuasiParticles* in: "*Hot Hadronic Matter*" [J. Rafelski et J. Letessier, eds.; Plenum Press, New York (1995)]

Havas P., *Some basic problems in the formulation of a relativistic statistical mechanics of interacting particles*, in: J. Meixner (ed.), *Statistical mechanics of equilibrium and non-equilibrium*, [North Holland Publ. Co.; Amsterdam (1965)]

Huang K., *Statistical mechanics* [Wiley;New York (1963)]
S. Ichimaru, *Basic Principles of Plasma Physics*, [Benjamin; Reading, Massachusetts (1973)
Israel W., *Relativistic kinetic theory of a simple gas*, J. Math. Phys., **4**, 1163 (1963)
Ivanov M.A., G.A. Schul'man, *Equilibrium in relation to pair formation in a quantizing magnetic field*, Soviet Phys. J., **20**, 1220 (1977)
Ivanov M.A., G.A. Schul'man, *On the physical conditions in the interior of superdense, magnetized, physical objects*, Soviet Astron., **23**, 197 (1979)
Ivanov M.A., G.A. Schul'man, *Kinetic equilibrium of beta processes in high-temperature, superdense matter with a strong frozen-in magnetic field*, Soviet Astron., **24**, 311 (1981)
Ivanov M.A., G.A. Schul'man, *Neutrino energy emission in β-interaction of electrons and positrons with nuclei of hot matter in the presence of a quantizing magnetic field*, Soviet Astron., **25**, 76 (1981)
Ivanov M.A., G.A. Schul'man, *Thermodynamic functions of equilibrium radiation and electron-positron pairs in quantizing magnetic fields*, Soviet Phys. J., (1978)
Ivanov, M. A.; S. S. Lipovetskij, ; V. S., Sekerzhitskij, *Neutronization of strongly magnetized matter*, Astronomicheskij Zhurnal, **70**, 531 (1993).
Jüttner F., *Das Maxwellsche Gesetz der Geschwindigkeitsverteilung in der Relativitätheorie*, Ann. Physik, **34**, 856 (1911)
Jüttner F., *Die Dynamik eines bewegten Gases in der Relativitätheorie*, Ann. Physik, **34**, 145 (1911)
Jüttner F., *Die relativistische Quantentheorie des idealen Gases*, Z. Physik, **47**, 542 (1928)
Kalman G., *Relativistic fermion gas interacting through a scalar field I. Hartree approximation*, Phys. Rev., **D9**, 1656 (1974)
Kandrup H.E., *Relativistic stellar dynamics as an example of relativistic statistical mechanics*, AstroPhys. Space Sci., **124**, 359 (1986)
Kandrup H.E., *Generalized Wigner functions in curved spaces: A new approach*, Phys. Rev., **D37**, 2165 (1988)
Kapusta J., *Finite-temperature field theory* [Cambridge University Press; Cambridge (1989)]
Khinchin A.I., *Mathematical foundations of statistical mechanics* [Dover; New York (1949)]
Klimontovich Yu. L., *Relativistic transport equations for a plasma I.*, Soviet Phys. J.E.T.P., **37**, 524 (1960)
Klimontovich Yu. L., *Relativistic transport equations for a plasma II.*, Soviet Phys. J.E.T.P., **37**, 876 (1960)
Klimov V.V., *Spectrum of elementary Fermi excitations in quark gluon plasma*, Soviet J. Nucl. Ener., **33**, 934 (1981)
Klimov V.V., *Collective oscillations in a hot quark-gluon plasma*, Soviet Phys. J.E.T.P., **55**, 199 (1982)
Kolb E.W., M.S., Turner, *Grand unified theories and the origin of the baryon asymmetry*, Ann. Rev. Nuclear Sci., **33**, 645 (1983)].
Kursunoglu B., *Relativistic plasma*, Nuclear Fusion, **1**, 213 (1960)
Kursunoglu B., *Relativistic plasma* in : Proceedings Fifth International Conference on Ionization Phenomena in Gases, Munich 1961 [North Holland; Amsterdam (1962)]
Kursunoglu B., *Relativistic plasma in a magnetic field*, Nuovo Cim., **43**, 209 (1966)
Kursunoglu B., *Symmetries of a relativistic plasma* in : *Relativistic Plasmas*, the Coral Gable Conference; O. Buneman, W.B. Pardo eds. [W. Benjamin; New York (1968)]
Le Bellac M., *Thermal field theory* [Cambridge University Press; Cambridge (2000)]
Lichnerowicz A., R. Marrot, *Propriétés statistiques des ensembles de particules en relativité restreinte*, C. R. Acad. Sci., **210**, 759 (1940)
Mangeney A., *Thesis*, Ann. Phys. (Paris), **10**, 191 (1965)
Marle C., *Sur l'établissement des équations de l'hydrodynamique des fluides relativistes dissipatifs I. L'équation de Boltzmann relativiste*, Ann. Inst. Henri Poincaré, **10A**, 67 (1969)
Marle C., *Sur l'établissement des équations de l'hydrodynamique des fluides relativistes dissipatifs II. Méthodes de résolution approchée de l'équation de Boltzmann relativiste*, Ann. Inst. Henri Poincaré, **10A**, 127 (1969)
Marmo G., N. Mukunda, E.C.G. Sudarshan, *Relativistic particle dynamics—Lagrangian proof of the no-interaction theorem*, Phys. Rev., **D30**, 2120 (1984)
Mornas L., *Transport coefficients of relativistic nuclear and neutron matter with in-medium effects*, Nuclear Phys, **A573**, 554 (1994)
Pauli W., *Theory of Relativity* (1921) [reprinted by Dover, New York (1981)].
Perez Rojas H., A.E. Shabad, *Thermodynamic potential in electron-positron system in a magnetic field*, Soviet Phys.- Lebedev Institute, n° 7, 13 (1976)
Perez Perez Rojas H., *Quantum statistics of an electron gas in an external constant uniform magnetic field (I). The Green's function and thermodynamic relations*, preprint Acad. Cien. Cuba n° 1 (1978)
Rojas H., *Polarization of a relativistic electron and positron gas in a strong magnetic field. Propagation of electromagnetic waves*, preprint Acad. Cien. Cuba n° 70 (1978)

Perez Rojas H., *Potencial termodinamico de un gas electr'on-positron en un campo magnetico constante y uniforme*, Ciencias Tec. Fis. Mat., n° 1, 59 (1981)
Perez Rojas H., A.E. Shabad, *Absorption and dispersion of electromagnetic eigenwaves of electron-positron plasma in a strong magnetic field*, Ann. Phys. (NY), **138**, 1 (1982)
Perez Rojas H., *Explicit evaluation of the thermodynamical potential for a gas of charged bosons at high temperatures*, Ciencias Tec. Fis. Mat., n°7, 60 (1987)
Perez Rojas H., R. Torres Rivero, *Bose-Einstein condensation with increasing temperature ?*, Phys. Lett., **A137**, 13 (1989)
Perez Rojas H., R., *Bose-Einstein condensation may occur in a constant magnetic field*, Physics Letters **B379**, 148 (1996)
Perez Rojas H., L. Villegas-Lelovski, *Bose-Einstein condensation in a constant magnetic field*, Braz. J. Phys., **30**, 410 (2000)
Pina E., R. Balescu, *Some aspects of relativistic statistical mechanics*, Acta Physica Austriaca, **28**, 309 (1968) 1985)
Remler E.A., *Use of the Wigner representation in scattering problems*, Ann. Phys. (NY), **95**, 455 (1975)
Remler E.A., *Some connections between relativistic classical mechanics, statistical mechanics and quantum field theory*, Phys. Rev., **D16**, 3464 (1977)
Remler E.A., *Particle propagation in quasiparticle models of relativistic heavy-ion collisions*, Phys. Rev. Lett., **54**, 989 (1985)
Rohrlich F., *Classical charged particles* [Addison-Wesley, Reading (Mass.), (1965)].
Rzewuski J., *Field Theory* (part I); PWN Publ. Warsaw (1964)
Scheidegger A.E., C.D. McKay, *Quantum statistics of fields*, Phys. Rev., **83**, 125 (1951)
Schönberg M., *Application of second quantization methods to the classical statistical mechanics*, Nuovo Cim., **9**, 1139 (1952); *A general theory of the second quantization methods*, ibid., **10**, 697 (1953).
Schulman G.A., Astrofiz., **10**, 542 (1974); Astrofiz., **11**, 89 (1975); Soviet Astron., **19**, 698 (1976); Soviet Astron., **20**, 425 (1976); Soviet Astron., **20**, 689 (1976); Astrofiz., **13**, 657 (1977); Soviet Astron., **21**, 590 (1977)
Schul'man G.A., *On the physical conditionsin the interiors of superdense, magnetized, astrophysical objects*, Soviet Astron., **23**, 197 (1979)
Serot, B.D., J.D. Walecka, *The relativistic nuclear many-body problem*, in : Adv.Nuclear Phys. 16 [J.W. Negele, E. Vogt, eds., Plenum; New York (1986)].
Silin V.P., *Collision integral for charged particles*, Physica, **81A**, 597 (1975)
Silin V.P., *On the electromagnetic properties of a relativistic plasma*, Soviet Phys. J.E.T.P., **11**, 1136 (1960)
Silin V.P., V.N. Ursov, *Kinetic theory of the quark-gluon plasma*, Soviet Phys. Doklady, **30**, 594 (1985)
Sivak H.D., *Fluctuations in the relativistic quantum plasma*, Ann. Phys. (NY), **159**, 351 (1985)
Sivak H.D., *Friedel oscillations in a relativistic quantum plasma*, Physica, **A129**, 408 (1985)
Sivak H.D., *Electromagnetic perturbations of the relativistic quantum plasma*, Phys. Rev., **A34**, 653 (1986)
Stewart J.M., *Non-equilibrium relativistic kinetic theory* [Springer; Berlin (1971)]
Stewart J.M., *Perturbations in an expanding universe of free particles*, Astrophys. J., **176**, 323 (1972)
Stewart J.M., *Relativistic thermodynamics and kinetic theory, with applications to cosmology* in : E. Schatzman (Ed.), *Cargèse lectures in physics* 6, 175 [Gordon and Breach, New-York (1973)]
Stewart J.M., *Fundamental problems in relativistic kinetic theory* in : Colloques Internationaux du Centre National de la Recherche Scientifique no. 236, *Théories cinétiques classiques et relativistes*, 151 [Editions du Centre National de la Recherche Scientifique, Paris, 1975)]
Synge J.L., *The relativistic gas* [North Holland; Amsterdam (1957)].
Tauber G.E., J.W. Weinberg, *Internal state of a gravitating gas*, Phys. Rev., **122**, 1342 (1961)
Tauber G.E., *On a general relativistic statistical mechanics of interacting particles*, in : B. Gal-Or (Ed.), *Modern developments in thermodynamics*, 247 [Israel Universities Press (Jerusalem) and Wiley (New-York) (1974)]
Tauber G.E., *Plasma kinetic equation in an expanding universe*, Preprint TAUP 1004-82 (1982 ?)
Titeica Serban, *O complementare statistica a teoriei electronice clasice*, Studii si Cercetari de fizica, **7**, 7 (1956)
van Dantzig D., *On relativistic thermodynamics*, Proc. Sect. Sci. Kon. Ned. Akad. Wetensch., **42**, 601 (1939)
van Dantzig D., *On the phenomenological thermodynamics of moving matter*, Physica, **6**, 673 (1939)
van Dantzig D., *Stress tensor and particle density in special relativity theory*, Nature, **143**, 855 (1939)
van Kampen N.G., *Lorentz-invariance of the distribution function in phase space*, Physica, **43**, 244 (1969)
van Weert Ch.G., W.P.H. de Boer, *Relativistic kinetic theory of quantum systems I. Wigner functions for a relativistic spin ½ system*, Physica, **81A**, 597 (1975)
van Weert Ch.G., *Wigner function and normal ordering*, Physica, **A100**, 641 (1980)
Walker A.G., *The Boltzmann equations in general relativity*, Proc. Edinburgh Math. Soc., **2**, 238 (1934)
Wigner E.P., *On the quantum correction for thermodynamic equilibrium*, Phys. Rev. 40, 749 (1932)

Winter J., *Covariant extension of the Wigner transformation to non-abelian Yang-Mills symmetries for a Vlasov equation approach to the quark-gluon plasma*, J. Phys. (Paris), **45**, C6 (1984)

Winter J., *Wigner transformation in curved space-time and the curvature correction of the Vlasov equation for semi-classical gravitating systems*, Phys. Rev., **D32**, 1871 (1985)

Current Issues in Numerical Relativistic (Magneto-)Hydrodynamics

J. M^a. Ibáñez

Departamento de Astronomía y Astrofísica, Universidad de Valencia, Valencia, Spain

Abstract. Astronomical observations have increased dramatically the number of astrophysical scenarios in which matter is evolving in presence of strong gravitational fields and/or reaches velocities near the speed of light. Stellar core collapse as a mechanism of hydrodynamical supernovae, relativistic jets inside collapsars or in the remnants of binary coalescing compact objects as a model of formation of gamma-ray bursts, relativistic jets associated to active galactic nuclei or microquasars, etc, are a few examples of these scenarios. Hence, modern Astrophysics is demanding the development of robust numerical multidimensional relativistic (magneto-)hydro-codes as tools to deep into the physical nature of those energetic phenomena.
  Here I review the present status on the field of Numerical Relativistic (Magneto-) Hydrodynamics (RMHD), paying particular attention to their characteristic structure. The current extensions of Riemann solvers (exact and approximate ones) to the relativistic regime are discussed.

Keywords: Numerical Methods: Hyperbolic Systems of Conservation Laws – Relativity – Jets – Accretion – Gamma-Ray Bursts
PACS: 04.25.Dm, 95.30.Qd, 97.60.-s

INTRODUCTION

Astrophysical scenarios involving relativistic flows have drawn the attention and efforts of many researchers since the pioneering studies of May & White [1] and Wilson [2]. To some extent, these papers can be considered as the basis of modern *numerical relativistic hydrodynamics*.

Many of the high-energetic phenomena occurring in our Universe are governed by relativistic (magneto-)hydrodynamical processes. The development during the nineties of new numerical techniques, i.e., the extension to the relativistic regime of the so-called *high-resolution shock-capturing schemes* (HRSC) able to solve the equations of relativistic (magneto-)hydrodynamics under extreme conditions (i.e., large flow Lorentz factors and strong shocks) revolutionized the fields, among others, of *extragalactic jets*, *gamma-ray bursts* (GRBs), and has contributed to deepen into the knowledge of the theory of *stellar core collapse*. They are examples of systems in which the evolution of matter is described within the framework of the theory of relativity (special or general).

1. Extragalactic Jets: Present simulations (in 2D and/or 3D, i.e, axisymmetry and/or without symmetries) cover all the relevant scales from the subparsec scale, at which jets are formed, and that remain hidden to observations, up to the megaparsec scale. Reader interested is addressed to [3] (and references therein) where an exhaustive study of the morphology and dynamics of relativistic magnetized jets is carried out, and the review [4].

2. Gamma-ray bursts: In the field of GRBs, simulations have also addressed every scale from their formation ($\simeq 10^7$ cm) to the late afterglow evolution ($\simeq 10^{18}$ cm). The use of GENESIS (a relativistic 3D-code, see [5]) has allowed to carry out simulations of relativistic outflows from remnants of compact object mergers and to study their viability as progenitors of short gamma-ray bursts (see [6]), and to validate the paradigm of jets from collapsars as a progenitor of long GRBs (see [7]).

3. Stellar core collapse: During the last years we have been witness of an outstanding progress in the field of general-relativistic stellar core collapse. Reader interested can address to the following references: [8] and [9], which are the first calculations of general-relativistic collapse of rotating configurations in the Conformally Flat approach of Einstein equations; [10] for an extended version (CFC+) of the above approach; [11] for an analysis of three-dimensional relativistic simulations of rotating neutron-star collapse to a Kerr black hole carried out with the **Whisky** code developed in the framework of an european collaboration (EU Network: Theoretical Foundations of Sources for Gravitational Wave Astronomy in the Next Century: Synergy between Supercomputer Simulations and Approximation Techniques, Contract HPRN-CT-2000-00137); [12] where the authors propose a new numerical strategy –the so-called *Mariage des Maillages* – based on the use of an appropriate matching of spectral methods (suitable for solving the elliptic constraints of the Einstein equations) with HRSC methods. All of these calculations are based on the HRSC numerical schemes, which were extended to the relativistic regime, in the beginning of nineties, by Martí, et al. in the reference [13].

Numerical studies in special relativistic magnetohydrodynamics (SRMHD) have been undertaken by a growing number of authors [14, 15, 16, 17, 3]. In particular, [14], [15], and [16] developed independent *upwind* high-resolution shock-capturing (HRSC) schemes (also referred to as Godunov-type schemes), providing the characteristic information of the corresponding system of equations, which is the crucial building block in such type of schemes. In addition, [14] and [15] proposed a comprehensive sample of tests to validate numerical MHD codes in special relativity (SR). Recently, [17] have developed a third order shock-capturing *central* scheme for SRMHD which sidesteps the use of *Riemann solvers* in the solution procedure (see, e.g. [18] for general definitions on HRSC schemes). Simulations of the morphology and dynamics of magnetized relativistic jets with Godunov-type schemes have been reported by [3]. In addition, the exact solution of the *Riemann problem* (i.e., an initial value problem with discontinuous data) in SRMHD, for some particular orientation of the magnetic field and the fluid velocity field, has been obtained by [19]. I have summarized in table 1 (see below) the current status of MHD codes in special and general relativity.

MATHEMATICAL FRAMEWORK

The Eulerian observer in the 3+1 formalism

In the 3+1 formalism the line element of the spacetime can be written as

$$ds^2 = -(\alpha^2 - \beta_i\beta^i)dt^2 + 2\beta_i dx^i dt + \gamma_{ij}dx^i dx^j, \tag{1}$$

where α (lapse function), β^i (shift vector) and γ_{ij} (spatial metric) are functions of the coordinates t, x^i. A natural observer associated with the 3+1 splitting is the one with four velocity **n** perpendicular to the hypersurfaces of constant t at each event in the spacetime. This is the so-called *Eulerian observer*. The contravariant and covariant components of **n** are given by

$$n^\mu = \frac{1}{\alpha}(1, -\beta^i), \tag{2}$$

and

$$n_\mu = (-\alpha, 0, 0, 0), \tag{3}$$

respectively. In spacetimes containing matter an additional natural observer is the one that follows the fluid during its motion, also called the *comoving observer*, with four-velocity **u**. With the standard definition, the three-velocity of the fluid as measured by the Eulerian observer can be expressed as

$$v^i \equiv \frac{h^i_\mu u^\mu}{-\mathbf{u}\cdot\mathbf{n}}, \tag{4}$$

where $-\mathbf{u}\cdot\mathbf{n} \equiv W$ is the relative Lorentz factor between **u** and **n**, while $h_{\mu\nu} = g_{\mu\nu} + n_\mu n_\nu$ is the the projector onto the hypersuface orthogonal to **n**, whose spatial terms are given by $h_{ij} = \gamma_{ij}$. From Eq. (4) it follows that

$$v^i = \frac{u^i}{\alpha u^t} + \frac{\beta^i}{\alpha}, \tag{5}$$

while $v_i = u_i/W$. Note that the Lorentz factor satisfies the relation $W = 1/\sqrt{(1-v^2)} = \alpha u^t$, where $v^2 = \gamma_{ij}v^i v^j$ is the squared modulus of the three-velocity of the fluid with respect to the Eulerian observer.

Magnetic field evolution

A complete description of the electromagnetic field in general relativity is provided by the Faraday electromagnetic tensor field $F^{\mu\nu}$. This tensor is related to the electric and magnetic field, E^μ and B^μ, measured by a generic observer with four-velocity U^μ, as follows,

$$F^{\mu\nu} = U^\mu E^\nu - U^\nu E^\mu - \eta^{\mu\nu\lambda\delta}U_\lambda B_\delta, \tag{6}$$

$\eta^{\mu\nu\lambda\delta}$ being the volume element,

$$\eta^{\mu\nu\lambda\delta} = \frac{1}{\sqrt{-g}}[\mu\nu\lambda\delta], \tag{7}$$

where g is the determinant of the 4-metric ($g = \det g_{\mu\nu}$) and $[\mu\nu\lambda\delta]$ is the completely antisymmetric Levi-Civita symbol. Both, **E** and **B** are orthogonal to **U**, $\mathbf{E}\cdot\mathbf{U} = \mathbf{B}\cdot\mathbf{U} = 0$. The dual of the electromagnetic tensor $^*F^{\mu\nu}$ is defined as

$$^*F^{\mu\nu} = \frac{1}{2}\eta^{\mu\nu\lambda\delta}F_{\lambda\delta}, \tag{8}$$

and in terms of the electric and magnetic field measured by the observer **U** is given by

$$^*F^{\mu\nu} = U^\mu B^\nu - U^\nu B^\mu + \eta^{\mu\nu\lambda\delta}U_\lambda E_\delta. \tag{9}$$

From these equations, **E** and **B** can be expressed in terms of the electromagnetic tensor and the four-velocity **U** as follows

$$E^\mu = F^{\mu\nu}U_\nu, \tag{10}$$
$$B^\mu = {^*F^{\mu\nu}}U_\nu. \tag{11}$$

In terms of the electromagnetic tensor, Maxwell's equations are written as follows,

$$\nabla_\nu {^*F^{\mu\nu}} = 0, \tag{12}$$
$$\nabla_\nu F^{\mu\nu} = 4\pi \mathscr{J}^\mu, \tag{13}$$

where ∇_ν stands for the covariant derivative and $\mathscr{J}^\mu$ is the electric four-current. According to Ohm's law, the latter can be in general expressed as

$$\mathscr{J}^\mu = \rho_q u^\mu + \sigma F^{\mu\nu}u_\nu, \tag{14}$$

where ρ_q is the proper charge density measured by the comoving observer and σ is the electric conductivity. Maxwell's equations can be further simplified if one assumes that the fluid is a perfect conductor. In this case the fluid has infinite conductivity and, in order to keep the current finite, the term proportional to the conduction current, $F^{\mu\nu}u_\nu$, must vanish, which means that the electric field measured by the comoving observer is zero. This case corresponds to the so-called ideal MHD condition. We can take advantage of this condition to express the electric field measured by the observer **U** as a function of the magnetic field **B** measured by the same observer and of the four-velocities U^μ and u^μ. Straightforward calculations give

$$E^\mu = \frac{1}{W}\eta^{\mu\nu\lambda\delta}u_\nu U_\lambda B_\delta. \tag{15}$$

If we choose **U** as the four-velocity of the Eulerian observer, $\mathbf{U} = \mathbf{n}$, Eq. (15) provides

$$E^0 = 0, \tag{16}$$
$$E^i = -\alpha\eta^{0ijk}v_j B_k, \tag{17}$$

or, in terms of three-vectors, $\vec{E} = -\vec{v} \times \vec{B}$, where the arrow means that the vector lies in the 'absolute space' and the cross product is defined using the induced volume element in the absolute space $\eta^{ijk} = \alpha \eta^{0ijk}$. Using the above relations, the dual of the electromagnetic field can be written in terms of the magnetic field only

$$^*F^{\mu\nu} = \frac{u^\mu B^\nu - u^\nu B^\mu}{W}, \tag{18}$$

and Maxwell's equations $\nabla^*_\nu F^{\mu\nu} = 0$ reduce to the divergence-free condition plus the induction equation for the evolution of the magnetic field

$$\frac{\partial(\sqrt{\gamma}B^i)}{\partial x^i} = 0, \tag{19}$$

$$\frac{1}{\sqrt{\gamma}}\frac{\partial}{\partial t}(\sqrt{\gamma}B^i) = \frac{1}{\sqrt{\gamma}}\frac{\partial}{\partial x^j}\{\sqrt{\gamma}[(\alpha v^i - \beta^i)B^j - (\alpha v^j - \beta^j)B^i]\}, \tag{20}$$

or, in terms of three-vectors,

$$\vec{\nabla} \cdot \vec{B} = 0 \tag{21}$$

$$\frac{1}{\sqrt{\gamma}}\frac{\partial}{\partial t}\left(\sqrt{\gamma}\vec{B}\right) = \vec{\nabla} \times \left[\left(\alpha\vec{v} - \vec{\beta}\right) \times \vec{B}\right]. \tag{22}$$

Conservation Equations

Once we have established the magnetic field evolution equation in the ideal MHD case, we need to obtain the evolution equations for the matter fields. These equations can be expressed as the local conservation laws of baryon number and energy-momentum. For the baryon number we have

$$\nabla_\nu J^\nu = 0, \tag{23}$$

J being the rest-mass current, $J^\mu = \rho u^\mu$, where ρ denotes the rest-mass density. The conservation of the energy-momentum is given by

$$\nabla_\nu T^{\mu\nu} = 0, \tag{24}$$

where $T^{\mu\nu}$ is the energy-momentum tensor. For a fluid endowed with a magnetic field, this tensor is obtained by adding the energy-momentum tensor of the fluid to that of the electromagnetic field:

$$T^{\mu\nu} = T^{\mu\nu}_{\text{Fluid}} + T^{\mu\nu}_{\text{EM}}. \tag{25}$$

When the fluid is assumed to be perfect, $T^{\mu\nu}_{\text{Fluid}}$ is given by

$$T^{\mu\nu}_{\text{Fluid}} = \rho h u^\mu u^\nu + p g^{\mu\nu}, \tag{26}$$

where $g_{\mu\nu}$ is the metric, p is the pressure, and h is the specific enthalpy, defined by $h = 1 + \varepsilon + p/\rho$, ε being the specific internal energy. The fluid is further assumed to be in local thermodynamic equilibrium, and there exists an equation of state of the form $p = p(\rho, \varepsilon)$ which relates the pressure with ρ and ε. On the other hand, the energy-momentum tensor $T_{EM}^{\mu\nu}$ of the electromagnetic field can be obtained from the electromagnetic tensor, $\mathbf{F}$, as follows

$$T_{EM}^{\mu\nu} = \frac{1}{4\pi}\left(F^{\mu\lambda}F^{\nu}{}_{\lambda} - \frac{1}{4}g^{\mu\nu}F^{\lambda\delta}F_{\lambda\delta}\right). \tag{27}$$

Furthermore, from Eq. (6) and exploiting the ideal MHD condition, the electromagnetic tensor can be expressed in terms of the magnetic field b^μ measured by the comoving observer as

$$F^{\mu\nu} = -\eta^{\mu\nu\lambda\delta}u_\lambda b_\delta, \tag{28}$$

and Eq. (27) can be rewritten as

$$T_{EM}^{\mu\nu} = \left(u^\mu u^\nu + \frac{1}{2}g^{\mu\nu}\right)b^2 - b^\mu b^\nu, \tag{29}$$

where $b^2 = b^\nu b_\nu$ and where the magnetic field four vector has been redefined by dividing it by the factor $\sqrt{4\pi}$. As a result, the total energy-momentum tensor, fluid plus electromagnetic field, is given by

$$T^{\mu\nu} = \rho h^* u^\mu u^\nu + p^* g^{\mu\nu} - b^\mu b^\nu. \tag{30}$$

where we have introduced the definitions $p^* = p + b^2/2$ and $h^* = h + b^2/\rho$. Note that if we consistently define $\varepsilon^* = \varepsilon + b^2/(2\rho)$, the following relation, $h^* = 1 + \varepsilon^* + p^*/\rho$, is fulfilled.

In order to write the evolution equations (23), (24) in a conservation form suitable for numerical applications, let us define a basis adapted to the Eulerian observer,

$$\mathbf{e}_{(\lambda)} = \{\mathbf{n}, \partial_i\}, \tag{31}$$

where ∂_i are the coordinate vectors that are tangent to the hypersurface t=const, and, therefore, $\mathbf{n} \cdot \partial_i = 0$. This allows us to define the following five 4-vectors $\mathscr{D}_{(A)}$:

$$\mathscr{D}_{(A)} = \{\mathbf{T}(\mathbf{e}_{(\lambda)}, \cdot), \mathbf{J}\}, \qquad A = 0, \ldots, 4. \tag{32}$$

Hence the above system of equations (23), (24) can be written as

$$\nabla_\nu \mathscr{D}_{(A)}^\nu = s_{(A)}, \tag{33}$$

where the five quantities $s_{(A)}$ on the right-hand side –the sources–, are

$$s_{(A)} = \{T^{\alpha\beta}\nabla_\mu e_{(\lambda)\nu}, 0\}. \tag{34}$$

The covariant derivatives of the basis vectors, $\nabla_\mu e_{(\lambda)\nu}$, are obtained in the usual manner as

$$\nabla_\mu e_{(\lambda)\nu} = \frac{\partial e_{(\lambda)\nu}}{\partial x^\mu} - \Gamma^\delta_{\nu\mu}e_{(\lambda)\delta}, \tag{35}$$

where $\Gamma^\delta_{\nu\mu}$ are the Christoffel symbols, and

$$e_{(0)\nu} = -\alpha\delta_{0\nu}, \quad e_{(k)\nu} = g_{k\nu} = (\beta_k, \gamma_{kj}). \tag{36}$$

In a similar way to the pure hydrodynamics regime [20], if we now define the following quantities measured by an Eulerian observer,

$$D \equiv -J_\nu n^\nu = \rho W \tag{37}$$
$$S_j \equiv -T(\mathbf{n}, \mathbf{e}_{(j)}) = \rho h^* W^2 v_j - \alpha b^0 b_j \tag{38}$$
$$\tau \equiv T(\mathbf{n}, \mathbf{n}) = \rho h^* W^2 - p^* - \alpha^2 (b^0)^2 - D \tag{39}$$

i.e. the rest-mass density, the momentum density of the magnetized fluid in the j-direction, and its total energy density (subtracting the rest-mass density in order to consistently recover the Newtonian limit), respectively, the system of GRMHD equations can be written explicitly in conservative form. Together with the equation for the evolution of the magnetic field as measured by the Eulerian observer, Eq. (20), the fundamental GRMHD system of equations can be written in the following general form

$$\frac{1}{\sqrt{-g}}\left(\frac{\partial\sqrt{\gamma}\mathbf{F}^0}{\partial x^0} + \frac{\partial\sqrt{-g}\mathbf{F}^i}{\partial x^i}\right) = \mathbf{S}, \tag{40}$$

where the quantities $\mathbf{F}^\mu$ ($\mathbf{F}^0$ being the state vector and $\mathbf{F}^i$ being the fluxes) are

$$\mathbf{F}^0 = \begin{bmatrix} D \\ S_j \\ \tau \\ B^k \end{bmatrix}, \tag{41}$$

$$\mathbf{F}^i = \begin{bmatrix} D\tilde{v}^i \\ S_j\tilde{v}^i + p^*\delta^i_j - b_j B^i/W \\ \tau\tilde{v}^i + p^* v^i - \alpha b^0 B^i/W \\ \tilde{v}^i B^k - \tilde{v}^k B^i \end{bmatrix} \tag{42}$$

with $\tilde{v}^i = v^i - \frac{\beta^i}{\alpha}$. The corresponding sources $\mathbf{S}$ are given by

$$\mathbf{S} = \begin{bmatrix} 0 \\ T^{\mu\nu}\left(\frac{\partial g_{\nu j}}{\partial x^\mu} - \Gamma^\delta_{\nu\mu} g_{\delta j}\right) \\ \alpha\left(T^{\mu 0}\frac{\partial \ln\alpha}{\partial x^\mu} - T^{\mu\nu}\Gamma^0_{\nu\mu}\right) \\ 0^k \end{bmatrix}, \tag{43}$$

where $0^k \equiv (0,0,0)^T$. Note that the following fundamental relations hold between the four components of the magnetic field in the comoving frame, b^μ, and the three vector components B^i measured by the Eulerian observer:

$$b^0 = \frac{WB^i v_i}{\alpha} \tag{44}$$

$$b^i = \frac{B^i + \alpha b^0 u^i}{W}. \tag{45}$$

Finally, the modulus of the magnetic field can be written as

$$b^2 = \frac{B^2 + \alpha^2 (b^0)^2}{W^2}, \tag{46}$$

where $B^2 = B^i B_i$.

HYPERBOLIC STRUCTURE

In Section we have written the GRMHD equations in conservative form anticipating the use of numerical methods specifically designed to solve conservation equations, as will be explained in the next Section. These methods strongly rely on the hyperbolic character of the equations and on the associated wave structure. Following [21], in order to analyze the hyperbolicity of the equations it is convenient to write them in a more suitable form. If we take the following set of variables, $\mathbf{V} = (u^\mu, b^\mu, p, s)$, where s is the specific entropy, the system of equations can be written as a quasi-linear system of the form

$$\mathscr{A}_B^{\mu A} \nabla_\mu V^B = 0, \tag{47}$$

where, A and B run from 0 to 9, as the number of variables, and the 10×10 matrices $\mathscr{A}^\mu$ are given by

$$\mathscr{A}^\mu = \begin{pmatrix} \mathscr{C} u^\mu \delta_\beta^\alpha & -b^\mu \delta_\beta^\alpha + P^{\alpha\mu} b_\beta & l^{\alpha\mu} & 0^{\alpha\mu} \\ b^\mu \delta_\beta^\alpha & -u^\mu \delta_\beta^\alpha & f^{\mu\alpha} & 0^{\alpha\mu} \\ \rho h \delta_\beta^\mu & 0_\beta^\mu & u^\mu/c_s^2 & 0^\mu \\ 0_\beta^\mu & 0_\beta^\mu & 0^\mu & u^\mu \end{pmatrix} \tag{48}$$

where c_s stands for the speed of sound

$$c_s^2 = \left(\frac{\partial p}{\partial e}\right)_s, \tag{49}$$

e being the mass-energy density of the fluid $e = \rho(1+\varepsilon)$. In Eq. (48) the following definitions are introduced:

$$\mathscr{C} = \rho h + b^2, \tag{50}$$
$$P^{\alpha\mu} = g^{\alpha\mu} + 2u^\alpha u^\mu, \tag{51}$$
$$l^{\mu\alpha} = (\rho h g^{\mu\alpha} + (\rho h - b^2/c_s^2) u^\mu u^\alpha)/\rho h, \tag{52}$$
$$f^{\mu\alpha} = (u^\alpha b^\mu/c_s^2 - u^\mu b^\alpha)/\rho h, \tag{53}$$

as well as the notation

$$0^\mu \equiv 0, \quad 0^{\alpha\mu} \equiv (0,0,0,0)^{\mathrm{T}}, \quad 0^\mu_\beta \equiv (0,0,0,0). \tag{54}$$

If $\phi(x^\mu) = 0$ defines a characteristic hypersurface of the above system (47), the characteristic matrix, given by $\mathscr{A}^\varepsilon \phi_\varepsilon$ can be written as

$$\mathscr{A}^\varepsilon \phi_\varepsilon = \begin{pmatrix} \mathscr{C} a \delta^\mu_\nu & m^\mu_\nu & l^\mu & 0^\mu \\ \mathscr{B} \delta^\mu_\nu & -a \delta^\mu_\nu & f^\mu & 0^\mu \\ \rho h \phi_\nu & 0_\nu & a/c_s^2 & 0 \\ 0_\nu & 0_\nu & 0 & a \end{pmatrix} \tag{55}$$

where $\phi_\mu = \nabla_\mu \phi$, $a = u^\mu \phi_\mu$, $\mathscr{B} = b^\mu \phi_\mu$, $l^\mu = l^{\mu\nu}\phi_\nu = \phi^\mu + (\rho h - b^2/c_s^2) a u^\mu / \rho h + \mathscr{B} b^\mu / \rho h$, $f^\mu = f^{\mu\nu}\phi_\nu = (a b^\mu / c_s^2 - \mathscr{B} u^\mu)/\rho h$, and $m^\mu_\nu = (\phi^\mu + 2 a u^\mu) b_\nu - \mathscr{B} \delta^\mu_\nu$. The determinant of the matrix (55) must vanish, i.e.

$$\det(\mathscr{A}^\mu \phi_\mu) = \mathscr{C} a^2 \mathscr{A}^2 \mathscr{N}_4 = 0, \tag{56}$$

where

$$\mathscr{A} = \mathscr{C} a^2 - \mathscr{B}^2, \tag{57}$$

$$\mathscr{N}_4 = \rho h \left(\frac{1}{c_s^2} - 1\right) a^4 - \left(\rho h + \frac{b^2}{c_s^2}\right) a^2 G + \mathscr{B}^2 G, \tag{58}$$

and $G = \phi^\mu \phi_\mu$. If we now consider a wave propagating in an arbitrary direction x with a speed λ, the normal to the characteristic hypersurface is given by the four-vector

$$\phi_\mu = (-\lambda, 1, 0, 0), \tag{59}$$

and by substituting Eq. (59) in Eq. (56) we obtain the so called *characteristic polynomial*, whose zeroes give the characteristic speed of the waves propagating in the x-direction. Three different kinds of waves can be obtained according to which factor in equation (56) becomes zero. For entropic waves $a = 0$, for Alfvén waves $\mathscr{A} = 0$, and for magnetosonic waves $\mathscr{N}_4 = 0$.

Let us next analyze in more detail the characteristic equation. First of all, since the four-vector ϕ_μ must be spacelike (this is a property of the RMHD system of equations [21]), it follows that $\phi^\mu \phi_\mu > 0$. In terms of the wave speed λ we obtain

$$-\alpha \sqrt{\gamma^{xx}} - \beta^x < \lambda < \alpha \sqrt{\gamma^{xx}} - \beta^x. \tag{60}$$

The characteristic speed λ of the entropic waves propagating in the x-direction, given by the solution of the equation $a = 0$, is the following

$$\lambda = \alpha v^x - \beta^x. \tag{61}$$

For Alfvén waves, given by $\mathscr{A} = 0$, there are two solutions corresponding, in general, to different speeds of the waves,

$$\lambda = \frac{b^x \pm \sqrt{\mathscr{C}} u^x}{b^0 \pm \sqrt{\mathscr{C}} u^t}. \tag{62}$$

In the case of magnetosonic waves it is however not possible, in general, to obtain explicit expressions for their speeds since they are given by the solutions of the quartic equation $\mathcal{N}_4 = 0$ with a, $\mathcal{B}$ and G explicitly written in terms of λ as

$$a = \frac{W}{\alpha}(-\lambda + \alpha v^x - \beta^x), \qquad (63)$$

$$\mathcal{B} = b^x - b^0 \lambda, \qquad (64)$$

$$G = \frac{1}{\alpha^2}(-(\lambda + \beta^x)^2 + \alpha^2 \gamma^{xx}). \qquad (65)$$

NUMERICAL ISSUES

There are many numerical difficulties involved in the long way towards building up a robust relativistic magneto-hydrodynamical code. Let me briefly comment some of them. Reader interested in details is addressed to Antón's Ph.D. Thesis [22] or the paper [23]:

1. **Recovering** the primitive variables from the conserved ones.
   The numerical procedure used to solve the GRMHD equations allows us to obtain the values of the conserved variables $\mathbf{F}^0$ at time $t + \Delta t$ from their values at time t. However, the values of the physical variables (i.e. ρ, ε, etc) are also needed at each time step in order to compute the fluxes. It is therefore necessary to solve the algebraic equations relating the conserved and the physical variables. For the classical MHD equations and an ideal gas equation of state the physical variables can be expressed as explicit functions of the conserved ones. Unfortunately, this cannot be done in GRMHD. Therefore, the resulting nonlinear algebraic system of equations has to be solved numerically (*at each numerical cell and timestep*). The procedure we follow is an extension to full general relativity of that developed by Komissarov [14] in the special relativistic regime. The basic idea of this procedure relies on the fact that it is not necessary to solve the system (37)-(39) for the three components of the momentum, but instead for its modulus $S^2 = S^i S_i$.

2. **Preserving** the divergenceless constraint.
   Among the methods designed to preserve the divergence of the magnetic field we use the *constrained transport method* designed by [24], first extended to HRSC methods by [25], and, recently used by, e.g., [26]. This scheme is based on the use of Stokes theorem after the integration of the induction equation on surfaces of constant t and x^i, Σ_{t,x^i}.

3. **Degeneracies** are present in classical (and relativistic) ideal magnetohydrodynamics. Just as in the classical case, the relativistic MHD equations have *degenerate states* in which two or more wavespeeds coincide, which breaks the strict hyperbolicity of the system. [14] has reviewed the properties of these degeneracies. In the fluid rest frame, the degeneracies in both classical and relativistic MHD are the same: either the slow and Alfvén waves have the same speed as the entropy wave when propagating perpendicularly to the magnetic field (Degeneracy I), or the slow or the fast wave (or both) have the same speed as the Alfvén wave when propagating in a direction aligned with the magnetic field (Degeneracy II). [22] have characterized

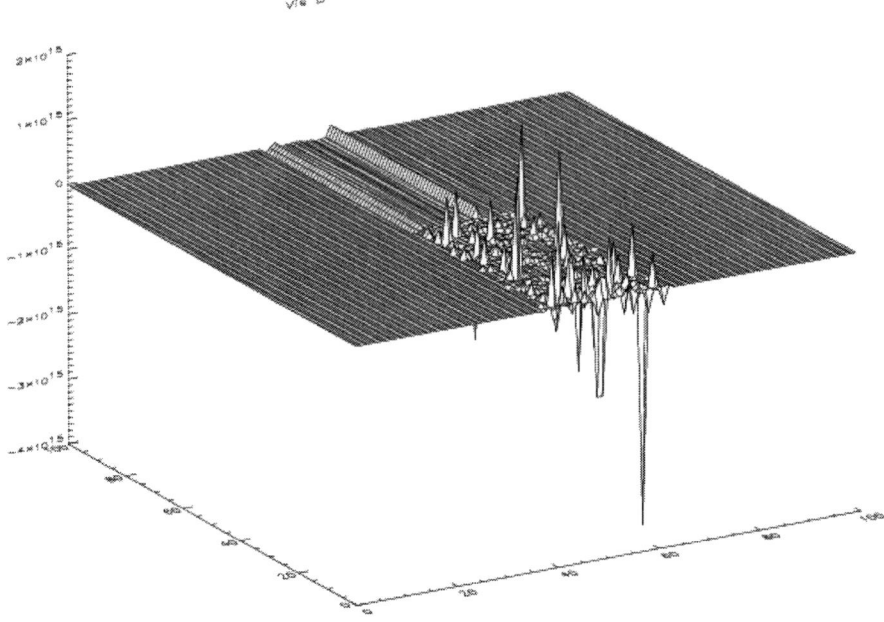

FIGURE 1. Transversal component of the magnetic field. Without renormalization

these degeneracies in terms of the components of the magnetic field four-vector normal and tangential to the Alfvén wavefront, $\mathbf{b}_n$, $\mathbf{b}_t$. When $\mathbf{b}_n = 0$, the system falls within Degeneracy I, while Degeneracy II is reached when $\mathbf{b}_t = 0$. Let us note that the previous characterization is covariant (i.e. defined in terms of four-vectors) and hence can be checked in any reference frame. In addition, [22] have also worked out a single set of right and left eigenvectors which are regular and span a complete basis in any physical state, including degenerate states. The *renormalization* procedure can be understood as a relativistic generalization of the work performed by [27] in classical MHD. This procedure avoids the ambiguity inherent to a change of basis when approaching a degeneracy, as done e.g. by [14].

The existence of degeneracies in the eigenvectors derived from the spectral decomposition of the Jacobian matrices associated to the fluxes of the RMHD system of equations makes it hazardous to implement linearized Riemann solvers based on the knowledge of that spectral decomposition. Nevertheless, we have succeeded in developing and implementing in the code a full-wave decomposition (Roe-type) Riemann solver based on a single, *renormalized set of right and left eigenvectors*, as discussed in detail in [22], which is regular for any physical state, including degeneracies. This Riemann solver is invoked in the code after a (local) linear coordinate transformation based on the procedure developed by [28] that allows to

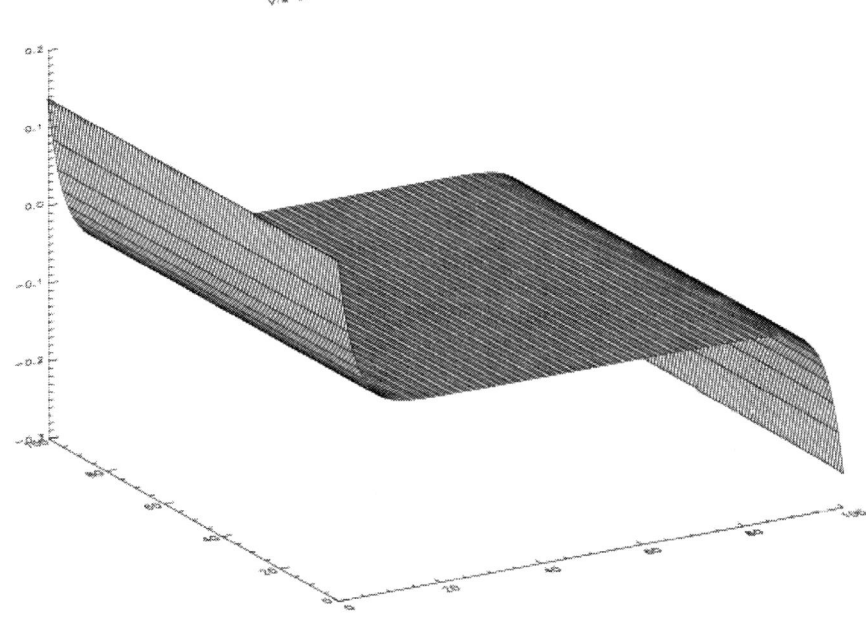

FIGURE 2. Transversal component of the magnetic field. With renormalization

use special relativistic Riemann solvers in general relativity, and which has been properly extended to include magnetic fields.

To illustrate the effects of the renormalization, we have considered a particular state of the system near a degenerate one, and computed the numerical fluxes (using a Roe-type linearized Riemann solver) corresponding to each one of the components of the vector of conserved variables. Figure 1 shows the numerical spurious oscillations in, e.g., the transversal component of the magnetic field spanning more than fifty orders of magnitud around its renormalized values which are displayed in Figure 2.

SUMMARY

Table 1 summarizes the main features of current RMHD-codes. All the codes in table 1, exception made of De Villiers & Hawley's, use a conservative formulation of the RMHD eqs. All the codes interpolate primitive variables (exception made of [15]). Basically, we can distinguish three different strategies followed by the authors (shown in table 1) in designing the different codes: i) RS-HRSC, i.e., the use of a high-resolution shock-capturing algorithm based on the solution of local Riemann problems. ii) Sym-HRSC, i.e, those based on some kind of central differencing (see, e.g., [37] and [38] for the

TABLE 1. Summary of current RMHD-codes

Code	Description
RS-HRSC	
Komissarov 1999, 2005 [14], [29]	Extension of FK96 to RMHD; right eigenvectors in primitive variables; no left eigenvectors (jumps in characteristic variables from physical conditions); switch between eigenvectors sets for degenerate and non-degenerate states; extra artificial viscosity and resistivity; 2nd order accuracy in space and time; extension to (test) GRMHD following the approach in [28]
Balsara 2001 [15]	Left eigenvectors from numerical inversion of right eigenvectors matrix; TVD interpolation on characteristic fields; 2nd order accuracy in space and time; 1D
Koldoba et al. 2002 [16]	Right and left eigenvectors in covariant variables directly used to obtain the fluxes of conserved variables; 1D
Antón et al. 2006 [22], [23]	Right and left eigenvectors in conserved variables from covariant ones; single (CONSISTENT) set of left/right eigenvectors for both degenerate and non-degenerate states; 2nd order accuracy in space (MINMOD) and time (RK); extension to (test) GRMHD following the approach in [28]
Sym-HRSC	
Koide et al. 1998 [30]	Symmetric TVD scheme with nonlinear numerical dissipation; 2nd order accuracy in space and time; (test) GRMHD
Del Zanna et al. 2003 [17]	HLL scheme for numerical fluxes; point-value representation of variables instead of cell averaged; 3rd order accuracy in space (CENO) and time (RK)
Gammie et al. 2003 [31]	HLL scheme for numerical fluxes; 2nd order in space and time; (test) GRMHD
Leismann et al. 2005 [3]	HLL scheme for numerical fluxes; 2nd order in space (MINMOD) and time (RK)
Duez et al. 2005 [32]	HLL scheme for numerical fluxes; 2nd order in space (MC, CENO, MINMOD) and time; full GRMHD
Shibata & Sekiguchi 2005 [33]	Symmetric TVD scheme with nonlinear numerical dissipation; 2nd order in space and time; full GRMHD
Qamar & Warnecke 2005 [34]	Combines a kinetic flux-splitting method and a 2nd order central scheme; SRMHD, 1D
Other approaches	
De Villiers & Hawley 2003 [35]	AV; extension of the code developed in [36] to (test-)GRMHD

relativistic regime). iii) The extension of relativistic codes based on the use of an artificial viscosity (AV) term to be used there were a shock is forming and aiming to smooth out the spurious oscillations due to Gibbs phenomenon. Reader interested can address to the excellent reviews [39], [40].

The knowledge of the characteristic structure of the system of equations of GRMHD, the use of the appropriate renormalization factors (to overcome the degenerate states, see [22]), and the recent advances in the exact solution of the Riemann Problem for the equations of SRMHD, combined with our proposal in [28] to extend to General-Relativity the solutions of special-relativistic Riemann problems, allows one to be optimistic in facing on the difficulties of an accurate description of complex astrophysical scenarios governed by relativistic magneto-hydrodynamical processes.

ACKNOWLEDGMENTS

The author thanks to the organizers of the Spanish Relativity Meeting, in particular J. Díaz Alonso and L. Mornas, for their extreme kindness and their contribution to create a friendly environment during all the days of the meeting. This contribution relies on extensive analytical and numerical work done in collaboration with M.A. Aloy, L. Antón, P. Cerdá-Durán, H. Dimmelmeier, G. Faye, J.A. Font, J.M$^{\underline{a}}$. Martí, J.A. Miralles, E. Müller, J. Novak, J.A. Pons, G. Schäfer, R. Romero, and O. Zanotti. This research has been supported by the Spanish Ministerio de Educación y Ciencia (grant AYA2004-08067-C03-01)

REFERENCES

1. M.M. May, and R.H. White, *Math. Comp. Phys.*, **7**, 219 (1967)
2. J.R. Wilson, *ApJ*, **173**, 431 (1972)
3. T. Leismann, L. Antón, L., M.A. Aloy, E. Müller, J. M$^{\underline{a}}$. Martí, J.A. Miralles, and J. M$^{\underline{a}}$. Ibáñez, *A&A*, **436**, 503 (2005)
4. M.A. Aloy, J. M$^{\underline{a}}$. Martí, and J. M$^{\underline{a}}$. Ibáñez, *Bulletin of the Spanish Astronomical Society*, **11**, 17 (2004)
5. M.A. Aloy, J. M$^{\underline{a}}$. Ibáñez, J. M$^{\underline{a}}$. Martí, and E. Müller, *ApJ Suppl.*, **122**, 151 (1999)
6. M.A. Aloy, H.-Th. Janka, and E. Müller, *A&A*, **436**, 273 (2005)
7. M.A. Aloy, E. Müller, J. M$^{\underline{a}}$. Ibáñez, J. M$^{\underline{a}}$. Martí, and A. MacFadyen, *ApJ Letters*, **531**, L119 (2000)
8. H. Dimmelmeier, J.A. Font, and E. Müller, *A&A*, **388**, 917 (2002)
9. H. Dimmelmeier, J.A. Font, and E. Müller, *A&A*, **393**, 523 (2002)
10. P. Cerdá-Durán, G. Faye, G., H. Dimmelmeier, J.A. Font, J.M$^{\underline{a}}$. Ibáñez, E. Müller, and G. Schäfer *A&A*, **439**, 1033 (2005)
11. L. Baiotti, I. Hawke, P.J. Montero, F. Loeffler, L. Rezzolla, N. Stergioulas, J.A. Font, and E. Seidel *Phys. Rev. D*, **71**, 024035 (2005)
12. H. Dimmelmeier, J. Novak, J., J.A. Font, J. M$^{\underline{a}}$. Ibáñez, and E. Müller, *Phys. Rev. D*, **71**, 064023 (2005)
13. J. M$^{\underline{a}}$. Martí, J. M$^{\underline{a}}$. Ibáñez, and J.A. Miralles, *Phys. Rev. D*, **43**, 3794 (1991)
14. S.S. Komissarov, *MNRAS*, **303**, 343 (1999)
15. D. Balsara, *ApJ Suppl.*, **132**, 83 (2001)
16. A. V. Koldoba, O. A. Kuznetsov, and G. V. Ustyugova, *MNRAS*, **333**, 932 (2002)
17. L. Del Zanna, N. Bucciantini, and P. Londrillo, *A&A*, **400**, 397 (2003)
18. E.F. Toro, *Riemann solvers and numerical methods for fluid dynamics*, Springer Verlag, Berlin, 1997
19. R. Romero, J. M$^{\underline{a}}$. Martí, J.A. Pons, J. M$^{\underline{a}}$. Ibáñez, and J.A. Miralles, *J. Fluid Mech.*, **544**, 323 (2005)

20. F. Banyuls, J.A. Font, J. M$^{\underline{a}}$. Ibáñez, J. M$^{\underline{a}}$. Martí, J.A. Miralles, *ApJ*, **476**, 221 (1997)
21. A.M. Anile, *Relativistic fluids and magneto-fluids*, Cambridge University Press, Cambridge, England, (1989)
22. L. Antón, *Ph.D. Thesis, University of Valencia* (2006)
23. L. Antón, O. Zanotti, J.A. Miralles, J. M$^{\underline{a}}$. Martí, J. M$^{\underline{a}}$. Ibáñez, J.A. Font, and J.A. Pons, *ApJ*, **637**, 296 (2006)
24. C. Evans, and J.F. Hawley, *ApJ*, **332**, 659 (1988)
25. D. Ryu, F. Miniati, T. W. Jones, and A. Frank, *ApJ*, **509**, 244 (1998)
26. P. Londrillo, and L. del Zanna, *J. Comput. Phys.*, **195**, 17 (2004)
27. M. Brio, and C. C. Wu, *J. Comput. Phys.*, **75**, 400 (1988)
28. J. A. Pons, J. A. Font, J. M$^{\underline{a}}$. Ibáñez, J. M$^{\underline{a}}$. Martí, and J. A. Miralles, *A&A*, **339**, 638 (1998)
29. S.S. Komissarov, *MNRAS*, **303**, 359 (2005)
30. S. Koide, K. Shibata, T. Kudoh, *ApJ Letters*, **495**, L63 (1998)
31. C. F. Gammie, J. C. McKinney, and G. Tóth, *ApJ*, **589**, 444 (2003)
32. M. D. Duez, Y. T. Liu, S. L. Shapiro, and B. C. Stephens, *ApJ*, **submitted**, (2005), (astro-ph/0503420)
33. M. Shibata, and Y. Sekiguchi, *Phys.Rev.D*, **submitted**, (2005)
34. S. Qamar, and G. Warnecke, *J. Comput. Phys.*, **205**, 182 (2005)
35. J. De Villiers, and J. F. Hawley, *ApJ*, **589**, 458 (2003)
36. J. F. Hawley, J. R. Wilson, and L. L. Smarr, *ApJ*, **277**, 296 (1984)
37. H. Nessyahu, and E. Tadmor, *SIAM J. Comp. Phys.*, **87**, 408 (1990)
38. A. Lucas-Serrano, J.A. Font, J. M$^{\underline{a}}$. Ibáñez, and J. M$^{\underline{a}}$. Martí, *A&A*, **428**, 703 (2004)
39. J. M$^{\underline{a}}$. Martí, and E. Müller, *Living Reviews in Relativity*, **6**, 7 (2003) http://www.livingreviews.org/Articles/
40. J.A. Font, *Living Reviews in Relativity*, **6**, 4 (2003) http://www.livingreviews.org/Articles/

The Shape of Space from Einstein to WMAP data

Jean-Pierre Luminet

*Laboratoire Univers et Théories, CNRS-UMR 8102,
Observatoire de Paris, F--92195 Meudon cédex, France*

Abstract. In this talk I review recent advances in cosmic topology since it has entered a new era of experimental tests. High redshift surveys of astronomical sources and accurate maps of the Cosmic Microwave Background radiation (CMB) are beginning to hint at the shape of the universe, or at least to limit the wide range of possibilities. Among those possibilites are surprising "wrap around" universe models in which space, whatever its curvature, may be smaller than the observable universe and generate topological lensing effects on a detectable cosmic scale. In particular, the recent analysis of CMB data provided by the WMAP satellite suggest a finite universe with the topology of the Poincaré dodecahedral spherical space. Such a model of a "small universe", the volume of which would represent only about 80% the volume of the observable universe, offers an observational signature in the form of a predictable topological lens effect on one hand, and rises new issues on the early universe physics on the other hand.

Keywords: General Relativity, Cosmology, Topology.
PACS: 98.80.Jk

THE SHAPE OF SPACE

The problem of the global shape of the universe can be decomposed into three intertwined questions.

First, what is the space curvature ? In homogeneous isotropic models of relativistic cosmology, there are only three possible answers. Three-dimensional space sections of spacetime may have zero curvature on the average – in such a case, two parallel lines keep a constant space separation and never meet, as in usual Euclidean space, sometimes called "flat space". Or space sections can be negatively curved, such as two any parallels diverge and never meet (such a space is the three-dimensional analogue of the Lobachevsky hyperbolic plane). Eventually, they can be positively curved, in which case all parallels reconverge and cross again (like on the two–dimensional surface of a sphere).

The property for physical space to correspond to one of these three possibilities depends on the way the total energy density of the Universe may counterbalance the kinetic energy of the expanding space. The normalized density parameter Ω, defined as the ratio of the actual density to the critical value that an Euclidean space would require, characterizes the present-day contents (matter and all forms of energy) of the Universe. If Ω is greater than 1, then space curvature is positive and geometry is spherical; if Ω_0 is smaller than 1 the curvature is negative and geometry is hyperbolic; eventually Ω_0 is strictly equal to 1 and space is Euclidean.

The second question about the shape of the Universe is to know whether space is finite or infinite – equivalent to know whether space contains a finite or an infinite amount of matter–energy, since the usual assumption of homogeneity implies a uniform distribution of matter and energy through space. From a purely geometrical point of view, all positively curved spaces (called spherical spaces whatever their topology) are finite, but the converse is not true : flat (Euclidean) or negatively curved (hyperbolic) spaces can have finite or infinite volumes, depending on their degree of connectedness (Ellis, 1971 ; Lachièze-Rey & Luminet, 1995). For instance, in a flat space with cubic torus topology, as soon as a particle or a light ray "exits" a given face of the fundamental cube, it "re-enters" from the opposite face, so that space is finite, although without a boundary.

From an observable point of view, it is necessary to distinguish between the "observable universe", which is the interior of a sphere centered on the observer and whose radius is that of the cosmological horizon (roughly the radius of the last scattering surface), and the physical space. Again there are only three logical possiblities. First, the physical space is infinite – like for instance the simply-connected Euclidean space. In this case, the observable universe is an infinitesimal patch of the full universe and, although it has long been the preferred model of many cosmologists, this is not a testable hypothesis. Second, physical space is finite (e.g. an hypersphere or a closed multiconnected space), but greater than the observable space. In that case, one easily figures out that if physical space is much greater that the observable one, no signature of its finitude will show in the observable data. But if space is not too large, or if space is not globally homogeneous (as is permitted in many space models with multiconnected topology) and if the observer occupies a special position, some imprints of the space finitude could be observable. Third, physical space is smaller than the observable universe. Such an apparently odd possibility is due to the fact that space can be multiconnected and have a small volume. There a lot of geometrical possibilites, whatever the curvature of space. As it is well-known, such "small universe" models may generate multiple images of light sources, in such a way that the hypothesis can be tested by astronomical observations.

The third question about the shape of the Universe deals with its global topological properties (see Luminet, 2001 for a non-technical book about all the aspects of topology and its applications to cosmology). It is interesting to point out that none of these global properties is given by Einstein's field equations, since they are partial differential equations describing only the local, metric structure of spacetime (Friedmann, 1924). The present-day topology and curvature of space take likely their origin in the early quantum conditions of the Universe, which also governed its time evolution.

The topological classification of homogeneous Riemannian 3-D spaces has made considerable progress during the last century. There are 18 Euclidean spaceforms (for a full description, see Riazuelo et al., 2004), a countable infinity of spherical spaceforms (see Gausmann et al, 2001) and a non-countable infinity of hyperbolic spaceforms (see Weeks, 1999.)

COSMIC CRYSTALLOGRAPHY

The topology and the curvature of space can be studied by using specific astronomical observations. For instance, from Einstein's field equations, the space curvature can be deduced from the experimental values of the total energy density and of the expansion rate. If the Universe was finite and small enough, we should be able to see « all around » it, because the photons might have crossed it once or more times. In such a case, any observer might identify multiple images of a same light source, although distributed in different directions of the sky and at various redshifts, or to detect specific statistical properties in the apparent distribution of faraway sources such as galaxy clusters. To do this, methods of « cosmic crystallography » have been devised (Lehoucq et al., 1996, 1999, 2000), and extensively studied by the Brazilian school of cosmic topology (Gomero et al., 2000, 2001a, 2002a,b, 2003; Fagundes & Gausmann, 1999) ; see also Marecki et al. (2005).

Basically, cosmic crystallography looks at the 3-dimensional apparent distribution of high redshift sources (e.g. galaxy clusters, quasars) in order to discover repeating patterns in the universal covering space, much like the repeating patterns of atoms observed in a crystal. « Pair Separation Histograms » (PSH) are in most cases able to detect a multiconnected topology of space, in the form of sharp spikes standing out above the noise distribution that is expected in the simply-connected case. Figures 1-3 visualize the « topological lens effect » generated by a multiconnected shape of space, and the way the topology can be determined by the PSH method.

However it was shown (Lehoucq et al., 2000; Gomero et al., 2002b) that PSH may provide a topological signal only when the holonomy group of space has Clifford translations, a property which excludes all hyperbolic spaces.

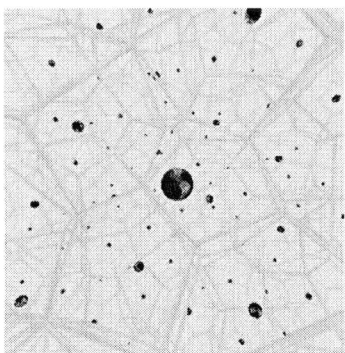

FIGURE 1. In a multi-connected Universe, the physical space is identified to a fundamental polyhedron, the duplicate images of which form the observable universe. Representing the structure of

apparent space is equivalent to representing its « crystalline » structure, each cell of which is a duplicate of the fundamental polyhedron. Here is depicted the closed hyperbolic Weeks space (only one celestial object is depicted, namely the Earth). As viewed from inside, it gives the illusion of a cellular space, tiled par polyhedra distorted with optical illusions (courtesy Jeffrey Weeks).

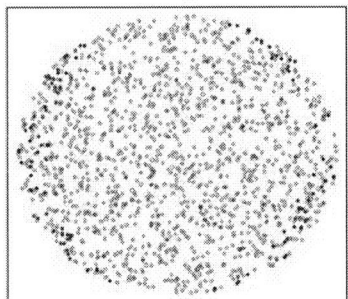

FIGURE 2. Sky map simulation in hypertorus flat space (left). The fundamental polyhedron is a cube with length = 60 % the horizon size and contains 100 « original » sources (dark dots). One observes 1939 topological images (light dots).

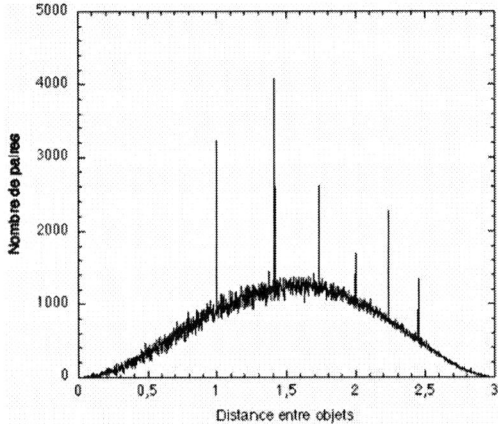

FIGURE 3. The Pair Separation Histogram corresponding to Figure 2 exhibits spikes which stand out at values and with amplitudes depending on the topological properties of space.

SPHERICAL LENSING

In the first investigations of cosmic topology, the search for the shape of space had focused on big bang models with flat or negatively curved spatial sections. Since 1999 however, a combination of astronomical (type Ia supernovae) and cosmological (temperature anisotropies of the CMB) observations suggest that the expansion of the universe is accelerating, and constrain the value of space curvature in a range which marginally favors a positively curved (i.e. spherical) model. As a consequence, spherical spaceforms have come back to the forefront of cosmology.

Gausmann et al. (2001) have investigated the full properties of spherical universes. The simplest case is the celebrated hypersphere, which is finite yet with no boundary. Actually there are an infinite number of spherical spaceforms, including lens spaces, prism spaces and polyhedral spaces. Gausmann et al. (2001) gave the construction and complete classification of such spaces, and discussed which topologies were likely to be detectable by crystallographic methods. They predicted the shapes of the pair separation histograms and they checked their predictions by computer simulations.

In addition, Weeks et al. (2003) and Gomero et al. (2001b) proved that the spherical topologies would be more easily detectable observationally than hyperbolic or flat ones. The reason is that, no matter how close space is to perfect flatness, only a finite number of spherical shapes are excluded by observational constraints. Due to the special structure of spherical spaces, topological imprints would be potentially detectable within the observable universe. Thus cosmologists are taking a renewed interest in spherical spaces as possible models for the physical universe.

THE UNIVERSE AS A DRUMHEAD

The main limitation of cosmic crystallography is that the presently available catalogs of observed sources at high redshift are not complete enough to perform convincing tests (Luminet and Roukema, 1998).

Fortunately, the topology of a small Universe may also be detected through its effects on such a "Rosetta stone" of cosmology as is the CMB fossil radiation (Levin, 2002 ; Riazuelo et al., 2004a).

If you sprinkle fine sand uniformly over a drumhead and then make it vibrate, the grains of sand will collect in characteristic spots and figures, called Chladni patterns. These patterns reveal much information about the size and the shape of the drum and the elasticity of its membrane. In particular, the distribution of spots depends not only on the way the drum vibrated initially but also on the global shape of the drum, because the waves will be reflected differently according to whether the edge of the drumhead is a circle, an ellipse, a square, or some other shape.

In cosmology, the early Universe was crossed by real acoustic waves generated soon after the big bang. Such vibrations left their imprints 380 000 years later as tiny density fluctuations in the primordial plasma. Hot and cold spots in the present-day 2.7 K CMB radiation reveal those density fluctuations. Thus the CMB temperature fluctuations look like Chladni patterns resulting from a complicated three-dimensional drumhead that vibrated for 380 000 years. They yield a wealth of information about the physical conditions that prevailed in the early Universe, as well as present geometrical properties like space curvature and topology. More precisely, density fluctuations may be expressed as combinations of the vibrational modes of space, just as the vibration of a drumhead may be expressed as a combination of the drumhead's harmonics. The shape of space can be heard in a unique way. Lehoucq et al. (2002) calculated the harmonics (the so-called "eigenmodes of the Laplace operator") for most of the spherical topologies, and Riazuelo et al. (2004b) did the same for all 18 Euclidean spaces. Then, starting from a set of initial conditions fixing how the universe originally vibrated (the so-called Harrison-Zeldovich spectrum), they evolved the harmonics forward in time to simulate realistic CMB maps for a number of flat and spherical topologies (Uzan et al., 2003a).

FIGURE 4. A multiconnected topology translates into the fact that any object in space may possess several copies of itself in the observable Universe. For an extended object like the region of emission of the CMB radiation we observe (the so-called last scattering surface) it can happen that it intersects with itself along pairs of circles. In this case, this is equivalent to say that an observer (located at the center of the last scattering surface) will see the same region of the Universe from different directions. As a consequence, the temperature fluctuations will match along the intersection of the last scattering surface with itself, as illustrated in the above figure. This CMB map is simulated for a multiconnected flat space – namely a cubic hypertorus whose length is 3.17 times smaller than the diameter of the last scattering surface. Only two duplicates are depicted.

The "concordance model" of cosmology describes the Universe as a flat infinite space in eternal expansion, accelerated under the effect of a repulsive "dark energy". The data collected by the NASA satellite WMAP (Bennett et al, 2003 ; Spergel et al., 2003) has recently produced a high resolution map of the CMB which showed the seeds of galaxies and galaxy clusters (figure 5) and allowed to check the validity of the dynamic part of the expansion model. However, combined with other astronomical data (Tonry et al., 2003), they suggest a value of the density parameter W0 = 1.02 ± 0.02 at the 1σ level. The result is marginally compatible with strictly flat space sections. Improved measurements could indeed lower the value of W0 closer to the critical value 1, or even below to the hyperbolic case. Presently however, taken at their face value, WMAP data favor a positively curved space, necessarily of finite volume since all spherical spaceforms possess this property. This provides (provisory) answers to the first two questions stated above.

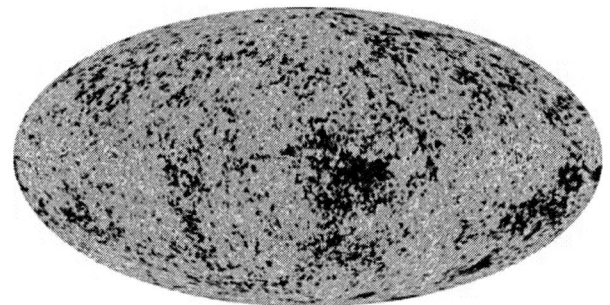

FIGURE 5. Map of temperature anisotropies of CMB as observed by WMAP telescope. WMAP Homepage : http://map.gsfc.nasa.gov

Now what about space topology ? There is an intriguing feature in WMAP data, already present in previous COBE mearurements (Hinshaw et al., 1996), although at a level of precision that was not significant enough to draw firm conclusions. The power spectrum of temperature anisotropies (figure 6) exhibits a set of "acoustic" peaks when anisotropy is measured on small and mean scales (i.e. concerning regions of the sky of relatively modest size). These peaks are remarkably consistent with the infinite flat space hypothesis. However, at large angular scale (for CMB spots typically separated by more than 60°), there is a strong loss of power which deviates significantly from the predictions of the concordance model. Thus it is necessary to look for an alternative.

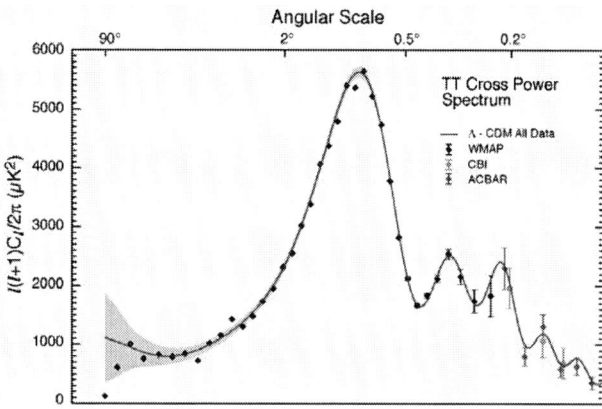

FIGURE 6. The CMB power spectrum depicts the minute temperature differences on the last scattering surface, depending on the angle of view. It shows a series of peaks corresponding to small angular separations (the position and amplitude of the main peak allows us to measure space curvature), but at larger angular scales, peaks disappear. According to the predictions of the concordance model (continuous curve), at such scales the power spectrum should follow the so-called "Sachs-Wolfe plateau". However, WMAP measurements in this region (black diamonds) fall well below the plateau for the quadrupole and the octopole moments (first two diamonds on the left). While the flat infinite

space model cannot explain this feature, multiconnected space models with a "well-proportioned" topology are remarkably consistent with such data. WMAP Homepage : http://map.gsfc.nasa.gov

CMB temperature anisotropies essentially result from density fluctuations of the primordial Universe : a photon coming from a denser region will loose a fraction of its energy to compete against gravity, and will reach us cooler. On the contrary, photons emitted from less dense regions will be received hotter. The density fluctuations result from the superposition of acoustic waves which propagated in the primordial plasma. Riazuelo et al. (2004a) have developed complex theoretical models to reproduce the amplitude of such fluctuations, which can be considered as vibrations of the Universe itself. In particular, they simulated high resolution CMB maps for various space topologies (Riazuelo et al., 2004b ; Uzan et al., 2003a) and were able to compare their results with real WMAP data. Depending on the underlying topology, the distribution of the fluctuations differs. For instance, in an infinite flat space, all wavelengths are allowed, and fluctuations must be present at all scales.

COSMIC HARMONICS

The CMB temperature fluctuations can be decomposed into a sum of *spherical harmonics*, much like the sound produced by a music instrument may be decomposed into ordinary harmonics. The "fundamental" fixes the height of the note (as for instance a 440 hertz acoustic frequency fixes the A of the pitch), whereas the relative amplitudes of each harmonics determine the tone quality (such as the A played by a piano differs from the A played by a harpsichord). Concerning the relic radiation, the relative amplitudes of each spherical harmonics determine the power spectrum, which is a signature of the geometry of space and of the physical conditions which prevailed at the time of CMB emission.

The first observable harmonics is the quadrupole (whose wavenumer is $l = 2$). WMAP has observed a value of the quadrupole 7 times weaker than expected in a flat infinite Universe. The probability that such a discrepancy occurs by chance has been estimated to 0.2 % only. The octopole (whose wavenumber is $l = 3$) is also weaker (72 % of the expected value). For larger wavenumbers up to $l = 900$ (which correspond to temperature fluctuations at small angular scales), observations are remarkably consistent with the standard cosmological model.

The unusually low quadrupole value means that long wavelengths are missing. Some cosmologists have proposed to explain the anomaly by still unknown physical laws of the early universe (Tsujikawa et al., 2003). A more natural explanation may be because space is not big enough to sustain long wavelengths. Such a situation may be compared to a vibrating string fixed at its two extremities, for which the maximum wavelength of an oscillation is twice the string length. On the contrary, in an infinite flat space, all the wavelengths are allowed, and fluctuations must be present at all scales. Thus this geometrical explanation relies on a model of finite space whose size *smaller* than the observable universe constrains the observable wavelengths below a maximum value.

WELL-PROPORTIONED SPACES

Such a property has been known for a long time, and was used to constrain the topology from COBE observations (Sokolov, 1993 ; Starobinsky, 1993). Preliminary oversimplified analyses (Stevens et al., 1993 ; de Oliveira-Costa & Smoot, 1995) suggested that any multi-connected topology in which space was finite in at least one space direction had the effect of lowering the power spectrum at large wavelengths. Weeks et al. (2004) reexamined the question and showed that indeed, some finite multiconnected topologies do lower the large--scale fluctuations whereas others may elevate them. In fact, the long wavelengths modes tend to be relatively lowered only in a special family of closed multiconnected spaces called "well-proportioned". Generally, among spaces whose characteristic lengths are comparable with the radius of the last scattering surface R_{lss} (a necessary condition for the topology to have an observable influence on the power spectrum), spaces with all dimensions of similar magnitude lower the quadrupole more heavily than the rest of the power spectrum. As soon as one of the characteristic lengths becomes significantly smaller or greater than the other two, the quadrupole is boosted in a way not compatible with WMAP data. The property was proved geometrically (Weeks et al., 2004), and checked out by numerical simulations (Riazuelo et al., 2004a). In the case of flat tori, they have varied their proportions and shown that a cubic torus lowers the quadrupole whereas an oblate or a prolate torus increase the quadrupole. They have also studied spherical spaces and shown that polyhedric spaces suppress the quadrupole whereas high order lens spaces (strongly anisotropic) boost the quadrupole. Thus, well-proportioned spaces match the WMAP data much better than the infinite flat space model.

THE POINCARÉ DODECAHEDRAL SPACE

Among the family of well-proportioned spaces, the best fit to the observed power spectrum is the *Poincaré Dodecahedral Space* (hereafter PDS) (Luminet et al., 2003).

PDS may be represented by a dodecahedron (a regular polyhedron with 12 pentagonal faces) whose opposite faces are glued after a 36° twist (figure 7). Such a space is positively curved, and is a multiconnected variant of the simply-connected hypersphere S^3, with a volume 120 times smaller. A rocket going out of the dodecahedron by crossing a given face immediately re-enters by the opposite face. Propagation of light rays is such that any observer whose line-of-sight intercepts one face has the illusion to see inside a copy of his own dodecahedron (figure 8).

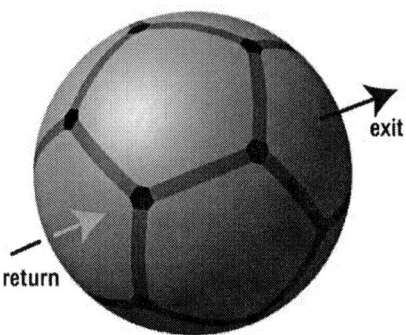

FIGURE 7. Poincaré Dodecahedral Space can be described as the interior of a spherical dodecahedron such that when one goes out from a pentagonal face, one comes back immediately inside the space from the opposite face, after a 36° rotation. Such a space is finite, although without edges or boundaries, so that one can indefinitely travel within it.

FIGURE 8. View from inside PDS perpendicularly to one pentagonal face. In such a direction, ten dodecahedra tile together with a 1/10th turn to tessellate the universal covering space S^3. Since the dodecahedron has 12 faces, 120 dodecahedra are necessary to tessellate the full hypersphere. Thus, an observer has the illusion to live in a space 120 times vaster, made of tiled doecahedra which duplicate like in a mirror hall (courtesy Jeffrey Weeks).

The associated power spectrum, namely the repartition of fluctuations as a function of their wavelengths corresponding to PDS, strongly depends on the value of the mass-energy density parameter. Luminet et al. (2003) computed the CMB multipoles for $l = 2, 3, 4$ and fitted the overall normalization factor to match the WMAP data at $l = 4$, and then examined their prediction for the quadrupole and the octopole as a function of Ω_0. There is a small interval of values within which the spectral fit is excellent, and in agreement with the value of the total density parameter deduced from WMAP data (1.02 ± 0.02). The best fit is obtained for $\Omega_0 = 1.016$ (figure 9). The result is quite remarkable because the Poincaré space has no degree of freedom. By contrast, a 3-dimensional torus, constructed by gluing together the opposite faces of a cube and which constitutes a possible topology for a finite Euclidean space, may be deformed into any parallelepiped : therefore its geometrical construction depends on 6 degrees of freedom.

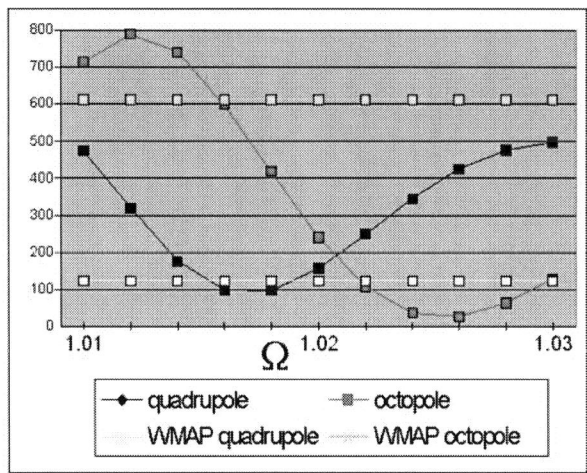

FIGURE 9. The values of the total mass-energy density parameter (assuming $\Omega_m = 0.28$) for which the Poincaré Dodecahedral Space fits the WMAP observations.

The values of the matter density Wm, of the dark energy density Ω_Λ and of the expansion rate H_0 fix the radius of the last scattering surface R_{lss} as well as the curvature radius of space R_c, thus dictate the possibility to detect the topology or not. For $\Omega_m = 0.28$, $\Omega_0 = 1.016$ and $H_0 = 62$ km/s/Mpc, $R_{lss} = 53$ Gpc and $R_c = 2.63$ R_{lss}. It is to be noticed that the curvature radius R_c is the same for the simply-connected universal covering space S^3 and for the multiconnected PDS. Incidently, the numbers above show that, contrary to a current opinion, a cosmological model with $\Omega_0 \sim 1.02$ is far from being "flat" (i.e. with $R_c = \infty$) ! For the same curvature radius, PDS has a volume 120 times smaller than S^3. Therefore, the smallest dimension of the fundamental dodecahedron is only 43 Gpc, and its volume about 80 % the volume of the observable universe (namely the volume of the last scattering surface). This implies that some points of the last scattering surface will have several copies. Such a lens effect is purely attributable to topology and can be precisely calculated in the framework of the PDS model. It provides a definite signature of PDS topology, whereas the shape of the power spectrum gives only a hint for a small, well-proportioned universe model.

To resume, the Poincaré Dodecahedral Space accounts for the low value of the quadrupole as observed by WMAP in the fluctuation spectrum, and provides a good value of the octopole. To be confirmed, the PDS model, which has been popularized as the "soccerball universe model", must satisfy two experimental tests :

1) A finer analysis of WMAP data, or new data from the future European satellite "Planck Surveyor" (scheduled 2007), will be able to determine the value of the energy density parameter with a precision of 1 %. A value lower than 1.009 will discard the Poincaré space as a model for cosmic space, in the sense that the size of the corresponding dodecahedron would become greater than the observable universe and would not leave any observable imprint on the CMB, whereas a value greater than 1.01 would strengthen its cosmological pertinence.

2) If space has a non trivial topology, there must be particular correlations in the CMB, namely pairs of "matched circles" along which temperature fluctuations should be the same (Cornish et al, 1998). The PDS model predicts 6 pairs of antipodal circles with an angular radius comprised between 5° and 55° (sensitively depending on the cosmological parameters).

Such circles have been searched in WMAP data by various teams, using various statistical indicators and massive computer calculations. On the one hand, Cornish et al. (2004) claimed to have found no matched circles on angular sizes greater than 25°, and thus rejected the PDS hypothesis. Moreover, they claimed that any reasonable topology smaller than the horizon was excluded. This is a wrong statement because they searched only for antipodal or nearly-antipodal matched circles. However Riazuelo et al. (2004b) have shown that for generic topologies (including the well-proportioned topologies which are good candidates for explaining the WMAP power spectrum), the matched circles are not back-to-back and space is not globally homogeneous, so that the positions of the matched circles depend on the observer's position in the fundamental polyhedron. The corresponding larger number of degrees of freedom for the circles search in the WMAP data generates a dramatic increase of the computer time, up to values which are out-of-reach of the present facilities.

On the other hand, Roukema et al. (2004) performed the same analysis for smaller circles, and found six pairs of matched circles distributed in a dodecahedral pattern, each circle on an angular size about 11°. This implies $\Omega_0 = 1.010 \pm 0.001$ for $\Omega_m = 0.28 \pm 0.02$, values which are perfectly consistent with the PDS model.

It follows that the debate about the pertinence of PDS as the best fit to reproduce CMB observations is fully open. Since then, the properties of PDS have been investigated in more details by various authors. Lachièze-Rey (2004) found an analytical expression of the eigenmodes of PDS, whereas Aurich et al. (2005) computed numerically the first 10 521 eigenfunctions up to the $l = 15$ mode and also supported the PDS hypothesis for explaining WMAP data. Eventually, the second-year WMAP data, originally expected by February 2004 but delayed for already two years due to unexpected surprises in the results, may soon bring additional support to a spherical multiconnected space model.

CONSEQUENCES FOR THE PHYSICS OF THE EARLY UNIVERSE

Finite well-proportioned spaces, and specially the Poincaré dodecahedral spherical space, open something like a "Pandora box" for the physics that prevailed in the early universe. The concordance model relies mostly on the hypothesis that the early universe underwent a phase of exponential expansion - the celebrated "inflationary process". Even without mentioning topological subtleties, it is good to recall that inflation theory gets into some troubles. In the simplest inflationary models, space is supposed to have become immensely larger than the observable universe after its phase of exponential growth. Therefore apositive curvature (i.e. $\Omega_0 > 1$), even weak, implies a finite space and sets strong constraints on the number of e-foldings that took place during an inflation phase. It is possible to build models of "low scale" inflation where the inflationary phase is short and leads to a detectable space curvature (Uzan et al., 2003b). It turns out that, if space is not flat, the possibility of a multiconnected topology is not in contradiction with the general idea of inflation, due a number of free and adjustable parameters in this kind of models. Yet, no convincing physical scenario has been proposed (see however Linde, 2003).

In most cosmological models, it is generally assumed that spatial homogeneity stays valid beyond the horizon scale. For instance, in the model of chaotic inflation (Linde et al., 1994 ; Guth, 2000), the universe could be very homogeneous but on scales much larger than the horizon scale. On this respect, the PDS model seems incompatible with chaotic inflation : it requires only one expanding bubble universe, of size sufficiently small to be entirely observable. In his seminal cosmological paper, Einstein (1917) had already emphasized that spatially closed universes had the advantage to eliminate boundary conditions (Wheeler, 1968). A small universe like the PDS or a well-proportioned one, in which the observer could have access to all the existing physical reality is still more advantageous (Ellis & Schreiber, 1986). It is the only type of model in which the astronomical future could be definitely predicted - such as the return of Halley's comet -, because only in such universes the observer could access to all the data in order to perform such predictions.

Maybe the most fundamental issue is to link the present-day topology of space to a quantum origin, since classical general relativity does not allow for topological changes during the course of cosmic evolution. Theories of quantum gravity could allow to address the problem of a quantum origin of space topology. For instance, in the approach of quantum cosmology, some simplified solutions of Wheeler-de Witt equations show that the sum over all topologies involved in the calculation of the wavefunction of the universe is dominated by spaces with small volumes and multiconnected topologies (Carlip, 1993 ; e Costa & Fagundes, 2001). In the approach of brane worlds (see Brax 2003 for a review), the extra--dimensions are often assumed to form a compact Calabi-Yau manifold ; in such a case, it would be strange that only the ordinary dimensions of our 3-brane would not be compact like the extra ones.

These are only heuristic indications on the way unified theories of gravity and quantum mechanics could "favor" multiconnected spaces. Whatsoever the fact that some particular multiconnected space models, such as PDS, may be refuted by future astronomical data, the question of cosmic topology will stay as a major question about the ultimate structure of our universe.

REFERENCES

1. W. Aurich, S. Lustig S. and F. Steiner, *Class. Quant. Grav.* **22**, 2061-2083 (2005).
2. C. L. Bennett et al, *Astrophys. J. Suppl. Ser.* **148**, 1-27 (2003).
3. P. Brax and C. Van de Bruck, *Class.Quant.Grav.* **20** R201-R232 (2003).
4. S. Carlip, *Class. Quant. Grav.* **10**, 207-218 (1993).
5. N. Cornish, D. Spergel and G. Starkman, *Class. Quant. Grav.* **15**, 2657-2670 (1998).
6. N. J. Cornish, D. N. Spergel, G. D. Starkman and E. Komatsu, *Phys. Rev. Lett.* **92**, 201302 (2004).
7. A. de Oliveira-Costa and G.F. Smoot, *Astrophys. J.* **448**, 447 (1995).
8. S. S. e Costa and H. V. Fagundes, *Gen.Rel.Grav.* **33**, 1489-1494 (2001).
9. A. Einstein, *Preuss. Akad.Wiss. Berlin Sitzber.* 142-152 (1917).
10. G. F. R. Ellis, *Gen. Rel. Grav.* **2** 7-21 (1971).
11. G. F. R. Ellis and W. Schreiber, *Phys. Lett.* **A115**, 97-107 (1986).
12. H. V. Fagundes and E. Gausmann, *Phys.Lett.* **A261**, 235-239 (1999).
13. A. Friedmann, *Z. Phys.* **21** 326-332 (1924).
14. E. Gausmann, R. Lehoucq, J.-P. Luminet, J.-P. Uzan and J. Weeks, *Class. Quant. Grav.* **18**, 5155-5188 (2001).
15. Gomero, G.I., *Class.Quant.Grav.* **20**, 4775-4784 (2003).
16. G.I. Gomero, M.J.Reboucas, and A.F.F. Teixeira, *Phys.Lett.* **A275**, 355-367 (2000); *Class.Quant.Grav.* **18**, 1885-1906 (2001a).
17. G.I.Gomero, M.J. Reboucas and R. Tavakol, *Class.Quant.Grav.* **18**, 4461-4476 (2001b) ; *Int.J.Mod.Phys.* **A17**, 4261-4272 (2002).
18. G.I. Gomero, A.F.F. Teixeira, M.J. Reboucas and A. Bernui, *Int.J.Mod.Phys.* **D11**, 869-892 (2002).
19. A. H.Guth, *Phys. Rep.* **333**, 555-574 (2000).
20. G. Hinshaw et al., *Astrophys. J. Lett.* **464**, L17-20 (1996).
21. M. Lachièze-Rey, *Class.Quant.Grav.* **21**, 2455-2464 (2004).
22. M. Lachièze-Rey and J. P. Luminet, *Phys. Rep.* **254**, 135-214 (1995).
23. R. Lehoucq, M. Lachièze-Rey and J.P. Luminet, *Astron. Astrophys.* **313**, 339-346 (1996).
24. R. Lehoucq, J.-P. Luminet and J.-P. Uzan, *Astron. Astrophys.* **344**, 735 (1999).
25. R. Lehoucq, J.-P. Uzan and J.-P. Luminet, *Astron. Astrophys.* **363**, 1 (2000).
26. R. Lehoucq, J.-P. Uzan and J. Weeks, *Class. Quant. Grav.* **20**, 1529-1542 (2003).
27. R. Lehoucq, J. Weeks, J.-P. Uzan, E. Gausmann and J.-P. Luminet, *Class. Quant. Grav.*, **19**, 4683-4708 (2002).
28. J. Levin, *Phys. Rep.* **365**, 251-333 (2002).
29. A. Linde, *JCAP* **0305**, 002 (2003).
30. A. Linde, D. and A. Mezhlumian, *Phys. Rev.* **D49**, 1783-1826 (1994).
31. J.- P. Luminet, *L'Univers chiffonné*, Fayard, Paris, 2005.
32. J.-P. Luminet and B. Roukema, in Proc. Cargese 98 summer school *Cosmology : The Universe at Large Scale*, edited by M. Lachièze-Rey, Kluwer Ac. Pub., NATO ASI 970491 (1998).
33. J.-P. Luminet, J. Weeks, A. Riazuelo, R. Lehoucq and J.-P. Uzan, *Nature* **425**, 593-595 (2003).
34. A. Marecki, B. Roukema and S. Bajtlik, *Astron. Astrophys.*, **435**, 427 (2005).
35. A. Riazuelo, J.-P. Uzan, R. Lehoucq and J. Weeks, *Phys.Rev.* **D69**, 103514 (2004a).
36. A. Riazuelo, J. Weeks, J.-P. Uzan, R. Lehoucq and J.-P. Luminet, *Phys Rev.* **D69**, 103518 (2004b).
37. B. F. Roukema, B. Lew, M. Cechowska, A. Marecki and S. Bajtlik, *Astron. Astrophy.* **423**, 821 (2004).
38. I.Y. Sokolov, *JETP Lett.* **57**, 617 (1993).
39. D. N. Spergel et al., *Astrophys. J. Suppl. Ser.* **148**,175-194 (2003).
40. A.A.Starobinsky, *JETP Lett.* **57**, 622 (1993).
41. D. Stevens, D. Scott and J. Silk, *Phys. Rev. Lett.* **71**, 20 (1993).
42. J. Tonry et al., *Astrophys.J.* **594**, 1-24 (2003).
43. S. Tsujikawa, R. Maartens and R. H. Brandenberger, *Phys. Lett.* **B574**, 141-148 (2003).
44. J.-P. Uzan, A. Riazuelo, R. Lehoucq and J. Weeks, *Phys. Rev.* **D69**, 043003 (2004).
45. J.-P. Uzan, U. Kirchner and G.F.R. Ellis, *Mon.Not.Roy.Astron.Soc.* **344** L65 (2003b).
46. J. Weeks, "SnapPea: A computer program for creating and studying hyperbolic 3-manifolds", available by anonymous ftp from http://geometrygames.org/SnapPea/
47. J. Weeks, R. Lehoucq and J.-P. Uzan, *Class.Quant.Grav.* **20**, 1529-1542 (2003).
48. J. Weeks, J.-P. Luminet, A. Riazuelo and R. Lehoucq, *Mon.Not.Roy.Astron.Soc.* **352**, 258-262 (2004).
49. J.-A.Wheeler, *Einstein's Vision*, Springer, Berlin, 1968.

Finding and using exact solutions of the Einstein equations

M.A.H. MacCallum

School of Mathematical Sciences, Queen Mary, University of London, Mile End Road, LONDON E1 4NS, U.K.
Email: m.a.h.maccallum@qmul.ac.uk

Abstract. The evolution of the methods used to find solutions of Einstein's field equations during the last 100 years is described. Early papers used assumptions on the coordinate forms of the metrics. Since the 1950s more invariant methods have been deployed in most new papers. The uses to which the solutions found have been put are discussed, and it is shown that they have played an important role in the development of many aspects, both mathematical and physical, of general relativity.

Keywords: General relativity;exact solutions;global properties;black holes;gravitational waves
PACS: 04.20.-q;04.20.Dw;04.20.Jb;04.25.-g;04.30.-w;04.40.Nr

1. INTRODUCTION

In an hour's talk such as this it is impossible to cover all that is known on the subject, especially since in some respects detail is of the essence in dealing with exact solutions. The three major recent reviews of rather general character [1, 2, 3] have a total of over 1200 pages and that is before moving to more specialized reviews such as [4, 5]. (The short summary of what is new in [1] as compared with the first edition, given below as an appendix, illustrates some ways in which the field has changed in the last 25 years.) Thus I shall be selective, dwelling at some length on particular solutions rather than trying to cram in as many as possible in the time. For example, I give particular attention to the Schwarzschild solution, the first solution known that did not have constant curvature. I apologize to the authors of the many excellent pieces of work I do not touch on.

Before learning the techniques for finding solutions of the Einstein equations and discovering their properties, one should ask "why this is a worthwhile endeavour?". Some colleagues do seem to regard it as something of a backwater in the theory. The reason it is still important stems from the nonlinearity of general relativity, one of its essential features. To understand the meaning of the theory, there are really three approaches. One can seek to prove global results, such as are described in [6] and were the subject of the Isaac Newton Institute programme in progress at the time of writing (see http://www.newton.cam.ac.uk/webseminars/pg+ws/2005/gmr/). One can try to use approximation, either in the form of iterated perturbation methods or numerical solutions. Finally, one can use exact solutions: as Mason and Woodhouse said, "they combine tractability with nonlinearity, so they make it possible to explore nonlinear phenomena while working with explicit solutions" [7], and as we shall see below, they have had considerable impact on the theory.

We also need to ask: "what is a solution?". We could assume a form for the metric, calculate its Einstein tensor, and so obtain a form for the energy-momentum through Einstein's equations. The pointlessness of such 'solutions' was made clear by Synge [8]. More generally, choosing a more complicated form of energy-momentum with a simple form of metric usually reduces the number of equations to be solved and makes the task of finding solutions easier. The vacuum case, and cases with an equivalent set of equations to actually solve, are more difficult.

Of course, exact solutions are very special cases. However, as Bicak [3] noted, Feynman said "The physicist is always interested in the special case. He is talking about something, he is not talking abstractly about anything. He wants to discuss the gravity law in three dimensions: he never wants the arbitrary force case in n dimensions. So a certain amount of reducing is necessary...". The special cases known exactly can be very useful as examples in guiding the global or approximative approaches.

Indeed without such uses finding exact solutions would be more like stamp-collecting than science. It requires ingenuity and the objects found may be beautiful, which is fine if you think relativity is a branch of pure mathematics, but if you think it is physics, that is not so good, despite Feynman's point. It is not the formulae for the solutions that are really of interest: "At present the main problem concerning solutions, in our opinion, is not to construct more but rather to understand more completely the known solutions with respect to *their local geometry, symmetries, singularities, sources, extensions, completeness, topology and stability*" (my emphasis). This remark by Ehlers and Kundt in 1962 [9] is as true now as when it was written, perhaps more so as a result of the large number of further solutions found since then. Part of my purpose here is to illustrate the crucial role such understanding of exact solutions has played in the development of our understanding of general relativity itself[1].

The illustrations of uses of specific solutions given below naturally focus on some of the best-known solutions. One might imagine that the very large number of other solutions known have little use. For many that may be true, at least so far, but many others have helped the exploration of the physics.

The effort required to find solutions is not trivial. Compared with Newtonian gravity, general relativity has one more independent variable, 9 more dependent ones (taking the metric approach), and equations of degree 8 in these variables rather than one, the result of these changes being that the general form of the field equations expands to 10 partial differential equations each with thousands of terms (in terms of the metric and coordinates). This complexity is the reason only some special solutions can be found, principally those with some special symmetry or algebraic property.

One might nevertheless hope for a general solution. The closest to this which I know of (the formalism of [10]) is however largely intractable. So I feel hopes for a useful general solution are slim and instead one has to make the simplifying assumptions already mentioned. Here the choices have developed over the last 90 years.

One disadvantage of those choices is that different ones may lead us to the same solution. I repeat a jeu d'esprit I first gave in a review some years ago [11], by giving a

---

[1] These aspects are covered at greater length in Bicak's review [3], which I found very useful when selecting points to cover below.

list of the most commonly-rediscovered solutions:

1. Flat space
2. The Schwarzschild solution
3. The Kasner solutions
4. Plane waves
5. Conformally flat perfect fluids
6. The Taub-NUT family of solutions
7. Static spherically symmetric perfect fluids
8. Cylindrically symmetric stationary electrovac solutions
9. Plane symmetric fluid and electrovac solutions
10. Spherically symmetric shearfree fluids

The 7th and subsequent places in this list are actually rather open to debate, for example the Harris Zund class are strong contenders. I will say something later about how to recognize known solutions.

I should add that in this review I am going to stick firmly to the number of dimensions that I know I exist in, i.e. 4, and avoid trespassing on the 5- and higher-dimensional work described by others. The development of exact solutions in such theories seems to me to be for the most part still at the stage of using only very simple metric forms.

2. THE PRE-EXISTING SOLUTIONS

Since the title of this meeting is 'A century of relativity', not the 90 of General Relativity, I am obliged to begin at the beginning, meaning the first solution, the spacetime of special relativity, Minkowski space:

$$ds^2 = dx^2 + dy^2 + dz^2 - dt^2. \tag{1}$$

It is flat, empty and is usually taken to have $\mathbb{R}^4$ topology.

The role of this solution has been considerable, for example:

- It provides the prototype for asymptotic flatness (see Ehlers' contribution to this volume)
- It is ideal for 'cutting and pasting', which was a technique used to great effect while the concepts of causal structure were being developed[2]
- It was the setting for the first work on acceleration horizons (the Unruh effect)
- In suitable coordinates, it is the Milne universe which is often used as the extreme Robertson-Walker model with $k = -1$. Apart from the use of this example in astrophysical predictions, this second choice of slicing of flat space, together with the three foliations of de Sitter space, helps to illustrate the fact that the 'open', 'flat' or 'closed' nature of spacetimes can depend on the slicing.

---

[2] As a student I had the pleasure of watching part of this work as member of Sciama's research group in Cambridge.

- It is a special case of many other models, whence its frequent rediscovery
- It is the main background for quantum field theory (QFT), and plays a role even for QFT in curved spaces.

The other solutions one could regard as in some sense known before GR was discovered were the other spaces of constant curvature, the de Sitter and anti-de Sitter spaces:

$$ds^2 = \frac{dx^2 + dy^2 + dz^2 - dt^2}{\left[1 + \frac{1}{4}K(x^2 + y^2 + z^2 - t^2)\right]^2}, \qquad (2)$$

where $K = \pm 1$. These have played a role in, for example,

- Inflation (for de Sitter)
- The AdS/CFT correspondence (anti de Sitter)
- The "no hair" theorems (de Sitter)
- The development of our understanding of particle horizons and event horizons.

Bicak [3] notes that stability against general non-linear vacuum perturbations has been proved for these three solutions. This is a 'robustness' result: it raises the question of how much of what we do is robust in this sense.

3. FINDING SOLUTIONS: THE FIRST PHASE

3.1. Methods used

Up to the 1950s the main method of finding new solutions started by postulating a coordinate form of the metric. Authors generally assumed one of:

- spherical symmetry,
- cylindrical symmetry,
- staticity or stationarity and axisymmetry, or
- a plane wave form.

In the large bibliography amassed as background for writing [1], only a part of which appears in the book itself, I found that nearly all papers before 1950 belonged to one of these groups, by far the largest group being the spherically symmetric cases. It should be noted that the metric forms were usually just written down, whereas now we would derive them from group-theoretic or other invariant assumptions.

As successful examples of this method, one can cite the work of Schwarzschild, Droste, Levi-Civita, Kasner, Chazy, Curzon, Brinkmann, Baldwin and Jeffrey, Lewis, van Stockum, Papapetrou, and Gödel, refs. [12] – [25], and of course the Robertson-Walker metrics. Many of these solutions have been important in later discussions, for example in establishing the reality of gravitational waves or elucidating the nature of directional singularities, but I will focus only on the two best-known.

3.2. The Schwarzschild solution

The original form of the Schwarzschild solution [12] used a radial coordinate r which had its origin at the horizon: for clarity I denote this r_0 below. Schwarzschild also gave (contrary to some statements in the literature) the form most often quoted now:

$$ds^2 = r^2(d\vartheta^2 + \sin^2\vartheta\, d\varphi^2) + A^{-1}dr^2 - A dt^2, \qquad (3)$$

where $A = 1 - 2m/r$. Here I have renamed Schwarzschild's R as r. Schwarzschild regarded this as an auxiliary form because it did not fulfil the condition $|\det(g)| = 1$ which Einstein had imposed in the initial formulation of General Relativity, a condition of course later discarded.

This solution initiated a discussion (see e.g. [26, 27]) on the meaning of the surface $r = 2m$ and the absence of a solution clearly analogous to a point mass in Newtonian theory. There is of course no alternative point mass solution to consider due to the uniqueness of (3) as the spherically symmetric vacuum solution (Birkhoff's theorem).

There are still authors who argue that Schwarzschild's original $r_0 = 0$ should be regarded as a singularity representing a point mass. The horizon clearly is not a regular point since the area of surrounding spheres has a limit $4\pi(2m)^2$ (which is part of the reason $r = 2m$ is now understood as a sphere, not a point). The argument can be made at various levels of sophistication, the simplest being that the metric components are singular (similar things could be said of the axis of spherical polars, but nobody argues this is singular!). To counter such arguments, it helps to note first that the horizon $r = 2m$ ($r_0 = 0$) is not in the coordinate patch for (3). (The waters here have been somewhat muddied by Hilbert's arguments that these coordinates could be continued to the interior, in which he overlooked that at the horizon the three coordinates (t, θ, φ) do not parametrize a three-dimensional manifold, as becomes immediately apparent on passing to the Kruskal-Szekeres picture.) The whole Schwarzschild patch $r > 2m$ is isometric to a region of the Kruskal-Szekeres solution (region I in the conformal diagram given as Fig. 1),

$$ds^2 = r^2(d\vartheta^2 + \sin^2\vartheta\, d\varphi^2) - 32m^3 r^{-1} e^{-r/2m} du\, dv, \qquad (4)$$

where r is defined implicitly by the equations giving u and v in terms of the previous t and r:

$$u = -(r/2m - 1)^{1/2} e^{r/4m} e^{-t/4m}, \quad v = (r/2m - 1)^{1/2} e^{r/4m} e^{t/4m}. \qquad (5)$$

By inspection of this form of the solution we see that considering the bounding surface $r = 2m$ at a given time as a point amounts to topologically identifying all the points on a sphere. One can produce an analogous effect by cutting and pasting flat space (in r_0-like coordinates).

One thing to note in these arguments is the ambivalence with which we treat coordinates. Introductions to General Relativity always emphasize the covariance of the equations, but practical examples often implicitly communicate the importance of particular coordinates. The problems this causes are seen at their worst when refereeing weak papers on exact solutions, where authors often refuse to accept that their solution is not new on the grounds its coordinate form is different from the known ones.

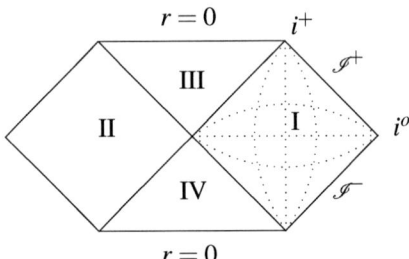

FIGURE 1. Conformal diagram of the Kruskal-Szekeres form of the Schwarzschild solution. Each point shown represents a two-sphere parametrized by θ and φ. Coordinates u and v are constant along lines at 45°. The 45° lines crossing in the centre of the figure are the horizon $r = 2m$. The 45° lines at the sides represent null infinity $\mathscr{I}$. Region I is isometric to the Schwarzschild region $r > 2m$. Region II is a second exterior. Region III is the black hole interior and region IV a white hole interior. The dotted curves running to i^o represent surfaces of constant t, and the ones through i^+ surfaces of constant r. Note that for all t, $r = 2m$ is the single point at the centre, i.e. a single sphere.

The full understanding of the Schwarzschild horizon was an important element in the general understanding of global properties of spacetimes which developed in the 1960s. Conformal pictures drawn on the same principles as Fig. 1 were also developed for the Schwarzschild solution's generalizations given by

$$A = 1 - 2m/r + e^2/r^2 - \tfrac{1}{3}\Lambda r^2,$$

the Reissner-Nordström solution being given by $\Lambda = 0 \neq e$ and the Köttler solution by $\Lambda \neq 0 = e$. Incidentally, these provide two illustrations of a 'metatheorem' that all named results have the wrong name: the Köttler solution is commonly called Schwarzschild-anti-de Sitter, though as far as I know neither Schwarzschild nor de Sitter gave this solution, and Weyl gave the general class to which the Reissner-Nordström metric belongs before Nordström's paper appeared. Other examples of this are the "Tolman-Bondi" solutions, due to Lemaître [28], whose paper Tolman [29] cited, and the "Bertotti-Robinson" solution, which Robinson himself noted was first found by Levi-Civita [14].

The Schwarzschild solution had a pivotal role in two other developments. Approximations to it were used in predictions of the classical tests of general relativity: those approximations were later generalized to the PPN formalism which provides the basis of analysis of solar system tests of relativity. It was also the setting in which Hawking made the link between the laws of black hole mechanics, quantum field theory and thermodynamics in his discovery of Hawking radiation.

The thermodynamic identification of black hole surface area with entropy is in general not fully understood in terms of microscopic states, unlike entropy in normal statistical mechanics (see [30]). However, for the 'extreme' charged case[3] (which is supersymmet-

---

[3] I am grateful to Ingemar Bengtsson and Gary Horowitz for correcting a misapprehension on my part about this work.

ric) or near-extreme charged black holes, a microscopic basis in string theory has been given [31]. Yet again the Schwarzschild solution's generalizations are playing a major role in advancing our understanding.

To summarize, the Schwarzschild solution assisted the development of our understanding of the following:

- there is no "point mass" solution in relativity, since there is no point centre;
- The role of coordinates has to be properly understood;
- global concepts such as
  black holes,
  event horizons,
  apparent horizons,
  trapped surfaces and the singularity theorems,
  cosmic censorship and naked singularities;
- PPN expansions; and
- Hawking radiation, QFT in curved spaces and black hole entropy.

3.3. The Robertson-Walker metric form: FLRW solutions

These are the solutions with the metric

$$ds^2 = -dt^2 + a^2(t)[dr^2 + \Sigma^2(r,k)(d\vartheta^2 + \sin^2\vartheta \, d\varphi^2)]. \tag{6}$$

where

$$\Sigma(r,k) = \sin r, r \text{ or } \sinh r, \text{respectively, when } k = 1, 0 \text{ or } -1. \tag{7}$$

It would be impossible to overemphasize the importance of these solutions in cosmology. They are fundamental to the inferences from observation of the presence on the cosmological scale of dark matter and "dark energy" (above the densities required by dynamics of galaxy clusters). These matters were discussed by other speakers at the meeting so I will not give details here.

There are many specific solutions due to Einstein himself, de Sitter, Friedman, Eddington, Lemaître, and so on (see Ch. 14 of [1]). The pivotal role of the ones due to Friedman and Lemaître led to those authors' names being coupled with the names of Robertson and Walker in the short name FLRW. The role of Robertson and Walker was that they independently elucidated the geometrical basis of the metric form in the 1930s.

One may also note the wide use of the Lemaître-Tolman-Bondi models, spherically symmetric inhomogeneous dust models, to describe collapse and voids, primordial black holes, and other inhomogeneities in cosmology (see [2]). For example, it has been shown that the number distribution of galaxies can be modelled by a non-evolving population in an LTB spacetime rather than an evolving population in FLRW.

4. FINDING MORE SOLUTIONS: THE SECOND PHASE

4.1. Two major themes

Two very big steps forward were made in the early 1950s, which provided the framework for many of the new solutions found up to now.

The first of these was Taub's 1951 paper [32] which contained the introduction of group-theoretic and differential geometric methods used in an essential way. (There was also some interaction with parallel work of Gödel.) This paper also discovered the Taub portion of the important Taub-NUT solution, and hinted at tetrad methods, though these latter were not fully developed until the 1960s, by Ellis [33, 34], Estabrook and Wahlquist [35], Newman and Penrose [36] and others (a development which I helped to codify [37] as far as the orthonormal tetrad version was concerned).

The second was Petrov's classification of the Weyl tensor ([38]). This, together with the solutions of the Kundt class, played an important part in the development of gravitational radiation theory and formed a big step on the road to invariant classification of solutions, discussed later.

One can briefly describe the Petrov classification as following from the fact that for any non-zero Weyl tensor C_{abcd}, as defined by

$$R^{ab}{}_{cd} = C^{ab}{}_{cd} - \tfrac{1}{3} R \delta^a_{[c} \delta^b_{d]} + 2 \delta^{[a}_{[c} R^{b]}_{d]}, \tag{8}$$

there are four null vectors k^a, the principal null directions, which satisfy the equation

$$k_{[e} C_{a]bc[d} k_{f]} k^b k^c = 0. \tag{9}$$

If these four vectors are distinct, one has the general case, known as Petrov type I. The remaining cases, where two or more coincide, are called algebraically special. If just one pair coincide, we have type II, if three coincide, type III, and if all four coincide, type N: the other possible degeneracy, two coinciding pairs, is called type D. The case of zero Weyl tensor, where the spacetime is conformally flat, is sometimes called type O.

These two developments led to the two main themes of the organization of the exact solutions book [1], i.e. classification by symmetry groups and algebraically special metrics.

Work on the latter area was much assisted by the development of the already-mentioned Newman-Penrose formalism [36], a calculation technique not based on coordinates but on a tetrad of null vectors, and therefore well adapted to the study of algebraically special spacetimes. These ideas led to the discovery of a number of important solutions, for example the following.

- The Taub-NUT solution (or rather its NUT part): [39].
- The Kerr solution, the rotating black hole: [40].
- The Robinson-Trautman class: [41].
- (Later) the Kerr-Schild ansatz and solution class: [42].

The excellent review of Ehlers and Kundt [9], perhaps the first modern general review of exact solutions, summarized the first decade of the resulting developments. I will now pause in describing methods for finding solutions and discuss the uses of three of these solutions or solution classes.

4.2. The Taub-NUT solution

The metric in this case can be given as

$$ds^2 = -U^{-1}d\tau^2 + (2\ell)^2 U (d\psi + \cos\theta \, d\phi)^2 + (\tau^2 + \ell^2)(d\theta^2 + \sin^2\theta \, d\phi^2), \quad (10)$$

where U, which is positive in the Taub region and negative in the NUT region, is given by

$$U(\tau) = -1 + 2\frac{m\tau + \ell^2}{\tau^2 + \ell^2}. \quad (11)$$

This played such a role that Misner described it as a "counterexample to almost anything" [43].

Papers by Misner [43] and Misner and Taub [44] established that the Taub and NUT regions can be joined, that the NUT region contains closed timelike lines and no sensible Cauchy surfaces, that there are two inequivalent maximal analytic extensions of the Taub region (or one non-Hausdorff manifold with both extensions), that Taub-NUT space is nonsingular in the sense of a curvature singularity, and that there are geodesics of finite affine parameter length.

To summarize, it has the following properties:

- the topology of group orbits changes at the horizon;
- there are closed timelike lines in the NUT region;
- the boundary of the Taub region has closed null geodesics;
- there is geodesic incompleteness at finite affine parameter without a curvature singularity; and
- there are inequivalent extensions, or a non-Hausdorff one.

The solutions also have applications and generalizations outside strict general relativity, e.g. in string theory or Euclidean quantum gravity.

The solution has thus had a great influence on studies of exact solutions and cosmological models which are spatially-homogeneous, and more generally on those which are hypersurface-homogeneous and self-similar (see e.g. the discussion in [45]), on cosmology in general, and on our understanding of global analysis and singularities in spacetimes.

4.3. pp-waves and plane waves

The plane waves were first found by Brinkmann [19] but their significance, showing among other things that gravitational waves were definitely not a coordinate effect, was not appreciated until the 1950s, in work of Bondi, Pirani and Robinson [46], Peres [47],

Hely [48] and others. These are the metrics

$$ds^2 = 2d\zeta d\bar{\zeta} - 2dudv - 2Hdu^2, \quad H = H(\zeta,\bar{\zeta},u), \tag{12}$$

with $H = A(u)\zeta^2 + \bar{A}(u)\bar{\zeta}^2 + B(u)\zeta\bar{\zeta}$. They are members of the more general pp-wave class, which are given, for electrovacuum cases, by

$$H = f(\zeta,u) + \bar{f}(\bar{\zeta},u) + \kappa_0 F(\zeta,u)\bar{F}(\bar{\zeta},u), \quad F_{ab} = 2k_{[a}F_{,b]}, \tag{13}$$

where the functions f and F are arbitrary functions analytic in ζ and dependent on the retarded time coordinate u. Here κ_0 is the constant in the Einstein equations.

Among their properties are:

- the wave front speed is the speed of light;
- they have a transverse character;
- there are focussing effects leading to caustics;
- there is no global Cauchy surface (i.e. no initial value problem).

These solutions are comprehensively discussed in the review of Ehlers and Kundt [9], so the treatment in the exact solutions book [1] adds virtually nothing. They have interesting singularity and horizon behaviour, and gave rise to a number of special cases of interest, such as sandwich waves and impulsive waves. One can also study collisions between them [49, 50, 51, 52] and the resulting metrics have been studied in the monograph of Griffiths [4]: see also Ch. 25 of [1].

They have more recently been used in understanding higher-dimensional theories and as examples in quantum field theory (they have the property of having no quantum corrections).

4.4. The Kerr solution

Together with the Robertson-Walker and Schwarzschild solutions, this is probably the best-known exact solution, because it represents the unique rotating vacuum black hole. It has a number of interesting mathematical properties, having indeed been found as the complement to the NUT investigation of Petrov type D vacuum solutions (the NUT solutions had initially been thought to be the only such solutions). The metric is

$$\begin{aligned}ds^2 &= \left(1 - \frac{2mr}{r^2 + a^2\cos^2\vartheta}\right)^{-1}\left[(r^2 - 2mr + a^2)\sin^2\vartheta d\varphi^2\right.\\ &\quad + \left(r^2 - 2mr + a^2\cos^2\vartheta\right)\left(d\vartheta^2 + \frac{dr^2}{r^2 - 2mr + a^2}\right)\bigg]\\ &\quad - \left(1 - \frac{2mr}{r^2 + a^2\cos^2\vartheta}\right)\left(dt + \frac{2mar\sin^2\vartheta d\varphi}{r^2 - 2mr + a^2\cos^2\vartheta}\right)^2.\end{aligned} \tag{14}$$

When the global structure of this metric was worked out, it emerged that it has a ring singularity (for $m > a$) through which further exterior regions can be connected: see e.g.

the summary in [53]. Moreover, the Hamilton-Jacobi and Klein-Gordon equations are separable in this metric, which is related to the fact that it has a non-trivial Killing tensor. It exhibits the phenomenon of an ergosphere, a region outside the black hole horizon but within which any particle has to corotate around the hole. There is a relation to work on characterization of stationary axisymmetric spacetimes by multipole moments: the Kerr solution has very specific relations between its moments which do not appear to be found in physical bodies of rotating fluid, posing a question about possible sources or the process of approach to the Kerr solution as the eventual black hole outcome of a collapse.

Work on its mathematical properties, however, is likely to be exceeded by the many papers on its astrophysical implications. For example, the ergosphere classically allows the Penrose process, in which a part of a body dividing within the ergosphere can emerge with more energy than the original body entered with. The wave version of this is superradiance, where the scattered wave has more energy than the incident wave, and this phenomenon was one of the stimuli to the laws of black hole mechanics later explained by Hawking. It is also related to the Blandford-Znajek mechanism [54], in which a magnetic field threading the black hole can extract rotational energy from the hole.

The most important of all the astrophysical uses of the Kerr solution is probably as a basis for accretion disk physics, thought, for example, to be responsible for the X-ray emission of X-ray binaries in the sky, and used in explanations of larger objects such as jets in active galactic nuclei. Observations of astronomical accretion disks are now suggesting the objects at their centres really are Kerr black holes, as the only way to explain both their short periods and other orbital data [55, 56]. Chandrasekhar remarked[4] "In my entire scientific life, extending over forty-five years, the most shattering experience has been the realization that an exact solution of Einstein's equations, discovered by the New Zealand mathematician Roy Kerr, provides an absolutely exact representation of untold numbers of massive black holes that populate the Universe..."

4.5. Finding more solutions: generating techniques

A third set of novel techniques appearing for the first time in papers in the 50s and early 60s (by Buchdahl, Ehlers, and Bonnor, for instance [57, 58, 59]) took rather longer to grow to maturity than the first two methods described in this section: this third topic is that of the generating techniques. Although they exist for metrics with one Killing vector, they are most used for stationary axisymmetric solutions, and the other classes with two commuting Killing vectors: cylindrical waves and colliding plane waves, boost-rotation symmetric spacetimes and cosmologies with two commuting spacelike Killing vectors. They work not only for vacuum, but for other forms of matter with characteristic propagation speed equal to the speed of light: massless scalar fields (or 'stiff fluid' in the case of a timelike gradient of the field), (massless) neutrinos, and electromagnetism.

---

[4] This remark came to my attention in Bicak's review [3].

This area exploded in the 1970s and 1980s in work of many people. The number of methods proliferated, and created a secondary industry of understanding the relations between them. At the same time the number of applications grew hugely. There is so much work that I shall not attempt to give references here but instead refer the reader to [1] (see also [4] and [5]). The best known formulation for the basic equations is that due to Ernst. For the stationary axisymmetric metrics

$$ds^2 = e^{-2U}(\gamma_{MN} dx^M dx^N + W^2 d\varphi^2) - e^{2U}(dt + A d\varphi)^2, \qquad (15)$$

where the metric functions U, γ_{MN}, W, A depend only on the coordinates $x^M = (x^1, x^2)$ which label the points on the 2-surfaces S_2 orthogonal to the orbits, and the electromagnetic field is given by a complex potential Φ, it reads

$$(\operatorname{Re}\mathscr{E} + \Phi\overline{\Phi}) W^{-1}(W\mathscr{E}_{,M})^{;M} = \mathscr{E}_{,M}(\mathscr{E}^{,M} + 2\overline{\Phi}\Phi^{,M}), \qquad (16)$$
$$(\operatorname{Re}\mathscr{E} + \Phi\overline{\Phi}) W^{-1}(W\Phi_{,M})^{;M} = \Phi_{,M}(\mathscr{E}^{,M} + 2\overline{\Phi}\Phi^{,M}). \qquad (17)$$

To recover the metric one needs the equations

$$A_{,M} = W e^{-4U} \varepsilon_{MN} \omega^N, \quad \omega_N = \operatorname{Im}\mathscr{E}_{,N} - i(\overline{\Phi}\Phi_{,N} - \Phi\overline{\Phi}_{,N}), \qquad (18)$$
$$e^{2U} = \operatorname{Re}\mathscr{E} + \overline{\Phi}\Phi, \qquad (19)$$

where ε_{MN} is the Levi-Civita tensor in the S_2.

Although I will not attempt a review here of the methods, the works of Geroch, Neugebauer and Kramer, Hoenselaers, Kinnersley and Xanthopoulos (HKX), Kinnersley and Chitre, Harrison, Belinski and Zakharov, Hauser and Ernst, Yamazaki, and Cosgrove were so influential their names should be mentioned.

These methods themselves can now be seen as embedded in an even more general context of symmetries of differential equations and integrable systems, involving concepts such as inverse scattering and Lax pairs, Bäcklund transformations, Riemann-Hilbert problems, prolongation and so on (see [7] or Ch. 10 of [1]). Thus the solutions have offshoots in mathematics and in other physical theories. They provide a unification of results on known solutions (for example, all stationary axisymmetric electrovacuum solutions in which a portion of the rotation axis is regular can be generated from flat space).

They also enable an infinite number of solutions to be obtained. At one stage there were a significant number of papers which exploited this possibility by exhibiting specific solutions, but it quickly became apparent that merely obtaining a new solution, when it can be done infinitely often (at least in principle), is pointless. Attention is now normally directed to ways to generate solutions with predetermined characteristics: for example one can ask what class of axis data gives a certain feature to the solution?

Of the solutions obtained or obtainable by these methods, a number have been of importance or interest, for example the Tomimatsu-Sato family, the double Kerr solution, the Neugebauer-Meinel dust disk, and a number of colliding wave and cosmological solutions. Brevity precludes detailed discussion and the reader is referred to the literature already cited.

5. FINDING AND USING MORE SOLUTIONS: RECENT DEVELOPMENTS

I now return to the fundamental question raised earlier: how can we compare solutions? In particular how can we test for local isometric equivalence? This is called the equivalence problem, and belongs to the general class of recognition problems. It has largely been answered, in theory and in practice, using ideas due to Cartan, Brans, Karlhede and others, with practical implementation and development by Åman, and later myself and my group, although in a formal sense its final step is undecideable. I will briefly describe the procedure now: a full description would be a lecture in itself. For a review see Ch. 9 of [1].

The key point is that scalar polynomial invariants, i.e. polynomials in the Riemann tensor and its derivatives in which all indices are contracted over, are insufficient. Instead we need curvature invariants of a more general sort, the 'Cartan invariants'. These are, for example, given by the components of the Weyl tensor referred to a tetrad chosen using the principal null directions.

These enable local characterization of the metric. The basic idea is to use invariantly defined tetrads, like the one from the principal null directions, and take components of the Riemann tensor and its derivatives in this frame. Thus the method has links with the Petrov classification and tetrad methods which were introduced in the 1950s and 1960s. Counting functionally independent invariants gives the dimension of the symmetry group, thus linking to the group theoretic ideas of that period, and additional information available can give the group structure.

These characterization methods can be used to check if solutions found are really new, or to search among known solutions for examples with desired local properties. They could in principle be used to find solutions, as well as classify known ones, and first examples of this method have been developed by Bradley, Karlhede and Marklund.

Other uses of this approach, i.e. the direct use of invariants, have been in understanding the limits of families of solutions without needing trial and error for the appropriate coordinate transformations (for example, studying the limits of the Schwarzschild family as $m \to \infty$); proving (non) existence of matchings by characterizing the geometries of the proposed matching surfaces (e.g. in work of my student Daniel Cox [60]); and in providing a method to 'unravel' directional singularities (in work of my student John Taylor [61]). The classifying quantities might also be used to give a topology on the space of solutions, which could even be of interest in numerical relativity.

Summarizing, the main methods for finding solutions in current use are still those outlined in Section 4, but the understanding and classification of these solutions can now be done in an invariant manner which should enable better use of the solutions found and may also provide a fresh and more invariant way to find more.

APPENDIX

The second edition of the exact solutions book contains about 400 pages of new material, covering hundreds of new solutions and references. In its preparation the authors read about 4000 new papers (as well as the 3000 read for the first edition). So there is far more material than I could talk about. Most of the new material is integrated into and expands existing chapters and sections: additional sections were added on the GHP formalism and other calculi, junction conditions and so on, and the chapter on solutions obtained by generating techniques was almost completely re-written.

The entirely new chapters or part-chapters were on:

- Homotheties
- Characterization by invariants
- Generating techniques themselves
- Dynamical systems methods
- Inhomogeneous solutions with two spacelike Killing vectors
- Colliding plane waves
- Special vector and tensor fields.

REFERENCES

1. H. Stephani, D. Kramer, M. MacCallum, C. Hoenselaers, and E. Herlt, *Exact solutions of Einstein's field equations, 2nd edition*, Cambridge University Press, Cambridge, 2003.
2. A. Krasiński, *Inhomogeneous cosmological models*, Cambridge University Press, Cambridge, 1997.
3. J. Bicak, "The role of exact solutions of Einstein's equations in the developments of general relativity and astrophysics: selected themes," in *Einstein's field equations and their physical implications*, Springer Verlag, Heidelberg, 2000, vol. 540 of *Lecture Notes in Physics*.
4. J. Griffiths, *Colliding plane waves in general relativity*, Oxford University Press, Oxford, 1991.
5. V. Belinski, and E. Verdaguer, *Gravitational solitons*, Cambridge University Press, Cambridge, 2001.
6. S. Hawking, and G. Ellis, *The large-scale structure of space-time*, Cambridge University Press, Cambridge, 1973.
7. L. Mason, and N. Woodhouse, *Integrability, Self-duality and twistor theory*, vol. 15 of *London Mathematical Society Monographs*, Oxford Science Publications, 1996.
8. J. Synge, *Relativity: the general theory*, North-Holland, Amsterdam, 1960.
9. J. Ehlers, and W. Kundt, "Exact solutions of the gravitational field equations," in *Gravitation: an introduction to current research*, edited by L. Witten, Wiley, New York and London, 1962, pp. 49–101.
10. D. Sciama, P. Waylen, and R. Gilman, *Phys. Rev. A* **187**, 1762–6 (1969).
11. M. MacCallum, "An overview of exact solutions of Einstein's equations and their classification," in *Highlights in gravitation and cosmology (Proceedings of the Goa conference 1987)*, edited by B. Iyer, A. Kembhavi, J. Narlikar, and C. Vishveshwara, Cambridge University Press, London and New York, 1989, pp. 3–14.
12. K. Schwarzschild, *Sitz. Preuss. Akad. Wiss.* pp. 189–196 (1916).
13. J. Droste, *Kon. Akad. Wetensch. Amsterdam, Proc. Sec. Sci.* **19**, 197–215 (1916-17).
14. T. Levi-Civita, *Rend. R. Accad. Lincei, Cl. sci. fis., mat. nat* **26**, 519–531 (1917).
15. H. Weyl, *Ann. Phys. (Germany)* **54**, 117 (1917).
16. E. Kasner, *Amer. J. Math.* **43**, 217 (1921).
17. J. Chazy, *Bull. Soc. Math. France* **52**, 17 (1924).
18. H. Curzon, *Proc. London Math. Soc.* **23**, 477 (1924).
19. H. Brinkmann, *Math. Ann.* **94**, 119–145 (1925).
20. O. Baldwin, and G. Jeffery, *Proc. Roy. Soc. Lond. A* **111**, 95 (1926).
21. T. Lewis, *Proc. Roy. Soc. Lond. A* **136**, 176 (1932).
22. W. van Stockum, *Proc. Roy. Soc. Edinburgh A* **57**, 135 (1937).
23. A. Papapetrou, *Proc. Roy. Irish Acad. A* **51**, 191 (1947).
24. A. Papapetrou, *Ann. Phys. (Germany)* **12**, 309 (1953).
25. K. Gödel, *Rev. Mod. Phys.* **21**, 447 (1949).

26. J. Eisenstaedt, *Arch. Hist. Exact Sci.* **27**, 157–198 (1982).
27. J. Eisenstaedt, *Arch. Hist. Exact Sci.* **37**, 275–357 (1987).
28. G. Lemaître, *Ann. Soc. Sci. Bruxelles A* **53**, 51 (1933), translation by M.A.H. MacCallum in Gen. Rel. Grav. **29**. 641-680 (1997).
29. R. Tolman, *Proc. Nat. Acad. Sci. (Wash.)* **20**, 169 (1934), reprinted in Gen. Rel. Grav. **29**, 935-943 (1997).
30. R. Wald, The thermodynamics of black holes, http://relativity.livingreviews.org/Articles/lrr-2001-6, Albert Einstein Institute (2001).
31. G. Horowitz, "Quantum States of Black Holes," in *Black Holes and Relativistic Stars*, edited by R. Wald, University of Chicago Press, Chicago, 1998, pp. 241–266.
32. A. Taub, *Ann. Math.* **53**, 472 (1951), reprinted, with editorial introduction by M.A.H. MacCallum, in Gen. Rel. Grav. **36**, 2699-2719 (2004).
33. G. Ellis, On general relativistic fluids and cosmological models, Ph.D. thesis, Cambridge University (1964).
34. G. Ellis, *J. Math. Phys.* **8**, 1171 (1967).
35. F. Estabrook, and H. Wahlquist, *J. Math. Phys.* **5**, 1629 (1964).
36. E. Newman, and R. Penrose, *J. Math. Phys.* **3**, 566 (1962).
37. M. MacCallum, "Cosmological models from the geometric point of view," in *Cargese Lectures in Physics, vol. 6*, edited by E. Schatzman, Gordon and Breach, New York, 1973, pp. 61–174.
38. A. Petrov, *Scientific Proceedings of Kazan State University (named after V.I. Ulyanov-Lenin), Jubilee (1804-1954) Collection* **114**, 55–69 (1954), translation by J. Jezierski and M.A.H. MacCallum, with introduction by M.A.H. MacCallum, Gen. Rel. Grav. **32**, 1661-1685 (2000).
39. E. Newman, L. Tamburino, and T. Unti, *J. Math. Phys.* **4**, 915 (1963).
40. R. Kerr, *Phys. Rev. Lett.* **11**, 237 (1963).
41. I. Robinson, and A. Trautman, *Proc. Roy. Soc. Lond. A* **265**, 463 (1962).
42. R. Kerr, and A. Schild, "A new class of vacuum solutions of the Einstein field equations," in *Atti del convegno sulla relatività generale; problemi dell'energia e ondi gravitationali.*, Barbèra, Firenze, 1965, p. 222.
43. C. Misner, "Taub-NUT space as a counterexample to almost anything," in *Relativity theory and astrophysics, vol. 1: Relativity and cosmology*, edited by J. Ehlers, Lectures in applied mathematics, volume 8, American Mathematical Society, Providence, R.I., 1963, pp. 160–169.
44. C. Misner, and A. Taub, *Zh. Eks. Teor. Fiz.* **55**, 233 (1968).
45. R. Jantzen, "Higher Dimensional Cosmological Models: the View from Above," in *Proc. 26th Liège Int. Astrophys. Colloq.*, edited by J. Demaret, Liège University Press, Liège, 1987, gr-qc/0309025.
46. H. Bondi, F. Pirani, and I. Robinson, *Proc. Roy. Soc. Lond. A* **251**, 519 (1959).
47. A. Peres, *Phys. Rev. Lett.* **3**, 571–572 (1959).
48. J. Hély, *Comptes Rendus Acad. Sci. (Paris)* **249**, 1867–1868 (1959).
49. R. Penrose, *Rev. Mod. Phys.* **37**, 215 (1965).
50. P. Szekeres, *Nature* **228**, 1183 (1970).
51. P. Szekeres, *J. Math. Phys.* **13**, 286 (1972).
52. K. Khan, and R. Penrose, *Nature* **229**, 185 (1971).
53. B. Carter, "Black hole equilibrium states," in *Black Holes (Les Houches Lectures)*, edited by B. DeWitt, and C. DeWitt, Gordon and Breach, New York, 1972, pp. 57–214.
54. R. Blandford, and R. Znajek, *Mon. Not. Roy. Astr. Soc.* **179**, 433 (1977).
55. R. Genzel, R. Schoedel, T. Ott, A. Eckart, T. Alexander, F. Lacombe, D. Rouan, and B. Aschenbach, *Nature* **425**, 934–937 (2003).
56. J. Miller, A. Fabian, C. Reynolds, M. Nowak, J. Homan, M. Freyberg, M. Ehle, T. Belloni, R. Wijnands, M. van der Klis, P. Charles, and W. Lewin, *Astrophys.J.* **606**, L131–L134 (2004).
57. H. Buchdahl, *Quart. J. Math. Oxford* **5**, 116 (1954).
58. J. Ehlers, Konstruktionen und charakterisierungen von lösungen der Einsteinschen gravitationsfeldgleichungen, Dissertation, Hamburg (1957).
59. W. Bonnor, *Z. Phys.* **161**, 439 (1961).
60. D. Cox, *Physical Review D* **68**, 124008 (2003).
61. J. Taylor, *Class. Quant. Grav.* **22**, 4961–4971 (2005).

Recent developments in the study of the Cosmic Microwave Background anisotropies

Enrique Martínez-González

Instituto de Física de Cantabria, Av. Los Castros s/n, 39005 Santander, Spain

Abstract.
We review recent developments in the study of the Cosmic Microwave Background anisotropies. This field has experienced a very strong evolution in recent years mainly due to the construction of very sensitive experiments allowing a very precise mapping of the microwave sky. The combination of the Cosmic Microwave Background with other cosmological data sets has provided for the first time an accurate picture of the universe in a consistent way, what is known as the *concordance model*. Here, we discuss the main observational facts supporting the concordance model as well as the problems interpreting some of the most important characteristics of this model.

Keywords: Cosmic Microwave Background, anisotropies, cosmology
PACS: 98.70.Vc, 98.80.-K, 98.80.Bp, 98.80.Es, 98.80.Ft

INTRODUCTION

The classical tests on which the standard cosmological model has been based are the expansion of the universe, the primordial nucleosinthesis and the Cosmic Microwave Background (CMB). The first observations showing that the universe was in fact expanding are due to Edwin Hubble [35], who at the end of the twenties of the last century realized that the galaxies around the Milky Way suffered a systematic redshift which increased with distance. Although problems with the distance calibrators, *the cepheids*, made Hubble derive a wrong value for the proportionality constant relating distance and velocity (the Hubble constant), however the expansion was out of question. About twenty years later, the ideas about the primordial origin of the lightest nuclei were established by George Gamow and collaborators [29]. They realized that all the elements could be formed in the core of the stars except for the lightest ones that had to be made in a dense and hot past in the early history of the expanding universe. Moreover, based on the same ideas a remanent radiation of that hot past was predicted, the CMB [4]. Finally, in 1965 the CMB was detected by Arno Penzias and Robert Wilson [66]. That finding was a definitive prove that the universe passed through a dense and hot phase, the *Big-Bang*, and also that the *Steady State* theory was not a valid description of the universe.

Those classical tests are still of fundamental importance to support the validity of the standard Big-Bang model at present. Besides, the large scale distribution of galaxies also represents a crucial cosmological test which complements very well the others. Its statistical properties have to match those of the initial matter fluctuations as derived from the CMB anisotropies. The classical tests, however, have been refined very much with the development of new sophisticated instruments. Thus, the expansion of the universe is now being tested with much better standard candles, supernovas of type Ia (SN Ia), which allow us to probe much larger distances (redshifts $z > 1$). This improvement in redshift implies a test not only for the Hubble parameter but also for the deceleration one $q = \Omega_m/2 - \Omega_\Lambda$. The latter is linked with the *cosmological constant* introduced by Albert Einstein in 1917 to find a static solution for the universe in the context of the recently formulated *General Theory of Relativity* [23], and only about ten years before Hubble established the expansion of the universe. As we will se below, the cosmological constant wrongly introduced to find a static solution plays paradoxically a crucial role in our present description of the universe.

The nucleosynthesis of the light elements one second after the Big-Bang has being tested not only in the vecinity of our Galaxy with the abundances of the elements ^4He [25] and ^7Li [9] (determined from observations of extragalactic HII regions in nearby galaxies and metal-poor stars in our own Galaxy, respectively) but also with the deuterium abundance derived from high-resolution spectra of high redshift quasars [44]. Those observations have imposed very strong constraints on the baryonic density which, on the other hand, are roughly consistent with the ones derived from the anisotropies of the CMB.

The CMB anisotropies, detected for the first time in 1992 with the COBE-DMR satellite [89], have become the most important data set to characterize our universe and, in particular, to impose strong constraints to the different cosmological parameters. In combination with large galaxy surveys and other cosmological data sets we are able to determine about twelve parameters with errors $\lesssim 10\%$, something unconceivable only ten years ago. Below, the present status about constraints on the cosmological parameters and the model of the universe is reviewed.

COSMIC MICROWAVE BACKGROUND ANISOTROPIES

The CMB anisotropies are produced by different physical effects acting on this radiation before the last-scattering surface, when the universe was some 380000 years old. These *primary anisotropies* contain very valuable information on the early history of the universe and can be observed today almost unaffected. After the temperature of the photons drop below some 3000 degrees they are not able to ionize the hydrogen atoms and propagate freely. The physical effects producing the primary anisotropies can be summarised by the following equation [60, 81]:

$$\frac{\Delta T}{T}(\vec{n}) \approx \frac{\phi_e(\vec{n})}{3} + 2\int_e^o \frac{\partial \phi}{\partial t}dt + \vec{n}\cdot(\vec{v}_o - \vec{v}_e) + \left(\frac{\Delta T}{T}(\vec{n})\right)_e \quad (1)$$

The gravitational redshift suffered by the CMB photons in their travel from the last-scattering surface to the observer is given by the first two terms in the r.h.s. of eq. 1. They are known as *Sachs-Wolfe (SW)* and *Integrated Sachs-Wolfe (ISW)* effects, respectively [80]. The velocity of the baryon-photon fluid at recombination generates a Doppler effect. The intrinsic temperature of the photons at recombination, represented by the fourth term in the r.h.s. of eq. 1, also contributes to the total anisotropy. Eq. 1 accounts for the anisotropies at recombination. Before recombination, changes in the gravitational potential, due to the imperfect baryon-photon coupling, can also produce a gravitational redshift (the early Integrated Sachs-Wolfe effect). The accurate computation of all the contributions requires to solve the linearized coupled Einstein-Boltzmann equations. Several codes have been developed to numerically solve those equations as a function of the cosmological parameters (about 10, see below), being some of them publically available (e.g. CMBFAST [85], CAMB [53]). A comparison among the results of different codes showed that the accuracy achieved is very good, reaching $\approx 0.1\%$ up to $\ell = 3000$ [84].

As will be explained below, a very relevant statistical quantity to compute is the 2-point correlation function of the temperature anisotropies or equivalently the correlation of the spherical harmonic coefficients:

$$< a_{\ell m} a^*_{\ell' m'} > = C_\ell \delta_{\ell \ell'} \delta_{mm'}, \qquad (2)$$

where C_ℓ is the anisotropy power spectrum and the a_{lm} are the coefficients of the spherical harmonic expansion

$$\frac{\Delta T}{T}(\vec{n}) = \sum_{\ell,m} a_{\ell m} Y_{\ell m}(\vec{n}), \qquad (3)$$

The homogeneity and isotropy of the universe have been assumed in eq. 2, i.e. the a_{lm} are not correlated for different l or m. In Fig. 1 the scales at which the different effects dominate the power spectrum C_ℓ are marked. As we can see, the gravitational effects dominate at the lower multipoles (large angular scales). At intermediate scales, $100 < \ell < 1000$, the spectrum is dominated by several oscillations, usually called *acoustic peaks*. These peaks appear as a consequence of the balance between the gravitational force and the radiation pressure. At the smaller scales, $\ell > 1000$, the C_ℓ are damped because of the width of recombination and the imperfections in the coupling of the photon-baryon fluid (*Silk effect*, [87]).

When the smaller scales in the matter distribution become nonlinear and their collapse give rise to the formation of the first stars and quasars, they start to reionize the surrounding matter. This ionized matter can interact again with the microwave photons and produce new anisotropies. *Secondary anisotropies* can be either produced by the scattering with free electrons or by the gravitational effect of the matter density evolution. The scattering can leave a very clear imprint when the photons happen to cross the core of rich galaxy clusters where the electron density and temperature are very high (of several 10^{-3}cm^{-3} and $\sim 10^8$K, respectively). In this case the electrons inject energy to the photons through inverse Compton scattering, producing a distorsion of the blackbody spectrum. This effect is called *the Sunyaev-Zeldovich effect (SZ)* [91].

The evolving gravitational wells in the large scale structure produce a gravitational redshift in the photons. This is known as the *late ISW effect* [80] or *the Rees-Sciama effect* when the evolution is non-linear [76]. Besides, the trajectory of the photons is lensed by the same gravitational wells producing a noticeble effect on the CMB anisotropies at arcmin angular scales. One important issue in the observation of the CMB anisotropies is therefore to distinguish between primary and secondary anisotropies. In the case of the SZ effect, as the frequency dependence is different from the Planckian one, it is possible to separate the effect from the intrincsic CMB anisotropies by using multifrequency

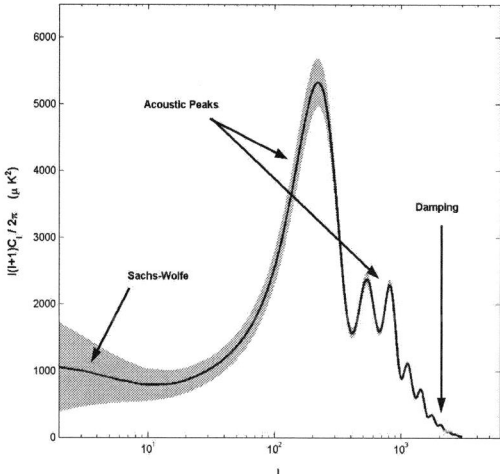

FIGURE 1. CMB power spectrum C_ℓ for the best fit model given in [6]. The gray band represents the cosmic variance. The spectrum has been computed with the CMBFAST code [85].

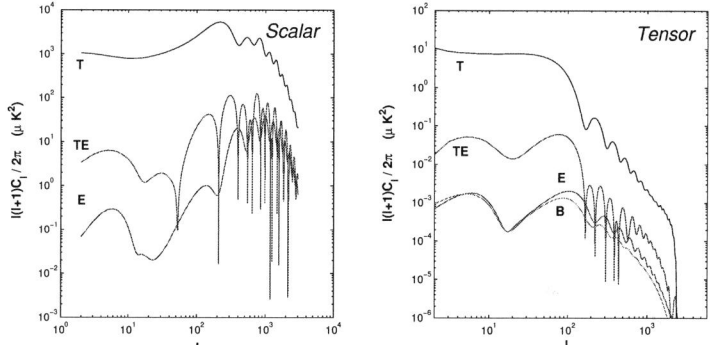

FIGURE 2. Angular power spectra for temperature and polarization. The C_ℓ's are plotted for scalar perturbations on the left and for tensor perturbations on the right, where a tensor/scalar ratio of $r = 0.01$ has been chosen. The rest of the parameters have been fixed to the best fit model of [6]. All the spectra have been computed with the CMBFAST code.

observations. However, for the lensing effect this is not possible and other properties, such as high-order correlations in the CMB temperature and polarization fields, have to be used (see e.g. [33, 30, 86]).

In the same manner that anisotropies in the temperature are expected in the gravitational instability scenario for structure formation, anisotropies in the polarization of the CMB are also expected. The physical effects giving rise to these anisotropies are, however, different. Linear polarization is generated by Thompson scattering at the end of recombination when the growth of the mean free path of the photons allows temperature anisotropies to grow. In particular a quadrupole is formed in the reference frame of each electron producing polarization after the scattering (see, e.g., [14]). The expected level of polarization is however small $\approx 5\%$. As shown by [104, 41] two rotationally invariant quantities E, B can be constructed from the Stokes parameters Q, U. The different behaviour of E, B under parity transformations implies that only three spectra are needed to characterise CMB polarization: $C_\ell^E, C_\ell^B, C_\ell^{TE}$.

In addition to the anisotropies generated by physical effects associated to the energy-matter density perturbations (*scalar perturbations*), new anisotropies can be generated by a background of primordial gravitational waves (*tensor perturbations*). This background is also a generic prediction of inflation. The primordial power spectra for both scalar and tensor perturbations are usually characterised by a scale-free law of the form:

$$P_s(k) = A_s(k/k_0)^{n_s}, P_t(k) = A_t(k/k_0)^{n_t}, k_0 = 0.05 Mpc^{-1}. \qquad (4)$$

There is no observational evidence of the primordial background of gravitational waves yet. Upper limits have been imposed by the combination of WMAP and other high resolution CMB data with large scale structure data implying a tensor-to-scalar amplitude ratio $r \equiv A_t/A_s < 0.9$ [6], see next sections. In any case, the amplitude of both temperature and polarization power spectra produced by tensor perturbations are believed to be several orders of magnitude below the ones corresponding to the scalar perturbations. However, the B-mode can only be generated by gravitational waves and thus its detection represents a unique evidence of the existence of this primordial background. This is the reason why there is now so much interest in planning new more sensitive CMB experiments to detect polarization (see e.g. [10]). The detection of the B-mode is expected to be much harder than the already detected E-mode because of its intrinsically smaller amplitude and the relatively larger foreground emissions (for a recent discussion of these isues see [97]). An additional complication comes from the lensing conversion of the E-mode to the B-mode [105] which is expected to dominate over the primordial B-mode at multipoles $\ell \gtrsim 100$ [46]. Several methods have been developed to reconstruct the gravitational lensing potential from the E and B-mode polarization correlations and remove the lensing contamination [34, 42, 83]. A more technical problem is the separation of the mixing of E and B-modes for observations with partial sky coverage for which other methods have been proposed [54, 12, 15].

Outside the standard inflationary scenario there are other possibilities to generate B-modes. For instance, cosmic strings or primordial magnetic fields would generate a vector and a vector plus tensor components of metric perturbations, respectively, contributing to the B-mode polarization (see e.g. [70, 55]). It is important to remark, however, that in the case of cosmic strings and other topological defects their role as seeds for the large scale structure formation is already very much constrained by the observed C_ℓ, see below.

In Fig. 2 the different temperature and polarization power spectra corresponding to the primary anisotropies (including the ISW effect) for the concordance model are plotted for the scalar and tensor perturbations assuming a value of $r = 0.01$. The power of the tensor perturbations is a few orders of magnitude below the corresponding scalar power spectra for $\ell \lesssim 100$ and decays strongly for larger multipoles.

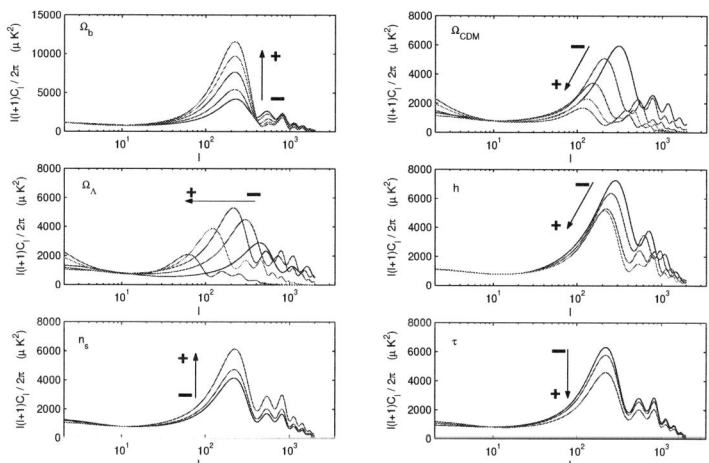

FIGURE 3. Dependence of the temperature power spectrum C_ℓ on some relevant cosmological parameters (the spectra have been computed with the CMBFAST code).

Many parameters are needed to characterize the standar cosmological model. Their variation changes the amplitude and shape of the temperature and polarization power spectra in many different ways (see Fig. 3 for their effect on the temperature power spectrum). The cosmological parameters can be classified depending on whether they characterize the background universe or the primordial power spectrum.

The background Friedmann-Robertson-Walker universe and its matter and energy content are determined by the following parameters: *physical baryonic density*, $w_b = \Omega_b h^2$; *physical matter density*, $w_m = \Omega_m h^2$, where Ω_m is the matter density parameter including the contributions from baryons, cold dark matter and neutrinos, $\Omega_m = \Omega_b + \Omega_{CDM} + \Omega_\nu$; *physical neutrino density*, $w_\nu = \Omega_\nu h^2$; *dark energy equation of state parameter*, $w \equiv p_{DE}/\rho_{DE}$; *dark energy density*, Ω_{DE} (in case $w = -1$ the dark energy takes the form of a cosmological constant and its energy contribution is denoted by Ω_Λ); *the Hubble constant*, $h \equiv H_0/100 \text{km s}^{-1} \text{Mpc}^{-1}$.

The scalar and tensor primordial power spectra are characterized by the following parameters: *amplitude of the primordial scalar power spectrum*, A_s, defined by $P_s(k) = A_s(k/k_0)^{n_s}$, where $k_0 = 0.05 \text{Mpc}^{-1}$; *scalar spectral index*, n_s; *running index*, $\alpha = dn_s/d\ln k$, normally determined at the scale $k_0 = 0.05 \text{Mpc}^{-1}$; *tensor-to-scalar ratio*, $r = A_t/A_s$; *tensor spectral index*, n_t, which is normally assumed to be $n_t = -r/8$ from the consistency relation of inflation (see e.g. [52]).

There is an extra parameter that accounts for the reionization history of the universe: *the optical depth*, $\tau = \sigma_T \int_{t_r}^{t_0} n_e(t) dt$, where σ_T is the Thompson cross-section and $n_e(t)$ is the electron number density as a function of time.

Besides, there are two possible types of matter density fluctuations: *adiabatic* which preserve the entropy per particle and *isocurvature* which preserve the total energy density. The standard inflationary scenario predicts fluctuations of the adiabatic type.

In Fig. 3 the dependence of the C_ℓ on some of the most relevant cosmological parameters

TABLE 1. Cosmological parameters using only WMAP first year data. In the fit the universe is assumed to be spatially flat and the value of the optical depth is constrained to $\tau < 0.3$ (from [90]).

Parameter	Values (68% CL)
w_b	0.024±0.001
w_m	0.14±0.02
h	0.72±0.05
A_s	0.9±0.1
τ	$0.166^{+0.076}_{-0.071}$
n_s	0.99±0.04

is shown. It is important to note that the manner in which some of the parameters enter in the calculation of the C_ℓ produce degeneracies. In particular, there is the well known geometrical degeneracy involving Ω_m, Ω_Λ and Ω_k, where Ω_k is the curvature density parameter, $\Omega_k \equiv 1 - \Omega_m - \Omega_\Lambda$ (an example of this degeneracy can be found in Fig. 5 of [62]). This fact makes clear the need to combine different cosmological data sets to break those degeneracies, as will be shown below.

COSMOLOGICAL CONSTRAINTS FROM THE CMB POWER SPECTRUM

Extraction of the cosmological parameters from the CMB data is performed using a maximum likelihood test. For the likelihood it is assumed that the temperature anisotropies are Gaussian. The most precise data for this analysis has been obtained with the space mission NASA WMAP, which has covered the whole sky from a priviledged position (the Lagrangian point 2 of the Sun-Earth system), saved from contaminating emissions coming from the Earth, Moon and Sun. In table 1 the values of the cosmological parameters determined with the first year data of that mission are shown. An interesting result that can be derived from this table is that the Einstein-de-Sitter model (i.e. flat geometry with null dark energy density) is many sigmas away from the best fit model (asuming a flat universe and an optical depth $\tau < 0.3$) [90].

The resolution of the WMAP experiment allows a good precision in the determination of the C_ℓ up to multipoles $\ell \lesssim 600$. Other experiments based on the ground like ACBAR [50], CBI [73], VSA [21] and balloon-borne like BOOMERANG [40] are able to reach higher resolutions covering small patches on the sky with good sensitivity. The data obtained with these experiments therefore complement very well the WMAP ones. In Fig. 4 the temperature anisotropy results from WMAP and the other high resolution experiments as a function of the multipole are shown. We can see that the concordance model follows quite well the data up to $\ell \approx 2000$. For higher multipoles only ACBAR and CBI experiments have enough resolution and the data seem to indicate an excess of power as compared to the model prediction. This excess can be interpreted as a contri-

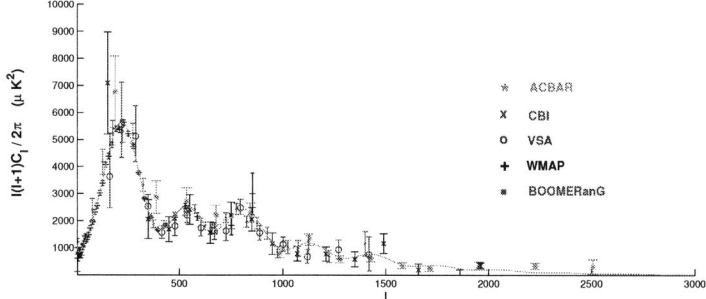

FIGURE 4. The temperature power spectrum C_ℓ measured by WMAP, ACBAR, CBI, VSA and BOOMERANG, compared to the best fit model given by [6].

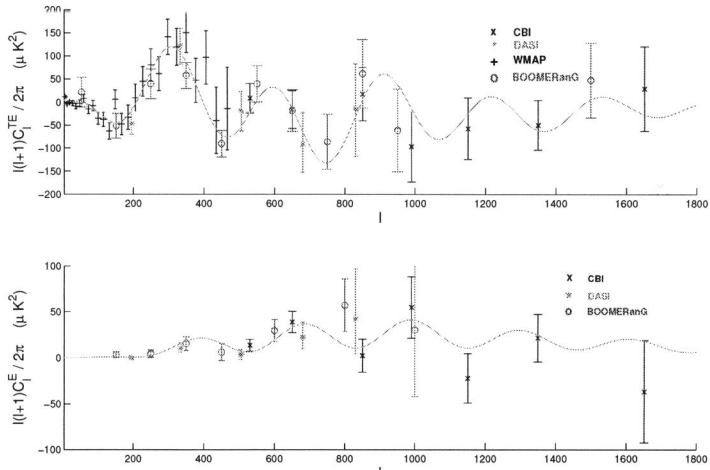

FIGURE 5. Polarization power spectra measured by DASI, WMAP, CBI and BOOBMERANG (TE) and DASI, CBI and BOOMERANG (E). Also plotted is the best fit model given by [6].

bution from secondary anisotropies coming from the SZ effect [8] or/and the emission of radio sources [94]. However, new multifrequency data at arcminutes resolution are necessary to confirm the excess and discriminate between possible causes.

Since the recent detection of polarization by the DASI experiment [49] several experiments have measured the TE cross-power spectrum C_ℓ^{TE} (WMAP [47], CBI [75] and BOOMERANG [69]) and the EE one C_ℓ^E (CBI [75] and BOOMERANG [63]). The sensitivity of the polarization data is not yet comparable to the temperature one. However, it is already sufficient to confirm the main features of the concordance model and, more specifically, the adiabatic nature of the primordial matter density fluctuations. Namely, that the peaks of the polarization power spectrum, C_ℓ^E, are out of phase with the temperature C_ℓ ones. The TE cross-power spectrum, C_ℓ^{TE}, and the E-mode polarization power spectrum, C_ℓ^E, are shown in Fig. 5.

The next challenge is the detection of the B-mode polarization. As commented in the previous section this detection would unambiguosly indicate the existence of a background of primordial gravitational waves. Moreover, it is the only chance that we have to know about its possible existence in the next decade since experiments aimed to directly detect gravitational waves are still not sensitive enough. The amplitude of this background, A_t, or equivalently the tensor-to-scalar ratio r, is proportional to the energy scale of inflation. Whereas the all-sky Planck space mission will be limitted by instrument sensitivity, being able to detect values down to $r \approx 0.05$, planned very sensitive ground-based experiments covering small patches of the sky and carrying arrays of 1000's of detectors like PolarBear [71], Clover[58], BRAIN [59] or QUIET [74], are expected to be limitted by the ability to remove the foregrounds and are expected to reach values of $r \approx 0.01$ in the best case. In order to significantly improve this limit we will have to wait for the next generation of space missions with $10^3 - 10^4$ detectors now under discussion by ESA, NASA and other space agencies [102, 10]. The combination of complete sky coverage, very sensitive multifrequency observations and an optimal control of systematics make the space missions the ultimate experiments to go down to values $r \sim 10^{-3} - 10^{-4}$ allowing to probe energy scales for inflation down to $\approx 5 \times 10^{15}$GeV [97].

COSMOLOGICAL CONSTRAINTS FROM OTHER DATA SETS

The first observations showing the existence of a dark energy dominating the dynamics of the universe were those based on the luminosity distance-redshift diagram determined with supernovae SN Ia [77, 68]. Those results were, however, taken with certain caution because of the assumption made that the low and high redshift SN Ia had the same light curve behaviour (in addition, there were many other possible systematics which raised some concern). In any case, SN Ia data complements very well the CMB anisotropies since for the former the dependence of the model prediction on the dark matter and dark energy density parameters enters approximately as $\Omega_m - \Omega_\Lambda$, and as $\Omega_m + \Omega_\Lambda$ for the latter (see Fig. 6).

The large galaxy surveys, *2-degree Field Galaxy Redshift Survey* 2dFGRS [67] and *Sloan Digital Sky Survey* SDSS [92] have recently reached ≈ 200000 measured redshifts on large fractions of the sky. A three-dimensional power spectrum $P(k)$ of density perturbations is estimated with each data set. The cosmological parameters enter in the prediction of $P(k)$ through the initial power spectrum (A_s, n_s) and the transfer function which linearly connects the initial and present spectra. Besides, an additional parameter, *the bias b*, is required to link the galaxy power spectrum to the matter one $P(k)$. Although this parameter will in principle depend on the scale, however, it is found that at large scales b is scale-independent (see e.g. [93]). There are still other problematics that need to be corrected before using the data for accurate parameter determination such as the redshift-space distorsions due to galaxy peculiar velocities, survey geometry effects, ... Once those effects are corrected the galaxy redshift surveys play a key role in breaking the CMB degeneracies. For instance, the degeneracies of the model predictions on both data sets on the plane formed by the spectral shape parameter $h\Omega_m$ and the baryon

TABLE 2. Cosmological parameters from WMAP, CBI, ACBAR and 2dFGRS combined data (from [6]).

Parameter	Values (68% CL)
w_b	0.0224 ± 0.0009
w_m	$0.135^{+0.008}_{-0.009}$
w_ν	< 0.0076 (95% CL)
w	< -0.78 (95% CL)
Ω_{DE}	0.73 ± 0.04
h	$0.71^{+0.04}_{-0.03}$
τ	0.17 ± 0.04
A_s	$0.833^{+0.086}_{-0.083}$
n_s	0.93 ± 0.03
α	$-0.031^{+0.016}_{-0.018}$
r	< 0.90 (95% CL)

fraction Ω_b/Ω_m are almost orthogonal.

Results on 11 free cosmological parameters combining CMB data (WMAP [32], CBI [73], ACBAR [50]) with the 2dFGRS galaxy redshift survey [67] are given in table 2. As can be seen from that table, the combination of different cosmological data sets has allowed for the first time an accuracy $\lesssim 10\%$ in the determination of most parameters. Results combining WMAP data with the SDSS galaxy redshift survey [92] have also produced similar values for the parameters [93].

In addition to CMB and galaxy surveys, other combinations including the HST key project value for the Hubble parameter [27], SN Ia magnitude-redshift data [78, 96], Lyα forest power spectrum [17, 57] or abundancies of reach clusters of galaxies can help to improve the results [90, 82, 72, 39]. Confidence contours in the plane $(\Omega_m, \Omega_\Lambda)$ for the combination CMB+SN Ia+cluster abundancies are given in Fig. 6. The complementarity of the three data sets to break degeneracies is clearly shown.

Recently, an independent piece of evidence that the universe is not Einstein-de Sitter (flat geometry with null dark energy density) has been found by cross-correlating the CMB map with galaxy survey maps. The same evolving gravitational potential wells which generate the large scale structure of the galaxy distribution also produce the gravitational redshift in the CMB photons a late times (the late ISW effect, see section on CMB anisotropies). The amplitude and sign of the cross-correlation depends on three parameters Ω_{DE}, Ω_k and h. For a flat universe, as indicated by many different observations as discussed above, a positive signal will unambiguously imply the presence of dark energy. This is the case found recently cross-correlating the WMAP map with different large scale structure surveys (the radiosource survey NVSS [16], the X-ray survey HEAO-1 [7], the optical SDSS [1], the near-infrared survey 2MASS [88]) [11, 26, 64, 3, 101]. In [101] three different methods were used to estimate the WMAP-NVSS cross-correlation: direct temperature anisotropy-galaxy number density, cross-power spectrum and covariance wavelet coefficients. A clear positive signal was found in the three cases using a maximum likelihood analysis, implying a value for the dark energy $\Omega_{DE} = 0.73^{+0.11}_{-0.14}$. The significance of a non-null dark energy reaches 3.5σ for

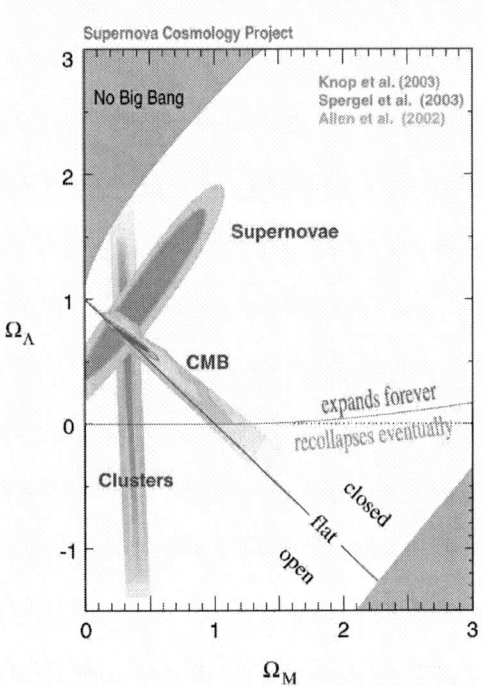

FIGURE 6. Confidence contours for the pair (Ω_m, Ω_Λ) combining SN Ia, CMB and cluster density data (taken from the Supernova Cosmology Project [45]).

the cross-power spectrum. This is also the maximum significance detection of the ISW effect up to date. In the same work, an analysis allowing variations in the dark energy equation of state parameter w also shows a prefered value close to -1. In other words, the dark energy is well described by a cosmological constant where the energy density remains constant with time.

THE CONCORDANCE MODEL

From the previous discussion about the cosmological implications of the large amounts of data already collected on the CMB anisotropies, galaxy redshift surveys, etc, maybe the most important result is the convergence of all those cosmological data sets towards the same model of the universe, *the concordance model*. Below we summarize the main characteristics of this model and other consequencies implied by the data:

- The geometry of the universe is very close to flat.
- The dynamics is dominated by a dark energy whose equation of state is almost that of a cosmological constant.

- Most of the matter content is in the form of cold dark matter.
- The large scale structure of the universe was seeded by quantum fluctuations in the very early universe that evolved via gravitational instability (the first evidence of this came from COBE-DMR [89]).
- The initial matter-energy density fluctuations were of the adiabatic type.
- Topological defects did not play a dominant role in the structure formation of the universe.
- Primordial gravitational waves do not appreciably contribute to neither the temperature nor the E-mode polarization anisotropies (up to the present sensitivities reached).
- The reionization of the universe happened at relatively early times, $z \gtrsim 10$.
- The initial fluctuations were close to a homogeneous and isotropic Gaussian random field (see however the next section).

BEYOND THE POWER SPECTRUM

There exist a richness of information in the distribution of the CMB temperature anisotropies. As we have seen in the previous sections, the C_ℓ contains very valuable information for constraining the cosmological parameters. Beyond the second-order moment, the overall n-point CMB temperature distribution can inform us about competing scenarios that might have taken place in the very early universe. If the temperature anisotropies are found to be consistent with a homogeneous and isotropic Gaussian random field then it would strongly support the standard inflationary model. On the contrary, deviations from Gaussianity would imply that alternative scenarios to the standard model are favoured. In particular, non-standard models of inflation and topological defects are expected to produce different levels of non-Gaussianity.

The situation is complicated by the effect of secondary anisotropies and foreground emissions. Both leave non-Gaussian signatures which are superimposed on the distribution of the primary anisotropies. The secondary anisotropies produced after recombination are caused by different physical processes. One of them is the SZ effect [91] generated when the microwave photons cross the hot gas present in the central cores of rich clusters, as discussed above. This effect produces a distortion in the black-body spectrum. The new anisotropies generated are therefore localized in the direction of rich clusters and subtend typical angular scales of a few arcminutes. A different effect is the one produced by the lensing of the microwave photons by the large scale structure of the universe. The angular scales more affected by lensing are below several arcminutes. Besides, the same gravitational potential wells which produce the lensing effect also produce the gravitational redshift known as the late ISW. This fact generates correlations between multipoles at very low and very high orders which clearly represent non-Gaussian features [33, 30, 86]. A last effect which also generates deviations from Gaussianity is the redshift suffered by the microwave photons when they cross evolving gravitational potentials generated by non-linear structures (*the Rees-Sciama effect* [76]). This effect is in general smaller than the other two and might be significant only for the most massive structures or largest voids in the universe.

Detailed studies of the distribution of the CMB temperature anisotropy in a systematic way started with COBE-DMR and have been intensified recently with WMAP. Most of the analyses of the data from those satellites and also other ground and balloon based experiments have concluded that the CMB anisotropy distribution is consistent with the inflationary prediction of a homogeneous and isotropic Gaussian random field (see e.g. [48]). In particular the unique all-sky data set provided by the first year of operation of WMAP has allowed a very precise test of normallity. From those data, however, several significant anomalies have been found whose interpretation is still unclear.

The first anomaly found is related to the amplitude and alignment of the lowest multipoles [6, 22, 20, 51]. The amplitudes of both quadrupole and octopole are small compared to the prediction by the standard model. The two multipoles are also aligned towards a direction close to the dipole one. If this anomaly is confirmed to be intrinsic to the CMB, and not due to systematics or foreground contamination, then a natural explanation would be a non-trivial topology for the universe [56, 79, 5]. Another possibility which has been proposed is the lensing of the dipole due to near-by structures [99]. In any case the lowest multipoles are very much affected by the Galactic mask which prevent us to get information from a significant fraction of the sky around the Galactic plane. The future Planck mission is expected to strongly reduce the area of that mask because of its higher sensitivity and resolution and wider frequency coverage. In addition the systematics will be quite different from that of WMAP due to both a different scanning strategy and geometry of the measurement and different instruments. Thus, it is expected that Planck can help in clarifying the origin of the lowest multipole anomaly.

A different kind of anomaly is the north-south asymmetry in the two and three-order moments maximized for a coordinate system with the poles close to the ecliptic ones [65, 24, 31]. It seems difficult to explain this anomaly with foreground contamination. Errors caused by an inacurate subtraction of the dipole might explain at least part of the asymmetry [28]. On the other hand, a good fit to this asymmetry and the cold spot (see below) (which also aliviates the low-multipole alignment and amplitude) can be obtained with a homogeneous and anisotropic Bianchi VII_h model with values of vorticity and shear consistent with observations [37]. However the value required for the matter content is too high [38].

A significant excess of kurtosis in the wavelet coefficients of the WMAP data at scales of several degrees has been detected [100]. That deviation is found to be due to the presence of a cold spot of $\approx 10°$ size located at Galactic coordinates ($b = -57°, l = 209°$) [18, 13] (see Fig. 7). This spot is unlikely to be due to foreground residuals [19]. Also it is very difficult to imagin systematics that can account for it. An intrinsic CMB origin is not discarded. A possibility in this direction is the gravitational redshift generated by the non-linear evolution of large scale structures [60, 61, 95] or by a non-Gaussian matter distribution [36]. Another possibility could be the effect of topological deffects in the form of textures [98]. [2] have suggested a model of the universe in the form of a finite spherical ball of dust and dark energy. A cold spot in the CMB is expected in this model if we are not too far from the dust-ball edge, and also suppression of the low multipoles.

Very recently, deviations with respect to the global isotropy of the universe have been found [103]. By assigning a prefered orientation to each pixel of the WMAP data using orientable wavelets some anomalous directions are found at the scale of several degrees (see Fig. 8). The most significant ones are very close to the ecliptic poles. Also they are

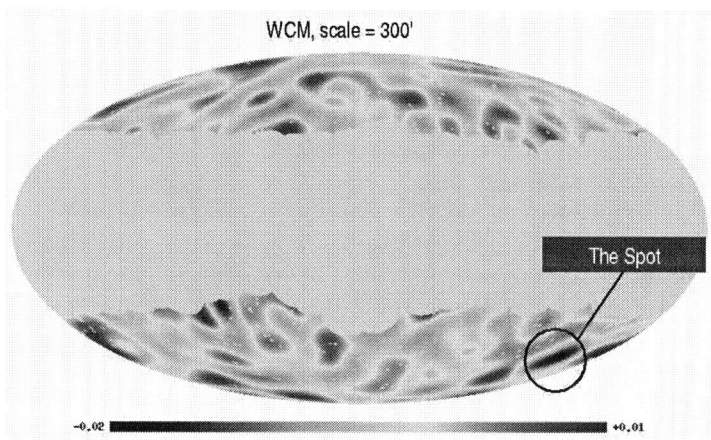

FIGURE 7. The Cold Spot found in the WMAP combined map after convolving with a Spherical Mexican Hat Wavelet with scale $300'$ (see [100, 18, 19])

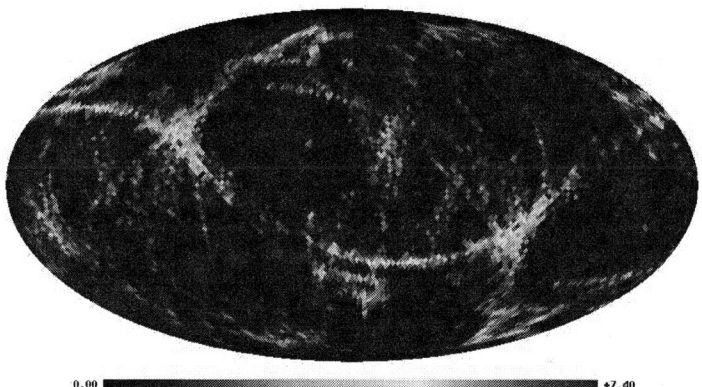

FIGURE 8. Excess of the number of times each pixel is seen by any other in σ units, where σ is the dispersion corresponding to Gaussian simulations (see [103])

ordered following a great circle in the sky whose perpendicular is close to the dipole direction. Those coincidences seem to indicate a systematic effect maybe associated to the scanning strategy of the mission. However, an intrinsic origin or foreground contamination cannot be discarded yet.

A more detailed analysis of the anomalies described above is necessary before any firm conclusion about their origin is reached. Maybe some of them could be a first hint of new physics.

CONCLUSIONS

The standard scenario of the universe is characterised by an almost flat geometry with its energy content dominated by a dark energy ($\approx 73\%$) whose equation of state is very close to that of a cosmological constant ($w \approx -1$). The matter content represents $\approx 27\%$ of which about 80% is in the form of cold dark matter and only 20% in the form of "normal" matter (baryons). Besides, the large scale structure of the universe originated from quantum fluctuations during an inflationary phase in the very early universe. This standard picture has been recently proved to be consistent with the large amounts of data already collected and which constitute the fundamental tests of the universe: the temperature and polarization anisotropies of the CMB, the large scale structure of the galaxy distribution, the magnitude-redshift diagram provided by SN Ia up to $z > 1$ and the primordial nucleosynthesis. This concordance of the standard model with all those different data sets makes it a solid framework to continue exploring our universe further.

There are, however, fundamental problems related to the nature of the dark energy and dark matter which completely dominates the dynamics of the universe. Although the dark energy is well represented by a cosmological constant, its natural explanation in terms of the vacuum energy implies a value around 100 orders of magnitude above the observed one. Regarding the dark matter, several candidates have been proposed but still we do not have any hint about neither its nature nor its evolutionary history.

Another fundamental problem is the origin of inflation and its dynamics. There is not any particle physics framework for inflation and for how the universe started to inflate in the early universe. Evenmore, we do not even know which is the specific model of inflation within, e.g., the classification of [43].

The anomalies found in the first year WMAP data are a matter for concern. Although some of them might be finally interpreted as systematics or foreground contamination, it is not clear that all of them can be interpreted that way. If proved to be intrinsic they could be a first hint of new physics. The next ESA Planck mission, scheduled for launch in a couple of years, will measure the temperature of the CMB over the whole sky with unprecedented sensitivity and resolution. It will also provide a map of polarization with better sensitivity than previous experiments. These new measurements will certainly have a profound impact on our understanding of the origin and evolution of the universe and, in particular, will help in clarifying the present anomalies found in the WMAP CMB data.

Finally, the major goal after Planck is to measure the CMB polarization with high sensitivity to search for the B-mode. This mode represents at present the best way to study the primordial background of gravitational waves expected in the standard model of inflation. The future programs "Beyond Einstein" and "Cosmic Vision" of NASA and ESA, respectively, include a polarization mission in their plans.

ACKNOWLEDGMENTS

I acknowledge financial support from the Spanish MEC project ESP2004-07067-C03-01. I also acknowledge the use of LAMBDA, support for which is provided by by the NASA Office of Space Science. I acknowledge the use of the software package CMB-

FAST (http://www.cmbfast.org) developed by Seljak and Zaldarriaga. The work has also used the software package HEALPix (http://www.eso.org/science/healpix) developed by K.M. Gòrski, E.F. Hivon, B.D. Wandelt, J. Banday, F.K. Hansen and M. Barthelmann.

REFERENCES

1. Abazajian K. et al., 2004, AJ, 128, 502
2. Adler R.J., Bjorken J.D. & Overduin J.M., 2006, gr-qc/0602102
3. Afshordi N., Loh Y.-S. & Strauss M. A., 2004, Phys. Rev. D, 69, 3524
4. Alpher R.A. and Herman R.C., 1948, Phys. Rev. 74, 1737
5. Aurich R., Lustig S. & Steiner F., 2005, astro-ph/0510847
6. Bennett C.L. et al., 2003, ApJS, 148, 1
7. Boldt E., 1987, Phys. Rev., 146, 215
8. Bond et al., 2005, ApJ, 626, 12
9. Bonifacio P. et al., 2002, A&A, 390, 91
10. Bouchet F.R., Benoît A., Camus Ph., Désert F.X., Piat M. & Ponthieu N., 2005: Charting the New Frontier of the Cosmic Microwave Background Polarization. In: *SF2A*, ed. by F. Casoli et al. (EDP Sciences 2004), astro-ph/0510423
11. Boughn S.P. & Crittenden R.G., 2004, Nature, 427, 45
12. Bunn E.F., Zaldarriaga M., Tegmark M. & de Oliveira-Costa A., 2003, Phys. Rev. D, 67, 023501
13. Cayón L., Jin L. & Treaster A., 2005, MNRAS, 362, 826
14. Challinor A., 2005, astro-ph/0502093
15. Challinor A. & Chon G., 2005, MNRAS, 360, 509
16. Condon J.J. et al., 1998, AJ, 115, 1693
17. Croft R.A.C. et al., 2002, ApJ, 581, 20
18. Cruz M., Martínez-González E., Vielva P. & Cayón L., 2005, MNRAS, 356, 29
19. Cruz M., Tucci M., Martínez-González E. & Vielva P., 2006, MNRAS, accepted
20. De Oliveira-Costa A., Tegmark M., Zaldarriaga M. & Hamilton A., 2004, Phys. Rev. D, 69, 063516
21. Dickinson C. et al., 2004, MNRAS, 353, 732
22. Efstathiou G., 2004, MNRAS, 348, 885
23. Einstein A., 1917, Sitz. Preuss. Akad. d. Wiss., 142
24. Eriksen H.K., Hansen F.K., Banday A.J., Gorski K.M. & Lilje P.B., 2004, ApJ, 605, 14
25. Fields B.D. & Olive K.A., 1998, ApJ, 506, 177
26. Fosalba P., Gaztañaga E. & Castander F., 2004, ApJL, 597, 89
27. Freedman W.L. et al., 2001, ApJ, 553, 47
28. Freeman, P.E., Genovese, C.R., Miller, C.J., Nichol, R.C. & Wasserman, L., 2006, ApJ, 638, 1
29. Gamow G., 1946, Phys. Rev., 70, 572
30. Goldberg D.M. & Spergel D.N., 1999, Phys. Rev. D59, 103002.
31. Hansen F.K., Banday A.J. & Gòrski K.M., 2004, MNRAS, 354, 641
32. Hinshaw G. et al., 2003, ApJ, 148, 135
33. Hu W., 2000, Phys. Rev. D, 62, 043007
34. Hu W. & Okamoto T., 2002, ApJ, 574, 566
35. Hubble E.P., 1929, Proc. Natl. Acad. Sci., 15, 168
36. Inoue K.T. & Silk J., 2006, astro-ph/0602478
37. Jaffe T.R., Banday A.J., Eriksen H.K., Gorski K.M. & Hansen F.K.,2005, ApJ, 629, L1, astro-ph/0503213
38. Jaffe T.R., Hervik S., Banday A.J., Gorski K.M., 2005, ApJ, submitted, astro-ph/0512433
39. Jassal H.K., Bagla J.S. & Padmanabhan T., 2005, astro-ph/0506748
40. Jones W.C. et al., 2005, ApJ, submitted, astro-ph/0507494
41. Kamionkowski M., Kosowsky A. & Sttebins A., 1997, Phys. Rev. D, 55, 7368
42. Kesden M., Cooray A. & Kamionkowski M., 2003, Phys. Rev. D, 67, 123507
43. Kinney W.H., 2002, Phys. Rev. D, 66, 083508
44. Kirkman, D., Tytler, D., Suzuki, N., O'Meara, J.M. & Lubin, D., 2003, ApJ, 149, 1
45. Knop R.A. *et al.*, 2003, ApJ, 598, 102

46. Knox L. & Song Y.-S., 2002, Phys. Rev. Lett., 89, 011303
47. Kogut A. et al., 2003, ApJS, 148, 161
48. Komatsu E. et al., 2003, ApJS, 148, 119
49. Kovac J.M. et al. 2002, Nature, 420, 772
50. Kuo C.L. et al., 2004, ApJ, 600, 32
51. Land K. & Magueijo J., 2005, Phys. Rev. Lett., 95, 071301
52. Liddle A.R. & Lyth D.H., 2000, *Cosmological Inflation and Large-Scale Structure*, (Cambridge University Press)
53. Lewis A., Challinor A. & Lasenby A., 2000, ApJ, 538, 473 (http://camb.info)
54. Lewis A., Challinor A. & Turok N., 2002, Phys. Rev. D, 65, 023505
55. Lewis A., 2004, Phys. Rev. D, 70, 043011
56. Luminet J.-P., Weeks J.R., Riazuelo A., Lehoucq R. & Uzan J.-P., 2003, Nature, 425, 593
57. McDonald P. et al., 2004, ApJ, submitted (astro-ph/0407377)
58. Maffei B. et al., 2005, EAS Publications Series, 14, 251-256
59. Masi S. et al., 2005, EAS Publications Series, 14, 87-92
60. Martínez-González E., Sanz J.L. & Silk J., 1990, ApJ, 335, 5
61. Martínez-González E. & Sanz J.L., 1990, MNRAS, 247, 473
62. Martínez-González E. & Vielva P., 2005: The Cosmic Microwave Background Anisotropies: Open Problems. In: *JENAM Workshop: The Many Scales in the Universe*, ed. by C. del Toro et al., (Kluwer), astro-ph/0510003
63. Montroy T.E. et al., 2005, ApJ, Submitted, astro-ph/0507514
64. Nolta M. R. et al., 2004, ApJ, 608, 10
65. Park C.-G., 2004, MNRAS, 349, 313
66. Penzias A.A. & Wilson R.W., 1965, ApJ, 142, 419
67. Percival W.J. et al., 2001, MNRAS, 327, 1297
68. Perlmutter S., 1999, ApJ, 517, 565
69. Piacentini F. et al., 2005, ApJ, submitted, astro-ph/0507507
70. Pogosian, L., Tye, S.-H.H., Wasserman, I. & Wyman, M., 2003, Phys. Rev. D, 68, 0235506
71. http://bolo.berkeley.edu/polarbear
72. Rapetti D., Steven S.W. & Weller J., 2005, MNRAS, 360, 555
73. Readhead A.C.S. et al., 2004, ApJ, 609, 498
74. http://quiet.uchicago.edu
75. Readhead A.C.S. et al., 2004, Science, 306, 836
76. Rees M.J. & Sciama D.W., 1968, Nature, 517, 611
77. Riess et al., 1998, ApJ, 116, 1009
78. Riess et al., 2001, ApJ, 560, 49
79. Roukema B.F., Lew B., Cechowska M., Marecki A. & Bajtlik S., 2004, astro-ph/0402608
80. Sachs R. K. & Wolfe A. M., 1967, ApJ, 147, 73
81. Sanz J.L., 1997, in The Cosmic Microwave Background, eds. C.H. Lineweaver et al., Kluwer Academic Publishers, p. 33
82. Seljak U. et al., 2005, Phys. Rev. D, 71, 103515
83. Seljak U. & Hirata C.M., 2004, Phys. Rev. D, 69, 043005
84. Seljak U., Sugiyama N., White M. & Zaldarriaga M., 2003, Phys. Rev. D, 68, 3507
85. Seljak U. & Zaldarriaga M., 1996, ApJ, 469, 437 (http://www.cmbfast.org)
86. Seljak U. & Zaldarriaga M., 1999, Phys. Rev. D60, 043504.
87. Silk J., 1968, ApJ, 151, 459
88. Skrutskie M.F. et al., 1997, in The Impact of Large Scale Near-IR Sky Survey, eds. Gazón F. et al., Kluwer, Dordrecht, p. 187
89. Smoot G. F. et al., 1992, ApJL, 396, L1
90. Spergel D. N. et al., 2003, ApJS, 148, 175
91. Sunyaev R.A. & Zeldovich Y.B., 1972, Comm. Astrophys. Space Phys., 4, 173
92. Tegmark M. et al., 2004, ApJ, 606, 702
93. Tegmark M. et al., 2004, Phys. Rev. D, 69, 103501
94. Toffolatti L., Negrello M., González-Nuevo J., de Zotti G., Silva L., Granato G.L. & Argüeso F., 2005, A&A, 438, 475, astro-ph/0410605
95. Tomita K., 2005, Phys. Rev. D, 72, 10

96. Tonry J.L. et al., 2003, ApJ, 594, 1
97. Tucci M., Martínez-González E., Vielva P. & Delabrouille J., 2005, MNRAS, 360, 935
98. Turok N. & Spergel D.N., 1990, Phys. Rev. Lett., 64, 2736
99. Vale C., 2005, APJL, submitted, astro-ph/0509039
100. Vielva P., Martínez-González E., Barreiro R.B., Sanz J.L. & Cayón L., 2004, ApJ, 609, 22
101. Vielva P., Martínez-González E. & Tucci M., 2006, MNRAS, 365, 891, astro-ph/0408252
102. White N.E., 2005, Adv. Space Res., 35, 96
103. Wiaux Y., Vielva P., Martínez-González E. & Vandergheynst P., 2006, Phys. Rev. Lett., submitted
104. Zaldarriaga M. & Seljak U., 1997, Phys. Rev. D, 55, 1830

Supersymmetry and the Supergravity Landscape

Tomás Ortín

Instituto de Física Teórica UAM/CSIC Facultad de Ciencias C-XVI, C.U. Cantoblanco, E-28049-Madrid, Spain

Abstract. In the recent times a lot of effort has been devoted to improve our knowledge about the space of string theory vacua ("the landscape") to find statistical grounds to justify how and why the theory selects its vacuum. Particularly interesting are those vacua that preserve some supersymmetry, which are always supersymmetric solutions of some supergravity theory. After an general introduction to how the pursuit of unification has lead to the vacuum selection problem, we are going to review some recent results on the problem of finding all the supersymmetric solutions of a supergravity theory. We will also review some interesting solutions that have been discovered using these methods and their relations.

Keywords: Supersymmetry, Supergravity, Landscape, Killing spinors,
PACS: 11.30.Pb,04.65.+e,11.25.-w

INTRODUCTION: UNIFICATION AND THE LANDSCAPE

Unification has been one of the most fruitful guiding principles in our search for the fundamental components and forces of the Universe. It is, however, more than just a wish or a prejudice that has produced important results for a while: it is indeed a logical necessity for the human mind to understand the Universe: the history of Physics could be written as the history of the process of unification of many different concepts, entities and phenomena into an ever smaller and more fundamental number of them. However, it was only in later times that we realized what we were doing and started doing it consciously, setting explicitly the unification of all forces and particles as our major goal.

It is this (sometimes feverish) pursuit of unification that has lead us to the vacuum selection problem in Superstring Theory and similar unification schemes that include gravity. If unification is a major goal, then, the vacuum selection problem is a major problem of Superstring Theory, perhaps the most important one.

In order to get some perspective over this problem we are going to review several instances of unification in Physics. We could go back to Archimedes or Newton but we will content ourselves with the *classical* period of unification that starts with Faraday and Maxwell, showing also that the process of unification underlies all the main advances in Theoretical Physics and is, in particular, strongly related to the symmetry principles on which many of our theories are based.

1. Electricity $\oplus$ Magnetism $\overset{\text{Faraday,Maxwell}}{\Longrightarrow}$ Electromagnetism

$$\vec{E}, \vec{B} \longrightarrow (F_{\mu\nu}) \equiv \begin{pmatrix} 0 & -\vec{E}^T \\ \vec{E} & \star\vec{B} \end{pmatrix}.$$

The unification of electricity and magnetism into a single interaction is the first paradigm of modern unification of interactions: the unification requires (or produces) a bigger group of symmetry because the equations of each field were invariant only under the Galilean group and the full set of Maxwell's equations are invariant under the Poincaré group. This had to be so: if two interactions are different manifestations of a single interaction, there must exist transformations that do not change the equations of the theory and transform one interaction into the other. Had the Special Theory of Relativity been proposed before Maxwell's equations, the latter could have been discovered by imposing Poincaré invariance on the incomplete equations of electricity and magnetism. However, the importance of symmetry principles was discovered much later.

Observe that the Principle of (Special) Relativity applied to Newtonian gravity implies the existence of gravitomagnetism and the combination of both into a single relativistic field of interaction. This interaction is not yet General Relativity, but contains its seeds.

2. **Space $\oplus$ Time $\overset{\text{Einstein,Minkowski}}{\Longrightarrow}$ Spacetime**

$$t, \vec{x} \longrightarrow (x^\mu) \equiv (ct, \vec{x}).$$

This is an example of unification of fundamental concepts (not interactions), although it is strongly related to our previous example because the increase in symmetry (from Galileo to Poincaré) is the same and the underlying mechanism is similar (if space and time are different aspects of spacetime, there must be transformations that take space into time and vice-versa). It is important to observe that the new symmetry is only apparent at high speeds, but it is never broken.

3. **Waves $\oplus$ Particles $\overset{\text{deBroglie}}{\Longrightarrow}$ Quantum particles**

This unification of two entities always believed to be distinct is required (and led to) Quantum Mechanics. It is, to this day, mysterious, perhaps because it is different from the other instances of unification: in this case there seems to be no underlying symmetry group transforming particles into waves and vice-versa.

4. **Gravity (GR) $\oplus$ Electromagnetism $\overset{\text{Kaluza, Klein, Einstein}}{\Longrightarrow}$ Higher – dimensional gravity**

$$g_{\mu\nu}, A_\mu \longrightarrow (\hat{g}_{\hat{\mu}\hat{\nu}}) \equiv \left(\begin{array}{c|c} k^2 & A_\nu \\ \hline A_\mu & g_{\mu\nu} \end{array} \right)$$

This attempt was unsuccessful (it was, may be, too early) but introduced many new ideas that have stayed around until now. In this theory there is also an increase of symmetry, but the scheme is more complicated: the vacuum of the theory (in modern parlance) could be 5-dimensional Minkowski spacetime, invariant under the 5-dimensional Poincaré group but this symmetry is spontaneously broken (again in modern parlance) to the 4-dimensional Poincaré group times $U(1)$ due to the (completely arbitrary) choice of vacuum (4-dimensional Minkowski spacetime times a

circle). General covariance implies that these symmetries are local in the resulting effective theory, a fact that can be formulated as the *Kaluza-Klein Principle*:
> Global invariances of the vacuum are local invariances of the theory.

An, originally unwanted, feature of the theory is that a new massless field is predicted: the *Kaluza-Klein scalar* (or *radion*) k. Its v.e.v., related to the radius of the internal circle, can also be fixed arbitrarily because there is no potential for this scalar. Fixing (*stabilizing*) the v.e.v. of scalars such as k that determine the size and shape of part of the vacuum spacetime (generically known as *moduli*) is nowadays known as the *moduli problem*. Explaining why the vacuum should be 4-dimensional Minkowski spacetime times a circle of all the possible classical solutions of 5-dimensional General Relativity is the simplest version of the *vacuum selection problem*.

5. **Quantum Mechanics $\oplus$ Relativistic Field Theory $\overset{\text{Many people}}{\Longrightarrow}$ QFT**
   A difficult but fruitful marriage.

6. **Weak interactions $\oplus$ Electromagnetism $\overset{\text{Glashow, Salam, Weinberg}}{\Longrightarrow}$ EW interaction**
   In this case, two Relativistic QFTs are unified.
   - Unification is achieved by an increase of *local* (Yang-Mills-type) symmetry, from $U(1)$ to $SU(2) \times U(1)$.
   - The symmetry is *spontaneously broken* by the Higgs mechanism: choice of vacuum by energetic reasons (minimization of the *ad hoc* Higgs potential). (This is the main difference with Kaluza-Klein and other theories including gravity in which different vacua are associated to different spacetimes and, therefore, different definitions of energy that cannot be compared.)
   - The spontaneous breaking of the symmetry renders the model renormalizable.
   - The symmetry is restored at high energies.
   - New massive particles are predicted associated to the enhanced symmetry (gauge bosons, found) and a new massless spin-0 particle is also predicted (Higgs boson, not yet found).

   This model, part of the Standard Model of Particle Physics, has had an extraordinary success and most unification schemes of relativistic QFTs have followed the same pattern. In particular

7. **Electroweak interaction $\oplus$ Strong interactions $\overset{\text{Many people}\cdots}{\Longrightarrow}$ Grand Unified Theory**
   This is an unsuccessful generalization of the electroweak unification scheme based on a semisimple gauge group ($SO(10), SU(5), \cdots$) spontaneously broken by a generalized Higgs mechanism to $SU(3) \times U(1)$. There are two main problems:
   - New massive and massless particles predicted may mediate proton disintegration (not observed).
   - Unification of coupling constants should occur at the energy at which the symmetry is restored, but this does not seems to work.

8. **Bosons $\oplus$ Fermions $\overset{\text{Golfand,Likhtman,Volkov,Akulov,Soroka,Wess and Zumino}}{\Longrightarrow}$ Superfields**

This is a new kind of unification based in an increase of (global spacetime) symmetry to *supersymmetry*, which should also be spontaneously broken by a yet unknown super-Higgs mechanism. It has many interesting properties:
- It is the most general extension of the Poincaré and Yang-Mills symmetries of the S-matrix (Haag-Lopuszanski-Sohnius theorem).
- This new symmetry can be combined with Yang-Mills-type symmetries (super-Yang-Mills theories) and with GUT models in which, in some cases, unification of coupling constants can be achieved.
- It can also be combined with g.c.t.'s, making it local (*supergravity* theories). We can have supergravity theories with Yang-Mills fields etc., but in most of these theories gravity is not unified with the other interactions since they belong to different supermultiplets.
- However, *extended* ($N > 1$) supergravities contain in the same supermultiplet of the graviton additional bosonic fields that may describe the other interactions. In this scheme all interactions would be described in a truly unified way. These extended supergravities can in general be obtained from compactification of simpler higher-dimensional supergravities. It was also discovered that many $N = 1$ supergravities coupled to Yang-Mills fields could also be obtained in the same way, by a careful choice of compact manifold (i.e. Kaluza-Klein vacuum). This lead to a new brand of unified theories which could describe everything (*Theories of Everything*). The first of these is

9. **Kaluza-Klein Supergravity [1, 2]**

    It is a combination of the Kaluza-Klein theories with supersymmetry. Now, a Kaluza-Klein vacuum is (arbitrarily) chosen that breaks spontaneously part of the (super)symmetries of the "original" vacuum (Minkowski spacetime for Poincaré supergravities and anti-De Sitter spacetime for aDS supergravities). Now, the rule of the game, the *supersymmetric Kaluza-Klein Principle*, is

    Global (super)symmetries of the vacuum are local (super)symmetries of the compactified theory.

    In general, the theories were based on compactifications of $N = 1, d = 11$ supergravity [3], the unique supergravity that can be constructed in the highest dimension in which a consistent supergravity can be constructed. It can accommodate the bosonic part of the Standard Model with minimal supersymmetry. However, these theories are anomalous and it is impossible to obtain the chiral structure of the Standard Model by compactification on *smooth* manifolds [4]. The vacuum of these theories was arbitrarily chosen to recover the Standard Model. The arbitrariness in the choice of vacuum replaces that of the choice of Higgs field and potential (and gauge interactions, dimensionality...). This makes these theories, conceptually, far superior, but raises to a very prominent place the vacuum selection problem.

    These problems and the advent of String Theory, in particular the Heterotic Superstring [5], which is anomaly-free and has chiral fermions, killed these theories, although they have been resurrected again by the same theory that killed them.

10. **Superstring Theories**

In these theories, all quantum particles are different vibration states of a single physical entity: the superstring. All known interactions could be described in this way. At low energies, one recovers an anomaly-free supergravity theory. However, there are still some problems:

- They are 10-dimensional, and require compactification. At low energies we are faced with 10-dimensional Kaluza-Klein supergravity and the vacuum selection problem.
- There are at least five superstring theories: Types IIA, Type IIB, Heterotic $SO(32)$, Heterotic $E(8) \times E(8)$ and Type I $SO(32)$. Which one should be considered?
- The theory seems to contain other extended objects besides strings: *D-branes* [6], NSNS-branes... Why should strings be fundamental [7]?

The answer to the last two questions lies on the *dualities* that related the different superstring theories and the different extended objects that occur in them [8]. Dualities are transformations that relate different theories: their spectra, interactions and coupling constants. Their existence allows the mapping of all scattering amplitudes of one theory into those of the other theory and vice-versa. In some cases, the mapping relates the coupling constant of one theory with the inverse coupling constant of the other theory and we talk about the non-perturbative S dualities. In other cases the non-trivial mapping affects only geometrical data of the compactification (moduli) and we talk about perturbative T dualities. These are characteristic of String Theories.

Dualities are, certainly, not symmetries of a single theory. Instead, they can be seen as symmetries in the space of theories. If two dual theories arise from two different compactifications (i.e. choices of vacua) of a given String Theory, then dualities can be seen as symmetries in the space of vacua. The extrapolation of this fact to the cases in which the theories are not known to originate from the same theory by different choices of vacuum is the basis of *M Theory*.

11. ⊕Superstring Theories $\overset{\text{Witten et al.}}{\Longrightarrow}$ M theory

In this (super-) unification scheme, all the superstring theories are understood as different duality-related vacua of an unknown theory called M Theory, whose low energy limit is $N = 1, d = 11$ supergravity, which was discovered by Witten [9] to be related to the strong-coupling limit (S duality) of the low-energy limit of Type IIA Superstring ($N = 2A, d = 10$ Supergravity).

Now we are back, in a sense, into the old Kaluza-Klein supergravity scenario, but all the Supergravity fields have got a String Theory meaning. It is amusing to see how, in this scheme, the low-energy limit of the Heterotic Superstring, which has chiral fermions, is related to the low-energy limit of M theory: 11-dimensional supergravity, which was apparently forbidden by Witten's no-go theorem [4]. The solution to the inconsistency is the use of non-smooth manifolds (orbifolds) [10], evading one the hypothesis of the theorem. This could have been done many years earlier, but, without Superstring Theory underlying the Supergravity theory other problems such as anomalies may never have been solved.

The unification scheme proposed by M theory is very attractive and could satisfy all our desires for unification: all particles and interactions may be explained in a unified way. Further, we no longer have different Superstring Theories to choose from. All the arbitrariness we had have disappeared, but only to be replaced by the arbitrariness in the choice of vacuum. Now there is only one theory and everything depends on that. But the theory seems to have nothing to tell us yet about how it chooses the vacuum and why our Universe is as we see it.

It has to be mentioned that, nowadays, we ask much more from a good candidate to *the* vacuum of our theory: it is not enough (but it is, certainly, a good starting point) that it gives the Standard Model of Particle Physics, but it should also explain the evolution of our Universe, that is, according to the most extended prejudices, it should give rise to an inflationary era and explain, in a fundamental way, dark energy.

With respect to this problem, there have been two main directions of work:

- Finding phenomenologically viable vacua (in the Particle Physics, e.g. [11] and/or cosmological, e.g. [12] sense).
- Find a vacuum-selection mechanism.

There has been no real progress in the second direction for many years.

The failure to solve the vacuum selection problem through some dynamical mechanism has favored recently a purely statistical approach in which one first has to explore and chart ("classify") the space of vacua a.k.a. *Landscape*. In this approach, our Universe is the way it is because the probability of this kind of Universe is overwhelming. Of course, this way of thinking can be combined with different forms of the Anthropic Principle.

Charting the superstring landscape is a very difficult problem and some simplifications have been suggested: for instance, one could consider all supersymmetric String Theory vacua, which correspond to different kinds of supergravities [13] or only the vacua with 4-dimensional Poincaré symmetry and a Calabi-Yau internal space, which correspond to $N=1, d=4$ supergravities and give Standard-Model-like theories [14]. One could also consider, as proposed by Van Proeyen [15], all possible supergravities, even if the stringy origin of many of them is unknown (the *supergravity landscape*).

In this talk, which is based on Refs. [16, 17, 18], we are going to review some recent general results on the classification of supersymmetric String Theory vacua and new techniques that can be used to find them, presenting some particular results on the classification of the supersymmetric vacua of the toroidally compactified Heterotic String Theory ($N=4, d=4$ SUGRA). First, we are going to define what is a supersymmetric configuration and its symmetry superalgebra, describing some useful special identities that they satisfy (*Killing spinor identities*). Then we will move on to define the problem of finding all the supersymmetric configurations of a given supergravity theory *Tod's problem* and we will explain the strategy to solve it in most (4-dimensional) cases. Finally, we will consider the case of $N=4, d=4$ supergravity.

SUPERSYMMETRIC CONFIGURATIONS AND SOLUTIONS

Supersymmetric configurations[1] (a.k.a. configurations with residual or unbroken or preserved supersymmetry) are classical bosonic configurations of supergravity (SUGRA) theories which are invariant under some supersymmetry transformations. Let us see what this definition implies.

Generically, the supersymmetry transformations take, schematically, the form

$$\delta_\varepsilon \phi^b \sim \bar\varepsilon \phi^f, \qquad \delta_\varepsilon \phi^f \sim \partial \varepsilon + \phi^b \varepsilon, \qquad (1)$$

where ϕ^b stands for bosonic fields (or products of an even number of fermionic fields) and ϕ^f for the fermionic fields (or products of an odd number of fermionic fields) and ε are the infinitesimal, local, parameters of the supersymmetry transformations, which are fermionic.

Then, a bosonic configuration (i.e. a configuration with vanishing fermionic fields $\phi^f = 0$) will be invariant under the infinitesimal supersymmetry transformation generated by the parameter $\varepsilon^\alpha(x)$ if it satisfies the *Killing spinor equations* (one equation for each ϕ^f), which have the generic form

$$\delta_\varepsilon \phi^f \sim \partial \varepsilon + \phi^b \varepsilon = 0. \qquad (2)$$

The concept of unbroken supersymmetry is a generalization of the concept of isometry, an infinitesimal general coordinate transformation generated by $\xi^\mu(x)$ that leaves the metric $g_{\mu\nu}$ invariant because it satisfies the *Killing (vector) equation*

$$\delta_\xi g_{\mu\nu} = 2\nabla_{(\mu} \xi_{\nu)} = 0. \qquad (3)$$

As it is well known, in this case, to each bosonic symmetry we associate a generator

$$\xi^\mu_{(I)}(x) \rightarrow P_I, \qquad (4)$$

of a symmetry algebra

$$[P_I, P_J] = f_{IJ}{}^K P_K, \quad \Leftrightarrow \quad [\xi_{(I)}, \xi_{(J)}] = f_{IJ}{}^K \xi_{(K)}, \qquad (5)$$

where the brackets in the right are Lie brackets of vector fields.

In our case, the unbroken supersymmetries are associated to the odd generators

$$\varepsilon^\alpha_{(n)}(x) \rightarrow \mathcal{Q}_n, \qquad (6)$$

of a superalgebra

$$[\mathcal{Q}_n, P_I] = f_{nI}{}^m \mathcal{Q}_m, \qquad \{\mathcal{Q}_n, \mathcal{Q}_m\} = f_{nm}{}^I P_I. \qquad (7)$$

---

[1] It will be very important for our discussion to distinguish between general field configurations and (classical) solutions of a given theory. General field configurations may or may not satisfy the classical equations of motion and, therefore, may or may not be classical solutions. As we are going to see, supersymmetry does not ensure that the equations of motion are satisfied.

The calculation of these commutators and anticommutators is explained in detail in Refs. [19, 20] and the consistency of the scheme was proven in [21]. According to the Kaluza-Klein principle we enunciated at the beginning, conveniently generalized to the supersymmetric case, this global supersymmetry algebra becomes the algebra of the local symmetries of the field theories constructed on this field configuration.

Of course, we do not want to construct field theories on just any field configuration but only on vacua of the theory. In general, for a field configuration to be considered a vacuum, we require that it is a classical solution of the equations of motion of the theory. Apart from this requirement, it is not clear what *a priori* characteristics a good vacuum must have except for classical and quantum stability, which are difficult to test in general, but which are, under certain conditions, guaranteed by the presence of unbroken supersymmetry. This is one of the reasons that makes supersymmetric vacua interesting. We also *prefer* highly symmetric vacua (such as Minkowski or anti-De Sitter space) since, on them, we can define a large number of conserved quantities, but it is uncertain why Nature should have the same prejudices.

Sometimes, when a vacuum solution has a clear (possibly warped) product structure, we can distinguish internal and spacetime (super-) symmetries and, if we choose this vacuum, our choice implies spontaneous compactification.

TOD'S PROBLEM

This is the problem of finding *all* the supersymmetric bosonic field configurations, i.e. all the bosonic field configurations ϕ^b for which a SUGRA's Killing spinor equations

$$\delta_\varepsilon \phi^f \Big|_{\phi^f=0} \sim \partial \varepsilon + \phi^b \varepsilon = 0, \tag{8}$$

have a solution ε, which includes all the possible supersymmetric vacua and compactifications.

Observe that, as we announced, not all supersymmetric bosonic field configurations satisfy the classical bosonic equations of motion for which we use the notations $\frac{\delta S}{\delta \phi^b}\Big|_{\phi^f=0} \equiv S_{,b}\Big|_{\phi^f=0} \equiv \mathcal{E}(\phi^b)$. Actually, the bosonic equations of motion of supersymmetric bosonic field configurations satisfy the so-called *Killing spinor identities* (KSIs) [16, 17] that relate different equations of motion of a supersymmetric theory. These identities can be derived as follows: The supersymmetry invariance of the action implies, for arbitrary local supersymmetry parameters ε

$$\delta_\varepsilon S = \int d^d x \, (S_{,b} \delta_\varepsilon \phi^b + S_{,f} \delta_\varepsilon \phi^f) = 0. \tag{9}$$

Taking the functional derivative w.r.t. the fermions and setting them to zero

$$\int d^d x \left[S_{,bf_1} \delta_\varepsilon \phi^b + S_{,b} (\delta_\varepsilon \phi^b)_{,f_1} + S_{,ff_1} \delta_\varepsilon \phi^f + S_{,f} (\delta_\varepsilon \phi^f)_{,f_1} \right] \Big|_{\phi^f=0} = 0, \tag{10}$$

The terms $\delta_\varepsilon \phi^b|_{\phi^f=0}$, $S_{,f}|_{\phi^f=0}$, $(\delta_\varepsilon \phi^f)_{,f_1}|_{\phi^f=0}$ vanish automatically because they are odd in fermion fields ϕ^f and so we are left with

$$\left\{ S_{,b}(\delta_\varepsilon \phi^b)_{,f_1} + S_{,ff_1} \delta_\varepsilon \phi^f \right\}\bigg|_{\phi^f=0} = 0. \tag{11}$$

This is valid for any fields ϕ^b and any supersymmetry parameter ε. For a supersymmetric field configuration ε is a Killing spinor $\delta_\varepsilon \phi^f|_{\phi^f=0}$ and we obtain the KSIs

$$\mathcal{E}(\phi^b)(\delta_\varepsilon \phi^b)_{,f_1}\bigg|_{\phi^f=0} = 0. \tag{12}$$

These non-trivial identities are linear relations between the bosonic equations of motion and can be used to solve Tod's problem, obtain BPS bounds etc. Let's see some examples.

Example: $N=1, d=4$ Supergravity.

This is the simplest supergravity theory. Its field content is $\{e^a{}_\mu, \psi_\mu\}$. The bosonic action (Einstein-Hilbert's) and the equations of motion (Einstein's) are

$$S|_{\psi_\mu=0} = \int d^4x \sqrt{|g|} R, \Rightarrow \mathcal{E}_a{}^\mu(e) \sim G_a{}^\mu. \tag{13}$$

The supersymmetry transformations of the graviton and gravitino are

$$\delta_\varepsilon e^a{}_\mu = -i\bar{\varepsilon}\gamma^a \psi_\mu, \qquad \delta_\varepsilon \psi_\mu = \nabla_\mu \varepsilon = \partial_\mu \varepsilon - \tfrac{1}{4}\omega_\mu{}^{ab}\gamma_{ab}\varepsilon. \tag{14}$$

The KSIs can be readily computed from the general formula Eq. (12) and simplified

$$-i\bar{\varepsilon}\gamma^a G_a{}^\mu = 0, \Rightarrow R = 0, -i\bar{\varepsilon}\gamma^a R_a{}^\mu = 0. \tag{15}$$

On the other hand, in trying to solve the Killing spinor equations (KSEs) which, here, take the form $\delta_\varepsilon \psi_\mu = \nabla_\mu \varepsilon = 0$, we can consider first their integrability conditions:

$$[\nabla_\mu, \nabla_\nu]\varepsilon = -\tfrac{1}{4}R_{\mu\nu}{}^{ab}\gamma_{ab}\varepsilon = 0, \Rightarrow R^\mu{}_a \gamma^a \varepsilon = 0. \tag{16}$$

Thus, at least in the case, the KSIs are contained in the integrability conditions. We will see later how to obtain more information from these identities.

Example: $N=2, d=4$ Supergravity.

This is the next simplest supergravity theory, if we do not consider adding matter supermultiplets to the $N=1$ theory. Its field content is $\{e^a{}_\mu, A_\mu, \psi_\mu\}$ (but now ψ_μ is a Dirac spinor, instead of a Majorana spinor as in the $N=1$ case). The bosonic action (Einstein-Maxwell's) and the equations of motion (Einstein's and Maxwell's) are

$$S|_{\psi_\mu=0} = \int d^4x \sqrt{|g|} \left[R - \tfrac{1}{4}F^2\right], \Rightarrow \begin{cases} \mathcal{E}_a{}^\mu(e) = -2\{G_a{}^\mu - \tfrac{1}{2}T_a{}^\mu\}, \\ \mathcal{E}^\mu(A) = \nabla_\alpha F^{\alpha\mu}. \end{cases} \quad (17)$$

The supersymmetry transformations are

$$\delta_\varepsilon e^a{}_\mu = -i\bar\varepsilon\gamma^a\psi_\mu + \text{c.c.}, \quad \delta_\varepsilon A_\mu = -2i\bar\varepsilon\psi_\mu + \text{c.c.}, \quad \delta_\varepsilon\psi_\mu = \nabla_\mu\varepsilon - \tfrac{1}{8}F^{ab}\gamma_{ab}\varepsilon \equiv \tilde{\mathcal{D}}_\mu\varepsilon. \quad (18)$$

Using the bosonic fields supersymmetry transformations, we find that the KSIs take the form

$$\bar\varepsilon\{\mathcal{E}_a{}^\mu(e)\gamma^a + 2\mathcal{E}^\mu(A)\} = 0. \quad (19)$$

On the other hand, the integrability conditions of the KSEs $\delta_\varepsilon\psi_\mu = \tilde{\mathcal{D}}_\mu\varepsilon = 0$ are

$$[\tilde{\mathcal{D}}_\mu, \tilde{\mathcal{D}}_\nu]\varepsilon = -\tfrac{1}{4}\left\{\left[R_{\mu\nu}{}^{ab} - e^a{}_{[\mu}T_{\nu]}{}^b\right]\gamma_{ab} + \nabla^a\left(F_{\mu\nu} + {}^*F_{\mu\nu}\gamma_5\right)\gamma_a\right\}\varepsilon = 0, \quad (20)$$

$$\Rightarrow \{\mathcal{E}_a{}^\mu(e)\gamma^a + 2[\mathcal{E}^\mu(A) + \mathcal{B}^\mu(A)\gamma_5]\}\varepsilon = 0.$$

In this case we get a more general formula from the integrability conditions, valid for the case in which the Bianchi identities are not satisfied. When they are satisfied we recover the KSIs, which is consistent since we have explicitly used the supersymmetry variations of the vector field in order to derive them, assuming, then, implicitly, that the Bianchi identities are satisfied.

The last formula (which we are also going to call KSI) has one important advantage over the original KSI: it is covariant under the $U(1)$ group of electric-magnetic duality rotations of the Maxwell and Bianchi identities that act as chiral rotations of the spinors.

SOLVING TOD'S PROBLEM

In 1983 showed in Ref. [22] that in $N=2, d=4$ SUGRA the problem could be completely solved using just integrability and consistency conditions. However, he used the Newman-Penrose formalism, unfamiliar to most particle physicists and suited only for $d=4$. Thus, there were no further results until 1995, when Tod, using again the same methods, solved partially the problem in $N=4, d=4$ SUGRA [23]. Then, in 2002, Gauntlett, Gutowski, Hull, Pakis and Reall proposed to translate the Killing spinor equation to tensor language and they solved the problem in minimal $N=1, d=5$ SUGRA [24]. This opened the gates to new results: in 2002 the problem was solved in *gauged* minimal $N=1, d=5$ SUGRA [25], in 2003 in minimal $N=(1,0), d=6$ SUGRA [26, 27] and gauged $N=2, d=4$ SUGRA [28], and in 2004 and 2005 in gauged minimal $N=1, d=5$ SUGRA coupled to Abelian vector multiplets [29, 30] and in $N=4, d=4$ SUGRA [18], completing the work started by Tod on this theory.

There is by now a well-defined recipe to attack this problem (at least in low dimensions) starting with only one assumption: the existence of one Killing spinor ε. The recipe consists in the following steps:

I Translate the Killing spinor equations and KSIs into tensorial equations.
   With the Killing spinor ε one can construct scalar, vector, and p- form bilinears $M \sim \bar{\varepsilon}\varepsilon$, $V_\mu \sim \bar{\varepsilon}\gamma_\mu\varepsilon, \cdots$ that are related by Fierz identities. These bilinears satisfy certain equations because they are made out of Killing spinors, for instance, if the KSE is of the general form

$$\delta_\varepsilon \psi_\mu = \tilde{\mathscr{D}}_\mu \varepsilon = [\nabla_\mu + \Omega_\mu]\varepsilon = 0, \Rightarrow \nabla_\mu M + 2\Omega_\mu M = 0, \quad (21)$$

The set of all such equations for the bilinears should be equivalent to the original spinorial equation or at least it should contain most of the information contained in it (but, certainly, not all of it).

II One of the vector bilinears (say V_μ) is always a Killing vector which can be timelike or null. These two cases are treated separately.

III One can get an expression of all the gauge field strengths of the theory using the Killing equation for those scalar bilinears: Ω_μ is usually of the form $F_{\mu\nu}V^\nu$ and, then Eq. (21) tells us that $F_{\mu\nu}V^\nu \sim \nabla_\mu \log M$. When V is timelike this determines completely F and, when it is null, it determines the general form of F. Of course, Eq. (21) is an oversimplified KSE and in real-life situations there are additional scalar factors, $SU(N)$ indices etc.

IV The Maxwell and Einstein equations and Bianchi identities are imposed on those field strengths F, getting second order equations for the scalar bilinears M.

V The KSIs guarantee that these three different sets of equations (plus the equations of the scalar fields, if any) are complicated combinations a a reduced number of simple equations involving a reduced number of scalar unknowns. Solving these equations for the scalar unknowns gives full solutions of the theory. The tricky part is, usually, identifying the right variables that satisfy simple equations and finding these equations as combinations of the Maxwell, Einstein etc. equations.

VI Finally, with the results obtained, the KSEs have to be solved, which may lead to additional conditions on the fields.

Let us see how this recipe works in the examples considered before.

Example: $N=1, d=4$ Supergravity.

With one (Majorana) Killing spinor ε the only bilinear that one can construct is a real vector bilinear V_μ which is always null. V_μ is also covariantly constant (i.e. it is a Killing vector and $V_\mu dx^\mu$ is an exact 1-form, which allows us to write $V_\mu dx^\mu = du$):

$$\delta_\varepsilon \psi_\mu = \nabla_\mu \varepsilon = 0, \Rightarrow \nabla_\mu V_\nu = 0, R^\mu{}_\nu V^\nu = 0, (\bar{\varepsilon} R^\mu{}_a \gamma^a \varepsilon = 0). \quad (22)$$

All the metrics with covariantly constant null vectors are Brinkmann pp-waves and have the form

$$ds^2 = 2du(dv + Kdu + A_{\underline{i}}dx^i) + \tilde{g}_{\underline{ij}}dx^i dx^j, \tag{23}$$

where all the components are independent of v, where v is defined by $V^\mu \partial_\mu \equiv \partial/\partial v$.

It can be checked that for all these metrics the KSE has solutions. These, then, are all the supersymmetric field configurations of $N=1, d=4$ SUGRA, but only those with $R_{\mu\nu} = 0$ are supersymmetric solutions.

Example: $N = 2, d = 4$ Supergravity.

With two Weyl spinors[2] ε^I one can construct the following independent bilinears

- A complex scalar $\bar{\varepsilon}^I \varepsilon^J \equiv M \varepsilon^{IJ}$
- A Hermitean matrix of null vectors $V^I{}_{J\mu} \equiv i\bar{\varepsilon}^I \gamma_\mu \varepsilon_J$

The KSEs imply the following equations for the bilinears:

$$\nabla_\mu M \sim F^+{}_{\mu\nu} V^I{}_I{}^\nu, \tag{24}$$

$$\nabla_\mu V^I{}_{J\nu} \sim \delta^I{}_J [MF^+{}_{\mu\nu} + M^* F^-{}_{\mu\nu}] - \Phi_{KJ(\mu}{}^\rho \varepsilon^{KI} F^-{}_{\nu)\rho} - \Phi^{JK}{}_{(\mu|}{}^\rho \varepsilon_{KJ} F^+{}_{|\nu)\rho} \tag{25}$$
$$\tag{26}$$

so the vector $V^\mu \equiv V^I{}_I{}^\mu$ is Killing and the other three are exact forms. The Fierz identities tell us that $V^\mu V_\mu \sim |M|^2 \geq 0$ can be timelike or null. When it is timelike, $V^\mu \partial_\mu \equiv \sqrt{2}\partial/\partial t$ and the metric can be put in the *conformastationary* form

$$ds^2 = |M|^2 (dt + \omega)^2 - |M|^{-2} d\vec{x}^2, \tag{27}$$

where, for consistency, the 1-form ω has to be related to M by

$$d\omega = i|M|^{-2\star}[MdM^* - \text{c.c.}]. \tag{28}$$

On the other hand, Eq. (24) gives

$$F^+ \sim |M|^{-2}\{V \wedge dM + i^\star[V \wedge dM]\}. \tag{29}$$

The KSIs are satisfied if Eq. (28) is satisfied. It can be seen that, then, any metric and 2-form field strength of the above form admit Killing spinors. On the other hand, all the equations of motion are combinations of the simple equation in 3-dimensional Euclidean space

$$\vec{\nabla}^2 M^{-1} = 0. \tag{30}$$

---

[2] In this theory one can use pairs of Majorana or Weyl spinors or single Dirac spinors. We now use, for convenience, pairs of Weyl spinors.

Thus, solving this equation for some M gives us a supersymmetric solution of all the equations of motion (all the fields are determined by M). These solutions of the Einstein-Maxwell theory are the Israel-Wilson-Perjés family [31, 32].

The case in which V is null is very similar to the $N=1$ case and we will not study it here in detail for lack of space.

TOD'S PROBLEM IN $N=4, D=4$ SUPERGRAVITY

This theory can be obtained by toroidal compactification on T^6 of $N=1, d=10$ SUGRA [33] (the effective field theory of the Heterotic String) and subsequent (consistent) truncation of the matter vector fields. The 10- and 4-dimensional fields are related as indicated in Fig. 1.

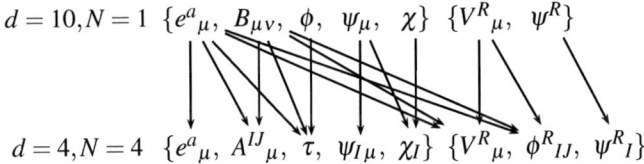

FIGURE 1. Relation between the fields of $N=1, d=10$ SUGRA $N=4, d=4$ SUGRA. The fields in curly brackets belong to the same supermultiplet. Both in $d=10$ and $d=4$ there a supergravity multiplet containing the graviton and vector supermultiplets, but the 4-dimensional vector supermultiplets originate from both the $d=10$ supergravity and vector supermultiplets. The $I, J = 1, \cdots, 4$ indices are $SU(4)$ indices related to the six internal dimensions using the isomorphism between $SO(6)$ and $SU(4)$. The $R, S = 1, \cdots, 22$ indices count the vector supermultiplets: 6 of them coming from the supergravity multiplet and 16 from 10-dimensional vector supermultiplets.

A special role is played by the *axidilaton* field $\tau = a + ie^{-\phi}$, where a is dual to the 4-dimensional Kalb-Ramond 2-form and plays the role of local θ parameter and ϕ is the 4-dimensional dilaton, which plays its usual role of local coupling constant.

It is convenient to start by studying the *pure* supergravity theory (without the vector supermultiplets) [34], for simplicity. The theory has global $SU(4)$ symmetry (duality) and, furthermore, only at the level of the equations of motion, an $SL(2,\mathbb{R})$ invariance (S duality) that rotates Maxwell equations into Bianchi identities and acts on the axidilaton according to

$$\tau' = \frac{\alpha\tau + \beta}{\gamma\tau + \delta}, \qquad \alpha\delta - \gamma\beta = 1. \qquad (31)$$

Observe that the $N=2$ and $N=1$ are included as truncations.

The bosonic action of the theory is

$$S = \int d^4x \sqrt{|g|} \left[R + \tfrac{1}{2}\frac{\partial_\mu \tau \partial^\mu \tau^*}{(\Im m\, \tau)^2} - \tfrac{1}{16}\Im m\, \tau F^{IJ\mu\nu} F_{IJ\mu\nu} - \tfrac{1}{16}\Re e\, \tau F^{IJ\mu\nu*} F_{IJ\mu\nu} \right]. \qquad (32)$$

It is convenient to denote the equations of motion by

$$\mathcal{E}_a{}^\mu \equiv -\frac{1}{2\sqrt{|g|}}\frac{\delta S}{\delta e^a{}_\mu}, \qquad \mathcal{E} \equiv -\frac{2\Im m\,\tau}{\sqrt{|g|}}\frac{\delta S}{\delta \tau}, \qquad \mathcal{E}^{IJ\mu} \equiv \frac{8}{\sqrt{|g|}}\frac{\delta S}{\delta A_{IJ\mu}}. \tag{33}$$

The Maxwell equation $\mathcal{E}^{IJ\mu}$ transforms as an $SL(2,\mathbb{R})$ doublet together with the Bianchi identity

$$\mathcal{B}^{IJ\mu} \equiv \nabla_\nu{}^\star F^{IJ\nu\mu}. \tag{34}$$

For vanishing fermions, the supersymmetry transformation rules of the gravitini and dilatini, generated by 4 spinors ε_I of negative chirality, are

$$\delta_\varepsilon \psi_{I\mu} = \mathcal{D}_\mu \varepsilon_I - \frac{i}{2\sqrt{2}}(\Im m\,\tau)^{1/2} F_{IJ}{}^+{}_{\mu\nu} \gamma^\nu \varepsilon^J, \tag{35}$$

$$\delta_\varepsilon \chi_I = \frac{1}{2\sqrt{2}}\frac{\slashed{\partial}\tau}{\Im m\,\tau}\varepsilon_I - \frac{1}{8}(\Im m\,\tau)^{1/2} \slashed{F}_{IJ}{}^- \varepsilon^J, \tag{36}$$

where $\mathcal{D}$ is the Lorentz plus $U(1)$ covariant derivative and where the $U(1)$ connection is given by

$$Q_\mu \equiv \frac{1}{4}\frac{\partial_\mu \Re e\,\tau}{\Im m\,\tau}. \tag{37}$$

The supersymmetry transformation rules of the bosonic fields take the form

$$\delta_\varepsilon e^a{}_\mu = -\frac{i}{4}(\bar{\varepsilon}^I \gamma^a \psi_{I\mu} + \bar{\varepsilon}_I \gamma^a \psi^I{}_\mu), \tag{38}$$

$$\delta_\varepsilon \tau = -\frac{i}{\sqrt{2}}\Im m\,\tau\, \bar{\varepsilon}^I \chi_I, \tag{39}$$

$$\delta_\varepsilon A_{IJ\mu} = \frac{\sqrt{2}}{(\Im m\,\tau)^{1/2}} \left[\bar{\varepsilon}_{[I}\psi_{J]\mu} + \frac{i}{\sqrt{2}}\bar{\varepsilon}_{[I}\gamma_\mu \chi_{J]} + \frac{1}{2}\varepsilon_{IJKL}\left(\bar{\varepsilon}^K \psi^L{}_\mu + \frac{i}{\sqrt{2}}\bar{\varepsilon}^K \gamma_\mu \chi^L\right)\right] \tag{40}$$

Given N chiral commuting spinors ε_I and their complex conjugates ε^I we can constructed the following independent bilinears:

1. A complex, antisymmetric, matrix of scalars

$$M_{IJ} \equiv \bar{\varepsilon}_I \varepsilon_J, \qquad M^{IJ} \equiv \bar{\varepsilon}^I \varepsilon^J = (M_{IJ})^*, \tag{41}$$

2. A complex matrix of vectors

$$V^I{}_{Ja} \equiv i\bar{\varepsilon}^I \gamma_a \varepsilon_J, \qquad V_I{}^J{}_a \equiv i\bar{\varepsilon}_I \gamma_a \varepsilon^J = (V^I{}_{Ja})^*, \tag{42}$$

which is Hermitean:

$$(V^I{}_{Ja})^* = V_I{}^J{}_a = V^J{}_{Ia} = (V^I{}_{Ja})^T. \tag{43}$$

Using the supersymmetry transformation rules of the bosonic fields, one can find the KSIs of this theory, associated to the gravitini and dilatini, respectively. However, just as in the $N=2, d=4$ example, since the Bianchi identities do not appear in these equations, they break S-duality covariance. This covariance can be restored by hand or re-deriving the KSIs from the KSEs integrability conditions. The result is

$$\mathcal{E}^\mu{}_a \gamma^a \varepsilon_I - \frac{i}{\sqrt{2}(\Im \tau)^{1/2}}(\mathcal{E}_{IJ}{}^\mu - \tau^* \mathcal{B}_{IJ}{}^\mu)\varepsilon^J = 0, \quad (44)$$

$$\mathcal{E}^* \varepsilon_I - \frac{1}{\sqrt{2}(\Im \tau)^{1/2}}(\mathcal{E}_{IJ} - \tau \mathcal{B}_{IJ})\varepsilon^J = 0. \quad (45)$$

It is useful to derive tensorial equations from these KSIs. Combining them we arrive to the following, which are chosen among the many possible tensorial KSIs by their interest. For timelike $V^a \equiv V^I{}_I$ we get

$$\mathcal{E}^{ab} - \tfrac{1}{2}\Im m\, \mathcal{E} V^a V^b - \frac{1}{\sqrt{2}}(\Im \tau)^{1/2} \Im m (M^{IJ} \mathcal{B}_{IJ}{}^a) V^b = 0, \quad (46)$$

$$\mathcal{E}^* V^a - \frac{i}{\sqrt{2}(\Im \tau)^{1/2}} M^{IJ}(\mathcal{E}_{IJ}{}^a - \tau \mathcal{B}_{IJ}{}^a) = 0, \quad (47)$$

$$\Im m [M^{IJ}(\mathcal{E}_{IJ}{}^a - \tau^* \mathcal{B}_{IJ}{}^a)] = 0. \quad (48)$$

Observe that the first equation implies the off-shell vanishing of all the Einstein equations with one or two spacelike components. Further, the Einstein equation is automatically satisfied when the Maxwell, Bianchi and complex scalar equations are satisfied and the scalar equation is automatically satisfied when the Maxwell and Bianchi are.

When V^a is null (we denote it by l^a), all the spinors ε_I are proportional and we can parametrize all of them by $\varepsilon_I = \phi_I \varepsilon$, where $\phi^I \phi_I = 1$. In order to construct tensor bilinears we define an auxiliary spinor η normalized by $\bar\varepsilon \eta = \tfrac{1}{2}$. With these two spinors we can construct a standard complex null tetrad

$$l_\mu = i\bar\varepsilon^* \gamma_\mu \varepsilon, \quad n_\mu = i\bar\eta^* \gamma_\mu \eta, \quad m_\mu = i\bar\varepsilon^* \gamma_\mu \eta = i\bar\eta \gamma_\mu \varepsilon^*, \quad m_\mu^* = i\bar\varepsilon \gamma_\mu \eta^* = i\bar\eta^* \gamma_\mu \varepsilon. \quad (49)$$

Then, in the null case, the KSIs take the form

$$(\mathcal{E}^\mu{}_a - \tfrac{1}{2} e_a{}^\mu \mathcal{E}^\rho{}_\rho) l^a = (\mathcal{E}^\mu{}_a - \tfrac{1}{2} e_a{}^\mu \mathcal{E}^\rho{}_\rho) m^a = 0, \quad (50)$$

$$\mathcal{E} = 0, \quad (51)$$

$$(\mathcal{E}_{IJ}{}^\mu - \tau^* \mathcal{B}_{IJ}{}^\mu)\phi^J = 0. \quad (52)$$

In this case supersymmetry implies that the scalar equations of motion are automatically satisfied. We are not going to work out here the null case, since it was treated completely in Ref. [23].

We are now ready to follow the recipe to find all the supersymmetric configurations of this theory. The first step consists in finding (Killing) equations for the spinor bilinears. From the vanishing of the gravitini supersymmetry transformation rule we find

$$\mathcal{D}_\mu M_{IJ} = \tfrac{1}{\sqrt{2}}(\Im\tau)^{1/2} F_{K[I}{}^+{}_{\mu\nu} V^K{}_{|J|}{}^\nu, \qquad (53)$$

$$\mathcal{D}_\mu V^I{}_{J\nu} = -\tfrac{1}{2\sqrt{2}}(\Im\tau)^{1/2} \big[M_{KJ} F^{KI-}{}_{\mu\nu} + M^{IK} F_{JK}{}^+{}_{\mu\nu}$$

$$- \Phi_{KJ(\mu}{}^\rho F^{KI-}{}_{\nu)\rho} - \Phi^{IK}{}_{(\mu|}{}^\rho F_{KI}{}^+{}_{|\nu)\rho} \big], \qquad (54)$$

and from that of the dilatini, we find

$$V^K{}_I \cdot \partial\tau - \tfrac{i}{2\sqrt{2}}(\Im\tau)^{3/2} F_{IJ}{}^- \cdot \Phi^{KJ} = 0, \qquad (55)$$

$$F_{IJ}{}^-{}_{\rho\sigma} V^J{}_K{}^\sigma + \tfrac{i}{\sqrt{2}}(\Im\tau)^{-3/2} \big(M_{IK}\partial_\rho\tau - \Phi_{IK\rho}{}^\mu \partial_\mu\tau \big) = 0. \qquad (56)$$

It is immediate to see that $V \equiv V^I{}_I$ is a Killing vector and that

$$V^\mu \partial_\mu \tau = 0. \qquad (57)$$

Further, using Eq. (53) and the antisymmetric part of Eq. (56) we find

$$F_{SR}{}^-{}_{\mu\nu} V^\nu = -\frac{\sqrt{2}i}{(\Im\tau)^{3/2}} M_{SR} \partial_\mu \tau - \frac{\sqrt{2}}{(\Im\tau)^{1/2}} \varepsilon_{SRIJ} \mathcal{D}_\mu M^{IJ}, \qquad (58)$$

which determines completely the vector field strengths in terms of the scalar bilinears, τ and the Killing vector V^a when this is timelike. In the null case, this equation gives us important constraints on the form of the field strengths, but does not completely determine them. From now on we will focus on the timelike case since it illustrates our procedure best. In this case we can write the metric in the conformastationary form Eq. (27), but, while in the $N=2, d=4$ case one could show that three of the vector bilinears where exact 1-forms and then the metric on the constant-time slices could be chosen to be Euclidean, in the $N=4, d-4$ case this is not possible and we have to live with a non-trivial 3-dimensional metric γ_{ij}. Thus

$$ds^2 = |M|^2 (dt+\omega)^2 - |M|^{-2} \gamma_{ij} dx^i dx^j, \qquad i,j=1,2,3, \qquad (59)$$

where ω has to satisfy the equation

$$d\omega = \tfrac{1}{\sqrt{2}}\Omega = \tfrac{i}{2\sqrt{2}} |M|^{-4} {}^\star\big[(M^{IJ} \mathcal{D} M_{IJ} - M_{IJ} \mathcal{D} M^{IJ}) \wedge \hat{V}\big]. \qquad (60)$$

Having the field strengths expressed in terms of the scalars M^{IJ}, τ, we move on to the next step and impose the Maxwell equations and Bianchi identities on them, to obtain equations that only involve those scalars. We also substitute the field strengths into the τ equation, obtaining another equation that only involves M^{IJ} and τ. Now comes the magic of supersymmetry: these three sets of equations are combinations of just two sets of much simpler equations in the 3-dimensional metric γ_{ij}:

$$n_{(3)}^{IJ} \equiv (\nabla_i + 4i\xi_i)\left(\frac{\partial^i N^{IJ}}{|N|^2}\right), \tag{61}$$

$$e_{(3)}^* \equiv (\nabla_i + 4i\xi_i)\left(\frac{\partial^i \tau}{|N|^2}\right), \tag{62}$$

where $N^{IJ} \equiv (\Im m\tau)^{1/2} M^{IJ}$ and ξ is defined by

$$\xi \equiv \tfrac{i}{4}|M|^{-2}(M_{IJ}dM^{IJ} - M^{IJ}dM_{IJ}), \tag{63}$$

and acts as a $U(1)$ connection.

In fact, we can write all the components of the equations of motion define above in terms of these two

$$\mathcal{E}_{00} = |M|^2\left[|M|^2 \Im m e_{(3)}^* - 2\Re e(N_{KL} n_{(3)}^{KL}) + \tfrac{1}{2} e_k{}^k\right], \tag{64}$$

$$\mathcal{E}_{0i} = 0, \tag{65}$$

$$\mathcal{E}_{ij} = |M|^2(e_{ij} - \tfrac{1}{2}\delta_{ij} e_k{}^k), \tag{66}$$

$$\mathcal{B}^{IJa} = -\sqrt{2}|M|^2 V^a \left\{\frac{N^{IJ} + \tilde{N}^{IJ}}{\Im m\tau} \Re e_{(3)} - i(n_{(3)}^{IJ} - \tilde{n}_{(3)}^{IJ})\right\}, \tag{67}$$

$$\mathcal{E}^{IJa} = -\sqrt{2}|M|^2 V^a \left\{\frac{N^{IJ} + \tilde{N}^{IJ}}{\Im m\tau} \Re(\tau e_{(3)}) - i(\tau^* n_{(3)}^{IJ} - \tau \tilde{n}_{(3)}^{IJ})\right\}. \tag{68}$$

$$\mathcal{E} = -|M|^2\left[|M|^2 e_{(3)} + 2i N_{KL} \tilde{n}_{(3)}^{KL}\right], \tag{69}$$

and a set of equations e_{ij} defined by

$$e_{ij} \equiv R_{ij}(\gamma) - 2\partial_{(i}\left(\frac{N^{IJ}}{|N|}\right)\partial_{j)}\left(\frac{N_{KL}}{|N|}\right)(\delta^{KL}{}_{IJ} - \mathcal{J}^K{}_I \mathcal{J}^L{}_J), \tag{70}$$

and which have to vanish in order to satisfy the KSIs and have supersymmetry[3]. These equations are conditions for the 3-dimensional metric γ_{ij}, but are not easy to solve

---

[3] The integrability condition of the equation for ω has to be satisfied as well in order to have supersymmetry. WE are going to discuss it later.

directly. We have to substitute our results into the original KSEs or into their integrability conditions. The solution one finds is that, in order to solve the $e_{ij} = 0$ equations have supersymmetry, the 3-dimensional metric has to take the form

$$\gamma_{i\underline{j}}dx^i dx^{\underline{j}} = dx^2 + 2e^{2U(z,z^*)}dzdz^*, \qquad (71)$$

and the connection ξ has to take the form

$$\xi = \pm\tfrac{i}{2}(\partial_{\underline{z}}Udz - \partial_{\underline{z}^*}Udz^*) + \tfrac{1}{2}d\lambda(x,z,z^*). \qquad (72)$$

Since ξ is defined in terms of the M^{IJ} scalars, this is a condition that these scalars have to fulfill, on top of Eqs. (61,62).

Further, to have supersymmetry, the integrability condition for the equation defining ω has to be satisfied as well. It takes the form

$$\nabla_{\underline{i}}\left(\frac{Q^i - \xi^i}{|M|^2}\right) = 0. \qquad (73)$$

The timelike case now has been completely solved. Let us put together the results: any supersymmetric configuration of $N = 4, d = 4$ supergravity in this class is given by a set of 7 complex functions M^{IJ}, τ which have to satisfy the following conditions:

1. $M^{[IJ}M^{K]L} = 0$. This is a condition that the scalar bilinears satisfy due to the Fierz identities.
2. $|M|^2 \neq 0$. We have assumed this, as definition of the timelike case ($V^2 \sim |M|^2 > 0$).
3. Eq. (73) has to be satisfied.
4. ξ has to take the form Eq. (72).

Given 7 complex functions satisfying these conditions, then, a supersymmetric field configuration of $N = 4, d = 4$ is given by the metric Eqs. (59,71) and the field strengths Eq. (58). These field configurations will be supersymmetric solutions if the expressions Eqs. (61,62) vanish.

This is the main result in the timelike case.

Now comes the problem of finding sets of 7 complex functions satisfying the above conditions, which is not an easy. We have been able to find two families of supersymmetric solutions based on the *Ansatz* for the M^{IJ}s

$$M_{IJ} = e^{i\lambda(x,z,z^*)}M(x,z,z^*)k_{IJ}(z), \quad M = M^*, \quad \lambda = \lambda^*, \quad k_{[IJ}k_{K]L} = 0. \qquad (74)$$

which give a connection ξ of the form Eq. (72) with

$$U = +\ln|k|, \qquad |k|^2 \equiv k^{IJ}(z^*)k_{IJ}(z). \qquad (75)$$

This *Ansatz* satisfies all the conditions except for Eq. (73). In the following two cases, at least, this last condition is also satisfied:

1. If the k_{IJ} are constants, then, normalizing $|k|^2 = 1$ for simplicity, $\xi = \tfrac{1}{2}d\lambda$ and $U = 0$. This case was considered by Tod in Ref. [23] and studied in detail in

Ref. [35]. Defining $\mathcal{H}_1 \equiv [(\Im m\,\tau)^{1/2} e^{-i\lambda} M]^{-1}$, and $\tau = \mathcal{H}_1/\mathcal{H}_2$ we get solutions if $\partial_{\underline{i}}\partial_{\underline{i}}\mathcal{H}_1 = \partial_{\underline{i}}\partial_{\underline{i}}\mathcal{H}_2 = 0$.

2. With $e^{i\lambda} = M = 1$ and constant τ we solve all constraints and all equations using the holomorphicity of the k_{IJ}s. The metric takes the form

$$ds^2 = |k|^2(dt+\omega)^2 - |k|^{-2}dx^2 - 2dzdz^*. \tag{76}$$

The metric and the supersymmetry projectors correspond to stationary strings lying along the coordinate x, in spite of the trivial axion field $\Re e\tau$. These solutions clearly deserve more study. Observe that this family is precisely the one that cannot be embedded in $N=2, d=4$ supergravity plus matter fields [36] and it is genuinely $N=4$.

CONCLUSIONS

The landscape approach offers an interesting, even if controversial, point of view over the vacuum selection problem. It also gives additional reasons to work on the problem of classification of supersymmetric solutions, whose 4-dimensional structure we have reviewed in this talk, emphasizing the difference between general supersymmetric configurations and solutions and showing how the KSIs can be used in this problem. We have applied the recipes to an interesting case: pure $N=4, d=4$ supergravity, but is should be clear that the same procedure could be used in more general contexts ($N=4, d=4$ coupled to matter, gauged etc. and other 4-dimensional theories [49, 50]). We also expect some of the techniques could also be of use in solving the much more complicated 11- and 10-dimensional problems [37, 48].

ACKNOWLEDGMENTS

This work has been supported in part by the Spanish grant BFM2003-01090.

REFERENCES

1. M. J. Duff, B. E. W. Nilsson and C. N. Pope, Phys. Rept. **130** (1986) 1.
2. T. Appelquist, A. Chodos and P.G.O. Freund, "Modern Kaluza-Klein Theories," Addison-Wesley, Reading (Massachusets, USA) (1987).
3. E. Cremmer, B. Julia and J. Scherk, Phys. Lett. B **76** (1978) 409.
4. E. Witten, in *Shelter Island Conference on Quantum Field Theory and the Fundamental Problems of Physics II*, MIT Press (1985).
5. D. J. Gross, J. A. Harvey, E. J. Martinec and R. Rohm, Nucl. Phys. B **256** (1985) 253.
6. J. Polchinski, Phys. Rev. Lett. **75** (1995) 4724 [arXiv:hep-th/9510017].
7. P. K. Townsend, arXiv:hep-th/9507048.
8. J. H. Schwarz, Nucl. Phys. Proc. Suppl. **55B** (1997) 1 [arXiv:hep-th/9607201].
9. E. Witten, Nucl. Phys. B **443** (1995) 85 [arXiv:hep-th/9503124].
10. P. Horava and E. Witten, Nucl. Phys. B **460** (1996) 506 [arXiv:hep-th/9510209].
11. L. E. Ibanez, F. Marchesano and R. Rabadan, JHEP **0111** (2001) 002 [arXiv:hep-th/0105155].
    M. Cvetic, G. Shiu and A. M. Uranga, Phys. Rev. Lett. **87** (2001) 201801 [arXiv:hep-th/0107143].

12. S. Kachru, R. Kallosh, A. Linde, J. Maldacena, L. McAllister and S. P. Trivedi, JCAP **0310**, 013 (2003) [arXiv:hep-th/0308055]. S. Kachru, R. Kallosh, A. Linde and S. P. Trivedi, Phys. Rev. D **68**, 046005 (2003) [arXiv:hep-th/0301240].
13. L. Susskind, arXiv:hep-th/0302219.
14. M. R. Douglas, JHEP **0305** (2003) 046 [arXiv:hep-th/0303194].
15. A. Van Proeyen, talk given at the *Workshop on Gravitational Aspects of String Theory*, Fields Institute, (Toronto, 2-6 May 2005)
16. R. Kallosh and T. Ortín, arXiv:hep-th/9306085.
17. J. Bellorín and T. Ortín, Phys. Lett. B **616** (2005) 118 [arXiv:hep-th/0501246].
18. J. Bellorin and T. Ortín, Nucl. Phys. B **726** (2005) 171 [arXiv:hep-th/0506056].
19. T. Ortín, Class. Quant. Grav. **19** (2002) L143 [arXiv:hep-th/0206159].
20. N. Alonso-Alberca and T. Ortín, Talk given at Spanish Relativity Meeting on Gravitation and Cosmology (ERE 2002), Mao, Menorca, Spain, 22-24 Sep 2002. arXiv:gr-qc/0210039.
21. J. Figueroa-O'Farrill, P. Meessen and S. Philip, Class. Quant. Grav. **22** (2005) 207 [arXiv:hep-th/0409170].
22. K. P. Tod, Phys. Lett. B **121** (1983) 241.
23. K. P. Tod, Class. Quant. Grav. **12** (1995) 1801.
24. J. P. Gauntlett, J. B. Gutowski, C. M. Hull, S. Pakis and H. S. Reall, Class. Quant. Grav. **20** (2003) 4587 [arXiv:hep-th/0209114].
25. J. P. Gauntlett and J. B. Gutowski, Phys. Rev. D **68** (2003) 105009 [Erratum-ibid. D **70** (2004) 089901] [arXiv:hep-th/0304064].
26. J. B. Gutowski, D. Martelli and H. S. Reall, Class. Quant. Grav. **20** (2003) 5049 [arXiv:hep-th/0306235].
27. A. Chamseddine, J. Figueroa-O'Farrill and W. Sabra, arXiv:hep-th/0306278.
28. M. M. Caldarelli and D. Klemm, JHEP **0309** (2003) 019 [arXiv:hep-th/0307022].
29. J. B. Gutowski and H. S. Reall, JHEP **0404** (2004) 048 [arXiv:hep-th/0401129].
30. J. B. Gutowski and W. Sabra, arXiv:hep-th/0505185.
31. W. Israel and G.A. Wilson, *J. Math. Phys.* **13**, (1972) 865.
32. Z. Perjés, *Phys. Rev. Lett.* **27** (1971) 1668.
    [33]
33. A. H. Chamseddine, Nucl. Phys. B **185** (1981) 403.
34. E. Cremmer, J. Scherk and S. Ferrara, Phys. Lett. B **74** (1978) 61.
35. E. Bergshoeff, R. Kallosh and T. Ortín, Nucl. Phys. B **478** (1996) 156 [arXiv:hep-th/9605059].
36. S. Ferrara, R. Kallosh and A. Strominger, Phys. Rev. D **52** (1995) 5412 [arXiv:hep-th/9508072].
37. J. P. Gauntlett and S. Pakis, JHEP **0304** (2003) 039 [arXiv:hep-th/0212008].
38. J. P. Gauntlett, J. B. Gutowski and S. Pakis, JHEP **0312** (2003) 049 [arXiv:hep-th/0311112].
39. J. P. Gauntlett, D. Martelli, J. Sparks and D. Waldram, Class. Quant. Grav. **21** (2004) 4335 [arXiv:hep-th/0402153].
40. O. A. P. Mac Conamhna, Phys. Rev. D **70** (2004) 105024 [arXiv:hep-th/0408203].
41. J. Gillard, U. Gran and G. Papadopoulos, Class. Quant. Grav. **22** (2005) 1033 [arXiv:hep-th/0410155].
42. M. Cariglia and O. A. P. Mac Conamhna, arXiv:hep-th/0411079.
43. M. Cariglia and O. A. P. MacConamhna, arXiv:hep-th/0412116.
44. U. Gran, J. Gutowski and G. Papadopoulos, arXiv:hep-th/0501177.
45. U. Gran, G. Papadopoulos and D. Roest, arXiv:hep-th/0503046.
46. O. A. P. Mac Conamhna, arXiv:hep-th/0504028.
47. U. Gran, J. Gutowski and G. Papadopoulos, arXiv:hep-th/0505074.
48. O. A. P. Mac Conamhna, arXiv:hep-th/0505230.
49. J. Bellorín, M. Hübscher and T. Ortín, in preparation.
50. P. Meessen and T. Ortín, in preparation.

Observing Dark Matter

Joseph Silk

Astrophysics, Denys Wilkinson Building, Keble Road, Oxford, OX1 3RH, UK

Abstract. Dark matter presents a challenge for gravitational theory. It is best attacked in two ways; by studying the confrontation of structure formation with observation and by direct and indirect searches. In this review, I will focus on those aspects of dark matter that are relevant for understanding galaxy formation, and describe the current issues that are of concern to theorists. I will review the status of baryonic dark matter, and describe the outlook for detecting the most elusive component, non-baryonic dark matter.

Keywords: Dark Matter
PACS: 95.35.+d

INTRODUCTION

The standard (or concordance) model of cosmology has a predominance of dark energy. Baryonic matter only amounts to 5% whereas non-baryonic matter is 30%. Dark energy is 65% of the mass energy today In contrast, luminous baryons (mostly in stars) constitute $\Omega_* = 0.005$ towards the total.

An important component of the standard model is the spectrum of primordial density fluctuations, measured in the linear regime via the temperature anisotropies of the CMB. This provides the initial conditions for large-scale structure and galaxy formation via gravitational instability once the universe is matter-dominated. Dark matter consequently provides the gravitational wells within which galaxies formed. The dark matter and galaxy formation paradigms are inextricably interdependent. Unfortunately we have not yet identified a dark matter candidate, nor do we understand the fundamental aspects of galaxy formation. Nevertheless, cosmologists have not been deterred, and have even been encouraged to develop novel probes and theories that seek to advance our understanding of these forefront issues.

Progress has been made on the baryonic dark matter front. Only about half of the baryons initially present in galaxies, or more precisely, on the comoving scales over which galaxies formed, are directly observed. We cannot predict with any certainty the mass fraction in dark baryons. Yet there are excellent candidates for the dark baryons, both compact and especially diffuse.

In contrast, we have at least one elegant and moderately compelling theory of particle physics, SUSY, that predicts the observed fraction of nonbaryonic dark matter. Unfortunately, we have no idea yet as to whether the required stable supersymmetric particles actually exist.

In this review talk, I will first describe the increasingly standard precision model of cosmology that will enable us to provide an inventory of cosmic baryons. I describe how nonbaryonic matter has been successfully used to provide an infrastructure for galaxy formation. I review the astrophysical issues and the current status of indirect detection experiments. I describe the outlook for future progress.

PRECISION COSMOLOGY

Dark matter must dominate over ordinary matter. Observations are compelling. Of course, by definition we do not observe matter if it is dark. Minimal theory is needed to take us from the observational plane to conclude that dark matter is required. This theory is that of gravity, and is well understood and verified over a wide range of scales. Gravity has been tested over scales that range from millimetres to megaparsecs. Newton's description of gravity is perfectly adequate, apart from generally small deviations due to the curvature of space near massive objects, such as stars, or more radically, black holes. Einstein's theory of gravity tells us that gravity curves space and measuring this effect was one of the great triumphs of 20th century physics. Nevertheless, pending its direct detection, dark matter remains a hypothesis that depends, inevitably, on our having the correct theory of gravitation. For the remainder of this review, however, I will assume the reality of dark matter dominance on scales from galactic to those spanning the entire universe.

The cosmological parameters

Modern cosmology has emphatically laid down a challenge to theorists. A combination of new experiments has unambiguously measured the key parameters of our cosmological model that describes the universe. These include the temperature fluctuations in the cosmic microwave background, the large galaxy redshift surveys, the studies of the intergalactic medium via the distribution of absorbing neutral clouds along different lines of sight and the use of distant Type Ia supernovae as standard candles. Cosmologists now mostly debate the error bars of the standard model parameters. The ingredients of the standard model in effect define the model. These most crucially are the FRW metric and the Friedmann-Lemaitre equations, and the contents of the universe: baryons, neutrinos, photons, baryons, dark matter and dark energy. Upon these constituents is superimposed a distribution of primordial adiabatic density (scalar) fluctuations characterised by a power spectrum of specified amplitude and spectral index. In addition, there may be a primordial gravity wave tensor mode of fluctuations. The number of free parameters in the standard model is 14, of which the most significant are: $H_0, \Omega_b, \Omega_m, \Omega_\Lambda, \Omega_\gamma, \Omega_\nu, \sigma_8, n_s, r, n_T$, and τ. One can also add an equation of state for dark energy parameter, $w = -p_\Lambda/\rho_\Lambda$, in effect really a function of redshift, and a rolling scalar (and possibly tensor) index, $dn_s/dlnk$.

No single observational set constrains all, or even most, of these parameters. There are well-known degeneracies, most notably between Ω_Λ and Ω_m, σ_8 and τ, and σ_8 and Ω_m. However use of multiple data sets helps to break these degeneracies. For example, CMB anisotropies fix the combination $\Omega_m + \Omega_\Lambda$, and $\Omega_b h^2$, and SNIa constrain $\Omega_m - \Omega_\Lambda$. Both weak lensing and peculiar velocity surveys specify the product $\Omega_m^{-0.6} \sigma_8$. Lyman alpha forest surveys extend the latter measurement to Mpc comoving scales, probing the currently nonlinear regime. Finally, baryon oscillations are providing a measure of Ω_m/Ω_b, independently of the CMB. Interpretation in terms of a standard model (Friedmann-Lemaitre plus adiabatic fluctuations) yields the concordance model with remarkably small error bars [1].

For example, the flatness of space is measured to be $\Omega_{total} = 1.02 \pm 0.02$. Dark energy in the form of a cosmological constant dominates the universe, with $\Omega_\Lambda = 0.72 \pm 0.02$. The dark energy equation of state is indistinguishable from that of a cosmological constant, with $w \equiv p_\Lambda/\rho_\Lambda c^2 = -0.99 \pm 0.1$, this uncertainty holding to $z \sim 0.5$. Even at $z \sim 1$, the claimed uncertainty around $w = -1$ is only 20 percent. Non-baryonic dark matter dominates over baryons with $\Omega_m = 0.27 \pm 0.02$ and $\Omega_b = 0.044 \pm 0.004$. Most of the baryons are non-luminous, since $\Omega_* = 0.005$.

The spectrum of primordial density fluctuations is unambiguously measured both in the CMB and in the large-scale galaxy distribution from deep redshift surveys, and found to be approximately scale-invariant, with scalar index $n_s = 0.98 \pm 0.02$. One can also constrain a possible relic gravitational wave background, a key prediction of inflationary cosmology by the tensor mode limit on relic gravitational waves: $T/S < 0.36$. It has been argued that a fundamental test of inflation requires sensitivity at a level $T/S \sim 0.01$ [2]. Neutrinos are known to have mass as a consequence of atmospheric (v_τ, v_μ) and solar (v_μ, v_e) oscillations, with a deduced mass in excess of 0.001 eV for the lightest neutrino. From the power spectrum of the density fluctuations, the inferred mass limit (on the sum of the 3 neutrino masses) is $\Sigma m_v < 0.4$eV.

However one note of caution should be added. These tight error bars all depend on adoption of a simple prior. If this extended, to allow for example for an admixture of generic primordial isocurvature fluctuations, the error bars on many of these parameters explode by up to an order of magnitude.

What lies ahead for cosmology?

Clearly, the devil is in the observational details. Popular models of inflation predict that $n \approx 0.97$. Space should be very close to being flat, with $\Omega = 1 + \mathcal{O}(10^{-5})$. The numbers of massive objects at high redshift should comply with predictions of any plausible biasing model based on the statistics of gaussian random fields. The universe as viewed in the CMB should be isotropic. Any deviations from these predictions would be immensely exciting.

Suppose deviations were to be found. This would allow all sorts of possible extensions to the standard model of cosmology. One might consider the signatures of string relics of superstrings or transplanckian features in $\delta T/T|_k$ [3]. Large-scale cosmology might be affected by compact topology or global anisotropy with observable signatures in CMB temperature and polarisation maps [4]. The initial conditions might involve primordial nongaussianity. Anthropically constrained landscape scenarios of the metauniverse prefer a slightly open universe [5]. Some of these features, and others, could be a consequence of compactification from higher dimensions.

THE GLOBAL BARYON INVENTORY

There are several independent approaches to obtaining the baryon abundance in the universe. At $z \sim 10^9$, primordial nucleosynthesis of the light elements yields $\Omega_b = 0.04 \pm 0.004$. At the epoch of matter-radiation decoupling, $z \sim 1000$, the ratios of odd and even CMB acoustic peak heights set $\Omega_b = 0.044 \pm 0.003$. At more recent epochs, Lyman alpha forest modelling of the intergalactic medium at $z \sim 3$ as viewed in absorption along different lines of sight towards high redshift quasars at $z \sim 3$ yields $\Omega_b \approx 0.04$. At the present epoch, on very large scales, of order 10 Mpc comoving linear regime equivalent, the intracluster baryon fraction measured via x-ray observations of massive galaxy clusters provides a baryon fraction of 15%. This translates into $\Omega_b \approx 0.04$. In summary, we infer that $\Omega_b = 0.04 \pm 0.005$ and $\Omega_b/\Omega_m = 0.15 \pm 0.02$.

One's immediate impression, at least until very recently, is that most of the baryons in the universe today are not accounted for. The reasoning is as follows. The luminous content in the form of stars sums to $\Omega_b \approx 0.004$ or 10% in spheroids, and $\Omega_b \approx 0.002$ or 5% in disks. There is also hot intracluster gas amounting to $\Omega_b \approx 0.002$ or 5%. Current epoch observations of the cold/warm photo-ionised IGM via the nearby Lyman alpha/beta forest at $10^4 - 10^5$K as well as CIII (at $z \sim 0$) yield a much larger baryonic reservoir of gas, $\Omega_b \approx 0.012$ or 30%. This gas is metal-poor, with an abundance of about 10% solar [6]. So far, this only accounts for 50% of current epoch baryons.

The probable breakthrough however has come with the recent detections of the warm-hot intergalactic medium at $T \lesssim 10^5 - 10^6$K at $z \sim 0$, observed in OVI absorption in the UV and especially via x-ray absorption via OVII and OVIII hydrogen-like transitions towards low redshift luminous AGN. Something like $\Omega_b \approx 0.012$ or 30% of the primordial baryon fraction appears to be in this form, enriched (in oxygen, at least) to about 10% of the solar value [7]. We now have $\gtrsim 80\%$ of the baryons accounted for today. The total baryon content sums to $\Omega_b = 0.032 \pm 0.005$. Given the measurement uncertainties, this would seem to remove any strong case for dark baryons being present.

However, the situation is not so simple. The Andromeda Galaxy and our own galaxy are especially well-studied regions, where dark matter and baryons can be probed in detail. In the Milky Way Galaxy, the virial mass out to 100 kpc is $M_{virial} \approx 10^{12} M_\odot$, whereas the baryonic mass, mostly in stars, is $M_* \approx 6-8 \times 10^{10} M_\odot$. The inferred baryon fraction is at most 8% [8]. This is in fact an upper limit as the dark mass estimate is a lower bound.

I infer that globally, there is no problem. Nevertheless the outstanding question is where are the galactic baryons? Most of the baryons are globally accounted for. But this is not the case for our own galaxy and most likely for all comparable galaxies. We cannot account for a mass in baryons comparable to that in stars. It is possible that up to 10% of all the baryons *may* be dark, and that the dark baryons are comparable in mass to the galactic stars.

The "missing" baryons

There are several possibilities for the "missing" baryons. Perhaps they never were present in the protogalaxy. Or they are in the outer galaxy. Or, finally, they have been ejected.

The first of these options seems very unlikely, especially given that modelling of disk formation generally requires an initial baryon fraction of at least $\sim 10\%$ in order for there to be sufficient cooling to form a cold, gravitationally unstable thin disk.

Consider the second option. The most likely candidates for dark baryons are massive baryonic objects or MACHOs. These are constrained by several gravitational microlensing experiments. The allowed mass range is between 10^{-8} and $10 M_\odot$, and the best current limit on the MACHO abundance is $\lesssim 20\%$ of the dark halo mass. In fact, one experiment, that of the MACHO Collaboration, claims a detection from some 20 events seen towards the LMC, most of which cannot be accounted for by star-star microlensing. The observed range of amplification time-scales specifies the mass of the lensing objects. The preferred MACHO mass is around $\sim 0.5 M_\odot$. This mass favours an interpretation in terms of old halo white dwarfs. Main sequence stars in this mass range can be excluded. Current searches for halo high velocity old white dwarfs utilise the predicted colours and proper motions as a discriminant from field dwarfs, and set a limit of $\lesssim 4\%$ of the dark halo mass on a possible old white dwarf component in the halo [9]. However even if this limit were to apply, an extreme star formation history and protogalactic IMF would be required. Observations at high redshift both of star-forming galaxies and of the diffuse extragalactic light background, combined with chemical evolution and SNIa constraints, make such a hypothesis extremely implausible.

If the empirical mass range constraint is relaxed, theory does not exclude either primordial brown dwarfs ($0.01 - 0.1 M_\odot$), primordial black holes (mass $\gtrsim 10^{-16} M_\odot$) or even cold dense H_2 clumps $\lesssim 1 M_\odot$. The latter have been invoked in the Milky Way halo in order to account for such phenomena as extreme scattering events [10] and unidentified submillimetre objects) [11]. However these possibilities seem to be truly acts of last resort in the absence of any more physical explanations.

There is indeed another possibility that seems far less ad hoc. The nearby intergalactic medium is enriched to about 10% of the solar metallicity, and contains of order 50% of the baryons in photo-ionised and collisionally ionised phases. This strongly suggests that ejection from galaxies via early winds must have occurred, and moreover would inevitably have expelled a substantial fraction of the baryons along with the heavy elements.

There are candidates for young galaxies undergoing extensive mass loss via winds. These are the Lyman break galaxies at $z \sim 2-4$. Observations of spectral line displacements of the interstellar gas relative to the stellar component as well as of line widths are indicative of early winds from L_* galaxies [12]. Studies of nearby starburst galaxies, essentially lower luminosity counterparts of the distant LBGs, show that the gas outflow rate in winds is of order the star formation rate. The ICM to $z \sim 1$ is enriched to about a third of the solar metallicity, again suggestive of massive early winds, in this case from early-type galaxies. Hence the "missing" baryons could be in the IGM, with about as much mass ejected in baryons as in stars remaining.

The ejection hypothesis however has to confront a theoretical difficulty. Winds from L_* galaxies cannot be reproduced by hydrodynamical simulations of forming galaxies [13]. The momentum source for gas expulsion appeals to supernovae. SN feedback works for dwarf galaxies and can explain the observed outflows in these systems. However an alternative feedback source is needed for massive galaxies. This most likely is associated with AGN, and the ubiquitous presence of central supermassive black holes in galaxy spheroids.

LARGE-SCALE STRUCTURE AND COLD DARK MATTER: THE ISSUES

The cold dark matter hypothesis has had remarkable successes in confronting observations of the large-scale structure of the universe. These have stemmed from predictions of the amplitude of the temperature fluctuations in the cosmic microwave background that are directly associated with the seeds of structure formation. The initial conditions for gravitational instability to operate in the expanding universe were measured. The formation of galaxies and galaxy clusters was explained, as was the filamentary nature of the large-scale structure of the galaxy distribution. Nor was only the amplitude confirmed as a prerequisite for structure formation. The Harrison-Zeldovich-Peebles ansatz of an initially scale-invariant fluctuation spectrum, later motivated by inflationary cosmology, has been confirmed over scales from 0.1 to 10000 Mpc, via a combination of CMB, large-scale galaxy distribution and IGM measurements.

Despite these stunning successes, difficulties remain in reconciling theory with observations. These centre on two aspects: the uncertainties in star formation physics that render any definitive predictions of observed galaxy properties unreliable, and the detailed nature of the dark matter distribution on small scales, where the simulations are also incomplete.

The former issues include such observables as the galaxy luminosity function, disk sizes and mass-to-light ratios, and the presence of old, red massive galaxies at high redshift. These difficulties in the confrontation of galaxy formation theory and observational data are plausibly resolved by improving the prescriptions for star formation and feedback, although there are as yet no definitive answers. The latter issues require high resolution dark matter simulations combined with hydrodynamic simulations of the baryons including star formation and feedback.

I will focus here on the dark matter conundrums, and in particular on the challenges posed by theoretical predictions of dark matter clumpiness, cuspiness and concentration. Implementation of numerical simulations of dark halos of galaxies in the context of hierarchical galaxy formation yields repeatable and reliable results at resolutions of up to $\sim 10^5 M_\odot$ in M_* halos. It is clear that the simulations predict an order of magnitude or more dwarf galaxy halos than the observed dwarf galaxies. It is more controversial but probably true that the dark halos of dwarf galaxies and of barred galaxies do not have the $\sim r^{-1}$ central cusps predicted by high resolution simulations. The dark matter concentration parameter, defined by the ratio of r_{200}, approximately the virial scale, to the scale length, within which the cusp profile is found, measures the cosmological density at virialisation, and hence should be substantially lower for galaxy clusters than for galaxies. This seems not to be the case in the best-studied examples of massive gravitationally lensed clusters, cf. [14].

Resurrection via fundamental physics

To resolve these conundrums, one is justified in asking whether there may be something fundamentally wrong with the standard CDM cosmology algorithm. For example, one may try to modify the nature of Newtonian gravity. It has been claimed that with relativistic variants of MOND, one can produce a self-consistent cosmology without recourse to any dark matter [15]. Another approach is to modify the nature of dark matter by introducing self-interactions. Finally, one can drop major assumptions about the nature of the initial conditions, such as the assumptions of adiabaticity and gaussianity.

Resurrection via astrophysics

There are at least two approaches to resolving the dark matter conundrums from astrophysics. One is by incorporating star and AGN formation inthe dense baryonic core that forms by gas dissipation. Massive gas outflows can effectively weaken the dark matter gravity, at least in the central cusp. These may include stellar feedback and massive winds, via supernovae, top-heavy IMF, hypernovae, or supermassive black hole-driven outflows Another mechanism that shows some promise in terms of generating an isothermal dark matter core is dynamical feedback, via a central massive rotating gas bar. Such bars may form generically and dissolve rapidly, but their dynamical impact on the dark matter has not been fully evaluated [16, 17, 18].

All of these are radical procedures, but some are more radical than others. Tinkering with fundamental physics, in essence, opens up a Pandora's box of phenomenology. It seems to me that one should first take the more conservative approach of examining the impact of astrophysics on the dark matter distribution before tinkering with more fundamental issues. Of course if one could learn about fundamental physics, such as a new theory of gravity or higher dimensional dark matter relics from dark matter modelling, this would represent an unprecedented and unique breakthrough. But this may be premature.

To proceed, one has to better understand when and how galaxies formed. Fundamental questions in galaxy formation theory remain unresolved. Why do massive galaxies assemble early? And how can their stars form rapidly, as inferred from the α/Fe abundance ratios? Where are the baryons? And if, as observations suggest, they are in the intergalactic medium, both the photo-ionised Lyman α forest and the collisionally ionised warm-hot intergalactic medium (WHIM), how and when is the intergalactic medium (IGM) enriched to 0.1 of the solar value? Why are disks as large as they are, and for that matter, what determines the core density and scalings of spheroids? Can the galaxy luminosity function be reconciled with the dark matter halo mass function? Does the predicted dark matter concentration allow a simultaneous explanation of both the Tully-Fisher relation and the luminosity function? And for that matter, is the dark matter consistent with barred galaxy and low surface brightness dwarf galaxy rotation curves?

AND SO TO GALAXY FORMATION

Many of these issues can be addressed and resolved, at the cost admittedly of introducing additional parameters. Consider the question of disk sizes. This has been a long-standing problem, a consequence of the loss of initial specific angular momentum from the baryons to the dark halo. There are two possible classes of solutions. One involves supernova feedback that imparts energy and momentum to the dissipating baryons. Another approach involves the formation of the disk by infall of gas-rich satellites which carry enough orbital angular momentum to account for the disk size and spin. In either case, low angular momentum material forms the spheroid. Recent simulations claim to reproduce the Tully-Fisher relation for disks, although the spheroids are generally too massive. Moreover, both approaches are too simplistic, since the age and chemical differences between disk and spheroid are not straightforwardly explained.

More success has been achieved by using supernova feedback to eject gas from forming dwarfs. The predicted excess of dwarfs is still present but most are dark. Outflows are achieved once the first few supernovae have formed, while the dwarf is still gas-rich, and are effective for dwarf escape velocities up to about 100 km/s. Observations of nearby starbursts find evidence for hot, metal-enriched outflows, with an outflow rate of order the star formation rate. These represent the most massive dwarfs, which have retained some gas. The far more numerous small dwarfs predicted by hierarchical CDM clustering formed very few stars before being wind-stripped, and so are virtually invisible.

On Milky Way scales, supernova feedback plays an important role in reducing the efficiency of star formation. Gas accretion adds a supply of cold gas to the disk, which is gravitationally unstable, if sufficiently cold, via the Toomre criterion. Stars form, some explode as supernovae and cold gas is heated and ejected in fountains into the halo. The disk stabilises pending a resupply of cold gas, both from gas cooling in the halo and from infall of gas-rich dwarfs. The instability recommences. Hence the star formation history of a disk may be depicted as a series of supernova bubble porosity-driven ministarbursts [21]. The gas consumption in star formation is inefficient, comparable with the observed value of about 2% for the star formation efficiency [22]. The disk remains supplied by gas for a Hubble time, although by today the gas fraction is only about 5%.

The baryon deficit in massive disks is understood in terms of mass outflows from the precursor dwarfs that formed the disks. The outflow rate is comparable to the star formation rate, thereby implying that of order half of the initial baryons have been ejected. The enriched ejecta fills the environment, in our case the Local Group. X-ray observations suggest that groups, at least those containing a massive early-type galaxy, are baryonically closed systems. For less dense environments the ejecta are likely to constitute the warm-hot intergalactic medium, pervasive diffuse gas at 10^5–10^6K that has recently been detected via absorption from highly ionized states such as OVI, OVII and OVIII.

Massive spheroids present a different problem. They are known to have formed stars with high efficiency in order to account for the observed high redshift population of massive red galaxies. Something other than supernova feedback is needed. The new ingredient comes from outflows from active galactic nuclei. These are powered by supermassive black holes. Until now, simulations have appealed to these outflows to eject gas and quench star formation. This is sufficient to obtain the red colours appropriate to an old stellar population. Motivation comes from the correlation between spheroid velocity dispersion (or mass) with SMBH mass over four decades, suggesting that both form contemporaneously. The protogalactic environment is ideal for growing the SMBH. Moreover, the empirical correlation is self-limiting: the momentum outflow in an Eddington wind drives out all residual diffuse gas once the SMBH mass attains the empirical value [20].

There is one more ingredient that is required, however. The high α/Fe ratio in massive spheroids means that the star formation time-scale is the inverse of that found in the usual hierarchy, wherein more massive systems form later and more slowly. I have suggested [23] that non-gravitational physics provides the key, and in particular the short time-scale associated with the jet-driven outflow can provide a means of inducing more efficient star formation than would be obtained with the cumulative ejecta of supernovae. Supernovae, even if their input was energetic enough via boosting the IMF, for example, could only operate globally on a dynamical time-scale. However, a fast jet, surrounded by its over-pressured cocoon, would frustrate itself in a clumpy protogalactic medium, driving turbulence and overtaking and compressing ambient clouds [19]. The more massive clouds would be triggered to form star formation over a time-scale between one and ten percent that of the gravitational crossing time, that is at an efficiency greater than that expected for supernovae by between one and two orders of magnitude. This is precisely what is needed to power an ultraluminous starburst and its associated outflow. The natural hierarchy is inverted via the impact of the SMBH.

Observations indeed show that both SMBH growth, as monitored by the AGN luminosity function, and spheroid assembly, as seen in the galaxy luminosity function, are anti-hierarchical, with growth peaking for both at $z \sim 2$ [24]. It remains to fully implement the AGN model into numerical simulations, but it is clear that it helps address many of the outstanding issues associated with massive galaxy formation. It has predictive power: for example the outflow rate should be proportional to the ratio of AGN to starburst luminosity.

OBSERVING COLD DARK MATTER

There is a motivated dark matter candidate, the lightest stable SUSY particle, under R parity conservation, or WIMP. As yet, direct detection experiments have not found any unambiguous evidence for its existence. The Milky Way halo provides a laboratory par excellence for indirect WIMP searches via annihilations into high energy particles and photons.

The relic WIMP freezes out at $n_x < \sigma_{ann} v > t_H \lesssim 1$, corresponding to a temperature $T \lesssim m_x/20k$. The resulting CDM density is $\Omega_x \sim \sigma_{weak}/\sigma_{ann}$. Halo annihilations of the LSSP occur into γ and ν, as well as $\bar{p}p$ and e^+e^- pairs. In fact, halo detectability may require clumpiness $<n^2>/<n>^2 \sim 100$. High resolution numerical simulations of dark halos actually reveal considerable clumpiness. This is inherent in the early freeze-out of massive particles and subsequent hierarchical clustering in the matter-dominated era. There are possible signals of anomalous gamma ray emission from the inner galaxy and a bump in the cosmic ray positron spectrum that may be indicators of dark matter annihilations. The uncertainties are large however, and improved data is urgently needed to assess this issue.

The 511 keV saga

The INTEGRAL/SPI satellite experiment has detected diffuse 511 keV line emission from throughout the galactic bulge. The 511 keV luminosity is 10^{43}s^{-1}. The line consists of a broad component with width approximately 5 keV and a dominant narrow component with width 1 keV [25]. The approximately spherically symmetric emission has a FWHM of 10 degrees.

There are no indications of any known radioactive source of e^+ such as SNII ejecta detected via the associated Al^{26} emission, or accelerated pairs such as might be associated with jets from massive x-r binaries The only plausible astrophysical e^+ sources associated with the old stellar population are low mass x-ray binaries and classical novae. However estimates of possible e^+ injection from these sources are low compared to the observed flux.

If dark matter annihilations are responsible for the observed flux, x-ray/γ ray constraints require that the DM consist of light (~ 20MeV) particles with, if thermal decoupling applies to give the relic density, $\sigma_{ann} \sim 10^{-5}$pb. Moreover the observed angular distribution of the 511 keV flux fits that expected from integrating over the square of the density for a dark matter profile with $\rho \overset{\infty}{\sim} r^{-0.5}$. Decaying dark matter does not fit the data [26].

Now the relic freeze-out condition sets $\sigma_{ann} \sim (0.2/\Omega_\chi)$pb. Reconciliation to the empirical cross-section is possible via S-wave suppression. In this case, $\sigma_{ann} \propto (m_\chi^2/m_U^4)(v/c)^2$, and the suppression factor relative to freeze-out is $(v/c)^2 \sim 10^{-5}$. There is one further adjustment for light dark matter to work: since $m_U \propto m_\chi^{1/2}$, the price to pay is a new light gauge boson m_U. A potential test involves exploiting the line-shape. The line-width is broadened in a hot interstellar medium, e.g. at $T \sim 10^6$K, as expected high above the galactic plane. By searching for 511 keV line emission from nearby dark-matter-dominated dwarf spheroidal galaxies, one would expect to detect a broader 511 keV line width.

The Galactic Centre as a CDM laboratory

Our galaxy formed around a core of CDM, the most massive of a distribution of clouds that clustered hierarchically to eventually form the dark halo. The CDM is weakly interacting and formed a central density concentration and density cusp, within which baryons condensed. By processes not fully understood, a supermassive black hole formed out of dissipative baryons at the centre of the galaxy, before most of the stars formed. The dissipative processes most likely involved shocked and compressed gas that lost angular momentum in the highly non-axisymmetric gravitational potential well of the forming galaxy as well as run-away collisional mergers of stars and seed black holes.

CDM reacted to the growth of the SMBH by developing a central density spike. The CDM cusp is steepened by adiabatic growth of a supermassive black hole from $\rho \propto r^{-\gamma}$, to $\rho \propto r^{-\gamma'}$, with $\gamma' = \frac{9-2\gamma}{4-\gamma}$ and $\gamma \sim 1$ is the initial cusp profile density index. The annihilation rate is amplified within a radius $GM_{bh}/\sigma^2 \sim 0.1(M_{BH}/3 \times 10^6 M_\odot)$pc. The supermassive black hole in the Galactic Centre could therefore provide a "smoking gun" where a spike of cold dark matter was retained [27]. Possible observables include high energy gamma rays and neutrinos. In fact, the enhanced neutralino annihilations measure CDM where Milky Way formation began, 12 Gyr ago.

Very high energy gamma rays from the Galactic Centre

Now the GC SMBH currently has a low accretion rate as measured by the x-ray flux from SgrA*, and current models do not predict a high flux of γ-rays. Detection of a high gamma ray flux would open up the possibility of radically new astrophysics. In fact, the HESS air Cerenkov gamma ray telescopes have measured a flux centered on SgrA* extending beyond 20 TeV [28] and that, if due to DM annihilations, would require $m_\chi \gtrsim 30$TeV and $\sigma_{ann} \sim 10$pb for an NFW profile. The data is consistent with an NFW profile if interpreted this way. However the inferred cross-section is far too high to give the relic density, but one can save the situation either if there is a DM cusp or if the relic neutralino is non-thermal, as expected for KK dark matter. If thermal freeze-out occurred, there is an additional problem, of providing a mass of at least 100 TeV. To get a viable thermal candidate with a large enough cross-section to give the relic density requires a substantial boost via co-annihilations. This is possible in extensions of the NMSSM model [29].

The case for intermediate mass black holes

Various processes, including black hole mergers and stellar heating may suppress the growth of a spike around the central SMBH. However plausible merger histories predict a distribution of IMBHs in the inner galaxy and halo [30]. In primordial clouds that cool via Lyman alpha emission, fragmentation is likely to be suppressed, facilitating the formation of an IMBH. Many of these pregalactic IMBHs should have retained their CDM spikes and be annihilation sources of hard gamma rays. They also accrete intergalactic gas and are possible candidates for ultraluminous x-ray sources in old stellar populations.

An early population of IMBH remnants would be sources of ionising photons and accelerators of cosmic rays via jet-driven outflows. Possible signatures of such energy input in the pregalactic era include early reionisation of the universe, preheating of the IGM, and a spallation origin of Li^6 in extremely metal-poor halo stars. All of these phenomena could be explained if of order 0.1% of the baryons was in the form of IMBHs.

WHERE NEXT?

One can envisage progress on a variety of fronts. In particle theory, one can readily imagine more than one DM candidate. Why not have 2 stable dark matter particles, one light, one heavy, as motivated by $N = 2$ SUSY? If one took the light dark matter and any of the possible heavy dark matter detections seriously, one could have a situation in which the light (a few MeV) spin-0 particle is subdominant but a $\sim 0.1 - 100\,\text{TeV}$ neutralino is the dominant relic [31].

Because a neutralino of mass $\gtrsim 1$ TeV is beyond the range of the LHC or even the ILC, astrophysical searches for DM merit serious consideration and modest funding. In direct detection, one might eventually hope to see a modulated signal, due to the effect of the Earth's motion through directed streams of CDM [32]. The streams are generic to tidal disruption of dark matter clumps. As for indirect detection, the prospects are exciting, because of the many complementary searches that are being launched. Evidence of neutralino annihilations may come from searches for γ, ν, e^+ and $\bar{p}$ signatures. Experiments under development include HESS2, MAGIC, VERITAS, GLAST (γ-rays), ICECUBE, ANTARES, KM3NET (ν), and PAMELA and AMS ($e^+, \bar{p}$). Targets include the Galactic Centre, the halo and even the sun, where neutralino annihilations in the solar core yield a potentially observable high energy neutrino flux [33].

In the area of galaxy formation theory, refined numerical simulations will explore the impacts of supernova and SMBH-driven outflows and bar evolution on the distribution and especially the concentration of CDM. A better understanding of IMBHs and the SMBH in the Galactic Centre could eventually provide a "smoking gun" where spikes of CDM were retained: the enhanced neutralino annihilations measure CDM where galaxy formation began, 12 Gyr ago. Fundamental physics could be probed: for example a higher dimensional signature, Kaluza-Klein dark matter, would have a spectral signature and branchings that are distinct from those of neutralinos. The prospect of multi-TeV dark matter is another tantalising probe. This provides a challenge for SUSY but is possibly a natural and fundamental scale for any stable relics surviving from n=3 extra dimensions.

REFERENCES

1. U. Seljak, Phys.Rev. **D71** (2005) 103515.
2. L. Boyle, P. Steinhardt, N. Turok, preprint astro-ph/0507455 (2005).
3. M. Gasperini and N. Nicotri, preprint hep-th/0511039 (2005).
4. A. Riazuelo et al., preprint astro-ph/0601433 (2005).
5. B. Freivogel et al., preprint hep-th/0505232 (2005).
6. C. Danforth, J. Shull, J. Rosenberg and J. Stocke, ApJ in press, astro-ph/0508656 (2005).
7. F. Nicastro et al., ApJ **629** (2005) 700.
8. A. Klypin, H. Zhao and R. Somerville, ApJ **573** (2002) 597.
9. M. Creze et al., A&A **426** (2004) 65.
10. M. Walker and M. Wardle, ApJ **498L** (1998) 125.

11. A. Lawrence, MNRAS **323** (2001) 147L.
12. K. Adelberger, C. Steidel, A. Shapley and M. Pettini, ApJ **584** (2003) 45.
13. V. Springel and L. Hernquist, MNRAS **339** (2003) 289.
14. M. Oguri et al., ApJ **632** (2005) 8410.
15. J. Bekenstein, PRD **70** (2004) 1502.
16. M. Weinberg and N. Katz, ApJ **580** (2002) 627.
17. O. Valenzuela and A. Klypin, MNRAS **345** (2003) 406.
18. E. Athanassoula, IAU Symposium **220**, eds. S. D. Ryder, D. J. Pisano, M. A. Walker, and K. C. Freeman. San Francisco: Astronomical Society of the Pacific (2004), p.273.
19. C. Saxton, G. Bicknell, R. Sutherland and S. Midgley, MNRAS **359** (2005) 781.
20. J. Silk and M. Rees, A&A **331L** (1998) 1.
21. J. Silk, MNRAS **343** (2003) 249.
22. R. Kennicutt, ApJ **498** (1998) 541.
23. J. Silk, MNRAS **364** (2005) 1337.
24. G. Hasinger, T. Miyaji and M. Schmidt, A&A **441** (2005) 447.
25. P. Jean et al., A&A **445** (2006) 579.
26. Y. Ascasibar, P. Jean, C. Boehm and J. Knoedlseder, A&A, submitted, astro-ph/0507142 (2005).
27. G. Bertone, G. Sigl and J. Silk, MNRAS **337** (2003) 98.
28. W. Hoffmann, in 29th International Cosmic Ray Conference Pune 101-104 (2005), available at http://icrc2005.tifr.res.in/htm.
29. S. Profumo, PRD **72** (2006) 103521.
30. R. Islam, J. Taylor and J. Silk, MNRAS **354** (2004) 427I.
31. C. Boehm, P. Fayet and J. Silk, PRD **69** (2004) 101302.
32. K. Freese, P. Gondolo and H. Newburg, PRD **71** (2005) 043516.
33. J. Silk, K. Olive and M. Srednicki, PRL **55**, (1986) 257.

Albert Einstein: A Man for the Millenium?

John Stachel

Department of Physics & Center for Einstein Studies Boston University

Keywords: History of science, Philosophy of science, Special Relativity, General Relativity and Gravitation
PACS: 01.65.+g, 01.70.+w, 03.304+p, 03.70+k, 04.,11.30+Cp

THE LONG VIEW OF HISTORY

True story. Henry Kissinger was in China in 1972, laying the groundwork for President Nixon's visit. At a meeting with Chinese prime minister Chou En-Lai, Mr. Kissinger asked the prime minister if he believed whether the 1789 French Revolution benefited humanity. After mulling over the question for a few minutes, Chou En-Lai replied, "It's too early to tell." (J. Lau, cited from <www.yellowbridge.com/humor/chinaamerica.html>)

FIGURE 1. Kissinger – Chou-En-Lai

Chou En-Lai, heir to a 5000-year old civilization, was obviously trying to "put in his place" the upstart from the barely-200-years-old United States. Yet his answer contains a good deal of wisdom. Often, the historical evaluation of the significance of some important event can change long after the event occurred.

Euclid　　　　Gauss　　　　Bolyai　　　　Lobachevski

FIGURE 2.

Here is an example from the history of science that is relevant to my topic: the case of Euclid, who flourished about 300 BCE. For over three millennia, if anyone asked the question "What was Euclid's major scientific contribution?", the answer was something like: "He codified *the* geometry of space." While I have placed emphasis on the singular, until the beginning of the 19th century there was no need to do so because there simply was no other geometry. As late as 1772, the renowned English philosopher David Hume wrote:

> Though there never were a circle or triangle in nature, the truths demonstrated by Euclid would forever retain their certainty and evidence. (*An Enquiry Concerning Human Understanding*, Section IV).

But, beginning with the work of Carl Friedrich Gauss, who coined the term "non-Euclidean geometry," it became clear that consistent alternative geometries could be developed that differed from Euclid's by negating his famous fifth or parallel postulate:

> If a straight line crossing two straight lines makes the interior angles on the same side less than two right angles, the two straight lines, if extended indefinitely, meet on that side on which are the angles less than the two right angles. (*Elements*, Book I)

It is easier to formulate the alternatives to this postulate if we use this equivalent form:

> Given any straight line and a point not on it, there exists one and only one straight line that passes through the point and never intersects the first line, no matter how far it is extended.

Gauss considered a geometry in which there could be *more than one* such parallel line, but did not publish his results

> for I fear the cry of the Bœotians [i.e., the philistines] which would arise should I express my whole view on this matter (letter to Bessel, 1829).

Results similar to Gauss' were soon published by János Bolyai (1831) and Nikolai Lobachevski (1829). Some years later, Bernhard Riemann described a second non-Euclidean geometry, in which there are *no parallel lines* (1854 lecture, 1868 posthumous publication).

Riemann   Poincaré   Minkowski   Lorentz

FIGURE 3.

As was soon discovered, two-dimensional Gauss-Bolyai-Lobachevski geometry can be interpreted as the geometry of a space of constant negative curvature (the surface of a hypersphere) embedded (locally) in three-dimensional Euclidean space; while two-dimensional Riemannian geometry can be interpreted as the geometry of a space of constant positive curvature (the surface of a sphere) embedded (globally) in three-dimensional Euclidean space. In both cases, "straight line" is to be interpreted as a geodesic (shortest curve between two points) of the surface.

In all three of these geometries, space is homogeneous and isotropic. Henri Poincaré developed what he called a "fourth geometry, as coherent as those of Euclid, Lobachevski and Riemann" (*Sur les hypothèses fondamentales de la géométrie*, 1887). The parallel postulate holds in this geometry and space is homogeneous; but it is no longer isotropic. In this two-dimensional version, the straight lines through any point fall into two classes; any line in one class can be "pseudo-rotated" into any other in the same class, but no "pseudo-rotation" can take a line of one class into a line of the other. The two classes are separated by a pair of straight lines, each of which is orthogonal to itself. With hindsight, we can see that this is a description of two-dimensional Minkowski space-time, the two classes consisting of the time-like and space-like lines, separated by a pair of null lines (see the next section); but no one seems to have realized this until long after the passing of Poincaré and Minkowski.

The four geometries mentioned were only the first of a host of new geometries invented since the floodgates were opened by Gauss. (Later, I shall discuss Weyl's definition of a geometry). Clearly, in the face of this profusion, the old answer to the question of Euclid's major contribution is unacceptable. A modern answer is given in *The Dictionary of Scientific Biography*:

> The *Elements* ... most remarkable feature is the arrangement of the matter so that one proposition follows on another in a strict logical order, with the minimum of assumptions and very little that is superfluous. ... The significance of Euclid's *Elements* in the history of thought is twofold. In the first place, it introduced into mathematical reasoning new standards of rigor which ... have been equaled again only in the past two centuries. In the second place, it marked a decisive step in the geometrization of mathematics ("Euclid," *DSB*, vol. IV).

Euclid's work has served as a model for many later attempts to logically organize other branches of science, and even philosophy (see Spinoza's *Ethica Ordine Geometrico Demonstrata* [*Ethics Demonstrated in Geometric Order*]

FIGURE 4. Einstein

What About Einstein?

Having adopted the long view of history, we are ready to consider the question: "How will Einstein be viewed at the end of the next millennium?" In 3005 (assuming humanity survives until then- and given the current state of the world, this is a big assumption), what will physicists regard as his major contribution?

Today, just once century after 1905, we can already see a sifting out of certain items of his total œuvre as most significant. Were we to list *all* his accomplishments, the list would be long indeed. Such a list would certainly include:

- His estimate of molecular size based on the change in viscosity of a liquid when particles are suspended in it.
- His demonstration, based on the first theory of a stochastic process, that microscopic fluctuation phenomena can be observed in Brownian motion.
- His development of a new kinematics in the special theory of relativity, and the deduction from it of such remarkable features as:
  - The path dependence of proper time intervals (twin-paradox")
  - The equivalence of mass and energy ("$E = mc^2$").
- His development of general relativity, still the best theory of gravitation that we have.

- His proposal of the light quantum hypothesis, which developed into the theory of the photon, the first elementary particle to be given a quantum treatment.
- His quantum theory of solids, which provided the basis for explaining the anomalous low-temperature behavior of crystalline solids.
- His explanation of Planck's law based on the introduction of the *A* & *B* coefficients, which placed the concept of transition probabilities at the center of atomic physics.
- The Einstein-Podolsky-Rosen "paradox," which highlighted the nature of the quantum entanglement of two or more systems.
- His work on Bose-Einstein statistics, leading to his prediction of the existence of Bose-Einstein condensates, only recently confirmed.
- The Einstein-Infeld-Hoffmann derivation of the equations of motion of massive bodies from the field equations of general relativity – the list could go on indefinitely.

But from the perspective of 2005, most physicists would probably agree that a list of the works that did the most to change the domain of physics must include:

- The light quantum hypothesis and quantum theory of solids, which ultimately led to the fulfillment of Einstein's early prediction that neither classical mechanics (including its special-relativistic modifications) nor classical electrodynamics could survive the onslaught of the quantum of action. Of course the form taken by the fulfillment– non-relativistic quantum mechanics and (special-)relativistic quantum field theory– left Einstein quite dissatisfied. This question is discussed in a later section.
- Special relativity (SR), which led to a realization that all of physics, including the future theory of elementary particles (and excluding only gravitation), would have to be reformulated in terms of representations of the Poincaré (or inhomogeneous Lorentz) group.
- General relativity (GR), which provides a theory of the inertio-gravitational field. It goes beyond the special theory by turning all space-time structures into dynamic fields. GR has survived 90 years of theoretical challenges and experimental tests, both local and astronomical, and forms the basis for current treatments of cosmology.

Given this 2005 perspective, perhaps it not excessive *hubris* to raise the question of how matters will look from the perspective of 3005. Clearly by then, most of the details mentioned above will have faded from sight; but I shall propose that Einstein's work on space-time structures provides clues suggesting a plausible guess about what will survive. However, before gazing into the crystal ball of prophecy, we need to look back at the history of the development of the concept of space-time in physics and discuss some philosophical controversies about its nature.

HISTORICAL SECTION

It has long been clear that space and time are intimately related in physics. Kinematics, the description of motion as change of place over time, involves both, as was already clear to Aristotle:

> Evidently time does not exist without a motion or change ... it must be something belonging to a motion ... "motion" in its most general and primary sense is change of place, which we call "locomotion" ... time is continuous because a motion is continuous (Aristotle, *Physica*, ca. 350 BCE, Book IV).

But the modern concept of a union of space and time in one abstract space, now called space-time, only developed in the 20th century (some 18th century anticipations are discussed below). In addition to the development of the various geometries discussed in the previous section, a number of other developments contributed to our current concept of space-time and its diagrammatic representation. I shall single out a few other crucial developments. Underlying the possibility of any further developments was:

0. The ability to create symbolic representations.
1. Representation of spatial intervals, and later other, non-spatial (concrete) magnitudes, by lengths (one dimensional diagrams).
2. Representation of (abstract) time intervals by lengths;
3. Combination of one-dimensional representations of spatial and temporal intervals in a single, two-dimensional diagram;
4. The representation of motion in two- and three-dimensional diagrams and coordinatization of two- and three-dimensional Euclidean space using mutually perpendicular axes
5. Recognition that two such coordinatizations are related by a coordinate transformation representing a rotation (orthogonal transformation).
6. Generalization of the concept of space beyond its use to describe three-dimensional physical space to higher-dimensional spaces, in particular the concept of time as a fourth dimension.
7. Formulation of the concept of affine spaces of arbitrary dimension and their use to formulate the principle of inertia.
8. The four-dimensional generalization of Poincaré's fourth geometry.
9. Formulation of the concept of Riemannian spaces of variable curvature, both in the metric sense (geodesics, Gaussian curvature) and in the affine sense (parallel transport, affine curvature);

I shall briefly indicate, to the best of my knowledge, when each of these concepts was introduced.

Perhaps the most extraordinary step on the road to the concept of space-time was step 2, the representation of a time interval by a length, which has been called the spatial representation of time by Henri Bergson (see *Durée et simultanéité*, 1922) and the spatialization of time by Émile Meyerson (see *La déduction relativiste*, 1925). It was the culmination of several preceding developments.

0) Underlying the possibility of any further developments is the ability to create symbolic representations. Until this uniquely human faculty to create pictorial and other lasting symbolic representations developed, no further progress was possible in the production of shared abstract concepts. The earliest preserved pictorial representations are the cave drawings and paintings, which are at most about 40,000 years old. Any cave art much older than that would have deteriorated beyond recognition, so it is hard to say just when the human ability to create such representations arose.

The earliest surviving examples include both abstract symbolic and naturalistic representational elements. Implicit in such drawings, especially in the naturalistic ones, is the concept of representation of the spatial dimensions and relations of external objects by lines in the drawing. This already manifests a considerable power of abstraction.

1A) In some of the abstract drawings, one finds geometrical patterns involving straight lines. In more naturalistic drawings, one finds representations of straight objects, such as spears or arrows, by straight lines.

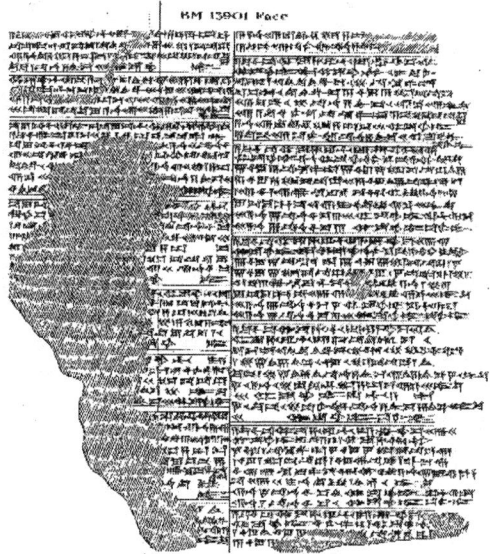

FIGURE 5. Cuneiform tablet

1B) I shall now jump forty millennia to something at an even higher level of abstraction: the first representation of a non-length by a length: Jens Høyrup has cited the earliest preserved instances of such representations from ca. 1800 BCE in clay tablets of the Mesopotamian scribal school. Here,

> a length is taken to represent something different from itself, viz. an area...[S]ince other (slightly later) Old Babylonian texts use lengths and widths to represent pure numbers, prices or complex arithmetical expressions, the step is real, no mere

accident. ...This is one of the great steps in the history of mathematics, one of the very greatest, and whoever feels a chill when faced with intellectual progress should feel it here (see *Lengths, Widths, Surfaces/A Portrait of Old Babylonian Algebra and Its Kin*, 2002).

2) Only a millennium-and-a-half later is this method of representation applied to the concept of time: "The application of the concept of continuum to time and process does not appear to have taken place before the middle of the fourth century BC" (Hans-Joachim Waschkies, *Von Eudoxos zu Aristoteles/Das Fortwirken der Eudoxischen Proportionentheorie in der Aristotelischen Lehre vom Kontinuum, 1977*).

It was Aristotle who, in his analysis of Zeno's paradoxes of motion, first represented a continuous time interval by a length. He utilized geometrical representations of both spatial and temporal intervals by one-dimensional line segments, placed parallel to one another, in order to compare the two (*Physica*, Book VI, Chapter 2). His example was followed by Archimedes, who employed similar diagrams (*On Spirals*, Propositions.I & II, ca. 225 BCE), an interesting case of a mathematician imitating a philosopher.

3A) Neither Aristotle, Archimedes, nor any of their successors for two millennia, combined the representations of temporal and spatial intervals into a single, two-dimensional diagram. The first to use time as part of a two-dimensional diagram appears to have been Nicholas Oresme in the mid-fourteenth century. To use somewhat anachronistic language, he plotted time against velocity, not distance (*Tractatus de configurationibus qualitatum et motuum*, ca. 1370; Marshall Clagett, ed. & transl., *Nicholas Oresme and the Medieval Geometry of Qualities and Motions*, 1968. For the claim that Giovanni di Casali preceded Oresme by a few years, see Marshall Clagett, *The Science of Mechanics in the Middle Ages*, 1961).

 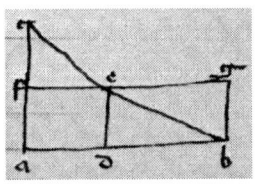

Nicholas Oresme (1321-1382)
"*Tractatus de configurationibus*"
The ordinate denotes the time, the abcissa denotes the velocity.

FIGURE 6.

4A) René Descartes used two-dimensional spatial diagrams, with the points of a curve referred to two orthogonal spatial axes, (*La Geométrie*, 1637), and when he had to represent motion in three dimensions, he did so by projecting it onto several two-dimensional diagrams.

6A) The conjunction of time and space in a two-dimensional diagram took another half-century. However, Descartes did give a definition of *dimension* that justifies the spatial representation of *any* quantifiable property of a system:

> By dimension, we understand nothing but the mode and reason, according to which some subject is considered to be measurable; so that not only length, breadth and depth are dimensions of a body, but in addition its gravity [i.e., weight] is a dimension, in accord with which subjects are weighed, its velocity is the dimension of motion, and an infinity of others of this type (*Regulae ad directionem ingenii* [*Rules for the direction of the understanding*], written between 1619-1628, first published in Dutch translation in 1684).

That for Descartes, time constitutes one such dimension is clear from the succeeding discussion.

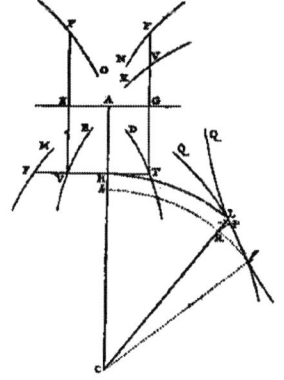

Pierre Varignon (1654-1722) actually plotted position, time and velocity on the same diagram.

FIGURE 7.

3B) The first two-dimensional graphic representation of a one-dimensional motion, plotting distance and time as orthogonal coordinates, was done by Pierre Varignon near the turn of the 18th century (*Règle générale pour toutes sortes de mouvements de vitesse quelconques variées à discrétion*, 1698). Varignon actually plotted position, time, and velocity on the same diagram. Indeed, he first defined the concept of instantaneous velocity, and so may be said to have introduced extended configuration space as well. He saw the conceptual problem raised by this work:

> Space and time being heterogeneous magnitudes, it is not properly they that are compared with each other in the relation called speed, but only the homogeneous magnitudes that express them; which here are, and will always be in what follows, either two lines, or two numbers, or two of any other homogeneous magnitudes that one wishes (*Des mouvements variés à volonté, comparés entre eux et avec les uniformes*, 1707).

Descartes   d'Alembert   Euler   Lagrange

FIGURE 8.

About fifty years later, this problem was discussed in more detail by Jean le Rond d'Alembert:

> One cannot compare with each other two things of a different nature, such as space and time; but one can compare the relation of portions of time with that of the portions of the space traversed. By its nature, time flows uniformly and mechanics assumes this uniformity. In addition, without knowing time in itself and without having a precise measure of it, we cannot represent the relation of its parts more clearly that by that of portions of an indefinite straight line. Now, the analogy that exists between the parts of such a line and that of the space traversed by a body that moves in any sort of way, can always be expressed by an equation: one may thus imagine a curve, the *abcissae* of which represent the portions of time that have elapsed since the start of the motion, the corresponding ordinates representing the spaces traversed during these portions of time: the equation of this curve will express, not the relation of the times to the spaces, but, if one may so put it, the relation of the relation that the parts of time have to their unit, to that [relation] that the parts of space have to their unit (D'Alembert, *Traité de Dynamique*, 1743).

4B) Leonhard Euler appears to have been the first to use three-dimensional diagrams to represent motions, and to resolve forces and motions into their components along three mutually perpendicular axes (*Recherches sur le mouvement des corps célestes en général*, 1749; *Découverte d'un nouveau principe de mécanique*, 1750)

5) Euler also realized that these three orthogonal axes could be chosen in many ways, each related to the other by a rigid rotation; and worked out the transformation between the two sets of Cartesian coordinates of a point (see, e.g., *Recherches sur la connoissance mécanique des corps*, 1758)

6B) D'Alembert was the first to discuss the concept of time as a fourth dimension:

> Above I said that it is not possible to conceive of more than three *dimensions*. A clever man of my acquaintance [d'Alembert himself?] believes that nevertheless one may regard duration as a fourth *dimension*, & that the product of time multiplied by solidity would be in some way a product of four *dimensions*. ("Dimension" in the *Encylopédie*, vol. 4, 1754)

Half a century later, Lagrange was less hesitant, affirming that:

One may regard mechanics as a four-dimensional geometry, and mechanical analysis [i.e., analytical mechanics] as an extension of geometrical analysis (*Théorie des fonctions analytiques*, 1797)

6C) Euler used six coordinates to treat the six degrees of freedom of a rigid body –three for translation, three for rotation (*Theoria motus corporum solidorum seu rigidorum* [*Theory of the motions of solid or rigid bodies*], 1765). Lagrange introduced the idea of treating all the degrees of freedom of a mechanical system, however many, as abstract dimensions (*Mécanique analytique*, 1788). But he prided himself on having no diagrams in his book.

Diagrams involving such quantities as pressure and volume were introduced in the nineteenth century. As so often in thermodynamics, the idea was adapted from engineering practice: James Watt and John Southern used indicator diagrams to calculate the work done by a steam engine. Originally regarded as a trade secret, such diagrams were not published until 1822. Émile Clapeyron used such a diagram to represent the Carnot cycle (*Mémoire sur la puissance motrice de la chaleur*, 1834) and the idea passed into general usage in thermodynamics.

Mathematicians and physicists thus became accustomed to Descartes' idea (see 6A above) that any magnitude may be treated as a dimension, and any number of such magnitudes represented by an abstract space of higher dimension.

7) Hermann Grassmann abstracted the concept of parallelism from its metrical associations in Euclidean geometry and developed the concept of an affine geometry, and applied it to any number of dimensions (*Die lineale Ausdehnungslehre*, 1844, 2nd ed. 1878). He applied this concept to a number of problems in mechanics; but only much later was it realized that a four-dimensional affine space is the proper geometric setting for the law of inertia (Hermann Weyl, *Raum-Zeit-Materie*, 1918).

8) In 1905, Poincaré introduced the concept of a four-dimensional representation of the Lorentz transformations (*Sur la dynamique de l'électron*), quite independently of his earlier work in 1887 (see the opening section). In 1907, Hermann Minkowski realized that such a four-dimensional unification of space and time is particularly suited to the visualization of the Lorentz transformations (*Das Relativitätsprinzip*). He adapted a four-dimensional coordinate system similar to Poincaré's, and carried its geometrical interpretation further. He introduced the term *space-time* and (rather pretentiously) named special-relativistic space-time "*die Welt*" [*the universe* or *world*], leading to such terms as world point, world line and world tube. Like Poincaré, he represented the temporal coordinate by an imaginary number, so that Lorentz transformations could be interpreted geometrically as rotations in a four-dimensional (but complex) Euclidean space. It is more common now to use a real time coordinate and pseudo-rotations in a real but non-Euclidean space-time.

9) Riemann generalized Gauss' theory of surfaces of variable curvature to spaces of any number of dimensions that are locally flat (Euclidean) but globally non-flat, i.e. having a curvature that varies from point to point. What does curvature mean here? Riemann defined a fourth rank tensor, now called the Riemann curvature tensor, as a generalization of the Gaussian curvature (*Habilitationsschrift*, 1854, posthumously published in 1868).

The concept of parallel transport in such a Riemannian space was not introduced until 1916 by Tullio Levi-Civita (*Nozione di parallelismo in una varietà qualunque e conseguente specificazione geometrica della curvatura Riemanniana*, 1917), in response to the formulation of general relativity. Hermann Weyl (*Reine Infinitesimalgeometrie*, 1918) soon generalized his work by defining the concept of a non-flat affine geometry. Locally it is an affine-flat space, but globally non-flat in a new sense: In such a space, parallel transport of a vector around a closed curve results in a different vector, the affine curvature tensor being a measure of the difference This concept led to a deeper understanding of the relation between affine connection and inertio-gravitational field in both Newtonian and general relativity theory (see below).

Space-Time Structures

Before turning to S-R space-time, I shall discuss the concept of Galilei-Newtonian (G-N) space-time, the space-time associated with the Galileian law of inertia and Newtonian dynamics. Logically it comes before SR space-time, although in actuality its four-dimensional version was only developed afterwards. One reason for this is that the mathematical structure of G-N space-time crucially involves the concept of an affine space (see above).

There are two distinct types of space-time structure inherent in both G-N and S-R space-time:

1) The *chrono-geometrical structure*, which determines the behavior of (ideal) measuring rods (geometry) and clocks (chronometry); and
2) The *inertial structure*, which governs the behavior of free particles (i.e., particles subject to no external forces);
3) there are also *compatibility conditions* between the two.

The S-R inertial structure (and later the inertio-gravitational structure in GR) is represented mathematically by an affine connection. This connection is usually derived from the chronogeometry, which is represented mathematically by a pseudo-metric, and treated as secondary if mentioned at all. But I shall emphasize the connection for two reasons:

1) Much recent progress in GR has come from emphasis on its primary role in the most fruitful formulations of the theory in preparation for canonical quantization;
2) It illuminates the connection (pun intended) as well as the contrast between GR and Yang-Mills gauge theories.

Geometry

In both G-N and S-R space-times, the geometry of the relative space of each inertial frame is Euclidean; it can be measured with (ideal) measuring rods at rest in that frame, for example. The distance along any spatial path depends on the path taken, and the *shortest distance* (geodesic) is along the *straightest path*. Mathematically an inertial frame is represented by a *fibration* of space-time; that is, a family of parallel time-like straight lines that fills the space-time, each line of which transvects the space-like hyperplanes of simultaneity (see below).

G-N Chronometry and Chrono-Geometry

In G-N kinematics, the chronometry is independent of the geometry: The time is *absolute* and *universal*, and space-time divides naturally into events that are *simultaneous* (i.e., occur at the same absolute time). Mathematically, the absolute time is represented by a *foliation* of space-time; that is, a family of parallel space-like hyper-planes of equal global time. A four-dimensional G-N *chrono-geometric* structure can be defined, which splits naturally into a unique chronometry (unique foliation) – time is absolute – and a three-parameter family of relative geometries (three-parameter family of fibrations) – space is relative to the choice of inertial frame.

S-R Chrono-Geometry

In contrast, S-R space and time are united in one absolute chrono-geometrical structure, represented mathematically by a flat four-dimensional pseudo-metric ("pseudo" because it has a non-definite signature, usually called Lorentzian), often called the Minkowski metric. This results in the existence at each point of space-time of a double null cone (consisting of a forward and backward cone) of events that have a null (i.e., zero) separation from the event in question. A *null separation* between two events is interpreted as the possibility of connecting them by a light signal (or any zero-rest mass particle); which way the signal can pass depends on which event is in the forward light cone of the other.

S-R Chronometry

Both *geometry* and *chronometry* are now *relative*. This results in a big difference between the two chronometries:

In G-N chronometry, as noted above, the absolute time along any path between two non-simultaneous events is independent of the path. All (ideal) clocks measure this absolute, universal time.

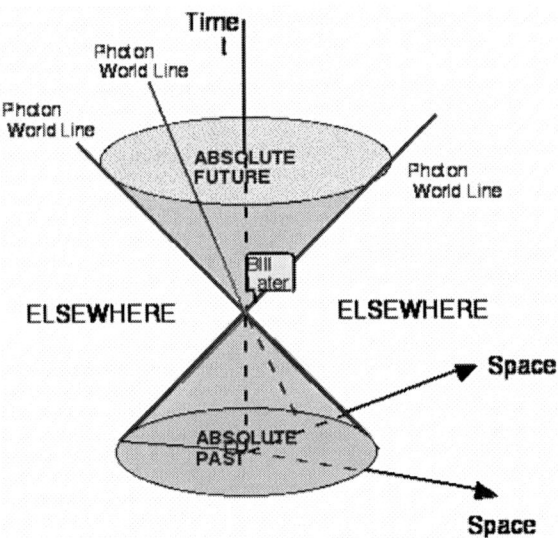

FIGURE 9. Minkowski space in 4 dimensions

In S-R chronometry, the time along a path between any two events with a time-like separation (i.e., each one is within the forward or backward light cone of the other), usually called the *proper time, depends on the time-like path* taken between them. In this respect, S-R time is more like space, but there is still a big difference: The *longest time interval* between two events is along the *straightest path* between them (this observation is the essence of the "twin paradox").

Inertial Structure

The use of the terms *straightest* and *parallel* actually encroaches upon the domain of the second type of space-time structure: the *inertial structure*, which determines the motion of *freely-falling* (i.e., net force-free) structureless bodies ("particles").

In both G-N and S-R space-times, such particles follow the time-like straightest inertial paths of space-time – straight lines for the flat space-times of both classical and special-relativistic physics, with the affine parameter coinciding with the absolute time in the first case, and with the proper time in the second. This is the mathematical expression of the law of inertia, common to both G-N and S-R space-times because it depends only on their common affine structure.

Affine Spaces

We shall only be concerned with torsion-free affine spaces.

Mathematically, to define the inertial structure, all we need is the concept of an affine space, for which parallelism and the ratio of parallel intervals are meaningful concepts. The affine structure defines the concept of parallelism for two vectors at neighboring points of space-time. In an affine space, a curve is *straight* in the sense that its tangent vector always remains parallel to itself as it is parallel-transported along the curve.

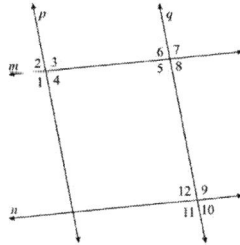

Affine space: Parallel intervals   Ratio of parallel vectors

FIGURE 10.

Compatibility Conditions

The two space-time structures – chrono-geometry and inertial structure – are *compatible* with each other. Mathematically this compatibility is expressed by the vanishing of the covariant derivative of the chrono-geometry. This has a number of kinematical consequences. For example:

1) The *extremal paths* (*geodesics*) as defined by the pseudo-metric (shortest paths for space-like curves, longest paths for time-like curves), coincide with the *straightest paths*, as defined by the inertial structure.
2) *Freely falling rods and clocks*, as defined by parallel transport with the inertial structure, continue to measure *proper space and time intervals* respectively, as defined by the chrono-geometry.

The Relativity Principle

Unless the relativity principle is taken into account, the combination of spatial and temporal dimensions in a single diagram shares a feature with the combination of such heterogeneous dimensions as pressure, volume and temperature in diagrams used to picture thermodynamic relations: While the three spatial coordinates in a given frame of reference (e.g. an inertial frame) can be mixed among themselves by rigid rotations (as noted above, a technique introduced in rigid body dynamics by Euler), the spatial and temporal coordinates of that frame can no more be mixed than can p, V and T.

Galilean Relativity

It follows from the laws of Newtonian mechanics, that no *mechanical* experiment can distinguish between any two inertial frames of reference. As long as it was believed that all physical phenomena could ultimately be reduced to mechanical interactions (the mechanical world view), this restriction seemed harmless.

Once this *Galilean relativity principle* is taken into account, the situation changes: Now, *the Galilei transformations*, which relate the Cartesian coordinates of an event with respect to two inertial frames of reference in relative motion with relative velocity $\mathbf{V}$, allow us to mix spatial and temporal coordinates:

$$\mathbf{r}' = \mathbf{r} - \mathbf{V}t,$$

where $\mathbf{r}'$, $\mathbf{r}$ are the Cartesian coordinate vectors relative to the origins of the respective inertial frames, and t is the absolute time (assuming the origins to coincide at time $t = 0$). Of course in classical (G-N) kinematics, time is absolute, and to emphasize this we must add

$$t' = t$$

to our transformation equations. Again one sees that space is relative to choice of an inertial frame of reference, but time remains universal and absolute.

Relativity Principle and Optics

With the rise of the wave theory of light and then Maxwell's explanation of light as a type of electromagnetic wave, the mechanical world view seemed to demand introduction of a mechanical medium-the ether- in which such waves would propagate. All attempts to detect the motion of the earth through the ether by optical or other electromagnetic phenomena failed.

The relativity principle seemed to apply to all these phenomena independently of the hypothetical ether. In 1874, Eleuthère Elie Nicolas Mascart formulated what we might call the Optical Principle of relativity, based on experimental tests of order (v/c), where v is the presumed velocity of the earth through the ether:

> No optical experiment can detect the motion of the earth through the ether. The earth's translational motion does not have a measurable influence on optical phenomena produced by a terrestrial source ... [T]hese phenomena do not provide us with a way to determine the absolute motion of a body and ... relative motions are

the only ones that we are able to determine. (*Modifications qu'éprouve la lumière par suite du mouvement de la source lumineuse et mouvement de l'observateur (deuxième partie)*, 1874).

Yet optical and later electromagnetic theory predicted the existence of such effects.

To explain this apparent paradox, Hendrik Antoon Lorentz and then Poincaré introduced the concept of local time and the length-contraction hypothesis, which they interpreted as dynamical "compensations" for the expected effects of motion through the ether. They introduced what Poincaré named the Lorentz transformations from the unprimed coordinates in the ether frame to the primed coordinates in the moving frame:

$$\mathbf{r}' = \gamma(\mathbf{r} - \mathbf{V}t) + (1 - \gamma)[\mathbf{r} - (\mathbf{V}\cdot\mathbf{r})\mathbf{V}/V^2]$$
$$t' = \gamma[t - (\mathbf{V}\cdot\mathbf{r})/c^2],$$

where $\gamma = [1 - (V/c)^2]^{1/2}$.

Neither Lorentz nor Poincaré realized the fundamental kinematical significance of these transformations; they interpreted them within the framework of N-G kinematics and the ether theory: Due to their motion through the ether, clocks *really* slow down and rigid rods *really* contract. There is a distinction between the "apparent," primed space and time coordinates of an event, as measured in a moving frame of reference by the slowed-down clocks and contracted measuring rods, and the "true," unprimed spatial and temporal coordinates as defined by clocks and rods at rest in the ether.

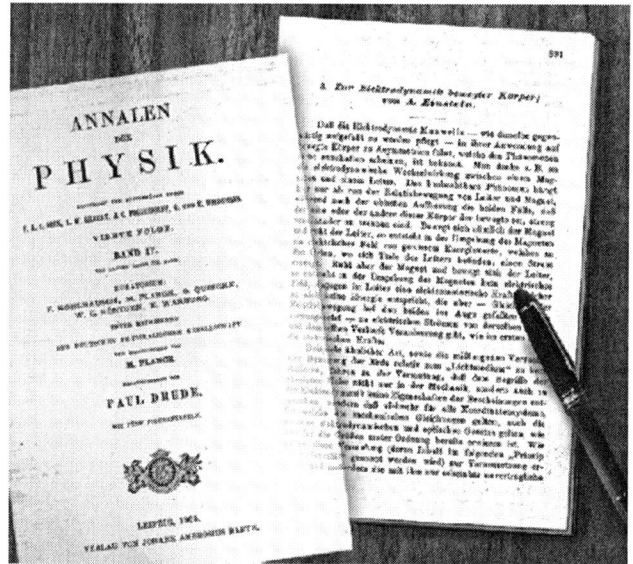

FIGURE 11. *"On the Electrodynamics of Moving Bodies"*

Special Relativity

It was Albert Einstein (*Zur Elektrodynamik bewegter Körper*, 1905) who first realized the need to replace such ideas, based on classical kinematics, with a new kinematics based on four key ideas:

- 1. Omit all reference to the hypothetical ether frame;
- 2. Take the failure of all attempts to detect absolute motion at face value, and postulate the relativity principle (all inertial frame of reference are equivalent) for all physical phenomena;
- 3. Add the well-tested postulate that the speed of light is independent of that of its source;
- 4. Combining 1, 2 and 3, one can derive the Lorentz transformations between any two inertial frames of reference. Interpret the measured spatial and temporal coordinates occurring in them as the "true" spatial and temporal coordinates of each inertial frame of reference; these transformations then form a group that does not single out any inertial frame.

The derivation of the Lorentz transformations requires that simultaneity of distant events be *defined* with respect to each inertial frame of reference in such a way as to make the speed of light the same in every inertial frame and independent of position and direction in that frame. It is important to realize that, if the concept of distant simultaneity is to be introduced at all, some definition always is needed. No physical result can depend on this definition; and it is even possible to dispense with such a definition. Bondi's K-calculus, for example, can treat the special theory without introducing such a definition (see, e.g., Hermann Bondi, *Relativity and Common Sense*, 1964).

In 1905 Einstein formulated his insights largely in ignorance of the most recent results of Lorentz and Poincaré, and treated space and time separately, rather than combined into space-time. But, since Einstein's new kinematics mixed both spatial *and* temporal coordinates in the transformation from one frame to another, the adoption of the space-time viewpoint, once suggested, was irresistible.

Global vs. Local Time: Newtonian Identity

The Newtonian absolute time is both *global* and *local*. It is:

Global, because it can be used for defining distant simultaneity in each inertial frame, and even universal, because this definition will give the same result for two events, no matter in which inertial frame the definition is used.

Local, because it provides the readings of any good clock along its world line, and absolute because the time difference read between any two events will be the same for all world lines.

Global vs. Local Time: Special-Relativistic Splitting

In SR, the global and local concepts of time, which coincide in Newtonian kinematics, split apart:

Global Time: No matters of fact can depend on the definition of global time (see above); but various definitions may be useful in different contexts. For example, the retarded time along the light cones emanating from some world line (as utilized in the K-calculus) will give the same global time for all world lines passing through any one event, but different global times for the same event as defined by different but parallel world lines. Since it depends only on a single world line, this definition may be extended to general relativity.

The Poincaré-Einstein convention is the most useful for an inertial frame of reference. It leads to different global times for the same event as defined in different inertial frames; but all world lines in the same inertial frame define the same global time for any event. Since it depends on distant parallelism that is independent of path, it cannot be extended to general relativity.

Local Time: The concept of local time is now the proper time along any time-like world line. It is *absolute* (i.e., frame-independent) like the Newtonian time, but unlike it in being *path-dependent*. As noted above the local time is more like the spatial distance: a good clock is more like a good pedometer than previously thought.

The Moral of This Tale

Loose talk about "space" and "time" being "relative" is just that, and often leads to serious philosophical misinterpretations. To sum up the moral again. In SR:

The *global space* (fibration of S-T) and *global time* (foliation of S-T) are both relative, but *nothing physically significant* depends on them.

Local space (integral along a space-like path) and *local time* (integral along a time-like path) are absolute, but both are *path dependent*

Curvature

In N-G and S-R space-times, both the chrono-geometrical and the inertial structures are *flat*, in the sense that:

For *chronogeometry*, there is no *Gaussian curvature*, defined by the pseudo-metric and associated with any of the two-sections through a point.

For *the inertial field*, there is no *affine curvature* associated with the parallel transport of a vector through space-time. Any vector parallel-transported around any closed curve coincides with itself when it returns to its starting point.

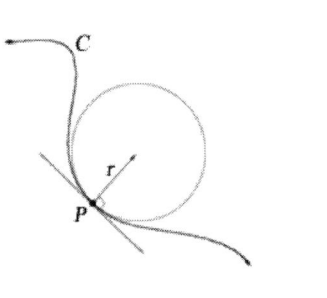
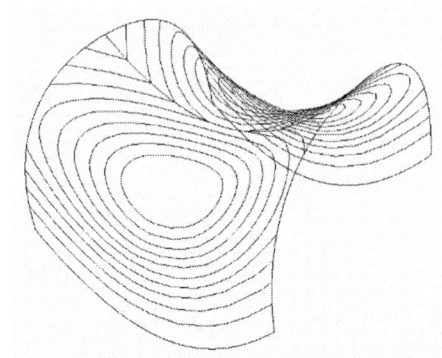

Osculating circle   Gaussian curvature

FIGURE 12.

Gaussian Curvature

The curvature of a plane curve at any of its points is the inverse of the radius of the osculating circle at that point.

At any point of a surface, each plane through that point intersects the surface in a plane curve. Take the maximum and minimum curvatures of these plane curves as the plane is varied. Their product is the Gaussian curvature of the surface at that point.

This definition depends on the embedding of the surface in Euclidean three-space. But Gauss showed that the Gaussian curvature is an intrinsic property of the surface. He proved that is can be expressed in terms of the metric components in the expression for the distance between two neighboring points of the surface in Gaussian (curvilinear) coordinates, the line element ds (*Theorema Egregium*):

$$ds^2 = g_{11}(dx^1)^2 + 2g_{12}(dx^1)(dx^2) + g_{22}(dx^2)^2.$$

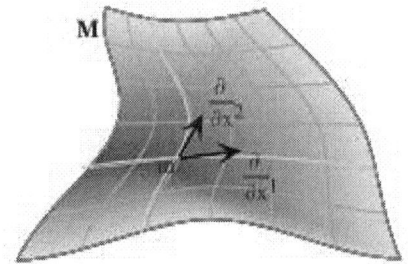

$$g_{11} = \frac{\partial}{\partial x^1} \cdot \frac{\partial}{\partial x^1} \qquad g_{12} = \frac{\partial}{\partial x^1} \cdot \frac{\partial}{\partial x^2}$$
$$g_{21} = \frac{\partial}{\partial x^2} \cdot \frac{\partial}{\partial x^1} \qquad g_{22} = \frac{\partial}{\partial x^2} \cdot \frac{\partial}{\partial x^2}$$

FIGURE 13.   Line element

Locally, this line element expresses Euclidean geometry. It is just Pythagoras' Theorem expressed in curvilinear coordinates. The Riemann curvature tensor generalizes Gaussian curvature to a space of any number of dimensions.

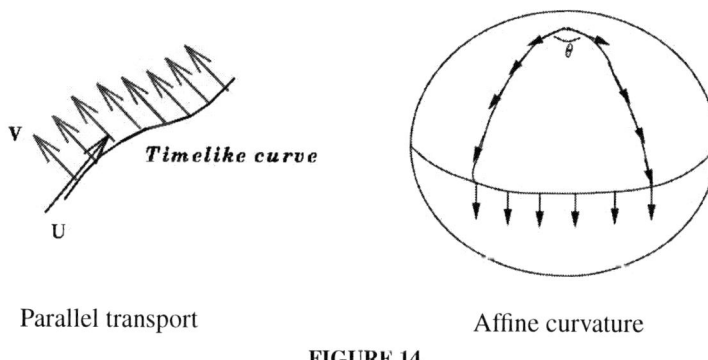

Parallel transport   Affine curvature

FIGURE 14.

Affine Curvature

Given a vector at any point in an affinely-connected space, the connection enables us to define the vector parallel to it at a neighboring point. By iterating this procedure, we may *parallel transport* a vector along any curve: What happens to a vector when it is transported parallel to itself around a closed curve? If there are any closed curves, for which the parallel-transported vector does not coincide with the original vector, one says the space is affinely curved. By taking a set of infinitesimal closed curves, one can define the components of the *affine curvature tensor*.

Equation of geodesic deviation

The affine curvature tensor has another application that is especially important for its physical interpretation. Consider an infinitesimal displacement vector connecting two neighboring affinely straight lines (i.e., curves such that the tangent vector field along the curve is parallel transported into itself). The affine curvature tensor is a measure of how this displacement vector changes as a function of the affine parameter as we proceed along the two straight lines. If the displacement vector change is accelerated, then the affine curvature has a non vanishing component related to the direction of the lines and of the displacement vector.

Physically, an affine straight line corresponds to the path of a freely falling body, and the equation of geodesic deviation measures whether there is any relative acceleration between two nearby freely-falling bodies. Mathematically, the amount of such relative acceleration in various directions is a measure of the components of the affine curvature tensor; physically, it is a measure of the gravitational tidal forces.

Newtonian Gravitation

Special-relativistic space-time proved sufficient for the analysis of all physical phenomena, for which gravitation may be neglected. But it must be modified to include gravitation, because of:

The Equivalence Principle

Because inertial and gravitational mass are equal, there is no (unique) way to separate the effects of inertia and gravitation on a "freely-falling" body. Once this is understood, even at the Newtonian level, gravitation can no longer be treated as an external force acting on bodies, but must be regarded as a modification of the hitherto fixed inertial structure of space-time. This structure now becomes dynamical, an *inertio-gravitational field*. While the inertial structures of both GN and SR space-times are associated with a flat affine connection, the inertio-gravitational field is associated with an affine structure that is no longer flat. The *affine curvature* associated with the Newtonian inertio-gravitational field describes *tidal gravitational forces*. This curvature obeys field equations that reduce to the field equations for the Newtonian gravitational "force" in any non-rotating frame of reference.

Although its symmetry group is enlarged to include all linearly-accelerated frames (this is the equivalence principle), the classical Newtonian chrono-geometrical structures are unmodified. The chrono-geometrical and inertio-gravitational structures remain compatible: ideal measuring rods and clocks still remain such in the presence of any inertio-gravitational field. But the compatibility conditions do not uniquely determine the inertio-gravitational field: Just enough freedom is left to introduce gravitational fields that reduce to the gradient of the Newtonian potential in non-rotating frames of reference.

General-Relativistic Space-Time

But special-relativistic chrono-geometry is no longer compatible with the dynamical inertio-gravitational field: The flat Minkowski metric is not compatible with the non-flat affine structure. To restore compatibility, the chrono-geometry must be modified: The pseudo-metric must become a non-flat, dynamical field that plays a dual role. In addition to determining the chrono-geometry, it also serves as the potentials that uniquely determine the inertio-gravitational field.

While the inertio-gravitational field traditionally was derived from the chrono-geometry, we favor the modern approach, which treats both as logically independent before the imposition of the field equations. One set of field equations then relates the inertio-gravitational field to all other matter and fields (the sources) by equating the contracted affine curvature tensor of the inertio-gravitational field to the stress-energy tensor of the sources. The other set of field equations are the compatibility conditions imposing the unique relation between chrono-geometry and inertio-gravitational field.

To succeed in formulating the special theory, Einstein had to attach physical significance to the coordinate system. To succeed in formulating general relativity, Einstein had to learn that coordinates have no inherent physical significance (see discusson below).

PHILOSOPHICAL SECTION

Two Concepts of Space: Absolute vs. Relational

Historically, since (at least) ancient Greek times, there has been a conflict between two views of the nature of space:

- The *absolute* concept: Space is a container, in which matter moves about. This view was espoused by Demokritos (and the Greek and Roman atomists):

> By convention are sweet and bitter, hot and cold, by convention is color; in truth are atoms and void" (Fr. 589, ca. 430 BCE. G. S. Kirk and J.E. Raven, *The Presocratic Philosophers*, 1957)

This is sometimes paraphrased as "Nothing exists but atoms and void. All else is mere opinion." Aristotle criticized the concept of a void:

> The believers in its reality present it to us as if it were some kind of receptacle or vessel, which may be regarded as full when it contains the bulk of which it is capable, and empty when it does not (*Physica*, Book VI).

- The *relational* concept: Space has no independent existence. It is just a certain set of positional relations between material entities. There cannot be a vacuum –the world is a *plenum*. Aristotle's doctrine is really a doctrine of *place* rather than *space*.

> The physicist must have a knowledge of Place ... because 'motion' in its most general and primary sense is change of place, which we call 'locomotion'... the motions of simple bodies (fire, earth, and so forth) show not only that place is something but that place has some kind of power [*dunamin*] (*Physica*, Book IV).

Aristotelianism triumphed and atomism vanished from the Western philosophical tradition for almost two millennia. With its revival in early modern times and subsequent adoption by Newton, the conflict between the absolute and relational concepts was renewed in the 17th and 18th centuries in the battle between Newtonianism and Cartesianism (the philosophy of Rene Descartes). As Voltaire wittily observed:

> A Frenchman who arrives in London, will find philosophy, like everything else,

| Demokritos | Aristotle | Voltaire | Newton | Leibniz |

FIGURE 15.

very much changed there. He had left the world a plenum, and he now finds it a vacuum (*Lettres philosophiques.*, ca. 1778, "Letter XIV, On Descartes and Sir Isaac Newton").

This time it was the absolute, Newtonian conception of space that triumphed in spite of the cogent arguments of Leibniz and Huygens against it:

> In fine, the better to resolve, if possible, every difficulty, he [Newton] proves, and even by experiments, that it is impossible there should be a plenum; and brings back the vacuum, which Aristotle and Descartes had banished from the world (*ibid.*, "Letter XV, On Attraction").

As Euler emphasized, absolute space seemed to be necessary if one wanted to use Newtonian dynamics (*Réflexions sur l'espace et le temps*, 1748).

Absolute versus Relational Concepts of Space and Time

Einstein summarized the situation in these words:

> Two concepts of space may be contrasted as follows:
> (a) space as positional quality of the world of material objects;
> (b) space as container of all material objects.
> In case (a), space without a material object is inconceivable. In case (b), a material object can only be conceived as existing in space; space then appears as a reality which in a certain sense is superior to the material world. ("Foreword" to Max Jammer, *Concepts of Space*, 1954).

Things versus Processes

The old emphasis on space and time favors the concept of *things*, which occupy regions of space at moments of time, but changing over (absolute) time. The new emphasis on space-time favors the concept of *processes*, which occupy regions of space-time, or even – with the development of the field concept – all of space-time. (*Events* are then defined as limiting case of processes, occupying vanishingly small regions of space-time.) We must now discuss the extension of our previous discussion of things in space to processes in space-time.

Two Concepts of Space-Time: Absolute vs. Relational

The *absolute*: Space-time is an independent container, in which processes take place. In addition to ponderable matter, such processes now include fields (e.g., the electromagnetic field) that may fill all space-time.

The *relational*: Space-time has no independent existence. It is just as certain set of relations between the elements of processes. There cannot be an empty space-time.

Pre-general Relativistic Situation

In the case of both Galilei-Newtonian and special-relativistic space-times, it was possible to hold either of these viewpoints although there were serious problems for the relational viewpoint:

1) The possible existence of regions of space-time that are devoid of all matter and fields.
2) The fact that the space-time structures remain the same, regardless of all the varying physical processes that can take place within them.
3) The fact that the space-time structures influence all physical processes (e.g., through the law of inertia), but are not influenced by them. This is a general problem with any fixed, background structures introduced into physics.

Fixed, Background Space-Time Structures

In the case of space-time, we refer to such structures as fixed, background space-time structures. Thus, we may sum up our previous discussion by saying that both GN and SR theories are based on fixed, background space-time structures.

Theories with background space-time structures have a kinematics that is logically prior to and independent of all dynamical physical theories. The slogan is: *Kinematics first, then dynamics!* The background space-time is a stage, upon which various dynamical dramas can be enacted.

Such background space-time structures are essential features of all current quantum theories:

GN space-time is the stage for the quantum mechanics of non-relativistic quantum systems.

SR space-time (Minkowski space) is the stage for relativistic quantum field theories as all thoughtful workers on the subject recognize:

> The basic concept[s] of the theory are quantum fields defined on space-time, not particles. Space-time is assumed to be a four-dimensional real vector space with given metrical properties and Einstein causality, such that the Poincaré group (constituted by translations and Lorentz transformations) is implied as a symmetry group. This space-time structure fixed in advance – called Minkowski space – forms the register for recording physical events. The predictions of a relativistic quantum field theory on the outcome of scattering processes are of probabilistic nature, in this respect similar to those of (non-relativistic) quantum mechanics. However, a novel feature occurs: in these processes particles can be created and annihilated. The quantum fields, in terms of which the theory is constructed, are operators that depend on space-time and act on the space of physical state vectors.(Hans Günther Dosch, Volkhard F. Müller and Norman Sieroka, "*Quantum Field Theory, Its Concepts Viewed from a Semiotic Perspective*", 2004).

In short, both the formalism of quantum field theory and the measurement processes that test its predictions presuppose the SR (Minkowski) space-time structure.

The General-Relativistic Revolution - The Triumph of Relationalism

As discussed above, in the general theory of relativity, both the inertio- gravitational and the chrono-geometrical structures are dynamical fields. We speak of such theories, which are free of any background space-time structures, as background free: In a background-free theory, with no non-dynamical structures, kinematics and dynamics cannot be separated. The slogan is: *No Kinematics Without Dynamics!!!!*

The problems for the relational viewpoint discussed above now disappear:

1) There are no "empty" regions of space-time: Wherever there is space and time (chrono-geometric structure), there is always (at least) an inertio-gravitational field (affine structure).
2) The space-time structures are not independent of the processes taking place within them. Chrono-geometry and inertio-gravitation are dynamical fields, obeying field equations that couple them to each other and to all other physical processes.
3) Thus, there is now reciprocal interaction between space-time and other processes. Physical processes do not take place *in space-time*. Space-time is just *an aspect of the totality of physical processes*.

General relativity more-or-less forces one to adopt the relational viewpoint.

> On the basis of the general theory of relativity ... space as opposed to 'what fills space' ... has no separate existence. If we imagine the gravitational field ... to be removed, there does not remain a space of the type [of the Minkowski space of SR], but absolutely nothing, not even a 'topological space' [i.e., a manifold]... There is no such thing as an empty space, i.e., a space without field. Space-time does not claim

Rosenfeld   Bronstein
FIGURE 16.

existence on its own, but only as a structural quality of the field (Einstein,"Relativity and the Problem of Space," in *Relativity: The Special and the General Theory*, 1952 edition).

THE PROBLEM OF QUANTUM GRAVITY

The greatest challenge to theoretical physics today is: How to invent a theoretical structure that encompasses both Quantum Field Theory (background-dependent) and General Relativity (background-independent)? "That is the Question."

Quantizing General Relativity

In 1916, Einstein stated that general relativity would require a quantum version for the same reason that electromagnetism did: A gravitationally bound system would ultimately radiate away all its energy unless it was quantized.

The earliest attempts to apply the methods of QFT, developed by Heisenberg and Pauli in the late 1920s, to GR came in the early 1930s, first by Leon Rosenfeld (see John Stachel, "The Early History of Quantum Gravity," in Bala Iyer and Biplap Bhawal, eds, *Black Holes, Gravitational Radiation and the Universe*, 1998, pp. 525-534). The basic philosophy behind this work was that only technical difficulties (the non-linearity of the field equations) stand in the way of application of standard methods of QFT to GR, and the way to begin was by quantizing the linearized approximation to the field equations.

In the 1930s, only one physicist realized that such attempts raised profound conceptual problems due to the unique features of gravitation as compared to electromagnetism: Matvei Petrovich Bronstein (see Gennady Gorelik, "First Steps of Quantum Gravity and the Planck Values," *Studies in the history of general relativity* [*Einstein Studies*, vol. 3], 1992, pp. 364-379). He was the only serious contender with Lev Davidovich Landau for leadership of Soviet theoretical physics. Both were imprisoned during the Stalinist purges of the mid-1930s: Landau survived, Bronstein perished.

In formal quantum electrodynamics, which does not take into consideration the structure of the elementary charge, there is no consideration limiting the increase of density .... With sufficiently high charge density in the test body, the measurement of the electrical field may be arbitrarily precise. In nature, there are probably limits to the density of the electrical charge... but formal quantum electrodynamics does not take these limits into account .... The quantum theory of gravitation represents a quite different case: it has to take into account the fact that the gravitational radius of the test body ... must be less than its linear dimensions ... The elimination of the logical inconsistencies connected with this requires a radical reconstruction of the theory, and in particular, the rejection of a Riemannian geometry dealing, as we see here, with values unobservable in principle, and perhaps also the rejection of our ordinary concepts of space and time, modifying them by some much deeper and nonevident concepts. *Wer's nicht glaubt, bezahlt einen Taler* ["*Let him who does not believe it pay a dollar*" – finale of a Grimm fable]. (Bronstein, *Quantentheorie schwacher Gravitationsfelder*, 1936)

There is no place here to say more about the history of quantum gravity (for the later history, see Carlo Rovelli, "Appendix B History" in *Quantum Gravity*, 2004).

But it is relevant to note that the conflict between those who see only technical problems in the application of existing techniques of QFT to general relativity and those who see profound conceptual issues in the reconciliation of quantum theory and general relativity, which started in the 1930s, continues to this day.

Background-Dependence versus Background-Independence

The first viewpoint is represented today mainly by people from the quantum field theory community. Their approach is basically to keep a background space-time (of however many dimensions), and somehow incorporate general relativity into the quantum formalism developed using this background structure. Currently, the strongest candidate put forward by advocates of this approach is string theory, or some variation or extension of it such as the elusive M-theory.

The second viewpoint is represented today mainly by people from the general relativity community. Their approach is to try to develop a background-independent formulation of quantum theory and apply it to general relativity. Currently, the strongest candidate put forward by advocates of this approach is loop quantum gravity (LQG), or some extension of it such as spin-foam theory. Without going into any details, I want to emphasize the importance of the connection in the LQG program. In contrast to previous approaches to canonical quantization, such as geometrodynamics, which took the metric as primary, the most important achievements of LQG are based on taking a particular form of the connection as primary.

Being from this community myself, it is natural that I favor the background-independent approach, but have a number of critical reservations about how it is currently carried out (see John Stachel, "Structure, Individuality and Quantum Gravity," Steven French,

Dean Rickles and Juha Saatsi, eds., *The Structural Foundations of Quantum Gravity*, to appear). But, as I shall emphasize, there are people in the string community who also favor this approach.

A New Formal Principle?

None of the current approaches has been completely successful in solving the basic problem of quantum gravity: the reconciliation of QFT with GR. In 1905, Einstein faced a similar situation in his attempts to reconcile Newtonian mechanics with Maxwell's electrodynamics. As he said much later:

> Gradually I despaired of the possibility of discovering the true laws by means of constructive efforts based on known facts. The longer and more desperately I tried, the more I came to the conviction that only the discovery of a universal formal principle could lead us to assured results (*Autobiographical Notes*, 1949)

Consideration of a striking common feature of QFT and GR has led me to propose a new formal principle that might serve as a guide in the further quest for a theory of quantum gravity, whatever direction(s) it may take.

From General Covariance to Permutation Invariance

What is the significance of the general covariance of the field equations of general relativity? If general covariance is given an active interpretation (as it should be – coordinate transformations can never have a direct physical significance), it requires invariance of the field equations under the diffeomorphism group acting on the underlying differentiable manifold M of space-time points. But what are diffeomorphisms? A little thought shows that they are just fancy permutations (automorphisms) of the homogeneous elements of M – permutations that are required to be continuous and differentiable because they act on the elements of a differentiable manifold – but permutations nevertheless.

I said above "the homogeneous elements of M", and this is an important part of the meaning of general covariance: the elements of M are not distinguished from each other unless and until some solution to the field equations is specified. Einstein's 1913 hole argument against general covariance was based on the tacit assumption that, just as in SR, the points of the space-time manifold could be individuated independently of the field, and it was only his realization in late 1915 that this assumption was untenable in a background-independent theory that enabled him to justify his adoption of generally covariant field equations (see John Stachel, "Einstein's Search for General Covariance, 1912-1915" in *Einstein and the History of General Relativity*, 1989, pp. 63-100). Indeed it is this question of individuation that distinguishes algebra from geometry.

Geometry vs Algebra

A *geometry* consists of a set of elements, together with some relations between them, such that all of the elements are homogeneous under the group of automorphisms (permutations) that preserves all the relations. In such a case, since the relations are primary, one may speak of "The things (elements) between the relations"

Example: Euclidean plane geometry, a manifold homeomorphic to R^2, together with the group of translations and rotations acting on the points of the manifold.

An *algebra* consists of a set of elements, together with some relations between them, such that each element is individuated independently of the relations between it and the other elements. In such a case, since the elements are primary, one may speak of "The relations between things (elements)."

Example: The plane rotation group, each element of which is characterized by an angle.

A *representation* of an abstract space (geometry) is called *algebraic* if it characterizes the space by means of some *coordinatization* of its elements (points). A coordinatization is a one-one correspondence between the elements of an algebra and those of a geometry. Since any one coordinatization individuates the otherwise homogeneous elements of a geometry, the only way to keep them homogeneous is to demand invariance of any geometrically significant result under *all* admissible coordinatizations. These concepts of geometry, and coordinatization are due to Hermann Weyl (see *The Classical Groups*, 1939). This concept of algebra is due to I. R. Shafarevich (see *Basic Notions of Algebra*, 1997).

The coordinatization of a differentiable manifold is generally a local operation, since usually, no one coordinatization can cover the entire manifold. One must carefully distinguish between coordinate transformations (re-coordinatizations of the differentiable manifold), which are local, passive mathematical operations, needed to ensure that the points of the manifold remain homogeneous, and having no physical significance; and the diffeomorphisms of the manifold, which are global, active point transformations of great potential physical significance as we shall see.

Background-Independent Theories and Diffeomorphisms

In a background-independent theory, there are no non-dynamical relations to be preserved on the set of space-time points; so all possible permutations of the points of space-time are permissible. If one adds the demand that these permutations be continuous (because space-time is a manifold) and differentiable (because it is a differentiable manifold), one gets the diffeomorphism group. As noted above, in GR the points of space-time have no inherent properties that individuate them. GR is a background independent theory, or in my terminology a "things-between relations" theory.

The Principle of Maximal Permutability

FIGURE 17.
S. MacLane
Courtesy of the University
of Chicago News Office

One can thus express the concept of general covariance in GR in the following form: The theory (GR) shall be invariant under all possible permutations of the basic entities of the theory (elements of space-time in GR) in the sense that any model of the theory (solution to the field equations of GR) shall be physically equivalent to any other model that results from it by such a permutation. In this form the principle for diffeomorphisms in general relativity can be both *generalized* and *abstracted*.

Generalization: "Generalization from cases refers to the way in which several specific prior results may be subsumed under a single more general theorem" (Saunders MacLane, *Mathematics, Form and Function*, 1986).

One can generalize the principle of diffeomorphism invariance from the pseudo-metric tensors and affine connections of general relativity to arbitrary geometric object fields, also called natural objects (see John Stachel and Mihaela Iftime, *Fibered Manifolds, Natural Bundles, Structured Sets, G-Sets and All That: The Hole Story From Space Time to Elementary Particles*, 2005; *The Hole Argument for Covariant Theories*, 2005)

Abstraction: "Abstraction by deletion ... One carefully omits parts of the data describing the mathematical concepts ... to obtain the more abstract concept" (Saunders MacLane, *ibid.*).

By dropping the assumptions of differentiability, one can extend the principle from theories based on differentiable manifolds to those based on *topological manifolds*; and by dropping the assumption of continuity the principle can be extended to theories based on *sets of discrete elements*.

Even if the concepts of space, time and space-time have to be greatly modified; or are themselves explained in terms of some more fundamental entities in some future theoretical advance, it is hard to believe that one would retreat from the relational to the absolute point of view concerning the fundamental entities, whatever their nature. This suggests adoption of the principle of *maximal permutability of the fundamental constituents* as a "universal formal principle" in Einstein's sense as a heuristic guide in the search for a theory of quantum gravity – and even beyond.

Elementary Particles, Field Quanta

The heuristic force of this principle is reinforced by the observation that, like the points of space-time, the particles of non-relativistic QM and the field quanta of special-relativistic QFT also lack inherent individuality and hence obey the principle. They are only individuated (to the extent that they are) by some process (Feynman's word) or phenomenon (Bohr's word), in which they are involved. In any quantum system in non-relativistic QM, both the bosons and the fermions of any species can be arbitrarily permuted among themselves without changing the probability amplitude for any process; so, like the points of space-time, they are also "things between relations". And in QFT, the field quanta in any Fock space state are also completely indistinguishable.

A Background-Independent String Theory?

As currently constituted, string theory is based on a fixed, background space time structure on a manifold of some number of dimensions higher than four. Regardless of the details of particular models, it is clear that the principle of maximal permutability is violated: Only diffeomorphisms that are symmetries of the fixed, background structure (usually a flat pseudo-metric) are permissible permutations of the elements of the manifold.

FIGURE 18. B. Greene
Photograph by Robert Birnbaum.

Many string theorists are aware of this problem. Brian Greene recently presented an appealing vision of how a background-free string theory might look, but he emphasized how far string theorists still are from realizing this vision (*The Fabric of the Cosmos – Space, Time, and the Texture of Reality*, 2004):

Since we speak of the "fabric" of spacetime, maybe spacetime is stitched out of strings much as a shirt is stitched out of thread. That is, much as joining numerous threads together in an appropriate pattern produces a shirt's fabric, maybe joining numerous strings together in an appropriate pattern produces what we commonly call spacetime's fabric. Matter, like you and me, would then amount to additional agglomerations of vibrating strings – like sonorous music played over a muted din, or an elaborate pattern embroidered on a plain piece of material – moving within the context stitched together by the strings of spacetime. ... [A]s yet no one has turned these words into a precise mathematical statement. As far as I can tell, the obstacles to doing so are far from trifling. .... [T]o make sense of this proposal, we would need a framework for describing strings that does not assume from the get-go that they are vibrating in a preexisting spacetime. We would need a fully spaceless and timeless formulation of string theory, in which spacetime emerges from the collective behavior of strings... Many researchers consider the development of a background-independent formulation to be the single greatest unsolved problem facing string theory.

Einstein's Greatest Contribution?

For reasons discussed above, I have been led to conjecture that, whatever form a future fundamental physical theory (such as some version of quantum gravity, or something even farther from our current conceptual framework) may take, there will be no absolute elements in it. Rather, its basic entities – whatever their nature – will be embedded in some discrete or continuous relational structure: The result will be a completely background-independent physics.

If I am proved right, then a millennium from now (assuming humanity still exists and has not relapsed into barbarism) then Einstein's greatest contribution to physics will be regarded as the development of the first, prototype background-independent physical theory!

As I indicated earlier, it is always dangerous to try to predict the future; still, I can draw wry comfort from the fact that – right or wrong – I shall not be around when my prophecy is finally tested. But I hope some of you may.

ACKNOWLEDGMENTS

I thank Lysiane Mornas for all the careful work put into the preparation of the written version of this paper.

Brane-world Cosmology

David Wands

Institute of Cosmology and Gravitation, University of Portsmouth, Mercantile House, Portsmouth P01 2EG, United Kingdom

Abstract. Brane-world models, where observers are restricted to a brane in a higher dimensional spacetime, offer a novel perspective on cosmology. I discuss some approaches to cosmology in extra dimensions and some interesting aspects of gravity and cosmology in brane-world models.

Keywords: Cosmology, extra dimensions, branes, gravity
PACS: 98.80.Cq, 98.80.Jk, 04.50.+h

COSMOLOGY AFTER EINSTEIN

A century after Einstein first proposed his theory of relativity, it has become a cornerstone of the physical sciences. Four dimensional spacetime provides the setting for describing physical processes and in particular provides the dynamical framework for cosmological models of our expanding Universe.

It was the general theory of relativity, proposed by Einstein in 1915, that for the first time provided equations with which to describe the dynamics of spacetime. Einstein's equation

$$G_{AB} + \Lambda g_{AB} = 8\pi G_N T_{AB}, \tag{1}$$

relates the intrinsic curvature, G_{AB}, of the metric, g_{AB}, to the local energy-momentum, T_{AB}, while allowing for the possibility of a non-zero cosmological constant, Λ. Consistency with Newtonian gravity in the weak-field, slow-motion limit is ensured by the appearance of Newton's constant, G_N, in the constant of proportionality.

In much of modern cosmology Einstein's tensor equation (1) conveniently reduces to the Friedmann constraint equation

$$3\left(H^2 + \frac{K}{a^2}\right) = \Lambda + 8\pi G_N \rho, \tag{2}$$

which relates the Hubble expansion, H, and spatial curvature K/a^2, of a homogeneous and isotropic Friedmann-Robertson-Walker (FRW) spacetime to the local energy density ρ.

Homogeneous and isotropic expansion has been used to build up a remarkably successful model for the evolution of our Universe starting with a hot Big Bang at a finite time in our past. This model has been tested not only by qualitative features such as the evolution of galaxy populations and the existence of a cosmic microwave background (CMB) radiation, but also quantitatively tested by comparing models of primordial nucleosynthesis with abundance of light elements.

One only needs to consider linear perturbations about a homogeneous and isotropic metric to build up a coherent picture of the formation of structure in our Universe. Small fluctuations, about one part in a hundred thousand, are observed in the temperature of the microwave background radiation and indicate the existence of small primordial perturbations in the distribution of matter and radiation in the early universe when the CMB last scattered, about 300,000 years after the Big Bang. These primordial density fluctuations provide the seeds around which the observed large-scale structure of our Universe can form simply by gravitational instability, in a cosmological model with appropriate contributions from radiation, baryonic matter, as well as cold dark matter and some form of dark energy, that behaves very much like Einstein's cosmological constant today. A wealth of observational data now enables cosmologists to put this basic picture to the test and attempt to measure parameters such as the density of different forms of matter, the nature of the primordial perturbations, and Einstein's gravitational laws.

At the same time fundamental questions remain unanswered. Why are there 3 large spatial dimensions (not 5 or 15)? why is the value of the cosmological constant so small? and what really happens at the initial Big Bang which represents a singular point at the start of our cosmological evolution? I cannot answer these questions in this talk, but I can show how brane-world models offer some novel and interesting perspectives on these issues. I should emphasize that this is a personal view and not intended to be a systematic review of all aspects of brane-world cosmology. For a more comprehensive review see [1].

EXTRA DIMENSIONS

Superstring theory is an attempt to unify gravity with the other fundamental interactions in a self-consistent quantum theory, based on strings (extended 1-dimensional objects) as the fundamental constituents of matter rather than point particles. In particular string theory should be finite and singularity free.

For example, the existence of a minimal length scale in the effective theory leads to a "T-duality" that relates expanding and contracting cosmological solutions and has been proposed as the basis for the pre-Big Bang scenario [2] that proposes a pre-Big Bang era preceding the hot Big Bang expansion. Unfortunately the nature of the transition from pre- to post-Big Bang is dependent on the nature higher-order, possibly non-perturbative, effects and remains elusive. This makes it hard to make robust predictions based on a pre-Big Bang phase.

It is not fair to say that string theory does not make any predictions. String theory does make a definite prediction for the number of spacetime dimensions. Spacetime should have 10 dimensions for a consistent, anomaly-free superstring theory [3]. This may not appear to be a huge success for the theory, but of we can only assert that there are four *observable* dimensions and it is quite possible that there exist *extra* dimensions that are very small and/or unobservable.

Only a few years after Einstein proposed his theory of dynamical four-dimensional spacetime, Kaluza began to consider the dynamical equations for a five-dimensional spacetime, realising that the degrees of freedom of the metric associated with the extra dimension could describe a vector field in our four-dimensional world [4]. If the extra

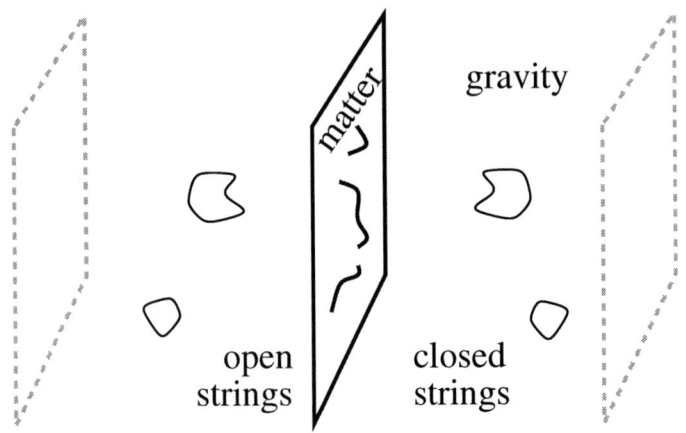

FIGURE 1. Matter fields are described by open strings confined to branes.

dimension is compact and very small, less than 10^{-19} m say, then only the zero-mode of the metric, or other fields, would be excited in terrestrial experiments. Higher harmonics in hidden dimension(s) correspond to very massive states, requiring large energies to excite them, and these can be consistently set to zero in a low-energy effective action.

In time it was realised that the size of the extra dimension was itself a scalar field and higher-dimensional models of gravity reduce to an effective scalar-tensor theory of gravity in four-dimensions at low energies. To avoid conflict with experimental tests of gravity the size of the extra dimensions must be fixed, but there has been little progress on how to stablise all the moduli fields describing the size and shape of hidden dimensions in string theory, until recently.

This "hosepipe view" of the extra dimensions being rolled up incredibly small and hence out of sight was almost universally adopted to deal with the embarrassment of extra dimensions in string theory until the 1990's. What changed in the mid 1990's was that realisation that other extended objects, higher-dimensional membranes, or "branes", should also play a fundamental role in string theory [5]. Branes opened up the possibility to related apparently different string theories, for instance string theories containing closed strings or those with open strings.

Branes can support open strings whose end-points lie on the brane. These open strings can describe matter fields which live on the brane. On the other hand perturbations of the higher dimensional bulk geometry are described by excitations of closed strings, such as the graviton. To a general relativist it should be clear that even if matter fields are restricted to a lower dimensional hypersurface, gravity as a dynamical theory of geometry must exist throughout the spacetime.

This lead several authors to consider the possibility that at least some of the extra dimensions could be far larger than had previously been imagined [6, 7]. They realised that while particle interactions are probed by high energy colliders on energies up to 1 TeV, and hence scales down to 10^{-19} m, gravity is barely tested on scales below 1 mm. If the extra dimensions were testable only via gravity then they might be relatively large. This offers a tantalising explanation for why gravity appears to be so weak when compared with the other interactions. The gravitational field of an object could leak out into the large but hidden dimensions and gravity in our four-dimensional world seems weaker.

To make this a little more precise, consider the gravitational field of a mass M in a D-dimensional spacetime. If we use Gauss's law to calculate the gravitational field strength g at a distance r then we find $g \propto G_D M/r^{D-2}$ for distances $r \ll R$, the radius of compactification of the hidden dimensions. But if $r \gg R$ then the gravitational field strength is given by

$$g = \frac{4\pi G_D M}{4\pi r^2 R^{D-4}}. \tag{3}$$

The effective value of Newton's constant in our apparently 4-dimensional world, G_4, can be identified as

$$G_4 \equiv \frac{G_D}{R^{D-4}}. \tag{4}$$

Given we observe only the four-dimensional effective gravitational coupling, from which we infer a very large effective Planck scale $M_4 = 1/\sqrt{G_4} = 10^{19}$ GeV, the true value of the Planck scale (the scale at which quantum gravity becomes important) could be much smaller in models with large extra dimensions.

For instance, Horava and Witten in 1996 [8] proposed a supergravity model in 11-dimensions with a fundamental Planck scale close to the Grand Unified (GUT) scale of 10^{16} GeV where one of the extra dimensions had a size considerably larger than the conventional Planck scale of 10^{-35} m. But the GUT scale is still far beyond terrestrial experiments and established particle physics models. What if quantum gravity was within reach of experiments like the LHC at CERN? If one hidden dimension was as large as 1 mm then the Planck scale could be as low as 10^8 GeV. With two large extra dimensions, the Planck scale could be as low as 1 TeV [7].

RANDALL-SUNDRUM MODEL

So far I have implicitly been discussing Minkowski branes in a higher dimensional Minkowski spacetime. This provides a good vacuum state for string theory but we need to go beyond flat spacetime to provide a cosmological model. Anti-de Sitter (AdS) spacetime, that is maximally symmetric space with a negative cosmological constant, $\Lambda^2 = -6k^2$, can also provide a useful vacuum state for string theory. This may not appear to be very promising for a cosmological model as a negative cosmological constant leads to a cosmological collapse and big crunch in homogeneous and isotropic cosmologies. However it turns out to be a fascinating spacetime in which to consider brane-world cosmology.

Randall and Sundrum produced two papers [9, 10] in 1999 which have had a huge impact in string theory and cosmology. They considered gravity on constant tension branes embedded in five-dimensional anti-de Sitter spacetime.

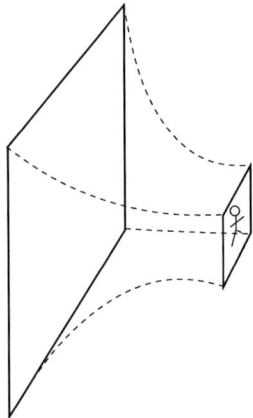

FIGURE 2. Randall Sundrum 1, where the hierarchy between the Planck scale and the TeV scale is due to the distance between two branes in compact AdS spacetime.

Branes can be embedded at fixed y-coordinate in a Gaussian normal coordinate system where the AdS_5 metric is written as

$$ds^2 = e^{-2k|y|}\eta_{\mu\nu}dx^\mu dx^\nu + dy^2. \qquad (5)$$

The exponential "warp factor" means that the volume of the extra dimensional space becomes small at large y. In their first paper [9] Randall and Sundrum showed that the large hierarchy between a fundamental TeV scale and the apparent Planck scale 10^{19} GeV could be "explained" by a large warp factor even if the size of the extra dimension (specifically the normal distance between branes) was relatively small. But in their second paper [10] they showed that even if there was no second brane, and the extra dimension extended to infinity, gravity remained effectively localised on a single brane as the integrated volume remained finite as $y \to \infty$. This they proposed as an "alternative to compactification".

The two-brane model [9], called RS1, is not so different from earlier attempts to compactify the hidden dimensions, other than that it operates in a curved bulk spacetime. It is still the large volume of the hidden space that makes gravity weaker on the brane than other forces. There is still a discrete spectrum of Kaluza-Klein states corresponding to higher harmonics on the hidden space, although the spectrum of eigenvalues is different in a curved space. And the size of the extra dimension, the distance between the two branes remains a scalar degree of freedom, known as the radion. It still leads to an effective scalar-tensor gravity in four dimensions at low energies [11, 12] which may be in conflict with experimental tests unless the radion is stabilised.

On the other hand, the one-brane model [10], inevitably known as RS2, offers a radically different model of dimensional reduction. The radion field in the RS1 model, decouples from gravity on the remaining brane in the limit that the second brane tends to spatial infinity. (The Brans-Dicke parameter $\omega \to \infty$ [11, 12].) And the discrete spectrum of KK modes is replaced with a continuum of bulk modes. However the lightest

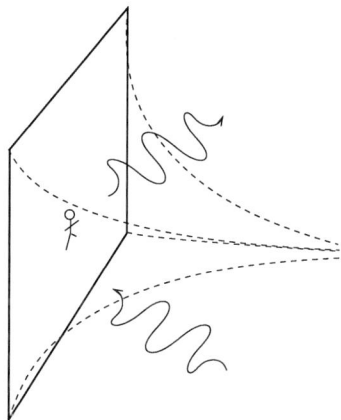

FIGURE 3. Randall Sundrum 2, where there is only one brane embedded in non-compact 5D spacetime.

modes are only weakly coupled to matter on the brane and gravity remains effectively four-dimensional on length scales greater than the AdS curvature scale, k^{-1}. More fundamentally though the single brane in AdS is an open system now where the initial state of matter on the brane (or branes) is not enough to determine the future evolution of the system. Instead one needs to specify initial data on a Cauchy hypersurface in the bulk. For example one might specify the AdS incoming vacuum state [13]. And fields on the brane can radiate into the bulk and information can escape to future null infinity.

In either of the RS models there is a simple and novel interpretation of our cosmological expansion. In the curved anti-de Sitter bulk spacetime (5) any motion of the brane, represented by a time-dependent trajectory $y = y_b(t)$, induces an FRW metric on the brane with scale factor $a = e^{-k|y_b|}$ [14, 15]. Cosmological expansion on the brane corresponds to motion in a curved bulk spacetime.

BRANE-WORLD GRAVITY

One way to understand the gravitational theory on a brane, such as the Randall-Sundrum branes in AdS, is to use the projected Einstein equations on the brane [16, 1]. Consider a codimension-one brane with unit normal vector n^A. The induced metric on the brane is then

$$g_{AB} = {}^{(5)}g_{AB} - n_A n_B, \qquad (6)$$

and the extrinsic curvature of the brane is

$$K_{AB} = g^{AC}\,{}^{(5)}\nabla_C n_B. \qquad (7)$$

The 4D Riemann tensor on the brane can be given in terms of the 5D Riemann tensor in the bulk and the brane's extrinsic curvature as [1]

$$R_{ABCD} = {}^{(5)}R_{EFGH}\, g_A^E g_B^F g_C^G g_D^H + 2K_{A[C}K_{D]B}. \qquad (8)$$

The higher-dimensional Einstein equations (1) determine the bulk Einstein tensor in terms of the bulk energy-momentum tensor. In the case of a vacuum bulk with only a cosmological constant we have

$$^{(5)}G_{AB} + \Lambda_5\,^{(5)}g_{AB} = 0. \tag{9}$$

The Israel-Darmois junction conditions determine the jump in the extrinsic curvature tensor across the brane in terms of the energy-momentum tensor localised on the brane where κ_5^2 is the gravitational coupling constant in 5-dimensions. In the Randall-Sundrum model the brane is a boundary of the bulk spacetime. This is equivalent to imposing a Z_2-symmetry across the brane so that $K_{AB} = K_{AB}^+ = -K_{AB}^-$ and hence

$$K_{AB} = -\frac{\kappa_5^2}{2}\left[T_{AB}^{\text{brane}} - \frac{1}{3}T^{\text{brane}} g_{AB}\right]. \tag{10}$$

This also occurs in the Horava-Witten model where the 10D boundary branes are fixed points of the orbifold S_1/Z_2. On the other hand the HW model also admits additional branes which can move within the bulk spacetime. In this case there is an additional freedom due to the averaged extrinsic curvature, $K_{AB}^+ + K_{AB}^-$, which is not directly constrained by the energy-momentum tensor on the brane [17], but for simplicity I will assume Z_2-symmetry across the brane in the following.

Finally putting all this together we can give an expression for the Einstein tensor for the induced metric on the brane [16]

$$G_{AB} + \Lambda g_{AB} = \kappa_4^2 T_{AB} + \kappa_5^4 \mathscr{S}_{AB} - \mathscr{E}_{AB}, \tag{11}$$

where (i) $\mathscr{S}_{AB}$ and (ii) $\mathscr{E}_{AB}$ represent modifications to the standard Einstein equations (1) due to (i) terms quadratic in the brane energy-momentum tensor and (ii) the 5D Weyl tensor projected on the brane.

The 4D intrinsic curvature (8) includes terms quadratic in the extrinsic curvature of the brane, and hence, via (10), the energy-momentum on the brane. Indeed we only recover a term linear in T_{AB} in Eq. (11) if the energy-momentum tensor on the brane contains a constant part due to a constant tension or vacuum energy density on the brane, σ, so that we split

$$T_{AB}^{\text{brane}} = \sigma g_{AB} + T_{AB}. \tag{12}$$

The effective 4D gravitational coupling constant for the renormalised energy-momentum tensor, T_{AB}, in the brane-world Einstein equations (11) is then given by

$$\kappa_4^2 = \frac{\kappa_5^4 \sigma}{6}. \tag{13}$$

The effect of terms quadratic in the matter energy-momentum tensor is given by

$$\mathscr{S}_{AB} = \frac{1}{12}T T_{AB} - \frac{1}{4}T_{AC}T_B^C + \frac{1}{24}g_{AB}\left(3T_{CD}T^{CD} - T^2\right). \tag{14}$$

It is represents a high-energy correction to the brane-world Einstein equations and is typically unimportant when the matter density is much less than the brane tension, $\rho \ll \sigma$.

The brane-world cosmological constant problem

The effective cosmological constant on the brane in Eq. (11) is given by

$$\Lambda = \frac{\Lambda_5}{2} + \frac{\kappa_5^4 \sigma^2}{12}. \tag{15}$$

In contrast to our usual 4D viewpoint that the vacuum energy density should simply vanish, or be very small, in the brane-world we require instead that there is a cancellation between the 4D and 5D contributions to the vacuum energy. An intriguing possibility in the brane-world is that 4D cosmological solutions might naturally seek out fixed points in an inhomogeneous 5D spacetime with small values of the cosmological constant – called self-tuning solutions [18].

A novel twist on the cosmological constant problem is provided by the model of Dvali, Gabadadze and Porrati (DGP) [19] who pointed out that quantum loop corrections to any classical model would be expected to induce terms in the effective energy-momentum tensor on the brane proportional to the brane Einstein tensor: $\Delta T_{AB}^{\text{brane}} = \alpha G_{AB}$. In a 4D model such corrections would simply renormalise the gravitational coupling κ_4^2. But substituted into (11) the brane-world Einstein equations become quadratic in the Einstein tensor. Thus in addition to the usual vacuum solution with $G_{AB} = 0$ when $\Lambda = 0$, there is a second (non-perturbative) solution with $G_{AB} \propto \alpha^{-2} g_{AB}$. The DGP model has sparked great interest as a novel explanation of the observed acceleration of our Universe [20], in terms of modified gravity rather than "dark energy", but there remain questions over whether the self-accelerating solutions admit unstable "ghost" modes [21].

Non-local brane gravity

Equation (11) leaves only the projected 5D Weyl tensor

$$\mathscr{E}_{AB} = {}^{(5)}C_{ECFD} g_A^E g_B^F n^C n^D. \tag{16}$$

undetermined by the local energy-momentum on or near the brane. This is the tidal part of the 5D gravitational field so is only determined when one has a solution to the full 5D Einstein equations with appropriate boundary conditions.

To the brane-bound observer it may be interpreted as an effective "Weyl fluid" with energy density $\tilde{\rho}$ and 4-velocity $\tilde{u}_A$ so that [1]

$$-\mathscr{E}_{AB} = \kappa_4^2 \left[\frac{\tilde{\rho}}{3} \left(g_{AB} + 4\tilde{u}_A \tilde{u}_B \right) + \tilde{\Pi}_{AB} \right]. \tag{17}$$

Because of the symmetries of the bulk Weyl tensor, $\mathscr{E}_{AB}$ is trace-free and hence the Weyl fluid is trace-free and has be interpreted as "dark radiation" [22]. This is consistent with the Maldacena's AdS-CFT conjecture [23] which implies that the higher-dimensional gravitational field is equivalent to a conformal field theory on the boundary [24].

The 4D Bianchi identities, $\nabla^A G_{AB} = 0$, imply from Eq. (11) that the Weyl fluid's energy $\tilde{\rho}$ and momentum $\tilde{\rho} \tilde{u}_A$ obey local conservation equations on the brane, driven by the quadratic energy-momentum tensor

$$\nabla^A \mathcal{E}_{AB} = \kappa_5^4 \nabla^A \mathcal{S}_{AB}. \tag{18}$$

The evolution of the Weyl anisotropic stress, $\tilde{\Pi}_{AB}$, however cannot in general be determined from initial conditions set solely on the brane. Thus while the projected equations, and the Weyl fluid description in particular, may be useful for interpreting 5D gravity as seen on the brane, it may be of limited use in deriving solutions. This intrinsic non-locality of 4D gravity in the brane-world is the reason why so many outstanding problems remain, including the nature of black hole solutions on the brane or anisotropic cosmologies, and require higher-dimensional solutions.

FRW COSMOLOGY ON A BRANE

One, important, case in which the projected field equations are sufficient is the behaviour of 4D homogeneous and isotropic (FRW) cosmologies. In this case the maximal symmetry of 3D space requires that the Weyl anisotropic stress vanishes and the general form of the modified Friedmann equation (2) on the brane is

$$3\left(H^2 + \frac{K}{a^2}\right) = \Lambda + \kappa_4^2 \rho \left(1 + \frac{\rho}{2\sigma}\right) + \frac{m}{a^4}, \tag{19}$$

where m is an integration constant on the brane set by the initial density of the Weyl fluid or dark radiation on the brane.

In fact a generalisation of Birkhoff theorem implies that the general 5D vacuum spacetime admitting an FRW brane cosmology is Schwarzschild-Anti-de Sitter (SAdS) [14, 15]. The integration constant m in (19) represents the mass of the black hole in the SAdS spacetime (though in a compact RS1 model the singularity may lie outside the physical region of the spacetime between the branes).

The cosmological expansion described by the modified Friedmann equation (19) can be interpreted as motion of the brane in a static, but curved bulk. The static bulk metric can be written as

$$ds^2 = -f(R)dT^2 + \frac{dR^2}{f(R)} + R^2 d\Omega_K^2, \tag{20}$$

where $d\Omega_K^2$ is the line element on a maximally symmetric 3-space, curvature K, and

$$f(R) = K + \left(\frac{R}{\ell}\right)^2 - \frac{m}{R^2}. \tag{21}$$

On the other hand if we choose a Gaussian normal coordinate in which the brane is at a fixed location $y = y_b$ then the line element becomes

$$ds^2 = -n^2(\tau,\chi)d\tau^2 + d\chi^2 + a^2(\tau,\chi)d\Omega_K^2, \tag{22}$$

where the explicit forms of n and a are given in Ref. [25]. The two coordinate systems are related by a pseudo-Lorentz transformation at the brane [27]

$$\begin{pmatrix} nd\tau \\ d\chi \end{pmatrix} = \Lambda(\theta) \begin{pmatrix} \sqrt{f}dT \\ dR/\sqrt{f} \end{pmatrix}, \tag{23}$$

where

$$\Lambda(\theta) \equiv \begin{pmatrix} \cosh\theta & \sinh\theta \\ \sinh\theta & \cosh\theta \end{pmatrix} \tag{24}$$

and the Lorentz factor due to the motion of the brane in the bulk coordinates is

$$\cosh\theta = \sqrt{1 + \frac{R^2 H^2}{f}}, \tag{25}$$

and H is the Hubble expansion rate (19).

This offers a novel perspective on 4D cosmology, not least the cosmological singularity problem. For instance, Garriga and Sasaki showed that an inflating brane and its SAdS bulk can be created "out of nothing" by a de Sitter-brane instanton [28]. Others have tried to describe the big bang singularity on the brane as a singular event within a regular higher-dimensional bulk, see for example Refs. [29, 30, 31, 32, 33].

Colliding FRW branes

One scenario that has attracted much attention is the ekpyrotic model [29, 30] where the Big Bang on the brane-world is identified as a collision between branes. In the original version a hidden brane traverses the bulk and when it hits the boundary brane its kinetic energy is released, heating our observable universe and initiating the hot Big Bang.

It turns out to be possible to give a complete description in general relativity of the collision of maximally symmetric codimension-one branes (or shells) in vacuum [26], similar to that envisaged in the original ekpyrotic model. Consider the simplest case of two incoming FRW brane-worlds (a and b) coalescing at a 3D collision surface to give one outgoing brane (c) as shown in Figure 4. The intervening regions (I, II and III) are necessarily SAdS in vacua. Thus the coordinate system on each brane (20) and in each bulk region (22) are related by pseudo-Lorentz transformations of the form given in (24).

There is a simple geometrical constraint that the product of all the Lorentz transformations (24), as one completes a circuit around the collision surface, is unity [34, 26]

$$\Pi_i \Lambda(\theta_i) = 1. \tag{26}$$

The junction conditions (10) enable one to relate the corresponding jump in the extrinsic curvature across each brane to the energy density on the brane, and hence one obtains a general relativistic version of the local conservation of energy-momentum at the collision [26]

$$\rho_c \cosh\theta_c = \rho_a \cosh\theta_a + \rho_b \cosh\theta_b. \tag{27}$$

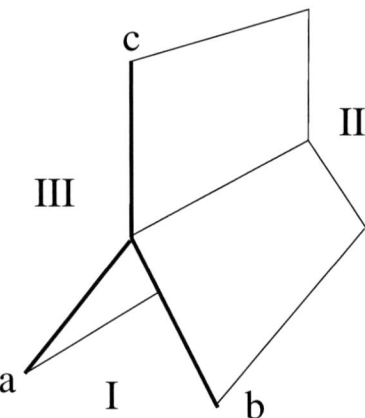

FIGURE 4. Collision between branes (a) and (b), separated by region I, resulting in single outgoing brane (c) between regions II and III.

It is remarkable that a geometrical identity (26) when combined with the general relativistic junction conditions (10) reduces to a formula (27) that is of exactly the same form as energy conservation in special relativity.

There is no curvature singularity at the brane collision and yet the collision does represent an abrupt change in the energy density and expansion rate on the brane.

In the original ekpyrotic scenario [29] the collision occurs between a moving bulk brane and a Z_2-symmetric boundary brane. The Z_2-symmetry at the boundary makes this equivalent to two symmetric bulk branes hitting the boundary at the same collision point. In principle one should also be able to calculate the spectrum of primordial perturbations inherited after the collision and thus test the model, see for instance Refs.[29, 35]. However, in the subsequent cyclic scenario [31] the collision takes place between two Z_2-symmetric boundary branes, which is equivalent to an infinite number of branes colliding at the same hypersurface with an infinite energy density. Equivalently the fifth dimension disappears at the collision. This means that the collision really is singular even from the five-dimensional viewpoint and one cannot calculate the properties of any brane-world that emerges from the collision in any conventional (e.g., general relativistic) theory. Instead one has to construct a non-singular "completion" of the theory in order to eliminate the diverges, see, for example Ref.[36].

Brane inflation

Branes have had a big impact in recent years upon attempts to construct models of inflation in the early universe within string theory. A long-standing problem has been the large number of light scalar fields, or moduli in the low-energy effective actions obtained from string theory. These would not only violate experimental tests of gravity today, as mentioned earlier, but also dilute the vacuum energy density required to drive inflation in the very early universe. But in recent years the inclusion of non-trivial antisymmetric form fields on the hidden dimensions, so-called flux compactifications [37], offer the possibility of stabilising all the moduli associated with the shape of the extra dimensions. The existence of branes plays a crucial role in providing a positive vacuum energy density in the low-energy effective theory.

Some of the resulting higher-dimensional geometries bear a close resemblance to the simple Randall-Sundrum brane-worlds. Near the branes the ten-dimensional spacetime may be distorted into a "throat" that is approximately described by $AdS_5 \times S_5$. This provides the setting, in the KKLT scenario [38], for brane-inflation [39] in which the motion of the test branes in this warped geometry plays the role of an inflaton field. The collision of the branes signals the end of inflation and can also lead to a symmetry breaking phase transition on the brane.

These brane inflation models are constructed from low-energy effective potentials in a four-dimensional effective action. The dynamics during inflation is similar to models of hybrid inflation in four dimensional Einstein gravity originally derived from supergravity models in the 1990's [40]. In particular the moving branes are test branes moving in a fixed background geometry, quite different from the self-gravitating branes in higher-dimensional gravity that I discussed earlier. It remains to be seen if it is possible to describe popular models of brane inflation in the covariant approach of Shiromizu, Maeda and Sasaki, and whether the gravitational backreaction of the branes can be taken into account, and whether it plays an important role. For tentative steps in this direction see [41, 42, 43].

Slow-roll inflaton on a brane

A simple way to study the effect of higher-dimensional gravity on inflation models is to consider inflation driven by the potential energy, V, of a four-dimensional inflaton field, ϕ, confined to a brane, embedded in a vacuum bulk [44].

At low energies, $\rho \ll \sigma$, we have seen that we can recover the conventional Friedmann equation for the cosmological expansion. On the other hand, at high energies the additional quadratic term in the modified Friedmann equation (19) leads to additional Hubble damping and actually assists slow-roll inflation. The conventional slow-roll parameters, which must be much smaller than unity for slow-roll inflation, become

$$\varepsilon = \frac{1}{16\pi G_N} \left(\frac{V'}{V}\right)^2 \to \left(\frac{4\sigma}{V}\right) \frac{1}{16\pi G_N} \left(\frac{V'}{V}\right)^2, \qquad (28)$$

$$\eta = \frac{1}{8\pi G_N} \left(\frac{V''}{V}\right) \to \left(\frac{2\sigma}{V}\right) \frac{1}{8\pi G_N} \left(\frac{V''}{V}\right), \qquad (29)$$

where a prime denotes derivatives with respect to ϕ. The slow-roll parameters are automatically suppressed at high energies [44], even for potentials that might otherwise be considered too steep to drive inflation [45].

To lowest order in the slow-roll approximation it is remarkably easy to calculate the primordial perturbation spectra expected to be generated by an inflaton on the brane.

Quantum fluctuations of a light scalar field ($|\eta| \ll 1$) in four-dimensional de Sitter spacetime ($\varepsilon \ll 1$) give the standard result for the power spectrum of inflaton fluctuations on the Hubble scale ($k = aH$)

$$\mathcal{P}_{\delta\phi} \simeq \left(\frac{H}{2\pi}\right)^2. \tag{30}$$

This determines the dimensionless curvature perturbation on uniform-density hypersurfaces [44]

$$\mathcal{P}_\zeta \simeq \left(\frac{H^2}{2\pi\dot\phi}\right)^2. \tag{31}$$

This is conserved on large (super-Hubble) scales for adiabatic density perturbations, even in brane-world gravity [46], simply as a consequence of local energy conservation [47, 48]. Thus the standard expression (31) gives the initial (primordial) density perturbations from slow-roll inflation driven by an inflaton on the brane. Of course the actual values for H and $\dot\phi$ in terms of the scalar field potential will differ due to the modified Friedmann equation at high energies.

To determine the amplitude of gravitational waves we do need to consider the five-dimensional metric perturbations. For a de Sitter brane in AdS$_5$ the wave equation is separable. We find a discrete zero-mode (as in the original Randall-Sundrum model) and a continuum of massive modes in the four-dimensional effective theory with $m^2 > 9H^2/4$ [28]. These massive modes are not excited by the de Sitter expansion – they remain in their vacuum state – and hence we need only consider the spectrum of tensor perturbations on super-Hubble scales from zero-mode vacuum fluctuations [27, 13]

$$\mathcal{P}_T = F(V/\sigma)64\pi G \left(\frac{H}{2\pi}\right)^2, \tag{32}$$

where $F(V/\sigma)$ represents the enhancement with respect to the standard four-dimensional result. At low-energies we have $F \to 1$ but at high energies we find $F \simeq 3V^2/4\sigma^2 \gg 1$. Nonetheless one can verify that the standard consistency relation between observables [49]

$$\frac{\mathcal{P}_T}{\mathcal{P}_\zeta} \simeq -2n_T, \tag{33}$$

still holds, where $n_T \equiv d\ln\mathcal{P}_T/d\ln k$ is the spectral tilt of the tensor power spectrum. It is quite unexpected that the non-trivial 5D modifications to the expected power spectra from slow-roll inflation leave the four-dimensional consistency relation unaltered [49, 50].

Quantum fluctuations on the brane

The above calculation would be expected hold to lowest order in the slow-roll approximation where one neglects the coupling between scalar field perturbations and the metric perturbations. On the other hand there is a small but finite coupling between inflaton field fluctuations and the metric even in slow-roll inflation. In 4D general relativity

Mukhanov [51] and Sasaki [52] showed how to consistently couple the linear field perturbations to the metric and hence one can derive small corrections to the scalar field fluctuations (30) during inflation.

One might expect something similar for an inflaton on the brane, but there are important differences. Although deviations from 4D gravity are small at low energies (indeed we have argued that for long-wavelength perturbations the curvature perturbation ζ is conserved as in general relativity), at high energies the brane perturbations are strongly coupled to bulk gravity. A perturbative calculation, introducing the coupling at first-order in the slow-roll parameters, shows that there may be a damping effect on the small scale (sub-horizon) modes on a single brane at high energies [53]. This is not so surprising as the single brane in AdS bulk is an open system and high frequency modes on the brane are coupled to bulk gravitons that can escape to future null infinity. This highlights the need to treat the coupled brane-bulk system in a consistent way in order to define the initial vacuum state on an inflating brane. This remains another unsolved problem in brane cosmology.

Remarkably little work seems to have been done to investigate the quantum field theory of a boundary (brane) field coupled to a higher-dimensional (bulk) field. Recently George [54] considered a linear oscillator in flat spacetime linearly coupled to another linear oscillator on its boundary. Above a critical coupling one finds an instability corresponding to a discrete tachyonic bound state. This has been extended [55], to the case of a Minkowski boundary in an AdS bulk (i.e., a Randall-Sundrum geometry). Again we found a critical coupling above which there exists an unstable bound state. But even for weak coupling one finds a small imaginary part in the frequency of resonant modes (quasi-normal modes [56]) of the coupled brane-bulk oscillators, which indicates the slow decay of oscillations on the brane. This is an example of an effect which would not be seen in a dimensionally-reduced low-energy effective theory which included only the zero-mode of the bulk field.

CONCLUSIONS

In summary, the brane-world has offered a novel perspective on many of the unsolved problems in cosmology including the number of spatial dimensions, the initial singularity and the cosmological constant problem.

There remain many unsolved problems even in the simplest case of a co-dimension one brane. There is no known solution for an isolated black hole on the brane. Indeed it has been conjectured that there is no static black hole solution on the brane [57]. Only a few special cases are known for anisotropic brane cosmologies [58]. Even linear perturbations about FRW branes are restricted to special cases or numerical solutions as the wave equation in the bulk is not in general separable in a Gaussian normal coordinate system.

Some of these problems are technical difficulties simply due to the extra complication of an inhomogeneous extra dimension, but some problems are also more fundamental. If we are seeking to describe a single brane in an infinite extra dimension that the brane alone does not form a closed system and we must specify initial data in the bulk to determine the subsequent evolution.

A century after Einstein introduced physicists to four-dimensional spacetime, braneworlds offer a big new higher-dimensional playground for relativists to explore. Cosmological models provide one area to study. If we can understand more about gravity in higher-dimensions then cosmology should also provide an opportunity to test these exotic ideas against observational data.

ACKNOWLEDGMENTS

I am grateful to the organisers for their hospitality in Oviedo and I thank my collaborators for the many collaborations upon which the work described in this article is based. This work is supported in part by PPARC grant PPA/G/S/2002/00576.

REFERENCES

1. R. Maartens, Living Rev. Rel. **7**, 7 (2004) [arXiv:gr-qc/0312059].
2. M. Gasperini and G. Veneziano, Phys. Rept. **373**, 1 (2003) [arXiv:hep-th/0207130].
3. M. B. Green, J. H. Schwarz and E. Witten, *Superstring Theory. Vol. 1: Introduction,* Cambridge University Press, Cambridge, 1987.
4. T. Kaluza, Sitzungsber. Preuss. Akad. Wiss. Berlin (Math. Phys.)) **1921**, 966 (1921).
5. J. Polchinski, *String theory. Vol. 2: Superstring theory and beyond,* Cambridge University Press, Cambridge, 1998.
6. I. Antoniadis, Phys. Lett. B **246**, 377 (1990);
7. N. Arkani-Hamed, S. Dimopoulos and G. R. Dvali, Phys. Lett. B **429**, 263 (1998) [arXiv:hep-ph/9803315]; I. Antoniadis, N. Arkani-Hamed, S. Dimopoulos and G. R. Dvali, Phys. Lett. B **436**, 257 (1998) [arXiv:hep-ph/9804398].
8. P. Horava and E. Witten, Nucl. Phys. B **460**, 506 (1996) [arXiv:hep-th/9510209].
9. L. Randall and R. Sundrum, Phys. Rev. Lett. **83**, 3370 (1999) [arXiv:hep-ph/9905221].
10. L. Randall and R. Sundrum, Phys. Rev. Lett. **83**, 4690 (1999) [arXiv:hep-th/9906064].
11. J. Garriga and T. Tanaka, Phys. Rev. Lett. **84**, 2778 (2000) [arXiv:hep-th/9911055].
12. S. Kanno and J. Soda, Phys. Rev. D **66**, 043526 (2002) [arXiv:hep-th/0205188].
13. D. S. Gorbunov, V. A. Rubakov and S. M. Sibiryakov, JHEP **0110**, 015 (2001) [arXiv:hep-th/0108017].
14. S. Mukohyama, T. Shiromizu and K. i. Maeda, Phys. Rev. D **62**, 024028 (2000) [Erratum-ibid. D **63**, 029901 (2001)] [arXiv:hep-th/9912287].
15. P. Bowcock, C. Charmousis and R. Gregory, Class. Quant. Grav. **17**, 4745 (2000) [arXiv:hep-th/0007177].
16. T. Shiromizu, K. i. Maeda and M. Sasaki, Phys. Rev. D **62**, 024012 (2000) [arXiv:gr-qc/9910076].
17. R. A. Battye, B. Carter, A. Mennim and J. P. Uzan, Phys. Rev. D **64**, 124007 (2001) [arXiv:hep-th/0105091].
18. N. Arkani-Hamed, S. Dimopoulos, N. Kaloper and R. Sundrum, Phys. Lett. B **480**, 193 (2000) [arXiv:hep-th/0001197]; S. Kachru, M. B. Schulz and E. Silverstein, Phys. Rev. D **62**, 045021 (2000) [arXiv:hep-th/0001206].
19. G. R. Dvali, G. Gabadadze and M. Porrati, Phys. Lett. B **485**, 208 (2000) [arXiv:hep-th/0005016].
20. C. Deffayet, G. R. Dvali and G. Gabadadze, Phys. Rev. D **65**, 044023 (2002) [arXiv:astro-ph/0105068].
21. D. Gorbunov, K. Koyama and S. Sibiryakov, arXiv:hep-th/0512097.
22. P. Kraus, JHEP **9912**, 011 (1999) [arXiv:hep-th/9910149].
23. J. M. Maldacena, Adv. Theor. Math. Phys. **2**, 231 (1998) [Int. J. Theor. Phys. **38**, 1113 (1999)] [arXiv:hep-th/9711200].
24. S. S. Gubser, Phys. Rev. D **63**, 084017 (2001) [arXiv:hep-th/9912001].

25. P. Binetruy, C. Deffayet, U. Ellwanger and D. Langlois, Phys. Lett. B **477**, 285 (2000) [arXiv:hep-th/9910219].
26. D. Langlois, K. i. Maeda and D. Wands, Phys. Rev. Lett. **88**, 181301 (2002) [arXiv:gr-qc/0111013].
27. D. Langlois, R. Maartens and D. Wands, Phys. Lett. B **489**, 259 (2000) [arXiv:hep-th/0006007].
28. J. Garriga and M. Sasaki, Phys. Rev. D **62**, 043523 (2000) [arXiv:hep-th/9912118].
29. J. Khoury, B. A. Ovrut, P. J. Steinhardt and N. Turok, Phys. Rev. D **64**, 123522 (2001) [arXiv:hep-th/0103239].
30. R. Kallosh, L. Kofman and A. D. Linde, Phys. Rev. D **64**, 123523 (2001) [arXiv:hep-th/0104073].
31. P. J. Steinhardt and N. Turok, Phys. Rev. D **65**, 126003 (2002) [arXiv:hep-th/0111098].
32. J. J. Blanco-Pillado and M. Bucher, Phys. Rev. D **65**, 083517 (2002) [arXiv:hep-th/0111089]; J. J. Blanco-Pillado, M. Bucher, S. Ghassemi and F. Glanois, Phys. Rev. D **69**, 103515 (2004) [arXiv:hep-th/0306151].
33. U. Gen, A. Ishibashi and T. Tanaka, Phys. Rev. D **66**, 023519 (2002) [arXiv:hep-th/0110286].
34. A. Neronov, JHEP **0111**, 007 (2001) [arXiv:hep-th/0109090].
35. D. H. Lyth, Phys. Lett. B **524**, 1 (2002) [arXiv:hep-ph/0106153]; J. Khoury, B. A. Ovrut, P. J. Steinhardt and N. Turok, Phys. Rev. D **66**, 046005 (2002) [arXiv:hep-th/0109050].
36. A. J. Tolley and N. Turok, Phys. Rev. D **66**, 106005 (2002) [arXiv:hep-th/0204091]; A. J. Tolley, N. Turok and P. J. Steinhardt, Phys. Rev. D **69**, 106005 (2004) [arXiv:hep-th/0306109].
37. S. B. Giddings, S. Kachru and J. Polchinski, Phys. Rev. D **66**, 106006 (2002) [arXiv:hep-th/0105097].
38. S. Kachru, R. Kallosh, A. Linde and S. P. Trivedi, Phys. Rev. D **68**, 046005 (2003) [arXiv:hep-th/0301240].
39. G. R. Dvali and S. H. H. Tye, Phys. Lett. B **450**, 72 (1999) [arXiv:hep-ph/9812483].
40. A. D. Linde, Phys. Rev. D **49**, 748 (1994) [arXiv:astro-ph/9307002]; E. J. Copeland, A. R. Liddle, D. H. Lyth, E. D. Stewart and D. Wands, Phys. Rev. D **49**, 6410 (1994) [arXiv:astro-ph/9401011].
41. Class. Quant. Grav. **22**, 3431 (2005) [arXiv:hep-th/0505256].
42. S. Mukohyama, Y. Sendouda, H. Yoshiguchi and S. Kinoshita, JCAP **0507**, 013 (2005) [arXiv:hep-th/0506050].
43. S. Kanno, J. Soda and D. Wands, JCAP **0508**, 002 (2005) [arXiv:hep-th/0506167].
44. R. Maartens, D. Wands, B. A. Bassett and I. Heard, Phys. Rev. D **62**, 041301 (2000) [arXiv:hep-ph/9912464].
45. E. J. Copeland, A. R. Liddle and J. E. Lidsey, Phys. Rev. D **64**, 023509 (2001) [arXiv:astro-ph/0006421].
46. D. Langlois, R. Maartens, M. Sasaki and D. Wands, Phys. Rev. D **63**, 084009 (2001) [arXiv:hep-th/0012044].
47. D. Wands, K. A. Malik, D. H. Lyth and A. R. Liddle, Phys. Rev. D **62**, 043527 (2000) [arXiv:astro-ph/0003278].
48. D. H. Lyth and D. Wands, Phys. Rev. D **68**, 103515 (2003) [arXiv:astro-ph/0306498].
49. G. Huey and J. E. Lidsey, Phys. Lett. B **514**, 217 (2001) [arXiv:astro-ph/0104006].
50. M. Bouhmadi-Lopez, R. Maartens and D. Wands, Phys. Rev. D **70**, 123519 (2004) [arXiv:hep-th/0407162].
51. V. F. Mukhanov, Sov. Phys. JETP **67**, 1297 (1988) [Zh. Eksp. Teor. Fiz. **94N7**, 1 (1988)].
52. M. Sasaki, Prog. Theor. Phys. **76**, 1036 (1986).
53. K. Koyama, S. Mizuno and D. Wands, JCAP **0508**, 009 (2005) [arXiv:hep-th/0506102].
54. A. George, J. Phys. A **38**, 7399 (2005) [arXiv:hep-th/0412067].
55. K. Koyama, A. Mennim and D. Wands, Phys. Rev. D **72**, 064001 (2005) [arXiv:hep-th/0504201].
56. S. S. Seahra, Phys. Rev. D **72**, 066002 (2005) [arXiv:hep-th/0501175].
57. R. Emparan, J. Garcia-Bellido and N. Kaloper, JHEP **0301**, 079 (2003) [arXiv:hep-th/0212132].
58. A. Fabbri, D. Langlois, D. A. Steer and R. Zegers, JHEP **0409**, 025 (2004) [arXiv:hep-th/0407262].

PART II: CONTRIBUTED PAPERS

Quantum Gravity and the fate of Lorentz invariance in the Standard Model

Jorge Alfaro

Dept. de Física Teórica C-XI, Facultad de Ciencias, Univ. Autónoma de Madrid, Cantoblanco, 28049, Madrid, Spain and
Facultad de Física, Pontificia Universidad Católica de Chile
Casilla 306, Santiago 22, Chile. [1]
jalfaro@puc.cl

Abstract. The most important problem of fundamental Physics is the quantization of the gravitational field. Experimental tests that discriminate among the theories proposed to quantize gravity are difficult due to the smallness of the new effects involved. So the violation of a well tested symmetry such as Lorentz invariance could be a good arena to explore the implications of the quantized gravitational field. Here we review recent work where we showed that the Standard Model(SM) itself contains tiny Lorentz invariance violation(LIV) terms coming from QG. All terms depend on one arbitrary parameter α that set the scale of QG effects. α can be estimated using the spectrum of Ultra High Energy Cosmic Rays.

Keywords: Quantum Gravity, Ultra High Energy Cosmic Rays
PACS: 04.60.Pp, 11.15.-q

INTRODUCTION

Recently various schemes have been advanced to select theories and predict new phenomena associated to the Quantum gravitational field [1, 2, 3, 4]. Most of the new phenomenology is related to some sort of Lorentz invariance violations(LIV's)[5, 6, 7]. Recently [8], this approach has been subjected to severe criticism.

In previous works [9], we asserted that the main effect of QG is to deform the measure of integration of Feynman graphs at large four momenta by a tiny LIV. The classical lagrangian is unchanged. In a similar manner, we can say that QG deforms the metric of space-time, introducing a tiny LIV proportional to $(d-4)\alpha$, d being the dimension of space time in Dimensional Regularization and α is the only arbitrary parameter in the model. The LIV could be due to quantum fluctuations of the metric of space-time produced by QG:virtual black holes as suggested in[1], D-branes as in [10], compactification of extra-dimensions or spin-foam anisotropies [11]. A precise derivation of α will have to wait for additional progress in the available theories of QG [2]

It is possible to have modified dispersion relations without a preferred frame(DSR)[12].

---

[1] Permanent Address.
[2] Such derivation must explain why the LIV parameter is so small. Progress in this direction is in [3, 4, 13]. There α appears as $(l_P/L)^2$, where l_P is Planck's lenght and L is defined by the semiclassical gravitational state in Loop Quantum Gravity. If $L \sim 10^{11} l_P$, an α of the right order is obtained

Notice, however, that in our case the classical lagrangian is invariant under usual linear Lorentz transformations but not under DSR. So our LIV is more akin to radiative breaking of usual Lorentz symmetry than to DSR. Moreover the regulator R defined below and the deformed metric (5) are given in a particular inertial frame, where spatial rotational symmetry is preserved. That is why, in this paper we are ascribing to the point of view of [6] which is widely used in the literature. The preferred frame is the one where the Cosmic Background Radiation is isotropic.

In the Standard Model, such LIV implies several remarkable effects, which are wholly determined up to one arbitrary parameter (α).The main effects are: The maximal attainable velocity for particles is not the speed of light, but depends on the specific couplings of the particles within the Standard Model. This is a necessary requirement to explain the Greisen[14],Zatsepin and Kuz'min[15](GZK) anomaly[6, 13, 16]. Since the Auger[17] experiment is expected to produce results in the near future, powerful tests of Lorentz invariance using the spectrum of UHECR will be available. Also, birrefringence occurs for charged leptons, but not for gauge bosons. In particular, photons and neutrinos have different maximum attainable velocities. This could be tested in the next generation of neutrino detectors such as NUBE[18, 19].

CUTOFF REGULATOR

To see what are the implications of the asymmetry in the measure for renormalizable theories, we will represent the Lorentz asymmetry of the measure by the replacement

$$\int d^d k -> \int d^d k R(\frac{k^2 + \alpha k_0^2}{\Lambda^2})$$

Here R is an arbitrary function, Λ is a cutoff with mass dimensions, that will go to infinity at the end of the calculation. We normalize $R(0) = 1$ to recover the original integral. $R(\infty) = 0$ to regulate the integral. α is a real parameter. This regulator has the property that for logarithmically divergent integrals, the divergent term is Lorentz invariant whereas when the cutoff goes to infinity a finite LIV part proportional to α remains.

ONE LOOP

Let D be the naive degree of divergence of a One Particle Irreducible (1PI) graph. The change in the measure induces modifications to the primitively log divergent integrals(D=0) In this case, the correction amounts to a finite LIV. The finite part of 1PI Green functions will not be affected.

Let us analyze the primitivily divergent 1PI graphs for bosons first.

Self energy: $\chi(p) = \chi(0) + A^{\mu\nu} p_\mu p_\nu + convergent$, $A^{\mu\nu} = \frac{1}{2}\partial_\mu \partial_\nu \chi(0)$. We have:

$$A^{\mu\nu} = c_2 \eta^{\mu\nu} + a^{\mu\nu}$$

c_2 is the log divergent wave function renormalization counterterm; $a^{\mu\nu}$ is a finite LIV. The on-shell condition is:
$$p^2 - m^2 - a^{\mu\nu}p_\mu p_\nu = 0$$
If spatial rotational invariance is preserved, the nonzero components of the matrix a are:
$$a^{00} = a_0; \quad a^{ii} = -a_1$$
So the maximum attainable velocity for this particle will be:
$$c_m = \sqrt{\frac{1-a_1}{1-a_0}} \sim 1 - (a_1 - a_0)/2 \tag{1}$$

For fermions, we have the self energy graph
$$\Sigma(p) = \Sigma(0) + s^{\mu\nu}\gamma_\nu p_\mu$$
$s^{\mu\nu}\gamma_\nu = \partial_\mu \Sigma(0)$. Moreover
$$s^{\mu\nu} = s\eta^{\mu\nu} + a^{\mu\nu}/2$$
s is a log divergent wave function renormalization counterterm; $a^{\mu\nu}$ is a finite LIV. The maximum attainable velocity of this particle will be given again by equation (1).

By doing explicit computations for all particles in the SM, we get definite predictions for the LIV, assuming a particular regulator R. However, the dependence on R amounts to a multiplicative factor. So ratios of LIV's are uniquely determined.

Gauge Bosons Consider the most general quadratic Lagrangian which is gauge invariant, but could permit LIV's [3]
$$L = c^{\mu\nu\alpha\beta} F_{\mu\nu} F_{\alpha\beta}$$

$c^{\mu\nu\alpha\beta}$ is antisymmetric in $\mu\nu$ and $\alpha\beta$ and symmetric by $(\alpha,\beta) <-> (\mu,\nu)$ It implies that the most general expression for the self-energy of the gauge boson will be
$$\Pi^{\nu\beta}(p) = c^{\mu\nu\alpha\beta} p_\alpha p_\mu \Pi(p) \tag{2}$$
We see that
$$p_\nu \Pi^{\nu\beta}(p) = 0$$
$c^{\mu\nu\alpha\beta}$ is given by a logarithmically divergent integral. We get:
$$c^{\mu\nu\alpha\beta} = c_2(\eta^{\mu\alpha}\eta^{\nu\beta} - \eta^{\mu\beta}\eta^{\nu\alpha}) + a^{\mu\nu\alpha\beta} \tag{3}$$
c_2 is a Lorentz invariant counterterm and $a^{\mu\nu\alpha\beta}$ is a LIV.

It is clear that the same argument applies to massive gauge bosons that got their mass by spontaneous gauge symmetry breaking as well as to the graviton in linearized gravity.

Explicit computations are simplified by using Dimensional Regularization as explained below.

---

[3] A Chern-Simons term is absent due to the symmetry $k_\mu -> -k_\mu$, which is preserved by the regulator.

LIV DIMENSIONAL REGULARIZATION

We generalize dimensional regularization to a d dimensional space with an arbitrary constant metric $g_{\mu\nu}$. We work with a positive definite metric first and then Wick rotate. We will illustrate the procedure with an example. Here $g = det(g_{\mu\nu})$ and $\Delta > 0$.

$$\sqrt{g}\int \frac{d^d k}{(2\pi)^d} \frac{k_\mu k_\nu}{(k^2+\Delta)^n} =$$
$$\frac{\sqrt{g}}{\Gamma(n)} \int_0^\infty dt\, t^{n-1} \int \frac{d^d k}{(2\pi)^d} k_\mu k_\nu e^{-t(g^{\alpha\beta}k_\alpha k_\beta + \Delta)} =$$
$$\frac{1}{(4\pi)^{d/2}} \frac{g_{\mu\nu}}{2} \frac{\Gamma(n-1-d/2)}{\Gamma(n)} \frac{1}{\Delta^{n-1-d/2}} \quad (4)$$

In the same manner, after Wick rotation, we obtain Appendix A4 of [18].

These definitions preserve gauge invariance, because the integration measure is invariant under shifts. To get a LIV measure, we assume that

$$g^{\mu\nu} = \eta^{\mu\nu} + (4\pi)^2 \alpha \eta^{\mu 0} \eta^{\nu 0} \varepsilon \quad (5)$$

where $\varepsilon = 2 - \frac{d}{2}$. A formerly divergent integral will have a pole at $\varepsilon = 0$, so when we take the physical limit, $\varepsilon -> 0$, the answer will contain a LIV term.

To define the counterterms, we used the minimal substraction scheme(MSS); that is we substract the poles in ε from the 1PI graphs.

Concrete examples will be provided in the following sections.

LIV Dimensional Regularization reinforces our claim that these tiny LIV's originates in Quantum Gravity. In fact the metric of space time changes by a correction of order ε from the Minkowsky metric and this is the origin of the effects studied above. Quantum Gravity is the strongest candidate to produce such effects because the gravitational field is precisely the metric of space-time and tiny LIV modifications to the flat Minkowsky metric may be produced by quantum fluctuations.

LIV IN THE STANDARD MODEL

We follow [20, 21] and use LIV Dimensional Regularization.

Photons: In the SM the photon self-energy can be written:

$$i\Pi^{\mu\nu} = i(q^2 g^{\mu\nu} - q^\mu q^\nu)\left(\frac{-23 e^2}{48\pi^2 \varepsilon} + finite\right) \quad (6)$$

so that the LIV photon self-energy in the SM is:

$$L\Pi^{\mu\nu}(q) = -\frac{23}{3} e^2 \alpha q_\alpha q_\beta$$
$$(\eta^{\alpha\beta}\delta_0^\mu \delta_0^\nu + \eta^{\mu\nu}\delta_0^\alpha \delta_0^\beta - \eta^{\nu\beta}\delta_0^\mu \delta_0^\alpha - \eta^{\mu\alpha}\delta_0^\nu \delta_0^\beta) \quad (7)$$

It follows that the maximal attainable velocity is

$$c_\gamma = 1 - \frac{23}{6}e^2\alpha \qquad (8)$$

We have included coupling to quarks and charged leptons as well as 3 generations and color.

Fermions: In the SM, the fermion self-energy is given by:

$$(4\pi^2)\Sigma(q) = -\frac{1}{\varepsilon}\slashed{q}\sum_{graphs}(|c_V+c_A|^2 P_L + |c_V-c_A|^2 P_R) + finite \qquad (9)$$

where the fermion-gauge boson vertex is:

$$l\gamma^k(c_V - c_A\gamma^5) \qquad (10)$$

and $P_L(P_R)$ are the L(R) helicity projectors.
Therefore

$$L\Sigma(q) = \frac{\alpha}{2}q_0\gamma^0\sum_{graphs}(|c_V+c_A|^2 P_L + |c_V-c_A|^2 P_R) \qquad (11)$$

We apply this last result to neutrinos and charged leptons below.

Neutrinos: The maximal attainable velocity is

$$c_\nu = 1 - (3+tan^2\theta_w)\frac{g^2\alpha}{8} \qquad (12)$$

In this scenario, we predict that neutrinos [19] emitted simultaneously with photons in gamma ray bursts will not arrive simultaneously to Earth. The time delay during a flight from a source situated at a distance D will be of the order of $(5 \times 10^{-23})D/c \sim 5 \times 10^{-6}$ s, assuming $D = 10^{10}$ light-years. No dependence of the time delay on the energy of high energy photons or neutrinos should be observed(contrast with [1]). Photons will arrive earlier since $\alpha < 0$(See below). These predictions could be tested in the next generation of neutrino detectors [20].

Using R_ξ-gauges we have checked that the LIV is gauge invariant. The gauge parameter affects the Lorentz invariant part only.

Electron self-energy in the Weinberg-Salam model. Birrefringence:
Define: $e_L = \frac{1-\gamma^5}{2}e$, $e_R = \frac{1+\gamma^5}{2}e$, where e is the electron field. We get

$$c_L = 1 - (\frac{g^2}{cos^2\theta_w}(sin^2\theta_w - 1/2)^2 + e^2 + g^2/2)\frac{\alpha}{2}; \qquad (13)$$

$$c_R = 1 - (e^2 + \frac{g^2 sin^4\theta_w}{cos^2\theta_w})\frac{\alpha}{2} \qquad (14)$$

The difference in maximal speed for the left and right helicities is $\sim (5 \times 10^{-24})$.

MESONS AND BARYONS

In order to apply our results to the computation of the UHECR spectrum and other phenomena, we must calculate the maximal attainable velocity of hadrons. As we mentioned before, the problem is hadronization. One way to get an estimation of the effect is using effective lagrangians.

We use the results of [22, 23] for the wave function renormalization of pions and nucleons in the chiral lagrangian and Heavy Baryon Chiral Perturbation Theory. They get:

$$Z_\pi^{-1} = 1 - \frac{4m_\pi^2}{3(4\pi)^2 F^2}\frac{1}{\varepsilon} + finite \qquad (15)$$

$$Z_N^{-1} = 1 - \frac{9g_A^2 m_\pi^2}{4(4\pi)^2 F^2}\frac{1}{\varepsilon} + finite \qquad (16)$$

Here, m_π is the renormalized pion mass, F is the renormalized decay constant of pions and g_A is the axial vector coupling constant, in the chiral limit.

Using the LIV metric, we can read off the maximal attainable velocities for pions and nucleons:

$$c_\pi = 1 + \frac{2m_\pi^2 \alpha}{3F^2}$$
$$c_N = 1 + \frac{9m_\pi^2 g_A^2 \alpha}{8F^2} \qquad (17)$$

REACTION THRESHOLDS

Knowing the LIV for nucleons, pions, photons and electrons, we proceed to study the reactions involved in the GZK cutoff:Photo-Pion Production $\gamma + p \to p + \pi$ and Pair Creation $\gamma + p \to p + e^+ + e^-$ [9].

Combining the two reactions and the standard values, $m_\pi = 139 Mev, g_A = 1.26, F = 92.4 Mev$, we get an upper and lower bound on α

$$2.2 \times 10^{-21} > -\alpha > 1.3 \times 10^{-24} \qquad (18)$$

First of all, we notice that $\alpha < 0$, in order to suppress the photopion production, thus removing the GZK cutoff. This implies that photons are the fastest particles and they arrive before neutrinos coming from the same source of GRB. Moreover, photons become unstable. They decay in a electron positron pair above an energy E_0[6]. See below.

Since $c_{photon} > c_{proton}$, the strong bound of [27] is avoided: Proton is stable under Cerenkov radiation in vacuum.

Photon unstability: It has been pointed out in [27, 6] that if $c_{photon} > c_{electron}$ then the process $\gamma \to e^+ + e^-$ is allowed above an energy E_0:

$$E_0 = m_e \sqrt{\frac{2}{\delta c}} \qquad (19)$$

where $\delta c = c_\gamma - c_e$.

In our case, we have:

$$\delta c_L = -\alpha(\frac{23}{6}e^2 - (\frac{g^2}{\cos^2\theta_w}(\sin^2\theta_w - 1/2)^2 + e^2 + g^2/2)/2) \qquad (20)$$

$$\delta c_R = -\alpha(\frac{23}{6}e^2 - (e^2 + \frac{g^2 \sin^4\theta_w}{\cos^2\theta_w})/2) \qquad (21)$$

Therefore, with

$$EL_0 = 2.3 \times 10^8 Gev$$
$$ER_0 = 1.9 \times 10^8 Gev \qquad (22)$$

So, we should not detect photons with energies above $2.3 \times 10^8 Gev$.

Neutral pion Stability: Following [6] we study the main decay process of neutral pion $\pi_0 \to \gamma + \gamma$. This is forbidden if $c_\gamma > c_\pi$ and above an energy

$$E_\pi = \frac{m_\pi}{\sqrt{2(c_\gamma - c_\pi)}} \qquad (23)$$

Using the bound $c_\gamma - c_\pi < 10^{-22}$ obtained in [28], we get

$$|\alpha| < 5.4 \times 10^{-23} \qquad (24)$$

In our numerical estimates we have chosen $\alpha = -5 \times 10^{-23}$.

We get $E_\pi = 10^{19} eV$. Therefore we expect that neutral pions above this energy are stable, so they could be a primary component of UHECR. Photons will be unstable above this energy by the same mechanism. Notice however that photons are unstable at a lower energy due to electron-positron pair creation (22).

CONCLUSIONS

In this paper we have reviewed the LIV induced by Quantum Gravity in the Standard Model. Studying several available processes, we found bounds on α:

From pair creation and absence of photopion creation: $2.2 \times 10^{-21} > -\alpha > 1.3 \times 10^{-24}$. From pion stability and the most stringent experimental bound found in [28]: $|\alpha| < 5.4 \times 10^{-23}$.

Then, several predictions are obtained: Photons are unstable above an energy $2.3 \times 10^8 Gev$. Neutral pions are stable above an energy $E_\pi = 10^{19} eV$; so they could be a primary component of UHECR, thus evading the GZK cutoff.

Moreover, in time of flight experiments, photons will arrive before neutrinos, assuming that they were emitted simultaneously at the source. No energy dependence of the time delay should be observed.

ACKNOWLEDGMENTS

The work of JA is partially supported by Secretaria de Estado de Universidades e Investigación SAB2003-0238(Spain). He wants to thank the organizers of ERE05 for a nice atmosphere during the conference; and to the Universities of Santiago de Compostela and Granada for a pleasent visit to them.

REFERENCES

1. Amelino-Camelia, G. et al., Nature 393, 763 (1998).
2. Gambini, R. and Pullin, J. , Phys. Rev. D 59, 124021 (1999).
3. Alfaro, J. Morales-Técotl, H.A. and Urrutia, L.F. , Phys. Rev. Lett. 84, 2318 (2000).
4. Alfaro, J. Morales-Técotl, H.A. and Urrutia, L.F., Phys. Rev. D 65, 103509(2002).
5. Colladay, D. and Kostelecky, V.A., Phys. Rev. D58, 116002(1998).
6. Coleman, S.and Glashow, S.L., Phys. Rev. D 59, 116008 (1999).
7. Bertolami, O., Colladay, Kostelecky, V.A. and Potting, R., Phys. Lett. B395(1997)178.
8. Collins, J. et al , Phys.Rev.Lett.93, 191301(2004).
9. Alfaro, J., Phys.Rev.Lett.94:221302,2005 ;Phys.Rev.D72:024027,2005.
10. J. Ellis, N.E. Mavromatos, D.V. Nanopoulos, G. Volkov, Gen.Rel.Grav. 32 (2000) 1777-1798.
11. For a comprehensive review on the loop quantum gravity framework see for example: Rovelli, C. , Quantum Gravity, Cambridge University Press, Cambridge (UK) 2004.
12. G. Amelino-Camelia, Int. J. Mod. Phys. **D 11** (2002) 35-60, Phys. Lett. **B 510** (2001) 255-263; N. Bruno, G. Amelino-Camelia, and J. Kowalski-Glikman, Phys. Lett. **B 522** (2001) 133; G. Amelino-Camelia, Nature **418** (2002) 34; G. Amelino-Camelia Int. J. Mod. Phys. **D 12** (2003) 1211;J. Magueijo and L. Smolin, Phys. Rev. Lett. **88** (2002) 190403; J. Magueijo and L. Smolin, Phys. Rev. **D 67** 044017 (2003).
13. Alfaro,J. and Palma,G., Phys. Rev. D 67,083003(2003);Phys. Rev. D **65**, 103516 (2002).
14. Greisen,K. Phys.Rev.Lett.16,748 (1966).
15. Zatsepin,G.T. and Kuz'min, V.A.,JETP Lett.4:78(1966), Pisma Zh.Eksp.Teor.Fiz.4:114(1966).
16. T.Kifune, Astroph. J. Lett. **518**, 21 (1999); G.Amelino-Camelia and T.Piran, Phys. Lett. B **497**, 265 (2001); J.Ellis, N.E.Mavromatos and D.V.Nanopoulos, Phys. Rev. D **63**, 124025 (2001); G.Amelino-Camelia and T.Piran, Phys. Rev. D **64**, 036005 (2001); G.Amelino-Camelia, Phys. Lett. B **528**, 181 (2002).
17. See http://www.auger.org/auger.html.
18. Waxman,E. and Bahcall, J.,Phys.Rev.Letts.78, 2292(1997).
19. Roy,M., Crawford,H.J. and Trattner, A., astro-ph/9903231.
20. Peskin, M. and Schroeder D., An introduction to Quantum Field Theory, Addison-Wesley Publishing Company, New York 1997.
21. Pokorski, S., Gauge Field Theories, Cambridge Monographs on Mathematical Physics, Cambridge University Press 2000.
22. G. Ecker and M. Mojzis, Phys. Lett. B 410(1997)266.
23. H. Fearing, R. Lewis, N. Mobed and S. Scherer, Phys. Rev. D 56 (1997) 1783.
24. V. Berezinsky, A.Z. Gazizov and S.I. Grigorieva, hep-ph/0107306; hep-ph/0204357.
25. F.W. Stecker and S.L. Glashow, Astropart.Phys. **16**, 97 (2001).
26. M. Takeda *et al.*, Phys. Rev. Lett. **81**, 1163 (1998). For an update see M. Takeda *et al.*, Astrophys. J. **522**, 225 (1999).
27. S. Coleman and S.L. Glashow, Phys. Lett. B 405, 249(1997).
28. E. Antonov et al.,JETP Letters 73(2001)446.

Ten-dimensional Supergravity Revisited

Eric Bergshoeff*, Mees de Roo*, Sven Kerstan* and Fabio Riccioni[†]

*Centre for Theoretical Physics, University of Groningen, Nijenborgh 4,
9747 AG Groningen, The Netherlands
[†]DAMTP, Centre for Mathematical Sciences, University of Cambridge,
Wilberforce Road, Cambridge CB3 0WA, UK

Abstract. We show that the exisiting supergravity theories in ten dimensions can be extended with extra gauge fields whose rank is equal to the spacetime dimension. These gauge fields have vanishing field strength but nevertheless play an important role in the coupling of supergravity to spacetime filling branes.

We discuss the role of these gauge fields in the construction of string theories with sixteen supercharges and mention their relation with a conjectured hyperbolic symmetry underlying string theory and M theory.

Keywords: Supergravity, Strings, Branes
PACS: 04.65.+e, 11.25.-w

1. INTRODUCTION

The Type II supergravity theories in ten dimensions form a starting point from which all lower dimensional maximal supergravities can be derived. The Type IIB [1, 2, 3] and IIA theory [4, 5, 6], with two supercharges of equal (opposite) chirality were both constructed around 1984. The Type IIA theory follows by dimensional reduction from $D = 11$ supergravity. It was extended in 1986 to include a massive parameter [7]. The IIB theory does not appear to have a higher dimensional origin. The bosonic fields of the two theories are

$$\text{IIA}: \quad g_{\mu\nu}, \phi, B_{(2)}, C_{(1)}, C_{(3)}, \tag{1}$$

$$\text{IIB}: \quad g_{\mu\nu}, \phi, B_{(2)}, C_{(0)}, C_{(2)}, C_{(4)}. \tag{2}$$

The subscripts (n) indicate the rank of an antisymmetric tensor gauge field, or n-form field. The IIB 4-form satisfies a self-duality relation, which prevents the construction of a covariant action.

A natural extension in both theories is the addition of duals of the n-form fields. In this way one can associate to every n-form field an $8-n$-form field ($n \geq 0$). The $n+1$-form curvatures are then related by a duality relation to the corresponding $9-n$-form curvatures. These forms therefore do not introduce new degrees of freedom, but instead provide a alternative way to view the role of these propagating fields. This is particularly profitable in the coupling of these fields to extended objects or branes. A p-brane, with p spatial extensions, couples in a natural way to a $p+1$-form field. The dual forms are therefore useful in studying the properties of p-branes with $p \geq 4$ (for $p = 3$ the brane couples to $C_{(4)}$, which is its own dual). The introduction of dual forms for the

RR potentials $C_{(n)}$ has led to a completely "democratic" formulation of IIA and IIB supergravity, where all RR forms appear simultaneously [8].

It is also possible to introduce n-form fields with rank $n \geq 9$. These do not carry propagating degrees of freedom, and are therefore not dual to the physical supergravity fields. Nevertheless, they also have interesting applications. In [9] the $C_{(9)}$ field in the massive IIA theory played an essential role in understanding the 8-brane domain wall. The dual of the curvature $G_{(10)}$ plays the role of a cosmological constant.

Ten-form fields couple to space-time filling branes. These are related to truncations of the IIB theory to $N = 1$ supersymmetric theories. A 9-brane charge is by itself inconsistent. This can be resolved by adding opposite charge on an orientifold plane, which triggers the truncation to a Type I string theory. The introduction of 10-form fields in IIB supergravity and the corresponding truncation to $N = 1$ were considered in [11]. There two 10-forms were obtained. One is an RR form $C_{(10)}$, which will reappear in our present work. The other, called $B_{(10)}$, has the wrong tension to be understood as the S-dual of $C_{(10)}$, so that these two fields could not arise from an $SU(1,1)$ (see Section 2) doublet. This problem will be resolved in this talk.

Since 9- and 10-forms do not carry physical degrees of freedom their number is not a priori limited. Of course they must be consistent with supersymmetry, and this turns out to lead to restrictions. The purpose of our work [10] is precisely to establish how many of these forms are possible in IIB supergravity, and to classify them in the correct $SU(1,1)$ representations. A similar investigation of IIA is presently under way [12].

In Section 2 we will review briefly the construction of [10] in Einstein frame. The coupling to branes and the brane tensions are discussed in Section 3. In Section 4 the truncation to $N = 1$ theories is treated. Our results turn out to have an intriguing relation to recent efforts to identify the symmetry group of M- and string theory. In Section 5 we comment on this correspondence.

2. SUPERSYMMETRY, $SU(1,1)$ AND FORM-FIELDS

The starting point of our analysis is the standard IIB supergravity as first formulated by [1, 2]. The theory exhibits an explicit $SU(1,1)$ symmetry, which acts on the two bosonic fields. The scalars parametrize an $SU(1,1)/U(1)$ coset. The scalars and fermions in the theory each has a charge associated with the local $U(1)$ symmetry, the gauge fields have zero charge. Under $SU(1,1)$ the fields $B_{(2)}$ and $C_{(2)}$ form a doublet $A_{(2)}^\alpha$ (satisfying $A_{(2)}^1 = (A_{(2)}^2)^*$, while $C_{(4)}$ (also written as $A_{(4)}$) corresponds to a singlet. The scalars are conveniently written as a matrix U:

$$U = \begin{pmatrix} V_-^1 & V_+^1 \\ V_-^2 & V_+^2 \end{pmatrix}. \tag{3}$$

Here $V_\pm^\alpha$, with charge ± 1 and with $\alpha = 1, 2$, form doublets of $SU(1,1)$. They are constrained by the relation

$$V_-^\alpha V_+^\beta - V_+^\alpha V_-^\beta = \varepsilon^{\alpha\beta}. \tag{4}$$

The supersymmetry transformations are, to terms bilinear in fermions:

$$\delta e_\mu{}^a = i\bar{\varepsilon}\gamma^a\psi_\mu + i\bar{\varepsilon}_C\gamma^a\psi_{\mu C} \ ,$$
$$\delta\psi_\mu = D_\mu\varepsilon + \tfrac{i}{480}F_{\mu\nu_1...\nu_4}\gamma^{\nu_1...\nu_4}\varepsilon + \tfrac{1}{96}G^{\nu\rho\sigma}\gamma_{\mu\nu\rho\sigma}\varepsilon_C - \tfrac{3}{32}G_{\mu\nu\rho}\gamma^{\nu\rho}\varepsilon_C \ ,$$
$$\delta A^\alpha_{\mu\nu} = V^\alpha_-\,\bar{\varepsilon}\gamma_{\mu\nu}\lambda + V^\alpha_+\,\bar{\varepsilon}_C\gamma_{\mu\nu}\lambda_C + 4iV^\alpha_-\,\bar{\varepsilon}_C\gamma_{[\mu}\psi_{\nu]} + 4iV^\alpha_+\,\bar{\varepsilon}\gamma_{[\mu}\psi_{\nu]C} \ ,$$
$$\delta A_{\mu\nu\rho\sigma} = \bar{\varepsilon}\gamma_{[\mu\nu\rho}\psi_{\sigma]} - \bar{\varepsilon}_C\gamma_{[\mu\nu\rho}\psi_{\sigma]C} - \tfrac{3i}{8}\varepsilon_{\alpha\beta}A^\alpha_{[\mu\nu}\delta A^\beta_{\rho\sigma]} \ ,$$
$$\delta\lambda = iP_\mu\gamma^\mu\varepsilon_C - \tfrac{i}{24}G_{\mu\nu\rho}\gamma^{\mu\nu\rho}\varepsilon \ ,$$
$$\delta V^\alpha_+ = V^\alpha_-\,\bar{\varepsilon}_C\lambda \ ,$$
$$\delta V^\alpha_- = V^\alpha_+\,\bar{\varepsilon}\lambda_C \ . \tag{5}$$

Here we have introduced

$$\begin{aligned}
P_\mu &= -\varepsilon_{\alpha\beta}V^\alpha_+\partial_\mu V^\beta_+ \ ,\\
Q_\mu &= -i\varepsilon_{\alpha\beta}V^\alpha_-\partial_\mu V^\beta_+ \ ,\\
G_{\mu\nu\rho} &= -\varepsilon_{\alpha\beta}V^\alpha_+ F^\beta_{\mu\nu\rho} \ ,\\
F^\alpha_{\mu\nu\rho} &= 3\partial_{[\mu}A^\alpha_{\nu\rho]} \ ,\\
F_{\mu\nu\rho\sigma\tau} &= 5\partial_{[\mu}A_{\nu\rho\sigma\tau]} + \tfrac{5i}{8}\varepsilon_{\alpha\beta}A^\alpha_{[\mu\nu}F^\beta_{\rho\sigma\tau]} \ .
\end{aligned} \tag{6}$$

Q_μ is the $U(1)$ gauge field which is implicitly present in the covariatizations in (5). For further details on notation we refer to [2]. An important property of these transformations is that the commutator of two supersymmetry transformations on the bosonic gauge fields closes on translations and gauge transformations. On the fermionic fields closure also requires supersymmetry transformations, local Lorentz transformations and the equations of motion, see [2] for details.

The way to obtain extensions of the supergravity multiplet above is to use this property of closure: we assume an initial form of the supersymmetry transformation of a proposed field, including free parameters, and determine these parameters by requiring closure. Since no new degrees of freedom can be introduced, closure will also require a relation between the additional and original fields. For the 6-form and 8-form fields this leads to a unique extension. For the 10-forms no relation with fields of the original IIB multiplet exists, so that the number of 10-forms is not determined a priori. We will come back to this point later in this section.

We find that the following fields can be introduced in IIB supergravity:

6-forms There is a doublet of 6-forms $A^a_{(6)}$, with $A^1_{(6)} = (A^2_{(6)})^*$, satisfying the duality relation

$$F^\alpha_{(7)\mu_1...\mu_7} = -\tfrac{i}{3!}\varepsilon_{\mu_1...\mu_7\mu\nu\rho}S^{\alpha\beta}\varepsilon_{\beta\gamma}F^{\gamma;\mu\nu\rho}_{(3)} \ , \tag{7}$$

where

$$S^{\alpha\beta} = V^\alpha_- V^\beta_+ + V^\alpha_+ V^\beta_- \ . \tag{8}$$

The algebra closes on these 6-forms, giving a translation and bosonic n-form gauge transformations, with $n = 1, 3, 5$.

8-forms There is a triplet of 8-forms $A_{(8)}^{\alpha\beta}$, symmetric in α,β, satisfying a reality condition

$$(A_{(8)}^{11})^* = A_{(8)}^{22}, \qquad (A_{(8)}^{12})^* = A_{(8)}^{12}, \tag{9}$$

and a duality relation

$$F_{(9)\,\mu_1\ldots\mu_9}^{\alpha\beta} = i\varepsilon_{\mu_1\ldots\mu_9}{}^{\sigma}\{V_+^{\alpha}V_+^{\beta}P_{\sigma}^* - V_-^{\alpha}V_-^{\beta}P_{\sigma}\}. \tag{10}$$

However, the three 8-forms are related to each other through a condition on the field-strengths,

$$\varepsilon_{\alpha\gamma}\varepsilon_{\beta\delta}V_+^{\alpha}V_-^{\beta}F_{(9)}^{\gamma\delta} = 0. \tag{11}$$

This implies that in the 8-form sector there are only two degrees of freedom, the 'duals' of the dilaton ϕ and the axion $C_{(0)}$. Note that the three potentials are not related by a local condition. On the 8-forms the algebra closes on translations, and on n-form gauge transformations, with $n = 1,3,5,7$. The existence of a triplet of 8-forms, and the relation between these forms, was also discussed in [13, 14].

10-forms There are 10-forms in two $SU(1,1)$ representations: a doublet $A_{(10)}^{\alpha}$, with the usual reality condition, and a quadruplet $A_{(10)}^{\alpha\beta\gamma}$, symmetric in α,β,γ, satisfying

$$(A^{111})^* = A^{222}, \qquad (A^{112})^* = A^{122}. \tag{12}$$

There are no conditions relating the 10-forms to other fields. However, the doublet and the quadruplet differ in an important respect. The doublet does not transform under n-form gauge transformations with $n < 9$, while the fields of the quadruplet transform under all lower rank gauge transformations. Of course the volume form is also a 10-form, and so is the product of the volume form with arbitrary functions of the scalar fields. However, the volume form is essentially a product of tenbeins, and therefore not a new field in the IIB supergravity theory

3. COUPLING TO BRANES

We now wish to investigate to which kind of branes the different n-form potentials couple. In particular we would like to know the tension of the corresponding branes. These tensions can be determined from the supersymmetry rules as follows. To be concrete let us consider the 8-form potentials. After gauge-fixing the generic supersymmetry rule of the 3 different potentials is as follows (in string frame):

$$\delta A_{(8)} \sim f(\tau,\bar\tau)\bar\varepsilon\gamma\bigl(a\psi_\mu + b\gamma_\mu\lambda\bigr) + \cdots, \tag{13}$$

where a,b are constants, the dots stand for other terms and the scalars have been expressed in terms of a complex scalar τ. The function $f(\tau,\bar\tau)$ can be expressed in terms of the dilaton ϕ and axion $C_{(0)}$ via the relation $\tau = C_{(0)} + ie^{-\phi}$. For our present purposes it is sufficient to consider the case of zero axion.

The same 8-form potentials may occur as Wess-Zumino terms in a supersymmetric 7-brane action as follows:

$$\underbrace{\mathscr{L}_{\text{brane}} \sim \quad e^{-\alpha\phi}}_{\text{brane tension at } C_{(0)}=0} \sqrt{-g} \; + \; A_{(8)} \; + \; \cdots \tag{14}$$

Before fixing kappa-symmetry the first, Nambu-Goto, term and the second, Wess-Zumino, term are separately supersymmetric. After gauge-fixing kappa-symmetry the (linear) supersymmetry variations of the two terms should cancel. Clearly, this is only possible if the function $f(\tau,\bar{\tau})$ is proportional to the brane tension $e^{-\alpha\phi}$. To achieve this one must consider particular combinations of the 8-form potentials. This enables us to read the brane tensions from the supersymmetry rules. This indeed works for two of the three 8-form potentials, leading to the combinations $C_{(8)}$ and $B_{(8)}$ in Table 1. They couple to the D7-brane and the S-dual $\widetilde{\text{D7}}$-brane (with exotic brane tension g_s^{-3}), respectively. However, for the third 8-form potential, called the combination $D_{(8)}$ in the Table, we find that $a=0$ in (13) and hence there is no corresponding supersymmetric brane action since the Nambu-Goto term always transforms to a term linear in the gravitino.

TABLE 1. The triplet of 8-form potentials and their 7-branes

potential	associated brane	tension
$C_{(8)}$	D7	g_s^{-1}
$D_{(8)}$	—	—
$B_{(8)}$	$\widetilde{\text{D7}}$	g_s^{-3}

A similar analysis can be performed for the 10-form potentials. The result is summarized in Tables 2 and 3. Note that, unlike $D_{(8)}$, the 10-form potentials $D_{(10)}$ and $E_{(10)}$ transform to the gravitino. Nevertheless we cannot associate a supersymmetric 9-brane action with these potentials since they do not transform to the correct combination of gravitino and dilatino to establish supersymmetry.

TABLE 2. The quadruplet of 10-form potentials and their 9-branes

potential	associated brane	tension
$C_{(10)}$	D9	g_s^{-1}
$D_{(10)}$	—	—
$E_{(10)}$	—	—
$B_{(10)}$	exotic	g_s^{-4}

4. TRUNCATION TO $N=1$

We can truncate our results for $N=2$ supergravity to find the $N=1$ algebra [10]. Since in $D=10$ the $N=1$ supergravity is unique, there is only one independent truncation, all others being related by field redefinitions. In spite of this, since there are two inequivalent

TABLE 3. The doublet of 10-form potentials and their 9-branes

potential	associated brane	tension
$\mathcal{D}_{(10)}$	solitonic	g_s^{-2}
$\mathcal{E}_{(10)}$	exotic	g_s^{-3}

$N = 1$ string theories, it is instructive to truncate the $N = 2$ theory in two different ways leading to the low energy limits of $D = 10$ heterotic and type I string theory. Hence we perform the "heterotic" and the "type I" truncations [11].
The heterotic truncation can be derived from the IIB algebra by setting

$$\varepsilon = \varepsilon_C. \tag{15}$$

This projects out the following fields from the IIB spectrum:

$$C_{(0)}, C_{(2)}, C_{(4)}, C_{(6)}, C_{(8)}, B_{(8)}, C_{(10)}, E_{(10)}, \mathcal{E}_{(10)}. \tag{16}$$

Further, $B_{(10)}$ and $\mathcal{D}_{(10)}$ turn out to be dependent fields of the form $e^{x\phi}\varepsilon_{(10)}$ (where $\varepsilon_{(10)}$ is the volume form) for some x in the truncated theory. Therefore, the field contents of the heterotic truncation of the $D = 10$, IIB supergravity is given by

$$\phi, B_{(2)}, B_{(6)}, D_{(8)}, D_{(10)}. \tag{17}$$

The supersymmetry algebra which is realised on these fields can easily be obtained by setting all truncated and dependent fields to zero in the IIB algebra as presented in [10]. The type I truncation can be derived from the IIB algebra by setting

$$\varepsilon = i\varepsilon_C. \tag{18}$$

This projects out the following fields from the IIB spectrum:

$$C_{(0)}, B_{(2)}, C_{(4)}, B_{(6)}, C_{(8)}, B_{(8)}, D_{(10)}, B_{(10)}, \mathcal{D}_{(10)}. \tag{19}$$

Similarly to the heterotic case, we find that $C_{(10)}$ and $\mathcal{E}_{(10)}$ turn out to be dependent fields of the form $e^{x\phi}\varepsilon_{(10)}$ for some x in the truncated theory. Therefore the field contents of the type I truncation is given by

$$\phi, C_{(2)}, C_{(6)}, D_{(8)}, E_{(10)}. \tag{20}$$

5. VERY EXTENDED SYMMETRY GROUPS

Collecting all n-form potentials of the IIB theory we find that they transform nonlinearly under the bosonic gauge transformations in the following generic form:

$$\delta A = d\Lambda + F \wedge \Lambda, \qquad F = dA + A \wedge F. \tag{21}$$

Since the gauge transformation rules only contain gauge-invariant curvatures, the bosonic gauge algebra is Abelian. Surprisingly, it turns out that the bosonic gauge transformations can all be rewritten in terms of

$$\Lambda_{(2n)} \equiv d\Lambda_{(2n-1)}. \tag{22}$$

After an appropriate (field-dependent) redefinition of the gauge fields and the parameters, all transformation rules become linear but the resulting bosonic gauge algebra is non-Abelian. Schematically, we thus obtain the following non-trivial commutators:

$$[\mathbf{2},\mathbf{2}] = \mathbf{4}, \qquad [\mathbf{2},\mathbf{4}] = \mathbf{6}, \qquad [\mathbf{2},\mathbf{6}] = \mathbf{8}, \cdots \tag{23}$$

We thus see an interesting structure arising: all gauge fields can be obtained by applying a number of times the basic **2** gauge transformation. This number is the so-called level of the gauge field. A similar structure arises in the IIA case where the basic building blocks are the RR 1-form **1** and the NS 2-form **2**:

$$[\mathbf{1},\mathbf{1}] = 0, \qquad [\mathbf{1},\mathbf{2}] = \mathbf{3}, \qquad [\mathbf{1},\mathbf{3}] = 0, \qquad [\mathbf{2},\mathbf{3}] = \mathbf{5}, \qquad [\mathbf{1},\mathbf{5}] = \mathbf{6}, \cdots \tag{24}$$

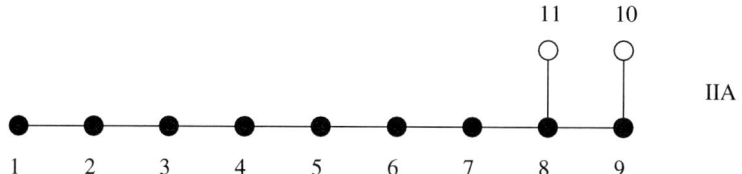

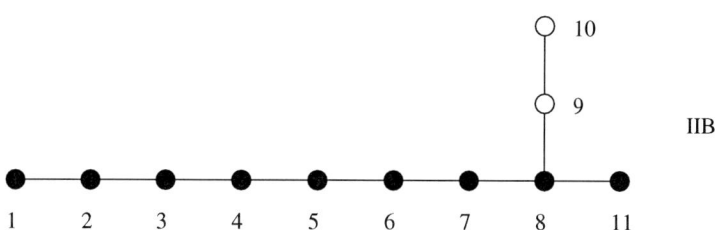

FIGURE 1. The Dynkin diagrams leading to IIA and IIB representations

The above is very reminiscent to recent work on a hyperbolic E_{11}-symmetry that might underly string and/or M-theory, see [10] for a list of references. In particular, in [15], representations of the E_{11} algebra are worked out for different embeddings of a bosonic $GL(10)$ subalgebra. This leads to the Dynkin diagrams of Figure 1. In these diagrams the horizontal line represents the $GL(10)$ subalgebra whereas the empty dots are related to our basic building blocks in the following way:

$$\text{IIA}: \quad 10 \leftrightarrow \mathbf{1}, \quad 11 \leftrightarrow \mathbf{2} \qquad (25)$$
$$\text{IIB}: \quad 9, 10 \leftrightarrow \mathbf{2}. \qquad (26)$$

It would be interesting to pursue this relationship further.

ACKNOWLEDGMENTS

One of us (E.B.) would like to thank the organizers of the XXVIII Spanish Relativity Meeting for a very pleasant atmosphere. We wish to thank Axel Kleinschmidt for useful discussions. E.B., S.K. and M. de R. are supported by the European Commission FP6 program MRTN-CT-2004-005104 in which E.B., S.K. and M. de R. are associated to Utrecht University. S.K. would like to thank the DAAD for financial support. Figure 1 is taken from [15]. The work of F.R. is supported by a European Commission Marie Curie Postdoctoral Fellowship, Contract MEIF-CT-2003-500308.

REFERENCES

1. J. H. Schwarz and P. C. West, *Symmetries And Transformations Of Chiral N=2 D = 10 Supergravity*, Phys. Lett. B **126** (1983) 301.
2. J. H. Schwarz, *Covariant Field Equations Of Chiral N=2 D = 10 Supergravity*, Nucl. Phys. B **226** (1983) 269.
3. P. S. Howe and P. C. West, *The Complete N=2, D = 10 Supergravity*, Nucl. Phys. B **238** (1984) 181.
4. F. Giani and M. Pernici, *N=2 Supergravity in Ten-Dimensions*, Phys. Rev. D**30** (1984) 325.
5. I. C. G. Campbell and P. C. West, *N=2 D=10 Nonchiral Supergravity and its Spontaneous Compactification*, Nucl. Phys. B**243** (1984) 112.
6. M. Huq and M. A. Namizie, *Kaluza-Klein Supergravity in Ten-Dimensions*, Class. Quant. Grav. **2** (1985) 293 [Erratum-ibid **2** (1985) 597].
7. L. J. Romans, *Massive N=2a Supergravity in Ten-Dimensions*, Phys. Lett. B**169** (1986) 374.
8. E. A. Bergshoeff, R. Kallosh, T. Ortín, D. Roest and A. Van Proeyen, *New formulations of D=10 supersymmetry and D8–O8 domain walls*, Class. Quant. Grav. **18** (2001) 3359 [arXiv:hep-th/0103233].
9. E. Bergshoeff, M. de Roo, M. B. Green, G. Papadopoulos and P. K. Townsend, *Duality of Type II 7-branes and 8-branes*, Nucl. Phys. B**470** (1996) 113-135 [arXiv:hep-th/9601150].
10. E. A. Bergshoeff, M. de Roo, S F. Kerstan and F. Riccioni, *IIB Supergravity Revisited*, JHEP 0508 (2005) 098 [arXiv:hep-th/0506013].
11. E. Bergshoeff, M. de Roo, B. Janssen and T. Ortín, *The super D9-brane and its truncations*, Nucl. Phys. B **550** (1999) 289 [arXiv:hep-th/9901055].
12. E. A. Bergshoeff, M. de Roo, S F. Kerstan, T. Ortín and F. Riccioni, *in preparation*.
13. P. Meessen and T. Ortín, *An Sl(2,Z) multiplet of nine-dimensional type II supergravity theories*, Nucl. Phys. B **541** (1999) 195 [arXiv:hep-th/9806120].
14. G. Dall'Agata, K. Lechner and M. Tonin, *D = 10, N = IIB supergravity: Lorentz-invariant actions and duality*, JHEP **9807** (1998) 017 [arXiv:hep-th/9806140].
15. A. Kleinschmidt, I. Schnakenburg and P. West, *Very-extended Kac-Moody algebras and their interpretation at low levels*, Class. Quant. Grav. **21** (2004) 2493 [arXiv:hep-th/0309198].

The Double Pulsar System J0737−3039A/B as Testbed for Relativistic Gravity

M. Burgay*, A. Possenti*, M. Kramer[†], R. N. Manchester**, N. D'Amico*,[‡], A. G. Lyne[†], M. A. McLaughlin[†], D. R. Lorimer[†], F. Camilo[§], I. H. Stairs[¶], P. C. C. Freire[∥] and B. C. Joshi[††]

*INAF – Osservatorio Astronomico di Cagliari, Loc. Poggio dei Pini, Strada 54, 09012 Capoterra (CA), Italy
[†]University of Manchester, Jodrell Bank Observatory, Macclesfield, Cheshire, SK11 9DL, UK
**Australia Telescope National Facility, CSIRO, P.O. Box 76, Epping, New South Wales 2121, Australia
[‡]Universit' degli Studi di Cagliari, Dipartimento di Fisica, SP Monserrato-Sestu km 0.7, 09042 Monserrato, Italy
[§]Columbia Astrophysics Laboratory, Columbia University, 550 West 120 th Street, New York 10027, USA
[¶]University of British Columbia, 6224 Agricultural Road Vancouver, BC V6T 1Z1, Canada
[∥]NAIC, Arecibo Observatory, Puerto Rico, US
[††]National Center for Radio Astrophysics, P.O. Bag 3, Ganeshkhind, Pune 411007, India

Abstract. The double pulsar system J0737−3039A/B is one of the most intriguing pulsar discoveries of the last decade. This binary system, with an orbital period of only 2.4-hr, provides a truly unique laboratory for relativistic gravity. Its discovery enhances of about an order of magnitude the estimate of the merger rate of double neutron stars systems, opening new possibilities for the current generation of gravitational wave detectors. In this contribution we summarize the present results and look at the prospects of future observations.

Keywords: Neutron star – Pulsar: individual (PSR J0737-3039A/B) – General Relativity
PACS: 97.60.Gb, 04.80.Cc

INTRODUCTION

The 22.7-ms binary ulsar PSR J0737-3039A (hereafter 'A') was discovered by our team in April 2003 [1] in the Parkes High-Latitude Pulsar Survey [2, in preparation]. Its short orbital period ($P_b = 2.4$ hrs), combined with a remarkably high value of the periastron advance ($\dot{\omega} = 16.9$ deg/yr), measurable after only few days of observations, identified it soon as a member of the most extreme relativistic binary system ever discovered. The compactness of the system, together with its short coalescence time ($T_{coal} = 85$ Myr) and low luminosity, boosts hopes to detect mergers of neutron stars with ground based gravitational wave detectors, increasing the estimates on the double neutron star coalescence rate by almost an order of magnitude [1, 3].

Analysis of follow-up observations, covering the entire orbit, led, in October 2003, to the discovery of the second pulsar in the system [4], the 2.8-s pulsar J0737-3039B (hereafter 'B'). The reason why the signal of pulsar A's companion was not detected earlier is that B is only bright in two short sections of the orbit; for the rest of the orbit the signal is very week or absent.

A closer inspection to the signals of both pulsar A and B reveals also other intriguing characteristics: pulsar A is eclipsed for ~ 30 s near superior conjunction and pulsar B shows variations in the pulse shape along the orbit [4]. More recently, also variations of the extent and location of B's bright phases and of the pulse shape on longer time scales have been observed [5]. These phenomena are probably related to the geodetic precession of pulsar A and B that are changing the geometry of the system and hence our view towards it.

In this contributions, we will concentrate on the description of the binary system J0737-3039A/B as test-ground for relativistic theories and on the implication of the discovery of this system on the probability to detect gravitational waves with ground based interferometers.

TEST OF GENERAL RELATIVITY

Due to their strong gravitational fields and rapid motions, the binary systems containing two neutron stars exhibit large relativistic effects [6]. When these are large enough, the system can be used for testing the predictions of theories of gravity in the strong-field limit. Tests can be performed when a number of relativistic corrections to the Keplerian description of an orbit, the so called, "post-Keplerian" (PK) parameters, can be measured. In each theory the PK parameters can be written as a function of the masses of the two stars and of the measurables Keplerian parameters. With the two masses as the only unknowns, the measurement of three or more PK parameters over-constrains the system hence providing tests for a given theory of gravity [7].

In General Relativity (GR) the post-Keplerian parameters can be written (at first post-Newtonian order, 1PN) as follows [6]:

$$\dot{\omega} = 3T_\odot^{2/3} \left(\frac{P_b}{2\pi}\right)^{-5/3} \frac{1}{1-e^2} (M_A+M_B)^{2/3},$$

$$\gamma = T_\odot^{2/3} \left(\frac{P_b}{2\pi}\right)^{1/3} e \frac{M_B(M_A+2M_B)}{(M_A+M_B)^{4/3}},$$

$$\dot{P}_b = -\frac{192\pi}{5} T_\odot^{5/3} \left(\frac{P_b}{2\pi}\right)^{-5/3} \frac{\left(1+\frac{73}{24}e^2+\frac{37}{96}e^4\right)}{(1-e^2)^{7/2}} \frac{M_A M_B}{(M_A+M_B)^{1/3}},$$

$$r = T_\odot M_B,$$

$$s = T_\odot^{-1/3} \left(\frac{P_b}{2\pi}\right)^{-2/3} x \frac{(M_A+M_B)^{2/3}}{M_B},$$

where P_b is the orbital period, e the eccentricity and x the projected semi-major axis of the orbit measured in light-s. The masses M_A and M_B of A and B respectively (or, in general, of the pulsar and its companion), are expressed in solar masses ($M_\odot$). We define the constant $T_\odot = GM_\odot/c^3 = 4.925490947 \mu s$ where G denotes the Newtonian constant of gravity and c the speed of light. The first PK parameter, $\dot{\omega}$, describes the relativistic advance of periastron. The parameter γ denotes the amplitude of delays in arrival times caused by the varying effects of the gravitational redshift and time dilation as the pulsar moves in its elliptical orbit at varying distances from the companion and with varying

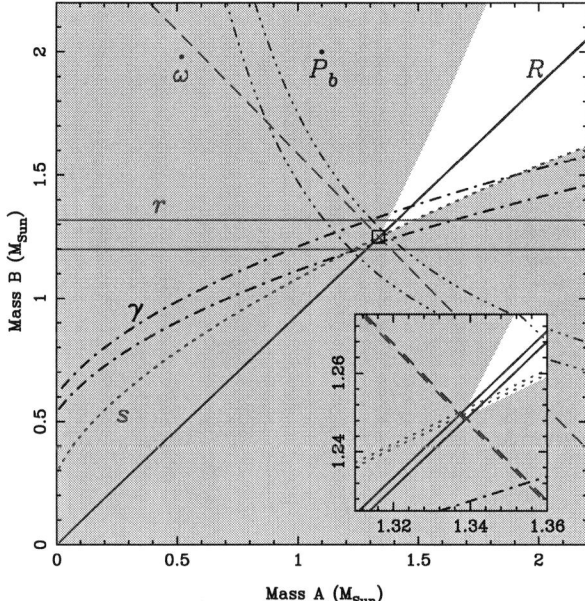

FIGURE 1. The observational constraints upon the masses M_A and M_B. The coloured regions are those which are excluded by the Keplerian mass functions of the two pulsars. Further constraints are shown as pairs of lines enclosing permitted regions as predicted by general relativity: (a) the measurement of the advance of periastron $\dot{\omega}$ (dashed lines); (b) the measurement of the mass ratio R (solid lines); (c) the measurement of the gravitational red-shift/time dilation parameter γ (dot-dash lines); (d) the measurement of Shapiro parameter r (solid horizontal lines) and Shapiro parameter s (dotted lines); (e) the measurement of the orbital decay (dot-dot-dot-dash lines). Inset is an enlarged view of the small square which encompasses the intersection of the three tightest constraints, with the scales increased by a factor of 16. The permitted regions are those between the pairs of parallel lines and we see that an area exists which is compatible with all constraints.

speeds. The decay of the orbit due to gravitational wave damping is expressed by the change in orbital period, $\dot{P}_b$. The other two parameters, r and s, are related to the Shapiro delay caused by the gravitational field of the companion.

The PK parameter can be plotted on a mass-mas diagram (see e.g. Fig. 1) and, if the theory tested is correct, the curves on the plane must intersect in a single point.

In this context, PSR J0737-3039A/B promises to be the most powerful instrument to test GR (and other theories) providing us with *two* pulsars, extremely stable clocks, in the same system. Timing measurements of pulsar A, infact, have already provided all 5 post-Keplerian parameters with high accuracy (see Table 1). Moreover, with the knowledge of the projected semimajor-axes for both A and B, we obtain a precise measurement of the mass ratio R of the two stars:

$$R \equiv M_A/M_B = x_B/x_A \qquad (1)$$

TABLE 1. Observed and derived parameters of PSRs J0737−3039A and B. Number in parentheses are standard errors on the last digit(s).

Pulsar	PSR J0737−3039A	PSR J0737−3039B
Pulse period P (ms)	22.699378556138(2)	2773.4607474(4)
Period derivative $\dot{P}$	$1.7596(2) \times 10^{-18}$	$0.88(13) \times 10^{-15}$
Epoch of period (MJD)	52870.0	
Right ascension α (J2000)	$07^h 37^m 51^s.24795(2)$	
Declination δ (J2000)	$-30°39'40''.7247(6)$	
Orbital period P_b (day)	0.1022515628(2)	
Eccentricity e	0.087778(2)	
Epoch of periastron T_0 (MJD)	52870.0120588(3)	
Advance of periastron $\dot{\omega}$ (deg yr^{-1})	16.900(2)	
Longitude of periastron ω (deg)	73.805(1)	73.805 + 180.0
Projected semi-major axis $x = a\sin i/c$ (sec)	1.415032(2)	1.513(4)
Gravitational redshift parameter γ (ms)	0.39(2)	
Shapiro delay parameter $s = \sin i$	0.9995(4)	
Shapiro delay parameter r (μs)	6.2(6)	
Orbital decay $\dot{P}_b$ (10^{-12})	−1.20(8)	
Mass ratio $R = M_A/M_B$	1.071(1)	

For every realistic theory of gravity, we can expect the mass ratio, R, to follow this simple relation [7], at least to 1PN order. Most importantly, the R value is not only theory-independent, but also independent of strong-field (self-field) effects which is not the case for PK-parameters. This provides a stringent and new constraint for tests of gravitational theories as any combination of masses derived from the PK-parameters *must* be consistent with the mass ratio. With five PK parameters already available, this additional constraint makes the double pulsar the most overdetermined system to date providing four possible tests for relativistic theories.

Since the precision with which we can measure the PK parameters increases with time, continued observation of J0737-3039A/B will eventually provide us with the necessity to include higher order corrections to the PK parameters. In particular, within few years, we could be able to measure the contribution of the spin-orbit coupling to the observed $\dot{\omega}$. This extra term in the periastron advance is related to the moment of inertia of the star which would be measured for the first time for a neutron star also providing tight constraints on the neutron stars' equation of state.

Another effect predicted by General Relativity is that, if the total angular momentum vector is not aligned with the spin axis, the latter will precess about the orbit normal. The predicted periods for geodetic precession for PSR J0737−3039A and B are only 75 and 71 yr respectively. Because of that the geometry of the system is expected to change in a short time scale and this should result in secular changes in the observed pulse shape, because the line of sight across the emission beam changes. Somewhat surprisingly, Parkes observations of pulsar A show no significant evidence for profile-shape variation over a three-year interval [8]. Near alignment of A's rotation axis with the orbit normal is a possible explanation for the observed lack of variations.

One of the many unique features of the double-pulsar system J0737−3039A/B, as mentioned above, is the dramatic orbital modulation of the pulsed emission from PSR J0737−3039B. Both the shape and intensity of the B pulse profile vary with orbital longitude, with two bright phases centered around longitudes (with respect to the ascending node) of $\sim 210°$ and $\sim 280°$, respectively. The pulse is weakly visible at other orbital longitudes except perhaps between $30°$ and $60°$.

This unprecedented behaviour has been interpreted as due to the interaction between radiation from pulsar J0737−3039A and B's magnetosphere [9, 10, 11, 12]. Direct evidence of the mutual interaction between the two pulsars is manifested by the drifting behaviour at A's period in the single pulses of B [13] and by the modulation at B's period of the emission from A during its eclipse [14].

Analysis of pulsar B over 20 months of observations showed also *secular* variations both in the pule profiles and in the orbital modulation pattern [5]. Pulse profiles are becoming single-peaked in both bright phases of the orbital modulation, although there is no clear variation in overall pulse width (Fig. 2, left). The shape of the orbital modulation is also varying systematically, with both bright phases shrinking in longitude (Fig. 2, right). The combined span of the two bright phases, however, is relatively constant and together they are shifting to higher longitudes at a rate of $\sim 3°\ \mathrm{yr}^{-1}$.

These effects can be ascribed to geodetic precession that, changing the geometry of the system, changes both our sight through it and the angle of impact of A's wind on B's open magnetic field lines, probably responsible for the orbital modulation of pulsar B signal.

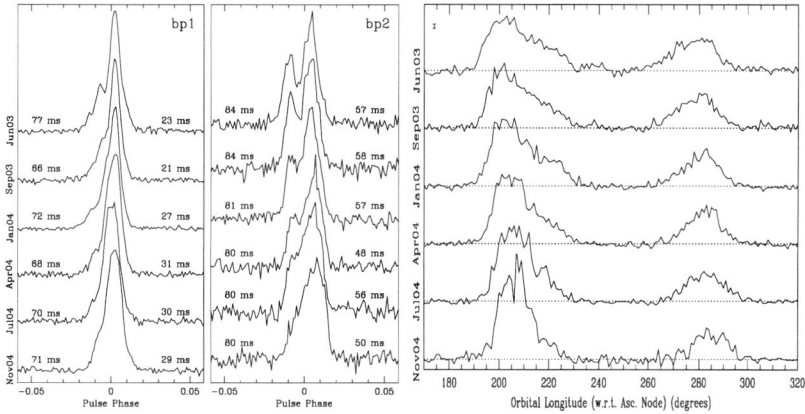

FIGURE 2. Left panel: mean pulse profiles of PSR J0737−3039B at 1390 MHz for the two bright phases, bp1 (left, orbital longitude $190°$ to $235°$) and bp2 (right, orbital longitude $260°$ to $300°$) at six epochs from 2003 June (top) to 2004 November (bottom). All profiles have been scaled to the same peak height. Numbers near each profile are the pulse widths at 10% (left) and 50% (right) of the pulse peak. Right panel: variation of pulse intensity as a function of orbital longitude for six epochs. The small bar represents the typical rms noise in the baseline.

A REVISED DOUBLE NEUTRON STAR COALESCENCE RATE.

The merging of a double-neutron-star system should produce a burst of emission of gravitational waves. Due to the energy budget and to the expected typical frequency of these events, they are among the primary targets for the current generation of ground-based gravitational waves detectors, which should be able to detect them up to a distance of about 20 Mpc. Hence, a key question is the occurrence rate of these double-neutron-star coalescences in a volume of universe of that radius. This rate can in turn be estimated on the basis of the rate of events in the Galaxy.

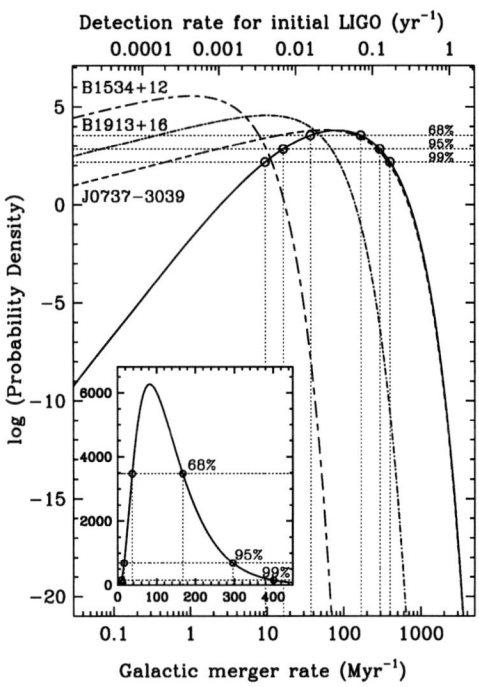

FIGURE 3. Probability density function of the actual double neutron star binary merger rate in the Galaxy (bottom axis) and the predicted initial LIGO detector rate (top axis). The solid line shows the total probability density. Dashed lines show those obtained for each of the three merging binary systems considered. Inset: Total probability density, and corresponding 68%, 95%, and 99% confidence limits, shown in a linear scale [3].

Among the double-neutron-star systems previously known, only three had tight enough orbits so that the two neutron stars will merge within a Hubble time. Two of them (PSR B1913+16 and PSR B1534+12) are located in the Galactic field, while the third (PSR B2127+11C) is found on the outskirts of a globular cluster. The contribution of globular cluster systems to the Galactic merger rate is estimated to be negligible [15]. Also, recent studies [16] have demonstrated that the current estimate of the Galactic merger rate $\mathscr{R}$ relies mostly on PSR B1913+16 characteristics. One can hence start by comparing the observed properties of B1913+16 and J0737−3039 systems. The latter will merge due to the emission of gravitational waves in ∼ 85 Myr, a time-scale that is a factor 3.5 shorter than that for PSR B1913+16. In addition, the estimated distance of PSR J0737−3039A/B (500 - 600 pc, based on the observed dispersion measure and a model for the distribution of ionized gas in the interstellar medium [17]) is an order of magnitude less than that of PSR B1913+16. These properties have a substantial effect on the prediction of the rate of merging events in the Galaxy.

For a given class k of binary pulsars in the Galaxy, apart from a beaming correction factor, the merger rate $\mathscr{R}_k$ is calculated as $\mathscr{R}_k = N_k/\tau_k$ [16]. Here τ_k is the binary pulsar lifetime and N_k is the scaling factor defined as the number of binaries in the Galaxy belonging to the given class. The shorter lifetime of J0737−3039 system [τ_{1913}/τ_{0737} = (365 Myr)/(185 Myr)] implies a doubling of the ratio $\mathscr{R}_{0737}/\mathscr{R}_{1913}$. A much more substantial increase results from the computation of the ratio of the scaling factors N_{0737}/N_{1913}. The luminosity L_{400}=30 mJy kpc^2 of PSR J0737−3039A is much lower than that of PSR B1913+16. For a planar homogeneous distribution of pulsars in the Galaxy, the ratio $N_{0737}/N_{1913} \sim L_{1913}/L_{0737} \sim 6$. Hence we obtain $\mathscr{R}_{0737}/\mathscr{R}_{1913} \sim 12$. Including the moderate contribution of the longer-lived PSR B1534+12 system to the total rate, we obtain an increase factor for the total merger rate of about an order of magnitude.

Extensive simulations [3] give results consistent with this simple estimate and show that the peak of the merger rate increase factor resulting from the discovery of J0737−3039 system lies in the range 5-7 (Fig. 3) and is largely independent of the adopted pulsar population model. For the reference model (model nr. 6 of [16]), the updated cosmic detection rate for first generation gravitational-wave detectors is about one every three years (for initial LIGO; two per year with advanced LIGO). Hence, with the discovery of PSR J0737−3039A/B the double-neutron-star coalescence rate estimates enter an astrophysical interesting regime. Within a few years of gravitational-wave detectors operations, it should be possible to directly test these predictions and, in turn, place better constraints on the cosmic population of double-neutron-star binaries.

ACKNOWLEDGMENTS

MB, AP and NDA received support from the Italian Ministry of University and Research (MIUR) under the national program *Cofin 2003*. The Parkes radio telescope is part of the Australia Telescope which is funded by the Commonwealth of Australia for operation as a National Facility managed by CSIRO. FC acknowledges support from NASA grant NNG05GA09G.

REFERENCES

1. M. Burgay, N. D'Amico, A. Possenti, R. N. Manchester, A. G. Lyne, B. C. Joshi, M. A. McLaughlin, M. Kramer, J. M. Sarkissian, F. Camilo, V. Kalogera, C. Kim, and D. R. Lorimer, *Nature* **426**, 531– 533 (2003).
2. M. Burgay et al. 2005, in preparation.
3. V. Kalogera, C. Kim, D. R. Lorimer, M. Burgay, N. D'Amico, A. Possenti, R. N. Manchester, A. G. Lyne, B. C. Joshi, M. A. McLaughlin, M. Kramer, J. M. Sarkissian, and F. Camilo, *ApJ* **601**, L179– L182 (2004).
4. A. G. Lyne, M. Burgay, M. Kramer, A. Possenti, R. N. Manchester, F. Camilo, M. McLaughlin, D. R. Lorimer, B. C. Joshi, J. E. Reynolds, and P. C. C. Freire, *Science* **303**, 1153–1157 (2004).
5. M. Burgay, A. Possenti, R. N. Manchester, M. Kramer, M. A. McLaughlin, D. R. Lorimer, I. H. Stairs, B. C. Joshi, A. G. Lyne, F. Camilo, N. D'Amico, P. C. C. Freire, J. M. Sarkissian, A. W.

Hotan, and G. B. Hobbs, *ApJ* **624**, L113–L116 (2005).
6. T. Damour, and N. Deruelle, *Ann. Inst. H. Poincaré (Physique Théorique)* **44**, 263–292 (1986).
7. T. Damour, and J. H. Taylor, *Phys. Rev. D* **45**, 1840–1868 (1992).
8. R. N. Manchester, and M. Kramer, *ApJ* (2005).
9. F. A. Jenet, and S. M. Ransom, *Nature* **428**, 919–921 (2004).
10. B. Zhang, and A. Loeb, *ApJ* **614**, L53–L56 (2004).
11. M. Lyutikov, *MNRAS* **353**, 1095–1106 (2004).
12. M. Lyutikov, *MNRAS* **362**, 1078–1084 (2005).
13. M. A. McLaughlin, M. Kramer, A. G. Lyne, D. R. Lorimer, I. H. Stairs, A. Possenti, R. N. Manchester, P. C. C. Freire, B. C. Joshi, M. Burgay, F. Camilo, and N. D'Amico, *ApJ* **613**, L57–L60 (2004).
14. M. A. McLaughlin, A. G. Lyne, D. R. Lorimer, A. Possenti, R. N. Manchester, F. Camilo, I. H. Stairs, M. Kramer, M. Burgay, N. D'Amico, P. C. C. Freire, B. C. Joshi, and N. D. R. Bhat, *ApJ* **616**, L131–L134 (2004).
15. E. S. Phinney, *ApJ* **380**, L17–L21 (1991).
16. C. Kim, V. Kalogera, and D. R. Lorimer, *ApJ* **584**, 985–995 (2003).
17. J. H. Taylor, and J. M. Cordes, *ApJ* **411**, 674–684 (1993).

The principle of general covariance has physical content

Alberto Chamorro

Department of Theoretical Physics, University of the Basque Country, 48080 Bilbao, Spain

Abstract. The issue of whether or not the Principle of General Covariance (GCP) has physical content has been matter of debate and confusion since the inception of General Relativity. In our view the physical meaning of coordinates is related to the question of the possible physical significance of that principle. We believe that the latter may be taken as an appropriate generalized principle of relativity with physical content. With the purpose of throwing light over the subject, after presenting our version of the GCP, we define and construct quasi-Minkowskian coordinates associated to the word-line of an observer who transports an orthonormal tetrad (QMCCω). We view the QMCCω as the coordinates that would be obtained by that observer by applying operational protocols valid in flat space-time to get the Lorentzian coordinates of an event. The set of all the QMCCω is in general an infinite family all of whose members collapse to the usual Lorentzian coordinates when the observer is in free fall, his or her space triad does not rotate ($\omega = 0$) and the curvature of space-time vanishes. This implements the idea that the set of all the operational protocols which are equivalent -in the sense of assigning the same numerical values- to obtain the Lorentzian coordinates of events in flat space-time split into inequivalent subsets of operational prescriptions under the presence of a gravitational field or when the observer is not inertial. Something similar must happen with all the physical quantities. Other considerations will be presented.

INTRODUCTION

Since first formulated nine decades ago the question of the possible physical content of the GCP has been a subject of polemic and confusion. Thus Kretschmann [1] in 1917 claimed the GCP to be devoid of physical content and that given enough mathematical ingenuity any theory could be set in a general covariant form. Einstein [2] begrudgingly accepted the objection stating however the heuristic value the GCP had in searching for a good theory and that that was a reason to prefer General Relativity to Newtonian gravitation which -in his opinion- would only be awkwardly casted into generally covariant form. Einstein was soon proved wrong as Cartan [3] in 1923 and Friedrichs [4] in 1927 found serviceable generally covariant formulations of Newtonian gravitation theory. See also Misner *et al*(1973, ch 12) [5]. In his excellent book Fock [6] makes interesting and critical remarks about the term "general relativity" adopted by Einstein to name his theory of gravitation and the connection of the term with general covariance that, in his view, is merely a logical requirement that is always satisfiable. Fock rightly points out that though Einstein had agreed with Kretschman objection as to the physical vacuity of the GCP his agreement was rather formal, because actually to the end of his life Einstein related the requirement of general covariance to the idea of some kind of "general relativity" and with the equivalence of all frames of reference. The subject has subsequently been addressed in several ways, for example, by Anderson [7](1967), Stachel [8](1986,

2002), Norton [9](1993) and Ellis and Matravers [10](1995). All these works while attemting to clarify the formulation and meaning of the GCP in our opinion fail to give it a specific expression susceptible of physical verification or contradiction. And certainly whatever the claim about the physical content of the GCP might be that should be subject to experimental test to be confirmed or refuted. Ellis and Matravers [10] point out how physicists and astrophysicists in fact almost always use preferred coordinate systems not merely to simplify the calculations but also to help define quantities of physical interest, and that this suggests that we should reconsider and perhaps refine the dogma od general covariance. In that spirit we present in this contribution a proposal for the GCP that may be proved wrong should that be the case, that is, that in principle may be falsifiable in the Popperian sense. The plan of the paper is as follows: In Section 2 we define a principle of general relativity and take it as the GCP. In Section 3 we construct a family of coordinates which are useful to endow with physical meaning the GCP. Some consequences of our formulation of the GCP and further work to be done to explore its potential will be indicated in Section 4.

THE GCP AS A PRINCIPLE OF GENERAL RELATIVITY

We will first formulate a principle of general relativity (GRP) as a physical principle subject to experimental test. To do that we need a few preliminay definitions: Let F be a physical quantity. Let $Q[F]$ be the set of all the different operational protocols -but equivalent in the sense that they yield the same values- that may be used to measure F in flat space-time when working in Lorentzian coordinates: $\eta_{\mu\nu} = diag(+1,+1,+1,-1)$. The physical meaning of F is given by $Q[F]$. Let S be an inertial frame and L the Lorentz group of transformations. Then if $\Lambda \in L$ one has that the action of Λ on S and F implies the following:

$$\Lambda : S \longrightarrow S', \ \Lambda : F \longrightarrow F' \Longrightarrow Q[F] = Q[F'] \ \forall \ \Lambda \in L \text{ and } \forall \ F.$$

Let O be an observer and C his world-line given by its equations $x^\lambda = f^\lambda(\tau)$, where the x^λ's, $\lambda = 0,1,2,3$, are any particular smooth but otherwise generic local coordinates and τ is O's proper time. O's four-velocity is $u^\lambda = \frac{dx^\lambda}{d\tau} = \dot{f}^\lambda(\tau)$, $u^\lambda u_\lambda = -c^2$. O transports an orthonormal tetrad $e_{(v)}$ along his world-line whose components verify

$$e^\lambda_{(v)} e_{(\mu)\lambda} = \eta_{v\mu} \ , \quad e^\mu_{(0)} = \frac{u^\mu}{c} . \tag{1}$$

The transportation law of the tetrad is given by the covariant derivatives of their components with respect to τ according to

$$\frac{De^\mu_{(\sigma)}}{d\tau} = \frac{1}{c^2}(u^\mu a^\nu - u^\nu a^\mu) e_{(\sigma)\nu} + \frac{1}{c} \omega_\alpha u_\beta \varepsilon^{\alpha\beta\mu\nu} e_{(\sigma)\nu}, \tag{2}$$

where

$$a^\nu = \frac{Du^\nu}{d\tau} , \quad \varepsilon_{\alpha\beta\gamma\delta} = (-g)^{\frac{1}{2}}[\alpha\beta\gamma\delta] , \quad \varepsilon^{\alpha\beta\gamma\delta} = -(-g)^{-\frac{1}{2}}[\alpha\beta\gamma\delta] ,$$

$$[\alpha\beta\gamma\delta] = \begin{cases} +1 & \text{if } \alpha\beta\gamma\delta \text{ is an even permutation of } 0123 \\ -1 & \text{if } \alpha\beta\gamma\delta \text{ is an odd permutation of } 0123 \\ 0 & \text{if } \alpha\beta\gamma\delta \text{ are not all different} \end{cases}$$

$$g = \det \|g_{\alpha\beta}\|,$$

and the $\omega'_\alpha s$ are the covariant components of a rotation pseudovector such that $u^\alpha \omega_\alpha = 0$.

Let us now introduce new coordinates, $\tilde{x}^\lambda = (c\tau, \tilde{x}^i) = (c\tilde{t}, \tilde{x}^i)$, $i = 1, 2, 3$, that verify the following conditions:

1. C is described in the $\tilde{x}^\lambda$ coordinates by: $\tilde{x}^i = 0$, $\tilde{t} = \tau$.
2. The restriction of the metric in the $\tilde{x}^\lambda$ coordinates on C is: $\tilde{g}_{\mu\nu}|_C = \eta_{\mu\nu}$, and $\frac{\partial \tilde{g}_{\mu\nu}}{\partial \tilde{x}^\lambda}|_C = 0$, when the four-acceleration of O and the four-rotation of the tetrad vanish: $\mathbf{a} = \omega = 0$.
3. $\mathbf{e}_{(\alpha)} = \frac{\partial}{\partial \tilde{x}^\alpha}|_C \iff e^\lambda_{(\alpha)}(\tau) = \frac{\partial x^\lambda}{\partial \tilde{x}^\alpha}|_C$.
4. The $\tilde{x}^\lambda$'s become the usual Lorentzian coordinates in a neighborhood of C when $\mathbf{a} = \omega = 0$ and the curvature tensor vanishes, $R_{\alpha\beta\mu\nu} = 0$, whithin that neighborhood.

The $\tilde{x}^\lambda$'s will be denoted henceforth by QMCCω (cuasi-Minkowskian coordinates relative to the world-line C and to the tetrad $\mathbf{e}_{(\alpha)}$ subject to the rotation ω).

Principle of General Relativity (GRP)

Let us have a generic smooth enough space-time $(\mathbf{M}, \mathbf{g})$, $\mathbf{M}$ and $\mathbf{g}$ respectively denoting the manifold and the metric. Let O be an observer in that space-time of world-line C that uses any type of QMCCω. Let F be any -generally multicomponent- physical quantity. And let us denote by $\tilde{F}$ the same quantity -or its components- in the QMCCω, $\tilde{x}^\lambda$'s.

We shall say that the Principle of General relativity (GRP) is verified when the following two conditions hold:

- The equations describing the behaviour of the physical quantities are covariant and they become their corresponding ones in Lorentzian coordinates in flat space-time $(\mathbf{M}, \eta)$ on C when they are expressed in terms of the $\tilde{x}^\lambda$'s and $\mathbf{a} = \omega = 0$. (This implies the Equivalence Principle in general, but curvature dependent terms may still appear if there is no minimal coupling or higher order derivatives are involved in the equations.)
- One has $Q[\tilde{F}] \subseteq Q[F]$, $\forall F$ and $\forall$ QMCCω.

Principle of General Covariance (GCP)

The GCP is the above GRP.

- It implies the Equivalence Principle.
- It is a principle of general relativity with physical content as previously defined.

Note that if both, $\tilde{x}^\lambda$ and $\tilde{\tilde{x}}^\lambda \in$ QMCCω, in general one will have that $Q[\tilde{x}] \neq Q[\tilde{\tilde{x}}]$, and that if $q[x] \in Q[x]$, where x stands for any set of Lorentzian coordinates in flat space-time, there will be QMCCω, $\tilde{x}^\lambda$, $\forall\, O\{C,\omega\}$, (observer of world-line C and reference tetrad with rotation ω), such that $q[x] \in Q[\tilde{x}] \in Q[x]$.

What is called for now is to verify that this interpretation of the GCP is coherent with the theory and agrees with experience. To that end we proceed to construct the QMCCω in the next Section.

CONSTRUCTION OF THE QMCCω'S

The $\tilde{x}^\lambda$'s will be constructed by expressing the x^λ's as power series of the $\tilde{x}^i$'s with coefficients depending on τ about C.

We get

$$x^\lambda = f^\lambda(\tau) + e^\lambda_{(k)}\tilde{x}^k + \frac{1}{2}\left(e^\lambda_{(\alpha)}\tilde{\Gamma}^\alpha_{ik}(\tau) - e^\rho_{(i)}e^\sigma_{(k)}\Gamma^\lambda_{\rho\sigma}(\tau)\right)\tilde{x}^i\tilde{x}^k + \Phi^\lambda(\tilde{x}) \,, \quad (3)$$

with

$$\tilde{\Gamma}^\alpha_{ik}(\tau) = A^\alpha_{ikv}(\tau)a^v + B^\alpha_{ikv}(\tau)\omega^v + O^\alpha_{ik}(2) \,, \quad (4)$$

where the A^α_{ikv} and B^α_{ikv} are arbitrary save by being sufficiently smooth and the constraints

$$A^\alpha_{ikv} = A^\alpha_{kiv} \,, \quad B^\alpha_{ikv} = B^\alpha_{kiv} \,, \quad A^\alpha_{ikv}e^v_{(0)} = B^\alpha_{ikv}e^v_{(0)} = 0 \,; \quad (5)$$

$O^\alpha_{ik}(2) = O^\alpha_{ki}(2)$, is an arbitrary smooth enough fuction of second order in the $\tilde{a}^i$'s and $\tilde{\omega}^i$'s, that is, of second order in the a^v's and ω^v's, otherwise $O^\alpha_{ik}(2)$ is a function of τ as so is $\Gamma^\lambda_{\rho\sigma}(\tau)$, as both are defined on C;

$$\Phi^\lambda(\tilde{x}) = \Psi^\lambda(\tilde{x}) + \Phi^\lambda_{(0)}(\tilde{x}) \,, \quad (6)$$

with $\Psi^\lambda(\tilde{x})$ verifying

$$\Psi^\lambda\,|_C = \frac{\partial \Psi^\lambda}{\partial \tilde{x}^i}\,|_C = \frac{\partial^2 \Psi^\lambda}{\partial \tilde{x}^i \partial \tilde{x}^j}\,|_C = 0 \,, \quad (7)$$

and vanishing whenever $\mathbf{a}, \omega$ and the curvature tensor, $\mathbf{R}$, in a finite neighborhood of C, all vanish; the latter is equivalent to the vanishing of the $\tilde{\Gamma}^\lambda_{\mu\nu}$'s in that neighborhood; otherwise the Ψ^λ's apart from being sufficiently smooth are arbitrary;

$$\Phi^\lambda_{(0)}(\tilde{x}) = \sum_{l,m,n} \frac{1}{l!m!n!} C^\lambda_{lmn}(\tau)(\tilde{x}^1)^l(\tilde{x}^2)^m(\tilde{x}^3)^n , \qquad (8)$$

with
$$l \geq 0, \ m \geq 0, \ n \geq 0, \ l+m+n \geq 3, \text{ and } l,m,n \text{ all being integers.}$$

The $C^\lambda_{lmn}(\tau)$'s may be systematically calculated by the following algorithm:
Consider the equation

$$\frac{\partial^2 \Phi^\lambda_{(0)}}{\partial \tilde{x}^i \partial \tilde{x}^k} = -\frac{\partial x^\rho}{\partial \tilde{x}^i} \frac{\partial x^\sigma}{\partial \tilde{x}^k} \Gamma^\lambda_{\rho\sigma}(\tilde{x}^\mu) + e^\beta_{(i)} e^\gamma_{(k)} \Gamma^\lambda_{\beta\gamma}(\tau) , \qquad (9)$$

where the first term on the rhs is taken as dependent, in general, on the $\tilde{x}^\mu$'s, while the second term only depends on τ as is evaluated on C. Eq. (9) is a consequence of considering the equation for the transformation of the Christoffel symbols on C and using eqs. (3), (6), and (7). The $C^\lambda_{lmn}(\tau)$'s are found by using the power series for $\Phi^\lambda_{(0)}$ given in eq. (8) and taking successive derivatives of eq. (9) with respect to the $\tilde{x}^j$'s, taking the result on C, and doing it all along as if $\tilde{\Gamma}^\alpha_{\mu\nu} = \Psi^\lambda = 0$ at all points.

So we get, for instance,

$$C^\lambda_{300}(\tau) = -\frac{\partial}{\partial \tilde{x}^1}\left(\frac{\partial x^\rho}{\partial \tilde{x}^1}\frac{\partial x^\sigma}{\partial \tilde{x}^1}\Gamma^\lambda_{\rho\sigma}(x)\right)\Big|_C = -4e^\sigma_{(1)}A^\rho_{11}\Gamma^\lambda_{\rho\sigma} - e^\rho_{(1)}e^\sigma_{(1)}\Gamma^\lambda_{\rho\sigma,\gamma}e^\gamma_{(1)},$$

with $A^\lambda_{ik} = -\frac{1}{2}e^\rho_{(i)}e^\sigma_{(k)}\Gamma^\lambda_{\rho\sigma}(\tau)$, all evaluated on C at the point corresponding to τ.
That way any $C^\lambda_{lmn}(\tau)$ may be expressed in terms of the $C^\lambda_{l'm'n'}$'s of lower order: $l'+m'+n' < l+m+n$, $l' \leq l$, $m' \leq m$, $n' \leq n$; the $e^\lambda_{(k)}(\tau)$, the $\Gamma^\alpha_{\mu\nu}(\tau)$'s, and the partial derivatives of the $\Gamma^\alpha_{\mu\nu}$'s with respect to the x^λ's up to order $\leq l+m+n-2$.

Fixing the Ψ^λ's, $A^\alpha_{ik\nu}$'s, $B^\alpha_{ik\nu}$'s and the $O^\alpha_{ik}(2)$'s uniquely determines a set of corresponding QMCCω. If space-time is flat and also $\mathbf{a} = \omega = 0$, it follows that $\Psi^\lambda = A^\alpha_{ik\nu} = B^\alpha_{ik\nu} = O^\alpha_{ik}(2) = 0$, and the entire family of the QMCCω's collapses to the unique usual Lorentzian coordinates corresponding to the choiced tetrad $e^\lambda_{(\nu)}$. This does not mean that if space-time is not flat and/or if $\mathbf{a} \neq 0$, or $\omega \neq 0$, by taking $\Psi^\lambda = A^\alpha_{ik\nu} = B^\alpha_{ik\nu} = O^\alpha_{ik}(2) = 0$, the resulting QMCC$\omega$ would be Lorentzian, as these do not symply exist for non-flat space-times or in non-inertial reference frames.

SOME CONSEQUENCES

The relationships of the generic given coordinates x^λ and two different sets of QMCCω, $\tilde{x}^\lambda$ and $\tilde{\tilde{x}}^\lambda$, may differ at most by terms of second order in the $\tilde{x}^i$'s and $\tilde{\tilde{x}}^i$'s if the observer is not in free fall (C is not a geodesic) and/or his/her choiced transported tetrad rotates, which corresponds to the freedom allowed to choose the $\tilde{\Gamma}^{\alpha}_{ik}(\tau)$'s via eq. (4), or by terms of third order in the same variables if the observer is in free fall and its reference tetrad is paralell transported, corresponding to the freedom allowed to choose the function Φ^{λ} when the space-time is not flat.

If the observer O is in free fall and his reference tetrad does not rotate the Equivalence Principle tells us that when he is at a point P at proper time τ_P, he should measure for the square, dl^2, of the spatial distance between P and any other very close point Q of QMCCω, $(c(\tau_P + \delta\tau_P), \delta\tilde{x}^i)$, corresponding to the given coordinates $x^{\lambda} = f^{\lambda}(\tau_P) + \delta x^{\lambda}$,

$$dl^2 = \delta_{ik}\delta\tilde{x}^j\delta\tilde{x}^k = e^{(k)}_{\alpha} e_{(k)\beta}\delta x^{\alpha}\delta x^{\beta} \tag{10}$$

that results from inverting the coordinates transformation in eq. (3). It is easy to see that one has at P

$$e^{(k)}_{\alpha} e_{(k)\beta} = g_{\alpha\beta} - \frac{g_{0\alpha}g_{0\beta}}{g_{00}}, \tag{11}$$

that yields the usually accepted result for dl^2 in eq. (10) [11].

It follows from eq. (3) that the values of tensor quantities measured on the worldline C corresponding to two different sets of QMCCω's -but with the same choiced reference tetrad- should be identical. This is clearly not so for their ordinary partial derivatives with respect to the spatial QMCCω coordinates. In future work this question will be considered in more detail and the general theory here presented will be applied to some concrete examples.

REFERENCES

1. E. Krestchmann, *Annalen der Physik*, **53**, 575 (1917).
2. A. Einstein, *Annalen der Physik*, **55**, 240 (1918).
3. E. Cartan, *Ann. Sci. ENS*, **40**, 325 (1923).
4. K. Friedrichs,*Matematische Annalen*, **98**,566 (1927).
5. C.W.Misner, K.S.Thorne and J.A.Wheeler,*Gravitation*(Freeman, San Francisco, 1973).
6. V. Fock,*The Theory of Space, Time and Gravitation*(Pergamon Press, 2nd Revised Ed. 1966), pp. 5-8, 178-182 and 392-396.
7. J.L. Anderson, *Principles of Relativity Physics*(Academic Press, New York,1967).
8. J. Stachel,"What a Physicist Can Learn from the History of Einstein's Discovery of General relativity", in *Proceedings of the Fourth Marcel Grossmann Meeting on General Relativity*, R. Ruffini, ed. (Elsevier,Amsterdam, 1986), pp. 1857-1862; "Einstein's Search for General Covariance, 1912-1915", in *Einstein from 'B' to 'Z'*(Birkhäuser,Boston, 2002), pp. 301-337.
9. J.D. Norton,*Rep. Prog. Phys.*, **56**, 791 (1993).
10. G.F.R.Ellis and D. R.Matravers, *Gen. Rel. Grav.*, **27**, 777 (1995).
11. See, for instance, C. Mœller, *The Theory of Relativity*(Oxford University Press, 2nd Ed. 1972).

Relativistic Positioning Systems

Bartolomé COLL

Systèmes de référence relativistes, SYRTE-CNRS, Observatoire de Paris, 75014 Paris, France
bartolome.coll@obspm.fr

Abstract. The theory of relativistic *location systems* is sketched. An interesting class of these systems are the relativistic *positioning systems*, which consists in sets of four clocks broadcasting their proper time. Among them, the more important ones are the *auto-located positioning systems*, in which every clock broadcasts non only its proper time but the proper times that it receives from the other three. At this level, no reference to any exterior system (the Earth surface, for example) or no synchronization are needed. Some properties are presented. In the SYPORT project, such a structure is proposed, eventually anchored to a classical reference system on the Earth surface, as the best relativistic structure for Global Navigation Satellite Systems.

Keywords: Coordinate systems, reference systems, positioning systems
PACS: 04.20.-q, 95.10.Jk

INTRODUCTION

In relativity, the physical space-time is modeled by a four-dimensional differential manifold. So, it admits, in general, an infinite variety of mathematical coordinate systems.

Among these abundant coordinate systems, only a few number of them is known which may be physically *interpreted*. This means that (some of) their ingredients, namely their coordinate lines or their coordinate (hyper-)surfaces, may be *imagined* as described by some physical objects, like point-like particles or clocks, dust, stretched strings or light signals.

Among this few number of physically interpretable coordinate systems, only one of them is known which, generically, may be physically *constructed*. This is the one based on the Poincaré-Einstein synchronization procedure, i.e. by means of two-way signals, sent by one observer equipped with a clock and returned by the events he want to locate. This system, of the observer's clock and the two-way signals, with the help of a theodolite, generates a four-dimensional coordinate system with spatial spherical coordinates around the observer. And this is the sole reasonable relativistic coordinate system that, up to now, one has been able to construct physically in generic, arbitrarily chosen, vacuum space-times. It is also often called *radar system*.

But this relativistic physical coordinate system suffers from an important default: the one of being *intrinsically retarded*. This means that the coordinates of every event in the (finite) neighborhood of the observer are necessarily known with an unavoidable delay not only by the observer, which, being separated from the event, expects such a delay, but also by the event itself which is constitutively present at the instant and place where it happens and its coordinates indicate.

Consequently, up for the very particular circumstances in which the observer, the events and the whole gravitational context are stationary, even the events are unable to know their proper coordinates in this system. So, in it, the physical properties of an event cannot be *experimentally* related to its position, still less without delay; in these coordinates, such relations between properties and positions cannot but be *calculated* and need for this purpose the help of a previous theory (often unknown) of their proper evolution.

Thus, the main problem is *how to construct physically good, not intrinsically retarded, coordinate systems*, i.e. systems such that every event in it be able to know its proper coordinates without delay? The class of such *positioning systems* is relatively restricted, and their paradigmatic representatives are systems of four clocks broadcasting their proper time.

Our purpose here is to introduce the basic concepts, to comment them and to present some qualitative aspects. This is organized as follows: in Section LOCATION SYSTEMS, some general notions and properties concerning the physical realizations of coordinate systems are explained; in Section POSITIONING SYSTEMS these systems are described and their principal properties presented, in particular the one of having a very good separation power for the space-time, and finally, in Section SYPOR PROJECT we shortly describe how these relativistic positioning systems should be used for primary reference and positioning of the Earth surroundings, replacing the at present Newtonian-relativistically-corrected conception of the Global Navigation Satellite Systems.

Details on these and other results are presented in this meeting by my collaborators [1], [2], [3].

LOCATION SYSTEMS

A coordinate system may be given in many different ways. But whatever they be, they are tantamount to give its (parameterized congruences of) *coordinate lines* or its (one-parameter families of) *coordinate (hyper)surfaces*. But lines and (hyper)surfaces may be physically constructed with many different materials and with many different protocols, giving raise to very different physical realizations of the same mathematical coordinate system. For this reason, it is convenient to distinguish by a different appellation coordinate systems ant their physical realizations. We call *location systems* the physical realizations of coordinate systems.

As physical realizations, location systems are physical objects and, consequently, able to be described in physical terms. For our purposes, the following physical description is sufficiently complete.

> • *In a region of the space-time, a* location system *is a* real *or* virtual, passive, *set of* physical fields, parameterized *in such a way that every event in the region be one-to-one characterized by the values of the parameters at the event.*

In this physical description, 'real' refers to the beforehand actual physical construction in all the domain of the whole set of physical fields[1], meanwhile 'virtual' refers to any other case; in particular when only the reference axes or surfaces are constructed

---

[1] For example, the Cartesian lines on a graph paper, at the millimetric scale.

beforehand, leaving afterward the construction of the sole axes or surfaces that contain the specific events of interest[2]

By 'passive' set of fields it is to be understood sufficiently weak physical fields so that their interaction with the events to be located may be considered negligible (a rigourous quantum field version of location systems would be necessarily 'active').

Finally, that a physical field is parameterized means that from the measure of some of its physical properties at every event of the domain, it is possible to extract a unique real number.

This description may be as well considered as the *physical definition* of location systems, so that alternatively, one can consider either locations systems as physical realizations of coordinate systems or coordinate systems as mathematical idealizations of location systems.

The use of location systems may respond to different needs or objectives; two of them are particularly important. As in astronomy, frequently the goal of some location systems is to allow one observer, generally considered at the origin, to locate with precision the events of his neighborhood. Location systems devoted to such a function are called (*relativistic*) *reference systems*. The goal of other location systems, like Global Navigation Satellite Systems (GNSS), is to indicate to every event of the region its own position. Location systems devoted to such a function are called (*relativistic*) *positioning systems*.[3]

In Newtonian theory, as far as the velocity of information is supposed to be infinite, both goals are *exchangeable* for any location system. But in general relativity this is no longer possible, and the goal of a location system strongly conditions its conception and its construction. In fact, one has a strong hierarchy between them: meanwhile it is impossible to construct a positioning system starting from a reference system by transmission of its data, it is always possible, and very easily, to construct a reference system starting from a positioning system (it is sufficient that every event send its coordinates to the observer). It is then evident that, whenever possible, it is a positioning system, and not a reference system, that has the most interest to be constructed. Of course, this it not always possible, as is the case, roughly speaking, for the space out of the Solar system[4], but for such regions, people conforms with slightly more than an

---

[2] For example, the Cartesian system defined on a white sheet of paper by two orthogonal lines is virtual, all the other lines of the congruence covering the sheet being not drawn beforehand; only those lines of the congruence crossing the points of interest are effectively and afterward constructed.

[3] Reference and positioning systems as defined here are *four-dimensional* objects, including time location. This is not still the common use, and so, the International Astronomical Union (IAU) considers separately time scales and (three-dimensional) reference systems. We believe that, from a relativistic point of view, it is imperative to gather them in a sole four-dimensional concept, if we want to adequate our points of view to the increasing and pressing presence of relativistic corrections. Also, the International Celestial Reference System (ICRS), in spite of its appellation, *is not* a reference system even in the three-dimensional sense, but only an *orientation system*; if at first glance on could consider these features as a simple matter of words, they induce to confusion students and professionals, delaying the construction of correctly conceived relativistic frames.

[4] A positioning system for the Solar system based on the signals of (basically) four millisecond pulsars has been proposed in [4].

orientation system [5], which is far from being the physical realization of a coordinate system (an orientation system is, in fact, nothing but a basis of the *tangent frame* to the observer for the light directions converging to him).

POSITIONING SYSTEMS

Positioning systems are here supposed to be generic, free and immediate.

A location system is *generic* (for a given class of space-times) if it can be constructed in any space-time (of the class). For example, Cartesian systems are not generic but for (the class of) Minkowski space-time, meanwhile harmonic systems are generic for (the class of) *all* space-times.

A location system is (gravity) *free* if its construction does not need the previous knowledge of the gravitational field[6]. For example, harmonic systems are not free; in fact, among the usual location systems, only the radar system is free.

A location system is *immediate* if every event of its domain may know its coordinates without delay. Immediate systems[7] are the antithesis of the already mentioned intrinsically retarded systems, to which belong the radar system. From the relativistic point of view, no one of the location systems known up no now are immediate.

The question is then if whether or not generic, free and immediate positioning systems can be constructed. Because involving real objects, the answer to this question is an *epistemic* answer, rather than a logical one, resulting from the analysis of at present methods, techniques and practical possibilities of physical construction of such systems. This analysis show that the set of generic, free and immediate relativistic positioning systems constitute a small class of location systems. As already mentioned, the paradigmatic representatives of this class are the location systems constituted by four clocks broadcasting their proper time.

In what sense four clocks broadcasting their proper time constitute a physical realization of a coordinate system? A coordinate system is defined by its coordinate lines, by its coordinate (hyper)surfaces or by a convenient set of these two ingredients. But there are obstructions to construct generic location systems by means of their parameterized (congruences of) lines. These obstructions are in part of structural or mathematical character, and in part of physical character. On one hand, in order that four congruences of lines be able to be parameterized in such a way that they constitute the coordinate lines of a coordinate system, they must obey constraint relations which are in general incompatible for pairs of generic congruences. On the other hand, even if one can imagine a dust of micro-clocks as one congruence of lines, four of such congruences will impose serious, in general insoluble, problems of time scale, synchronization and indi-

---

[5] See the end of next to last footnote.

[6] A location system is a physical object that lives in the physical space-time. In it, even if we do not know the metric, such objects as test particles, light rays or signals follow specific paths which, a priori, may allow constructing a location system.

[7] From the Late Latin 'immediatus', 'without anything between'.

vidual accelerations of the clocks, the problem being more serious for light beams [5]. In short, only in particular space-times and under particular conditions location systems may be constructed by means of their coordinate lines. Consequently, the physical fields able to construct generic location systems are those defining one-parameter families of hypersurfaces. This is because four one-parameter families of hypersurfaces constitute generically the coordinate hypersurfaces of a coordinate system. Now, one clock broadcasting its proper time describes in the space-time a time-like line of which every event is the vertex of the future light cone formed by the electromagnetic signal broadcasting the time of the event, so that the set of these cones constitute a one parametric (proper time) family of (null hyper)surfaces. So, the four clocks broadcasting their proper time construct physically the coordinate hypersurfaces of a coordinate system.

At every event in the domain of such a coordinate system, a receiver able to read the value of the proper time coded by every one of the four signals (light cones) reaching the event will obtain the four times $\{\tau^1, \tau^2, \tau^3, \tau^4\}$ that constitute the coordinates of the event.

What about the coordinate lines of such a positioning system? The coordinate lines of a coordinate system are the locus of points where all but one of the coordinate hypersurfaces cut together. A light cone contains either light-like or space-like directions, the first ones being the generators, so that the intersection of three non-coincident light cones cannot but be a space-like curve. Consequently, the coordinate lines of a positioning system constitute four parameterized congruences of space-like lines.

As we see, in spite of the fact that they are physically well defined, positioning systems constitute the physical realizations of coordinate systems which are unusual for us. Thus, meanwhile we are accustomed to coordinate systems on which, a point-like object may have, along its evolution, up to three constant (adapted) coordinates, the four emission coordinates of a positioning system necessarily change whatever the point-like material object be. Nevertheless, the incomparable operational character of such systems is worthy of an effort to better understand them and, on the way, to liberate ourselves of unjustified Newtonian prejudices about the space-time.

In a *grid* of parameters $\{\tau^1, \tau^2, \tau^3, \tau^4\}$, any user receiving continuously his coordinates may draw his trajectory. An important class of positioning systems are the *auto-located positioning systems*, which allow the user to know also the trajectories of the emitting clocks in the grid.

The necessary and sufficient condition for a positioning system to be auto-located is that every clock broadcast the proper time that it directly receives from the other clocks.

This is because, joint to its proper time, the three times that the clock receives constitute in fact its proper coordinates. More precisely, the clocks are at the border of the coordinate domain that they generate, because one of the light cone coordinate hypersurfaces of the domain is not differentiable at the positions of every clock, but the coordinates themselves are continuous along the world line of the clocks.

In an auto-located positioning system, a user receives at every instant sixteen times, $\{\tau^{ij}\}$, where $\tau^{ii} \equiv \tau^i$ is the proper time of the clock i, and τ^{ij}, $i \neq j$, are the times received by the clock i, from the clocks j. Then, $\{\tau^i\}$ are the coordinates of the user and $\{\tau^i, \tau^{ij}\}$, $i \neq j$, are the coordinates of the clock i.

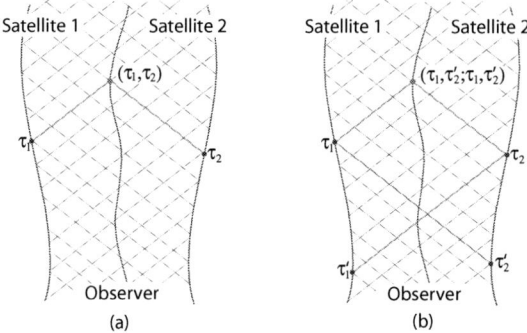

FIGURE 1. (a) Two-dimensional positioning system. (b) Auto-located two-dimensional positioning system.

The simplest examples of positioning systems are found in a two-dimensional space-time, as shown in Figure 1(a). In the internal region delimited by the trajectories of two clocks, the positioning system, constituted by the radiated electromagnetic fields broadcasting the proper time of the clocks, is well defined (the exterior regions correspond to the shadow of every clock for the signal of the other, and are characterized by a vanishing Jacobian). The emission coordinates of a user are the data (τ^1, τ^2) that he receives.

Because of the linearity of the light cones, the analysis of the two-dimensional case is particularly easy, and allow to understand with no much effort interesting features which, for the most part, remain qualitatively valid in higher dimensions. Nevertheless, dimension two is singular for some properties, in part due to the fact that coordinate lines and coordinate surfaces coincide.

Figure 1(b) shows an auto-located two-dimensional positioning system. Here, every user receives the four data $(\tau^1, \tau^{12}; \tau^2, \tau^{21})$ from which he can extracts his proper emission coodinates (τ^1, τ^2) and also the emission coordinates (τ^1, τ^{12}) and (τ^2, τ^{21}) of the satellites 1 and 2 respectively.

Some basic properties of the two-dimensional case are presented in this same meeting by Ferrando [1].

An important property of positioning systems is that, for a given precision of the clocks, they improve the separation power of the events with respect to Cartesian location systems. The definition of the meter being based in that of the second, we can consider a Cartesian location system as constituted, at best, by four clocks of a given *precision*, i.e. of a given *separation power* of instants, one of them devoted to the measure of the time coordinate of the events, and the other three assigned to the construction of the three Cartesian space-like distance coordinates of them. The interesting result is that, if *same* precision clocks are assigned for the construction of the emission coordinates of a positioning system, there always exists some space-time regions where the separation power of space-time events in these emission coordinates is three times greater than that obtained by a Cartesian protocol.

Other interesting basic properties of positioning system in the four-dimensional case are presented in this same meeting by Pozo [3].

SYPOR PROJECT

The Global Positioning System (GPS) is a classical positioning system based on a set of satellites around the Earth, that allows their users, receiving the signals of some of the satellites, to know their positions with respect to the Earth surface (nominally in the Word Geodetic System, WGS 84). The information about the trajectory of the satellites with respect to the WGS 84 is centralized in the Master Control Monitory Station, at Colorado Springs, where the equations for the predicted trajectories are elaborated and sent to every satellite. Other Global Navigation Satellite Systems (GNSS), actual or in progress (GLONASS, Galileo) are based in a similar structure.

The Global Positioning System (GPS) is a jewel of the military engineering, has largely fulfilled the schedule of conditions specifying delivering time and performances and offers an almost unlimited area of civil applications. It is clear that:

- as a *technological object*, of engineering interest, the only possible improvements seem to be those derived from the technical improvements of its components, but...
- as a *physical object*, of scientific interest, it seems nevertheless to admit radical conceptual innovations...

The need for these conceptual innovations is based in the following facts. It starts with an incorrect theory, the Newtonian one, that must be corrected with 'relativistic terms'. From the beginning, it constraints times and synchronizations with respect to the GPS time, a sort of advanced TAI (International Atomic Time), which is a global conventional time, not a local physical one. It uses the satellites not as the best supports for standard clocks (microgravity), but as (unfortunately (!) moving!) beacons controlled from the Earth. Control and positioning are made with espect to the WGS 84, a reference system (not a positioning one!) which, in addition, is a virtual one (see above).

The aim of the project SYPOR (SYstèmes de POsitionnement Relativistes) is to construct a *complete relativistic theory* of GNSS, i.e. a theory in which *only relativistic concepts* are involved, irrespective of the acceptable *numerical* simplifications that error bars and weakness of some quantities can justify. Such a theory will be *convergent* for increasing precision, meanwhile the present one is clearly *divergent*, and strongly mixed in its basis to Newtonian conventional protocols.

It is true that, for purposes of geodesy and positioning, the Earth may be frequently considered as a Newtonian system, i.e. a physical system correctly described by Newtonian theory. But a constellation of satellite-borne clocks interchanging their proper time around the Earth, is a *relativistic system* on its own (principally because the importance of the gravitational and Doppler correction terms). Consequently, the best, shortest and clearest way to improve present GNSS is to directly use the best concepts of relativity theory.

For this reason, the project SYPOR proposes to uncouple GNSS in two hierarchical systems:

- A primary system, Earth-surface independent, constituted by the constellation of satellites acting as an atlas (union of sets of four neighboring satellites) of primary auto-located relativistic positioning systems, related only to the mass content of the Earth. Its physical realization implies an Inter Satellite Link (ILS) between neighboring satellites, a device on every satellite to send to the Earth the links directly received by every satellite, and a device on some of the satellites of the system to connect the constellation of satellites to the orientation system ICRS.

- A secondary system, Earth-surface dependent, coupling the virtual and intrinsically retarded Earth reference system (WGS 84 or ITRF) to the real and immediate primary main system.

A *space agency* could, or even should, limit its task to realize the primary system, and delegate to global or local Earth agencies the task of attaching (secondary) terrestrial reference systems to it.

Let us remark that if the primary system alone does not allow the users to situate with respect to the Earth surface, it nevertheless allow every user to situate with respect to the constellation (or with respect to any more conventional reference system *deduced* from it) and two or more users to know their relative position.

A final remark. The TAI may be improved by satellited clocks, as is contemplate by the project ACES (Atomic Clock Ensemble in Space). But with the notion of relativistic position system in mind, it becomes clear that four or more satellited atomic clocks not only are able to supply an International Atomic Time, but also to constitute an International Atomic Coordinate System (IACS).

Such a system would reconcile us more deeply and faithfully with relativity theory than the most part of the relativistic chattering that we have contemplate this year all over the Earth.

ACKNOWLEDGMENTS

Long time ago, this subject covered a corner of my private garden of thoughts for weekends and holidays. But every flower that sprouted in it, every idea, I showed it to my friends Joan FERRANDO, Juan Antonio MORALES, Albert TARANTOLA and José Maria POZO, who watered it carefully. For this reason, it is a pleasure for me to acknowledge their important contribution to the subject.

REFERENCES

1. J. J. Ferrando, "Coll positioning systems: a two-dimensional approach," in *these proceedings* (2005).
2. J. A. Morales, "Coordinates and frames from the causal point of view," in *these proceedings* (2005).
3. J. M. Pozo, "Emission coordinates and the central observer," in *these proceedings* (2005).
4. B. Coll and A. Tarantola, "A Galactic positioning system," in *Journées Systèmes de Référence Spatio-Temporels*, St. Petersburg, 22-25 September (2003), edited by A. Finkelstein and N. Capitaine, I.A.A., Russian Academy of Sciences and Observatoire de Paris, 2004, pp. 333-334. See also http://coll.cc .
5. B. Coll, "Coordenadas luz en relatividad," in *Spanish Relativity Meeting ERE 85*, edited by Pub. Servei de Publications de l'ETSEIB, Barcelona, 1985, pp. 29-38. See also http://coll.cc for an English translation.

Geometry and experience: Einstein's 1921 paper and Hilbert's axiomatic system.

François De Gandt

Université Lille III, UMR 8163 "Savoirs et Textes", BP 60149, 59653 Villeneuve d'Ascq, France

Abstract. In his 1921 paper Geometrie und Erfahrung, Einstein decribes the new epistemological status of geometry, divorced from any intuitive or a priori content. He calls that "axiomatics", following Hilbert's theoretical developments on axiomatic systems, which started with the stimulus given by a talk by Hermann Wiener in 1891 and progressed until the Foundations of geometry in 1899. Difficult questions arise : how is a theoretical system related to an intuitive empirical content?

Keywords: geometry physics axioms experience intuition space euclidean idealization
PACS: 01.70.+w

I shall concentrate on a paper [1] which Einstein published in 1921. In the title *Geometrie und Erfahrung*, the word *Erfahrung* must be understood in a large sense, not as experiment, but as experience, that is our contact with given reality, and the question is: how is geometry related to natural objects, or to real things, perceptively given ? what are the ties or absence of ties between mathematics and intuitive knowledge ?

The paper, which is seven pages long in the first version, deals first with the status of geometry, then contains a discussion of Poincaré's point of view, and ends with some conjectures about the possible extensions of Einstein's theories to the « submolecular » level and to the cosmic scale, discussing the interest of a repulsive constant. I comment here only the first pages about geometry.

A NEW CONCEPTION OF GEOMETRY

There is a privilege which mathematics enjoys, namely its certainty and rigour, which the thinkers of other domains can only admire. But this would not be a subject for jealousy from other disciplines, if in addition mathematics were not also useful, if it were a void deductive arrangement and a pure game. But it is not, since physical science owes its high degree of certainty to the use of mathematics. How is this fusion possible ? Is it possible to understand the way in which the perfection of mathematics radiates into natural science ? Einstein then says he will offer the unique admissible answer, but in fact he simply makes the difficulty even more insolvable: if geometry is certain, it is not physical, if it is physical, it cannot be certain. Certainty and factual content exclude each other.

This clearcut opposition is a recent feature, a modern way of thinking. Einstein characterizes it by the keyword « axiomatics ». Modern geometry is axiomatically conceived and build, it is therefore independent of any intuive content. The axiomatic way consists in separating strictly logico-formal elements from the intuitive ones. Yesterday a geometrical truth was founded on a small stock of first truths, warranted by their evidence; the principles of geometry were facts, given by some sort of direct knowledge. Einstein proposes the example: knowing what a straight line is, and what a point, a traditional geometer can assert with evidence « by two points in space passes one and only one straight line ».

Now things have changed: a geometry is a free creation, presupposing no intuitive knowledge. A historical remark must be made here: it is a new conception, but not entirely divorced from the traditional view of geometry. Since the early beginnings of greek science, geometry had acquired a remarkable independence from natural truth, from intuitive content: the sequence of propositions was tied together through reasoning, deductive dependence, and only the first principles were considered as borrowed from experience. Euclid starts with definitions, axioms and postulates, and the rest of geometrical discourse must be generated according to logical or general rules (equals added to equals give equals, etc.) and operations permitted at the beginning (to join points by segments, to construct a circle from a given center with a given radius). But greek geometry was not completely independent of experience. The very beginning is related (in a way which is difficult to make precise and rigorous) to some factual knowledge, to some familiarity with the spatial objects in our world. The notions of point, of line, of angle, have to be assumed as well-known, even if the definition makes the notion more strictly determined (« a point is that which has no parts », « the boundary of a surface is a line »)[1]. The first principles must be taken for granted on the basis of some intuitive acquaintance.

The modern status of mathematics, for Einstein, is summed up in the word axiomatics. Today, the new freedom which runs under the name of axiomatics presupposes no acquaintance with the spatial behaviour of objects, no previous knowledge of what are a point or a line. Geometry deals with certain objects, but the names « point » or « straight line » are irrelevant, since we do not assume any knowledge or intuition of these objects, and, adds Einstein, they can even be considered as defined by the axioms (he quotes Moritz Schlick in this respect).

HILBERTIAN AXIOMATICS

The paradigm of this formal procedure had been given by Hilbert in his *Grundlagen der Geometrie* in 1899: the exposition of geometry begins with the supposition of three sorts of objects, totally unspecified, and devoid of any empirical content, but which we may name points, lines and planes - if we want to give them names, for the sake of commodity or heuristic interest. In speaking of axiomatics, Einstein places himself in

---

[1] A definition in the euclidean sense is something different from what it means today in a formal theory: our theories do not start with definitions as Euclid does, they start with primitive terms and axioms, and if there are « definitions » at all, they arrive later in the systematic sequence and are only abreviations for certain combinations of primitive terms. The first exposition of this modern conception can be found in Pascal's *De l'esprit géométrique*: a theory begins with primitive terms and primitive propositions (none of which is a definition).

the spirit of Hilbert[2] and the Italian school of mathematics (Peano, Padoa, Pieri, et al.). Other important mathematicians of the same or previous generations (Poincaré, Cantor, Frege, Brouwer) would probably not agree with this description of axiomatics as the essence of modern mathematics, they would insist more on the task of describing ideal facts.

The hilbertian flavour of Einstein's paper could be made more striking if we have in mind the celebrated statement by Hilbert: geometry is a totally abstract thing, it does not speak of points, lines and planes, it must be equally true of any sorts of objects, like tables, chairs and beer-mugs[3]. This provocative formula comes to us through oral tradition, but we can find in some writings of Hilbert equivalent statements. For instance, to Frege Hilbert writes that a mathematical theory is a purely formal structure, a « scaffolding of concepts », and that it must equally hold of systems of points, or of the system of love, or the system of law, or even of the system of chimney-sweeping (« Liebe, Gesetz, Schornsteinfeger »)[4].

WIENER'S TALK IN 1891

The sentence about tables and chairs is reported by a student of Hilbert, Blumenthal, and is supposed to have been uttered in the waiting-room of a railway station - presumably in front of a beerglass -. We know the circumstances[5], and they shed some light on the occasion, scope and intention of the sentence. In 1891, in Halle[6], the young David Hilbert attended a lecture at the annual meeting of the recently founded Union of German Mathematicians; of the lecture, given by Hermann Wiener, we possess only a short version in the Annual Report [4]. The subject was: *On the foundations and construction of geometry*. The author explains that the mathematician has as one of his tasks to reduce and simplify the mathematical theories, that is to say, to find the simplest objects and operations as starting points for his theories. This goal is easily reached in arithmetics, but it could be interesting to try to do the same for geometry, by a regression to the simplest objects. And Wiener proposes to follow the pattern offered by projective geometry, and to consider two sorts of « elements » or « objects », say objects of a sort *a* and of a sort *b*, such that two objects of the sort *a* « give » (ergeben) an object of the sort *b*, and vice versa. Objects are thus related by « operations ». This amounts to a formal translation of what happens in projective geometry, where two points determine a straight line, and two straight lines determine a point. Wiener calls this mutual generating process a « *Verknüpfung* », a term which encompasses both *Verbinden* and *Schneiden*

---

[2] See for instance, three years before, an important paper by D. Hilbert, *Axiomatisches Denken* [2]
[3] « man muss jederzeit an Stelle von Punkten, Geraden, Ebenen, Tische, Stühle, Bierseidel, sagen können ».
[4] Letter to Frege, 29 december 1899
[5] The biography by Constance Reid is to be completed by the study of Michael-Markus Toepell, *Uber die Enstehung von David Hilberts « Grundlagen der Geometrie »* [3] (see especially page 42).
[6] At that time, Cantor and Husserl held teaching positions in the University of Halle, and it is highly probable that they could listen to the lecture given by Wiener. Hilbert was 30 and he had a position in Königsberg, far away in Ostpreussen.

(to link and to cut).

But in his 1891 paper Wiener refrains from a completely formal attitude, as if he were frightened by his own boldness. He adds that, of course, we could generate a long sequence of propositions by operating on such quasi-objects, we could in this way create a theory which runs parallel to the propositions of geometry; but it would not be geometry properly speaking, since geometry rests upon our intuition of space. The decisive sentence is as follows in the shorter version of the talk:

« From these {objects and operations} it is possible to construct an abstract science, which will be independent of the axioms of geometry, but the propositions of which are parallel, step by step, to the propositions of geometry. » [7]

Along this abstract construction you imitate geometry, but it is not geometry, it is just parallel. Wiener thus maintains a gap between a purely formal theory and real and genuine geometry, which rests upon certain axioms (axiom must here be understood in the traditional sense, with the meaning of evident truth).

There comes precisely the provocation of the young Hilbert. He tells later that the talk struck him very deeply, and gave him a sort of lasting stimulus. Travelling back from Halle to Königsberg, through Berlin, he proposed to some friend (or to himself), in reaction to Wiener's talk, a daring step : after all, could'nt we accept to treat both things on the same footing, and decide that geometry is nothing else than this formal sequence of abstract objects and operations ? We then suppress the distinction, which Hermann Wiener maintained, between the science of empirical space and an abstract scaffolding of concepts. The roots which Euclid and all the tradition sought in the intuition of spatial relations are then cut. We may continue speaking of points, lines, and planes, but any family of three-sorted objects (any « system » with three ranks of membership) would fit as well, provided that they obey the relations stipulated by the axioms.

FOUNDATIONS OF GEOMETRY 1899

The stimulus was very profitable, since Hilbert went on and developed a long study of formal systems of geometry, especially focussed on the choice of axiom systems for projective geometry and on the discussion of various non-euclidean geometries. All this culminated in the publication of the *Foundations of geometry* in 1899. There he gave a classification of geometric axioms in five series: incidence (*Verknüpfung*), order, parallelism, congruence and continuity; he studied the independence of axioms, showing what happens if one removes a given axiom or group of axioms.

Now what remains of the old geometry ? In other words, what is the relation of this new study of various systems of axioms and the traditional « science » of empirical space ? Hilbert is not very clear on that philosophical point. His book, or more precisely the successive editions of his book show hesitations and ambiguities. For instance, in the first edition he spoke of the first principles as « Grundtatsachen » (fundamental facts) and

---

[7] « aus diesen eine abstrakte Wissenschaft aufbauen kann, die von den Axiomen der Geometrie unabhängig ist, deren Sätze aber Schritt für Schritt mit den Sätzen der Geometrie parallel sind. » [4], page 46

later he replaced the phrase by « Grundsätze » (fundamental propositions). A sentence in his Introduction is highly ambiguous and can be translated in two opposite ways: he writes that the study (establishing the axioms and investigating their mutual dependence) « flows into » the logical analysis of our spatial intuition[8]. The verb can be made to mean either « amounts to » or « opens the way to » , a difference which is philosophically very significative: in one case he supposes that his work is the analysis of our intuition, in the second case, he simply alludes to a further study of our intuition of space, a study which could then be psychological or philosophical, and in which the preparatory axiomatic study could be useful but not final. We do not how Hilbert conceived the adjustment of axiomatic system and spatial intuition, a question on which many efforts were spent between 1870 and 1920, in the years of the fashion of non-euclidean geometry, without arriving at established results and well-founded agreement.

WHAT IS PRACTICAL GEOMETRY ?

Let us come back to Einstein and his 1921 paper. He goes beyond the strict duality asserted in the beginning, and proposes to transform pure geometry into « practical geometry » , by enuntiating a separate proposition, which serves as a link between the formal system and the empirical world. Then the « conceptual schemes » as he calls them, which are used in the formal construction, would be given a content: it suffices to stipulate that the three-dimensional objects of euclidean geometry are the solids of our experience, or more precisely, that they behave in the same way, as far as their « positions » are concerned. Now geometry becomes an inductive science under the name of « practical geometry » , says Einstein, and it is permitted to try to test the truth of such a theory. For instance we can ask whether the geometry of the world is euclidean or not, the question possesses a definite meaning.

The exact status of such an extra proposition or axiom is difficult to make precise, it is in any case totally different from the status of the other axioms. It serves as a « meta-axiom », but its scope and acceptability are dubious. The difficulty which we encountered in the beginning is not solved: this axiom speaks of the objects in our world, the objects which fall under our empirical intuition, and claims that they are the same, with respect to position, as the objects of the euclidean formal system.

The strategy proposed in the paper seems straightforward and simple, but the appearence could be disappointing. It is not enough to suppose that the empirical solids « correspond » to the three-dimensional entities of the formal system. You have for instance to add: « with respect to position », presupposing that you know what a position is, and what it means to consider a physical solid only as a positional object, well defined in space. Does it presuppose some sort of idealization ? How are we supposed to gain the knowledge of positions and of their relations in space ?

---

[8] « Die bezeichnete Aufgabe läuft auf die logische Analyse unserer räumlichen Anschauung hinaus. » The two french translations have chosen the first case: it amounts to (cela revient à, c'est). Rossier 1971: « Ce problème est celui ... » ; Laugel 1900: « Ce problème revient à » . In french you have the ambiguity between two very close renderings: « revenir à » and « déboucher sur » .

Einstein here mentions the operation of measuring and the relation with the trajectory of light, supposed to be rectilinear. He does not recall some other conditions which he discussed in the seminal paper on special relativity and elsewhere : the possibility of having access, practically and effectively, to the magnitude to be measured, for instance, the permission to superpose the same rod several times. There are pre-conditions to the act of relating geometry and experience, of which Einstein mentions only a part.

A more complete discussion of this line of thought would lead us much too far, in particular it would involve a commentary of the debate between Poincaré and Einstein, sketched in the subsequent pages of Einstein's 1921 paper. Let us only mention a reproach expressed by Einstein: Poincaré has cut « the original and immediate relation which links geometry and physical reality ». This reproach sounds surprising, since the link between reality and geometry had already been abandoned at the very beginning of the paper, and by Einstein himself, following Hilbert and others. But the idea is that with the help of the extra proposition mentioned above, the abstract geometry can receive an empirical content, whereas according to Poincaré the operation of testing the truth of a geometrical system against the empirical behaviour of objects has no sense at all.

Poincaré is not the sole interesting competitor, and other thinkers could be usefully invoked in this respect, for instance Bridgman [5] and his ideas about the effectuability of measurements, or Husserl [6] and his long critical efforts about idealization in the modern science of nature.

REFERENCES

1. Albert Einstein, *Geometrie und Erfahrung*, in *Sitzungsberichte der Preussischen Akademie der Wissenschaften*, **1** 123-130 (1921)
2. D. Hilbert, *Axiomatisches Denken*, Mathematische Annalen, **78**, 405-415 (1918)
3. Michael-Markus Toepell, *Uber die Enstehung von David Hilberts « Grundlagen der Geometrie »*, Vandenhoek und Ruprecht, Göttingen (1986)
4. Hermann Wiener, *Ueber Grundlagen und Aufbau der Geometrie* (Kurzfassung), Jahresbericht der Deutschen Mathematiker Vereinigung **1**, 45-48 (1891)
5. Maila Walters, *Science and cultural crisis, an intellectual biography of Percy Williams Bridgman*, Stanford (1990).
6. F. De Gandt, *Husserl et Galilée, sur la crise des sciences européennes*, Paris Vrin (2004)

A weighted de Rham operator leading to local potentials for Riemann and Weyl tensors

S. Brian Edgar * and José M.M. Senovilla[†]

*Matematiska institutionen, Linköpings universitet, Linköping, Sweden S-581 83.
e-mail: bredg@mai.liu.se
[†]Física Teórica, Universidad del País Vasco, Apartado 644, 48080 Bilbao, Spain.
e-mail: josemm.senovilla@ehu.es

Abstract. We introduce a *weighted de Rham operator* which acts on arbitrary tensor fields by considering their structure as r-fold forms. We can thereby define *associated superpotentials* for all tensor fields in all dimensions and, from any of these superpotentials, we can deduce in a straightforward and natural manner the existence of $2r$ potentials for any tensor field, where r is its form-structure number. By specialising this result to *symmetric* double forms, we are able to obtain a pair of potentials for the Riemann tensor, and a single $(2,3)$-form potential for the Weyl tensor due to its tracelessness. This latter potential is the n-dimensional version of the double dual of the classical four dimensional $(2,1)$-form Lanczos potential.

Keywords: de Rham Laplacian, tensor-valued differential forms, local potentials, curvature tensors
PACS: 02.40.Ky, 02.40.Vh, 04.20.Cv

NOTATION AND STANDARD RESULT FOR P-FORMS

We begin by formulating familiar results for p-forms in a manner from which we can generalise to more general tensors,

The graded algebra of exterior forms is Λ, with Λ^p denoting the set of exterior p-forms. The standard operations on the exterior algebra are the

exterior differential (also called *curl*), $d : \Lambda^p \longrightarrow \Lambda^{p+1}$

co-differential (also called *divergence*), $\delta : \Lambda^p \longrightarrow \Lambda^{p-1}$.

de Rham Laplacian operator, $\Delta : \Lambda^p \longrightarrow \Lambda^p$: $\Delta \equiv d\delta + \delta d$

In index notation these operators are given by

$$(d\Sigma)_{a_1...a_{p+1}} \equiv (p+1)\nabla_{[a_1}\Sigma_{a_2...a_{p+1}]} = (p+1)\Sigma_{[a_2...a_{p+1};a_1]}$$

$$(\delta\Sigma)_{a_2...a_p} \equiv -\nabla^i\Sigma_{ia_2...a_p} = -\Sigma_{ia_2...a_p}{}^{;i}$$

$$-(\Delta\Sigma)_{a_1...a_p} = \nabla^i\nabla_i\Sigma_{a_1...a_p} - pR_{i[a_1}\Sigma^i{}_{a_2...a_p]} + \frac{p(p-1)}{2}R_{ij[a_1a_2}\Sigma^{ij}{}_{a_3...a_p]}$$

for all $\Sigma \in \Lambda^p$.

Result 1 (Hodge decomposition (local))

Given any p-form Σ, there always exists a local superpotential $\overset{o}{\Sigma}$ of the same type, given by $\Delta \overset{o}{\Sigma} = \Sigma$.

Furthermore, there always exist a pair of local potentials (Ψ, Γ) with $\Psi \in \Lambda^{p-1}$ and $\Gamma \in \Lambda^{p+1}$ such that

$$\Sigma = d\Psi + \delta\Gamma$$

where $\Psi = \delta\overset{o}{\Sigma}$ and $\Gamma = d\overset{o}{\Sigma}$.

NOTATION FOR DOUBLE FORMS

We believe that the best way to extend the principles of the previous section to arbitrary tensors is to consider them as *r-fold forms*. This terminology was extensively considered in [10]. For simplicity of presentation we will consider only 2-fold (double) (q, p)-forms.

$$T^{a_1 \ldots a_q}{}_{b_1 \ldots b_p} \equiv T^{[a_1 \ldots a_q]}{}_{[b_1 \ldots b_p]}$$

$$(d_1 T)^{a_1 \ldots a_{q+1}}{}_{b_1 \ldots b_p} = (q+1) T^{[a_2 \ldots a_{q+1}}{}_{b_1 \ldots b_p}{}^{;a_1]}$$

$$(\delta_1 T)^{a_1 \ldots a_{q-1}}{}_{b_1 \ldots b_p} = -T^{i a_1 \ldots a_{q-1}}{}_{b_1 \ldots b_p; i}$$

$$(d_2 T)^{a_1 \ldots a_q}{}_{b_1 \ldots b_{p+1}} = (p+1) T^{a_1 \ldots a_q}{}_{[b_2 \ldots b_{p+1}; b_1]}$$

$$(\delta_2 T)^{a_1 \ldots a_q}{}_{b_1 \ldots b_{p-1}} = -T^{a_1 \ldots a_q}{}_{i b_1 \ldots b_{p-1}}{}^{;i}$$

$$\Delta_1 = d_1 \delta_1 + \delta_1 d_1$$

$$\Delta_2 = d_2 \delta_2 + \delta_2 d_2$$

We emphasise that the covariant derivative acts on *all* indices, i.e., the extra tensor indices as well as the explicit form indices. Although the definition for d was extended to tensor-valued forms in [2] and is well known, as far as we are aware *all three types of operators*, in the form given above, were first introduced in [1], and are less familiar.

Of course, one can mix the operators Δ_1 and Δ_2, weighting them, and obtain new Laplace-type operators with similar or better properties. A particularly good one is,

Theorem 1 *The operator $\bar{\Delta}$ given by*

$$\bar{\Delta} \equiv \frac{1}{2}(\Delta_1 + \Delta_2) \tag{1}$$

is linear, self-adjoint, respects all index symmetry properties, and commutes with all trace operations when acting on double (q, p)-forms.

Note,
$$\bar{\Delta} = \frac{1}{2}(\Delta_L + \nabla^2)$$
where Δ_L is the familiar Lichnerowicz operator [8, 9].
($\bar{\Delta}$ is equivalent to the Lichnerowicz operator Δ_L only for p-forms)

So we find that although Δ_L and $\bar{\Delta}$ do not coincide, they are closely related, and indeed $\bar{\Delta}$ has most of the useful properties of Δ_L; but crucially $\bar{\Delta}$ in addition has direct links with d_i and δ_i. Hence we believe that $\bar{\Delta}$ is a more powerful alternative than the classical Lichnerowicz operator Δ_L since it is so well adapted to dealing with the r-fold form structure of the tensors.

RESULTS FOR DOUBLE FORMS

So we emphasise that an extremely important consequence of this new operator $\bar{\Delta}$ is that for any given tensor field T, from the Cauchy-Kovalewski theorem, we can deduce that there exists an *associated* superpotential $\overset{o}{T}$, by which we mean a superpotential, not just with the same form-structure number and block ranks, but also *with the same index symmetries and trace properties* as T. Moreover, from this associated superpotential, in the case of double (q,p)-forms, a set of fout potentials can be obtained in a natural and straightforward manner from the definition of $\bar{\Delta}$

$$\bar{\Delta}\overset{o}{T} \equiv \frac{1}{2}(\Delta_1 + \Delta_2)\overset{o}{T} = \frac{1}{2}(\delta_1 d_1 + d_1 \delta_1 + \delta_2 d_2 + d_2 \delta_2)\overset{o}{T}$$

Explicitly, we have the following result,

Theorem 2 *Given any double (q,p)-form $T^{a_1...a_q}{}_{b_1...b_p}$ there always exists a local superpotential $\overset{o}{T}{}^{a_1...a_q}{}_{b_1...b_p}$ of exactly the same type as $T^{a_1...a_q}{}_{b_1...b_p}$, such that*

$$\bar{\Delta}\overset{o}{T} = T.$$

Furthermore, there always exist local potentials $Y_{(1)}, Y_{(2)}, Z_{(1)}, Z_{(2)}$ such that

$$T = \frac{1}{2}\left(\delta_1 Y_{(1)} + d_1 Z_{(1)} + \delta_2 Y_{(2)} + d_2 Z_{(2)}\right)$$

where
the double $(q+1, p)$-form $Y_{(1)} = d_1 \overset{o}{T}$,
the double $(q, p+1)$-form $Y_{(2)} = d_2 \overset{o}{T}$,
the double $(q-1, p)$-form $Z_{(1)} = \delta_1 \overset{o}{T}$,
the double $(q, p-1)$-form $Z_{(2)} = \delta_2 \overset{o}{T}$.

When the double form is symmetric $T^{a_1...a_p}{}_{b_1...b_p} = T_{b_1...b_p}{}^{a_1...a_p}$, the number of potentials is halved to two,

Theorem 3 *Given any double symmetric (p,p)-form $T^{a_1...a_p}{}_{b_1...b_p}$ there always exists a local superpotential $\overset{o}{T}{}^{a_1...a_p}{}_{b_1...b_p}$ of exactly the same type as $T^{a_1...a_p}{}_{b_1...b_p}$, such that*

$$\bar{\Delta}\overset{o}{T} = T \ .$$

Furthermore, there always exist local potentials Y, Z such that

$$T^{a_1...a_p}{}_{b_1...b_p} = \frac{1}{2}\left(Y^{a_1...a_p}{}_{b_1...b_p i}{}^{;i} + Y_{b_1...b_p}{}^{a_1...a_p i}{}_{;i}\right.$$
$$\left. + pZ^{a_1...a_p}{}_{[b_1...b_{p-1};b_p]} + pZ_{b_1...b_p}{}^{[a_1...a_{p-1};a_p]}\right)$$

where
  the double $(p, p+1)$-form $Y^{a_1...a_p}{}_{b_1...b_{p+1}} = (p+1)\overset{o}{T}{}^{a_1...a_p}{}_{[b_1...b_p;b_{p+1}]}$
  the double $(p,p-1)$-form $Z^{a_1...a_p}{}_{b_1...b_{p-1}} = -\overset{o}{T}{}^{a_1...a_p}{}_{ib_1...b_{p-1}}{}^{;i}$.

N DIMENSIONS: RESULTS FOR CURVATURE TENSORS

We can specialise the previous theorem to the case of Riemann candidates, that is tensors with the algebraic properties of a Riemann curvature tensor.

Theorem 4 *Any Riemann candidate tensor $\mathscr{R}^{ab}{}_{cd}$ has <u>a pair</u> of local potentials, the double $(2,3)$-form $Y^{ab}{}_{cde}$ and the double $(2,1)$-form $Z^{ab}{}_c$, such that*

$$\mathscr{R}^{ab}{}_{cd} = \frac{1}{2}\left(Y^{ab}{}_{cdi}{}^{;i} + Y_{cd}{}^{abi}{}_{;i} - 2Z^{ab}{}_{[c;d]} - 2Z_{cd}{}^{[a;b]}\right)$$

where the potentials satisfy

$$Y_{a[bcde]} = 0, \qquad Z_{[abc]} = 0.$$

The potentials themselves can be given in terms of a Riemann candidate superpotential $\overset{o}{\mathscr{R}}{}^{ab}{}_{cd}$, where $\bar{\Delta}\overset{o}{\mathscr{R}} = \mathscr{R}$.

A further specialisation can be made to the case of Weyl candidates, that is tensors with the algebraic properties of a Weyl curvature tensor.

Theorem 5 *Any Weyl candidate tensor $\mathscr{W}^{ab}{}_{cd}$ has <u>one</u> double $(2,3)$-form local potential $P^{ab}{}_{cde}$ such that*

$$\mathscr{W}^{ab}{}_{cd} = \frac{1}{2}\left(P^{ab}{}_{cdi}{}^{;i} + P_{cd}{}^{abi}{}_{;i} - 2P_{i[c}{}^{abi}{}_{;d]} - 2P^{i[a}{}_{cdi}{}^{;b]}\right)$$

where the potential satisfies

$$P_{a[bcde]} = 0, \qquad P^{ab}{}_{abc} = 0 \ .$$

The potential itself can be given in terms of a Weyl candidate superpotential $\overset{o}{\mathscr{W}}{}^{ab}{}_{cd}$ where $\bar{\Delta}\overset{o}{\mathscr{W}} = \mathscr{W}$.

This result was originally obtained, by a more direct route, in [3]. The operator $\bar{\Delta}$ which we have used in this theorem is easily seen to coincide with the operator which we constructed for the superpotential of the Weyl candidate tensor in [3].

FOUR DIMENSIONS: THE LANCZOS POTENTIAL

In *four* dimensions all Weyl candidates $\mathscr{W}^{ab}{}_{cd}$ have a double $(2,1)$-form potential, called the Lanczos potential, $H^{ab}{}_c = H^{[ab]}{}_c$ such that [7]

$$\mathscr{W}^{ab}{}_{cd} = 2H^{ab}{}_{[c;d]} + 2H_{cd}{}^{[a;b]} - 2\delta^{[a}_{[c}\left(H^{b]i}{}_{d];i} + H_{d]i}{}^{b];i}\right)$$

where the potential satisfies

$$H_{[abc]} = 0, \qquad H^{ab}{}_b = 0.$$

In *four* dimensions, the double dual of a double $(2,3)$-form is of course a double $(2,1)$-form, and it is straightforward to establish the direct relationship with the new potential in the previous section,

$$H^{ab}{}_c = -(*P*)^{ab}{}_c.$$

Therefore, we have recovered the Lanczos potential $H^{ab}{}_c$ for the Weyl tensor, showing that *in four dimensions, the double dual of the new potential $P^{ab}{}_{cde}$ is the classical Lanczos potential $H^{ab}{}_c$*.

SUMMARY AND DISCUSSION

The inspiration for this paper was the result — that there exists a single potential for any Weyl candidate in any dimension — which we obtained in [3] in a rather pragmatic manner; here, that result has been shown to be a special case of a much more general result, which itself is a consequence of a significant generalisation and innovation in formalism. The underlying approach has been the systematic consideration of tensors as r-fold forms which has been explained and discussed at length in [10].

The generalisation of the differential form approach to tensor-valued forms and the extension of the use of the exterior differential d to such quantities is well known; we have emphasised the extension of the use of δ and Δ also, and highlighted how an r-fold form can be thought of as r different tensor-valued forms, by taking each of the r sets of antisymmetric indices as the form indices, and defining the three pair of operators d_i, δ_i and Δ_i, $i = 1, 2$, associated with each in turn. However, we believe the explicit introduction of the *weighted de Rham operator* $\bar{\Delta}$ as defined in this paper adds an important new ingredient. Two crucial properties of this operator are that it enables us, via a simple Laplace-like equation, to define an *associated superpotential* of exactly the same tensor type, and in addition to define *potentials in a very natural manner*; other generalised Laplacian operators for tensor-valued forms lack one or both of these properties. Hence we believe that $\bar{\Delta}$ is a more powerful operator than Δ_L, and so will be very useful in the type of formal investigations where Δ_L has been used previously [8], [9], and as a powerful alternative in investigations of harmonic tensors.

As an application of the general result for all tensors, we considered the Riemann curvature tensor showing that it can be written in terms of a pair of potentials. The usefulness of this pair of potentials, and in particular, the implications which the Bianchi equations impose on their relationship, need further study.

As emphasised in [3], we now have, in all dimensions, an explicit potential for the Weyl tensor which supplies a tensor which is an 'integral' of the Weyl tensor at the level of the connection, and whose 'square' has units L^{-2} which are precisely the units we would expect for gravitational 'energies'. This suggests the usefulness of the wave equation and the super-energy tensor of this potential, *in all dimensions with Lorentz signature*, as an alternative to the Bel-Robinson tensor for such mathematical investigations as positivity properties, stability and the Cauchy problem for the Einstein equations.

We have built our results on second order Laplace-like equations, with appeals to the Cauchy-Kovalewski theorem so that our results are local for analytic pseudo-Riemannian metrics. However, we expect that these results can be generalised in the usual way; from the point of view of general relativity we can appeal to stronger theorems [5], [6] when we specialise to spaces with Lorentz signature. However, of course the potentials are by definition first order, and now having established their existence it would be more natural to consider them in a first order system. In particular, the single potential for the Weyl tensor is an attractive candidate for deeper analysis in this context; preliminary investigations indicate that more direct and powerful results can be obtained by treating this definition as part of a first order symmetric hyperbolic system.

When we choose a potential we know that it is not unique, and a full understanding of the role of gauge, which is more complicated for tensor-valued forms than for single p-forms, will be very important for further work. In particular we will need to have a set of explicit gauge equations to complete the first order symmetric hyperbolic system for the Weyl tensor. Furthermore, for the Weyl tensor, we would hope for a second-order linear equation for its potential with principal part of type $\nabla^h \nabla_h P^{ab}{}_{cde}$, so that this will give an elliptic equation for positive-definite metrics and a wave equation for Lorentzian signature; in order to obtain such a simple version of the second-order linear equation for the potential we will need to exploit the gauge freedom.

The ideas outlined in this talk are developed in more detail in [4].

ACKNOWLEDGEMENTS

JMMS gratefully acknowledges financial support from the Wenner-Gren Foundation, Sweden, and from grants FIS2004-01626 of the Spanish CICyT and no. 9/UPV 00172.310-14456/2002 of the University of the Basque Country. JMMS thanks the Matematiska institutionen, Linköpings universitet, where this work was partly carried out, for hospitality.

SBE acknowledges travel support from the G S Magnusons fond, The Royal Swedish Academy of Sciences.

REFERENCES

1. Bel, L. *Étude de certains opérateurs definis sur les formes tensorielles (r,s)*, Ann. di Mat. Pura ed Applicata, **51**, serie 4, 171-192. (1963).
2. Cartan, É. *Lecons sur la géométrie des espaces de Riemann*, 2nd edn. Gauthier-Villars, Paris. (1946).
3. Edgar S. B. and Senovilla J. M. M. *A local potential for the Weyl tensor in all dimensions*, Class. Quantum Grav. **21**, L133 - L137 (2004).
4. Edgar S. B. and Senovilla J. M. M. *A weighted de Rham operator acting on arbitrary tensor fields and their local potentials*, http://arxiv.org/abs/math.DG/0505538. To be published in J. Geom. Phys.
5. Friedlander, F. G. *The Wave Equation on a Curved Spacetime*, Cambridge University Press. (1975).
6. Hawking, S. W. and Ellis, G. F. R. *The large scale structure of space-time*, Cambridge Monographs on Mathematical Physics, No. 1. Cambridge University Press, London-New York. (1973).
7. Lanczos, C. *The splitting of the Riemann tensor*, Rev. Mod. Phys., **34**, 379-389. (1962).
8. Lichnerowicz, A. *Propagateurs et commutateurs en relativité générale*, Publ. Scient. des Hautes études scientifiques, No.10, Hermann, Paris. (1961)
9. Lichnerowicz, A. *Propagateurs, commutateurs, et anticommutateurs en relativité générale*, in "Relativity, Groups and Topology", eds. C. DeWitt and B.S. DeWitt, Gordon and Breach, New York. (1964).
10. Senovilla, J. M. M. *Super-energy Tensors*, Class. Quantum Grav. **17**, 2799 (2000).

Kac-Moody algebras in gravity and M-theories

Laurent Houart

Service de Physique Théorique et Mathématique, Université Libre de Bruxelles,
and
The International Solvay Instiitutes,
Campus Plaine C.P. 231
Boulevard du Triomphe, B-1050 Bruxelles, Belgium

Abstract.
The formulation of gravity and M-theories as very-extended Kac-Moody invariant theories is reviewed. Exact solutions describing intersecting extremal brane configurations smeared in all directions but one are presented. The intersection rules characterising these solutions are neatly encoded in the algebra. The existence of dualities for all $\mathscr{G}^{+++}$ and their group theoretical-origin are discussed.

Keywords: Space-Time Symmetries, p-branes, String Dualities, Classical Theories of Gravity
PACS: 04.20.Cv, 04.50.+h, 11.25.-w, 11.25.Yb

A theory containing gravity suitably coupled to forms and dilatons may exhibit upon dimensional reduction down to three dimensions a simple Lie group $\mathscr{G}$ symmetry non-linearly realised. The scalars of the dimensionally reduced theory live in a coset $\mathscr{G}/\mathscr{H}$ where $\mathscr{G}$ is in its maximally non-compact form and $\mathscr{H}$ is the maximal compact subgroup of $\mathscr{G}$. A maximally oxidised theory is such a Lagrangian theory defined in the highest possible space-time dimension D namely a theory which is itself not obtained by dimensional reduction. These maximally oxidised actions have been constructed for all $\mathscr{G}$ [1] and they include in particular pure gravity in D dimensions and the low energy effective actions of the bosonic string and of M-theory.

It has been conjectured that these theories, or some extensions of them, possess the much larger very-extended Kac-Moody symmetry $\mathscr{G}^{+++}$. $\mathscr{G}^{+++}$ algebras are defined by the Dynkin diagrams depicted in Fig.1, obtained from those of $\mathscr{G}$ by adding three nodes [2]. One first adds the affine node, labelled 3 in the figure, then a second node, 2, connected to it by a single line and defining the overextended $\mathscr{G}^{++}$ algebra[1], then a third one, 1, connected by a single line to the overextended node. Such $\mathscr{G}^{+++}$ symmetries were first conjectured in the aforementioned particular cases [4, 5] and the extension to all $\mathscr{G}^{+++}$ was proposed in [6]. In a different development, the study of the properties of cosmological solutions in the vicinity of a space-like singularity, known as cosmological billiards [7], revealed an overextended symmetry $\mathscr{G}^{++}$ for all maximally oxidised theories [8, 9].

The possible existence of this Kac-Moody symmetry $\mathscr{G}^{+++}$ motivates the construction of a Lagrangian formulation explicitly invariant under $\mathscr{G}^{+++}$ [10]. The action $S_{\mathscr{G}^{+++}}$

---

[1] In the context of dimensional reduction, the appearance of $E_8^{++} = E_{10}$ in one dimension has been first conjectured by B. Julia [3].

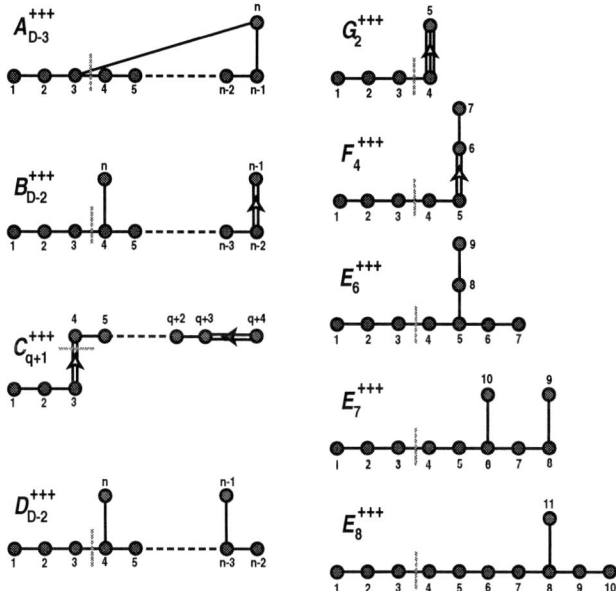

FIGURE 1. The nodes labelled 1,2,3 define the Kac-Moody extensions of the Lie algebras. The horizontal line starting at 1 defines the 'gravity line', which is the Dynkin diagram of a A_{D-1} subalgebra.

is defined in a reparametrisation invariant way on a world-line, a priori unrelated to space-time, in terms of fields $\phi(\xi)$ living in a coset $\mathscr{G}^{+++}/K^{+++}$ where ξ spans the world-line. A level decomposition of $\mathscr{G}^{+++}$ with respect to the subalgebra A_{D-1} of its gravity line (see Fig. 1) is performed where D is identified to the space-time dimension[2]. The subalgebra K^{+++} is invariant under a 'temporal' involution which ensures that the action is $SO(1,D-1)$ invariant at each level where the index 1 of A_{D-1} is identified to a time coordinate.

Each $\mathscr{G}^{+++}$ contains indeed a subalgebra $GL(D)$ such that $SL(D)(=A_{D-1}) \subset GL(D) \subset \mathscr{G}^{+++}$. The generators of the $GL(D)$ subalgebra are taken to be $K^a_{\ b}$ $(a,b=1,2,\ldots,D)$ with commutation relations

$$[K^a_{\ b}, K^c_{\ d}] = \delta^c_b K^a_{\ d} - \delta^a_d K^c_{\ b}. \qquad (1)$$

The $K^a_{\ b}$ along with abelian generators R_u ($u = 1 \ldots q$), which are present when the corresponding maximally oxidised action $S_{\mathscr{G}}$ has q dilatons[3], are the level zero generators.

---

[2] Level expansions of very-extended algebras in terms of the subalgebra A_{D-1} have been considered in [11, 12, 13].

[3] All the maximally oxidised theories have at most one dilaton except the C_{q+1}-series characterised by q dilatons. In the rest of the paper we omit the u index.

The step operators of level greater than zero are tensors $R_{d_1\ldots d_s}{}^{c_1\ldots c_r}$ of the A_{D-1} subalgebra. Each tensor forms an irreducible representation of A_{D-1} characterised by some Dynkin labels. In principle it is possible to determine the irreducible representations present at each level [12, 13]. The lowest levels contain antisymmetric tensor step operators $R^{a_1 a_2 \ldots a_r}$ associated to electric and magnetic roots arising from the dimensional reduction of field strength forms in the corresponding maximally oxidised theory. They satisfy the tensor and scaling relations

$$[K^a_b, R^{a_1\ldots a_r}] = \delta^{a_1}_b R^{aa_2\ldots a_r} + \ldots + \delta^{a_r}_b R^{a_1\ldots a_{r-1} a}, \tag{2}$$

$$[R, R^{a_1\ldots a_r}] = -\frac{\varepsilon_A a_A}{2} R^{a_1\ldots a_r}, \tag{3}$$

where a_A is the dilaton coupling constant to the field strength form and ε_A is $+1(-1)$ for an electric (magnetic) root [6]. The temporal involution Ω_1 generalises the Chevalley involution to allow identification of the index 1 to a time coordinate in $SO(1, D-1)$. It is defined by

$$K^a_b \stackrel{\Omega_1}{\mapsto} -\varepsilon_a \varepsilon_b K^b_a \quad R \stackrel{\Omega_1}{\mapsto} -R \quad , \quad R_{d_1\ldots d_s}{}^{c_1\ldots c_r} \stackrel{\Omega_1}{\mapsto} -\varepsilon_{c_1}\ldots \varepsilon_{c_r} \varepsilon_{d_1}\ldots \varepsilon_{d_s} \bar{R}_{c_1\ldots c_r}{}^{d_1\ldots d_s}, \tag{4}$$

with $\varepsilon_a = -1$ if $a = 1$ and $\varepsilon_a = +1$ otherwise. It leaves invariant a subalgebra K^{+++} of $\mathscr{G}^{+++}$. The fields $\varphi(\xi)$ living in the coset space $\mathscr{G}^{+++}/K^{+++}$ parametrise the Borel group built out of Cartan and positive step operators in $\mathscr{G}^{+++}$. Its elements $\mathscr{V}$ are written as

$$\mathscr{V}(\xi) = \exp(\sum_{a \geq b} h^a_b(\xi) K^b_a - \phi(\xi) R) \exp(\sum \frac{1}{r!s!} A_{b_1\ldots b_s}{}^{a_1\ldots a_r}(\xi) R_{a_1\ldots a_r}{}^{b_1\ldots b_s} + \ldots), \tag{5}$$

where the first exponential contains only level zero operators and the second one the positive step operators of levels strictly greater than zero. Defining

$$dv(\xi) = d\mathscr{V} \mathscr{V}^{-1} \quad d\tilde{v}(\xi) = -\Omega_1 dv(\xi) \quad dv_{sym} = \frac{1}{2}(dv + d\tilde{v}), \tag{6}$$

one obtains, in terms of the ξ-dependent fields, an action $S_{\mathscr{G}^{+++}}$ invariant under global $\mathscr{G}^{+++}$ transformations, defined on the coset $\mathscr{G}^{+++}/K^{+++}$

$$S_{\mathscr{G}^{+++}} = \int d\xi \frac{1}{n(\xi)} \langle (\frac{dv_{sym}(\xi)}{d\xi})^2 \rangle, \tag{7}$$

where $n(\xi)$ is an arbitrary lapse function ensuring reparametrisation invariance on the world-line and $<,>$ is the invariant bilinear form.

Writing

$$S_{\mathscr{G}^{+++}} = S^{(0)}_{\mathscr{G}^{+++}} + \sum_A S^{(A)}_{\mathscr{G}^{+++}}, \tag{8}$$

where $S^{(0)}_{\mathscr{G}^{+++}}$ contains all level zero contributions, one obtains

$$S^{(0)}_{\mathscr{G}^{+++}} = \frac{1}{2}\int d\xi \frac{1}{n(\xi)} \left[\frac{1}{2}(g^{\mu\nu} g^{\sigma\tau} - \frac{1}{2}g^{\mu\sigma} g^{\nu\tau})\frac{dg_{\mu\sigma}}{d\xi}\frac{dg_{\nu\tau}}{d\xi} + \frac{d\phi}{d\xi}\frac{d\phi}{d\xi}\right], \tag{9}$$

$$S^{(A)}_{\mathscr{G}^{+++}} = \frac{1}{2r!s!} \int d\xi \frac{e^{-2\lambda\phi}}{n(\xi)} \left[\frac{DA^{v_1...v_s}_{\mu_1...\mu_r}}{d\xi} g^{\mu_1\mu'_1}...g^{\mu_r\mu'_r} g_{v_1 v'_1}...g_{v_s v'_s} \frac{DA^{v'_1...v'_s}_{\mu'_1...\mu'_r}}{d\xi} \right]. \quad (10)$$

The ξ-dependent fields $g_{\mu\nu}$ are defined as $g_{\mu\nu} = e^a_\mu e^b_\nu \eta_{ab}$ where $e^a_\mu = (e^{-h(\xi)})^a_\mu$. The appearance of the Lorentz metric η_{ab} with $\eta_{11} = -1$ is a consequence of the temporal involution Ω_1. The metric $g_{\mu\nu}$ allows a switch from the Lorentz indices (a,b) of the fields appearing in Eq.(5) to $GL(D)$ indices (μ,ν). $D/D\xi$ is a covariant derivative generalising $d/d\xi$ through non-linear terms arising from non-vanishing commutators between positive step operators and λ is the generalisation of the scale parameter $-\varepsilon_A a_A/2$ to all roots.

The $\mathscr{G}^{+++}$-invariant actions $S_{\mathscr{G}^{+++}}$ leads to two distinct actions invariant under the overextended Kac-Moody algebra $\mathscr{G}^{++}$. The first one $S_{\mathscr{G}^{++}_C}$ is constructed from $S_{\mathscr{G}^{+++}}$ by performing a consistent truncation. The corresponding $\mathscr{G}^{++}$ algebra is obtained from $\mathscr{G}^{+++}$ by deleting the node labelled 1 from the Dynkin diagram of $\mathscr{G}^{+++}$ depicted in Fig. 1. The truncation is achieved by putting to zero in the coset representative the field multiplying the Chevalley generator H_1 and all the fields multiplying the positive step operators associated to roots whose decomposition in terms of simple roots contains the deleted root α_1. This theory carries a Euclidean signature and is the generalisation to all $\mathscr{G}^{++}$ of the $E_8^{++} = E_{10}$ invariant action first proposed in reference [14] in the context of M-theory and cosmological billiards. The parameter ξ is then identified with the time coordinate and the action restricted to a defined number of lowest levels is equal to the corresponding maximally oxidised theory in which the fields depend only on this time coordinate. A second $\mathscr{G}^{++}$-invariant action $S_{\mathscr{G}^{++}_B}$ is obtained from $S_{\mathscr{G}^{+++}}$ by performing the same consistent truncation *after* conjugation by the Weyl reflection in the hyperplane perpendicular to the simple root corresponding to the node 1 of figure 1. The non-commutativity of the temporal involution with the Weyl reflection [15, 16] implies that after Weyl reflection the index 2 of A_{D-1} is now identified to the time coordinate. Consequently the second action $S_{\mathscr{G}^{++}_B}$ is inequivalent to the first one [17].

In $S_{\mathscr{G}^{++}_B}$, ξ is identified with a space-like direction and the action is characterised by a Lorentzian signature $(1, D-2)$. This theory admits exact solutions which are identical to those of the corresponding maximally oxidised theory describing intersecting extremal brane configurations smeared in all directions but one [17, 10, 18]. For each of the $A = 1...\mathscr{N}$ branes present in the intersecting brane configuration and characterised by $\lambda_1...\lambda_{q_A}$ longitudinal spacelike directions, one has one non-zero field component corresponding to one positive step operator associated with one positive real root α_A.

$$A_{2\lambda_1...\lambda_{q_A}} = \varepsilon_{2\lambda_1...\lambda_{q_A}} \left[\frac{2(D-2)}{\Delta_A}\right]^{1/2} H_A^{-1}(\xi) \qquad A = 1...\mathscr{N}, \quad (11)$$

and

$$p^a = \sum_{A=1}^{\mathscr{N}} p_A^a = \sum_{A=1}^{\mathscr{N}} \frac{\eta_A^a}{\Delta_A} \ln H_A(\xi) \qquad a = 2,3,...,D \quad (12)$$

$$\phi = \sum_{A=1}^{\mathcal{N}} \phi_A = \sum_{A=1}^{\mathcal{N}} \frac{D-2}{\Delta_A} \varepsilon_A a_A \ln H_A(\xi), \tag{13}$$

where $p^a \equiv -h_a{}^a$ and $h_a^b = 0$ if $a \neq b$. Here $\eta_A^a = q_A + 1$ or $-(D-3-q_A)$ depending on whether the direction a is perpendicular or parallel to the q_A-brane and $\Delta_A = (q_A + 1)(D-3-q_A) + \frac{1}{2}a_A^2(D-2)$. The factor ε_A is $+1$ for an electric brane and -1 for a magnetic one. Each of the branes in the configuration is thus described as electrically charged and is characterised by one positive harmonic function in ξ-space, namely one has

$$\frac{d^2 H_A(\xi)}{d\xi^2} = 0 \qquad A = 1 \ldots \mathcal{N}. \tag{14}$$

The consistent troncation yields for the spatial direction 1 the result

$$p^1 = \sum_{A=1}^{\mathcal{N}} \frac{q_A + 1}{\Delta_A} \ln H_A(\xi), \tag{15}$$

identifying it to a direction transverse to all branes. The Eqs. (12), (13) and (15) are solutions provided the intersection rules [19]

$$\bar{q} + 1 = \frac{(q_A + 1)(q_B + 1)}{D - 2} - \frac{1}{2} \varepsilon_A a_A \varepsilon_B a_B \tag{16}$$

are satisfied.

The intersection rules Eq.(16) are neatly and elegantly encoded in the group structure [18]. They can indeed be expressed as an orthogonality condition between the real positive roots of $\mathscr{G}_B^{++}$ (and $\mathscr{G}^{+++}$) for all branes present in the configuration [18] namely

$$\alpha_A \cdot \alpha_B = 0 \qquad A \neq B = 1 \ldots \mathcal{N} \tag{17}$$

When $\mathscr{G}$ is not simply laced there is an additional condition in order to have a solution of $S_{\mathscr{G}_B^{++}}$, namely for each pair α_A, α_B with $A \neq B$ one must have

$$\alpha_A + \alpha_B \neq \text{root}. \tag{18}$$

The conditions Eqs. (17) and (18) are in fact the input that permits the derivation of the exact solutions by allowing a reduction of $S_{\mathscr{G}_B^{++}}$ to quadratic terms. Furthermore there is a one-to-one correspondence between the exact solutions of $S_{\mathscr{G}_B^{++}}$ and the space-time intersecting extremal brane solutions of the corresponding maximally oxidised theory. Indeed, the configurations satisfying Eq. (17) and not Eq.(18) correspond in the maximally oxidised theory to configurations which are not solutions of the equations of motion because of the presence of Chern-Simons terms in the space-time action [18].

In the particular case of $\mathscr{G}^{++} = E_8^{++}$, it is well-known that the Weyl reflection generated by the root α_{11} has an interpretation in terms of type IIA string duality. It correspond to a double T-duality in the direction 9 and 10 followed by an exchange of the two radii [20, 21, 6]. Furthermore the change of signatures which occur as a consequence of the non-commutativity of the temporal involution and the Weyl reflections [15, 16] are in agreement with the exotic phases of M-theory discussed in [22, 23].

The action of Weyl reflections generated by simple roots not belonging to the gravity line on the exact extremal brane solutions has been studied for *all* $\mathscr{G}^{+++}$-theory [10]. The existence of Weyl orbits of extremal brane solutions similar to the U-duality orbits existing in M-theory has been discovered. This fact strongly suggests a general group-theoretical origin of 'dualities' for all $\mathscr{G}^{+++}$-theories transcending string theories and supersymmetry. Furthermore, exotic phases of all the M-theories (all the $\mathscr{G}_B^{++}$ theories) related by 'duality' Weyl transformations to the conventional phase[4] characterised by a signature $(1, D-2)$ have been uncovered and classified [24].

We exemplify the existence of U-duality-like Weyl orbits of extremal branes in the case $\mathscr{G}_B^{++} = E_7^{++}$ whose corresponding maximally oxidised theory is gravity coupled to a 4- and a 2- form field strength with one dilaton in 9 space-time dimensions [1]. The Dynkin diagram of E_7^{+++} is depicted in Fig.1, which exhibits the two simple electric roots α_{10} and α_9 corresponding respectively to the step operators R^{789} and R^9 which couple to the electric potentials A_{789} and A_9.

We take as input the electric extremal 2-brane $\mathbf{e}_{(8,9)}$ in the directions $(8,9)$ associated with the 4-form field strength whose corresponding potential[5] is A_{289} and submit it to the non trivial Weyl reflection W_{10} associated with the electric root α_{10} of Fig.1. We display below, both for $\mathbf{e}_{(8,9)}$ and its transform, the vielbein components $p^a \equiv -h_a{}^a$ with $a = 2\ldots 9$ and the the dilaton value ϕ, of the brane solution Eqs.(12) and (13) as a nine-dimensional vector where the last component is the dilaton. We also indicate the transform of the step operator R^{289} under the Weyl transformation. We obtain

$$(-4,3,3,3,3,3,-4,-4;2\sqrt{7})\frac{\ln H(\xi)}{14} \quad \mathbf{e}_{(8,9)} \quad R^{289} \tag{19}$$
$$\downarrow W_{10}$$
$$(-7,0,0,0,0,7,0,0;0)\frac{\ln H(\xi)}{14} \quad \mathbf{kk}_{\mathbf{e}\,(7)} \quad K_7^2 \tag{20}$$

The 2-brane transforms into a KK-wave in the direction 7 characterised by a non-zero K_7^2 [10]. This is reminiscent of a double T-duality in M-theory.

We now move the electric brane through Weyl reflections associated with roots of the gravity line to $\mathbf{e}_{(5,9)}$ and submit it to the Weyl reflection W_{10}. We now find that the brane $\mathbf{e}_{(5,9)}$ is invariant but moving it to the position $\mathbf{e}_{(5,6)}$, we get

$$(-4,3,3,-4,-4,3,3,3;2\sqrt{7})\frac{\ln H(\xi)}{14} \quad \mathbf{e}_{(5,6)} \quad R^{256} \tag{21}$$
$$\downarrow W_{10}$$
$$(-1,6,6,-1,-1,-1,-1,-1;4\sqrt{7})\frac{\ln H(\xi)}{14} \quad \mathbf{m}_{(5,6,7,8,9)} \quad R^{256789} \tag{22}$$

---

[4] The other orbits have been discussed in [25].
[5] We recall that in $\mathscr{G}_B^{++}$ the index 2 is identified to the time coordinate

This is a magnetic 5-brane in the directions $(5,6,7,8,9)$ associated to the 2-form field strength ! It is expressed in terms of its dual potential A_{256789}. Submit instead $\mathbf{e}_{(5,9)}$ to to the Weyl reflection W_9 associated with the electric root α_9 of Fig.1. The 2-brane $\mathbf{e}_{(5,9)}$ is again invariant, but moving it to to the position $\mathbf{e}_{(5,6)}$, we now get

$$(-4,3,3,-4,-4,3,3,3;2\sqrt{7})\frac{\ln H(\xi)}{14} \qquad \mathbf{e}_{(5,6)} \qquad R^{256}$$
$$\downarrow W_9$$
$$(-3,4,4,-3,-3,4,4,-3;-2\sqrt{7})\frac{\ln H(\xi)}{14} \qquad \mathbf{m}_{(5,6,9)} \qquad R^{2569} \qquad (23)$$

This is a magnetic 3-brane in the directions $(5,6,9)$ associated to the 4-form field strength, expressed in terms of its dual potential A_{2569}.

Finally, let us submit the magnetic 5-brane $\mathbf{m}_{(5,6,7,8,9)}$ obtained in Eq.(22) to the Weyl reflection W_9. One obtains

$$(-1,6,6,-1,-1,-1,-1,-1;4\sqrt{7})\frac{\ln H(\xi)}{14} \qquad \mathbf{m}_{(5,6,7,8,9)} \qquad R^{256789}$$
$$\downarrow W_9$$
$$(0,7,7,0,0,0,0,-7;0)\frac{\ln H(\xi)}{14} \qquad \mathbf{kk}_{\mathbf{m}\,(1,3,4;9)} \qquad R^{256789,9} \qquad (24)$$

Eq.(24) describes, as in M-theory, a purely gravitational configuration, namely a KK-monopole with transverse directions (1,3,4) and Taub-NUT direction 9 in terms of a dual gravity tensor $h_{256789,9}$ [10].

The approach based on Kac-Moody algebras constitutes certainly a very-exciting and innovative attempt to understand gravitational theories encompassing string theories which could lead to a completely new formulation of gravitational interactions where the structure of space-time is hidden somewhere in these huge algebras [14, 17, 26] or even huger ones [27].

ACKNOWLEDGMENTS

I would like to warmly thank my collaborator and friend François Englert with whom almost all the original results presented here have been derived. I am also grateful to Sophie de Buyl, Marc Henneaux and Nassiba Tabti for enjoyable collaborations.

REFERENCES

1. E. Cremmer, B. Julia, H. Lü and C. N. Pope, *Higher dimensional origin of D=3 coset symmetries*, hep-th/9909099.
2. M. Gaberdiel, D. Olive and P. West, *A class of Lorentzian Kac-Moody algebras*, Nucl. Phys. **B645** (2002) 403, hep-th/0205068.
3. B. Julia, *Kac-Moody Symmetry of gravitation and Supergravity Theories*, Proc. AMS-SIAM Chicago meeting July 1982, Lecture in Applied Mathematics **21** (1985) 355; B. Julia, *Group Disintegrations*,

in *Superspace and Supergravity* edited by S.W. Hawking and M. Rocek, Cambridge University Press (1981).
4. P. West, E_{11} *and M theory*, Class. Quant. Grav. **18** (2001) 4443, `hep-th/0104081`.
5. N. D. Lambert and P. C. West, *Coset symmetries in dimensionally reduced bosonic string theory*, Nucl. Phys. **B615** (2001) 117, `hep-th/0107209`.
6. F. Englert, L. Houart, A. Taormina and P. West, *The symmetry of M-theories*, J. High Energy Phys. **09** (2003) 020, `hep-th/0304206`.
7. T. Damour, M. Henneaux and H. Nicolai, *Cosmological billiards*, Class. and Quant. Grav. **20** (2003) R145, `hep-th/0212256`, and references therein.
8. T. Damour, M. Henneaux, *E(10), BE(10) and arithmetical chaos in superstring cosmology*, Phys. Rev.Lett. **86** (2001) 4749, `hep-th/0012172`; T. Damour, M. Henneaux, B. Julia and H. Nicolai, *Hyperbolic Kac-Moody algebras and chaos in Kaluza-Klein models*, Phys. Lett. **B509** (2001) 323, `hep-th/0103094`.
9. T. Damour, S. de Buyl, M. Henneaux and C. Schomblond, *Einstein billiards and overextensions of finite-dimensional simple Lie algebras*, J. High Energy Phys. **08** (2002) 030, `hep-th/0206125`.
10. F. Englert and L. Houart, $\mathcal{G}^{+++}$ *invariant formulation of gravity and M-theories:exact BPS solutions*, J. High Energy Phys. **01** (2004) 002, `hep-th/0311255`.
11. P. West, *Very Extended E8 and A8 at low levels, gravity and supergravity*, Class. Quant. Grav. **20** (2003) 2393, `hep-th/0212291`.
12. H. Nicolai and T. Fischbacher, *Low level representations of E10 and E11*, in: *Kac-Moody Lie algebra and related topics*, eds. N. Sthanumoorthy and K.C. Misra, Contempary Mathematics 343, American Mathematical Society, 2004, `hep-th/0301017`.
13. A. Kleinschmidt, I. Schnakenburg and P. West, *Very-extended Kac-Moody algebras and their interpretation at low levels*, Class. Quant. Grav. **21** (2004) 2493, `hep-th/0309198`.
14. T. Damour, M. Henneaux and H. Nicolai, E_{10} *and a small tension expansion of M-theory*, Phys. Rev. Lett. **89** (2002) 221601, `hep-th/0207267`.
15. A. Keurentjes, E_{11}: *Sign of the times*, Nucl. Phys. **697** (2004) 302, `hep-th/0402090`.
16. A. Keurentjes, *Time-like T-duality algebra*, J. High. Energy Phys. **11** (2004) 34, `hep-th/0404174`.
17. F. Englert, M. Henneaux and L. Houart, *From very-extended to overextended gravity and M-theories*, J. High Energy Phys. **02** (2005) 070, `hep-th/0412184`.
18. F. Englert and L. Houart, $\mathcal{G}^{+++}$ *invariant formulation of gravity and M-theories:exact intersecting brane solutions*, J. High Energy Phys. **05** (2004) 059, `hep-th/0405082`.
19. R. Argurio, F. Englert and L. Houart, *Intersection rules for p-branes*, Phys. Lett. **B398** (1997) 61, `hep-th/9701042`.
20. S. Elizur, A. Giveon, D. Kutasov and E. Rabinovici, *Algebraic aspects of matrix theory on* T^d, Nucl. Phys. **B509** (1998) 122, `hep-th/9707217`; N. A. Obers and P. Pioline, *U-duality and M-theory*, Phys. Rept. **318** (1999) 113-225 `hep-th/9812139`.
21. T. Banks, W. Fischler and L. Motl, *Dualities versus singularities*, J. High Energy Phys. **01** (1999) 019, `hep-th/9811194`.
22. C. M. Hull, *Timelike T-duality, de Sitter space, large N gauge theories and topological field theory*, J. High Energy Phys **07** (1998) 021, `hep-th/9806146`; C. M. Hull and R. R. Khuri, *Branes, times and dualities*, Nucl. Phys. **B536** (1998) 219, `hep-th/9808069`.
23. C. M Hull, *Duality and the signature of space-time*, J. High Energy Phys. **11** (1998) 017, `hep-th/9807127`.
24. S. de Buyl, L. Houart and N. Tabti, *Dualities and signatures of* $\mathcal{G}^{++}$-*invariant theories*, J. High Energy Phys. **06** (2005) 084, `hep-th/0505199`.
25. A. Keurentjes, *Poincaré Duality and* $\mathcal{G}^{+++}$ *algebra's*, `hep-th/0510212`.
26. T. Damour and H. Nicolai, *Higher order M-theory corrections and the Kac-Moody algebra* E_{10}, Class. Quant. Grav. **22** (2005) 2849, `hep-th/0504153`.
27. A. Kleinschmidt and P. West, *Representations of* $\mathcal{G}^{+++}$ *and the role of space-time*, J. High Energy Phys. **04** (2004) 033, `hep-th/0312247`.

Hyperboloidal data and evolution

S. Husa*, C. Schneemann[†], T. Vogel[†] and A. Zenginoğlu[†]

*Friedrich Schiller University Jena
[†]MPI for Gravitational Physics Potsdam

Abstract. We discuss the hyperboloidal evolution problem in general relativity from a numerical perspective, and present some new results. Families of initial data which are the hyperboloidal analogue of Brill waves are constructed numerically, and a systematic search for apparent horizons is performed. Schwarzschild-Kruskal spacetime is discussed as a first application of Friedrich's general conformal field equations in spherical symmetry, and the Maxwell equations are discussed on a nontrivial background as a toy model for continuum instabilities.

Keywords: Numerical relativity, continuum instabilities, initial data, conformal compactification
PACS: 04.25.Dm, 04.20.Ha, 04.30.-w

INTRODUCTION

In this paper we consider algorithms for numerical relativity (NR) based on hyperboloidal slices – spacelike hypersurfaces characterized by a mean extrinsic curvature χ that approaches a finite value in the limit $r \to \infty$. Correspondingly such slices are not asymptotically euclidean, but rather reach out to null infinity, and thus provide an alternative to null surfaces for tracking radiation signals to large distances from their source, e.g. in order to predict signals in a gravitational wave detector (see e.g. [1, 2]). Being spacelike, hyperboloidal slices are in some sense more flexible than null surfaces, and thus interesting for constructing numerical relativity codes aimed at gravitational wave physics. In this paper we briefly discuss our general ideas about the design of numerical codes for hyperboloidal evolution and some preliminary results from two perspectives we believe to be of key importance: the need to control continuum instabilities and fitness to accurately resolve gravitational wave signals. We also list a few new results.

As is the case with other disciplines of computational physics, an essential part of the art of NR is to make the physical continuum features manifest in the discrete system that is then solved by a computer. In general relativity (GR) diffeomorphism invariance gives rise to a variety of problems not familiar from other theories and which are not yet understood in sufficient depth to provide a fully satisfactory basis for numerical simulations. A typical evolution scheme has many more computational than physical degrees of freedom, the extra degrees of freedom correspond to gauge choice and the presence of constraints – one should therefore not be surprised to find a generic tendency for instabilities in the excess degrees of freedom, and indeed the multitude of formulations of the Einstein equations are typically plagued by instabilities whose precise causes have often remained elusive, and we expect much further work to be necessary in order to understand the dos and don'ts of NR. As a simple (linear) example for the type of problems that have to be expected, we will consider hyperboloidal evolution of an electromagnetic field on Minkowski background.

At least from an observational point of view it is clearly desirable to design numerical codes with accurate GW signal prediction in mind. This is difficult for various reasons. First, in all physically relevant scenarios gravitational radiation is only a relatively small effect in the energy balance of the system. Second, in GR such fundamental quantities as energy, momentum, or emitted gravitational radiation energy can only be defined unambiguously in terms of asymptotic limits. Consequently, it also becomes particularly difficult to formulate physically motivated boundary conditions along the lines of "outgoing radiation boundary conditions" at finite distance from the sources.

Conformal compactification, originally suggested by Penrose [3] allows to discuss asymptotics in terms of local differential geometry and has provided a very fruitful framework to approach many problems in mathematical relativity. Naturally, it also raises hopes for a consistent notion and quantitative treatment of GW signals. However, since the conformal framework is extremely flexible, it does not by itself determine a strategy for NR, and additional physical intuition and practical insights are necessary to bring this technique to fruition in numerical simulations. In the following, we will briefly review the connection between asymptotics, conformal compactification and gravitational wave (GW) signals before sketching our strategy to develop codes for the hyperboloidal evolution problem. We then present some new numerical results concerning Friedrich's general conformal field equations in spherical symmetry as a simple window into the interplay of spatial and null infinity, the Maxwell equations on a nontrivial background as a toy model for continuum instabilities, and initial data that generalize Brill waves to the hyperboloidal context.

CONFORMAL COMPACTIFICATION AND RESCALING

A key idea behind conformal rescaling is to compute "order unity" quantities, e.g. for a massless scalar field Φ a rescaling of the type $\Psi := r\Phi$, which asymptotically just cancels the known fall-off of the radiation from an isolated source. This allows one to work with quantities that are finite even asymptotically. Such a procedure can furthermore improve the numerical conditioning of radiation problems. Generally, it is useful in computational work to factor out what is already known. The idea of conformal compactification is to perform a conformal transformation on the metric $g_{ab} = \Omega^2 \tilde{g}_{ab}$ and view the physical space-time $\tilde{\mathcal{M}}$ as a submanifold $\tilde{\mathcal{M}} = \{p \in \mathcal{M} \,|\, \Omega(p) > 0\}$ of some manifold $\mathcal{M}$ completed by boundary points $\partial \tilde{\mathcal{M}} = \{p \in \mathcal{M} \,|\, \Omega(p) = 0\}$ lying "at infinity" with respect to $\tilde{g}_{ab}$. The definition of a certain type of asymptotics, like asymptotic flatness, then proceeds in terms of asymptotic properties of the conformal factor, which define a desired physical fall-off behavior (see e.g. [1]). Note that in a relativistic theory, we need to deal with three types of directions toward infinity: timelike ($\iota^\pm$), spacelike (ι^0) and null ($\mathscr{I}^\pm$), and these limits have very different physical significance. In particular, observers situated at "astronomical" distances (e.g. GW detectors) can be modeled through geometric objects at future null infinity [4]. Clearly, a thorough physical understanding of the problem of consistently modeling GW sources and detectors in a single picture is very desirable.

However, writing the Einstein tensor in terms of the rescaled metric makes it immediately clear that taking this concept to the level of the field equations can not be straightforward:

$$\tilde{G}_{ab}[\Omega^{-2}g] = G_{ab}[g] - \frac{2}{\Omega}(\nabla_a \nabla_b \Omega - g_{ab}\nabla_c \nabla^c \Omega) - \frac{3}{\Omega^2}g_{ab}(\nabla_c \Omega)\nabla^c \Omega.$$

In the new variables the equations are formally singular at $\partial \tilde{\mathcal{M}}$ whereas multiplication by Ω^2 leads to a degenerate principal part for $\Omega = 0$. A very general prescription for regularizing the rescaled Einstein equations has been obtained by Friedrich through the formulation of the regular conformal field equations [5]. The fact that this is actually possible for the Einstein equations, is a nontrivial result and may certainly seem surprising. Unfortunately, it is achieved at a high price of introducing a large number of new evolution variables, which complicates the numerical implementation and increases the risk of triggering continuum instabilities (for numerical results see [2]).

Compactification techniques have been used in NR for quite some time, but have often been based on less general regularization techniques, e.g. through restriction to a special class of gauges. Compactification in null directions has been very successful in the characteristic approach (see e.g. [6] and [7] for recent results) and is well understood. Compactification of spacelike infinity has not only been used to construct initial data (see e.g. [8, 9, 10]), but encouraging results have also been obtained in the time evolution problem [11], where black holes are modeled as "internal asymptotic ends", often referred to as punctures, and recently also to get rid of the boundary problem in NR [12]. In the evolution context, however, some open questions remain, e.g. because compactification at i^0 leads to a "piling up" of waves. At $\mathscr{I}^+$ this effect does not appear – waves leave the physical spacetime through the boundary $\mathscr{I}^+$. Also, regularity issues of the equations at spatial infinity are not yet fully understood, although much progress has been made with Friedrich's general conformal field equations [13], for which we discuss a simple application below.

A code that utilizes hyperboloidal slices to compactify null infinity can profit from all the flexibility in gauge that a Cauchy approach offers. However, following the idea to factor out what is already known and making the physical continuum features also manifest in the numerical code leads to the problem of making manifest the rigid structure of null infinity in addition to the fall-off of the "gravitational field". Particularly important seem the shear-free property of null infinity and the existence of a natural class of time coordinates associated with affine parameters of the null geodesic generators of $\mathscr{I}$, known as Bondi time. It is this time coordinate which corresponds to the proper time of distant observers [1], and which thus corresponds to an "undistorted signal", as in Fig. 1. We suggest to use the gauge freedom to make the rigid structure of $\mathscr{I}$ manifest and freeze it to a fixed coordinate sphere as discussed in detail by Andersson [14] (in particular here the connection between the 3+1 split and the Bondi gauge is discussed, and essentially the same recommendation to use such a gauge as starting point for regularization is given). Fixing $\mathscr{I}^+$ to a coordinate sphere, it is natural to identify it also with the boundary of the computational domain, and thus to restrict oneself to the physical part of the spacetime. First experiments along these lines with scalar fields on a Schwarzschild background have yielded the ringdown results in Fig. 1.

As an example for the type of coordinates we have in mind, consider computing just the domain of dependence of a piece of Minkowski space with initial data given on a ball. Appropriate coordinates are those which are also adapted to self-similarity:

$$ds^2 = \frac{e^{-2\tau}}{R^2} \left[-(R^2 - r^2) d\tau^2 - 2r dr d\tau + dr^2 + r^2 \left(d\theta^2 + \sin\theta^2 d\varphi^2 \right) \right].$$

Using the same type of coordinates in the compactified spacetime with R identified with the initial location of $\mathscr{I}$ yields the picture in Fig. 1. Freezing $\mathscr{I}$ to a coordinate sphere essentially corresponds to a choice of the shift vector on $\mathscr{I}$, which leads to two problems: First, the prescription of shift needs to be compatible with a well-posed evolution system, and second, one also needs to choose well for the shift vector away from $\mathscr{I}$, in order not to distort the geometry in the interior of the spacetime.

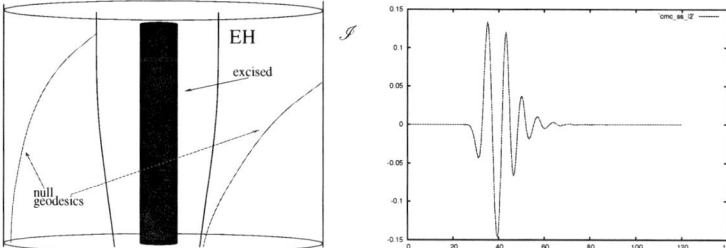

FIGURE 1. Left: A sketch depicting a situation with $\mathscr{I}^+$ frozen to a coordinate sphere. Right: The ringdown of a scalar field with angular momentum number $l = 2$ on a compactified Schwarzschild background with CMC-slices and $\mathscr{I}^+$ frozen to a coordinate sphere.

Our strategy to develop codes for the hyperboloidal initial value problem has thus been threefold: First, we have developed a computational infrastructure that allows us to confront equations as complex as the conformal field equations without tying us down to a particular form of the equations. To this end, the Kranc code generation and tensor manipulation package [15] has been developed. Second, it has proven very fruitful to learn as much as possible from the characteristic approach, which is less general, but works well. Third, we have started a number of smaller projects that allow us test what works and what does not in simplified situations, and actually start a mathematical analysis of the properties of our algorithms. We present some preliminary results below.

GENERAL CONFORMAL FIELD EQUATIONS

The approach suggested above and sketched in Fig. (1) is well adapted to computing gravitational wave signals, but can not reproduce a global representation of the spacetime, which includes spacelike infinity ι^0. From the point of view of an observer at $\mathscr{I}^+$, ι^0 represents the infinite past, which is clearly relevant for certain questions, e.g. a quasi-stationary solution may have persisted for a very long time, before violent dynamics sets in. Friedrich's general conformal field equations [13], which rely on the conformal Gauss gauge, allow for a global treatment, in which different asymptotic regions can be handled with one system of equations. Using this method can provide initial data for a hyperboloidal code that is actually determined from a Cauchy surface. Also, numerical experiments with this system might give rise to a better understanding of the regularity issues around spatial infinity.

As a first step we have used the general conformal field equations to construct the Schwarzschild-Kruskal solution with initial data specified on a Cauchy surface. Using the conformal Gauss gauge, in which by spherical symmetry all equations become ordinary differential equations, it was possible for the first time to cover the entire Schwarzschild-Kruskal spacetime including spacelike, null and timelike infinity and the domain close to the singularity (Fig. 2). These results can also be seen as a feasibility study of the conformal Gauss gauge. Current work is directed to the numerical solution of the general conformal field equations for non-spherically symmetric initial data.

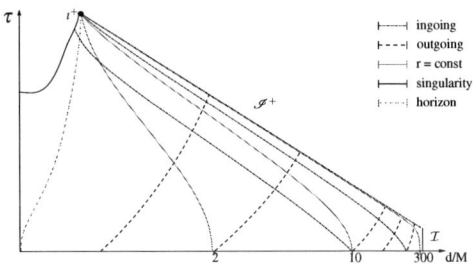

FIGURE 2. Schwarzschild-Kruskal spacetime in a conformal Gauss gauge.

MINKOWSKI SPACE

A natural first exercise when entering uncharted territory in NR is to consider Minkowski space. A comparison of evolutions of Minkowski space in various gauges with the full conformal field equations has been reported in [2], a particularly interesting case is, when $\mathscr{I}^+$ is identified with a fixed coordinate sphere and the conformal geometry is chosen stationary. In this case a constraint violating continuum instability is found. Inspection of the equations suggests the instability to be due to the type of effect described in this section. Recently, the mathematical tools to clarify this have been discussed by Frauendiener and Vogel [16].

As a simple exercise for this type of problem, we have analyzed the case of a Maxwell field (E^a, B^a) on Minkowski space sliced by non-trivial hypersurfaces:

$$\partial_t E^a - \beta^b \partial_b E^a - \alpha \, \varepsilon^{abc} h_{cd} \partial_b B^d = \alpha(\varepsilon^{abc} h_{cd} \chi_b B^d - \chi E^a + \varepsilon^{abc} h_{cd} \Gamma^d_{eb} B^e) - E^b \partial_b \beta^a,$$

$$\partial_t B^a - \beta^b \partial_b B^a + \alpha \, \varepsilon^{abc} h_{cd} \partial_b E^d = -\alpha(\varepsilon^{abc} h_{cd} \chi_b E^d + \chi B^a + \varepsilon^{abc} h_{cd} \Gamma^d_{eb} E^e) - B^b \partial_b \beta^a$$

The constraints and constraint propagation equations then are

$$0 = \mathcal{E} := \partial_b E^b + \Gamma^b_{cb} E^c, \qquad \partial_t \mathcal{E} - \beta^a \partial_a \mathcal{E} = -\alpha \chi \mathcal{E},$$

$$0 = \mathcal{B} := \partial_b B^b + \Gamma^b_{cb} B^c, \qquad \partial_t \mathcal{B} - \beta^a \partial_a \mathcal{B} = -\alpha \chi \mathcal{B}.$$

Here, h_{ab}, χ, χ_a, Γ^a_{bc}, α and β^a are 3-metric, mean extrinsic curvature, acceleration, Christoffel symbol of h_{ab}, lapse and shift respectively. From the constraint propagation equations one directly reads off a stability prognosis in the spirit of Frauendiener and Vogel [16]: if $\chi < 0$, a constraint violating continuum instability has to be expected, whereas $\chi > 0$ should result in constraint damping. Both effects will be demonstrated below for a simple class of slices in Minkowski space with a fixed sign of χ. Note that densitizing the evolved fields can change the sign of the χ factors. This example thus demonstrates that one needs to be aware of a subtle interplay between the evolution system, choice of variables and gauge. Continuum instabilities of this type are essentially an ODE effect in the sense that they are determined by lower order source terms rather than spatial derivatives. Consequently, it is important to realize that for numerical purposes, analyzing the principal part is only a starting point. In general, lower order terms have to be carefully analyzed and a formulation of the theory has to be chosen that avoids instabilities. Clearly, this process benefits from avoiding excess baggage when formulating the equations one starts with.

In nonlinear situations, the decay of the fields is delayed by nonlinear interactions, and the ODE effects have an even stronger influence than in linear situations. In order to monitor these effects over a considerable amount of time, we consider a finite box with ideally conducting walls, i.e. a cavity in which the field excitation is reflected back and forth. We foliate Minkowski space with simple hyperboloids that are bent only in x-direction and are flat in yz-directions: $\{t = \text{const}\}$-surfaces with $t(T,X,Y,Z) = T - \kappa(\sqrt{1+X^2} - 1)$. For initial data we use analytically known eigenmodes of the cavity, transformed appropriately from standard Minkowski to the curved coordinates. The results of our experiments are presented in fig. 3.

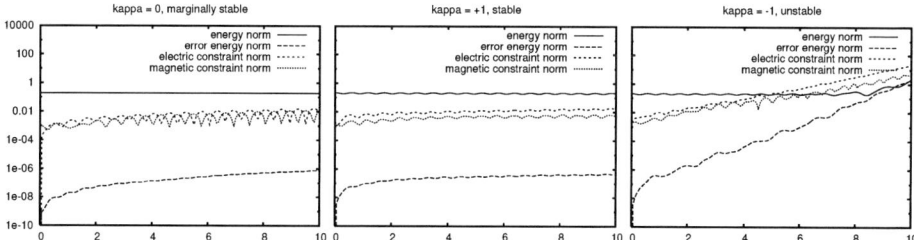

FIGURE 3. (a) Evolution in standard Minkowski coordinates; behaves nicely, energy conserved, linear drift away from exact solution (error energy norm depends linearly on time). Oscillations in the constraints due to lowered accuracy of constraint calculation at the boundary (stencil limitation). (b) Evolution in stable foliation ($\kappa = +1$); behaves nicely, energy conserved, drift away from exact solution *better* than linear. (c) Evolution in unstable foliation ($\kappa = -1$), behaves very badly, exponential growth of both constraints, exponential deviation from exact solution, finally triggers exponential growth of the energy.

SOLUTION OF THE CONSTRAINTS

Solving the constraints is interesting from two perspectives: first it is a necessary prerequisite for evolutions, and second, since a general procedure for solving the regular conformal constraints is not known, it provides an interesting example of a more ad-hoc regularization procedure for the Einstein equations. We consider an isotropic initial hypersurface, i.e. $\tilde{\chi}_{ab} = \tilde{\chi}\tilde{h}_{ab}/3$ with $\tilde{\chi} = \text{const}$. This ansatz solves the momentum constraint and is in some sense analogous to time symmetry for asymptotically euclidean slices. By applying the Lichnerowicz-York procedure to the *rescaled* metric $\Omega^2 \tilde{h}_{\mu\nu} = \phi^4 h_{\mu\nu}$, the Hamiltonian constraint is converted into the Yamabe equation

$$4\Omega^2 D^\mu D_\mu \phi - 4\Omega D^\mu \Omega D_\mu \phi - \left(\frac{R}{2}\Omega^2 + 2\Omega D^\mu D_\mu \Omega - 3D^\mu \Omega D_\mu \Omega\right)\phi = \frac{1}{3}\tilde{\chi}^2 \phi^5,$$

where D_μ denotes the spatial covariant derivative and R its Ricci scalar. For $\Omega \neq 0$ this is a semilinear elliptic equation, but its principal part vanishes on the conformal boundary and standard elliptic theory cannot be applied. The existence of smooth solutions ϕ has been proven in [17] under the condition that the extrinsic 2-curvature induced on the initial cut of $\mathscr{I}$ by the *free metric* is pure trace. The Yamabe equation then also determines the boundary values to be $\phi^2 = 3|\tilde{\chi}|^{-1}\sqrt{D^\mu \Omega D_\mu \Omega}$ on $\mathscr{I}$.

As an example, we consider the simple *axisymmetric* Brill ansatz

$$d\sigma^2 = e^{aq(\rho,z)}\left(d\rho^2 + dz^2\right) + \rho^2 d\varphi^2, \quad q(\rho,z) = \rho^2 e^{-(\rho^2+z^2)}.$$

Such data are well studied in the asymptotically euclidean regime where it is known that for small amplitudes a the waves eventually disperse, leaving flat space behind, whereas for large values of a the waves collapse and form trapped surfaces (in particular we have used such data to test our code against known results [18]). In the hyperboloidal case, the problem becomes nonlinear due to the non-vanishing of $\tilde{\chi}$, which we set to unity without restricting generality. Choosing the conformal gauge as $\Omega = 1 - r^2$ puts $\mathscr{I}$ to $r = 1$ and makes the regularity condition on the extrinsic 2-curvature of $\mathscr{I}$ be identically satisfied. The resulting nonlinear boundary value problem can be simply discretized with 2nd order finite differences and solved through a preconditioned GMRES method [19].

For the physical interpretation of the data it is interesting to search for marginally trapped surfaces, i.e. surfaces on which the null expansion Θ_+ vanishes. Note that since $\mathscr{I}^+$ is a "surface at infinity", the expansions take the unique values $\Theta_+ = \frac{4}{3}\tilde{\chi}$, $\Theta_- = 0$ there. Since the geometry is euclidean in the vicinity of the axis, the expansions have their flat space behavior $\Theta_\pm \to \pm\infty$ for $r \to 0$. Marginal surfaces can now develop if there exist values of the amplitude a for which Θ_+ becomes non-positive in between. Surprisingly, while in the asymptotically euclidean case this happens generically, for the classes of data, we have studied, Θ_+ remains strictly positive and no trapped surfaces exist even for extremely high amplitudes.

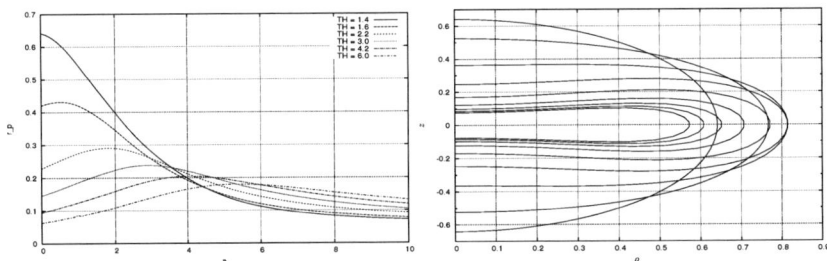

FIGURE 4. Left: Polar radii of equiexpansion surfaces for different amplitudes. Right: Shape of the surface $\Theta_+ = 1.4$ for $a = 0$ (innermost), $3, 6, \ldots, 24$ (outermost).

CONCLUSIONS

The prime motivation to study evolutions based on hyperboloidal slicings is that they enable us to reach null infinity with the flexibility of Cauchy codes. Using the example of the Maxwell equations we have also discussed that hyperboloidal hypersurfaces may tend to create either strong constraint damping or growth, which makes them interesting both as a model for what can go wrong and as a potential remedy. The general conformal field equations allow us to treat null and spacelike infinity in a unified picture, which we hope to help understand the physical significance of the idealizations one makes when using the compactified picture. In order to develop hyperboloidal codes that can handle physically interesting situations involving dynamical black holes and gravitational radiation, we believe it will be fruitful to obtain a fresh perspective on the compactification problem and consider adapted gauges as a starting point for regularizing equations rather than proceeding in the opposite direction.

ACKNOWLEDGMENTS

The authors thank C. Lechner for her work on computer algebra tools, especially those for the Maxwell equations, and I. Hinder for his continuing development work on Kranc. This work was supported in part by the SFB/Transregio 7 "Gravitational Wave Astronomy" of the German Science Foundation.

REFERENCES

1. J. Frauendiener, *Living Rev. Relativity* **7** (2004).
2. S. Husa, *Lect. Notes Phys.* **617**, 159–192 (2003), gr-qc/0204057.
3. R. Penrose, *Phys. Rev. Lett.* **10**, 66–68 (1963).
4. J. Frauendiener, *Class.Quant.Grav.* **17**, 373–387 (2000), gr-qc/9808072.
5. H. Friedrich, *Proc. Roy. Soc. Lond.* **A375**, 169–184 (1981).
6. N. T. Bishop, R. Gomez, L. Lehner, M. Maharaj, and J. Winicour, *Phys. Rev.* **D56**, 6298–6309 (1997), gr-qc/9708065.
7. Y. Zlochower, R. Gomez, S. Husa, L. Lehner, and J. Winicour, *Phys. Rev.* **D68**, 084014 (2003).
8. S. Husa, *Phys. Rev.* **D54**, 7311–7321 (1996), gr-qc/9606042.
9. S. Brandt, and B. Brügmann, *Phys. Rev. Lett.* **78**, 3606–3609 (1997), gr-qc/9703066.
10. S. Husa, *Asymptotically flat initial data for gravitational wave spacetimes, conformal compactification and conformal symmetry*, Ph.D. thesis, University of Vienna (1998).
11. B. Brügmann, *Int. J. Mod. Phys.* **D8**, 85–100 (1999), gr-qc/9708035.
12. F. Pretorius, *Phys. Rev. Lett.* **95**, 211101 (2005), gr-qc/0507014.
13. H. Friedrich, *50 years of the Cauchy problem in general relativity* (2004), gr-qc/0304003.
14. L. Andersson, *Lect. Notes Phys.* **604**, 183–194 (2002), gr-qc/0205083.
15. S. Husa, I. Hinder, and C. Lechner (2004), gr-qc/0404023.
16. J. Frauendiener, and T. Vogel, *Class. Quant. Grav.* **22**, 1769–1793 (2005), gr-qc/0410100.
17. L. Andersson, P. T. Chruściel, and H. Friedrich, *Comm. Math. Phys.* **149**, 587–612 (1992).
18. M. Alcubierre, S. Brandt, B. Brügmann, C. Gundlach, J. Massó, E. Seidel, and P. Walker, *Class. Quant. Grav.* **17**, 2159–2190 (2000), gr-qc/9809004.
19. Y. Saad, *Iterative Methods for Sparse Linear Systems*, SIAM, 2003, second edn.

Stochastic inflation with coloured noise

Heinz-Peter Breuer* and Kerstin E. Kunze[†]

*Physikalisches Institut, Universität Freiburg, Hermann-Herder-Str. 3, D-79104 Freiburg, Germany
[†]Departamento de Física Fundamental, Universidad de Salamanca, Plaza de la Merced s/n, E-37008 Salamanca, Spain

Abstract. In most approaches to stochastic inflation the noise term driving the dynamics is taken to be a white noise process. This white noise is closely linked to a step-function cutoff used to coarse-grain the inflaton field. Taking a different cutoff naturally leads to a coloured noise process. We study the effect of coloured noise on the dynamics of the inflaton, in particular, the resulting probability distributions. Furthermore, the corresponding stochastic differential equation of the inflaton is solved in a perturbative way.

Keywords: cosmology, models of inflation
PACS: 98.80.Cq

STOCHASTIC INFLATION

Inflation explains the observed flatness and homogeneity on large scales. Furthermore, during the course of inflation, quantum fluctuations of the inflaton are stretched beyond the horizon and turned into classical density fluctuations that serve as seeds of structure formation in the beginning of the epoch of the standard model of cosmology.

Modes leaving the horizon and becoming classical have an effect on the value of the coarse grained inflaton. This backreaction is modelled in the stochastic approach to inflation [1, 2]. The effect of quantum fluctuations entering the coarse graining domain, which is larger than the horizon, is modelled by a classical noise term. Thus the evolution of the inflaton is described by a stochastic differential equation.

The nature of the noise term depends on the way of filtering between modes below and above the coarse graining scale. In the original approach, the window function used to filter these modes was taken to be a step function. This led to a white noise in the equation of motion determining the evolution of the coarse grained scalar field [3].

However, it was realized that this is an effect of choosing a sharp cutoff. In general it is expected that the cutoff is smooth rather than sharp. A smooth cutoff leads in general to a coloured noise term in the equation of motion of the inflaton (see for example, [4]). In the following the effect of a coloured noise term will be discussed.

Basic equations

The dynamics of the inflaton ϕ in slow roll inflation is described by the equation

$$\dot{\phi} \simeq -\frac{V'}{3H}, \qquad (1)$$

where $V = V(\phi)$ is the potential and $' \equiv \frac{d}{d\phi}$. Furthermore, the Friedmann equation is given by

$$H^2 \simeq \frac{8\pi}{3M_P^2} V. \qquad (2)$$

In the evolution of the inflaton in its potential there are two competing effects [2]. On the one hand there is the change in the value of ϕ due to its classical rolling down the potential. This is given by

$$\Delta \phi \simeq \dot{\phi} \Delta t. \qquad (3)$$

On the other hand, there is the change due to the quantum fluctuations which become classical outside the horizon. These imply a change

$$\delta \phi = \pm \frac{H}{2\pi}. \qquad (4)$$

During inflation the horizon has a characterisitc size of H^{-1}, where H in a de Sitter space-time is strictly constant and in slow roll inflation slowly varies with time. Thus a causal domain has a volume of the order H^{-3}. During a time interval H^{-1} there are e^3 new domains appearing. Each of them contains an almost homogeneous scalar field given by $\phi - \Delta \phi + \delta \phi$. There is a critical value ϕ_s for which the change due to the quantum fluctuations is larger than the classical change due to the inflaton rolling down its potential. Thus there is the possibility that the inflaton "walks up" its potential. This critical value ϕ_s is determined by $\Delta \phi = \delta \phi$ leading to

$$\frac{2\pi}{3} \frac{V'}{H^3} \bigg|_{\phi_s} = 1. \qquad (5)$$

In a small fraction of domains the field value will continue to grow. However, once the Planck boundary at $V(\phi_{M_P}) = M_P^4$ is reached inflation will stop. Furthermore, there will be domains in which ϕ never becomes small enough in order to stop inflation. Thus there are eternally inflating regions. Of course, our universe evolved from a domain in which inflation stopped at some point. Thus the global structure of the universe as a whole (not just our observable part/universe) becomes very complicated [2]. Our observable universe corresponds only to part of the whole universe.

Effective description of stochastic inflation

The scalar field is divided into two parts as follows (see, for example, [5]),

$$\phi(x) = \varphi(x) + \psi(x). \qquad (6)$$

φ is the "system field" containing modes whose physical wavelengths are larger than the coarse graining scale, which is larger than the horizon. ψ is the "environment field" consisting of field modes whose physical wavelengths are smaller than the coarse graining scale.

The aim is to determine the dynamics of the coarse grained field φ. This can be written as (see, for example, [6])

$$\phi(x) = \varphi(x) + \frac{1}{(2\pi)^{\frac{3}{2}}} \int d^3k W(k - \sigma a H) \left[a_{\mathbf{k}} \phi_{\mathbf{k}}(\tau) e^{-i\mathbf{k}\cdot\mathbf{x}} + a_{\mathbf{k}}^\dagger \phi_{\mathbf{k}}^*(\tau) e^{i\mathbf{k}\cdot\mathbf{x}} \right], \qquad (7)$$

where $W(u)$ is a window function, σ a parameter characterizing the size of the coarse graining domain and $a_{\mathbf{k}}$ and $a_{\mathbf{k}}^\dagger$ are the usual annihilation and creation operators.

In the original approach to stochastic inflation the window function is chosen to be a step function $W(u) = \theta(u)$ [3]. The equations of motion lead to a stochastic differential equation for the coarse grained field $\varphi(t)$,

$$\frac{d\varphi}{dt} + \frac{1}{3H}\frac{dV}{d\varphi} = \frac{H^{\frac{3}{2}}}{2\pi}\eta(t), \qquad (8)$$

where $\eta(t)$ is a white noise, that is its two-point correlation function is given by

$$\langle \eta(t)\eta(t')\rangle = \delta(t - t'). \qquad (9)$$

According to equation (8) the evolution of φ is described by a nonlinear diffusion process. The corresponding probability distribution is determined by a Fokker-Planck equation

$$\frac{\partial P_c}{\partial t} = \frac{\partial}{\partial \varphi}\left(\frac{1}{2}D^{1-\beta}\frac{\partial(D^\beta P_c)}{\partial \varphi} + \kappa \frac{dV}{d\varphi}P_c\right), \qquad (10)$$

where $D = \frac{H^3}{4\pi^2}$ is the diffusion coefficient, $\kappa = \frac{1}{3H(\varphi)}$ is the mobility coefficient and the parameter β determines whether the Itô ($\beta = 1$) or Stratonovich ($\beta = \frac{1}{2}$) definition of the stochastic differential equation (8) is used.

The stationary solution ($\frac{\partial P_c}{\partial t} = 0$) of equation (10) is given by

$$P_c \sim \exp\left(\frac{3M_P^4}{8V(\varphi)}\right). \qquad (11)$$

This corresponds to the probability for creation of the universe [7, 2].

The stochastic differential equation (8) can be written more generally as

$$\frac{d\varphi}{dt} = g(\varphi(t)) + \sqrt{D(\varphi(t))}x(t), \qquad (12)$$

where g is the deterministic drift contribution given as before by

$$g = -\frac{1}{3H}\frac{dV}{d\varphi}. \qquad (13)$$

Furthermore, D is the diffusion coefficient given by

$$D(\varphi) = \frac{H^3}{4\pi^2}. \tag{14}$$

Finally, $x(t)$ is the noise term. In the case of a step function cutoff we have white noise, $x(t) = \eta(t)$. However, it seems more natural to expect that the cutoff is not exactly sharp, but is a smooth cutoff. Furthermore, white noise is not realized in nature. It is an idealization. The choice of a cutoff different from a step function leads naturally to a coloured noise term. The power spectrum $S(\omega)$ of a given stochastic process is defined by the Fourier transform of its autocorrelation function. Accordingly, the spectrum of a process with a δ-shaped autocorrelation is independent of the frequency ω. This is the reason why it is called white noise. A coloured noise process is characterized by a power spectrum which does depend on frequency [8]. Thus, our aim is to study the stochastic differential equation (12) with a coloured noise term, but leaving the drift term and diffusion coefficient unchanged, that is using equations (13) and (14). Furthermore, it is assumed that the noise process has zero mean, $\langle x(t) \rangle = 0$.

COUPLING TO AN ORNSTEIN-UHLENBECK PROCESS

A Markovian process is a stochastic process $\xi(t)$ with short memory time [9]. To give a precise definition one considers the conditional probability density $p(\xi, t | \xi_1, t_1; \xi_2, t_2; \ldots; \xi_n, t_n)$. This is defined to be the probability density for the process to take the value ξ at time t, under the condition that it assumed the values $\xi_1, \xi_2, \ldots, \xi_n$ at corresponding previous times $t_1 > t_2 > \ldots > t_n$. The process is then said to be Markovian if this conditional probability density only depends on the value ξ_1 the process assumed at the latest time t_1, i. e.: $p(\xi, t | \xi_1, t_1; \xi_2, t_2; \ldots; \xi_n, t_n) = p(\xi, t | \xi_1, t_1)$. This property leads to a great simplification of the mathematical description because it allows to reconstruct the whole hierarchy of joint probability distributions from the conditional density $p(\xi, t | \xi_1, t_1)$. In the case of a diffusion process, for example, the latter is determined by the Fokker-Planck equation. The mathematical description and analysis of non-Markovian processes is generally much more involved. However, there are certain types of non-Markovian processes that can be described by Markovian processes through the introduction of appropriate auxiliary variables. An example of this type will be discussed in the following.

The time-dependent probability density of a diffusion process is governed by a Fokker-Planck equation (see, for example, equation (10)). A typical example is given by the famous Wiener process. The Wiener process is a Markovian and Gaussian process, but it is not stationary. Another well-known example is the Ornstein-Uhlenbeck process $x(t)$. This process is essentially the only process which is Markovian, Gaussian and stationary. It has an exponentially decaying autocorrelation function

$$\langle x(t)x(t') \rangle = \frac{\sigma^2}{2\gamma} e^{-\gamma|t-t'|}, \tag{15}$$

and obeys the stochastic differential equation

$$\frac{d}{dt}x(t) = -\gamma x(t) + \sigma \eta(t), \tag{16}$$

where $\eta(t)$ describes white noise. The quantity γ is the relaxation rate and $\tau_c = \gamma^{-1}$ represents the autocorrelation time, while σ measures the strength of the fluctuations.

The aim is to study the effects of a coloured noise in the equation of motion of the inflaton (cf. equation (12)). Strictly speaking the form of the noise process is determined by the choice of a particular window function in equation (7) which in turn is determined by the coarse graining procedure. The coarse graining procedure is influenced by the model of decoherence of the quantum fluctuations leaving the horizon and becoming classical. However, since, to our knowledge, there is no preferred choice of the window function, we take a phenomenological approach and try to investigate in general the effects of a coloured noise process on the dynamics of the coarse grained scalar field φ.

The Ornstein-Uhlenbeck process is an example of a noise process that is coloured but still accessible to both the analysis with a Fokker-Planck type equation and direct solution of the corresponding stochastic differential equation. Coupling the inflaton to an Ornstein-Uhlenbeck process leads to the following set of stochastic differential equations,

$$\frac{d}{dt}\varphi(t) = g(\varphi(t)) + \sqrt{D(\varphi(t))}x(t) \tag{17}$$

$$\frac{d}{dt}x(t) = -\gamma x(t) + \gamma \eta(t), \tag{18}$$

where the drift coefficient g and diffusion coefficient D are defined as before. In equation (18) the parameter σ which measures the strength of the fluctuations of $x(t)$ was taken to be equal to the relaxation rate γ. This leads to the autocorrelation function

$$\langle x(t)x(t')\rangle = \frac{\gamma}{2}e^{-\gamma|t-t'|}, \tag{19}$$

such that in the limit of zero autocorrelation time, $\tau_c \to 0$, one recovers the δ-function correlation of white noise. The power spectrum of this process is given by

$$S(\omega) = \int d\tau e^{i\omega\tau}\langle x(t+\tau)x(t)\rangle = \frac{\gamma^2}{\omega^2 + \gamma^2}. \tag{20}$$

Thus, for a finite autocorrelation time the course grained scalar field $\varphi(t)$ follows a non-Markovian dynamics because it is coupled to a coloured noise process $x(t)$ involving a high-frequency cutoff of the order of the relaxation rate. One might argue that the correlation time $\tau_c = \gamma^{-1}$ of the process is of the order of the decoherence time that it takes the quantum modes to become classical. A decoherence time scale is given by the Hubble time, thus one might choose

$$\gamma = H. \tag{21}$$

With this choice, γ becomes a function of time, though slowly varying in slow roll inflation. Equation (18) then describes some kind of generalized Ornstein-Uhlenbeck process, since the process is no longer stationary.

The joint process $(\varphi(t), x(t))$ is Markovian. Therefore, it is possible to write down a Fokker-Planck equation for the corresponding probability distribution $P = P(\varphi, x, t)$, given by [10]

$$\frac{\partial P(\varphi, x, t)}{\partial t} = -\frac{\partial}{\partial \varphi}\left[\left(g + \sqrt{D}x\right)P\right] + \frac{\partial}{\partial x}[\gamma x P] + \frac{1}{2}\frac{\partial^2}{\partial x^2}\left[\gamma^2 P\right]. \tag{22}$$

Integrating out the auxiliary variable x gives the reduced probability distribution,

$$Q(\varphi, t) = \int dx P(\varphi, x, t). \tag{23}$$

Although the process $\varphi(t)$ is in general non-Markovian one can construct for constant γ an effective Fokker-Planck type equation for the reduced probability distribution $Q(\varphi, t)$ [10],

$$\frac{\partial Q(\varphi, t)}{\partial t} = -\frac{\partial}{\partial \varphi}gQ + \frac{1}{2}\left(1 - e^{-\gamma t}\right)\frac{\partial}{\partial \varphi}\sqrt{D}\frac{\partial}{\partial \varphi}\sqrt{D}Q. \tag{24}$$

The explicit time dependence of the diffusion coefficient is due to the non-Markovian property of the coarse-grained field $\varphi(t)$. Furthermore, for times much larger than the correlation time, $t \gg \gamma^{-1}$, equation (24) determining $Q(\varphi, t)$ approaches the Fokker-Planck equation in the case where the inflaton is coupled to a white noise process (cf. equation (10)). Moreover, it is the Fokker-Planck equation (10) in the Stratonovich interpretation.

Compared with the coupling to a white noise process, the dynamics of the probability distribution for the coarse grained field φ coupled to an Ornstein-Uhlenbeck process is significantly altered for times shorter than the correlation time. Diffusion of the coarse grained field φ is highly suppressed for times shorter than the correlation time.

PERTURBATIVE SOLUTIONS

The stochastic differential equation (12) depends on the particular choice of the noise process $x(t)$. In the last section a particular choice, namely an Ornstein-Uhlenbeck process has been discussed. It was used that for this particular type of coloured noise it is possible to discuss the probability distribution of the coarse grained field $\varphi(t)$ using a Fokker-Planck type equation. In general, however, it is not possible to find Fokker-Planck type equations for the probability distribution of $\varphi(t)$. For an arbitrary form of the coloured noise process the stochastic differential equation (12) has to be solved. This of course is a difficult endeavour. However, if deviations from the deterministic back ground value are assumed to be small, it is possible to solve the stochastic differential equation for the coarse grained scalar field $\varphi(t)$ by a perturbation expansion.

Assume that the coarse grained field $\varphi(t)$ can be expanded as follows [10],

$$\varphi(t) = \varphi_c(t) + \alpha \varphi_1(t) + \alpha^2 \varphi_2(t) + ..., \tag{25}$$

where $\varphi_c(t)$ is a solution of the deterministic equation $\dot{\varphi} = g(\varphi)$. α is a parameter that controls the order of involvement of the noise term, namely, the term proportional to α corresponds to the linear noise approximation. Using the perturbation expansion (25) in the stochastic differential equation

$$\frac{d\varphi}{dt} = g(\varphi) + \sqrt{D(\varphi)}\alpha x(t) \qquad (26)$$

leads to a set of stochastic differential equations. Up to second order in α [10]

$$\frac{d\varphi_c}{dt} = -\frac{M_P^2}{4\pi} H'(\varphi_c) \qquad (27)$$

$$\frac{d\varphi_1}{dt} = -\frac{M_P^2}{4\pi} H''(\varphi_c)\varphi_1 + \frac{H^{\frac{3}{2}}(\varphi_c)}{2\pi} x(t) \qquad (28)$$

$$\frac{d\varphi_2}{dt} = -\frac{M_P^2}{4\pi} H''(\varphi_c)\varphi_2 - \frac{M_P^2}{8\pi} H'''(\varphi_c)\varphi_1^2 + \frac{3}{4\pi} H^{\frac{1}{2}}(\varphi_c)H'(\varphi_c)\varphi_1 x(t), \qquad (29)$$

where $' = \frac{d}{d\varphi_c}$. The solution for φ_1 is given by

$$\varphi_1(t) = \frac{H'(\varphi_c(t))}{2\pi} \int_0^t d\tau \frac{H^{\frac{3}{2}}(\varphi_c(\tau))}{H'(\varphi_c(\tau))} x(\tau), \qquad (30)$$

assuming $\varphi_1(0) = 0$. Since $\langle x(t) \rangle = 0$, it follows that $\langle \varphi_1(t) \rangle = 0$. Furthermore,

$$\langle \varphi_1^2(t) \rangle = \left(\frac{H'(\varphi_c(t))}{2\pi}\right)^2 \int_0^t d\tau \int_0^t d\sigma \frac{H^{\frac{3}{2}}(\varphi_c(\tau))}{H'(\varphi_c(\tau))} \frac{H^{\frac{3}{2}}(\varphi_c(\sigma))}{H'(\varphi_c(\sigma))} \langle x(\tau)x(\sigma) \rangle. \qquad (31)$$

And the expectation value of $\varphi_2(t)$ is given by

$$\langle \varphi_2(t) \rangle = H'(\varphi_c(t)) \int_0^t d\sigma \left[-\frac{M_P^2}{8\pi} \frac{H'''(\varphi_c(\sigma))}{H'(\varphi_c(\sigma))} \langle \varphi_1^2(\sigma) \rangle \right.$$
$$\left. + \frac{3}{8\pi^2} H^{\frac{1}{2}}(\varphi_c(\sigma))H'(\varphi_c(\sigma)) \int_0^\sigma d\tau \frac{H^{\frac{3}{2}}(\varphi_c(\tau))}{H'(\varphi_c(\tau))} \langle x(\tau)x(\sigma) \rangle \right]. \qquad (32)$$

Thus the autocorrelation function of the noise determines these expectation values. These expressions can be used to calculate the curvature perturbation spectrum resulting in this type of stochastic inflationary models. This has been done in [10].

CONCLUSIONS

Replacing the step function cutoff in the original approach to stochastic inflation by any other type of window function leads naturally to a coloured noise term in the equation of motion of the coarse grained scalar field. Here the effect of such a coloured noise was studied in two different settings. For the particular example of coupling the inflaton to an Ornstein-Uhlenbeck process it is possible to find a Fokker-Planck type equation for the probability distribution of the coarse grained scalar field. For a general type of coloured noise this is not possible. However, assuming that fluctuations around the deterministic solution of the course grained scalar field remain small, the stochastic differential equation can be solved through a perturbation expansion. This leads to explicit expressions for the dynamics of all statistical quantities in terms of the correlation functions of the coloured noise. In particular, for any given model the procedure yields specific predictions about many quantities of interest, e. g., about the power spectrum, the spectral index and the non-Gaussian character of the fluctuations.

ACKNOWLEDGMENTS

K.E.K. is supported in part by the programme "Ramón y Cajal" of the M.E.C. (Spain) and Spanish Science Ministry Grants FPA 2002-02037, FPA2005-04823 and BFM 2003-02121.

REFERENCES

1. A. D. Linde, Phys. Lett. B **175** (1986) 395; H. E. Kandrup, Phys. Rev. D **39** (1989) 2245; A. D. Linde and A. Mezhlumian, Phys. Lett. B **307** (1993) 25;
2. A. D. Linde, D. A. Linde and A. Mezhlumian, Phys. Rev. D **49** (1994) 1783;
3. A. A. Starobinsky in *Field Theory, Quantum Gravity and Strings* ed. H. J. de Vega and N. Sanchez; A. S. Goncharov and A. D. Linde, Sov. Phys. JETP **65** (1987) 635.
4. S. Winitzki and A. Vilenkin, Phys. Rev. D **61** (2000) 084008.
5. E. Calzetta and B. L. Hu, Phys. Rev. D **52** (1995) 6770.
6. J. Martin and M. A. Musso, Phys. Rev. D **71** (2005) 063514.
7. J. B. Hartle and S. W. Hawking, Phys. Rev. D **28** (1983) 2960.
8. H. Risken, *The Fokker-Planck Equation* (Springer-Verlag, Heidelberg, Germany, 1996).
9. H. P. Breuer and F. Petruccione, *The Theory of Open Quantum Systems* (Oxford University Press, Oxford, 2002).
10. H. P. Breuer and K. E. Kunze, *in preparation*.

On fixed points of quantum gravity

Daniel Litim

School of Physics and Astronomy
University of Southampton, Southampton SO17 1BJ, U.K.
and CERN, Theory Group, CH – 1211 Geneva 23

Abstract. We study the short distance behaviour of euclidean quantum gravity in the light of Weinberg's asymptotic safety scenario. Implications of a non-trivial ultraviolet fixed point are reviewed. Based on an optimised renormalisation group, we provide analytical flow equations in the Einstein-Hilbert truncation. A non-trivial ultraviolet fixed point is found for arbitrary dimension. We discuss a bifurcation pattern in the spectrum of eigenvalues at criticality, and the large dimensional limit of quantum gravity. Implications for quantum gravity in higher dimensions are indicated.

Keywords: Quantum gravity, renormalisation group, fixed points, higher dimensions
Preprint: SHEP-0542, CERN-PH/TH-2005-256

INTRODUCTION

Gravitational interactions at distances sufficiently large compared to the Planck length are described by the classical theory of general relativity. At smaller length scales, quantum effects are expected to become important. The quantisation of general relativity, however, still poses problems. It is known since long that four-dimensional quantum gravity is perturbatively non-renormalisable, meaning that an infinite number of parameters have to be fixed to renormalise standard perturbation theory. One may wonder whether a quantum theory of gravity in terms of the metric degrees of freedom can exist as a well-defined, non-trivial and cutoff-independent local theory down to arbitrarily small distances. It is generally believed that the above requirements imply the existence of a non-trivial ultraviolet (UV) fixed point under the renormalisation group (RG), governing the short-distance physics. Then it would suffice to adjust a finite number of parameters, ideally taken from experiment, to make the theory asymptotically safe. The corresponding short distance fixed point action would then provide a valid microscopic starting point to access classical general relativity as a "low energy phenomenon" of a local quantum field theory in the metric field.

For quantum gravity, this asymptotic safety scenario has been introduced by Weinberg [1]. In the vicinity of two dimensions, a non-trivial fixed point has been identified within perturbation theory, to leading [1]-[3] and subleading order [4] in $\varepsilon = d - 2 \ll 1$. In the last couple of years, non-perturbative renormalisation group studies have been performed in the four-dimensional case. A non-trivial fixed point has been detected within various renormalisation group studies, also including higher dimensional operators or non-interacting scalar, vector and matter fields [5]-[16]. Additional indications for the existence of a non-trivial fixed point in four dimensions have been provided through lattice simulations within both Regge's simplicial lattice formulation [17] and the causal dynamical triangulations approach, *e.g.* [18].

From a renormalisation group point of view, the "critical" dimension of quantum gravity is $d_{\rm cr}=2$.[1] In two dimensions, the gravitational coupling has vanishing canonical dimension, and standard perturbation theory is applicable. In turn, for any $d > d_{\rm cr}$ the gravitational coupling has negative mass dimension, indicating that the theory is perturbatively non-renormalisable. Hence, one expects that the local renormalisation group properties of quantum theories of gravity for different dimensions with $d > d_{\rm cr}$ share qualitative properties. We conclude that the case of four dimensions, from a quantum gravity point of view, is by no means distinguished. Continuity in the dimension, together with the indications for a non-trivial fixed point in four dimensions, suggest that a non-trivial fixed point should persist, to the least, in some vicinity of four dimensions. We emphasize that this heuristic line of reasoning is solely based on local RG properties of the theory, and insensitive against global properties within specific dimensions.

In this contribution, we discuss the asymptotic safety scenario in the context of quantum gravity. Based on a wilsonian renormalisation group, we provide unique analytical fixed point solution in the Einstein-Hilbert truncation for any dimensions $d > d_{\rm cr}$ [12]. The approach is related to the integrating-out of momentum modes from a path integral representation of the theory [19], ammended by an appropriate optimisation [20, 21]. Consequently, the reliability of results based on optimised flows is enhanced [22]. Results and implications for quantum gravity in higher dimensions are discussed.

ASYMPTOTIC SAFETY

We recall a few general requirements and implications of the asymptotic safety scenario and a non-trivial fixed point in quantum gravity [1]. First of all, the asymptotic safety scenario relies on the existence of a (non-trivial) fixed point at short distances. This generalises a pattern observed for perturbatively renormalisable theories, which often are related to a non-interacting UV fixed point, *e.g.* asymptotic freedom in QCD. Secondly, it is mandatory that the short-distance fixed point is connected with the long-distance behaviour of the theory by a well-defined renormalisation group trajectory. Elsewise the putative fixed point would remain disconnected from the known physics at large distances. Finally, it is required that the UV fixed point displays at most a finite number of (infrared) unstable directions. Elsewise, the predictive power is spoiled, because an infinite number of unstable directions would require the fine-tuning of infinitely many parameters in order to reach the IR limit. Then, the fixed point together with the RG trajectory serve as a fundamental definition of the theory.

We proceed with a discussion of the renormalisation group flow for the gravitational coupling G in d dimensions. We introduce the renormalised dimensionless coupling as $g = \mu^{d-2} Z_G^{-1}(\mu) G$ and the graviton wave function renormalisation factor $Z_G(\mu)$, normalised as $Z_G(\mu_0) = 1$ at $\mu = \mu_0$. The momentum scale μ denotes the renormalisation scale, and can be thought of as an energy scale E, or momentum transfer p, or, more formally, as some wilsonian momentum cutoff k as used below. The graviton anomalous

---

[1] This is different from most particle physics theories, where the critical dimension of the relevant couplings is $d = 4$, *e.g.* QED, QCD.

dimension, given by $\eta = -\mu\partial_\mu \ln Z_G$, is a non-trivial function of g and other couplings parametrising the theory. Then the RG flow reads

$$\mu\partial_\mu g = (d-2+\eta)g. \tag{1}$$

From (1) we conclude that the gravitational coupling displays a non-interacting (gaussian) fixed point $g_* = 0$. This entails $\eta = 0$, i.e. classical scaling. On the other hand, non-trivial RG fixed points, if they exist, correspond to the implicit solutions of

$$\eta_* = 2 - d. \tag{2}$$

Hence, a non-trivial fixed point of quantum gravity in $d > 2$ implies a negative integer value for the graviton anomalous dimension, precisely counter-balancing the canonical dimension of G.

Integer values for anomalous dimensions are well-known from other gauge theories at criticality: in the d-dimensional abelian higgs theory, for example, the abelian charge e^2 has mass dimension $[e^2] = 4 - d$, whence $\beta_{e^2} = (d - 4 + \eta) e^2$. In three dimensions, the theory displays a non-perturbative infrared fixed point at $e_*^2 \neq 0$, leading to $\eta = 1$ [23]. The fixed point belongs to the universality class of standard superconductors with the charged scalar field describing the Cooper pair. The integer value $\eta = 1$ implies that the magnetic field penetration depth and the Cooper pair correlation length scale with the same universal critical exponent at the phase transition [23, 24]. In turn, scalar theories at criticality, not protected by a local gauge symmetry, often display non-integer anomalous dimensions characterising the universality class.

In quantum gravity, a non-trivial fixed point behaviour leads to two important implications. First of all, the dimensionful gravitational coupling constant scales as $G(\mu) \to g_*/\mu^{d-2}$ at the fixed point. In the case of an ultraviolet fixed point for diverging of μ, the coupling G becomes arbitrarily small. This indicates that gravity might become asymptotically free at short distances, similar to QCD. Conversely, in case of an infrared fixed point for vanishing μ, the dimensionful coupling grows large. This behaviour implies non-trivial long distance modifications of gravity, e.g. [25]. Secondly, the scalar part of the renormalised graviton propagator scales $\sim p^{-2+\eta}$. Here we have identified μ with the momentum scale p. At an UV fixed point, this leads to an increased momentum decay $\sim p^{-d}$, which, in position space, corresponds to a logarithmic behaviour $\sim \ln(|x-y|\mu)$ of the propagator, reminiscent from two dimensions. Hence, the integer value of (2), on the level of the graviton propagator, implies a dimensional crossover from d-dimensional behaviour in the perturbative regime to an effectively two-dimensional behaviour in the vicinity of a non-trivial fixed point. A similar result is found for the spectral dimension of quantum gravity space-times at short distances [26, 18].

RENORMALISATION GROUP

Whether or not the non-trivial fixed point is realised in quantum gravity can only be assessed by an explicit renormalisation group study. We apply a wilsonian renormalisation group, based on a momentum cutoff for the propagting degrees of freedom (for reviews, see [19]). It is based on a scale-dependent action functional Γ_k of the mean gravitational field $\langle g_{\mu\nu}\rangle_k$, where k denotes a wilsonian momentum cutoff scale. The wilsonian flow equation describes the change of Γ_k under an infinitesimal variation of k. By construction, Γ_k comprises all quantum fluctuations down to the momentum scale $q^2 \approx k^2$. The flow interpolates between some microscopic action at short distances $k \to \infty$ and the full quantum effective action at large distances $k \to 0$. Diffeomorphism invariance under local coordinate transformations is controlled by modified Ward identities [5], similar to those known for non-Abelian gauge theories [27]. In its modern formulation, the flow with respect to the logarithmic scale parameter $t = \ln k$ is given by

$$\partial_t \Gamma_k = \frac{1}{2} \text{Tr} \frac{1}{\Gamma_k^{(2)} + R_k} \partial_t R_k. \tag{3}$$

Here, the trace stands for a momentum integration and a sum over indices and fields, and R_k denotes an appropriate wilsonian momentum cutoff at momentum scale $q^2 \approx k^2$. For quantum gravity, we consider the flow (3) for Γ_k in the Einstein-Hilbert truncation

$$\Gamma_k = \frac{1}{16\pi G_k} \int d^d x \sqrt{g} \left[-R(g) + 2\bar{\lambda}_k \right] + \text{classical gauge fixing}, \tag{4}$$

retaining the volume element and the Ricci scalar as independent operators. In (4) we have introduced the gravitational coupling constant G_k and the cosmological constant $\bar{\lambda}_k$. The Ansatz (4) differs from the standard Einstein-Hilbert action in d Euclidean dimensions by the fact that the gravitational coupling and the cosmological constant have turned into running couplings. We introduce dimensionless renormalised gravitational and cosmological constants g_k and λ_k as

$$g_k = k^{d-2} G_k \equiv k^{d-2} Z_{N,k}^{-1} \bar{G}, \qquad \lambda_k = k^{-2} \bar{\lambda}_k. \tag{5}$$

and $Z_{N,k}$ denotes the wave function renormalisation factor for the Newtonian coupling. Their flows follow from (3) by an appropriate projection onto the operators in (4). For explicit constructions of a momentum cutoff in the background field gauge with gauge fixing parameter α, see [5, 7]. Closely related flows have also been derived in [13] within a proper-time approximation [28]. Here, we use the approach of [5, 7] for the tensorial structure of the momentum cutoff in addition with an optimised momentum cutoff [21] for its scalar part. This leads to an analytical flow equation

$$\partial_t \lambda \equiv \beta_\lambda = \frac{P_1}{P_2 + 4(d+2)g} \qquad \partial_t g \equiv \beta_g = \frac{(d-2)g P_2}{P_2 + 4(d+2)g} \tag{6}$$

$$P_1 = -16\lambda^3 + 4\lambda^2(4 - 10dg - 3d^2g + d^3g) + 4\lambda(10dg + d^2g - d^3g - 1)$$
$$+ d(2+d)(d - 16g + 8dg - 3)g \tag{7}$$

$$P_2 = 8(\lambda^2 - \lambda - dg) + 2. \tag{8}$$

For convenience, a factor $1/\alpha$ is absorbed into the definition (5), and a factor $c_d = (4\pi)^{d/2-1}\Gamma(d/2+2)$ is absorbed into the definition of g. Furthermore, we have performed the limit $\alpha \to \infty$ in (6), where the results take their simplest form. The flow for arbitrary α is given in [15]. The flow of g vanishes identically in two dimensions. For the anomalous dimension, we find

$$\eta = \frac{(d+2)g}{g - g_{\text{bound}}}, \qquad g_{\text{bound}}(\lambda) = \frac{(1-2\lambda)^2}{2(d-2)}. \qquad (9)$$

The anomalous dimension diverges at $g = g_{\text{bound}}(\lambda)$, which limits the domain of validity. For all real λ and $d > 2$, we have $g_{\text{bound}} \geq 0$. The anomalous dimension vanishes for $g = 0$, and in $d = 2$ dimensions.

ANALYTICAL FIXED POINTS

Next we identify the non-trivial fixed points of $\beta_g(\lambda_*, g_*) = 0 = \beta_\lambda(\lambda_*, g_*)$. From $P_2 = 0$, (8), we deduce that $g_* = g_{\text{bound}}(\lambda_*) \times (d-2)/(2d)$. With (9), we conclude that the gravitational coupling fixed point is positive for any real fixed point λ_*. Inserting this result into (7), equating $P_1 = 0$, and factoring-out a common factor $(1-2\lambda)^2$ leads to a simple quadratic equation with two real solutions, as long as $d \geq (1+\sqrt{17})/2 \approx 2.56$. The physical fixed point obeys $\lambda < \frac{1}{2}$ and reads

$$\lambda_* = \frac{d^2 - d - 4 - \sqrt{2d(d^2-d-4)}}{2(d-4)(d+1)}, \qquad g_* = \frac{\Gamma(d/2+2)}{(4\pi)^{1-d/2}} \frac{(\sqrt{d^2-d-4} - \sqrt{2d})^2}{2(d-4)^2(d+1)^2}. \quad (10)$$

Here, we have reinserted the factor c_d in g. The solution (10) is continuous and well-defined for all $d \geq 2.56$ and becomes complex for lower dimensions. For general $\alpha < \infty$, the fixed point extends down to $d = 2$.

Fixed points are non-universal quantities and may depend on unphysical parameters. In turn, the rates at which small perturbations about the fixed point grow with scale are universal. These are denoted as $-\theta$, or $-1/\nu$ in the statistical physics literature, and given by the eigenvalues of the stability matrix at criticality. In four dimensions, and reinserting c_d, we find $\theta = \theta' + i\theta''$, with

$$\lambda_* = \frac{1}{4}, \quad g_* = \frac{3\pi}{8}, \quad \theta' = \frac{5}{3}, \quad \theta'' = \frac{\sqrt{167}}{3}, \quad |\theta| = \frac{8}{\sqrt{3}}. \qquad (11)$$

This compares well with the lattice result $\nu \approx 1/3$ of [17]. In Fig. 1, the trajectory connecting the UV fixed point with the gaussian one is given, together with the running couplings along the separatrix. The complex eigenvalue is reflected by a rotating trajectory at the UV fixed point. It originates from a strong mixing of the scaling of the volume operator and the Ricci scalar. Along the separatrix, the anomalous dimension displays a cross-over from a perturbatively small value towards the non-trivial fixed point value (2). The fixed point and the flow pattern persists for arbitrary gauge fixing parameter.

HIGHER DIMENSIONS

The fixed point (10), and the corresponding solution for arbitrary gauge fixing parameter [12], is found independently of the dimension. The fixed point is unique, real, positive,

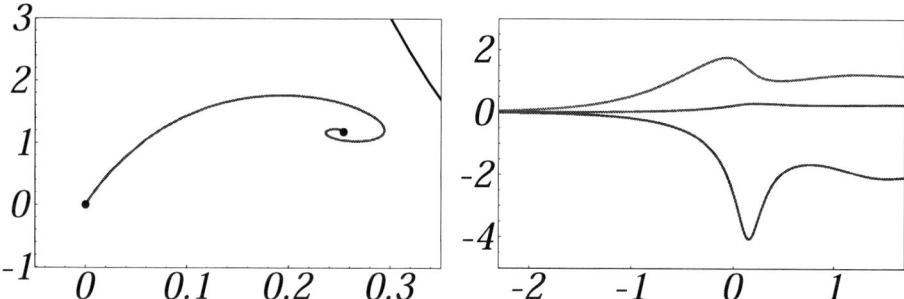

FIGURE 1. Phase diagram of quantum gravity in four dimensions. Left panel: The separatrix $g(\lambda)$ in four dimensions (red line), connecting the non-trivial UV fixed point (right dot) with the gaussian fixed point (left dot). The thin full line in the upper right corner denotes $g = g_{\text{bound}}$, see (9). Right panel: running couplings along the separatrix. From top to bottom: the dimensionless gravitational coupling (violet line), the dimensionless cosmological constant (red line) and the anomalous dimension (blue line) as a function of $\ln(k/M_{\text{Pl}})$ in four dimensions. The strong correlation between the scaling of the volume element $\sqrt{g}$ and the Ricci scalar $\sqrt{g}R$ in the vicinity of the fixed point is responsible for the rotation of the separatrix, and for the oscillatory behaviour of the trajectories at scales $k > M_{\text{Pl}}$.

and continuously connected to the perturbatively known fixed point in two dimensions. This is quite remarkable, also in view of the fact that the Einstein-Hilbert truncation is expected to be more sensitive to higher dimensional operators in higher than in lower dimensions. For a detailed study of the general cutoff and gauge fixing independence of this fixed point, see [16].

Here, we point out three noteworthy aspects of the fixed point for general dimension. For more details, see [15]. First of all, we notice a non-trivial bifurcation in the eigenvalue spectrum, as a function of dimension. For $2 < d_{\text{low}} < d < d_{\text{up}} < \infty$, the universal eigenvalues of the stability matrix are complex, and real otherwise (see Tab. 1 for numerical values). Complex eigenvalues indicate that the operators $\sqrt{g}$ and $\sqrt{g}R$ scale with a similar strength, and that they are subject to strong mixing effects, parametrised by the off-diagonal elements of the Jacobi matrix. This is the case in four dimensions. In turn, real eigenvalues indicate that the scaling behaviour at the fixed point is dominated by the volume element, while the Ricci scalar remains subleading. This behaviour is well-known in the vicinity of two dimensions. The interesting new result is that a similar behaviour persists for sufficiently large dimension. The large dimensional limit, in this light, shares an important similarity with the limit of small dimensions. If this structure persists in the full theory, we expect that an expansion about two dimensions (in inverse dimension) has a finite radius of convergence, given by the lower (upper) critical dimension.

Secondly, it is possible to perform the large dimensional limit. For any $\alpha \in [0,1]$ (and similarily for other α), one finds the universal eigenvalues $\theta_1 = d^3/156$ and $\theta_2 = 24d/13$, related to the scaling of the $\sqrt{g}$ and $\sqrt{g}R$ operator, respectively. The index $\nu = 1/\theta_2$ agrees very well with the result $\nu = 1/(2d)$ as obtained previously for vanishing cosmological constant $\lambda = 0$ [12]. This should also be compared with $\nu = 1/(d-1)$ based on geometrical considerations [29], with $\nu \approx 1.9/d$ based on an extrapolation of

TABLE 1. Lower and upper critical dimensions as functions of the gauge fixing parameter. The data is obtained using a wilsonian momentum cutoff with tensorial momentum structure as in [5] (for $\alpha = 1_A$) or [7] (for general $\alpha \geq 0$), and a scalar momentum structure as in [21].

α	0	$\frac{1}{2}$	1	1_A	2	∞
d_{low}	2.900	2.901	2.872	2.756	2.765	2.562
d_{up}	23.727	23.672	24.282	23.985	17.394	21.381

low dimensional lattice results [30], and $v = 1/(d-2)$ as obtained within a perturbative expansion about two dimensions [2, 3].

Finally, a non-trivial fixed point of quantum gravity in higher dimensions is of immediate interest for phenomenological applications. It has been suggested that the fundamental Planck scale may be as low as the electroweak scale, if gravity propagates in a $d = 4 + n$ dimensional "bulk" with n "extra" spatial dimensions, while standard model particles only propagate in a four-dimensional "brane"[31, 32]. Then high-energy particle physics experiments at LHC are potentially sensitive to the fundamental Planck scale and quantum gravitational effects. Up to now, phenomenological effects due to higher dimensional gravity have been studied using classical propagators and vertices, ammended by an ultraviolet cutoff, *e.g.* [33]. In turn, a fixed point scenario as described here is ultraviolet finite and would not require an additional UV regularisation.

CONCLUSIONS

We have discussed basic implications of an asymptotic safety scenario for quantum gravity. It has been emphasized that the case of four space-time dimensions, from a renormalisation group point of view, is not distinguished. This is reflected by the non-trivial fixed point structure in the Einstein Hilbert theory, which displays an UV fixed point for arbitrary dimension, *e.g.* (10). Maximal reliability in the present truncation is guaranteed by the underlying optimisation. It was pointed out that the large dimensional limit – in view of its scaling properties at criticality – shares qualitative similarities with the low dimensional limit, the vicinity of $d = 2$. In contrast, in the neighbourhood of four dimensions, scaling at criticality is characterised by strong operator mixing. This has lead to an interesting bifurcation pattern in the eigenvalue spectrum.

In four dimensions and below, our results are fully consistent with previous renormalisation group studies. The fixed point is remarkably stable, with only a mild dependence on the gauge fixing parameter. The phase diagram is equally robust. Furthermore, it is encouraging that the qualitative picture achieved so far is backed-up by lattice simulations. The fixed point consistently extends to higher dimensions, a region which previously has not been accessible. Hence it is likely that the fixed point exists in the full theory. We expect that the analytical form of the flow, crucial for the present analysis, is equally useful in extended truncations. If the above picture persists in these cases, quantum gravity exists as a well-defined local quantum field theory in the metric field down to arbitrarily short distances.

ACKNOWLEDGMENTS

This work is supported by an EPSRC Advanced Fellowship.

REFERENCES

1. S. Weinberg, in *General Relativity: An Einstein centenary survey*, Eds. S.W. Hawking and W. Israel, Cambridge University Press (1979), p. 790.
2. R. Gastmans, R. Kallosh and C. Truffin, Nucl. Phys. B **133** (1978) 417.
3. S. M. Christensen and M. J. Duff, Phys. Lett. B **79** (1978) 213.
4. T. Aida and Y. Kitazawa, Nucl. Phys. B **491** (1997) 427 [hep-th/9609077].
5. M. Reuter, Phys. Rev. D **57** (1998) 971 [hep-th/9605030].
6. W. Souma, Prog. Theor. Phys. **102** (1999) 181 [hep-th/9907027].
7. O. Lauscher and M. Reuter, Phys. Rev. D **65** (2002) 025013 [hep-th/0108040].
8. O. Lauscher and M. Reuter, Class. Quant. Grav. **19** (2002) 483 [hep-th/0110021]
   O. Lauscher and M. Reuter, Phys. Rev. D **66** (2002) 025026 [hep-th/0205062].
9. M. Reuter and F. Saueressig, Phys. Rev. D **65** (2002) 065016 [hep-th/0110054].
10. R. Percacci and D. Perini, Phys. Rev. D **67** (2003) 081503 [hep-th/0207033]
    R. Percacci and D. Perini, Phys. Rev. D **68** (2003) 044018 [hep-th/0304222].
11. P. Forgacs and M. Niedermaier, hep-th/0207028
    M. Niedermaier, JHEP **0212** (2002) 066 [hep-th/0207143],
    M. Niedermaier, Nucl. Phys. B **673** (2003) 131 [hep-th/0304117].
12. D. F. Litim, Phys. Rev. Lett. **92** (2004) 201301 [hep-th/0312114].
13. A. Bonanno and M. Reuter, JHEP **0502** (2005) 035 [hep-th/0410191].
14. R. Percacci, hep-th/0511177.
15. D. F. Litim, to appear.
16. P. Fischer, D. F. Litim, to appear.
17. H. W. Hamber, Phys. Rev. D **61** (2000) 124008, [hep-th/9912246],
    H. W. Hamber, Phys. Rev. D **45** (1992) 507.
18. J. Ambjorn, J. Jurkiewicz and R. Loll, Phys. Rev. Lett. **95** (2005) 171301 [hep-th/0505113],
    J. Ambjorn, J. Jurkiewicz and R. Loll, Phys. Rev. D **72** (2005) 064014 [hep-th/0505154].
19. J. Berges, N. Tetradis and C. Wetterich, Phys. Rept. **363** (2002) 223 [hep-ph/0005122],
    J. Polonyi, Centr.Eur.Sci.J.Phys. 1 (2002) 1, [hep-th/0110026],
    D. F. Litim and J. M. Pawlowski, hep-th/9901063.
20. D. F. Litim, Phys. Lett. **B486** (2000) 92 [hep-th/0005245].
21. D. F. Litim, Phys. Rev. D **64** (2001) 105007 [hep-th/0103195],
    D. F. Litim, Int. J. Mod. Phys. A **16** (2001) 208 [hep-th/0104221].
22. D. F. Litim, Nucl. Phys. B **631** (2002) 128 [hep-th/0203006], JHEP **0111** (2001) 059 [hep-th/0111159], Acta Phys. Slov.**52** (2002) 635 [hep-th/0208117].
23. B. Bergerhoff, F. Freire, D. Litim, S. Lola and C. Wetterich, Phys. Rev. B **53** (1996) 5734,
    B. Bergerhoff, D. Litim, S. Lola and C. Wetterich, Int. J. Mod. Phys. A **11** (1996) 4273.
24. I. F. Herbut and Z. Tesanovic, Phys. Rev. Lett. **76** (1996) 4588 [cond-mat/9605185].
25. E. Bentivegna, A. Bonanno and M. Reuter, JCAP **0401** (2004) 001 [astro-ph/0303150].
26. O. Lauscher and M. Reuter, JHEP **0510** (2005) 050 [hep-th/0508202].
27. M. Reuter and C. Wetterich, Phys. Rev. D **56** (1997) 7893 [hep-th/9708051],
    F. Freire, D. F. Litim and J. M. Pawlowski, Phys. Lett. B **495** (2000) 256 [hep-th/0009110].
28. D. F. Litim and J. M. Pawlowski, Phys. Lett. B **516** (2001) 197 [hep-th/0107020], Phys. Rev. D **65** (2002) 081701 [hep-th/0111191], Phys. Rev. D **66** (2002) 025030 [hep-th/0202188].
29. H. W. Hamber and R. M. Williams, Phys. Rev. D **70** (2004) 124007 [hep-th/0407039].
30. H. W. Hamber and R. M. Williams, hep-th/0512003.
31. N. Arkani-Hamed, S. Dimopoulos and G. R. Dvali, Phys. Lett. B **429** (1998) 263 [hep-ph/9803315].
32. I. Antoniadis, Phys. Lett. B **246** (1990) 377.
33. G. F. Giudice, R. Rattazzi and J. D. Wells, Nucl. Phys. B **544** (1999) 3 [hep-ph/9811291],
    G. F. Giudice, T. Plehn and A. Strumia, Nucl. Phys. B **706** (2005) 455 [hep-ph/0408320].

The Objectivity of Spacetime: Dirac Observables and Gauge Variables for the Gravitational Field.

Luca Lusanna

Sezione INFN di Firenze, Polo Scientifico, Via Sansone 1, 50019 Sesto Fiorentino (FI), Italy

Abstract. In special relativity the chronogeometrical structure is absolute and there is no notion of instantaneous 3-space: distant clock synchronization is based on a convention, which changes going from inertial to non-inertial frames. On the contrary in Einstein general relativity the metric tensor has a double role: 1) it is the mediator of the gravitational interaction; 2) it determines the dynamical chronogeometrical structure of spacetime. As a consequence, it "teaches relativistic causality" to to every other field with the following implications: a) the conventions for distant clock synchronization in non-inertial frames are dynamically determined; b) the 8 arbitrary gauge variables hidden in the metric tensor describe "inertial effects" in non-inertial frames; c) the Dirac observables of the gravitational field describe "tidal effects" in non-inertial frames. A Hamiltonian reading of Bergmann-Komar intrinsic pseudo-coordinates, in connection with Einstein's Hole Argument, leads to a identification of the notions of "spacetime" and "gravitational field", which must be taken into account in any approach to quantum gravity.

Keywords: special relativity, general relativity, canonical formalism and constraint theory, clock synchronization, philosophy of science
PACS: 03.30+p, 04.20.-q, 04.20Cv, 04.20Fy

In 1914 Einstein, during his researches for developing general relativity, faced the problem arising from the fact that the requirement of general covariance would involve a threat to the physical objectivity of the points of space-time M^4, which in classical field theories are usually assumed to have a well defined individuality. He formulated the Hole Argument, according to which to each active diffeomorphisms interchanging space-time points is associated a different solution: determinism can be reobtained only *abandoning the physical objectivity of space-time points*. Einstein avoided the problem with the pragmatic *point-coincidence argument*: the only real world-occurrences are the (coordinate-independent) space-time coincidences (like the intersection of two world-lines). However, the problem was reopened by Stachel [1] and then by Earman and Norton [2] and this started a rich philosophical debate that is still alive today. Stachel suggested that a physical individuation of the point-events of M^4 could be done only by using *four individuating fields depending on the 4-metric on M^4*, namely that a tensor field on M^4 is needed to identify the points of M^4.

To clarify the problematic underlying the Hole Argument, we must review of the chrono-geometrical structure of special and general relativity with a special emphasis on the role of non-inertial frames and of the conventions for the synchronization of distant clocks.

In special relativity the chrono-geometrical structure of Minkowski space-time is non dynamical. It replaces the notions of absolute time and instantaneous Euclidean 3-space of the Galilei space-time underlying Newtonian physics. The Galilei relativity principle is replaced by the relativistic one. In both cases ideal inertial observers with their associated inertial frames, employing Cartesian coordinates, and connected by a kinematical group (either Galilei or Poincare') of transformations are privileged. The light postulates state that the two-way (or round trip) velocity of light c (only one clock is needed in its definition) is constant and isotropic. The Lorentz signature of Minkowski 4-metric tensor implies that every time-like observer can identify the light-cone (the conformal structure, i.e. the locus of the trajectories of light rays) in each point of the world-line. However there is *no notion of an instantaneous 3-space, of a spatial distance and of a one-way velocity of light between two observers* (the problem of the synchronization of distant clocks). Since the relativity principle privileges inertial observers and Cartesian coordinates $x^\mu = (x^o = ct; \vec{x})$ with the time axis centered on them (inertial frames), the $x^o = const.$ hyper-planes of inertial frames are usually taken as Euclidean instantaneous 3-spaces, on which all the clocks are synchronized. Indeed they can be selected as simultaneity 3-spaces with *Einstein's convention* for the synchronization of distant clocks to the clock of an inertial observer: this implies that the two-way and one-way velocities of light coincide.

However, real observers are never inertial and for them Einstein's convention for the synchronization of clocks is not able to identify globally defined simultaneity 3-surfaces, which could also be used as Cauchy surfaces for Maxwell equations. The 1+3 *point of view* tries to solve this problem starting from the local properties of an accelerated observer, whose world-line is assumed to be the time axis of some frame. Since only the observer 4-velocity is given, this only allows to identify the tangent plane of the vectors orthogonal to this 4-velocity in each point of the world-line. Then, both in special and general relativity, this tangent plane is identified with an instantaneous 3-space and 3-geodesic Fermi coordinates are defined on it and used to define a notion of spatial distance. However this construction leads to coordinate singularities, because the tangent planes in different points of the world-line will intersect each other at distances from the world-line of the order of the (linear and rotational) *acceleration radii* of the observer. Another type of coordinate singularity arises in all the proposed uniformly rotating coordinate systems: if ω is the constant angular velocity, then at a distance r from the rotation axis such that $\omega r = c$, the ${}^4g_{oo}$ component of the induced 4-metric vanishes. See Ref.[3] for a review of these topics and for the *locality hypothesis* (standard clocks and rods do not feel acceleration and at each instant the detectors of the instantaneously comoving inertial observer give the correct data).

This state of affairs and the need of predictability (a well-posed Cauchy problem for field theory) lead to the necessity of abandoning the 1+3 point of view and to shift to the 3+1 one. In this point of view, besides the world-line of an arbitrary time-like observer, it is given a 3+1 splitting of Minkowski space-time, namely a foliation of it whose leaves are space-like hyper-surfaces. Each leaf is both a Cauchy surface for the description of physical systems and an instantaneous (in general Riemannian) 3-space, namely a notion of simultaneity implied by a clock synchronization convention different from Einstein's one. The extra structure of the 3+1 splitting of Minkowski space-time allows to enlarge its atlas of 4-coordinate systems with the definition of *Lorentz-scalar observer-dependent radar 4-coordinates* $\sigma^A = (\tau; \sigma^r)$, $A = \tau, r$. Here τ is either the proper time of the accelerated observer or any monotonically increasing function of it, and is used to label the simultaneity leaves Σ_τ of the foliation. On each leaf Σ_τ the point of intersection with the world-line of the accelerated observer is taken as the origin of curvilinear 3-coordinates σ^r, which can be assumed to be globally defined since each Σ_τ is diffeomorphic to R^3. To the coordinate transformation $x^\mu \mapsto \sigma^A$ (x^μ are the standard Cartesian coordinates) is associated an inverse transformation $\sigma^A \mapsto x^\mu = z^\mu(\tau, \sigma^r)$, where the functions $z^\mu(\tau, \sigma^r)$ describe the embedding of the simultaneity surfaces Σ_τ into Minkowski space-time. The 3+1 splitting leads to the following induced 4-metric (a functional of the embedding):

$$^4g_{AB}(\tau,\sigma^r) = \frac{\partial z^\mu(\sigma)}{\partial \sigma^A} {}^4\eta_{\mu\nu} \frac{\partial z^\nu(\sigma)}{\partial \sigma^B} = {}^4g_{AB}[z(\sigma)],$$

where $^4\eta_{\mu\nu} = \varepsilon(+---)$ with $\varepsilon = \pm 1$ according to particle physics or general relativity convention respectively. The quantities

$$z^\mu_A(\sigma) = \frac{\partial z^\mu(\sigma)}{\partial \sigma^A}$$

are cotetrad fields on Minkowski space-time. An admissible 3+1 splitting of Minkowski space-time must have the embeddings $z^\mu(\tau, \sigma^r)$ of the space-like leaves Σ_τ of the associated foliation satisfying certain Mller conditions (to avoid the quoted coordinate singularities) and each simultaneity surface Σ_τ must tend to a space-like hyper-plane at spatial infinity.

As a consequence, any admissible 3+1 splitting leads to the definition of a *non-inertial frame centered on the given time-like observer* and coordinatized with Lorentz-scalar observer-dependent radar 4-coordinates. It turns out that Mller conditions forbid uniformly rotating non-inertial frames: only differentially rotating ones are allowed (the ones used by astrophysicists in the modern description of rotating stars). In Ref.[3] there is a detailed discussion of this topic and there is the simplest example of 3+1 splittings whose leaves are space-like hyper-planes carrying admissible differentially rotating 3-coordinates.

The description of physical systems in non-inertial frames is done by means of *parametrized Minkowski theories* (see Ref.[4]). Given any isolated system (particles, strings, fields, fluids) admitting a Lagrangian description, one makes the coupling of the system to an external gravitational field and then replaces the 4-metric $^4g_{\mu\nu}(x)$ with the induced metric $^4g_{AB}[z(\tau, \sigma^r)]$ associated to an arbitrary admissible 3+1 splitting. The Lagrangian now depends not only on the matter configurational variables but also on the embedding variables $z^\mu(\tau, \sigma^r)$. Since the action principle turns out to be invariant under frame-preserving diffeomorphisms ($\tau \mapsto \tau'(\tau, \sigma^r)$, $\sigma^r \mapsto \sigma'^r(\sigma^s)$), at the Hamiltonian level there are four first-class constraints. As a consequence, Dirac's theory of constraints implies that the configuration variables $z^\mu(\tau, \sigma^r)$ are arbitrary *gauge variables* (their fixation identifies a non-inertial frame) describing the *spatio-temporal appearances* of the phenomena in non-inertial frames, which, in turn, are associated to extended physical laboratories using a metrology for their measurements compatible with the notion of simultaneity of the non-inertial frame (think to the description of the

Earth given by GPS). Therefore, all the admissible 3+1 splittings, namely all the admissible conventions for clock synchronization, and all the admissible non-inertial frames centered on time-like observers are *gauge equivalent*. The resulting effective Hamiltonian for the τ-evolution turns out to contain the potentials of the *relativistic inertial forces* present in the given non-inertial frame. We see that already in special relativity *non-inertial Hamiltonians are coordinate-dependent quantities* like the notion of energy density in general relativity.

Inertial frames centered on inertial observers are a special case of gauge fixing in parametrized Minkowski theories. For each configuration of an isolated system there is an special 3+1 splitting associated to it: the foliation with space-like hyper-planes orthogonal to the conserved time-like 4-momentum of the isolated system. This identifies an intrinsic inertial frame, the *rest-frame*, centered on a suitable inertial observer (the Fokker-Pryce center of inertia of the isolated system) and allows to define the *Wigner-covariant rest-frame instant form of dynamics* for every isolated system. Let us remark that parametrized Minkowski theories allow to have a description of positive-energy relativistic particles on the simultaneity surfaces Σ_τ: this eliminates the problem of relative times, which for a long time has been an obstruction to the theory of relativistic bound states and to relativistic statistical mechanics. This framework made possible to develop a coherent formalism for all the aspects of relativistic kinematics both for N particle systems and continuous bodies and fields: i) the classification of the intrinsic notions of collective variables (canonical non-covariant center of mass; covariant non-canonical Fokker-Pryce center of inertia; non-covariant non-canonical Møller center of energy); ii) canonical bases of center-of-mass and relative variables; iii) canonical spin bases and dynamical body-frames for the rotational kinematics of deformable systems; iv) multipolar expansions for isolated and open systems; v) the relativistic theory of orbits; vi) the Møller radius (a classical unit of length identifying the region of non-covariance of the canonical center of mass of a spinning system around the covariant Fokker-Pryce center of inertia; it is an effect induced by the Lorentz signature of the 4-metric; it could be used as a physical ultraviolet cutoff in quantization). See Ref.[5] for a comprehensive review and the references to the main related papers.

In Ref.[6] there is the quantization of relativistic scalar and spinning particles in a class of non-inertial frames, whose simultaneity surfaces Σ_τ are space-like hyper-planes with arbitrary admissible linear acceleration and carrying arbitrary admissible differentially rotating 3-coordinates. It is based on a multi-temporal quantization scheme for systems with first-class constraints, in which only the particle degrees of freedom are quantized. The gauge variables, describing the appearances (inertial effects) of the motion in non-inertial frames, are treated as c-numbers (like the time in the Schroedinger equation with a time-dependent Hamiltonian) and the physical scalar product does not depend on them. The main open problem is the quantization of the scalar Klein-Gordon field in non-inertial frames, due to the Torre and Varadarajan no-go theorem [7], according to which in general the evolution from an initial space-like hyper-surface to a final one is *not unitary* in the Tomonaga-Schwinger formulation of quantum field theory. From the 3+1 point of view there is evolution only among the leaves of an admissible foliation and the possible way out from the theorem lies in the determination of all the admissible 3+1 splittings of Minkowski space-time satisfying the following requirements: i) existence of an instantaneous Fock space on each simultaneity surface Σ_τ (i.e. the Σ_τ's must admit a generalized Fourier transform); ii) unitary equivalence of the Fock spaces on Σ_{τ_1} and Σ_{τ_2} belonging to the same foliation (the associated Bogoliubov transformation must be Hilbert-Schmidt), so that the non-inertial Hamiltonian is a Hermitean operator; iii) unitary gauge equivalence of the 3+1 splittings with the Hilbert-Schmidt property. The overcoming of the no-go theorem would help also in quantum field theory in curved space-times and in condensed matter (here the non-unitarity implies non-Hermitean Hamiltonians and negative energies).

In the years 1913-16 Einstein developed general relativity relying on the equivalence principle (equality of inertial and gravitational masses of bodies in free fall). This led to the geometrization of the gravitational interaction. The principle of general covariance, at the basis of the tensorial nature of Einstein's equations, has the two following consequences: i) the invariance of the Hilbert action under *passive* diffeomorphisms (the coordinate transformations in the pseudo-Riemannian space-time M^4), so that the second Noether theorem implies the existence of first-class constraints at the Hamiltonian level; ii) the mapping of solutions of Einstein's equations among themselves under the action of *active* diffeomorphisms of M^4 extended to the tensors over M^4 (dynamical symmetries of Einstein's equations). The basic field of metric gravity is the 4-metric tensor with components ${}^4g_{\mu\nu}(x)$ in an arbitrary coordinate system of M^4. The peculiarity of gravity is that the 4-metric field, differently from the fields of electromagnetic, weak and strong interactions and from the matter fields, has a *double role*: i) it is the mediator of the gravitational interaction (in analogy to all the other gauge fields); ii) it determines the chrono-geometric structure of the space-time M^4 in a dynamical way through the line element $ds^2 = {}^4g_{\mu\nu}(x)\,dx^\mu\,dx^\nu$. As a consequence, the gravitational field *teaches relativistic causality* to all the other fields: for instance it tells to classical rays of light and to quantum photons and gluons which are the allowed trajectories for massless particles in each point of M^4.

Let us remark that in all known formulations particle and nuclear physics are a chapter of the theory of representations of the Poincare' group in inertial frames in the spatially non-compact Minkowski space-time. As a consequence, if one looks at general relativity from the point of view of particle physics, the main problem to get a unified theory is how to reconcile the Poincare' group (the kinematical group of the transformations connecting inertial frames) with the diffeomorphism group implying the non-existence of global inertial frames in general relativity (special relativity holds only in a small neighborhood of a body in free fall). To this end let us consider the ADM formulation of metric gravity and its extension [${}^4g_{\mu\nu}(x) = {}^4E^{(\alpha)}_\mu(x)\,{}^4\eta_{(\alpha)(\beta)}\,{}^4E^{(\beta)}_\nu(x)$, (α) are flat indices] to tetrad gravity (needed to describe the coupling of gravity to fermions; it is a theory of time-like observers endowed with a tetrad field, whose time-like axis is the unit 4-velocity and whose spatial axes are associated to a choice of three gyroscopes). Then, after having restricted the model to globally hyperbolic, topologically trivial, spatially non-compact space-times (admitting a global notion of time), let us introduce a 3+1 splitting of the space-time M^4 and let choose the world-line of a time-like observer and the associated radar 4-coordinates like in special relativity. The leaves Σ_τ (assumed to be Riemannian 3-manifolds diffeomorphic to R^3) are both Cauchy surfaces and simultaneity surfaces corresponding to a convention for clock synchronization. For the induced 4-metric we get

$$\begin{aligned}{}^4g_{AB}(\sigma) &= \frac{\partial z^\mu(\sigma)}{\partial \sigma^A} {}^4g_{\mu\nu}(x) \frac{\partial z^\nu(\sigma)}{\partial \sigma^B} \\ &= {}^4E_A^{(\alpha)} {}^4\eta_{(\alpha)(\beta)} {}^4E_B^{(\beta)} \\ &= \varepsilon \begin{pmatrix} (N^2 - {}^3g_{rs} N^r N^s) & -{}^3g_{su} N^u \\ -{}^3g_{ru} N^u & -{}^3g_{rs} \end{pmatrix}(\sigma). \end{aligned}$$

Here ${}^4E_A^{(\alpha)}(\tau, \sigma^r)$ are adapted cotetrad fields (in general relativity they replace the embeddings of special relativity as configuration variables), $N(\tau, \sigma^r)$ and $N^r(\tau, \sigma^r)$ the lapse and shift functions and ${}^3g_{rs}(\tau, \sigma^r)$ the 3-metric on Σ_τ with signature $(+++)$.

Let us try to identify a class of space-times and an associated suitable family of admissible 3+1 splittings able to incorporate particle physics and giving a model for the solar system or our galaxy (and hopefully allowing an extension to the cosmological context) with the following further requirements (see Ref.[8]): 1) M^4 must be asymptotically flat at spatial infinity and the 4-metric must tend asymptotically at spatial infinity to the Minkowski 4-metric in every coordinate system. Therefore, in these space-times there is an *asymptotic background 4-metric* and this will allow to avoid the decomposition ${}^4g_{\mu\nu} = {}^4\eta_{\mu\nu} + {}^4h_{\mu\nu}$ in the bulk. 2) The boundary conditions on each leaf Σ_τ of the admissible 3+1 splittings must be such to reduce the Spi group of asymptotic symmetries to the ADM Poincare' group. This is possible only if the admissible 3+1 splittings have all the leaves Σ_τ tending to Minkowski space-like hyper-planes orthogonal to the ADM 4-momentum at spatial infinity [8]. In turn this implies that every Σ_τ is the rest frame of the instantaneous 3-universe and that there are asymptotic inertial observers to be identified with the *fixed stars* (in a future extension to the cosmological context they could be identified with the privileged observers at rest with respect to the background cosmic radiation). In absence of matter the Christodoulou and Klainermann [9] space-times are good candidates. Since the simultaneity leaves Σ_τ are the rest frame of the instantaneous 3-universe, at the Hamiltonian level it is possible to define the rest-frame instant form of metric and tetrad gravity [8, 10, 11]. If matters is present, the limit of this description for vanishing Newton constant will produce the rest-frame instant form description of the same matter in the framework of parametrized Minkowski theories and the ADM Poincare' generators will tend to the kinematical Poincare' generators of special relativity (*deparametrization of general relativity to special relativity*).

In ADM tetrad gravity the 16 cotetrads fields may be replaced by i) 3 boost parameters $\varphi(\tau, \sigma^r)$ (adapting the cotetrads to Σ_τ), ii) cotriads ${}^3e_{(a)r}(\tau, \sigma^r)$ on Σ_τ, iii) lapse and shift functions $N(\tau, \sigma^r)$, $N_{(a)}(\tau, \sigma^r) = [{}^3e^s_{(a)} N_s](\tau, \sigma^r)$. The local invariances of the ADM action imply the existence of 14 first-class constraints (10 primary and 4 secondary): i) $\pi_N(\tau, \sigma^r) \approx 0$ implying the secondary super-hamiltonian constraint $\mathcal{H}(\tau, \sigma^r) \approx 0$; ii) $\pi_{\tilde{N}(a)}(\tau, \sigma^r) \approx 0$ implying the secondary super-momentum constraints $\mathcal{H}_{(a)}(\tau, \sigma^r) \approx 0$; iii) $\pi_{\vec{\varphi}(a)}(\tau, \sigma^r) \approx 0$; iv) three constraints $M_{(a)}(\tau, \sigma^r) \approx 0$ generating rotations of the cotriads. As a consequence there are 14 gauge variables describing the *generalized inertial effects* in the non-inertial frame defined by the chosen admissible 3+1 splitting of

M^4 centered on an arbitrary time-like observer. The remaining independent "two + two" degrees of freedom are the gauge invariant Dirac observables (DO) of the gravitational field describing *generalized tidal effects*.

In the canonical approach it is possible to make a separation of the gauge variables from the DO by means of a Shanmugadhasan canonical transformation (see Ref.[12]). These transformations define a canonical basis adapted to the existing first-class constraints. Since no-one knows how to solve the super-hamiltonian constraint (except that in the post-Newtonian approximation), the best we can do is to look for a quasi-Shanmugadhasan canonical transformation adapted to the other 13 first-class constraints (the only constraints to be Abelianized are $M_{(a)}(\tau,\sigma^r) \approx 0$ and $\mathscr{H}_{(a)}(\tau,\sigma^r) \approx 0$) [9]:

$\varphi^{(a)}$	N	N_r	$^3e_{(a)r}$
≈ 0	≈ 0	≈ 0	$^3\tilde{\pi}^r_{(a)}$

$\longrightarrow$

$\varphi^{(a)}$	N	$N_{(a)}$	$\alpha_{(a)}$	ξ^r	ϕ	$r_{\bar{a}}$
≈ 0	≈ 0	≈ 0	≈ 0	≈ 0	π_ϕ	$\pi_{\bar{a}}$

Here, $\alpha_{(a)}(\tau,\sigma^r)$ are three Euler angles and $\xi^r(\tau,\sigma^r)$ are three parameters giving a coordinatization of the action of 3-diffeomorphisms on the cotriads $^3e_{(a)r}(\tau,\sigma^r)$. The configuration variable $\phi(\tau,\sigma^r) = (det\,^3g(\tau,\sigma^r))^{1/12}$ is the conformal factor of the 3-metric: it can be shown that it is the unknown in the super-hamiltonian constraint (also named the Lichnerowicz equation). The gauge variables are N, $N_{(a)}$, $\varphi_{(a)}$, $\alpha_{(a)}$, ξ^r and π_ϕ, while $r_{\bar{a}}$, $\pi_{\bar{a}}$, $\bar{a} = 1,2$, are the DO of the gravitational field (in general they are not tensorial quantities). Even if we do not know the expression of the final variables in terms of the original ones, we note that this a *point* canonical transformation with known inverse [9], so that the old momenta are linear functionals of the new ones. The first-class constraints are the generators of the Hamiltonian gauge transformations, under which the ADM action is quasi-invariant (second Noether theorem). In particular those generated by the super-hamiltonian constraint $\mathscr{H}(\tau,\sigma^r) \approx 0$ transform an admissible 3+1 splitting into another admissible one by realizing a normal deformation of the simultaneity surfaces Σ_τ. As a consequence, all the conventions about clock synchronization are gauge equivalent as in special relativity.

In spatially compact space-times without boundary the Dirac Hamiltonian H_D is weakly zero, being a linear combination of the primary constraints (*frozen picture*) and the super-hamiltonian constraint is assumed to give a local evolution in some internal time (*Wheeler-DeWitt interpretation*). On the contrary, in spatially non-compact space-times the needed addition of the *DeWitt surface term* gives a non-vanishing Dirac Hamiltonism: in the rest-frame instant form we get [8] $H_D = \check{E}_{ADM} + (constraints) = E_{ADM} + (constraints) \approx E_{ADM}$. Here $\check{E}_{ADM}$ is the *strong ADM energy*, a surface term analogous to the one defining the electric charge as the flux of the electric field through the surface at spatial infinity in electromagnetism. Since we have $\check{E}_{ADM} = E_{ADM} + (constraints)$, we see that the non-vanishing part of the Dirac Hamiltonian is the *weak ADM energy* $E_{ADM} = \int d^3\sigma\, \mathscr{E}_{ADM}(\tau,\sigma^r)$, namely the integral over Σ_τ of the ADM energy density (in electromagnetism this corresponds to the definition of the electric charge as the volume integral of matter charge density). However, the ADM energy density $\mathscr{E}_{ADM}(\tau,\sigma^r)$ is a *coordinate-dependent quantity* because it depends on the gauge variables (namely on the inertial effects present in the non-inertial frame): this is the *problem of energy* in general relativity. As a consequence, to get a deterministic evolution for the DO we must fix the gauge completely, that is we have to add 14 gauge-fixing constraints and to pass to Dirac brackets. In this way all the gauge variables are fixed to be either numerical functions or well determined functions of the DO. As a consequence, in a completely fixed gauge (i.e. in a non-inertial frame centered on a time-like observer and with its pattern of inertial forces, corresponding to an extended physical laboratory with fixed metrological conventions) the ADM energy density $\mathcal{E}_{ADM}(\tau,\sigma^r)$ becomes a well defined function only of the DO and the Hamilton equations for them with E_{ADM} as Hamiltonian are a hyperbolic system of partial differential equations for their determination. For each choice of Cauchy data for the DO on a Σ_τ, we obtain a solution of Einstein's equations in the radar 4-coordinate system associated to the chosen 3+1 splitting of M^4.

A universe M^4 (a 4-geometry) is the equivalence class of all the completely fixed gauges with gauge equivalent Cauchy data for the DO on the associated Cauchy and simultaneity surfaces Σ_τ. In each gauge we find the solution for the DO in that gauge (the tidal effects) and then the explicit form of the gauge variables (the inertial effects). Moreover, also the extrinsic curvature of the simultaneity surfaces Σ_τ is determined. Since the simultaneity surfaces are asymptotically flat, it is possible to determine their embeddings $z^\mu(\tau,\sigma^r)$ in M^4. As a consequence, differently from special relativity, the conventions for clock synchronization and the whole chrono-geometrical structure of M^4 (gravito-magnetism, 3-geodesic spatial distance on Σ_τ, trajectories of light rays in each point of M^4, one-way velocity of light) are *dynamically determined*.

A first application of this formalism [13] has been the determination of post-Minkowskian background-independent gravitational waves in a completely fixed non-harmonic 3-orthogonal gauge with diagonal 3-metric. The weak field approximation $r_{\bar{a}}(\tau,\sigma^r) << 1$, $\pi_{\bar{a}}(\tau,\sigma^r) << 1$, based on a Hamiltonian linearization scheme, allows to get a solution of linearized Einstein's equations, in which the configurational DO $r_{\bar{a}}(\tau,\sigma^r)$ play the role of the two polarizations of the gravitational wave and we can evaluate the embedding $z^\mu(\tau,\sigma^r)$ of the simultaneity surfaces of this gauge explicitly.

Regarding the Hole Argument it can be shown [14] that the role of active diffeomorphisms is equivalent to Hamiltonian gauge transformations restricted to solutions of Einstein's equations. Moreover, the *intrinsic pseudo-coordinates* of Bergmann and Komar [15], four scalar functions $F^A[w_\lambda]$, $A,\lambda = 1,..,4$, of the four eigenvalues $w_\lambda({}^4g,\partial{}^4g)$ of the Weyl tensor, can be used as individuating fields. As shown in Ref.[14] the individuation of point-events can be done by considering an arbitrary admissible 3+1 splitting of M^4 with a given time-like observer and the associated radar 4-coordinates σ^A and by imposing the gauge fixings $\chi^A(\tau,\sigma^r) = \sigma^A - F^A[w_\lambda] \approx 0$. After having fixed the other gauge freedoms of tetrad gravity, we arrive at a completely fixed gauge in which, after the transition to Dirac brackets, we get $\sigma^A \equiv \tilde{F}^A[r_{\bar{a}}(\sigma),\pi_{\bar{a}}(\sigma)]$, namely that the radar 4-coordinates of a point in M^4_{3+1}, the copy of M^4 coordinatized with the chosen non-inertial frame, are determined *off-shell* by the four DO of that gauge: in other words the individuating fields are the genuine tidal effects of the gravitational field. Some consequences of this identification of the point-events of M^4 are: 1) The space-time M^4 and the gravitational field are essentially the same entity. The presence of matter modifies the solutions of Einstein equations, i.e. M^4, but does not play any role in this identification. Instead matter is fundamental for establishing a (still lacking) dynamical theory of measurement not using test objects. 2) The reduced phase space of this model of general relativity is the space of abstract DO (pure tidal effects without inertial effects), which can be thought as four fields on an abstract space-time $\tilde{M}^4 = \{equivalence\ class\ of\ all\ the\ admissible\ non-inertial\ frames\ M^4_{3+1}\ containing\ the\ associated\ inertial\ effects\}$. 3) Each radar 4-coordinate system of an admissible non-inertial frame M^4_{3+1} has an associated *non-commutative structure*, determined by the Dirac brackets of the functions $\tilde{F}^A[r_{\bar{a}}(\sigma), \pi_{\bar{a}}(\sigma)]$ determining the gauge. 4) Conjecture: there should exist privileged Shanmugadhasan canonical bases of phase space, in which the DO (the tidal effects) are also *Bergmann observables* [16], namely coordinate-independent scalar tidal effects. As a final remark, let us note that these results on the identification of point-events are *model dependent*. In spatially compact space-times without boundary, the DO are *constants of the motion* due to the frozen picture. As a consequence, the gauge fixings $\chi^A(\tau, \sigma^r) \approx 0$ (in particular χ^τ) cannot be used to rebuild the temporal dimension: probably only the instantaneous 3-space of a 3+1 splitting can be individuated in this way.

For an extended version of the talk see Ref.[17], where there is a list of the problems now under investigation in this approach, from the Hamiltonian reformulation of Newman-Penrose formalism for the search of Bergmann observables till the inclusion of matter and the treatment of the two-body problem. A strategy for the quantization is also delineated.

REFERENCES

1. Stachel,J. *Einstein's Search for General Covariance, 1912–1915*. Ninth International Conference on General Relativity and Gravitation, Jena (1980), ed. E.Schmutzer (Cambridge Univ.Press, Cambridge, 1983).
2. Earman,J. and Norton,J. *What Price Spacetime Substantivalism? The Hole Story*, British Journal for the Philosophy of Science **38**, 515–525 (1987).
3. Alba,D. and Lusanna, L. *Simultaneity, Radar 4-Coordinates and the 3+1 Point of View about Accelerated Observers in Special Relativity* (2003) (gr-qc/0311058); *Generalized Radar 4-Coordinates and Equal-Time Cauchy Surfaces for Arbitrary Accelerated Observers* (2005), submitted to Gen.Rel.Grav. (gr-qc/0501090).
4. Lusanna, L. *The N- and 1-Time Classical Description of N-Body Relativistic Kinematics and the Electromagnetic Interaction*, Int. J. Mod. Phys. **A12**, 645-722 (1997); *The Chronogeometrical Structure of Special and General Relativity: towards a Background-Independent Description of the Gravitational Field and Elementary Particles* (2004), invited contribution to the book *Progress in General Relativity and Quantum Cosmology* (Nova Science) (gr-qc/0404122).
5. Alba, D., Lusanna, L. and Pauri, M. *New Directions in Non-Relativistic and Relativistic Rotational and Multipole Kinematics for N-Body and Continuous Systems* (2005), invited contribution for the book *Atomic and Molecular Clusters: New Research* (Nova Science) (hep-th/0505005).
6. Alba, D. and Lusanna, L. *Quantum Mechanics in Non-Inertial Frames with a Multi-Temporal Quantization Scheme: I) Relativistic Particles* (hep-th/0502060); II) Non-Relativistic Particles (hep-th/0504060), to appear in Int.J.Mod.Phys..
7. Torre, C.G. and Varadarajan, M. *Functional Evolution of Free Quantum Fields*, Clas. Quantum Grav. **16**, 2651-2668 (1999).
8. Lusanna, L. *The Rest-Frame Instant Form of Metric Gravity*, Gen.Rel.Grav. **33**, 1579-1696 (2001) (gr-qc/0101048).

9. Christodoulou, D., and Klainerman, S. *The Global Nonlinear Stability of the Minkowski Space*. (Princeton University Press, Princeton, 1993).
10. De Pietri,R., Lusanna,L., Martucci,L. and Russo,S. *Dirac's Observables for the Rest-Frame Instant Form of Tetrad Gravity in a Completely Fixed 3-Orthogonal Gauge*, Gen.Rel.Grav. **34**, 877-1033 (2002) (gr-qc/0105084).
11. Lusanna,L. and Russo,S. *A New Parametrization for Tetrad Gravity*, Gen.Rel.Grav. **34**, 189-242 (2002) (gr-qc/0102074).
12. Lusanna, L. *The Shanmugadhasan Canonical Transformation, Function Groups and the Second Noether Theorem*, Inter.J.Mod.Phys. **A8**, 4193-4233 (1993).
13. Agresti,J., De Pietri,R., Lusanna,L. and Martucci,L. *Hamiltonian Linearization of the Rest-Frame Instant Form of Tetrad Gravity in a Completely Fixed 3-Orthogonal Gauge: a Radiation Gauge for Background-Independent Gravitational Waves in a Post-Minkowskian Einstein Space-Time*, Gen.Rel.Grav. **36**, 1055-1134 (2004) (gr-qc/0302084).
14. Lusanna, L. and Pauri, M. *General Covariance and the Objectivity of Space-Time Point-Events*, talk at the Oxford Conference on Spacetime Theory (2004), to appear in History and Philosophy of Modern Physics (gr-qc/0503069); *The Physical Role of Gravitational and Gauge Degrees of Freedom in General Relativity. I: Dynamical Synchronization and Generalized Inertial Effects; II: Dirac versus Bergmann Observables and the Objectivity of Space-Time*, to appear in Gen.Rel.Grav. (gr-qc/0403081 and 0407007).
15. Bergmann, P.G. and Komar, A. *Poisson Brackets between Locally Defined Observables in General Relativity*, Phys.Rev.Lett. **4**, 432-433 (1960).
16. Bergmann, P.G. *Observables in General Relativity*, Rev.Mod.Phys. **33**, 510-514 (1961).
17. Lusanna, L. *General Covariance and its Implications for Einstein's Space-Times* (gr-qc/0510024).

Deformed Einstein Gravity

Frank Meyer

Max-Planck Institute for Physics, Föhringer Ring 6, D-80805 Munich, Germany

Abstract. We introduce the necessary concepts for an algebraic construction of a gravity theory on noncommutative spaces. The θ-deformed diffeomorphisms are studied and a tensor calculus is defined. This leads to a deformed Einstein-Hilbert action which is invariant with respect to deformed diffeomorphisms. The dynamical variable is the vierbein field. The deformed action is a deformation of the usual Einstein-Hilbert action and reduces to it in the limit where the noncommutativity vanishes. This contribution is based on joint work with P. Aschieri, C. Blohmann, M. Dimitrijevi ć, P. Schupp and J. Wess.

Keywords: Noncommutative Geometry, Deformed Diffeomorphisms, Deformed Gravity
PACS: 02.40.Gh, 02.20.Uw, 04.20.-q, 04.60.-m, 11.10.Nx

NONCOMMUTATIVE SPACES

It is expected that in order to obtain a better understanding of physics at short distances and in order to cure the problems occuring when trying to quantize gravity one has to change the nature of space-time in a fundamental way. One way to do so is to implement noncommutativity by taking coordinates which satisfy the commutation relations

$$[\hat{x}^\mu, \hat{x}^\nu] = C^{\mu\nu}(\hat{x}) \neq 0. \tag{1}$$

The function $C^{\mu\nu}(\hat{x})$ is unknown. For physical reasons it should be a function that vanishes at large distances where we experience the commutative world and may be determined by experiments [1]. We denote the algebra generated by noncommutative coordinates $\hat{x}^\mu$ which are subject to the relations (1) by $\mathscr{A}$ (*algebra of noncommutative functions*). In what follows we will exclusively consider the θ-deformed case which may at very short distances provide a reasonable approximation for $C^{\mu\nu}(\hat{x})$

$$[\hat{x}^\mu, \hat{x}^\nu] = i\theta^{\mu\nu} = \text{const.} \tag{2}$$

but we note that the algebraic construction presented here can be generalized to more complicated noncommutative structures of the above type which possess the Poincaré-Birkhoff-Witt (PBW) property.

SYMMETRIES ON DEFORMED SPACES

In general the commutation relations (1) are not covariant with respect to undeformed symmetries. For example the canonical commutation relations (2) break Lorentz symmetry if we assume that the noncommutativity parameters $\theta^{\mu\nu}$ do not transform.

The question arises whether we can *deform* the symmetry in such a way that it acts consistently on the deformed space (i.e. leaves the deformed space invariant) and such that it reduces to the undeformed symmetry in the commutative limit. The answer is yes: Lie algebras can be deformed in the category of Hopf algebras (Hopf algebras coming from a Lie algebra are also called Quantum Groups)[1]. Quantum group symmetries lead to new features of field theories on noncommutative spaces. Because of its simplicity, θ-deformed spaces are very well-suited to study those.

In the following we will construct explicitly a θ-deformed version of diffeomorphisms which consistently act on the noncommutative space (2). Then we present a gravity theory which is invariant with respect to this deformed diffeomorphisms [2, 3, 4].

DIFFEOMORPHISMS

Gravity is a theory invariant with respect to diffeomorphisms. However, to generalize the Einstein formalism to noncommutative spaces in order to establish a gravity theory, it is important to first understand that diffeomorphisms possess more mathematical structure than the algebraic one: They are naturally equipped with a Hopf algebra structure. In the common formulations of physical theories this additional Hopf structure is hidden and does not play a crucial role. It is our aim to deform the algebra of diffeomorphism in such a way that it acts consistently on a noncommutative space. This can be done by exploiting the full Hopf structure. In this section we first introduce the concept of diffeomorphisms as Hopf algebra in the undeformed setting.

Diffeomorphisms are generated by vector-fields ξ. Acting on functions, vector-fields are represented as linear differential operators $\xi = \xi^\mu \partial_\mu$. Vector-fields form a Lie algebra Ξ with the Lie bracket given by

$$[\xi, \eta] = \xi \times \eta$$

where $\xi \times \eta$ is defined by its action on functions

$$(\xi \times \eta)(f) = (\xi^\mu (\partial_\mu \eta^\nu) \partial_\nu - \eta^\mu (\partial_\mu \xi^\nu) \partial_\nu)(f).$$

The Lie algebra of *infinitesimal diffeomorphisms* Ξ can be embedded into its universal enveloping algebra which we want to denote by $\mathscr{U}(\Xi)$. The universal enveloping algebra is an associative algebra and possesses a natural Hopf algebra structure. The coproduct is defined as follows on the generators[2]:

$$\begin{aligned} \Delta : \mathscr{U}(\Xi) &\to \mathscr{U}(\Xi) \otimes \mathscr{U}(\Xi) \\ \Xi \ni \xi &\mapsto \Delta(\xi) := \xi \otimes 1 + 1 \otimes \xi. \end{aligned} \qquad (3)$$

---

[1] To be more precise the universal enveloping algebra of a Lie algebra can be deformed. The universal enveloping algebra of any Lie algebra is a Hopf algebra and this gives rise to deformations in the category of Hopf algebras.

[2] The structure maps are defined on the generators $\xi \in \Xi$ and the universal property of the universal enveloping algebra $\mathscr{U}(\Xi)$ assures that they can be uniquely extended as algebra homomorphisms (respectively anti-algebra homomorphism in case of the antipode S) to the whole algebra $\mathscr{U}(\Xi)$.

For a precise definition and more details on Hopf algebras we refer the reader to text books [5]. For our purposes it shall be sufficient to note that the coproduct implements how the Hopf algebra acts on a product in a representation algebra (Leibniz-rule). Scalar fields are defined by their transformation property with respect to infinitesimal coordinate transformations:

$$\delta_\xi \phi = -\xi\phi = -\xi^\mu(\partial_\mu \phi). \tag{4}$$

The product of two scalar fields is transformed using the Leibniz-rule

$$\delta_\xi(\phi\psi) = (\delta_\xi\phi)\psi + \phi(\delta_\xi\psi) = -\xi^\mu(\partial_\mu\phi\psi) \tag{5}$$

such that the product of two scalar fields transforms again as a scalar. The above Leibniz-rule can be understood in mathematical terms as follows: The Hopf algebra $\mathcal{U}(\Xi)$ is represented on the space of scalar fields by infinitesimal coordinate transformations δ_ξ. On scalar fields the action of δ_ξ is explicitly given by the differential operator $-\xi^\mu \partial_\mu$. Of course, the space of scalar fields is not only a vector space - it possesses also an algebra structure - such as $\mathcal{U}(\Xi)$ is not only an algebra but also a Hopf algebra - it possesses in addition the co-structure maps defined above. We say that a Hopf algebra H acts on an algebra A (or more precisely we say that A is a left H-module algebra) if A is a module with respect to the algebra H and if in addition for all $h \in H$ and $a, b \in A$

$$h(ab) = \mu \circ \Delta h(a \otimes b) \tag{6}$$
$$h(1) = \varepsilon(h). \tag{7}$$

Here μ is the multiplication map defined by $\mu(a \otimes b) = ab$. In our concrete example where $H = \mathcal{U}(\Xi)$ and A is the algebra of scalar fields we indeed have that the algebra of scalar fields is a $\mathcal{U}(\Xi)$-module algebra. This can be seen easily if we rewrite (5) using (3) for the generators $\xi \in \Xi$ for $\mathcal{U}(\Xi)$:

$$\delta_\xi(\phi\psi) = (\delta_\xi\phi)\psi + \phi(\delta_\xi\psi) = \mu \circ \Delta\xi(\phi \otimes \psi).$$

It is also evident that

$$\delta_\xi 1 = 0 = \varepsilon(\xi)1.$$

Now we are in the right mathematical framework: We study a Lie algebra (here infinitesimal diffeomorphisms Ξ) and embed it in its universal enveloping algebra (here $\mathcal{U}(\Xi)$). This universal enveloping algebra is a Hopf algebra via a natural Hopf structure induced by (3).

Physical quantities live in representations of this Hopf algebras. For instance, the algebra of scalar fields is a $\mathcal{U}(\Xi)$-module algebra. The action of $\mathcal{U}(\Xi)$ on scalar fields is given in terms of infinitesimal coordinate transformations δ_ξ.

Similarly one studies tensor representations of $\mathcal{U}(\Xi)$. For example vector fields are introduced by the transformation property

$$\delta_\xi V_\alpha = -\xi^\mu(\partial_\mu V_\alpha) - (\partial_\alpha \xi^\mu)V_\mu$$
$$\delta_\xi V^\alpha = -\xi^\mu(\partial_\mu V^\alpha) + (\partial_\mu \xi^\alpha)V^\mu.$$

The generalization to arbitrary tensor fields is straight forward:

$$\delta_\xi T^{\mu_1\cdots\mu_n}_{\nu_1\cdots\nu_n} = -\xi^\mu(\partial_\mu T^{\mu_1\cdots\mu_n}_{\nu_1\cdots\nu_n}) + (\partial_\mu \xi^{\mu_1}) T^{\mu\cdots\mu_n}_{\nu_1\cdots\nu_n} + \cdots + (\partial_\mu \xi^{\mu_n}) T^{\mu_1\cdots\mu}_{\nu_1\cdots\nu_n}$$
$$-(\partial_{\nu_1}\xi^\nu) T^{\mu_1\cdots\mu_n}_{\nu\cdots\nu_n} - \cdots - (\partial_{\nu_n}\xi^\nu) T^{\mu_1\cdots\mu_n}_{\nu_1\cdots\nu}.$$

As for scalar fields, we also find that the product of two tensors transforms like a tensor. Summarizing, we have seen that scalar fields, vector fields and tensor fields are representations of the Hopf algebra $\mathscr{U}(\Xi)$, the universal enveloping algebra of infinitesimal diffeomorphisms. The Hopf algebra $\mathscr{U}(\Xi)$ acts via *infinitesimal coordinate transformations* δ_ξ which are subject to the relations:

$$[\delta_\xi, \delta_\eta] = \delta_{\xi \times \eta} \tag{8}$$

$$\Delta \delta_\xi = \delta_\xi \otimes 1 + 1 \otimes \delta_\xi . \tag{9}$$

The transformation operator δ_ξ is explicitly given by differential operators which depend on the representation under consideration. In case of scalar fields this differential operator is given by $-\xi^\mu \partial_\mu$.

DEFORMED DIFFEOMORPHISMS

The concepts introduced in the previous subsection can be deformed in order to establish a consistent tensor calculus on the noncommutative space-time algebra (2). In this context it is necessary to account the full Hopf algebra structure of the universal enveloping algebra $\mathscr{U}(\Xi)$.

In our setting the algebra $\hat{\mathscr{A}}$ possesses a noncommutative product defined by

$$[\hat{x}^\mu, \hat{x}^\nu] = i\theta^{\mu\nu}. \tag{10}$$

We want to deform the structure maps (9) of the Hopf algebra $\mathscr{U}(\Xi)$ in such a way that the resulting deformed Hopf algebra which we denote by $\mathscr{U}(\hat{\Xi})$ consistently acts on $\hat{\mathscr{A}}$. In the language introduced in the previous section this means that we want $\hat{\mathscr{A}}$ to be a $\mathscr{U}(\hat{\Xi})$-module algebra. We claim that the following deformation of $\mathscr{U}(\Xi)$ does the job. Let $\mathscr{U}(\hat{\Xi})$ be generated as algebra by elements $\hat{\delta}_\xi$, $\xi \in \Xi$. We leave the algebra relation undeformed

$$[\hat{\delta}_\xi, \hat{\delta}_\eta] = \hat{\delta}_{\xi \times \eta} \tag{11}$$

but we deform the co-sector

$$\Delta \hat{\delta}_\xi = e^{-\frac{i}{2}\hbar\theta^{\rho\sigma}\hat{\partial}_\rho \otimes \hat{\partial}_\sigma}(\hat{\delta}_\xi \otimes 1 + 1 \otimes \hat{\delta}_\xi) e^{\frac{i}{2}\hbar\theta^{\rho\sigma}\hat{\partial}_\rho \otimes \hat{\partial}_\sigma}, \tag{12}$$

where $[\hat{\partial}_\rho, \hat{\delta}_\xi] = \hat{\delta}_{(\partial_\rho \xi)}$. The deformed coproduct (12) reduces to the undeformed one (9) in the limit $\theta \to 0$. We have to check whether the above deformation is a good one in the sense that it leads to a consistent action on $\hat{\mathscr{A}}$. First we need a differential operator acting on fields in $\mathscr{A}$ which represents the algebra (11). Let us consider the differential operator

$$\hat{X}_\xi := \sum_{n=0}^{\infty} \frac{1}{n!}(-\frac{i}{2})^n \theta^{\rho_1 \sigma_1} \cdots \theta^{\rho_n \sigma_n} (\hat{\partial}_{\rho_1} \cdots \hat{\partial}_{\rho_n} \xi^\mu) \hat{\partial}_\mu \hat{\partial}_{\sigma_1} \cdots \hat{\partial}_{\sigma_n}. \tag{13}$$

Then indeed we have

$$[\hat{X}_\xi, \hat{X}_\eta] = \hat{X}_{\xi \times \eta}. \tag{14}$$

It is therefore reasonable to introduce scalar fields $\hat{\phi} \in \mathscr{A}$ by the transformation property

$$\hat{\delta}_\xi \hat{\phi} = -(\hat{X}_\xi \hat{\phi}).$$

The next step is to work out the action of the differential operators $\hat{X}_\xi$ on the product of two fields. A calculation [2] shows that

$$(\hat{X}_\xi(\hat{\phi}\hat{\psi})) = \mu \circ (e^{-\frac{i}{2}h\theta^{\rho\sigma}\hat{\partial}_\rho \otimes \hat{\partial}_\sigma}(\hat{X}_\xi \otimes 1 + 1 \otimes \hat{X}_\xi)e^{\frac{i}{2}h\theta^{\rho\sigma}\hat{\partial}_\rho \otimes \hat{\partial}_\sigma} \hat{\phi} \otimes \hat{\psi}).$$

This means that the differential operators $\hat{X}_\xi$ act via a *deformed Leibniz rule* on the product of two fields. Comparing with (12) we see that the deformed Leibniz rule of the differential operator $\hat{X}_\xi$ is exactly the one induced by the deformed coproduct (12):

$$\hat{\delta}_\xi(\hat{\phi}\hat{\psi}) = e^{-\frac{i}{2}h\theta^{\rho\sigma}\hat{\partial}_\rho \otimes \hat{\partial}_\sigma}(\hat{\delta}_\xi \otimes 1 + 1 \otimes \hat{\delta}_\xi)e^{\frac{i}{2}h\theta^{\rho\sigma}\hat{\partial}_\rho \otimes \hat{\partial}_\sigma}(\hat{\phi}\hat{\psi}) = -\hat{X}_\xi \triangleright (\hat{\phi}\hat{\psi}).$$

Hence, the deformed Hopf algebra $\mathscr{U}(\hat{\Xi})$ is indeed represented on scalar fields $\hat{\phi} \in \mathscr{A}$ by the differential operator $\hat{X}_\xi$. The scalar fields form a $\mathscr{U}(\hat{\Xi})$-module algebra.

In analogy to the previous section we can introduce vector and tensor fields as representations of the Hopf algebra $\mathscr{U}(\hat{\Xi})$. The transformation property for an arbitrary tensor reads

$$\hat{\delta}_\xi \hat{T}^{\mu_1 \cdots \mu_r}_{\nu_1 \cdots \nu_s} = -(\hat{X}_\xi \hat{T}^{\mu_1 \cdots \mu_n}_{\nu_1 \cdots \nu_n}) + (\hat{X}_{(\partial_\mu \xi^{\mu_1})} \hat{T}^{\mu \cdots \mu_n}_{\nu_1 \cdots \nu_n}) + \cdots + (\hat{X}_{(\partial_\mu \xi^{\mu_n})} \hat{T}^{\mu_1 \cdots \mu}_{\nu_1 \cdots \nu_n})$$
$$- (\hat{X}_{(\partial_{\nu_1} \xi^\nu)} \hat{T}^{\mu_1 \cdots \mu_n}_{\nu \cdots \nu_n}) - \cdots - (\hat{X}_{(\partial_{\nu_n} \xi^\nu)} \hat{T}^{\mu_1 \cdots \mu_n}_{\nu_1 \cdots \nu}).$$

Up to now we have seen the following:

- Diffeomorphisms are generated by vector-fields $\xi \in \Xi$ and the universal enveloping algebra $\mathscr{U}(\Xi)$ of the Lie algebra Ξ of vector-fields possesses a natural Hopf algebra structure defined by (9).
- The algebra of scalar fields $\phi \in \mathscr{A}$ is a $\mathscr{U}(\Xi)$-module algebra.
- The universal enveloping algebra $\mathscr{U}(\Xi)$ can be deformed to a Hopf algebra $\mathscr{U}(\hat{\Xi})$ defined in (11,12).
- $\mathscr{U}(\hat{\Xi})$ consistently acts on the algebra of noncommutative functions $\hat{\mathscr{A}}$, i.e. the algebra of noncommutative functions is a $\mathscr{U}(\hat{\Xi})$-module algebra.
- Regarding $\mathscr{U}(\hat{\Xi})$ as the underlying "symmetry" of the gravity theory to be built on the noncommutative space $\hat{\mathscr{A}}$, we established a full tensor calculus as representations of the Hopf algebra $\mathscr{U}(\hat{\Xi})$.

NONCOMMUTATIVE GEOMETRY

The deformed algebra of infinitesimal diffeomorphisms and the tensor calculus covariant with respect to it is the fundamental building-block for the definition of a noncommutative geometry on θ-deformed spaces. In this section we sketch the important steps towards a deformed Einstein-Hilbert action [2]. A first ingredient is the *covariant derivative* $\hat{D}_\mu$. Algebraically, it can be defined by demanding that acting on a vector-field it produces a tensor-field

$$\hat{\delta}_\xi \hat{D}_\mu \hat{V}_\nu \stackrel{!}{=} -(\hat{X}_\xi \hat{D}_\mu \hat{V}_\nu) - (\hat{X}_{(\partial_\mu \xi^\alpha)} \hat{D}_\alpha \hat{V}_\nu) - (\hat{X}_{(\partial_\nu \xi^\alpha)} \hat{D}_\mu \hat{V}_\alpha) \tag{15}$$

The covariant derivative is given by a *connection* $\hat{\Gamma}_{\mu\nu}{}^\rho$

$$\hat{D}_\mu \hat{V}_\nu = \hat{\partial}_\mu \hat{V}_\nu - \hat{\Gamma}_{\mu\nu}{}^\rho \hat{V}_\rho.$$

From (15) it is possible to deduce the transformation property of $\hat{\Gamma}_{\mu\nu}{}^\rho$

$$\hat{\delta}_\xi \hat{\Gamma}_{\mu\nu}{}^\rho = (\hat{X}_\xi \hat{\Gamma}_{\mu\nu}{}^\rho) - (\hat{X}_{(\partial_\mu \xi^\alpha)} \hat{\Gamma}_{\alpha\nu}{}^\rho) - (\hat{X}_{(\partial_\nu \xi^\alpha)} \hat{\Gamma}_{\mu\alpha}{}^\rho) + (\hat{X}_{(\partial_\alpha \xi^\rho)} \hat{\Gamma}_{\mu\nu}{}^\alpha) - (\hat{\partial}_\mu \hat{\partial}_\nu \hat{\xi}^\rho).$$

The *metric* $\hat{G}_{\mu\nu}$ is defined as a symmetric tensor of rank two. It can be obtained for example by a set of vector-fields $\hat{E}_\mu{}^a$, $a = 0,\ldots,3$, where a is to be understood as a mere label. These vector-fields are called *vierbeins*. Then the symmetrized product of those vector-fields is indeed a symmetric tensor of rank two

$$\hat{G}_{\mu\nu} := \frac{1}{2}(\hat{E}_\mu{}^a \hat{E}_\nu{}^b + \hat{E}_\nu{}^b \hat{E}_\mu{}^a) \eta_{ab}.$$

Here η_{ab} stands for the usual flat Minkowski space metric. Let us assume that we can choose the vierbeins $\hat{E}_\mu{}^a$ such that they reduce in the commutative limit to the usual vierbeins $e_\mu{}^a$. Then also the metric $\hat{G}_{\mu\nu}$ reduces to the usual, undeformed metric $g_{\mu\nu}$.

The inverse metric tensor we denote by upper indices

$$\hat{G}_{\mu\nu} \hat{G}^{\nu\rho} = \delta_\mu^\rho.$$

We use $\hat{G}_{\mu\nu}$ respectively $\hat{G}^{\mu\nu}$ to raise and lower indices.

The curvature and torsion tensors are obtained by taking the commutator of two covariant derivatives[3]

$$[\hat{D}_\mu, \hat{D}_\nu] \hat{V}_\rho = \hat{R}_{\mu\nu\rho}{}^\alpha \hat{V}_\alpha + \hat{T}_{\mu\nu}{}^\alpha \hat{D}_\alpha \hat{V}_\rho$$

which leads to the expressions

$$\hat{R}_{\mu\nu\rho}{}^\sigma = \hat{\partial}_\nu \hat{\Gamma}_{\mu\rho}{}^\sigma - \hat{\partial}_\mu \hat{\Gamma}_{\nu\rho}{}^\sigma + \hat{\Gamma}_{\nu\rho}{}^\beta \hat{\Gamma}_{\mu\beta}{}^\sigma - \hat{\Gamma}_{\mu\rho}{}^\beta \hat{\Gamma}_{\nu\beta}{}^\sigma$$

$$\hat{T}_{\mu\nu}{}^\alpha = \hat{\Gamma}_{\nu\mu}{}^\alpha - \hat{\Gamma}_{\nu\mu}{}^\alpha.$$

---

[3] The generalization of covariant derivatives acting on tensors is straight forward [2].

If we assume the *torsion-free* case, i.e.

$$\hat{\Gamma}_{\mu\nu}{}^\sigma = \hat{\Gamma}_{\nu\mu}{}^\sigma,$$

we find an unique expression for the metric connection (Christoffel symbol) defined by

$$\hat{D}_\alpha \hat{G}_{\beta\gamma} \stackrel{!}{=} 0$$

in terms of the metric and its inverse[4]

$$\hat{\Gamma}_{\alpha\beta}{}^\sigma = \frac{1}{2}(\hat{\partial}_\alpha \hat{G}_{\beta\gamma} + \hat{\partial}_\beta \hat{G}_{\alpha\gamma} - \hat{\partial}_\gamma \hat{G}_{\alpha\beta})\hat{G}^{\gamma\sigma}.$$

From the curvature tensor $\hat{R}_{\mu\nu\rho}{}^\sigma$ we get the curvature scalar by contracting the indices

$$\hat{R} := \hat{G}^{\mu\nu}\hat{R}_{\nu\mu\rho}{}^\rho.$$

$\hat{R}$ indeed transforms as a scalar which may be checked explicitly by taking the deformed coproduct (12) into account.

To obtain an integral which is invariant with respect to the Hopf algebra of deformed infinitesimal diffeomorphisms we need a measure function $\hat{E}$. We demand the transformation property

$$\hat{\delta}_\xi \hat{E} = -\hat{X}_\xi \hat{E} - \hat{X}_{(\partial_\mu \xi^\mu)} \hat{E}. \tag{16}$$

Then it follows with the deformed coproduct (12) that for any scalar field $\hat{S}$

$$\hat{\delta}_\xi \hat{E}\hat{S} = -\hat{\partial}_\mu(\hat{X}_{\xi^\mu}(\hat{E}\hat{S})).$$

Hence, transforming the product of an arbitrary scalar field with a measure function $\hat{E}$ we obtain a total derivative which vanishes under the integral. A suitable measure function with the desired transformation property (16) is for instance given by the determinant of the vierbein $\hat{E}_\mu{}^a$

$$\hat{E} = \det(\hat{E}_\mu{}^a) := \frac{1}{4!}\varepsilon^{\mu_1\cdots\mu_4}\varepsilon_{a_1\cdots a_4}\hat{E}_{\mu_1}{}^{a_1}\hat{E}_{\mu_2}{}^{a_2}\hat{E}_{\mu_3}{}^{a_3}\hat{E}_{\mu_4}{}^{a_4}.$$

Now we have all ingredients to write down the *Einstein-Hilbert action* on $\mathcal{A}$ as

$$\hat{S}_{\text{EH}} := \int \det(\hat{E}_\mu{}^a)\hat{R} + \text{complex conj.}.$$

It is by construction invariant with respect to deformed diffeomorphisms meaning that

$$\hat{\delta}_\xi \hat{S}_{\text{EH}} = 0.$$

---

[4] We don't introduce a new symbol for the metric connection.

In this section we have presented the fundamentals of a noncommutative geometry on the algebra $\mathcal{A}$ and defined an invariant Einstein-Hilbert action. It is a deformation of the usual Einstein-Hilbert action. Using the star-product formalism it is possible to map the algebraic quantities to functions depending on commutative variables. Then it is possible to study explicitly deviations of the undeformed theory in orders of a deformation parameter [4, 2]. Very interesting is also to study a generalization of the above concepts to a more general class of noncommutative structures given by a twist [3]. This class contains in particular lattice-like spacetime algebras which may indeed provide a regularization of the field theory under consideration.

ACKNOWLEDGMENTS

I would like to thank the organizers of ERE2005 for a nice and interesting conference. The results presented here were obtained in collaboration with P. Aschieri, C. Blohmann, M. Dimitrijević, P. Schupp and J. Wess and would like to thank them for a fruitful collaboration.

REFERENCES

1. J. Wess, *Fortsch.Phys.* **49**, 377-385 (2001).
2. P. Aschieri, C. Blohmann, M. Dimitrijević, F. Meyer, P. Schupp and J. Wess, *Class. Quant. Grav.* **22**, 3511 (2005).
3. P. Aschieri, M. Dimitrijević, F. Meyer and J. Wess, *Noncommutative geometry and gravity*, hep-th/0510059.
4. F. Meyer, *Noncommutative Spaces and Gravity*, hep-th/0510188.
5. Chari, V. and Pressley, A., *A Guide to Quantum Groups*, Cambridge: University Press (1995).

Do Evanescent Modes Violate Relativistic Causality?

Günter Nimtz

II. Physikalisches Institut, Universität zu Köln, Zülpicher Str. 77, 50937 Köln, Germany

Abstract. Experiments have demonstrated that evanescent modes (photonic tunneling) violate the relativistic causality. Detectors made click receiving a tunneled signal in advance to receiving the vacuum traveled one. The primitive causality claiming only that effect follows cause is not violated. It is shown that evanescent modes are near field phenomena. Apparently, this phenomenon represents the exception of the rule of relativistic causality.

Keywords: Superluminal propagation, causality
PACS: 42.50.-p,03.65.Xp

INTRODUCTION

Einstein's II. postulate states that in vacuum c is the uppermost (signal) velocity. (We are interested only in the signal velocity which is in charge of the propagation of the cause to the subsequent effect.) Einstein wrote in his paper Ref. [1]: *and also introduce another postulate, which is only apparently irreconcilable with the former, namely, that light is always propagated in empty space with a definite velocity c which is independent of the state of motion of the emitting body.* Initiated by much ado about this postulate Sommerfeld and Brillouin Ref. [2] studied wave propagation in polarizable media. The quintessence of their investigations was that in such media the signal velocity does never exceed c. From now on it was assumed that Einstein's II. postulate valid for vacuum is representative for all environments, including evanescent modes and potential barriers Refs. [3, 4], for instance. However, recent experiments revealed that the near field phenomena evanescent modes perform superluminal velocities between cause and effect. Superluminal (faster than c) signal transmission by evanescent modes was shown by Enders and Nimtz already 1992 Ref. [5] and reproduced in different experiments later Ref. [6, 7, 8]. Evanescent modes are solutions of the Helmholtz equation and mathematical equivalent to the tunneling solutions of the Schrödinger equation.

WAVENUMBER AND SIGNAL VELOCITY

First some elementary quantities and relations of wave propagation are recalled. The refractive index n with which we are familiar from Snellius' law and the wave number k describing the wave propagation in space are given by the relations

$$n(\omega) = n'(\omega) - in''(\omega) \qquad (1)$$

$$k(\omega) = k_0 \cdot n(\omega) \qquad (2)$$
$$k_0 = 2\pi/\lambda_0 = 2\pi/(\lambda \cdot n), \qquad (3)$$

where n' and n'' are the real and imaginary parts of the refractive index due to polarization of matter, ω is the angular frequency and λ_0 is the vacuum wavelength. Both quantities, k and n, are in general complex functions of frequency. The imaginary parts describe the attenuation and amplification of waves. The attenuation may be caused either by dissipation or by reflection. In the case of an evanescent mode the refractive index $n(\omega)$ and the wave number $k(n)$ are purely imaginary in charge of the reflection.

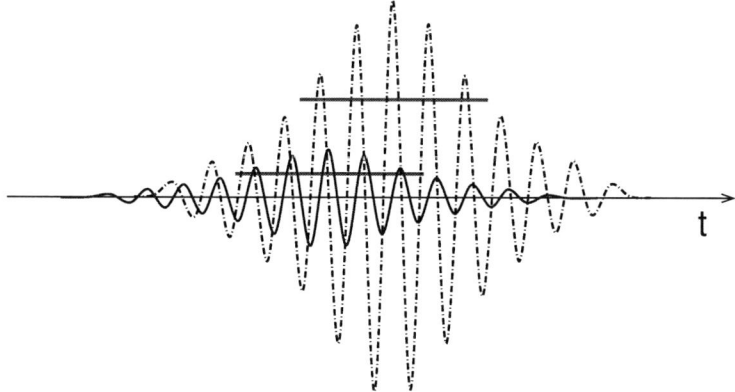

FIGURE 1. Sketch of the oscillations of two wave packets (i.e. pulses) vs time. The larger packet has traveled slower than the attenuated one. The horizontal bars indicate the half width of the packets, which do not depend on the packet's magnitude. The figure illustrates the gradual beginning of physical wave packets.

Now let us go to shortly discuss the velocities of waves. We are exclusively interested in the propagation of a cause, which is given by the signal velocity. The different quantities of propagation are made plausible by the sketch of two traveling wave packets displayed in Fig. 1. A signal cannot be sent with a harmonic traveling wave involving only a single frequency. If you want to send a message, you must modulate the wave. A signal can be decoded by a change in frequency, in amplitude, or in duration. The sketched waves are amplitude modulated packets representative for digital voltage pulses. The voltage oscillates with a frequency ν within a frequency band $\Delta \nu$. The pulses begin and end gradually with time. In consequence a physical signal has no well defined front. A well defined front and tail of a signal would presuppose an infinite frequency band width and then an infinite energy in consequence of $h\nu$. The general relationship that holds between the frequency bandwidth and the time duration is given by (see e.g. Ref. [9])

$$\Delta \nu \cdot \Delta t \geq 1, \qquad (4)$$

where $\Delta \nu$ and $\Delta t \ll \infty$. The phase velocity is given by the motion of a fixed point stuck to the oscillations. The group velocity is given by the speed of the maximum of the pulse

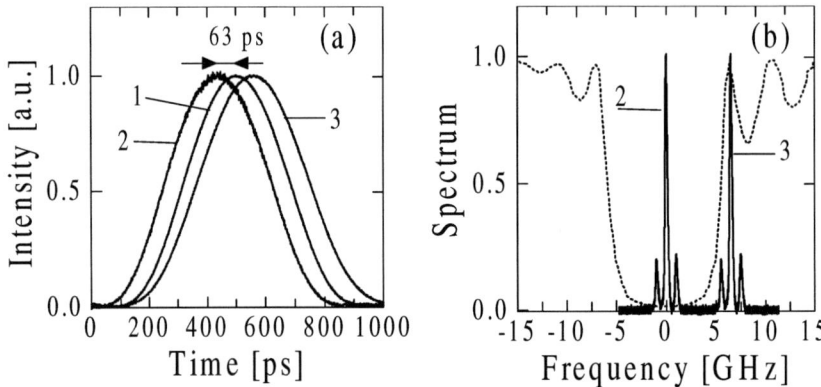

FIGURE 2. (a) Measured delay time of three digital signals [6] and spectrum of the photonic lattice transmission. Pulse trace 1 was recorded in vacuum. Pulse 2 traversed a photonic lattice in the center of the frequency band gap (see spectrum in part (b) of the figure), and pulse 3 was recorded for the pulse traveling through the fiber outside the forbidden band gap. The tunneling barrier was a photonic lattice of a quarter wavelength periodic dielectric hetero-structure fiber. The frequency zero point in part (b) corresponds to the infrared signal carrier frequency of $2 \cdot 10^{14}$ Hz and to the mid frequency of the forbidden gap of the lattice.

or of an other modulation. Phase and group velocities are equal in vacuum, but they may differ if traveling through interacting matter like glass. The signal velocity equals the group velocity as long as the dispersion is negligible, i.e. as long as the modulation is not reshaped. In strongly dispersive media with $n = n(\omega)$ group and signal velocities can be different and the signal may have lost its information, i.e. the cause. Here we are interested in the problem of causality, in cause and subsequent effect. Incidentally, a signal and then an effect can only be detected by its energy. In this respect signal and energy velocities are equal.

The notions on wave propagation are presented in text books, see Refs. [2, 9, 10], for instance. The group time delay is given by the derivative of the wave's phase φ

$$t_{group} = dx/dv_{group} = d(k \cdot x)/d\omega = d\varphi/d\omega \tag{5}$$

with the group velocity

$$v_{group} = \frac{x}{t_{group}} = \frac{x}{d\varphi/d\omega} = d\omega/dk, \tag{6}$$

The group time delay gives the time delay of a signal for traversing a distance x in the case of a negligible dispersion. In Fig. 2 an evanescent digital signal is shown which traveled along a fiber with a superluminal signal velocity of $2 \cdot c$.

MAXWELL AND SCHRÖDINGER EQUATIONS

For electromagnetic waves the propagation of waves are described by the Maxwell equations and for massive particles in the non–relativistic regime, by the Schrödinger equation. The Maxwell equations in media characterized by some refractive index $n = \sqrt{\mu\varepsilon}$ where μ and ε are the relative permeability and the relative permittivity, respectively, lead to the wave equation

$$-\nabla^2 \phi(\vec{x},t) + \frac{n^2}{c^2}\frac{\partial^2}{\partial t^2}\phi(\vec{x},t) = 0, \qquad (7)$$

ϕ being any component of the electrical and the magnetic fields. In vacuum characterized by $n = 1$ waves propagate with the velocity $c = (\mu_0\varepsilon_0)^{-1/2}$, where μ_0 and ε_0 are the permeability and the permittivity, respectively. If we describe phenomena periodic in time with frequency $\nu = \omega/(2\pi)$,

$$\phi(\vec{x},t) = \phi_x(\vec{x})e^{i\omega t}, \qquad (8)$$

then the wave equation reduces to the Helmholtz equation

$$\nabla^2 \phi_x(\vec{x}) + \frac{n^2\omega^2}{c^2}\phi_x(\vec{x}) = 0. \qquad (9)$$

As usual, this equation will be solved by a plane wave ansatz

$$\phi_a(\vec{x}) = \phi_0 e^{-i\vec{k}\cdot\vec{x}}, \qquad (10)$$

what leads to a relation between the wave number and the refractive index

$$k^2 = \frac{n^2\omega^2}{c^2} = k_0^2 n^2 = k_0^2 \varepsilon\mu. \qquad (11)$$

If k and then n are imaginary numbers the solution is called an evanescent mode. The imaginary wave number is usually expressed by the Greek letter κ.

Similar features can be found for the stationary Schrödinger equation

$$\nabla^2 \psi(\vec{x}) + \frac{2m}{\hbar^2}(E - U(\vec{x}))\psi(\vec{x}) = 0. \qquad (12)$$

where E is the energy of the stationary state, m is the mass of the particle and $U(\vec{x})$ is a position–dependent potential, the barrier potential, for example. This relation is mathematically equivalent to the Helmholtz equation Again, a plane wave ansatz yields for the wave number k

$$k^2 = \frac{2m}{\hbar^2}(E - U) = k_0^2 - \frac{2mU}{\hbar^2}, \qquad (13)$$

Particles in regions for which $E < U$, that is, inside the potential barrier, are quantum analogues of electromagnetic evanescent modes.

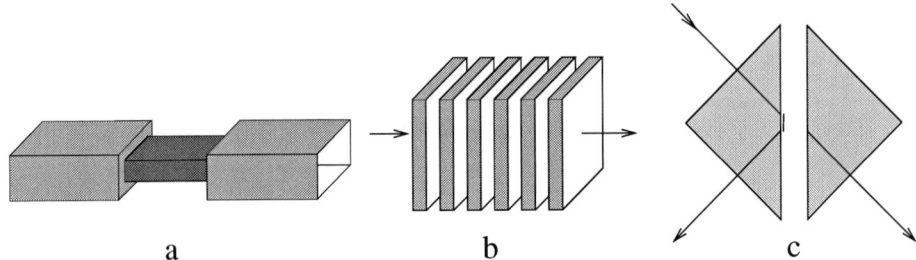

FIGURE 3. Sketch of three prominent photonic barriers. a) illustrates an undersized wave guide (the central part of the wave guide has a cross-section being smaller than half the wavelength in both directions perpendicular to propagation), b) a 1-dimensional photonic lattice (periodic dielectric hetero structure), and c) the frustrated total internal reflection of a double prism, where total reflection takes place at the boundary from a denser (the first prism with refractive index n_1) to a lesser dielectric medium (with refractive index n_2).

PHOTONIC BARRIERS: EXAMPLES OF EVANESCENT MODES

Prominent examples of evanescent modes are found a) in undersized wave guides (both dimensions of the guide cross section are smaller than half the vacuum wavelength), b) in the forbidden frequency bands of periodic dielectric hetero–structures (photonic lattice), and c) with double prisms in the case of frustrated total internal reflection (FTIR) [8, 11]. The examples are illustrated in Fig. 3. Dielectric lattices are analogous to electronic lattices of semiconductors with forbidden energy gaps. The square number of the imaginary refractive index n''^2 corresponds to a negative effective potential $E - U$ in the Schrödinger equation.

EVANESCENT MODES: A NEAR FIELD PHENOMENON

According to text books superluminal signal velocities are violating Einstein causality, implying that cause and effect can be interchanged [3, 4]. Actually, it can be shown for frequency band *unlimited groups* that the mathematically constructed signal front travels always at a velocity $\leq c$, and only the peak of the group has traveled with superluminal velocity Ref. [12]. The tunneled signal having non-evanescent frequency components is reshaped and its front tail has propagated at luminal velocity. Such an approach does not describe physical signals as those displayed in Figs. 1, 2 for instance. In this physical case the signal has gradually formed a front tail. A pulse reshaping did not happen due to the narrow frequency band and the complete envelope of the signal traveled at superluminal velocity. A physical signal can not be described by a Gauss function, because a signal is frequency band and time duration limited [9, 13]. As shown above for all physical signals the relation Eq. 4 holds, which, incidentally, is proportional to the information content of a signal as was shown by Shannon [13].

A signal can be detected only if its power is above the Johnson noise P_{JN}. The thermal noise was observed and measured by Johnson in 1928 and is theoretically elaborated by the Nyquist Theorem, see for instance [14]. The theorem is of great importance in experimental physics and electronics. It is concerned with the spontaneous thermal fluctuations of voltage across an electric circuit element. The theorem gives a quantitative expression for thermal noise power generated by a resistor in thermal equilibrium.

$$P_{JN} = kT\Delta f, \tag{14}$$

The relation gives an estimate for the near field extension of evanescent modes. The power $P(x)$ of a signal has to be detected. Then superluminal signal propagation is limited by the relationship, which gives the minimum tunneled signal power:

$$P(x) = P_0 e^{-2\kappa x} \geq kT\Delta f, \tag{15}$$

where P_0 is the incident power of the evanescent mode, κ is the imaginary wave number of the evanescent mode, x the length of the evanescent region, k the Boltzmann constant, T the temperature, and Δf the frequency range of the signal. For example an infrared signal source of 1 mW power, a carrier frequency of $2\ 10^{14}\ Hz$ (1.5 μ m wavelength), and an imaginary wave number in the barrier $\kappa = 115\,\text{m}^{-1}$ at a temperature of $T = 300\,\text{K}$. Thus the Johnson noise with $\approx 1\,\mu\text{W}$ limits a detectable near field up to 0.03 m, corresponding to about 20 000 wavelengths of this infrared digital signal and this special photonic barrier.

SUPERLUMINAL SIGNALS DO NOT VIOLATE PRIMITIVE CAUSALITY

Does superluminal signal velocity violate the principle of causality actually? The line of arguments showing how to manipulate the past in the case of superluminal signal velocities is illustrated in Fig. 4. There are displayed two frames of reference. In the first one lottery numbers are presented as points on the time coordinate with zero time duration. At $t = 0$ the counters are closed. Mary (A) sends the lottery numbers to her girl friend Susan (B) with a signal velocity of 4 c. Susan, moving in the second inertial system at a relative speed of 0.75 c, sends the numbers back at a speed of 2 c, to arrive in the first system of Mary at $t = -1$ s, thus in time to deliver the correct lottery numbers before the counters close at $t = 0$.

The time shift of a point on the time axis of reference system A into the past is given by [4, 15],

$$t_A = -\frac{L}{c} \cdot \frac{(v_r - c^2/v_s - c^2/v'_s + c^2 v_r/v_s v'_s)}{(c - cv_r/v'_s)}, \tag{16}$$

where L is the transmission length of the signal, v_r is the velocity between the two inertial systems A and B. The condition for the change of chronological order is $t_A < 0$, the time shift between the systems A and B. This interpretation assumes, however, a signal to be of zero time neglecting its duration. In the example with the lottery data, the signal was assumed to be a point in space-time. However, a physical signal has a finite duration like the pulses sketched along the time axis in Fig. 5. The general relationship for the bandwidth-time interval product of a signal, i.e. a packet of oscillations is given by Eq. 4. A zero time duration of a signal would require an infinite frequency bandwidth.

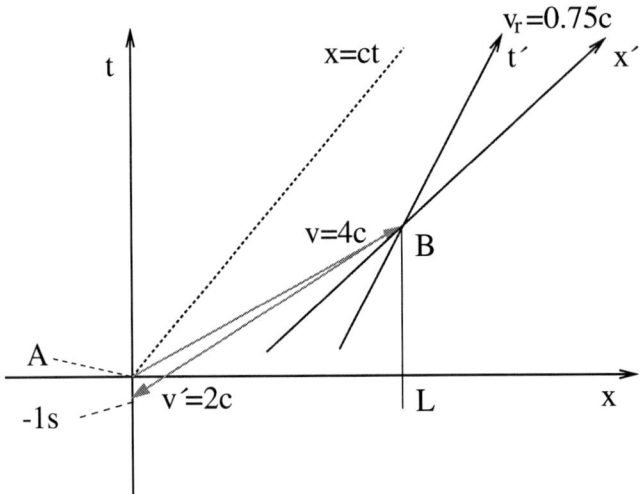

FIGURE 4. Coordinates of two inertial observers **A** $(0,0)$ and **B** with $O(x,t)$ and $O'(x',t')$ moving with a relative velocity of $0.75 \cdot c$. The distance L between **A** and **B** is 2 000 000 km. **A** makes use of a signal velocity $v_s = 4 \cdot c$ and **B** makes use of $v'_s = 2 \cdot c$ (in the sketch is $v \equiv v_s$). The numbers in the example are chosen arbitrarily. The signal returns -1 s in the past in **A**.

Taking into consideration the dispersion of the transmission of tunneling barriers, the frequency band of a signal has to be narrow in order to suppress signal reshaping. Assuming a signal duration of 4 s the complete information is obtained with superluminal signal velocity at 3 s at a positive time as illustrated in Fig. 5. The compulsory finite duration of all signals is the reason that a superluminal velocity does not violate the principle of causality. A shorter signal with the same information content would have an equivalently broader frequency bandwidth, compare Eq.(4). As a consequence, an increase of v_s or v'_s cannot violate the principle of causality.

For instance, the dispersion relation of FTIR $k \propto \omega$ elucidates this universal behavior: Assuming a wavelength $\lambda_0 = c/v$, a tunneling time $\tau = T = 1/v$, and a tunneling gap between the prisms $d = j\lambda_0$ ($j = 1, 2, 3, ...$) the superluminal signal velocity is $v_s = jc$, (remember the tunneling time is independent of barrier length). However, with increasing v_s the bandwidth Δv (that is the tolerated imaginary wave number width $\Delta \kappa$) of the signal decreases $\propto 1/d$ in order to guarantee the same amplitude distribution of all frequency components of the signal. In spite of an increasing superluminal signal velocity $v_s \to \infty$ the general causality cannot be violated because the signal time duration increases analogously $\Delta t \to \infty$ (Eq. 4).

SUMMING-UP

Experimental studies with evanescent modes displayed superluminal velocities between cause and effect thus violating the relativistic causality, but not the primitive causality. Evanescent modes and tunneling present a near-field effect.

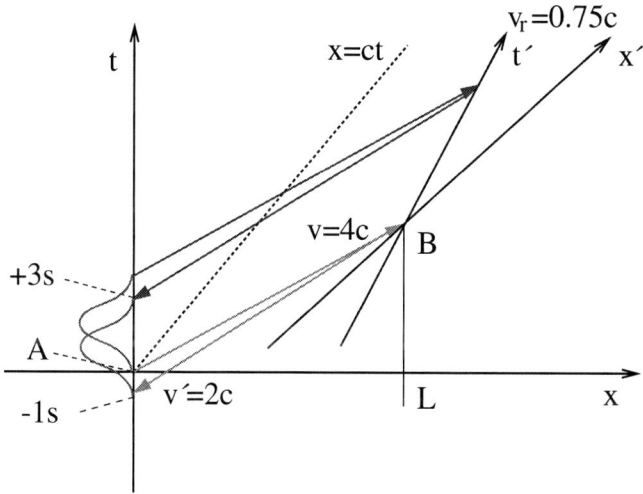

FIGURE 5. In contrast to Fig. 4 the pulse–like signal has now a finite duration of 4 s. This data is used for a clear demonstration of the effect. In all superluminal experiments, the signal length is long compared with the measured negative time shift. In this sketch the signal envelope ends in the future with 3 s (in the sketch is $v \equiv v_s$).

ACKNOWLEDGMENTS

The author thanks C. Laemmerzahl and A.A. Stahlhofen for useful discussions.

REFERENCES

1. A. Einstein, *Ann. Phys. (Leipzig)*, **17**, 891 (1905)
2. L. Brillouin, *Wave propagation and group velocity*, Academic Press, New York (1960)
3. M. Fayngold, *Special relativity and motions faster than light*, Wiley-VCH, Weinheim, (2002)
4. R. U. Sexl and H. K. Urbantke, *Relativity, Groups, Particles*, Springer, Wien, NewYork (2001)
5. A. Enders and G. Nimtz, *J.Phys. I, France*, **2**, 1693 (1992)
6. S. Longhi, M. Marano, P. Laporta, and M. Belmonte, *Phys. Rev.* **E64**, 055602 (2001)
7. S. Longhi, P. Laporta, M. Belmonte, and E. Recami, *Phys. Rev.* **E65**, 046610 (2002)
8. G. Nimtz, *Prog. Quantum Electronics* **27**, 417 (2003)
9. *berkeley physics course*, **3**, chap.6, McGraw-Hill, New York and London (1968)
10. A. Papoulis, *The Fourier Integral And Its Applications*, McGraw-Hill, New York, Secs. 7.5 and 7.6 (1962)
11. A. Haibel, G. Nimtz, and A. A. Stahlhofen, *Phys. Rev.* **E 63**, 047601 (2001)
12. Th. Emig, *Phys.Rev.* **E 54**, 5780 (1996)
13. C. E. Shannon, *Bell Sys. Tech. J.* **27**, 379, and 623 (1948)
14. C. Kittel, *Thermal Physics*, John Wiley & Sons, New York (1968), pp. 402-405
15. P. Mittelstaedt, *Eur. Phys. J* **B 13**, 353 (2000)

Holographic Dark Energy and Present Cosmic Acceleration

Diego Pavón[*] and Winfried Zimdahl[†]

[*]*Departamento de Física, Universidad Autónoma de Barcelona, 08193 Bellaterra, Spain*
[†]*Institut für Theoretische Physik, Universität zu Köln, D-50937 Köln, Germany*

Abstract. We review the notion of holographic dark energy and assess its significance in the light of the well documented cosmic acceleration at the present time. We next propose a model of holographic dark energy in which the infrared cutoff is set by the Hubble scale. The model accounts for the aforesaid acceleration and, by construction, is free of the cosmic coincidence problem.

Keywords: Cosmology, Holography, Late accelerated expansion, Dark energy
PACS: 98.80.Jk

INTRODUCTION

There is a growing conviction among cosmologists that the Universe is currently experiencing a stage of accelerated expansion not compatible with the up to now favored Einstein-de Sitter model [1]. According to the latter, the Universe should be now decelerating its expansion. This conviction is deeply rooted in observational grounds, mainly in the low brightness of high redshift supernovae type Ia which are fainter than allowed by the aforesaid model but consistent with accelerated models [2], as well as in other cosmological data. These include the position of the first acoustic peak of cosmic microwave background radiation (CMBR), which suggests that the Universe is spatially flat or nearly flat [3], combined with estimations of the amount of mass at cosmological scales -see e.g. [4]-, and correlations of the anisotropies of CMBR with large scale structures [5]. Overall, the data strongly hint at a Universe dominated by some form of energy -the so called, "dark energy"- that would contribute about 70 percent to the total energy density and nonrelativistic matter (dust) which would contribute the remaining 30 percent.

The trouble with dark energy is that we can only guess about its nature. To begin with, it must possess a huge negative pressure, at least high enough to violate the strong energy condition, something required (within general relativity) to drive accelerated expansion, and should cluster only at the highest accessible scales. The straightforward candidate is the cosmological constant, Λ, whose equation of state is simply $p_\Lambda = -\rho_\Lambda$, and whose energy is evenly distributed. Yet, it faces two serious drawbacks. On the one hand its quantum field theoretical value is about 123 orders of magnitude larger than observed; on the other hand, it entails the *coincidence problem*, namely: "Why are the vacuum and dust energy densities of precisely the same order today?" [6]. (Bear in mind that the energy density of dust redshifts with expansion as a^{-3}, where a denotes the scale factor of the Robertson–Walker metric). This is why a large variety of candidates -quintessence and tachyon fields, Chaplygin gas, phantom fields, etc.-, of varying plausibility, have been proposed in the last years -see Ref. [7] for reviews. Unluckily, however, there is not a clear winner in sight.

Recently, a new form of dark energy based on the holography notion and related to the existence of some or other cosmic horizon has been proposed [8]. Here, we present a specific model of holographic dark energy that accounts for the current stage of cosmic acceleration and is free from the coincidence problem that besets so many models of late acceleration [9]. The outline of this work is as follows. We first recall the notion of holography which is receiving growing attention and discuss possible choices for the infrared cutoff. Then we present our model of holographic dark energy. The last section is devoted to the conclusions and final remarks.

HOLOGRAPHY

We begin by recalling the notion of holography as introduced by 't Hooft [10] and Susskind [11]. Consider the world as three-dimensional lattice of spin-like degrees of freedom and assume that the distance between every two neighboring sites is some small length ℓ. Each spin can be in one of two sates. In a region of volume L^3 the number of quantum states will be $N(L^3) = 2^n$, with $n = (L/\ell)^3$ the number of sites in the volume, whence the entropy will be $S \propto (L/\ell)^3 \ln 2$. One would expect that if the energy density does not diverge, the maximum entropy varies as L^3, i.e., $S \sim L^3 \Lambda^3$, where $\Lambda \equiv \ell^{-1}$ is to be identified with the ultraviolet cutoff. However, the energy of most states described by this formula would be so large that they will collapse to a black hole of size in excess of L^3. Therefore, a reasonable guess is that in the quantum theory of gravity the maximum entropy should be proportional to the area, not the volume, of the system under consideration. (Bear in mind that the Bekenstein–Hawking entropy is $S_{BH} = A/(4\ell_{Pl}^2)$, where A is the area of the black hole horizon).

Consider now a system of volume L^3 of energy slightly below that of a black hole of the same size but with entropy larger than that of the black hole. By throwing in a very small amount of energy a black hole would result but with smaller entropy than the original system thus violating the second law of thermodynamics. As a consequence, Bekenstein proposed that the maximum entropy of the system should be proportional to its area rather than to its volume [12]. In keeping with this, 't Hooft conjectured that it should be possible to describe all phenomena within a volume by a set of degrees of freedom which reside on the surface bounding it. The number of degrees of freedom should be not larger than that of a two-dimensional lattice with about one binary degree of freedom per Planck area.

Holographic energy interpreted as dark energy

Inspired by these ideas, Cohen et al. [8] argued that an effective field theory that saturates the inequality

$$L^3 \Lambda^3 \leq S_{BH}, \tag{1}$$

necessarily includes many states with $R_s > L$, where R_s is the Schwarzschild radius of the system under consideration. Indeed, a conventional effective quantum field theory is expected to describe a system at temperature T provided that $T \leq \Lambda$. So long as $T \gg L^{-1}$, the energy and entropy will correspond to those of radiation ($E \simeq L^3 T^4$, and $S \simeq L^3 T^3$). When (1) is saturated (by setting $T = \Lambda$ in (1)) at $T \simeq m_{Pl}^{2/3} L^{-1/3}$, the Schwarzschild radius becomes $R_s \sim m_{Pl}^{2/3} L^{5/3} \gg L$.

Therefore it appears reasonable to propose a stronger constraint on the infrared (IR) cutoff L that excludes all states lying within R_s, namely:

$$L^3 \Lambda^4 \leq m_{Pl}^2 L \qquad (2)$$

(obviously, Λ^4 is the zero-point energy density associated to the short-distance cutoff). So, we conclude that $L \sim \Lambda^{-2}$ and $S_{max} \simeq S_{BH}^{3/4}$.

By saturating the inequality (2) -which is not compelling at all- and identifying Λ^4 with the holographic dark energy density we have

$$\rho_x = 3c^2 M_p^2 / L^2, \qquad (3)$$

where c^2 is a dimensionless constant and $M_p^2 \equiv (8\pi G)^{-1}$.

The infrared cutoff

Before building a cosmological model of late acceleration on the above ideas the IR cutoff must be specified. All the proposals in the literature identify L with the radius of one or another cosmic horizon. The simplest (and most natural) choice is the Hubble radius, H^{-1}. However, as shown by Hsu [13], this faces the following difficulty. For an isotropic, homogeneous and spatially flat universe dominated by nonrelativistic matter and dark energy the Friedmann equation $\rho_m + \rho_x = 3M_p^2 H^2$ together with $\rho_m \propto a^{-3}$ implies that ρ_x also redshifts as a^{-3}. In virtue of the conservation equation $\dot{\rho}_x + 3H(\rho_x + p_x) = 0$ it follows that p_x vanishes, i.e., there is no acceleration. So, this first choice seems doomed.

Two other, not so natural choices, are:
(i) $L = R_{ph}$ [14, 15], where

$$R_{ph} = a(t) \int_0^t \frac{dt'}{a(t')}$$

is the particle horizon. Yet, this option does not fare much better. Assuming the dark energy to dominate the expansion, Friedmann's equation reduces to $HR_{ph} = c$. Therefore, $H \propto a^{-(1+\frac{1}{c})}$ and consequently the equation of state parameter of the dark energy $w \equiv p_x/\rho_x = -(1/3) + (2/3c)$ is found to be larger than $-1/3$ whence this dark energy candidate does not violate the strong energy condition and cannot drive late acceleration either.

(ii) $L = R_H$ [16], where

$$R_H = a(t) \int_t^\infty \frac{dt'}{a(t')}$$

is the radius of the future event horizon, i.e., the boundary of the volume a given observer may eventually see. Assuming again the dark energy to dominate the expansion it is found that $w = -(1/3) - (2/3c) < -1/3$. Thus, this choice is compatible with accelerated expansion.

INTERACTING DARK ENERGY

This section focuses on our recent model of late acceleration based on three main assumptions, namely, (i) the dark energy density is given by Eq. (3), (ii) $L = H^{-1}$, and (iii) matter and holographic dark energy do not conserve separately but the latter decays into the former with rate Γ, i.e.,

$$\dot{\rho}_m + 3H\rho_m = \Gamma \rho_x, \qquad (4)$$
$$\dot{\rho}_x + 3H(1+w)\rho_x = -\Gamma \rho_x. \qquad (5)$$

As it can be checked, there is a relation connecting w to the ratio between the energy densities, $r \equiv \rho_m/\rho_x$, and Γ, namely, $w = -(1+r)\Gamma/(3rH)$, such that any decay of the dark energy $\Gamma > 0$ into pressureless matter implies a negative equation of state parameter, $w < 0$. It also follows that the ratio of the energy densities is a constant, $r_0 = (1-c^2)/c^2$, whatever Γ -see Ref. [9] for details.

In the particular case that $\Gamma \propto H$ one has $\rho_m, \rho_x \propto a^{-3m}$ and $a \propto t^n$ with $m = (1 + r_0 + w)/(1 + r_0)$ and $n = 2/(3m)$. Hence, there will be acceleration for $w < -(1+r_0)/3$. In consequence, the interaction is key to simultaneously solve the coincidence problem and have late acceleration. For $\Gamma = 0$ the choice $L = H^{-1}$ does not lead to acceleration. Before going any further, we wish to emphasize that models in which matter and ark energy interact with each other are well known in the literature -see [17] and references therein- and presently they are being contrasted with cosmological data [18].

Obviously, prior to the current epoch of accelerated expansion (during the radiation and matter dominated epochs) r must not have been constant but decreasing toward its current value r_0, otherwise the standard picture of cosmic structure formation would be irremediably spoiled (as usual, a subindex zero means present time). To incorporate this we must allow the parameter c^2 to vary with time. Hence, we now have

$$\dot{\rho}_x = -3H\left[1 + \frac{w}{1+r}\right]\rho_x + \frac{(c^2)^{\cdot}}{c^2}\rho_x. \qquad (6)$$

Combining it with the conservation equation (5) and contrasting the resulting expression with the evolution equation for r, namely,

$$\dot{r} = 3Hr\left[w + \frac{1+r}{r}\frac{\Gamma}{3H}\right], \tag{7}$$

yields $(c^2)^{\cdot}/c^2 = -\dot{r}/(1+r)$, whose solution is

$$c^2(t) = \frac{1}{1+r(t)}. \tag{8}$$

At sufficiently long times, $r \to r_0$ whence $c^2 \to c_0^2$.
In this scenario w depends also on c^2 according to

$$w = -\left(1 + \frac{1}{r}\right)\left[\frac{\Gamma}{3H} + \frac{(c^2)^{\cdot}}{3Hc^2}\right]. \tag{9}$$

Since the holographic dark energy must fulfil the dominant energy condition (and therefore it is not compatible with "phantom energy") [19], the restriction $w \geq -1$ sets constraints on Γ and c^2.

DISCUSSION AND CONCLUSIONS

The holographic dark energy seems to be a simple, reasonable and elegant alternative (within general relativity) to account for the present state of cosmic accelerated expansion. It can solve the coincidence problem provided that matter and holographic energy do not conserve separately. In this connection, it was pointed out by Das et al. [20], that because the interaction modifies the dependence of matter density on the scale factor the observers who endeavor to fit the observational data under the assumption of noninteracting matter will likely infer an effective w lower than -1. Therefore, most of the claims in favor of phantom energy may be considered as lending support to the dark energy–matter interaction. Yet, models of holographic dark energy must be further constrained by observations.

It should be noted that, contrary to what one may think, the infrared cutoff does not necessarily change when c^2 is varied. Indeed, the holographic bound can be expressed as $\rho_x \leq 3c^2 M_p^2/L^2$. Now, we first considered that it was saturated (i.e., the equality sign was assumed in the above expression) and that $L = H^{-1}$. Since the saturation of the bound is not at all compelling, and the "constant" $c^2(t)$ augments with expansion (as r decreases) up to attaining the constant value $(1 + r_0)^{-1}$, the expression $\rho_x = 3c^2(t)M_p^2H^2$, in reality, does not entail a modification of the infrared cutoff, which still is $L = H^{-1}$. What happens is that, as $c^2(t)$ grows, the bound gets progressively saturated up to full saturation when, asymptotically, c^2 becomes a constant. Put another way, the infrared cutoff stays $L = H^{-1}$ always, what changes is the degree of saturation of the holographic bound.

Before closing we would like to stress that there is no guarantee that the present accelerated epoch will not be followed by subsequent period of decelerated expansion. Models to that effect, partly motivated by string theory demands, have been advanced,

-see, e.g. [21]. If, indeed, the present epoch is followed by a decelerated one, then the future cosmic horizon will simply not exist and models of holographic dark energy based on that choice of the IR cutoff will be seen as essentially flawed.

ACKNOWLEDGMENTS

We thank the organizers of the XVIIIth edition of the "Encuentros Relativistas Españoles" for this opportunity. Our research was partly supported by the Spanish Ministry of Science and Technology under Grants BFM2003-06033 and BFM2000-1322.

REFERENCES

1. P.J.E. Peebles, *Principles of Physical Cosmology*, Princeton University Press, Princeton, 1993.
2. A.G. Riess, *et al.*, *Astron. J.*, **116**, 1009 (1998); S. Perlmutter, *et al.*, *Astrophys. J.*, **517**, 565 (1999); A.G. Riess, *et al.*, *Astrophys. J.*, **607**, 665 (2004).
3. D.N. Spergel *et al.*, *Astrophys. J. Suppl. Ser.*, **148**, 175 (2003).
4. P.J.E. Peebles, "Testing General Relativity on the Scales of Cosmology", in *Proceedings of the 17th International Conference on General Relativity and Gravitation* (in press), astro-ph/0410284; M. Tegmark, et al., *Phys. Rev. D*, **69** 103501 (2004); *ibid.*, *Astrophys. J.*, **606**, 702 (2004).
5. S. Boughn, and R. Crittenden, *Nature*, **427**, 45 (2003).
6. P.J. Steinhardt, "Cosmological Challenges for the 21st Century" in *Critical Problems in Physics*, edited by V.L. Fitch and D.R. Marlow, Princeton University Press, Priceton, 1997, pp. 123-144.
7. P.J.E. Peebles, *Rev. Mod. Phys.* **75**, 559 (2003); S. Carroll, "Why is the Universe Accelerating?" in *Measuring and Modelling the Universe*, Carnegie Observatory, Astrophysics Series, Vol. 2, edited by W.L. Freedman Cambridge University Press, Cambridge, 2004; T. Padmanbhan, Phys. Reports, **380**, 235 (2003); J.A.S. Lima, Braz. J. Phys., **34**, 194 (2004); V. Sahni, astro-ph/0403324; Proceedings of the I.A.P. Conference *On the Nature of Dark Energy*, edited by P. Brax, J. Martin and J-P. Uzan, Frontier Group, Paris, 2002; Proceedings of the IVth Marseille Cosmology Conference *Where Cosmology and Fundamental Physics Meet*, edited by V. Lebrun, S. Basa and A. Mazure, Frontier Group, Paris, 2004.
8. A.G. Cohen, D.B. Kaplan, and A.E. Nelson, *Phys. Rev. Lett.*, **82**, 4971 (1999).
9. D. Pavón and W. Zimdahl, *Phys. Lett. B*, **268**, 206 (2005).
10. G. 't Hooft, "Dimensional Reduction in Quantum Gravity", gr-qc/9311026.
11. L. Susskind, *J. Math. Phys. (N.Y.)*, **36**, 6377 (1995).
12. J.D. Bekenstein, *Phys. Rev. D*, **49**, 1912 (1994).
13. S.D.H. Hsu, *Phys. Lett. B*, **594**, 13 (2004).
14. W. Fischler and L. Susskind, "Holography and Cosmology", hep-th/9806039.
15. R. Bousso, *JHEP*, **9907**, 004 (1999).
16. M. Li, *Phys. Lett. B*, **603**, 1 (2004); Q.G. Huang, and M. Li, *JCAP* 08(2004)013.
17. L. Amendola, *Phys. Rev. D*, **62**, 043511 (2000); L.P. Chimento, A.S. Jakubi and D. Pavón, *Phys. Rev. D*, **62**, 062508 (2000); W. Zimdahl, D. Pavón and L.P.Chimento, *Phys. Lett. B*, **521**, 133 (2001); L.P. Chimento, A.S. Jakubi, D. Pavón and W. Zimdahl, *Phys. Rev. D*, **67**, 083513 (2003); G.R. Farrar, and P.J.E. Peebles, *Astrophys. J.* **604**, 1 (2004); S. del Campo, R. Herrera, and D. Pavón, *Phys. Rev. D*, **70**, 043540 (2004); S. del Campo, R. Herrera and D. Pavón, *Phys. Rev. D* **71**, 123529 (2005).
18. D. Pavón, S. Sen, and W. Zimdahl, *JCAP* 05(2004)009; G. Olivares, F. Atrio, and D. Pavón, *Phys. Rev. D*, **71**, 063523 (2005).
19. D. Bak and S-J. Rey, *Class. Quantum Grav.* **17**, L83 (2000); E.E. Flanagan, D. Marolf, and R.M. Wald, *Phys. Rev. D* **62**, 084035 (2000).
20. S. Das, P.S. Corasaniti, and J. Khoury, astro-ph/0510628.
21. M. Sami, and T. Padmanabhan, *Phys. Rev. D*, **67**, 083509 (2003); V. Sahni and Y. Shtanov, *JCAP* 11(2003)014; N. Bilić, G.B. Tupper and R. Viollier, *JCAP* 10(2005)003.

Quantum Gravity as Theory of "Superfluidity"

B.M. Barbashov*, V.N. Pervushin*, A.F. Zakharov[†,]* and V.A. Zinchuk*

*Joint Institute for Nuclear Research, 141980, Dubna, Russia
[†]National Astronomical Observatories of Chinese Academy of Sciences, Beijing 100012, China

Abstract. A version of the cosmological perturbation theory in general relativity (GR) is developed, where the cosmological scale factor is identified with spatial averaging of the metric determinant logarithm and the cosmic evolution acquires the pattern of a superfluid motion: the absence of "friction-type" interaction, the London-type wave function, and the Bogoliubov condensation of quantum universes. This identification keeps the number of variables of GR and leads to a new type of potential perturbations. A set of arguments is given in favor of that this "superfluid" version of GR is in agreement with the observational data.

Keywords: Cosmology, Quantum Gravity, CMB Radiation, Wheeler – De-Witt Equation, Bogoliubov Transformations
PACS: 04.60.-m, 98.80-k, 95.35+d, 9536+x, 98.80Qc

INTRODUCTION

Separation of the cosmological scale factor from metrics in general relativity (GR) is well known as the cosmological perturbation theory proposed by Lifshitz [1] and applied as a basic tool for analysis of modern observational data in astrophysics and cosmology [2]. However, the number of variables of the Lifshitz theory differs from GR. In the present paper, a new version of the cosmological perturbation theory [3] is discussed, where the logarithm of cosmological scale factor coincides with spatial averaging the metric determinant logarithm, so that the number of variables of GR is conserved. The conservation of the number of variables allows us to solve the energy constraint, fulfil the Dirac Hamiltonian reduction and quantization. This Quantum Gravity has some attributes of the theory of superfluidity of the type of Landau's "superfluid" dynamics [4], London's unique wave function [5], and Bogoliubov's squeezed condensate [6] and gives us a new possibility to explain the "CMBR primordial power spectrum" and other topical problems of modern cosmology.

"SUPERFLUID" MOTION AND POTENTIAL PERTURBATIONS

GR was given by Einstein and Hilbert 90 years ago with the *"dynamic"* action

$$S = \int d^4x \sqrt{-g} \left[-\frac{\varphi_0^2}{6} R(g) + \mathscr{L}_{(M)} \right], \tag{1}$$

where $\varphi_0^2 = \frac{3}{8\pi} M_{\text{Planck}}^2$ is the Newton constant, $\mathscr{L}_{(M)}$ is the matter Lagrangian, and an *"interval"* $g_{\mu\nu} dx^\mu dx^\nu \equiv \omega_{(\alpha)} \omega_{(\alpha)} = \omega_{(0)} \omega_{(0)} - \omega_{(1)} \omega_{(1)} - \omega_{(2)} \omega_{(2)} - \omega_{(3)} \omega_{(3)}$ presented here by the components of an orthogonal simplex of reference $\omega_{(\alpha)}$.

These simplex components in the reference frame of the Dirac – ADM Hamiltonian approach to GR [7] take the form

$$\omega_{(0)} = \psi^6 N_d dx^0, \qquad \omega_{(b)} = \psi^2 \mathbf{e}_{(b)i}(dx^i + N^i dx^0); \qquad (2)$$

here $\mathbf{e}_{(a)i}$ are triads of the spatial metrics with $\det|\mathbf{e}| = 1$, N_d is the Dirac lapse function, N^i is shift vector, and ψ is a determinant of the spatial metric. All these components were considered by Dirac as the potentials except of two transverse triads distinguished by the constraints $\partial_j e^j_{(a)} = 0$ and the zero determinant momentum $p_\psi = 0$. Latter contradicts to Friedmann-type evolution in the homogeneous approximation of GR, where the determinant component is identified with the cosmological scale factor ($\psi^2 \simeq a(x^0)$) and its momentum coincides with the Hubble parameter. The question appears about the status of the Hubble parameter in the Dirac Hamiltonian approach [7].

To answer this question we consider the cosmological perturbation theory [1, 2] $g_{\mu\nu} = a^2(x^0)\widetilde{g}_{\mu\nu}$ in terms of the Dirac – ADM variables, where one can get the new lapse function $\widetilde{N}_d = [\sqrt{-\widetilde{g}}\, \widetilde{g}^{00}]^{-1} = a^2 N_d$ and spatial determinant $\widetilde{\psi} = (\sqrt{a})^{-1}\psi$.

After the separation of the spatial metric determinant in the curvature $\sqrt{-g}R(g) = a^2\sqrt{-\widetilde{g}}R(\widetilde{g}) - 6a\partial_0\left[\widetilde{N}_d^{-1}\partial_0 a\right]$ the determinant part of the Lagrangian takes the form

$$L = -\varphi_0^2 \int d^3x \widetilde{N}_d \left[4a^2 (\overline{v_\psi})^2 + \left(\frac{\partial_0 a}{\widetilde{N}_d}\right)^2 \right] - \boxed{2\varphi_0^2 \partial_0 a^2 \int d^3x \overline{v_\psi}} + \ldots, \qquad (3)$$

where $\overline{v_\psi} = \left[(\partial_0 - N^l \partial_l)\log \widetilde{\psi} - \frac{1}{6}\partial_l N^l\right]/\widetilde{N}_d$ is a local velocity. One can see that this Lagrangian contains the velocity-velocity interaction distinguished by the box in Eq. (3). This interaction mixes the canonical momenta

$$P_a \equiv \frac{\partial L}{\partial(\partial_0 a)} = -\varphi_0^2 \int d^3x \left[\boxed{4a\, \overline{v_\psi}} + 2\frac{\partial_0 a}{\widetilde{N}_d}\right], \qquad (4)$$

$$\overline{p_\psi} = \frac{\partial \mathscr{L}}{\partial(\partial_0 \log \widetilde{\psi})} = -2\varphi_0^2 a \left[4a\overline{v_\psi} + \boxed{2\frac{\partial_0 a}{\widetilde{N}_d}}\right], \qquad (5)$$

so that the integral of the local momentum $\overline{p_\psi}$ coincides with the one of cosmic motion

$$\int d^3x \overline{p_\psi} = -2\varphi_0^2 a \int d^3x \left[4a\overline{v_\psi} + \boxed{2\frac{\partial_0 a}{\widetilde{N}_d}}\right] \equiv \boxed{2aP_a}, \qquad (6)$$

and the Hamiltonian approach is failure. The reason of that is the increase of the number of variables after the separation $a(x^0)$.

In order to keep the number of variables, we identify $\log\sqrt{a}$ with the spatial volume "averaging" of $\log\psi$: $\log\sqrt{a} = \langle\log\psi\rangle \equiv \int d^3x \log\psi / V_0$, where $V_0 = \int d^3x < \infty$ is a finite volume. In this case, the new determinant variable $\widetilde{\psi}$ and its velocity satisfy the identities

$$\int d^3x \log \widetilde{\psi} = \int d^3x [\log\psi - \langle\log\psi\rangle] \equiv 0, \qquad \int d^3x \overline{v_\psi} \equiv 0. \qquad (7)$$

This means that all the box terms in Eqs (3)– (6) disappear together with the "friction-type" velocity-velocity interaction. The momenta (4) and (5) are completely separated

$$P_a \equiv = -2V_0\varphi_0^2 \partial_0 a \langle (\widetilde{N}_d)^{-1} \rangle = -2V_0\varphi_0^2 \frac{da}{d\zeta} \equiv -2V_0\varphi_0^2 a', \qquad (8)$$

$$\overline{p_\psi} = \frac{\partial \mathscr{L}}{\partial(\partial_0 \log \widetilde{\psi})} = -8\varphi_0^2 a^2 \overline{v_\psi}, \qquad (9)$$

here the averaging $\langle(\widetilde{N}_d)^{-1}\rangle$ determines the time interval $d\zeta = \langle(\widetilde{N}_d)^{-1}\rangle^{-1} dx^0$ that is invariant with respect to reparametrizations of the coordinate evolution parameter x^0. In this case, the scale transformation $g_{\mu\nu} = a^2(x^0) \widetilde{g}_{\mu\nu}$ converts the GR action (1) into the "friction-free" one

$$S[\varphi_0] = \widetilde{S}[\varphi] - \int dx^0 (\partial_0 \varphi)^2 \int \frac{d^3x}{\widetilde{N}_d}, \qquad (10)$$

where $\widetilde{S}[\varphi]$ is the action (1) in terms of metrics $\widetilde{g}$ and the running scale $\varphi(x^0) = \varphi_0 a(x^0)$ of all masses including the Planck one φ_0. The energy constraint $\delta S[\varphi_0]/\delta \widetilde{N}_d = 0$ takes algebraic form [9] with respect to the invariant local lapse function $N_{\text{inv}} \equiv \widetilde{N}_d \langle (\widetilde{N}_d)^{-1} \rangle$

$$-\frac{\delta \widetilde{S}[\varphi]}{\delta \widetilde{N}_d} \equiv \widetilde{T}_0^0 = \left[\frac{\partial_0 \varphi}{\widetilde{N}_d}\right]^2 = \frac{1}{N_{\text{inv}}^2}\left[\frac{d\varphi}{d\zeta}\right]^2 = \frac{\varphi'^2}{N_{\text{inv}}^2}, \qquad (11)$$

where $\widetilde{T}_0^0$ is the local energy density. This equation has the resolution in both the local sector

$$N_{\text{inv}} \equiv \widetilde{N}_d \langle (\widetilde{N}_d)^{-1} \rangle = \left\langle \sqrt{\widetilde{T}_0^0} \right\rangle \left(\sqrt{\widetilde{T}_0^0}\right)^{-1} \qquad (12)$$

and the global one

$$\zeta(\varphi_0|\varphi) \equiv \int dx^0 \langle (\widetilde{N}_d)^{-1} \rangle^{-1} = \pm \int_\varphi^{\varphi_0} d\widetilde{\varphi} \left\langle \sqrt{\widetilde{T}_0^0(\widetilde{\varphi})} \right\rangle^{-1} \qquad (13)$$

where the invariant time ζ is connected with the scale factor by the Hubble-like law.
The explicit dependence of $\widetilde{T}_0^0$ on $\widetilde{\psi}$ was given by Lichnerowicz [10]

$$\widetilde{T}_0^0 = \widetilde{\psi}^7 \hat{\triangle} \widetilde{\psi} + \sum_I \widetilde{\psi}^I a^{\frac{I}{2}-2} \tau_I, \qquad \tau_I \equiv \langle \tau_I \rangle + \overline{\tau}_I, \qquad (14)$$

where $\hat{\triangle} \widetilde{\psi} \equiv \frac{4\varphi^2}{3} \partial_{(b)} \partial_{(b)} \widetilde{\psi}$ is the Laplace operator and τ_I is partial energy density marked by the index I running a set of values $I = 0$ (stiff), 4 (radiation), 6 (mass), 8 (curvature), 12 (Λ-term) in correspondence with a type of matter field contributions. The negative contribution $-(16/\varphi^2)\overline{p_\psi}^2$ of the spatial determinant momentum in the energy density $\tau_{I=0}$ can be removed by the Dirac constraint [7] of the zero velocity of the spatial volume element

$$\overline{p_\psi} = -8\varphi^2 \frac{1}{\widetilde{N}_d}\left[(\partial_0 - N^l \partial_l) \ln \widetilde{\psi} - \frac{1}{6}\partial_l N^l\right] = -8\varphi^2 \frac{\partial_0 \widetilde{\psi}^6 - \partial_l[\widetilde{\psi}^6 N^l]}{\widetilde{\psi}^6 \widetilde{N}_d} = 0. \qquad (15)$$

In the class of functions $\overline{F} = F - \langle F \rangle$, the classical equation $\delta S/\delta \log \widetilde{\psi} = 0$ takes the form

$$\widetilde{N}_d \widetilde{\psi} \frac{\partial \widetilde{T}_0^0}{\partial \widetilde{\psi}} + \widetilde{\psi} \triangle \left[\frac{\partial \widetilde{T}_0^0}{\partial \triangle \widetilde{\psi}} \widetilde{N}_d \right] = 0 \rightarrow \overline{7 N_{\text{inv}} \widetilde{\psi}^7 \widehat{\triangle} \widetilde{\psi}} + \overline{\widetilde{\psi} \widehat{\triangle} [N_{\text{inv}} \widetilde{\psi}^7]} + \sum_I \overline{I \widetilde{\psi}^I a^{\frac{I}{2} - 2} \tau_I} = 0$$

and gives both the determinant $\widetilde{\psi} = 1 + \overline{\mu}$ and the lapse function (12) $N_{\text{inv}} \widetilde{\psi}^7 = 1 - \overline{\nu}$. The first order of this equation determines $\overline{\mu}$ and $\overline{\nu}$ in the form of a sum [3]

$$\overline{\mu} = \frac{1}{2} \int d^3 y \left[D_{(+)}(x,y) \overline{T}_{(+)}^{(\mu)}(y) + D_{(-)}(x,y) \overline{T}_{(-)}^{(\mu)}(y) \right], \quad (16)$$

$$\overline{\nu} = \frac{1}{2} \int d^3 y \left[D_{(+)}(x,y) \overline{T}_{(+)}^{(\nu)}(y) + D_{(-)}(x,y) \overline{T}_{(-)}^{(\nu)}(y) \right], \quad (17)$$

where

$$\beta = \sqrt{1 + \lceil \langle \tau_{(2)} \rangle - 14 \langle \tau_{(1)} \rangle \rceil / (98 \langle \tau_{(0)} \rangle)},$$

$$\overline{T}_{(\pm)}^{(\mu)} = [1 \pm 49\beta] \overline{\tau_{(0)}} \mp 7\beta \overline{\tau_{(1)}}, \quad \overline{T}_{(\pm)}^{(\nu)} = [7 \pm (14\beta)^{-1}] \overline{\tau_{(0)}} - \overline{\tau_{(1)}} \quad (18)$$

are the local currents, $D_{(\pm)}(x,y)$ are the Green functions satisfying the equations

$$[\pm \widehat{m}_{(\pm)}^2 - \widehat{\triangle}] D_{(\pm)}(x,y) = \delta^3(x-y), \quad (19)$$

where $\widehat{m}_{(\pm)}^2 = 14(\beta \pm 1)\langle \tau_{(0)} \rangle \mp \langle \tau_{(1)} \rangle$, $\tau_{(n)} = \sum_I I^n a^{\frac{I}{2}-2} \tau_I$. In the case of point mass distribution in a finite volume V_0 with the zero pressure and the density $\overline{\tau_{(0)}}(x) = \overline{\tau_{(1)}}(x)/6 \equiv M \left[\delta^3(x-y) - 1/V_0 \right]$, solutions (16), (17) take a very interesting form

$$\widetilde{\psi} = 1 + \overline{\mu}(x) = 1 + \frac{r_g}{4r} \left[\gamma_1 e^{-m_{(+)}(z)r} + (1-\gamma_1) \cos m_{(-)}(z) r \right], \quad (20)$$

$$N_{\text{inv}} \widetilde{\psi}^7 = 1 - \overline{\nu}(x) = 1 - \frac{r_g}{4r} \left[(1-\gamma_2) e^{-m_{(+)}(z)r} + \gamma_2 \cos m_{(-)}(z) r \right], \quad (21)$$

where $\gamma_1 = \frac{1+7\beta}{2}$, $\gamma_2 = \frac{14\beta-1}{28\beta}$, $r_g = \frac{3M}{4\pi \varphi^2}$, $r = |x-y|$. The zero volume velocity (15) gives the diffeo-invariant shift of the coordinate origin

$$\langle (\widetilde{N}_d)^{-1} \rangle N^i = \left(\frac{x^i}{r} \right) \left(\frac{\partial_\zeta V}{\partial_r V} \right), \quad V(\zeta, r) = \int^r d\widetilde{r} \, \widetilde{r}^2 \widetilde{\psi}^6(\zeta, \widetilde{r}). \quad (22)$$

In the infinite volume limit $\langle \tau_{(n)} \rangle = 0$, $a = 1$ solutions (20) and (21) coincide with the isotropic version of the Schwarzschild solutions: $\widetilde{\psi} = 1 + \frac{r_g}{4r}$, $N_{\text{inv}} \widetilde{\psi}^7 = 1 - \frac{r_g}{4r}$, $N^k = 0$.

The main differences of the superfluid-type version of GR from the Lifshitz version [1, 2] are the potential perturbations of the scalar components $N_{\text{inv}}, \widetilde{\psi}$ instead of the kinetic ones and the nonzero shift vector $N^k \neq 0$ determined by Eq. (15). Recall that just the kinetic perturbations are responsible for the "primordial power spectrum" in the inflationary model [2]. The problem appears to describe CMBR by the potential perturbations given by Eqs. (16) – (22).

LONDON WAVE FUNCTION & BOGOLUIBOV CONDENSATION

The main consequence of the separation of the cosmological scale factor is the London-type globalization of the energy constraint (12). It fixes only the scale momentum P_φ

$$P_\varphi^2 \equiv [P_a/\varphi_0]^2 = E_\varphi^2 \equiv \left[2\int d^3x\sqrt{\widetilde{T}_0^0}\right]^2, \qquad (23)$$

in contrast to the energy constraint in GR without the cosmological scale factor. The "reduced" action can be obtained as values of the Hamiltonian action for the energy constraint $P_\varphi^2 = E_\varphi^2$ [9]:

$$S[\varphi_I|\varphi_0]|_{P_\varphi=\pm E_\varphi} = \int_{\varphi_I}^{\varphi_0} d\widetilde{\varphi} \left\{\int d^3x \left[\sum_F P_F \partial_\varphi F + C \mp 2\sqrt{\widetilde{T}_0^0(\widetilde{\varphi})}\right]\right\}, \qquad (24)$$

where C is the sum of all Dirac constraints [7, 9]. Momenta $P_{\varphi\pm} = \pm E_\varphi$ become the generator of evolution of all variables with respect to the evolution parameter φ [9] forward and backward, respectively. The negative energy problem can be solved by the primary quantization of the energy constraint $[P_\varphi^2 - E_\varphi^2]\Psi_u = 0$ and the secondary quantization $\Psi_u = (1/\sqrt{2E_\varphi})[A^+ + A^-]$ by the Bogoliubov transformation $A^+ = \alpha B^+ + \beta^* B^-$, in order to diagonalize the equations of motion by the condensation of "universes" $< 0|\frac{i}{2}[A^+A^+ - A^-A^-]|0> = R(\varphi)$ and describe cosmological creation of a "number" of universes $< 0|A^+A^-|0> = N(\varphi)$ from the stable Bogoliubov vacuum $B^-|0> = 0$. Vacuum postulate $B^-|0> = 0$ leads to an arrow of the invariant time $\zeta \geq 0$ (13) and its absolute point of reference $\zeta = 0$ at the moment of creation $\varphi = \varphi_I$ [3, 9]; whereas the Planck value of the running mass scale $\varphi_0 = \varphi(\zeta = \zeta_0)$ belongs to the present day moment ζ_0. The reduced action (24) shows us that the initial data at the beginning $\varphi = \varphi_I$ are independent of the present-day ones at $\varphi = \varphi_0$, therefore the proposal about an existence of the Planck epoch $\varphi = \varphi_0$ at the beginning [2] looks very doubtful.

W-,Z- FACTORY VERSUS THE PLANCK EPOCH

The low-energy expansion of the *"reduced action"* (24) over the field density T_{sm}

$$2d\varphi\sqrt{\widetilde{T}_0^0} = 2d\varphi\sqrt{\rho_0(\varphi) + T_{sm}} = d\varphi\left[2\sqrt{\rho_0(\varphi)} + T_{sm}/\sqrt{\rho_0(\varphi)}\right] + \ldots$$

gives the sum: $S^{(+)}|_{constraint} = S^{(+)}_{cosmic} + S^{(+)}_{field} + \ldots$, where $S^{(+)}_{cosmic}[\varphi_I|\varphi_0] = -2V_0 \int_{\varphi_I}^{\varphi_0} d\varphi\sqrt{\rho_0(\varphi)}$ is the reduced cosmological action and

$$S^{(+)}_{field} = \int_{\eta_I}^{\eta_0} d\eta \int d^3x \left[\sum_F P_F \partial_\eta F - T_{sm}\right] \qquad (25)$$

is the standard field action in terms of the conformal time: $d\zeta = d\eta = d\varphi/\sqrt{\rho_0(\varphi)}$ in the conformal flat space–time with running masses $m(\eta) = a(\eta)m_0$.

This low-energy expansion identifies the "conformal quantities" with the observable ones including the conformal time $d\eta$, instead of $dt = a(\eta)d\eta$, the coordinate distance r, instead of Friedmann one $R = a(\eta)r$, and the conformal temperature $T_c = Ta(\eta)$, instead of the standard one T. In this case, the cosmological redshift of the spectral lines

$$\frac{E_{\text{emission}}}{E_0} = \frac{m_{\text{atom}}(\eta_0 - r)}{m_{\text{atom}}(\eta_0)} \equiv \frac{\varphi(\eta_0 - r)}{\varphi_0} = a(\eta_0 - r) = \frac{1}{1+z}$$

is explained by the running masses $m = a(\eta)m_0$ in action (25) [11, 12, 13].

The conformal observable distance r loses the factor a, in comparison with the nonconformal one $R = ar$. In the units of "conformal quantities" the Supernova data [15, 16] are consistent with the dominance of the stiff (rigid) state, $\Omega_{\text{Rigid}} \simeq 0.85 \pm 0.15$, $\Omega_{\text{Matter}} = 0.15 \pm 0.15$ [12, 13]. If $\Omega_{\text{Rigid}} = 1$, we have the square root dependence of the scale factor on conformal time $a(\eta) = \sqrt{1 + 2H_0(\eta - \eta_0)}$. Just this time dependence of the scale factor on the measurable time (here – conformal one) is used for description of the primordial nucleosynthesis [17]. This stiff state is formed by a free scalar field when $E_\varphi = 2V_0\sqrt{\rho_0} = Q/\varphi$ and $Q/2V_0 = H_0\varphi_0^2 = H_I\varphi_I^2$ is an integral of motion.

These initial data φ_I and H_I can be determined by the parameters of matter cosmologically created from the vacuum at the beginning of a universe $\eta \simeq 0$.

The Standard Model (SM) density T_{sm} in action (25) shows us that W-, Z- vector bosons have maximal probability of this cosmological creation due to their mass singularity [14]. One can introduce the notion of a particle in a universe if the Compton length of a particle defined by its inverse mass $M_I^{-1} = (a_I M_W)^{-1}$ is less than the universe horizon defined by the inverse Hubble parameter $H_I^{-1} = a_I^2(H_0)^{-1}$ in the stiff state. Equating these quantities $M_I = H_I$ one can estimate the initial data of the scale factor $a_I^2 = (H_0/M_W)^{2/3} = 10^{-29}$ and the primordial Hubble parameter $H_I = 10^{29}H_0 \sim 1$ mm$^{-1} \sim 3K$. Just at this moment there is an effect of intensive cosmological creation of the vector bosons described in [14]; in particular, the distribution functions of the longitudinal vector bosons demonstrate us a large contribution of relativistic momenta. Their conformal (i.e. observable) temperature T_c (appearing as a consequence of collision and scattering of these bosons) can be estimated from the equation in the kinetic theory for the time of establishment of this temperature $\eta_{\text{relaxation}}^{-1} \sim n(T_c) \times \sigma \sim H$, where $n(T_c) \sim T_c^3$ and $\sigma \sim 1/M^2$ is the cross-section. This kinetic equation and values of the initial data $M_I = H_I$ give the temperature of relativistic bosons

$$T_c \sim (M_I^2 H_I)^{1/3} = (M_0^2 H_0)^{1/3} \sim 3K \tag{26}$$

as a conserved number of cosmic evolution compatible with the Supernova data [12, 15, 16]. We can see that this value is surprisingly close to the observed temperature of the CMB radiation $T_c = T_{\text{CMB}} = 2.73$ K.

The primordial mesons before their decays polarize the Dirac fermion vacuum (as the origin of axial anomaly) and give the baryon asymmetry frozen by the CP-violation. The value of the baryon–antibaryon asymmetry of the universe following from this axial anomaly was estimated in [14] in terms of the coupling constant of the superweak-interaction

$$n_b/n_\gamma \sim X_{CP} = 10^{-9}. \tag{27}$$

The boson life-times $\tau_W = 2H_I\eta_W \simeq (2/\alpha_W)^{2/3} \simeq 16$, $\tau_Z \sim 2^{2/3}\tau_W \sim 25$ determine the present-day visible baryon density

$$\Omega_b \sim \alpha_W = \alpha_{QED}/\sin^2\theta_W \sim 0.03. \tag{28}$$

All these results (26) – (28) testify to that all visible matter can be a product of decays of primordial bosons, and the observational data on CMBR can reflect parameters of the primordial bosons, but not the matter at the time of recombination. In particular, the length of the semi-circle on the surface of the last emission of photons at the life-time of W-bosons in terms of the length of an emitter (i.e. $M_W^{-1}(\eta_L) = (\alpha_W/2)^{1/3}(T_c)^{-1}$) is $\pi \cdot 2/\alpha_W$. It is close to $l_{min} \sim 210$ of CMBR, whereas $(\triangle T/T)$ is proportional to the inverse number of emitters $(\alpha_W)^3 \sim 10^{-5}$.

The temperature history of the expanding universe copied in the "conformal quantities" looks like the history of evolution of masses of elementary particles in the cold universe with the conformal temperature $T_c = a(\eta)T$ of the cosmic microwave background. In the superfluid version of cosmology [12], the dominance of stiff state $\Omega_{Stiff} \sim 1$ determines the parameter of spatial oscillations $\hat{m}_{(-)}^2 = \frac{6}{7}H_0^2[\Omega_R(z+1)^2 + \frac{9}{2}\Omega_{Mass}(z+1)]$. The values of red shift in the recombination epoch $z_r \sim 1100$ and the clusterization parameter $r_{clustering} = \dfrac{\pi}{\hat{m}_{(-)}} \sim \dfrac{\pi}{H_0\Omega_R^{1/2}(1+z_r)} \sim 130\,\text{Mpc}$ recently discovered in studies of a large scale periodicity in redshift distribution [18] lead to a reasonable value of the radiation-type density $10^{-4} < \Omega_R \sim 3 \cdot 10^{-3} < 5 \cdot 10^{-2}$ at the time of this epoch.

CONCLUSIONS

The conservation of the number of variables of GR after the separation of the cosmological scale factor from all fields leads to the Hamiltonian version of GR where the cosmic evolution acquires the pattern of a superfluid motion without the friction-type interaction. In this version the scale factor is an evolution parameter in the "field space of events" $[\varphi|\widetilde{F}^{(n)}]$, and its canonical momentum (i.e. the Hubble "parameter") plays the role of the generator of evolution of the fields $\widetilde{F}^{(n)}$. The values of the scale factor momentum for solutions of the equations of motion can be called the "reduced energies".

The solution of the problem of the "negative reduce energy" by the primary quantization and the secondary one (on the analogy of the pathway passed by QFT in the 20th century) reveals in GR other attributes of the theory of superfluid quantum liquid: London-type WDW wave function and Bogoliubov-type condensate of quantum universes. The postulate of the quantum Bogoliubov vacuum as the state with the minimal "energy" leads to the absolute beginning of geometric time.

The Hamiltonian approach leads to potential perturbations of the scalar metric components in contrast to the standard cosmological perturbation theory [1] keeping only the kinetic perturbations which are responsible for the "primordial power spectrum" in the inflationary model [2]. The Quantum Gravity considered as the theory of superfluidity gives us a possibility to explain this "spectrum" and other topical problems of cosmology by the cosmological creation of the primordial W-, Z- bosons from vacuum, when their Compton length coincides with the universe horizon.

The correspondence principle as the low-energy expansion of the reduced action identifies the conformal quantities with the "measurable" ones, and the uncertainty principle establishes the point of the beginning of the cosmological creation of the primordial W-, Z- bosons from vacuum due to their mass singularity at the moment $a_I^2 \simeq 10^{-29}, H_I^{-1} \simeq 1$ mm. In this case, the equations describing the longitudinal vector bosons in SM are close to the equations of the inflationary model used for description of the "power primordial spectrum" of the CMB radiation [2]. We listed the set of theoretical and observational arguments in favor of that the CMB radiation can be a final product of primordial vector W-, Z- bosons cosmologically created from the vacuum.

ACKNOWLEDGMENTS

The authors are grateful to A.A. Gusev, A.V. Efremov, E.A. Kuraev, V.V. Nesterenko, V.B. Priezzhev, and S.I. Vinitsky for interesting and critical discussions. AFZ is grateful to the National Natural Science Foundation of China (NNSFC) (Grant # 10233050) for a partial financial support.

REFERENCES

1. E.M. Lifshitz, *ZhETF*, **16**, 587–602 (1946); J.M. Bardeen, *Phys.Rev. D*, **22**, 1882–1905 (1980).
2. V. F. Mukhanov, H. A. Feldman, and R. H. Brandenberger, *Phys. Rep.* **215**, 206–333 (1992).
3. V.N. Pervushin and V.A. Zinchuk, gr-qc/0504123;
   B.M. Barbashov, V.N. Pervushin, A.F. Zakharov, and V.A. Zinchuk, gr-qc/0509006.
4. L.D. Landau, *ZhETF*, **11**, 592–614 (1941).
5. F. London, *Nature*, **141**, 643–645 (1938).
6. N.N. Bogoliubov, *J. Phys.*, **11**, 23–32 (1947).
7. P.A.M. Dirac, *Proc. Roy. Soc.*, **A 246**, 333–344 (1958); P.A.M. Dirac, *Phys. Rev.*, **114**, 924–930 (1959); R. Arnovitt, S. Deser, and C.W. Misner, *Phys. Rev.*, **117**, 1595–1602 (1960).
8. A.L. Zelmanov, *Dokl. AN USSR*, **107**, 815–818 (1956); *Dokl. AN USSR*, **209**, 822–825 (1973).
9. M. Pawlowski and V.N. Pervushin, *Int. J. Mod. Phys.*, **16**, 1715–1742 (2001); [hep-th/0006116]; B.M. Barbashov, V.N. Pervushin, and D.V. Proskurin, *Theor. Math.Phys.*, **132**, 1045–1058 (2002).
10. A. Lichnerowicz, *Journ. Math. Pure and Appl.*, **23**, 37 – 47 (1944).
11. J.V. Narlikar, *Introduction to Cosmology*, Jones and Bartlett, Boston, 1983.
12. D. Behnke, D.B. Blaschke, V.N. Pervushin, and D.V. Proskurin, *Phys. Lett. B*, **530**, 20–26 (2002); [gr-qc/0102039].
13. D. Behnke, *PhD Thesis*, Rostock Report MPG-VT-UR 248/04 (2004)
14. D.B. Blaschke *et al.*, *Physics of Atomic Nuclei*, **67**, 1050 – 1062 (2004); [hep-ph/0504225].
15. A.G. Riess *et al.*, *Astron. J.*, **116**, 1009–1038 (1998); S. Perlmutter *et al.*, *Astrophys. J.*, **517**, 565–586 (1999); A.G. Riess *et al.*, *Astrophys. J.*, **607**, 665–687 (2004).
16. A.G. Riess *et al.*, *Astrophys. J.*, **560**, 49–71 (2001).
17. S. Weinberg, *First Three Minutes. A Modern View of the Origin of the Universe*, Basic Books, Inc., Publishers, New-York, 1977.
18. W.J. Cocke and W.G. Tifft, *Astrophys. J.*, **368**, 383–389 (1991).

Second-Order Symmetric Lorentzian Manifolds

José M. M. Senovilla

Física Teórica, Universidad del País Vasco, Apartado 644, 48080 Bilbao, Spain.

Abstract. Spacetimes with vanishing second covariant derivative of the Riemann tensor are studied. Their existence, classification and explicit local expression are considered. Related issues and open questions are briefly commented.

Keywords: Symmetric spaces, curvature invariants, parallel null vector fields, Mp-waves.
PACS: 02.40.Ky, 04.20.-q, 04.50.+h

INTRODUCTION

Our aim is to characterize, as well as to give a full list of, the n-dimensional manifolds $\mathscr{V}$ with a metric g of Lorentzian signature such that the Riemann tensor $R^\alpha{}_{\beta\gamma\delta}$ of $(\mathscr{V}, g)$ *locally* satisfies the second-order condition

$$\nabla_\mu \nabla_\nu R^\alpha{}_{\beta\gamma\delta} = 0. \tag{1}$$

It is quite surprising that, hitherto, despite their simple definition, this type of Lorentzian manifolds have been hardly considered in the literature. Probably this is due to the classical results concerning these manifols in the proper Riemannian case, to the difficulties arising in other signatures, and to the little reward: only very special cases survive.

Apart from their obvious mathematical interest, from a physical point of view they are relevant in several respects: as a second local approximation to any spacetime (using for instance expansions in normal coordinates); as examples with a finite number of terms in Lagrangians; as interesting exact solutions for supergravity/superstring or M-theories; for invariant classifications; for solutions with parallel vector fields or spinors.

A more complete treatment, with a full list of references, is given in [23].

SYMMETRIC SPACES AND ITS GENERALIZATIONS

Semi-Riemannian manifolds satisfying (1) are a direct generalization of the classical locally *symmetric* spaces which satisfy

$$\nabla_\mu R^\alpha{}_{\beta\gamma\delta} = 0. \tag{2}$$

These were introduced, studied and classified by E. Cartan [10] in the proper Riemannian case[1], see e.g. [11, 16, 15], and later in [6, 9, 7] for the Lorentzian and general semi-

---
[1] With a positive-definite metric.

TABLE 1. The hierarchy of conditions on the Riemann tensor

$R^\alpha{}_{\beta\gamma\nu} \propto \delta^\alpha_\gamma g_{\beta\nu} - \delta^\alpha_\nu g_{\beta\gamma}$	$\nabla_\mu R^\alpha{}_{\beta\gamma\delta} = 0$	$\nabla_\mu \nabla_\nu R^\alpha{}_{\beta\gamma\delta} = 0$	$\nabla_{[\mu} \nabla_{\nu]} R^\alpha{}_{\beta\gamma\delta} = 0$
constant curvature	symmetric	2-symmetric	semisymmetric

Riemannian cases—see e.g. [8, 19] and references therein. They are themselves generalizations of the constant curvature spaces and, actually, there is a hierarchy of conditions, shown in Table 1, that can be placed on the curvature tensor. In the table, the restrictions on the curvature tensor decrease towards the right and each class is *strictly* contained in the following ones. The table has been stopped at the level of semi-symmetric spaces, defined by the condition[2] $\nabla_{[\mu} \nabla_{\nu]} R^\alpha{}_{\beta\gamma\delta} = 0$, which were introduced also by Cartan [11] and studied in [24, 25] as the natural generalization of symmetric spaces for the proper Riemannian case—see also [4] and references therein.

Why was semisymmetry considered to be the natural generalization of local symmetry? And, why not going further on to higher derivatives of the Riemann tensor? The answer to both questions is actually the same: a classical theorem [17, 18, 27] states that in any proper Riemannian manifold

$$\nabla_{\mu_1} \ldots \nabla_{\mu_k} R^\alpha{}_{\beta\gamma\delta} = 0 \iff \nabla_\mu R^\alpha{}_{\beta\gamma\delta} = 0 \quad (3)$$

for any $k \geq 1$ so that, in particular, (1) is strictly equivalent to (2) in proper Riemannian spaces. This may well be the reason why there seems to be no name for the condition (1) in the literature. However, an analogous condition has certainly been used for the so-called *k*-recurrent spaces [26, 12]; thus, I will call the spaces satisfying (1) *second-order symmetric*, or in short *2-symmetric*, and more generally *k-symmetric* when the left condition in (3) holds—see [23] for further details.

Results at *generic* points

As a matter of fact, the equivalence (3) holds as well in "generic" cases of semi-Riemannian manifolds of any signature. For some results on this one can consult [27, 12]. By "generic point" the following is meant: any point $p \in \mathscr{V}$ where the matrix $(R^{\alpha\beta}{}_{\gamma\delta})|_p$ of the Riemann tensor, considered as an endomorphism on the space of 2-forms $\Lambda_2(p)$, is non-singular. Then, for instance one can prove the following general result, see [23] for a proof.

Proposition 1 *For any tensor field T, and at* generic *points, one has*

$$\overbrace{\nabla \cdots \cdots \nabla}^{k} T = 0 \iff \nabla T = 0$$

for any $k \geq 1$.

---

[2] (Square) round brackets enclosing indices indicate (anti-)symmetrization, respectively.

Of course, these results apply in particular to the Riemann tensor, and in fact sometimes even stronger results can be proven. For instance, one can prove a conjecture in [12], namely, that all k-symmetric (and also all k-recurrent) spaces are necessarily of constant curvature on a neighbourhood of any generic $p \in \mathcal{V}$. As a matter of fact, a slightly more general result is proven in [23]:

Theorem 1 *All semi-symmetric spaces are of constant curvature at generic points.*

Therefore, there is little room for spaces (necessarily of non-Euclidean signature) which are k-symmetric but *not* symmetric nor of constant curvature. It is remarkable that there have been many studies on 2-recurrent spaces, but surprisingly enough the assumption that they are *not* 2-symmetric has always been, either implicitly or explicitly, made. The paper [23] tries to fill in this gap for the case of 2-symmetry and Lorentzian signature.

LORENTZIAN 2-SYMMETRY

To deal with the problem of k-symmetric and k-recurrent spaces one needs to combine several different techniques. Among them (i) pure classical standard tensor calculus by using the Ricci and Bianchi identities; (ii) study of *parallel* (also called covariantly constant) tensor and vector fields, and their implications on the manifold holonomy structure; and (iii) consequences on the curvature invariants. I now present the main points and results needed to reach the sought results. It turns out that the so-called "superenergy" and causal tensors [22, 3] are very useful, providing positive quantities associated to tensors that can be used to replace the ordinary positive-definite metric available in proper Riemannian cases.

Identities in 2-symmetric semi-Riemannian manifolds

Of course, some tensor calculation is obviously needed, mainly to prove some helpful quadratic identities. To start with, one needs a generalization of Proposition 1 to the case of non-generic points.

Lemma 1 *Let $(\mathcal{V}, g)$ be an n-dimensional 2-symmetric semi-Riemannian manifold of any signature. If $\nabla_\lambda \nabla_\mu T_{\mu_1 \ldots \mu_q} = 0$ then*

$$\sum_{i=1}^{q} \nabla_\nu R^\rho{}_{\alpha_i \lambda \mu} T_{\alpha_1 \ldots \alpha_{i-1} \rho \alpha_{i+1} \ldots \alpha_q} - R^\rho{}_{\nu \lambda \mu} \nabla_\rho T_{\alpha_1 \ldots \alpha_q} = 0, \quad (4)$$

$$(\nabla_\nu R^\rho{}_{\tau \lambda \mu} + \nabla_\tau R^\rho{}_{\nu \lambda \mu}) \nabla_\rho T_{\mu_1 \ldots \mu_q} = 0, \quad (5)$$

$$(\nabla_\nu R^\rho{}_\mu - \nabla_\mu R^\rho{}_\nu) \nabla_\rho T_{\mu_1 \ldots \mu_q} = 0, \quad (\nabla^\rho R_{\mu\nu} - 2\nabla_\nu R^\rho{}_\mu) \nabla_\rho T_{\mu_1 \ldots \mu_q} = 0. \quad (6)$$

By using the decomposition of the Riemann tensor,

$$R_{\alpha\beta\lambda\mu} = C_{\alpha\beta\lambda\mu} + \frac{2}{n-2}\left(R_{\alpha[\lambda}g_{\mu]\beta} - R_{\beta[\lambda}g_{\mu]\alpha}\right) - \frac{R}{(n-1)(n-2)}\left(g_{\alpha\lambda}g_{\beta\mu} - g_{\alpha\mu}g_{\beta\lambda}\right) \quad (7)$$

a selection of the formulas satisfied in 2-symmetric manifolds are given next

Lemma 2 *The Riemann, Ricci and Weyl tensors of any n-dimensional 2-symmetric semi-Riemannian manifold of any signature satisfy*

$$R^{\rho}{}_{\alpha\lambda\mu}R_{\rho\beta\gamma\delta} + R^{\rho}{}_{\beta\lambda\mu}R_{\alpha\rho\gamma\delta} + R^{\rho}{}_{\gamma\lambda\mu}R_{\alpha\beta\rho\delta} + R^{\rho}{}_{\delta\lambda\mu}R_{\alpha\beta\gamma\rho} = 0 \quad (8)$$

$$R^{\rho}{}_{\nu\lambda\mu}\nabla_{\rho}R_{\alpha\beta\gamma\delta} + R^{\rho}{}_{\alpha\lambda\mu}\nabla_{\nu}R_{\rho\beta\gamma\delta} + R^{\rho}{}_{\beta\lambda\mu}\nabla_{\nu}R_{\alpha\rho\gamma\delta} +$$
$$+ R^{\rho}{}_{\gamma\lambda\mu}\nabla_{\nu}R_{\alpha\beta\rho\delta} + R^{\rho}{}_{\delta\lambda\mu}\nabla_{\nu}R_{\alpha\beta\gamma\rho} = 0 \quad (9)$$

$$\nabla_{(\tau}R^{\rho}{}_{\nu)\lambda\mu}\nabla_{\rho}R_{\alpha\beta\gamma\delta} = 0, \; \nabla_{(\tau}R^{\rho}{}_{\nu)\lambda\mu}\nabla_{\rho}C_{\alpha\beta\gamma\delta} = 0, \; \nabla_{(\tau}R^{\rho}{}_{\nu)\lambda\mu}\nabla_{\rho}R_{\alpha\beta} = 0, \quad (10)$$

$$R_{\rho(\mu}R^{\rho}{}_{\nu)\alpha\beta} = 0, \; R^{\rho}{}_{\mu[\alpha\beta}R_{\gamma]\rho} = 0, \; C^{\rho}{}_{\mu[\alpha\beta}R_{\gamma]\rho} = 0, \; R^{\rho\sigma}R_{\rho\mu\sigma\nu} = R_{\mu}{}^{\rho}R_{\rho\nu}, \quad (11)$$

$$R^{\rho}{}_{\alpha\lambda\mu}C_{\rho\beta\gamma\delta} + R^{\rho}{}_{\beta\lambda\mu}C_{\alpha\rho\gamma\delta} + R^{\rho}{}_{\gamma\lambda\mu}C_{\alpha\beta\rho\delta} + R^{\rho}{}_{\delta\lambda\mu}C_{\alpha\beta\gamma\rho} = 0, \quad (12)$$

$$(n-2)\left(C_{\rho[\alpha}{}^{\lambda\mu}C^{\rho}{}_{\beta]\gamma\delta} + C_{\rho[\gamma}{}^{\lambda\mu}C^{\rho}{}_{\delta]\alpha\beta}\right) - 2\left(R_{[\alpha}{}^{[\lambda}C^{\mu]}{}_{\beta]\gamma\delta} + R_{[\gamma}{}^{[\lambda}C^{\mu]}{}_{\delta]\alpha\beta}\right) -$$
$$-2\left(R_{\rho}{}^{[\lambda}\delta^{\mu]}_{[\alpha}C^{\rho}{}_{\beta]\gamma\delta} + R_{\rho}{}^{[\lambda}\delta^{\mu]}_{[\gamma}C^{\rho}{}_{\delta]\alpha\beta}\right) + 2\frac{R}{n-1}\left(\delta^{[\lambda}_{[\alpha}C^{\mu]}{}_{\beta]\gamma\delta} + \delta^{[\lambda}_{[\gamma}C^{\mu]}{}_{\delta]\alpha\beta}\right) = 0 \quad (13)$$

and their non-written traces, such as the appropriate specializations of (6). Actually, (8) and (11-13) are valid in arbitrary semi-symmetric *spaces.*

Holonomy and reducibility in Lorentzian manifolds

Some basic lemmas on local holonomy structure are also essential. The classical result here is the de Rham decomposition theorem [20, 16] for positive-definite metrics. However, this theorem does not hold as such for other signatures, and one has to introduce the so-called *non-degenerate reducibility* [28, 29, 30]. See also [1] for the particular case of Lorentzian signature. To fix ideas, recall that the holonomy group [16] of $(\mathscr{V},g)$ is called reducible (when acting on the tangent spaces) if it leaves a non-trivial subspace of $T_p\mathscr{V}$ invariant. And it is called non-degenerately reducible if it leaves a non-degenerate subspace (that is, such that the restriction of the metric is non-degenerate) invariant.

Only a simple result is needed. This relates the existence of parallel tensor fields to the holonomy group of the manifold in the case of Lorentzian signature. It is a synthesis (adapted to our purposes) of the results in [14] but generalized to arbitrary dimension n (see [23] for a proof):

Lemma 3 *Let $D \subset \mathscr{V}$ be a simply connected domain of an n-dimensional Lorentzian manifold $(\mathscr{V},g)$ and assume that there exists a non-zero parallel symmetric tensor field $h_{\mu\nu}$ not proportional to the metric. Then (D,g) is reducible, and further it is not non-degenerately reducible only if there exists a null parallel vector field which is the unique parallel vector field (up to a constant of proportionality).*

Some important remarks are in order here:

1. If there is a parallel 1-form v_μ, then so is obviously $h_{\mu\nu} = v_\mu v_\nu$ and the manifold (arbitrary signature) is reducible, the Span of v^μ being invariant by the holonomy

group. If v_μ is *not* null, then $(\mathscr{V}, g)$ is actually *non-degenerately* reducible. In this case, the metric can be decomposed into two orthogonal parts as $g_{\mu\nu} = c v_\mu v_\nu + (g_{\mu\nu} - c v_\mu v_\nu)$, where $c = 1/(v^\mu v_\mu)$ is constant. Thus, necessarily $g_{\mu\nu}$ is a *flat extension* [21] of a $(n-1)$-dimensional non-degenerate metric $g_{\mu\nu} - c v_\mu v_\nu$.

2. If there is a parallel non-symmetric tensor $H_{\mu\nu}$, then its symmetric part is also parallel, so that one can put $h_{\mu\nu} = H_{(\mu\nu)}$ in the lemma. In the case that $H_{\mu\nu} = H_{[\mu\nu]} \neq 0$ is antisymmetric, then in fact one can define $H_{\mu\rho} H_\nu{}^\rho = h_{\mu\nu}$, which is symmetric, parallel, non-zero and *not* proportional to the metric if $n > 2$. For these last two statements, see e.g. [3].

3. Actually, the above can also be generalized to an arbitrary parallel p-form $\Sigma_{\mu_1 \ldots \mu_p}$ by defining $h_{\mu\nu} = \Sigma_{\mu\rho_2\ldots\rho_p} \Sigma_\nu{}^{\rho_2\ldots\rho_p}$.

Curvature invariants in 2-symmetric Lorentzian manifolds

Recall that a curvature scalar invariant [13] is a scalar constructed polynomially from the Riemann tensor, the metric, the covariant derivative and possibly the volume element n-form of $(\mathscr{V}, g)$. They are called linear, quadratic, cubic, etcetera if they are linear, quadratic, cubic, and so on, on the Riemann tensor. This defines its *degree*. The *order* can be defined for homogeneous invariants, that is, so that they have the same number of covariant derivatives in all its terms. This number is the order of the scalar invariant. Of course, all non-homogeneous invariants can be broken into their respective homogeneous pieces, and therefore in what follows only the homogeneous ones will be considered. Similarly, one can define curvature 1-form invariants, or more generally, curvature *rank* $-r$ invariants in the same way but leaving $1, \ldots, r$ free indices [13].

A simple but very useful lemma is the following [23]

Lemma 4 *Let (D, g) be as before with arbitrary signature. Any 1-form curvature invariant which is parallel must be necessarily null (possibly zero).*

It follows that, in 2-symmetric spaces, either R is constant or $\nabla_\mu R$ is null and parallel. This is a particular example of the following general important result [23].

Proposition 2 *Let $D \subset \mathscr{V}$ be a simply connected domain of an n-dimensional 2-symmetric Lorentzian manifold $(\mathscr{V}, g)$. Then either*

- *all (homogeneous) scalar invariants of the Riemann tensor of order m and degree up to $m+2$ are constant on D; or*
- *there is a parallel null vector field on D.*

(Observe also that there will be no non-zero invariants involving derivatives of order higher than one. Then, the degree is necessarily greater or equal than the order.)

The previous proposition has immediate consequences providing more information about curvature invariants. For instance [23]

Corollary 1 *Under the conditions of Proposition 2, either there is a parallel null vector field on D or the following statements hold*

1. *All curvature scalar invariants of any order and degree formed as functions of the homogeneous ones of order m and degree up to $m+2$ are constant on D;*
2. *All 1-form curvature invariants of order m and degree up to $m+1$ are zero.*
3. *All scalar invariants with order equal to degree vanish.*
4. *All rank-2 tensor invariants with order equal to degree are zero.*

Remark: Of course, it can happen that the mentioned curvature invariants vanish *and* there is a parallel null vector field too.

There is a very long list of vanishing curvature invariants as a result of this Corollary—if there is no null parallel vector field—. The list of the quadratic ones is (only an independent set [13] is given, omitting those contaning $\nabla_\mu R = 0$):

$$R^{\mu\nu}\nabla_\alpha R_{\mu\nu} = 0, \ R^{\mu\nu}\nabla_\mu R_{\nu\alpha} = 0, \quad (14)$$

$$R^{\mu\nu\rho\alpha}\nabla_\mu R_{\nu\rho} = 0, \ R^{\mu\nu\rho\sigma}\nabla_\mu R_{\nu\rho\sigma\alpha} = 0 = R^{\mu\nu\rho\sigma}\nabla_\alpha R_{\mu\nu\rho\sigma}, \quad (15)$$

$$\nabla_\alpha R^{\mu\nu}\nabla_\beta R_{\mu\nu} = \nabla_\mu R_{\nu\beta}\nabla_\alpha R^{\mu\nu} = \nabla_\mu R_{\nu\alpha}\nabla^\mu R^\nu{}_\beta = \nabla_\mu R_{\nu\alpha}\nabla^\nu R^\mu{}_\beta = 0, \quad (16)$$

$$\nabla^\mu R^{\nu\rho}\nabla_\alpha R_{\beta\rho\mu\nu} = \nabla^\mu R^{\nu\rho}\nabla_\mu R_{\alpha\nu\beta\rho} = 0, \quad (17)$$

$$\nabla_\alpha R^{\mu\nu\rho\sigma}\nabla_\beta R_{\mu\nu\rho\sigma} = \nabla^\sigma R^{\mu\nu\rho\alpha}\nabla_\sigma R_{\mu\nu\rho\beta} = 0 \quad (18)$$

where of course the traces of (16-18) vanish, and one could also write the same expressions using the Weyl tensor instead of the Riemann tensor.

MAIN RESULTS

All necessary results to prove the main theorems have now been gathered. Then, by using the so-called future tensors and "superenergy" techniques [22, 3] one can prove the following[3] [23]

Theorem 2 *Let $D \subset \mathscr{V}$ be a simply connected domain of an n-dimensional 2-symmetric Lorentzian manifold $(\mathscr{V}, g)$. Then, if there is no null parallel vector field on D, (D,g) is either Ricci-flat (i.e. $R_{\mu\nu} = 0$) or locally symmetric.*

Finally, one can at last prove that the narrow space left between locally symmetric and 2-symmetric Lorentzian manifolds can only be filled by spaces with a parallel null vector field.

Theorem 3 *Let $D \subset \mathscr{V}$ be a simply connected domain of an n-dimensional 2-symmetric Lorentzian manifold $(\mathscr{V}, g)$. Then, if there is no null parallel vector field on D, (D,g) is in fact locally symmetric.*

---

[3] It must be stressed that this proof is only valid for Lorentzian manifolds, as the definition of future tensors requires this signature.

Thus we have arrived at

Theorem 4 *Let $D \subset \mathcal{V}$ be a simply connected domain of an n-dimensional 2-symmetric Lorentzian manifold $(\mathcal{V}, g)$. Then, the line element on D is (possibly a flat extension of) the direct product of a certain number of locally symmetric proper Riemannian manifolds times either*

1. *a Lorentzian locally symmetric spacetime (in which case the whole (D,g) is locally symmetric), or*
2. *a Lorentzian manifold with a parallel null vector field so that its metric tensor can be expressed locally as an appropriately restricted case of formula (19) below.*

Again, the following remarks are important:

1. Of course, the number of proper Riemannian symmetric manifolds can be zero, so that the whole 2-symmetric spacetime, if not locally symmetric, is given just by a line-element of the form (19) restricted to be 2-symmetric.
2. Although mentioned explicitly for the sake of clarity, it is obvious that the block added in any flat extension can also be considered as a particular case of a locally symmetric part building up the whole space.
3. This theorem provides a full characterization of the 2-symmetric spaces using the classical results on the symmetric ones: their original classification (for the semisimple case) was given in [2], and the general problem was solved for Lorentzian signature in [9]. Combining these results with those for proper Riemannian metrics [10, 11, 15], a complete classification is achieved.

Thus, the only 2-symmetric non-symmetric Lorentzian manifolds contain a parallel null vector field. The most general local line-element for such a spacetime was discovered by Brinkmann [5] by studying the Einstein spaces which can be mapped conformally to each other. In appropriate local coordinates $\{x^0, x^1, x^i\} = \{u, v, x^i\}$, $(i, j, k, \ldots = 2, \ldots, n-1)$ the line-element reads

$$ds^2 = -2du(dv + Hdu + W_i dx^i) + g_{ij} dx^i dx^j \tag{19}$$

where the functions H, W_i and $g_{ij} = g_{ji}$ are independent of v, otherwise arbitrary, and the parallel null vector field is given by

$$k_\mu dx^\mu = -du, \qquad k^\mu \partial_\mu = \partial_v. \tag{20}$$

It is now a simple matter of calculation to identify which manifolds among (19) are actually 2-symmetric. Using Theorem 4 and its remarks, this will provide —by direct product with proper Riemannian symmetric manifolds if adequate— all possible *non-symmetric* 2-symmetric spacetimes. By doing so [23] one finds, among other results, that (i) the g_{ij} are a one-parameter family, depending on u, of locally symmetric proper Riemannian metrics[4]; (ii) for a given choice of g_{ij} in agreement with the previous point,

---

[4] As these are classified in e.g. [11, 15], the part g_{ij} of the metric is completely determined. For an explicit formula, one only has to take any of them from the list and let any arbitrary constants appearing there to be functions of u.

the integrability conditions provide the explicit form of the functions H and W_j; (iii) finally, the scalar curvature coincides with the corresponding scalar curvature $\bar{R}$ of g_{ij}: $R = \bar{R}$. Due to (i), the function $\bar{R}$ depends only on u, and thus the 2-symmetry implies

$$R = \bar{R}(u) = au + b \tag{21}$$

where a and b are constants. In particular, $\nabla_\mu R = -ak_\mu$. Thus, we see that given any locally symmetric proper Riemannian g_{ij} and letting the constants appearing there to be functions of u is too general, and these functions are restricted by the 2-symmetry so that, for example, (21) holds.

REFERENCES

1. L. Bérard-Bergery and A. Ikemakhen, "On the holonomy of Lorentzian manifolds", in *Differential geometry: geometry in mathematical physics and related topics, Los Angeles, CA, 1990* pp. 27–40, *Proc. Sympos. Pure Math.*, **54**, Part 2, (Amer. Math. Soc., Providence, R.I., 1993)
2. M. Berger, *Ann. Sci. École Norm. Sup.* **74** 85–177 (1957)
3. G. Bergqvist and J.M.M. Senovilla, *Class. Quantum Grav.* **18** 5299-5325 (2001)
4. E. Boeckx, O. Kowalski, L. Vanhecke, *Riemannian manifolds of conullity two* World Sci. Singapore 1996
5. H. W. Brinkmann, *Math. Ann.* **94** 119-145 (1925)
6. M. Cahen and R. McLenaghan, *C. R. Acad. Sci. Paris Sér. A-B* **266** A1125–A1128 (1968)
7. M. Cahen and M. Parker, *Bull. Soc. Math. Belg.* **22** 339–354 (1970)
8. M. Cahen and M. Parker, *Mem. Amer. Math. Soc.* **24** no. 229 (1980)
9. M. Cahen and N. Wallach, *Bull. Amer. Math. Soc.* **76** 585–591 (1970)
10. É. Cartan, *Bull. Soc. Math. France* **54** 214-264 (1926); **55** 114-134 (1927)
11. É. Cartan, *Leçons sur la Géométrie des Espaces de Riemann*, 2nd ed., Gauthier-Villars, Paris 1946
12. C.D. Collinson and F. Söler, *Tensor (N.S.)* **30** 87–88 (1976)
13. S.A. Fulling, R.C. King, B.G. Wybourne, C.J. Cummins, *Class. Quantum Grav.* **9** 1151–1197 (1992)
14. G.S. Hall, *J. Math. Phys.* **32** 181–187 (1991)
15. S. Helgason, *Differential Geometry, Lie groups, and Symmetric Spaces*, Academic Press, New York 1978
16. S. Kobayashi and K. Nomizu, *Foundations of Differential Geometry*, Wiley, Interscience, New York, vol.I 1963, vol.II 1969
17. A. Lichnerowicz, Courbure, nombres de Betti, et espaces symétriques, *Proceedings of the International Congress of Mathematicians, Cambridge, Mass., 1950* vol. 2, 216–223 (Amer. Math. Soc., Providence, R. I., 1952)
18. K. Nomizu and H. Ozeki, *Proc. Nat. Acad. Sci. USA* **48** 206-207 (1962)
19. B. O'Neill, *Semi-Riemannian Geometry*, Academic Press, New York 1983
20. G. de Rham, *Comment. Math. Helv.* **26** 328–344 (1952)
21. H.S. Ruse, A.G. Walker, T.J. Willmore, *Harmonic spaces*, (Consiglio Nazionale delle Ricerche Monografie Matematiche, 8) Edizioni Cremonese, Rome 1961
22. J.M.M. Senovilla, *Class. Quantum Grav.* **17** 2799-2841 (2000)
23. J. M. M. Senovilla, preprint (2005) (available upon request).
24. Z.I. Szabó, *J. Diff. Geom.* **17** 531-582 (1982)
25. Z.I. Szabó, *Geom. Dedicata* **19** 65-108 (1985)
26. H. Takeno, *Tensor (N.S.)* **27** 309–318 (1973)
27. S. Tanno, *Ann. Mat. Pura Appl. (4)* **96** 233–241 (1972)
28. H. Wu, *Illinois J. Math.* **8** 291–311 (1964)
29. H. Wu, *Bull. Amer. Math. Soc.* **70** 610–617 (1964)
30. H. Wu, *Pacific J. Math.* **20** 351–392 (1967)

Searches for continuous gravitational wave sources with LIGO and GEO

Alicia M. Sintes for the LIGO Scientific Collaboration

Departament de Física, Universitat de les Illes Balears, Cra. Valldemossa Km. 7.5, E-07122 Palma de Mallorca, Spain
Max-Planck-Institut für Gravitationsphysik, Albert Einstein Institut, Am Mühlenberg 1, D-14476 Golm, Germany

Abstract. An overview of the searches for continuous gravitational wave signals in LIGO and GEO performed on different recent science runs and results are presented. This includes both searching for gravitational waves from known pulsars as well as blind searches over a wide parameter space.

Keywords: gravitational waves, interferometers, pulsars, neutron stars
PACS: 04.80.Nn, 95.55.Ym, 97.60.Gb, 07.05.Kf

INTRODUCTION

Construction of the LIGO [1, 2] and GEO [3] instruments began in the mid-1990s. When the construction phase of the project was completed, the LIGO instruments were officially inaugurated on November 1999. Since then the commissioning of the instruments has proceeded in a sequence of engineering and science runs at increasing sensitivity. The four science runs to date are:

S1: August 23 - September 9, 2002

S2: February 14 - April 14, 2003

S3: October 31, 2003 - January 9, 2004

S4: February 22 - March 23, 2005

Both LIGO and GEO are to begin a full science run in November, with the aim of gathering data continuously for 18 months. During the previous science runs although these instruments were still to reach their design sensitivity, their performances were sufficiently good to justify a serious test of our search algorithms on real interferometer data, in particular, to search for continuous gravitational waves from very dense, rapidly-spinning stars, such as neutron or quark stars.

Rapidly rotating neutron stars are the most likely sources of periodic, persistent gravitational waves in the frequency band between ~ 100 and ~ 1000 Hz. These objects generate gravitational waves through a variety of mechanisms, including non-axisymmetric distortions of the star, velocity perturbations in the star's fluid, and free precession. Regardless of the specific mechanism, the emitted signal is a quasi-periodic continuous wave whose frequency changes slowly during the observation time due to the intrinsic frequency drift induced by the energy loss through gravitational wave emission (and possibly other mechanisms) and the motion of the detector with respect to the source. As the intrinsic amplitude of gravitational waves from this class of sources is several orders of magnitudes smaller than the typical root-mean-square value of the noise, detection can only be achieved by means of long integration times, of the order of weeks-to-months.

Code to search for this kind of sources has been under development within the LIGO Scientific collaboration (LSC) since the mid- to late 1990s. For S1-S4 the LSC pulsar upper limits group (PULG) has developed several methods to search and set upper limits on signals from radio pulsars as well as to perform an all-sky search for unknown neutron stars [4, 5, 6, 7, 8, 9, 10]. So far none of the searches conducted provided a detection but upper limits were set on the gravitational wave emission and these results are summarized in this paper.

SEARCH METHODS AND RESULTS

In the S1 analysis [4], two techniques were used to set upper limits on gravitational wave emission from pulsar J1939+2134 (the fastest rotating known millisecond pulsar): a Bayesian time-domain method [11, 12] and a classical frequency-domain method. The main result from this S1 analysis was an upper limit on signals from pulsar J1939+2134 of $h_0 < 1.4 \times 10^{-22}$ with 95% confidence.

For the LIGO S2 run, the time-domain search (which is more suitable for targeted sources) was expanded to include all well-known isolated pulsars with putative gravitational wave frequencies above 40 Hz [5]. Using the S2 data, multi-detector upper limits were set on gravitational wave emission from 28 pulsars including J1939+2134 and the Crab pulsar. The tightest limit on gravitational wave strain came from pulsar J1910-5959D with a 95% upper limit that $h_0 < 1.7 \times 10^{-24}$. At that time this was the lowest upper limit, for an astrophysical source, ever set by an interferometric gravitational wave detector (see Fig. 1). The same method is currently applied to analyze S3-S4 LIGO and GEO data [9]. The main change in this search since S2 has been the addition of pulsars in binary systems. Currently 93 pulsars are being search of which 60 are binaries and 33 are isolated. The improved sensitivity of the detectors in S3-S4 promise to give interesting results for several sources. For the Crab pulsar, we should be within a factor of a few of the spin-down based upper limit. Moreover, for S5, expectations for the Crab would be that within a year we would beat the spin-down limit.

The other frequency-domain statistical technique used in the S1 analysis is currently being used for broad all-sky searches for unknown sources. In fact, this search has been the test-bench for the core science analysis that the *Einstein@home* [10] project is carrying out. *Einstein@home* is a public distributed computing project that the LSC has been operating since February 2005, built using the Berkeley Open Infrastructure for Network Computing (BOINC). Members of the general public can quickly and easily install the software on their Windows, Macintosh, or Linux personal computer. When otherwise idle, their computer downloads data from *Einstein@home*, searches it for pulsar signals, then uploads information about any candidates.

A similar technique is used in [8] which uses S2 data from two of the LIGO detectors to perform two different searches: (i) for signals from isolated sources over the whole sky and the frequency band 160–728.8 Hz and (ii) for a signal from the Low-Mass X-ray Binary Scorpius X-1 over orbital parameters and in the frequency bands 464–484 Hz

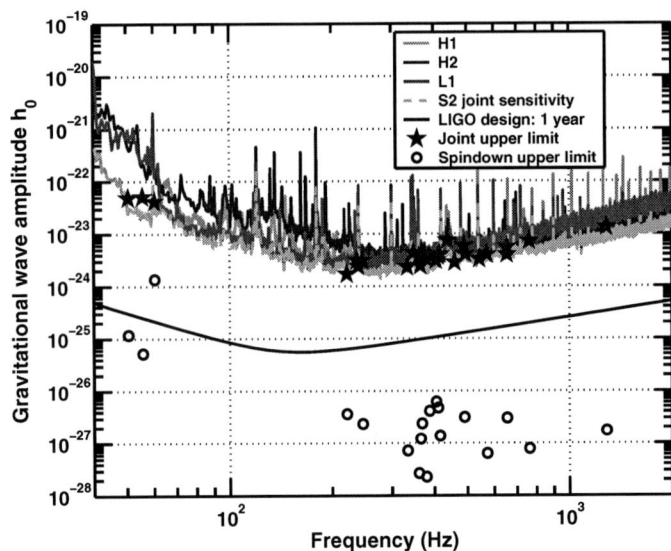

FIGURE 1. Upper curves: h_0 amplitudes detectable from a known generic source with a 1% false alarm rate and 10% false dismissal rate, as given by Eq. (2.2) in [4] for single detector analyzes and for a joint detector analysis. All the curves use typical S2 sensitivities and observation times. H1 and H2 are the 4 km arm and the 2 km arm detectors located in Hanford WA. L1 is the 4 km arm detector situated the in Livingston Parish LA. Lower curve: LIGO design sensitivity for 1 yr of data. Stars: upper limits found using the S2 data for 28 known pulsars. Circles: spin-down upper limits for the pulsars with negative frequency derivative values if *all* the measured rotational energy loss were due to gravitational waves and assuming a moment of inertia of 10^{45} g cm^2.

and 604–624 Hz. It is the first time that a coherent analysis is carried out over such a wide frequency band, using data in coincidence and (in one case) for a rotating neutron star in a binary system.

Future continuous wave searches will involve searching longer data stretches (or order months to years) for unknown sources over a large frequency band, vast portions of the sky and spin-down parameter values. It is well known that the computational cost of coherent techniques (as those applied in the S1 analysis) for searches of this type is absolutely prohibitive, thus hierarchical methods have been proposed [13, 14, 15, 16], where coherent and incoherent search stages are alternated in order to identify efficiently statistically significant candidates.

An essential step towards the actual implementation of a "hierarchical pipeline" for production analysis is the thorough investigation and characterization of its building blocks – the coherent and incoherent stages – over a large parameter space and on actual data sets; in fact the optimal sensitivity can be ultimately achieved through careful tuning of a variety of search parameters that are difficult to determine on pure theoretical grounds, including the choice of thresholds at each stage, the different tilings of the parameter space, the quality cuts in the data and the choice of coincidence windows.

In [7] we report results obtained by applying for the first time a coherent analysis

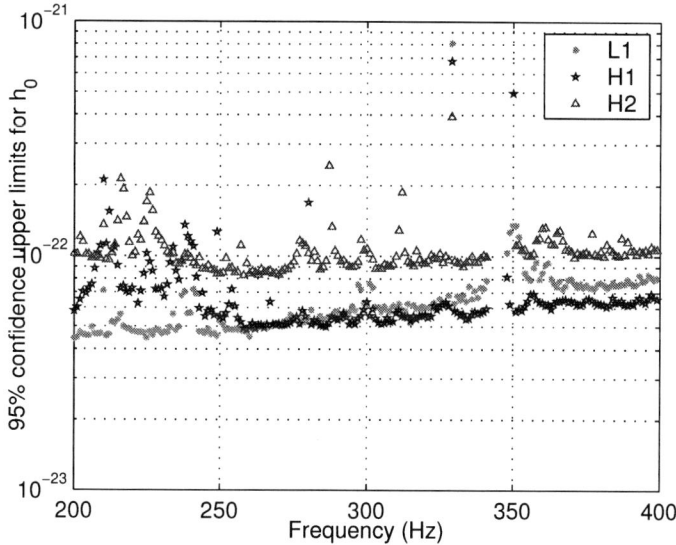

FIGURE 2. The 95% confidence upper limits on h_0 over the whole sky and different spin-down values in 1 Hz bands using the Hough transform technique on the S2 data set.

and an incoherent analysis, respectively, to the data collected during the S2 run. The search method is based on the Hough transform, which is a computationally efficient and robust pattern recognition technique. We apply this technique to perform an all-sky search for isolated spinning neutron stars using the two months of data. The main results of this paper are all-sky upper limits on the strength of gravitational waves emitted by unknown isolated neutron stars on a set of narrow frequency bands in the range 200–400 Hz. Our best 95% frequentist upper limit that we obtain in this frequency range is $h_0 < 4.43 \times 10^{-23}$ (see Fig. 2). Based on the statistics of neutron stat population with optimistic assumptions, this upper limit is about 1 order of magnitude larger than the amplitude of the strongest expected signal, but with 1 yr of data at design sensitivity for initial LIGO, we should gain about 1 order of magnitude in sensitivity, thus enabling us to detect signals smaller that what is predicted by the statistical argument mentioned above.

Other incoherent techniques, such as "stack-slide" [14] or "power-flux" as well as the Hough transform [17], are used by the PULG to analyze S4 data. All of them use, in some way, the power from the Fourier transforms of short stretches of data, which are added in a way that compensates for the Earth's motion and the pulsar's spin-down during the observation period. These analyzes provide us with the first thorough understanding and characterization of such approaches and allow us to place upper-limits on regions of the parameter space that have never been explored before.

The development of sophisticated analysis techniques, such as the one described here, together with the computing power of Einstein@home will allow the deepest pulsar searches.

ACKNOWLEDGMENTS

The authors gratefully acknowledge the support of the United States National Science Foundation for the construction and operation of the LIGO Laboratory and the Particle Physics and Astronomy Research Council of the United Kingdom, the Max-Planck-Society and the State of Niedersachsen/Germany for support of the construction and operation of the GEO600 detector. The authors also gratefully acknowledge the support of the research by these agencies and by the Australian Research Council, the Natural Sciences and Engineering Research Council of Canada, the Council of Scientific and Industrial Research of India, the Department of Science and Technology of India, the Spanish Ministerio de Educación y Ciencia, the John Simon Guggenheim Foundation, the David and Lucile Packard Foundation, the Research Corporation, and the Alfred P. Sloan Foundation.

REFERENCES

1. Abramovici A *et al*. 1992 *Science* **256**, 325
2. Barish B and Weiss R 1999 *Phys. Today* **52**, 44
3. Willke B *et al*. 2002, *Class. Quant. Grav.* **19** 1377
4. Abbott B *et al*. (The LIGO Scientific Collaboration) 2004 *Phys. Rev. D* **69** 082004
5. Abbott B *et al*. (The LIGO Scientific Collaboration) 2005 *Phys. Rev. Lett.* **94** 181103
6. Krishnan B for the LIGO Scientific Collaboration 2005 *Class. Quant. Grav.* **22** S1265-S1276
7. Abbott B *et al*. (The LIGO Scientific Collaboration) 2005, gr-qc/0508065 [*Phys. Rev. D* (to be published)]
8. Abbott B *et al*. (The LIGO Scientific Collaboration) 2005, Report No. LIGO-P050008-00-Z
9. Dupuis R J for the LIGO Scientific collaboration 2005, gr-qc/0509018 [*Class. Quant. Grav.* (to be published)]
10. http://einstein.phys.uwm.edu/
11. Dupuis R J and Woan G 2005, gr-qc/0508096 [*Phys. Rev. D* (to be published)]
12. Dupuis R J 2004, *Ph.D. thesis*, University of Glasgow
13. Brady P, Creighton T, Cutler C and Schutz B F 1998 *Phys. Rev. D* **57** 2101
14. Brady P R and Creighton T 2000 *Phys. Rev. D* **61** 082001
15. Cutler C Gholami I and Krishnan B 2005, *Phys. Rev. D* **72** 042004
16. Papa M A, Schutz B F and Sintes A M 2001, in *Gravitational waves: A challenge to theoretical astrophysics*, ICTP Lecture Notes Series, Vol. III, edited by V. Ferrari, J.C. Miller, L. Rezzolla. p. 431.
17. Krishnan B, Sintes A M, Papa M A, Schutz B F, Frasca S and Palomba C 2004 *Phys. Rev. D* **70** 082001

PART III: SHORT COMMUNICATIONS

Black Hole Entanglement Entropy

Juan Manuel Tejeiro-Sarmiento* and José Robel Arenas-Salazar *

Observatorio Astronómico Nacional, Universidad Nacional de Colombia, Bogotá, Colombia

Abstract. Entanglement entropy is calculated for an extension of Bombeli et al. black hole entropy flat model to maximally extended Schwarzschild spacetime in order to present an integrated model with thermofield-dynamical description. Local analysis shows that entanglement entropy is strongly concentrated near the horizon.

Keywords: Black hole entropy, Entanglement entropy, Black hole thermodynamics
PACS: 04.70.Dy

INTRODUCTION

The statistical nature of the Bekenstein-Hawking entropy S_{BH}, assigned to a black hole of surface area A, remains unknown. Different derivations of S_{BH} have been proposed to provide a microscopic explanation of the Bekenstein-Hawking Entropy [1, 2, 3]. Maybe the most promising and appropriate approach is that S_{BH} is entanglement entropy, associated with modes and correlations hidden from outside observers by the presence of a horizon. A program of this type was first clearly formulated by Bombelli et al. [4]. It was independently re-initiated by M. Srednicki [5] and by V. P. Frolov and I. D. Novikov [6]. From this approach can be obtained the entanglement entropy, which is proportional to the area of dividing wall [7].

On the other hand, to judge whether this speculation is true or not in thermodynamics, it is necessary to investigate the whole structure of thermodynamics obtained from the entanglement entropy, because thermodynamic entropy acquires a physical significance only when it is related to the energy and the temperature of a system. But, the black hole entanglement thermodynamics directly constructed from Bombelli et al.'s approach leads to problems and ambiguities [8, 9, 10].

Entanglement approach has some problems [11], but our main purpose in this paper is to propose a model to calculate effectively S_{BH} as a finite and thermal entanglement entropy in the context of the thermofield dynamics of black holes.

Entanglement entropy only arises artificially when it is pretended that modes beneath the horizon can't be seen and we form a density matrix for parochial observers in right or left sectors of the Schwarzschild spacetime maximally extended. Then, it is required to research carefully the distribution of energy density and pressure for the Hartle-Hawking and Boulware states, and to find out precisely where the entropy is located. Thus, by resorting to thermofield dynamics of black holes [12, 13, 14], it is shown below that entanglement entropy is strongly concentrated near the horizon.

ENTANGLEMENT LOCAL ENTROPY

In order to connect up the Bombelli et al. approach to the thermofield dynamics of black holes, the clue is to realize that the scalar field Hamiltonian is negative in left Kruskal sector. This is in fact the unique and only choice for an eternal black hole that makes the action regular and gives a conserved energy [15].

Consider the Hartle-Hawking state $|0\rangle_H$ defined over the complete Kruskal space:

$$|0\rangle_H = Z^{-\frac{1}{2}} \sum_n e^{-\frac{1}{2}\beta E_n} |n\rangle_{BL} |n\rangle_{BR}, \tag{1}$$

with

$$Z = \sum_n e^{-\beta E_n}, \quad \beta^{-1} = T_H, \tag{2}$$

where T_H is Hawking temperature and, $|n\rangle_{BL}$ and $|n\rangle_{BR}$ are excitations of the parochial Boulware states $|0\rangle_{BL}$ and $|0\rangle_{BR}$, respectively.

To calculate the entanglement entropy S consider the density matrix ρ_H obtained from (1) and the reduced density matrix $\rho^{(+)}$, given by

$$\rho_H = |0\rangle_{HH}\langle 0|, \quad \rho^{(+)} = tr_{B(-)} \rho_H, \tag{3}$$

where the trace is taken over the degrees of freedom corresponding to left Kruskal sector. Thus, we may calculate entropy as

$$S = -tr(\rho^{(+)} \ln \rho^{(+)}). \tag{4}$$

Although, according to equation (3), $\rho^{(+)}$ is a thermal reduced entanglement density matrix, it is necessary to describe previously the thermal nature of S, i.e., the distribution of energy density and pressure for the Boulware and Hartle-Hawking states, and to determine precisely where the entropy is located. Without this thermal characterization, the general procedure to calculate it would be purely formal.

To describe where the entropy is located and its thermal ambient, consider the expectation value for the component T_{00} of the stress-energy tensor with respect to Boulware and Hartle-Hawking states for a scalar field. In general, for a scalar field, the expectation value of the stress-energy tensor $T_{\alpha\beta}(x)$ is given by

$$\langle T_{\alpha\beta}(x,x')\rangle = \mathscr{D}_{\alpha\beta'} W(x,x'), \tag{5}$$

where $W(x,x')$ is the Wightman function and

$$\mathscr{D}_{\alpha\beta'} = \partial_{(\alpha} \partial_{\beta')} - \frac{1}{2} g_{\alpha\beta} \left(\partial^\gamma \partial_{\gamma'} + m^2 \right). \tag{6}$$

Then,

$$\lim_{x \to x'} \langle T_0^0(x,x') \rangle_{H\text{-}B} = \partial^0 \partial_{0'} W_{H\text{-}B} = -\int_0^\infty \frac{E}{e^{\frac{E}{T}} - 1} \frac{4\pi p^2 dp}{(2\pi)^3}, \tag{7}$$

where W_{HB} is the difference between Wightman functions associated to Boulware and Hartle-Hawking states, and it has been defined local proper energy per mode E by $E = \frac{\omega}{\sqrt{g_{00}}}$ and $p\,dp = \frac{\omega\,d\omega}{g_{00}} = E\,dE$.

The equation (7) is the expected thermal expression for energy density of a hot scalar field, calculated under WKB approximation, which is good near the horizon. So, the entropy S can be written as the volume integral of local density. Thus, we can argue that it represents an approximate localization of entropy, strongly concentrated near the horizon. This result means that there is a well-defined and finite entropy density near the horizon.

EFFECTIVE ENTANGLEMENT ENTROPY

The thermal system described above and associated to entanglement entropy, it actually allow us to think that entropy arises physically located near the horizon. Thus, we can calculate it from the expression (2) in terms of the partition function Z:

$$S = -\beta \frac{\partial}{\partial \beta} \ln Z + \ln Z, \tag{8}$$

where

$$\ln Z = \sum_{l,n,\eta=\pm} (2l+1) \ln\left(\frac{1}{1-e^{-\beta \omega_{ln}}}\right). \tag{9}$$

Hence, to get S finite we need to restrict the discrete sets $\{l = 0, 1, 2, \cdots\}$ and $\{n = 1, 2, \cdots\}$.

Since the thermal description, by the expression (7) summarized, is the same as the one established for the modified brick wall model [16], we can use this model to do an effective restriction to the sets $\{l\}$ and $\{n\}$.

Thus,

$$S = \int_{r_0+\varepsilon}^{R} 4\pi r^2 \frac{dr}{\sqrt{g_{00}}} s(r), \tag{10}$$

where

$$s(r) = \frac{1}{3T^2} \int_0^{\infty} \frac{p^2 e^{\frac{E}{T}}}{(e^{\frac{E}{T}}-1)^2} \frac{4\pi p^2 \, dp}{h^3}. \tag{11}$$

The integral (10) is dominated by two contributions, for large $r = R$ and for small $r_0 + \varepsilon$. The former corresponds to a volume term, proportional to $\frac{4}{3}\pi r^3$, which represents the entropy and energy of a homogeneous quantum gas in a flat space at a uniform temperature T_H. The latter is the contribution of gas near the inner wall $r = R_0$. According to this, the wall contribution to the total entropy is obtained

$$S_{\text{wall}} = \frac{N}{90\pi \alpha^2} \frac{1}{4} A, \tag{12}$$

where N accounts for helicities and the number of particle species, A is the wall area and α is the proper altitude of the inner wall above the horizon of the exterior geometry.

S_{wall} diverges in the limit $\alpha \to 0$, but introducing a brick wall cutoff we can obtain the Bekenstein-Hawking entropy from (12)

$$S_{wall} = S_{BH}, \qquad (13)$$

where α has been adjusted by invoking quantum gravity effects.

DISCUSSION

We have shown that S_{BH} can be considered as entanglement entropy, which is well defined near the horizon and presents a thermal nature according to thermofield dynamics of black holes.

With a Hamiltonian that is negative in left sector of the Schwarzschild spacetime maximally extended, in the context of the thermofield dynamics of black holes, we can solve the problems and ambiguities risen from the original approach with the considerations of "entanglement energy" and "entanglement temperature." In particular, from this formalism the Bombelli et al.'s Lagrangian can be generalized, to be applied to curved spacetime in order to include horizons, necessary condition to derive a thermal entanglement density matrix [15].

ACKNOWLEDGMENTS

We are indebted to Werner Israel, for his stimulating discussions and helpful comments on the manuscript.

REFERENCES

1. J. D. Bekenstein, "Do we understand black hole entropy," in *Proceedings of the Seventh Marcel Grossmann Meeting on General Relativity, Stanford, USA, 1994*, edited by R. T. Jantzen and G. M. Keiser, World Scientific, Singapore, 1996, p. 39, gr-qc/9409015.
2. S. Mukohyama, *Ph.D. thesis*, Kyoto University, 1998, gr-qc/9812079.
3. V. P. Frolov and D. V. Fursaev, *Class. Quantum Grav.*, **15**, 2041-2074 (1998).
4. L. Bombelli, R. K. Koul, J. Lee, and R. D. Sorkin, *Phys. Rev. D*, **34**, 373-383 (1986).
5. M. Srednicki, *Phys. Rev. Lett.*, **71**, 666-669 (1993).
6. V. Frolov and I. Novikov, *Phys. Rev. D*, **48**, 4545-4551 (1993).
7. H. Terashima, *Phys. Rev. D* **61**, 104016-1-11 (2000).
8. S. Mukohyama, M. Seriu and H. Kodama, *Phys. Rev. D*, **55**, 7666-7679 (1997).
9. S. Mukohyama, M. Seriu and H. Kodama, *Phys. Rev. D*, **58**, 064001 (1998).
10. J. R. Arenas and J. M. Tejeiro, in *Proceedings of the Ninth Marcel Grossmann Meeting on General Relativity, Rome, Italy, 2000*, edited by V. G. Gurzadyan, R. T. Jantzen and R. Ruffini, World Scientific, Singapore, 2002, Part B. pp. 1509-1510.
11. D. V. Fursaev, *Phys. Part. Nucl.*, **36**, 81-99 (2005)
12. Y. Takahashi and H. Umezawa, *Collective Phenomena*, **2**, 55-80 (1975).
13. W. Israel, *Phys. Lett. A*, **57**, 107-110 (1976).
14. H. Umezawa, *Advanced Field Theory*, AIP Press, New York, 1993.
15. J. R. Arenas and J. M. Tejeiro, "Entanglement entropy of black holes", in preparation.
16. S. Mukohyama and W. Israel, *Phys. Rev. D*, **58**, 104005-1-10 (1998).

Is there Any Evidence for Integrated Sachs-Wolfe Signal in WMAP First Year Data?

C. Hernández–Monteagudo[*], R. Génova–Santos[†] and F. Atrio–Barandela[**]

[*]*Department of Physics and Astronomy. University of Pennsylvania. Philadelphia, USA and Max Planck Institut für Astrophysik (MPA) Garching, Germany*
[†]*Instituto de Astrofísica de Canarias. Tenerife, Spain.*
[**]*Física Teórica. Universidad de Salamanca. Spain.*

Abstract. We introduce a pixel-to-pixel comparison method to detect the temperature anisotropies on the Cosmic Microwave Background induced by the time variation of gravitational potentials along the line of sight. We demonstrate it to be more sensitive than the cross-correlation method used in previous studies. We compare the recent WMAP data with templates constructed from galaxy catalogues. A positive cross-correlation between both data sets will be a signature of an accelerated expansion of the Universe. Contrary to other authors, we fail to detect any signal, except those coming from foreground residuals. Either the effect of an accelerated expansion is not present on the WMAP data or the galaxy catalogues at present do not trace the evolution of the large scale gravitational field.

Keywords: cosmic microwave background, cosmological constant, Cosmology, Relativity.
PACS: 98.80.Es, 98.70.Vc, 98.65.Dx

INTRODUCTION

Observations of high redshift supernovae [1] and of the Cosmic Microwave Background (CMB) temperature anisotropies by the WMAP satellite [2] indicate the Universe is expanding with positive acceleration. In this period, the growth of matter density perturbations slows down and variations of the gravitational potential along the line of sight induce a late time temperature anisotropy on CMB photons. We write the perturbed metric element as: $ds^2 = a^2[-(1+2\Psi)d\eta^2 + (1-2\Phi)\gamma_{ij}dx^i dx^j]$ where Ψ, Φ are the Bardeen potentials that give the gravitational time dilation and the perturbation to the 3-space curvature. The temperature pattern of a photon gas of wavenumber $k\hat{k}$ in the direction of observation $\hat{n}$ is ($\mu = \hat{k}\hat{n}$):

$$\Theta(\eta_o, k, \mu) = e^{ik\mu(\eta_{dec}-\eta_o)}\left[\Theta^{SW} + \Theta^{Dopl} + \Theta^{ISW}\right] \quad (1)$$

$$\Theta^{SW} = \left[\frac{1}{4}D_{intrinsic,\gamma} + \Psi\right](\eta_{dec}, k)$$

$$\Theta^{Dopl} = -i\mu k V_\gamma(\eta_{dec}, k)$$

$$\Theta^{ISW} = \int_{\eta_{dec}}^{\eta_o} d\eta\, e^{ik\mu(\eta-\eta_o)}(\dot{\Psi}+\dot{\Phi})(\eta, k).$$

The Integrated Sachs-Wolfe (ISW) effect [3, 4] can be detected by cross-correlating the distribution of matter density perturbations with the pattern of temperature anisotropies

TABLE 1. Results of the pixel-to-pixel comparison method.

BAND	Q		V		W	
FWHM	1°	2°	1°	2°	1°	2°
2MASS	-	-1.8± 0.3	-	-1.15 ± 0.3	-	-0.9 ± 0.3
NVSS-I	1.0±0.14	0.9±0.2	0.5±0.14	0.4±0.2	0.1±0.14	0.13±0.2
NVSS-II	0.9±0.15	-	0.4 ± 0.15	0.4± 0.2	0.18 ± 0.15	-
NVSS-III	0.8±0.17	-	0.4 ± 0.17	-	0.2 ± 0.17	-

on the sky. Several groups have carried out this analysis using the latest CMB data and galaxy catalogues [5] and found positive correlations at the $\approx$ 2–3 σ confidence level.

THE PIXEL-TO-PIXEL COMPARISON METHOD.

Up to date, positive detections of the ISW contribution to CMB data were based on computing different variants of the 2-point cross-correlation of galaxy templates and CMB data. We shall introduce a different method, the pixel-to-pixel comparison, that has been succesfully apply to trace the contribution of hot gas to the CMB data [6]. In this method, we assume that the CMB temperature anisotropy in a given direction of the sky has several components: intrinsic $\mathbf{T}_{cmb}$ of cosmological origin, noise $\mathbf{N}$, foreground residuals $\mathbf{F}$, and an additional component that can be traced by a template constructed from the matter distribution $\mathbf{M}$; i.e. $\mathbf{T} = \mathbf{T}_{cmb} + \tilde{\alpha}\mathbf{M} + \mathbf{N} + \mathbf{F}$, where $\tilde{\alpha}$ gives the matter contribution (in units of temperature) to the CMB anisotropies. The pixel-to-pixel comparison method is a minimum variance estimator of $\tilde{\alpha}$ with error bar σ_α:

$$\alpha = \frac{\mathbf{T}\mathscr{C}^{-1}\mathbf{M}^T}{\mathbf{M}\mathscr{C}^{-1}\mathbf{M}^T}; \quad \sigma_\alpha = \sqrt{\frac{1}{\mathbf{M}\mathscr{C}^{-1}\mathbf{M}^T}}, \quad (2)$$

Due to the large number of data points, the covariance matrix C of the CMB data can not be inverted for the whole sky. Therefore we divide the sky in patches of the same number of N_{pix} compute α^β for each patch and compute a weighted average α with error bar σ_α

$$\alpha = \frac{\sum_\beta \alpha^\beta/(\sigma_\alpha^\beta)^2}{\sum_\beta 1/(\sigma_\alpha^\beta)^2}, \sigma_\alpha^2 = \frac{1}{\sum_\beta 1/(\sigma_\alpha^\beta)^2}\left[1 + \frac{2\sum_{\beta_1<\beta_2} \mathbf{M}_{\beta_1}\mathscr{C}^{-1}_{\beta_1\beta_1}\mathscr{C}^{\beta_1\beta_2}\mathscr{C}^{-1}_{\beta_2\beta_2}\mathbf{M}^T_{\beta_2}}{\sum_\beta 1/(\sigma_\alpha^\beta)^2}\right]. \quad (3)$$

$\mathscr{C}^{\beta_1,\beta_2}$ is the covariance matrix between pixels of two different patches β_1, β_2.

We tested the efficiency of our method by generating 500 Monte Carlo realizations of the CMB sky. At each realization we added a map of temperature anisotropies of rms 1.5μK distributed like a galaxy template. We estimate $\tilde{\alpha}$ using the cross-correlation function and the pixel-to-pixel comparison methods. In either method, the mean value of α was 1.5μK, the input value, but the dispersion was 6 times smaller in the pixel-to-pixel comparison method than in the correlation method.

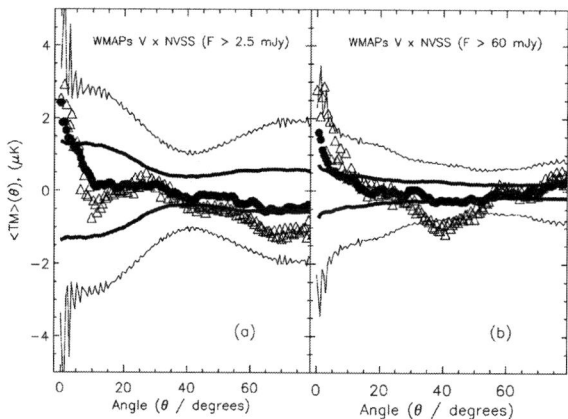

FIGURE 1. (a) Cross correlation function of the WMAP V band with the template based on NVSS-I. Templates and data were convolved with a gaussian beam FWHM equal to 1° (filled circles) and 4° (triangles). Error bars are given by the thick and thin solid lines, respectively. (b) As in (a), but for the template based on NVSS-II sources of flux greater than 60 mJy.

CMB DATA AND GALAXY CATALOGUE TEMPLATES

We constructed templates from galaxy catalogues assignig to each pixel a value equal to the number of galaxies within that pixel. We assumed that the depth of gravitational potential wells was proportional to the number of collapsed halos at each location. Although this hypothesis is correct in linear theory and on large scales, is not clear how well our galaxy samples trace the real halo population at every redshift. Sources located at redshifts where there is no significant ISW contribution could degrade our templates as potential ISW tracers. We constructed different templates containing galaxies at different redshift intervals to test this effect. To describe the low redshift Universe, we used the 2MASS catalog [7]. To probe the high redshift universe, we used the NVSS survey of extragalactic sources [8]. We constructed 3 templates: containing all sources brighter than 2.5 mJy (NVSS-I), brighter than 60 mJy (NVSS-II) and containing only unresolved sources (NVSS-III). The templates were convolved with beams of width 1 and 2 degrees.

RESULTS AND DISCUSSION

We compared the WMAP data and the different templates using both the cross-correlation and the pixel-to-pixel comparison method. In Table 1 we show our results of comparing WMAP V-band with all the templates. The analysis was carried out in the V-band and only when their was a positive detection the analysis was carried out at different frequencies. The data indicates an anticorrelation of the 2MASS template with the CMB dada. This anticorrelation desappeared after rotating the template a beam size, indicating the signal was associated to point sources and has a Sunyaev-Zeldovich spectrum. It corresponds to the same signal found at 12 arcmin scales [6]. We repeated the analysis using the cross-correlation techniques. With respect to the results for the NVSS catalogues, the correlation was posivite but depends on frequency, as would do foreground residuals associated to an unresolved population of radio galaxies. Also in this case the correlation desapeared after rotating one beam size.

In Fig. 1 we give the cross-correlation function for the NVSS-I and NVSS-II templates. The results are compatible with those of the pixel-to-pixel method: there was no evidence for an ISW signal in the 2MASS based template. For NVSS templates, the pixel-to-pixel comparison and cross correlation detected a positive signal of amplitude $\sim$ 0.5–1 μK. Comparing with observations from other bands, we found the amplitude depended on frequency. We checked that the correlation disappear when the template was rotated 0.5 degrees, suggesting the contribution was coming from a foreground residual. Since the pixel-to-pixel comparison method is more sensitive that the cross-correlation method, we conclude that what other authors could have taken a positive detection was no more than contribution from foreground residuals. To conclude, our results suggest that there is no evidence of ISW signatures in first-year WMAP data, suggesting that galaxy catalogues do not properly trace the time evolution of the large scale gravitational field. Since the galaxy catalogues used here have been also used to claim detection of the ISW effect in CMB data, they are in clear contradiction with previous studies.

ACKNOWLEDGMENTS

CHM was supported by the European Community through the Human Potential Programme under contract HPRN-CT-2002-00124 (CMBNET). FAB acknowledges financial support from the Spanish Ministerio de Educación y Ciencia (projects BFM2000-1322 and AYA2000-2465-E) and from the Junta de Castilla y León (projects SA002/03 and SA010C05).

REFERENCES

1. Perlmutter, S., et al. 1997, ApJ, 483, 565
2. Bennett, C. L., et al. 2003, ApJS, 148, 1; Spergel, D.N. et.al., 2003, ApJS, 148, 175.
3. Sachs, R. K., & Wolfe, A. M. 1967, ApJ, 147, 73.
4. Hu, W. & Sugiyama, N. (1995) Phys. Rev. D 51, 2599.
5. Afshordi, N. 2004, Phys Rev D70, 083536; Boughn, S. P., & Crittenden, R. G. 2005, MNRAS, 360, 1013; Fosalba, P., & Gaztañaga, E. 2004, MNRAS, 350, L37; Fosalba, P., Gaztañaga, E., & Castander, F. J. 2003, ApJL, 597, L89; Nolta, M. R., et al. 2004, ApJ, 608, 10; Scranton, R. et al. 2003, Preprint, astro-ph/0307335; Afshordi, N., Loh, Y., & Strauss, M. A. 2004, PhRvD, 69, 083524; Vielva et al. 2005, MNRAS, to be published.
6. Hernández-Monteagudo, C. & Rubiño-Martín, J. A. 2004, MNRAS, 347, 403; Hernández-Monteagudo, C., Genova-Santos, R. & Atrio-Barandela, F. (2004) ApJ, 613, L89.
7. Jarrett, T. H., et al. 2003, AJ, 125, 525
8. Condon, J.J. et al., 1998, AJ, 115, 1693

Static axisymmetric spacetimes with prescribed multipole moments

Thomas Bäckdahl

Matematiska Institutionen, Linköpings Universitet, SE-581 83 Linköping, Sweden.
e-mail: thbac@mai.liu.se

Abstract. We present a new method of finding the static axisymmetric spacetime corresponding to any given set of multipole moments. In addition to an implicit algebraic form for the general solution, we also give a power series expression for all finite sets of multipole moments.

Keywords: static, axisymmetric, multipole moments
PACS: 04.20.Ha, 04.20.Jb

MULTIPOLE MOMENTS FROM SPACETIMES

There are several techniques available to obtain solutions to Einstein's vacuum field equations. It is however often difficult to find physical interpretations for these solutions. Relativistic multipole moments aid the interpretation since they can be compared with their classical counterpart. For the static asymptotically flat case Geroch [2] gave a purely geometrical definition for these moments. In the axisymmetric case Herberthson [3] showed that this definition can be considerably simplified. This simplification makes it possible to answer the natural and important question: Given a set of multipole moments, what is the corresponding spacetime? Thus, can we find exact solutions with the physical properties prescribed beforehand?

We start by stating Geroch's definition [2]. Let V be a 3-surface orthogonal to the timelike Killing vector ξ^a and $\tilde{V}$ its conformal compactification. One then defines a potential $\psi = 1 - \sqrt{-\xi_a \xi^a}$. Let $P = \tilde{\psi}$. Define the sequence $P, P_{a_1}, P_{a_1 a_2}, \ldots$ of tensor fields recursively:

$$P_{a_1 \ldots a_n} = C[\nabla_{a_1} P_{a_2 \ldots a_n} - \frac{(n-1)(2n-3)}{2} \tilde{R}_{a_1 a_2} P_{a_3 \ldots a_n}] \quad (1)$$

where $C[\cdot]$ stands for taking the totally symmetric and trace-free part. The multipole moments are then given by the tensors $P, P_{a_1}, P_{a_1 a_2}, \ldots$ at space-like infinity.

If we now consider the axisymmetric case, the metric can be written

$$ds^2 = -e^{2\alpha} dt^2 + e^{2(\beta-\alpha)}(dR^2 + dZ^2) + R^2 e^{-2\alpha} d\phi^2, \quad (2)$$

where $\alpha = \frac{1}{2}\ln|\xi_a\xi^a|$, which vanishes at infinity, is axisymmetric and flat-harmonic with respect to the cylindrical coordinates R, Z and ϕ. With a simple coordinate change, the conformal compactification $\tilde{V}$ has the metric

$$d\tilde{s}^2 = e^{-2\beta}r^2\sin^2\theta d\phi^2 + dr^2 + r^2 d\theta^2. \quad (3)$$

Henceforth we use the scalars m_n to define the moments $m_n C[z_{a_1} z_{a_2} \ldots z_{a_n}]$, where $z_a = (dz)_a$ and $z = r\cos\theta$.

The fact that $\tilde{\alpha} = -\alpha/r$ is harmonic makes it possible to write

$$\tilde{\alpha}(r,\theta) = \sum_{n=0}^{\infty} a_n r^n P_n(\cos\theta). \tag{4}$$

Using the coefficients a_n we can then form the function

$$Y(r) = \tilde{\alpha}_L(r) = \sum_{n=0}^{\infty} a_n r^n \frac{(2n)!}{2^n n!^2}. \tag{5}$$

The simplification given by Herbethson [3] then consists of the following theorem (with minor changes):

Theorem 1 *Suppose that a static axisymmetric asymptotically flat spacetime M is given by the flat-harmonic function α, which after conformal rescaling is given by (4). Let Y be given by (5), β_L by*

$$\beta_L(r) = \int_0^r r \left(Y + 2r \frac{dY}{dr} \right)^2 dr \tag{6}$$

and define κ_L through

$$\kappa_L(r) = -\ln\left(1 - r \int_0^r \frac{e^{2\beta(r)} - 1}{r^2} dr - rC\right) + \beta_L(r). \tag{7}$$

Put $\rho(r) = re^{\kappa_L(r) - \beta_L(r)}$ and define $y: \mathbb{R}^+ \cup 0 \to \mathbb{R}$ implicitly by $y(\rho) = e^{(\beta_L(r) - \kappa_L(r))/2} Y(r)$. Then the multipole moments $m_0, m_1, m_2, \ldots$ of M are given by $m_n = \frac{d^n y}{d\rho^n}(0)$.

In the definition of κ_L, there appears a constant C. This constant affects $\kappa'(0)$. In particular, one can make $\kappa'(0) = 0$.

SPACETIMES FROM MULTIPOLE MOMENTS

Now we consider the problem of finding the appropriate spacetime for a given set of multipole moments. Use the moments $\{m_n\}$ to define the function

$$y = y(\rho) = \sum_{n=0}^{\infty} \frac{m_n \rho^n}{n!} \tag{8}$$

as in Theorem 1. We assume that the series converges in some neighbourhood of $\rho = 0$. The space-time is then determined if we know $Y(r) = \sqrt{\frac{\rho(r)}{r}} y(\rho(r))$, since this gives α. Thus we only need to know ρ as a function of r. This is given implicitly by the following theorem.

Theorem 2 *If y is given by (8) and ρ, κ are given implicitly by the relations in Theorem 1, then*

$$0 = -\frac{\rho}{r} + \rho \int_0^\rho \frac{1}{\sigma^2} \int_0^\sigma 2\rho (y + 2\rho y_\rho)^2 \, d\rho \, d\sigma + 1 + \kappa'(0)\rho. \tag{9}$$

If the number of non-zero multipole moments are finite, the equation reduces to an algebraic equation of the form:

$$a_n \rho^n + a_{n-1} \rho^{n-1} + \cdots + a_1 \rho + a_0 = 0. \tag{10}$$

We can then express Y in terms of powers of a solution to this equation. Solving general algebraic equations are usually difficult but Sturmfels [4] showed that the solutions can be written in terms of power series. Powers of series are in general quite cumbersome, but in our case we can write them down using the following theorem:

Theorem 3 *Let ρ be the solution of (10) with the appropriate asymptotics and assume that $\gamma \in \mathbb{R}^+$. Then*

$$\rho^\gamma = \left(\frac{-a_0}{a_1}\right)^\gamma \sum_{v_2 \ldots v_n \geq 0} \frac{\gamma a_0^{\sum_{j=2}^n (j-1) v_j} \Gamma(\gamma + \sum_{j=2}^n j v_j)}{(-a_1)^{\sum_{j=2}^n j v_j} \Gamma(1 + \gamma + \sum_{j=2}^n (j-1) v_j)} \prod_{j=2}^n \frac{a_j^{v_j}}{v_j!}. \tag{11}$$

With these tools it is possible to write down the potential for the spacetime with any desired finite set of multipole moments.

EXAMPLE

For the pure 2^n-pole we have $y(\rho) = q\frac{\rho^n}{n!}$, so that equation (9) reduces to

$$\frac{(1+2n)q^2}{n!(n+1)!} \rho^{2n+2} - \frac{\rho}{r} + 1 = 0.$$

Theorem 3 helps us to write down $Y(r) = \frac{q\rho^{n+1/2}}{\sqrt{r}n!}$ as a power series:

$$Y(r) = \frac{r^n q}{n!} \sum_{i=0}^\infty \frac{(2n+1)\Gamma(2(n+1)i + n + 1/2) r^{2(n+1)i} c^i}{2\Gamma((2n+1)i + n + 3/2) i!},$$

where $c = \frac{(1+2n)q^2}{n!(n+1)!}$. The potential $\tilde{\alpha}$ is then easily obtained

$$\tilde{\alpha}(r, \theta) = \frac{(2n+1)q}{n!} \sum_{i=0}^\infty \frac{(2(n+1)i + n)! \sqrt{\pi} c^i r^{2(n+1)i+n} P_{2(n+1)i+n}(\cos\theta)}{\Gamma((2n+1)i + n + 3/2) i! 2^{2(n+1)i+n+1}}. \tag{12}$$

This gives the spacetime.

REFERENCES

1. T. Bäckdahl, and M. Herberthson, *Static axisymmetric spacetimes with prescribed multipole moments*, Class. Quantum Grav. **22**, 1607–1621 (2005).
2. R. Geroch, *Multipole moments: II. Curved space*, J. Math. Phys. **11**, 2580–2588 (1970).
3. M. Herberthson, *The gravitational dipole and explicit multipole moments of static axisymmetric spactimes*, Class. Quantum Grav. **21**, 5121–5138 (2004).
4. B. Sturmfels, *Solving algebraic equations in terms of $\mathscr{A}$-hypergeometric series*, Discrete Math. **210**, 171–181 (2000)

Bigravity and Massive Gravity

D. Blas

Departament de Física Fonamental, Universitat de Barcelona,
Diagonal 647, 08028 Barcelona, Spain.

Abstract. We discuss some issues concerning the global structure of spherically symmetric solutions of bigravity. We propose maximal extensions of manifolds where two causal structure coexist. Besides we make some comments about the perturbations of these solutions and their relation to massive gravity and the cosmological constant problem.

Keywords: Bigravity, massive gravity, cosmological constant.
PACS: 04.50.+h, 11.30.Cp, 95.36.+x

INTRODUCTION

Massive gravity, in the broad sense, has been recently proposed as a possible alternative to dark energy (see e.g. [1] and references therein). Within this framework, the currently observed acceleration of the universe is not caused by any sort of exotic matter but is due to the dynamic of gravity in the infrared regime. This dynamic differ from the usual one of General Relativity by the addition of a mass term to the graviton, which means a weaker gravitational force at long distances (in comparison with the mass). At the linear level, one can simply add to the well-known Fierz-Pauli massless lagrangian, which describes particles of spin-2 propagating in a Minkowski background, a general mass term which breaks the gauge invariance of the theory:

$$\mathscr{L}_{FP} = \frac{1}{4}\partial_\mu h^{\nu\rho}\partial^\mu h_{\nu\rho} - \frac{1}{2}\partial_\mu h^{\mu\rho}\partial_\nu h^\nu_\rho + \frac{1}{2}\partial^\mu h \partial^\rho h_{\mu\rho} - \frac{1}{4}\partial_\mu h \partial^\mu h - \frac{1}{4}m^2(h^{\mu\nu}h_{\mu\nu} - rh^2) \tag{1}$$

It is well known that for $r = 1$ this lagrangian describes only the spin-2 degrees of freedom of the symmetric tensor $h_{\mu\nu}$ and that they have a proper behaviour (i.e. they are not neither ghosts or tachyons). Otherwise there is a ghost scalar degree of freedom which propagates [2, 3]. This seems to point to $r = 1$ as the only consistent choice; however if one calculates the gravitational interaction for massive bodies and compares it with the deflection of light in this theory, the massless limit does not coincide with the $m = 0$ calculation, and indeed is ruled out by observations in the Solar System [4, 5]. This discontinuity[1], known as van Dam-Veltman-Zakharov (vDVZ) discontinuity, has been argued to disappear in the non-linear regime [7] but the problem is then how to provide a theory of non-linear massive gravity. It is obvious that a covariant theory involving a single metric can not give rise to such mass terms for the metric as $g^{\mu\nu}g_{\nu\alpha} = \delta^\mu_\alpha$ and we need to include new fields in the theory. A concrete example is provided by the Kaluza-Klein reductions of models in higher dimensions where infinite new fields are added from the four dimensional point

---

[1] Recently there has been some interest in Lorentz breaking mass term where ghost and tachyon free theories with well defined massless limits can be defined [6].

of view. The vDVZ discontinuity has been shown to disappear at the non-linear level in certain models [8]. In principle a simpler choice would be to consider just the coupling of the metric to fields which condensate. The most straightforward choice (which can be seen as a first step in the higher dimensional settings) is thus to consider the existence of two interacting metrics (see [9] for further motivations for these theories).

f-g BIGRAVITY AND SPHERICAL SOLUTIONS

The simplest model of bigravity which we may think of is a theory given by two metrics each of which coupled to a particular kind of matter (*weakly coupled worlds* in the terminology of [9]) and where the metrics interact through a interaction term. That is, the action is given by

$$S = \int dx^4 \sqrt{-g}\left(\frac{-R_g}{2\kappa_g} + L_g\right) + \int dx^4 \sqrt{-f}\left(\frac{-R_f}{2\kappa_f} + L_f\right) + S_{int}[f,g] \quad (2)$$

where L_f and L_g refer to the lagrangian of the matter in each of the sectors which we will reduce to a vaccum energy contribution ($L_x = -\rho_x$). $S_{int}[f,g]$ is a general interaction term to which we demand to yield a Fierz-Pauli form when expanding one of the metric around Minkowski. For definiteness we will work with the form

$$S_{int}[f,g] = -\frac{\zeta}{4}\int d^4 (-g)^u (-f)^v (f^{\mu\nu} - g^{\mu\nu})(f^{\sigma\tau} - g^{\sigma\tau})(g_{\mu\sigma}g_{\nu\tau} - g_{\mu\nu}g_{\sigma\tau}) \quad (3)$$

with $2(u+v) = 1$ [10]. By a suitable coordinate choice we can write the most general static and spherical solution as[2]

$$g_{\mu\nu}dx^\mu dx^\nu = J dt^2 - K dr^2 - r^2\left(d\theta^2 + \sin^2\theta\, d\phi^2\right) \quad (4)$$
$$f_{\mu\nu}dx^\mu dx^\nu = C dt^2 - 2D dt dr - A dr^2 - B\left(d\theta^2 + \sin^2\theta\, d\phi^2\right) \quad (5)$$

The most generic solution for $D(r) \neq 0$ is given by two Schwarzschild-(Anti)de Sitter solutions with cosmological constants [11]

$$\frac{\Lambda_f}{\kappa_f} = \frac{\zeta}{4}\left(\frac{3}{2}\right)^{4u}\beta^u\{3v + 9\beta(1-v)\} + \rho_f, \quad (6)$$

$$\frac{\Lambda_g}{\kappa_g} = \frac{\zeta}{4}\left(\frac{2}{3}\right)^{4v}\beta^{-v}\{3u - 9\beta(1+u)\} + \rho_g. \quad (7)$$

where β is an integration constant. Thus we see that not only do we recover known solutions of ordinary GR without any trace of the vDVZ discontinuity, but also the cosmological constants depend on an integration parameter which can be fine tuned to give any desired value to one of them.

For $D(r) = 0$ the most general solution is not known but in [1] we proved that if one of the metric is solution of the Einstein's equations with an arbitrary cosmological constant,

---

[2] We are assuming that both metrics have the same signature and share the timelike direction.

the most general solution is given by two proportional metrics. These solutions are also important because, as claimed in [1], when two metrics with non compatible causal structure coexist, Cauchy horizons tend to appear and one can question the stability of these solutions. More concretely the solution is given by two Schwarzschild-(Anti)de Sitter metrics $f_{\mu\nu} = \gamma g_{\mu\nu}$ with

$$\Lambda_g = -6\kappa_g \zeta \gamma^{4\nu}(-1+\gamma)(-1-2u+2\gamma u)/(2\gamma^2) + \kappa_g \rho_g, \tag{8}$$
$$\Lambda_f = -6\kappa_f \zeta \gamma^{-4u}(-1+\gamma)(1-2v+2\gamma v)/(2\gamma^2) + \kappa_f \rho_f. \tag{9}$$

which γ satisfying $\Lambda_g = \gamma \Lambda_f$. Again, we find known solution of GR and now the cosmological constants depend on a parameter γ which in contrast to the previous situation must satisfy a dynamical equation which can give rise to the off-loading of one of the cosmological constants to the other sector, yielding a small Λ even if the matter contribution is large [12].

CAUSAL STRUCTURE

In general, when dealing with bigravity solutions we will face with two coexisting non compatible causal structures, which means that problems with global hyperbolicity, closed time-like curves (CTC) or geodesic completeness for both metrics simultaneously may appear. One way to study these issues is through mapping the null-geodesics of one of the metrics in the conformal diagram of the other one. As an example we can consider the solution with a Minkowski and a de Sitter metric coexisting in a manifold which is geodesically complete for Minkowski but not for de Sitter (Fig. 1). From Fig. 1, there is a

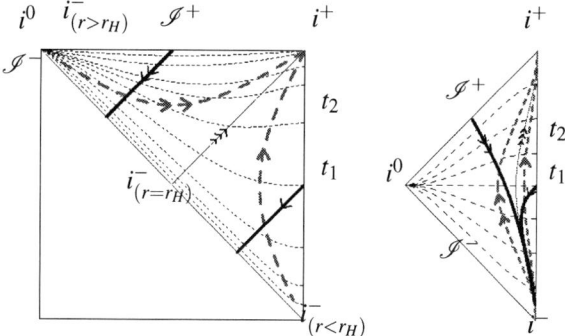

FIGURE 1. Causal diagram for de Sitter with Minkowski. We plot de Sitter null geodesics in the Minkowski causal structure. See [1] for further explanations.

region of the geodesically complete manifold for both metrics which is not visible for one of them (a sort of Cauchy horizon for the Minkowski observer). However, by interaction we may go from the already geodesically complete manifold for Minkowski to a new region of the space-time where again we have two metrics interacting. The simplest thing we can do is to add another Minkowski space time and find a space-time with a causal structure showed in Fig. 2. From Fig. 2 we can also see a behaviour which is characteristic of the presence of a Killing horizon for only one of the metrics: once we impose geodesic completeness, there is not a common Cauchy surface for both metrics. This is easily seen

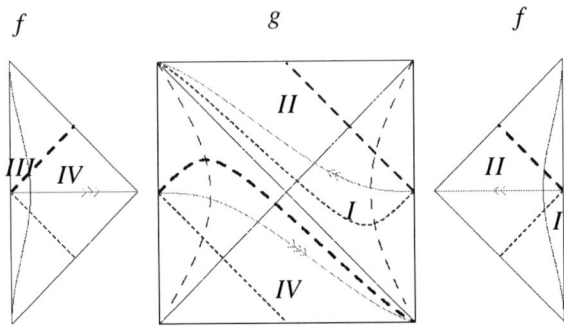

FIGURE 2. Geodesically Complete diagram for both Minkowski and de Sitter. We plot Minkowski geodesics.

in Fig. 2 by studying null cones for the Minkowski space (f) close to the horizon. As the Cauchy surface must be space-like and cross the horizon, a null geodesic for f which is close enough to the horizon will cross any Cauchy surface for g twice. This result can be generalized to general bigravity settings [12].

We can study more complicated solutions where both metrics have horizons which do not coincide. In such a case the manifold must be extended in both metrics, and this extension gives rise to a kind of "stair" [1].

Concerning CTC, we may imagine that using both causal structures to propagate a signal we could construct a CTC even if both metrics are causally well-behaved. Nevertheless, we have proved that these curves are absent for all the known solutions of the $f - g$ bigravity theories (cfr. [1]).

MASSIVE GRAVITY IN BIGRAVITY AND DE SITTER SPACE

Finally, some words are in order concerning the perturbations of these solutions. For $D \neq 0$ solutions the interaction term breaks the symmetry of any of the solutions (even if it preserves a $SO(3)$). In particular, considering perturbations around two Minkowski metrics which do not share a Lorentz symmetry we find a spectrum with a massive graviton with a mass term breaking the Lorentz symmetry [12]. As we mentioned before this kind of behaviour can give rise to a well behaved massive gravity theory (see [6, 13]). However, as expected, we find a tachyonic behaviour in this case where there is not a common causal structure.

Concerning the solutions with two proportional metrics, perturbations to these solutions amounts to perturbations of two ordinary Schwarzschild-(Anti)de Sitter, one of them including a generic mass term which depends on the interaction term. We have studied the most generic mass term for de Sitter space with a lagrangian similar to (1) but in this curved background and found that for

$$0 \leq m^2(1-r) \leq \frac{3H^2}{4\kappa_-} \tag{10}$$

where $H^2 = \Lambda_g/3$ and $\kappa_- = (\kappa_f + \gamma \kappa_g)\gamma^{-1}$ we find ghost free solutions [12]. Nevertheless, these solutions have tachyons for high momentum in comparison to the cosmological scale except for the Fierz-Pauli case $r = 1$.

I would like to thank C. Deffayet and J. Garriga with whom all this work has been done.

REFERENCES

1. D. Blas, C. Deffayet and J. Garriga, arXiv:hep-th/0508163.
2. M. Fierz and W. Pauli, Proc. Roy. Soc. Lond. A **173** (1939) 211.
3. P. Van Nieuwenhuizen, Nucl. Phys. B **60**, 478 (1973).
4. H. van Dam and M. J. G. Veltman, Nucl. Phys. B **22** (1970) 397.
5. V. I. Zakharov, JETP Letters (Sov. Phys.) **12** (1970) 312
6. V. A. Rubakov, arXiv:hep-th/0407104.
7. A. I. Vainshtein, Phys. Lett. B **39** (1972) 393.
8. C. Deffayet, G. R. Dvali, G. Gabadadze and A. I. Vainshtein, Phys. Rev. D **65** (2002) 044026 [arXiv:hep-th/0106001].
9. T. Damour and I. I. Kogan, Phys. Rev. D **66** (2002) 104024 [arXiv:hep-th/0206042].
10. C. J. Isham, A. Salam and J. A. Strathdee, Phys. Rev. D **3** (1971) 867.
11. A. Salam and J. A. Strathdee, Phys. Rev. D **16** (1977) 2668.
12. D. Blas, C .Deffayet, J. Garriga, *in preparation*.
13. N. Arkani-Hamed, H. C. Cheng, M. A. Luty, S. Mukohyama and T. Wiseman, arXiv:hep-ph/0507120.

Nuclear matter equation of state in relativistic nonlinear models: results and applications

J. Pastor, J.C. Caillon, J. Labarsouque

Centre d'Etudes Nucléaires de Bordeaux Gradignan,
Université Bordeaux I, IN2P3,
le Haut Vigneau, BP 120, 33175 Gradignan Cedex, France

Abstract.
We have determined the equation of state of nuclear matter according to relativistic non-linear models. In particular, we are interested in regions of high density and/or high temperature, in which the thermodynamic functions have different behaviours depending on which model one used. As applications, we have determined the maximal mass of neutron stars and studied the process of two-pion annihilation into e^+e^- pairs in dense and hot matter. We have found that these two observables are strongly sensitive to the nonlinear self-coupling terms of the Lagragian.

Keywords: Equation of state of nuclear matter, Maximal mass of neutron stars, Dilepton production rate
PACS: 21.65.+f,24.10.Jv;25.75.Dw

INTRODUCTION

A better knowledge of the nuclear matter equation of state would be very helpful for a better understanding of many physical and astrophysical phenomena. For example, the properties of nuclear matter at finite temperature and at densities far away from the saturation one are an essential input to calculate the dilepton production rate in nucleus-nucleus collisions. Concerning astrophysical systems, the equation of state of asymmetric nuclear matter plays a major role in determining the mass of neutron stars or the final evolution of the supernova collapse. The determination of the nuclear matter properties as functions of density and temperature is thus a very fundamental problem. However, up to now, the nuclear matter equation of state is still largely undetermined, more particularly, in the regions of high density and temperature.

Since a description of the properties of a nuclear system starting from QCD is not yet available, a number of effective theories have been developed. Among them, relativistic nonlinear models are effective theories which have gained more and more success in the description of nuclear matter and ground state properties of finite nuclei. In these models, the Lagrangian includes usual linear nucleon-meson couplings (as in the original Walecka model[1]) as well as couplings between the mesonic fields. The nucleons interact self-consistently through meson fields exchange in mean field approximation. Among many models, the models NL1[2], NL-SH[3] and NL3[4] contain only σ self-coupling terms in the Lagrangian while TM1 (for medium and heavy nuclei) and TM2[5] (for light nuclei) include in addition an ω self-coupling term. It has also been proposed two parameter sets G1 and G2[6], coming from an effective field theory which allows all scalar-vector coupling terms up to fourth order in the Lagrangian.

All of these models lead to results which compare rather well with the ground state properties of finite nuclei, but we may wonder if they also can provide a realistic description of dense and/or hot nuclear matter. An indication may be provided, for example, by the measurements of both astrophysical quantities and dilepton production in relativistic heavy-ion collisions. For this later, experiments from CERES collaboration[7] have shown that there is an excess of dileptons over those expected in the low invariant mass region, excess which cannot be explained by uncertainties and errors in the normalization procedure. Thus, in this work, we have first compared the equations of state for symmetric nuclear matter obtained in the various nonlinear models mentioned above. Then, as applications, we have considered the maximum mass of neutron stars as well as the process of two-pion annihilation into e^+e^- pairs in dense and hot matter which provides the main source of the observed dileptons in the low mass region.

EQUATION OF STATE

In Fig.1 we have chosen to show only the isotherms of the nuclear matter equation of state obtained in the models NL1 and G1 but all the nonlinear models previously cited have been considered. In the low density and temperature regime, the curves exhibit a typical Van der Waals-like interaction (liquid-gas phase transition) practically identical in all the models studied. At the opposite, the high density limit is strongly model dependent. Indeed, for models without ω self-coupling like NL1, NL-SH and NL3, we have obtained $p \to \mathcal{E}$, as it can be seen in Fig.1 (left panel). On the other hand, for models with an ω self-coupling like TM1, TM2 and G1, the limit $p \to \frac{\mathcal{E}}{3}$ has been found (as it can be seen in the right panel of Fig.1) implying that the thermodynamic speed of sound approaches one third of the velocity of light. In fact, we have shown[8] that such a limit is obtained whatever the nonlinear model used provided that an ω self-coupling term be included in the Lagrangian. Since these two limits are reached rather quickly, i.e. for densities close to five times the saturation one, any process involving nuclear matter at such densities should be strongly sensitive to these two different behaviors.

APPLICATIONS

As a first application, we have determined the maximal mass of neutron stars using the Tolman-Oppenheimer-Volkoff equations. We have considered the equation of state for pure neutron matter at zero temperature obtained in the relativistic nonlinear models previously discussed. We have plotted, in Fig.2 (left panel), the maximal mass of neutron stars as a function of the central density in the star. Our results show two different behaviors related to the two high density limits obtained. Indeed, for models without ω self-coupling like NL1, NL-SH and NL3, we have found a value of the maximal mass close to 2.1 times the solar mass while for models with an ω self-coupling like TM1, TM2 and G1, this value is then 2.8.

As a second application, we have determined[9] to what extent the nonlinear couplings (of the more elaborated models G1 and G2) are able to increase the dilepton pro-

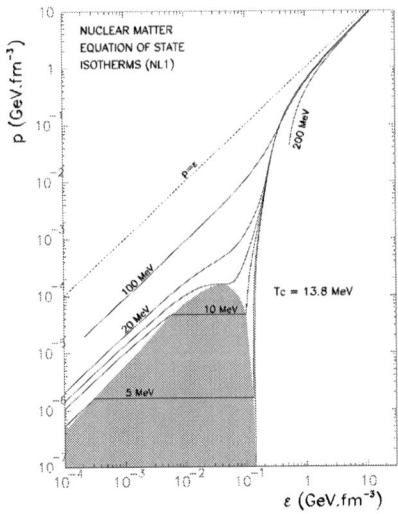 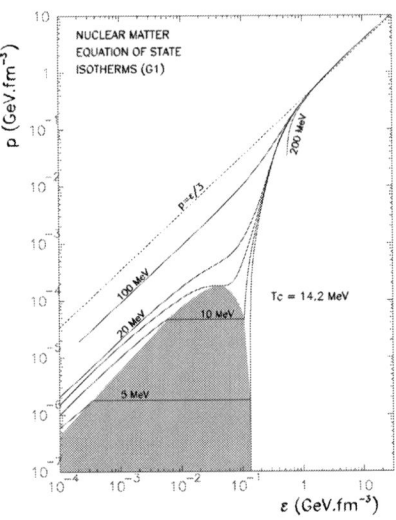

FIGURE 1. Equations of state for symmetric nuclear matter in models NL1[2] (left panel) and G1[6] (right panel). The region of liquid-gas phase coexistence is shown by the shaded area.

duction in the low invariant mass regime. We have considered the two-pion annihilation process $\pi^+\pi^- \to \rho \to e^+e^-$ which provides the main source of the observed dileptons in the low mass region. Therefore, the in-medium ρ meson propagator has been taken into account in this process. We have plotted, in Fig.2 (right panel), the dilepton production rates at baryon density equal to twice the saturation one and for a temperature of 100 MeV. In this figure, the solid curve represents the results using the G1 model, while for the long dashed curve, the G2 model has been used. For comparison, we have also displayed the result obtained using the free space ρ meson propagator (given by the short dashed curve). As we can see, we have obtained an enhancement of the dilepton production rate using the G1 model (in comparison with the free space result) for invariant masses below 700 MeV. At the opposite, the G2 model leads to a strong decrease for invariant masses below 800 MeV. We have shown that such a difference arises from the opposite sign of the $\sigma\rho\rho$ coupling constant in the G1 and G2 models.

CONCLUSION

We have determined the equation of state of nuclear matter according to relativistic nonlinear models. We have obtained two different high density limits related to the presence or not of an ω self-coupling term in the Lagragian. As applications, we have determined the maximal mass of neutron stars and studied the process of two-pion annihilation into e^+e^- pairs in dense and hot matter. We have found a value of the

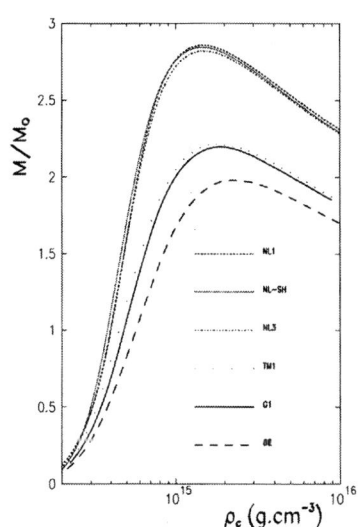 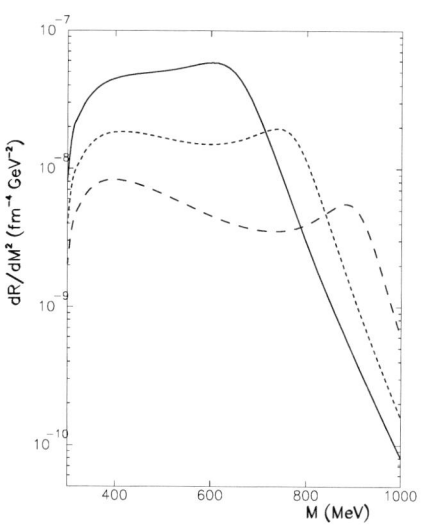

FIGURE 2. Left panel : Maximal mass of neutron stars as a function of the central density in the star. Right panel : The $\pi^+\pi^- \to \rho \to e^+e^-$ dilepton production rate at baryon density $\rho_B = 2\rho_0$ and temperature $T = 100$ MeV (see text).

maximal mass of neutron stars strongly dependent on the presence or not of an ω self-coupling term in the Lagrangian. However, the astrophysical mesurements do not allow a precise discrimination of the nonlinear models used. On the same way, the dilepton spectra obtained is strongly sensitive to the $\sigma\rho\rho$ nonlinear term. In particular, the G1 model gives an enhancement of the dilepton production rate in the low invariant mass region in good agreement with the invariant mass dependence of the data obtained in relativistic heavy ions collisions at CERN/SPS energies. Thus, this result indicates that the G1 model seems realistic enough to describe dense and hot nuclear matter such that produced in relativistic heavy-ion collisions.

REFERENCES

1. B.D. Serot and J.D. Walecka, Advances in Nuclear Physics **16** (1986) 1
2. P.G. Reinhard, M. Rufa, J. Maruhn, W. Greiner and J. Friedrich, Z. Phys. **A 323** (1986) 13
3. M.M. Sharma, M.A. Nagarajan and P. Ring, Phys. Lett. **B 312** (1993) 377
4. G.A. Lalazissis, J. König and P. Ring, Phys. Rev. **C55** (1997) 540
5. Y. Sugahara and H. Toki, Nucl. Phys. **A 579** (1994) 557
6. R.J. Furnstahl, B.D. Serot, H.-B. Tang, Nucl. Phys. **A 615** (1997) 441
7. P. Wurm for the CERES Collaboration, Nucl. Phys. **A 590**(1995) 103c
8. J. Pastor, J. C. Caillon, J. Labarsouque, Nucl. Phys. **A 735** (2004) 125
9. J. Pastor, J.C. Caillon, J. Labarsouque, Nucl. Phys. **A 757** (2005) 456

Neutrino mixing and dark energy

M. Blasone[*], A. Capolupo[*], S. Capozziello[†] and G. Vitiello[*]

[*]Dipartimento di Fisica "E.R. Caianiello" and INFN, Università di Salerno, I-84100 Salerno, Italy
[†]Dipartimento di Scienze Fisiche, Università di Napoli "Federico II" and INFN Sez. di Napoli, Compl. Univ. Monte S. Angelo, Ed.N, Via Cinthia, I-80126 Napoli, Italy

Abstract. We report on the recent result that the non–perturbative vacuum structure associated with neutrino mixing leads to a non–zero contribution to the value of the dark energy.

1. INTRODUCTION

In the context of Quantum Field Theory (QFT), the vacuum for neutrinos with definite mass is not invariant under the field mixing transformation and in the infinite volume limit it is unitarily inequivalent to the vacuum for the neutrinos with definite flavor [1]-[12]. This phenomenon affects the oscillation formulas which turns out to be different from the usual Quantum Mechanics ones [13] and it is also crucial in order to obtain a non–zero contribution to the dark energy [14].

In this report we show that the vacuum energy induced by the neutrino mixing may contribute to the value of the dark energy and we estimate its value by using the cut-off proportional to the natural scale of the neutrino mixing phenomenon.

The formalism of the fermion mixing in QFT is introduced in Section 2. In Section 3 the neutrino mixing contribution to the dark energy is computed and the conclusions are drown.

2. FERMION MIXING

For simplicity, we consider two Dirac neutrino fields. The Pontecorvo mixing transformations are [13]

$$v_e(x) = v_1(x)\cos\theta + v_2(x)\sin\theta$$
$$v_\mu(x) = -v_1(x)\sin\theta + v_2(x)\cos\theta, \qquad (1)$$

where $v_e(x)$ and $v_\mu(x)$ are the fields with definite flavors, θ is the mixing angle and v_1 and v_2 are the fields with definite masses $m_1 \neq m_2$. Explicitly v_1 and v_2 are given by

$$v_j(x) = \frac{1}{\sqrt{V}}\sum_{\mathbf{k},r}\left[u^r_{\mathbf{k},j}(t)\,\alpha^r_{\mathbf{k},j} + v^r_{-\mathbf{k},j}(t)\,\beta^{r\dagger}_{-\mathbf{k},j}\right]e^{i\mathbf{k}\cdot\mathbf{x}}, \qquad j=1,2, \qquad (2)$$

with $u^r_{\mathbf{k},j}(t) = e^{-i\omega_{k,j}t} u^r_{\mathbf{k},j}(0)$, $v^r_{\mathbf{k},j}(t) = e^{i\omega_{k,j}t} v^r_{\mathbf{k},j}(0)$ and $\omega_{k,j} = \sqrt{\mathbf{k}^2 + m_j^2}$. The vacuum for the α_j and β_j operators is denoted by $|0\rangle_{1,2}$: $\alpha^r_{\mathbf{k},j}|0\rangle_{12} = \beta^r_{\mathbf{k},j}|0\rangle_{12} = 0$.

The anticommutation relations are the usual ones (see Ref.[1]). The orthonormality and completeness relations are: $u^{r\dagger}_{\mathbf{k},j} u^s_{\mathbf{k},j} = v^{r\dagger}_{\mathbf{k},j} v^s_{\mathbf{k},j} = \delta_{rs}$, $u^{r\dagger}_{\mathbf{k},j} v^s_{-\mathbf{k},j} = v^{r\dagger}_{-\mathbf{k},j} u^s_{\mathbf{k},j} = 0$, and $\sum_r (u^r_{\mathbf{k},j} u^{r\dagger}_{\mathbf{k},j} + v^r_{-\mathbf{k},j} v^{r\dagger}_{-\mathbf{k},j}) = 1$.

The flavor fields can be expanded as (we use $(\sigma, j) = (e, 1), (\mu, 2)$):

$$\begin{aligned} \nu_\sigma(x) &= G_\theta^{-1}(t) \nu_j(x) G_\theta(t) \\ &= \frac{1}{\sqrt{V}} \sum_{\mathbf{k},r} \left[u^r_{\mathbf{k},j}(t) \alpha^r_{\mathbf{k},\nu_\sigma}(t) + v^r_{-\mathbf{k},j}(t) \beta^{r\dagger}_{-\mathbf{k},\nu_\sigma}(t) \right] e^{i\mathbf{k}\cdot\mathbf{x}}, \end{aligned} \quad (3)$$

where $G_\theta(t)$ is the generator of the mixing transformations (1) given by

$$G_\theta(t) = \exp\left[\theta \int d^3\mathbf{x} \left(\nu_1^\dagger(x) \nu_2(x) - \nu_2^\dagger(x) \nu_1(x) \right) \right]. \quad (4)$$

The flavor fields ν_e and ν_μ in Eq.(3) are expanded in the same basis of ν_1 and ν_2. The flavor annihilation operators and the flavor vacuum are defined as

$$\begin{pmatrix} \alpha^r_{\mathbf{k},\nu_\sigma}(t) \\ \beta^{r\dagger}_{-\mathbf{k},\nu_\sigma}(t) \end{pmatrix} = G_\theta^{-1}(t) \begin{pmatrix} \alpha^r_{\mathbf{k},i} \\ \beta^{r\dagger}_{-\mathbf{k},i} \end{pmatrix} G_\theta(t). \quad (5)$$

$$|0(t)\rangle_{e,\mu} \equiv G_\theta^{-1}(t) |0\rangle_{1,2}. \quad (6)$$

$|0(t)\rangle_{e,\mu}$ turns out to be orthogonal to the vacuum for the mass eigenstates $|0\rangle_{1,2}$ in the infinite volume limit. Note the time dependence of $|0(t)\rangle_{e,\mu}$. In the following we will denote the flavor vacuum state at the reference time $t = 0$ as $|0\rangle_{e,\mu}$.

The explicit expression of the flavor annihilation/creation operators for $\mathbf{k} = (0, 0, |\mathbf{k}|)$ is:

$$\begin{aligned} \alpha^r_{\mathbf{k},\nu_e}(t) &= \cos\theta\, \alpha^r_{\mathbf{k},1} + \sin\theta \left(U^*_\mathbf{k}(t) \alpha^r_{\mathbf{k},2} + \varepsilon^r V_\mathbf{k}(t) \beta^{r\dagger}_{-\mathbf{k},2} \right) \\ \alpha^r_{\mathbf{k},\nu_\mu}(t) &= \cos\theta\, \alpha^r_{\mathbf{k},2} - \sin\theta \left(U_\mathbf{k}(t) \alpha^r_{\mathbf{k},1} - \varepsilon^r V_\mathbf{k}(t) \beta^{r\dagger}_{-\mathbf{k},1} \right) \\ \beta^r_{-\mathbf{k},\nu_e}(t) &= \cos\theta\, \beta^r_{-\mathbf{k},1} + \sin\theta \left(U^*_\mathbf{k}(t) \beta^r_{-\mathbf{k},2} - \varepsilon^r V_\mathbf{k}(t) \alpha^{r\dagger}_{\mathbf{k},2} \right) \\ \beta^r_{-\mathbf{k},\nu_\mu}(t) &= \cos\theta\, \beta^r_{-\mathbf{k},2} - \sin\theta \left(U_\mathbf{k}(t) \beta^r_{-\mathbf{k},1} + \varepsilon^r V_\mathbf{k}(t) \alpha^{r\dagger}_{\mathbf{k},1} \right), \end{aligned} \quad (7)$$

with $\varepsilon^r = (-1)^r$ and

$$U_\mathbf{k}(t) \equiv u^{r\dagger}_{\mathbf{k},2}(t) u^r_{\mathbf{k},1}(t) = v^{r\dagger}_{-\mathbf{k},1}(t) v^r_{-\mathbf{k},2}(t) \quad (8)$$

$$V_\mathbf{k}(t) \equiv \varepsilon^r u^{r\dagger}_{\mathbf{k},1}(t) v^r_{-\mathbf{k},2}(t) = -\varepsilon^r u^{r\dagger}_{\mathbf{k},2}(t) v^r_{-\mathbf{k},1}(t). \quad (9)$$

We have:

$$U_\mathbf{k}(t) = |U_\mathbf{k}| e^{i(\omega_{k,2} - \omega_{k,1})t}, \qquad V_\mathbf{k}(t) = |V_\mathbf{k}| e^{i(\omega_{k,2} + \omega_{k,1})t} \quad (10)$$

$$|U_{\mathbf{k}}| = \left(\frac{\omega_{k,1}+m_1}{2\omega_{k,1}}\right)^{\frac{1}{2}} \left(\frac{\omega_{k,2}+m_2}{2\omega_{k,2}}\right)^{\frac{1}{2}} \left(1+\frac{|\mathbf{k}|^2}{(\omega_{k,1}+m_1)(\omega_{k,2}+m_2)}\right), \quad (11)$$

$$|V_{\mathbf{k}}| = \left(\frac{\omega_{k,1}+m_1}{2\omega_{k,1}}\right)^{\frac{1}{2}} \left(\frac{\omega_{k,2}+m_2}{2\omega_{k,2}}\right)^{\frac{1}{2}} \left(\frac{|\mathbf{k}|}{(\omega_{k,2}+m_2)} - \frac{|\mathbf{k}|}{(\omega_{k,1}+m_1)}\right), \quad (12)$$

$$|U_{\mathbf{k}}|^2 + |V_{\mathbf{k}}|^2 = 1. \quad (13)$$

The condensation density is given by

$$_{e,\mu}\langle 0|\alpha_{\mathbf{k},i}^{r\dagger}\alpha_{\mathbf{k},i}^{r}|0\rangle_{e,\mu} = {}_{e,\mu}\langle 0|\beta_{\mathbf{k},i}^{r\dagger}\beta_{\mathbf{k},i}^{r}|0\rangle_{e,\mu} = \sin^2\theta \, |V_{\mathbf{k}}|^2, \quad i=1,2. \quad (14)$$

$|V_{\mathbf{k}}|^2$ is zero for $m_1 = m_2$, it has a maximum at $|\mathbf{k}| = \sqrt{m_1 m_2}$. For $|\mathbf{k}| \gg \sqrt{m_1 m_2}$, it goes like $|V_{\mathbf{k}}|^2 \simeq (m_2 - m_1)^2/(4|\mathbf{k}|^2)$.

3. NEUTRINO MIXING CONTRIBUTION TO THE DARK ENERGY

We use the formalism presented above to derive the neutrino mixing contribution to the dark energy. The calculation is performed in a Minkowski spacetime but it can be easily extended to curved space-times. In the Minkowski metric, the $(0,0)$ component of the energy momentum tensor density $\mathcal{T}_{00}$ is

$$\mathcal{T}_{00}(x) = \frac{i}{2}\sum_{\sigma=e,\mu} : \left(\bar{\nu}_\sigma(x)\gamma_0 \overleftrightarrow{\partial}_0 \nu_\sigma(x)\right) := \frac{i}{2}\sum_{j=1,2} : \left(\bar{\nu}_j(x)\gamma_0 \overleftrightarrow{\partial}_0 \nu_j(x)\right): \quad (15)$$

where $:\ldots:$ denotes the customary normal ordering with respect to the mass vacuum in the flat space-time. In terms of the annihilation and creation operators of fields ν_1 and ν_2, the energy-momentum tensor $T_{00} = \int d^3x \mathcal{T}_{00}(x)$ is given by

$$T_{00} = \sum_{r,j}\int d^3\mathbf{k}\, \omega_{k,j} \left(\alpha_{\mathbf{k},j}^{r\dagger}\alpha_{\mathbf{k},j}^{r} + \beta_{-\mathbf{k},j}^{r\dagger}\beta_{-\mathbf{k},j}^{r}\right). \quad (16)$$

Note that T_{00} is time independent.

The expectation value of T_{00} in the flavor vacuum $|0\rangle_{e,\mu}$ gives the contribution $\langle \rho_{vac}^{mix}\rangle$ of the neutrino mixing to the vacuum energy density:

$$_{e,\mu}\langle 0|T_{00}|0\rangle_{e,\mu} = \langle \rho_{vac}^{mix}\rangle \eta_{00}. \quad (17)$$

Within the QFT formalism for neutrino mixing we have $_{e,\mu}\langle 0|T_{00}|0\rangle_{e,\mu} = {}_{e,\mu}\langle 0(t)|T_{00}|0(t)\rangle_{e,\mu}$ for any t. We then obtain

$$_{e,\mu}\langle 0|T_{00}|0\rangle_{e,\mu} = \sum_{r,j}\int d^3\mathbf{k}\, \omega_{k,j}\left({}_{e,\mu}\langle 0|\alpha_{\mathbf{k},j}^{r\dagger}\alpha_{\mathbf{k},j}^{r}|0\rangle_{e,\mu} + {}_{e,\mu}\langle 0|\beta_{\mathbf{k},j}^{r\dagger}\beta_{\mathbf{k},j}^{r}|0\rangle_{e,\mu}\right),$$

and

$$_{e,\mu}\langle 0|T_{00}|0\rangle_{e,\mu} = 8\sin^2\theta \int d^3\mathbf{k}\, (\omega_{k,1}+\omega_{k,2})\,|V_\mathbf{k}|^2 = \langle \rho_{vac}^{mix}\rangle \eta_{00}, \qquad (18)$$

i.e.

$$\langle \rho_{vac}^{mix}\rangle = 32\pi^2 \sin^2\theta \int_0^K dk\, k^2 (\omega_{k,1}+\omega_{k,2})|V_\mathbf{k}|^2, \qquad (19)$$

where the cut-off K has been introduced. Eq.(19) shows that the dark energy gets a non-zero contribution induced purely from the neutrino mixing [14]. Notice that such a contribution is indeed zero in the no-mixing limit ($\theta = 0$ and/or $m_1 = m_2$). Moreover, the contribution is absent in the traditional phenomenological (Pontecorvo) mixing treatment.

Choosing the cut-off proportional to the natural scale appearing in the mixing phenomenon, $K \simeq \sqrt{m_1 m_2}$ or the cut-off scale given by the sum of the two neutrino masses, $K = m_1 + m_2$ [15], and using $m_1 = 7 \times 10^{-3} eV$, $m_2 = 5 \times 10^{-2} eV$ and $\sin^2\theta \simeq 0.3$, we have $\langle \rho_{vac}^{mix}\rangle = 0.4 \times 10^{-47} GeV^4$, which is in agreement with the estimated value of the dark energy.

In conclusion, the QFT treatment of the neutrino mixing leads to a non zero contribution to the dark energy [14]. Our result discloses a new possible, non-perturbative mechanism contributing to the dark energy value.

ACKNOWLEDGEMENTS

We thank the organizers of the XXVIII Spanish Relativity Meeting E.R.E. 2005, Oviedo 2005, Spain. We also acknowledge the ESF network COSLAB, INFN and MIUR.

REFERENCES

1. M. Blasone and G. Vitiello, Annals Phys. **244**, 283 (1995).
2. M. Blasone, P.A. Henning and G. Vitiello, Phys. Lett. B **451**, 140 (1999).
3. M. Blasone, P. Jizba and G. Vitiello, Phys. Lett. B **517**, 471 (2001).
4. M. Blasone and G. Vitiello, Phys. Rev. D **60**, 111302 (1999).
5. K. Fujii, C. Habe and T. Yabuki, Phys. Rev. D **59**, 113003 (1999); Phys. Rev. D **64**, 013011 (2001).
6. M. Binger and C.R. Ji, Phys. Rev. D **60**, 056005 (1999).
7. M. Blasone, A. Capolupo, O. Romei and G. Vitiello, Phys. Rev. D **63**, 125015 (2001).
8. C.R. Ji, Y. Mishchenko, Phys. Rev. D **64**, 076004 (2001); Phys. Rev. D **65**, 096015 (2002).
9. M. Blasone, A. Capolupo and G. Vitiello, Phys. Rev. D **66**, 025033 (2002).
10. A. Capolupo, Ph.D. Thesis [hep-th/0408228].
11. A. Capolupo, C. R. Ji, Y. Mishchenko and G. Vitiello, Phys. Lett. B **594**, 135 (2004).
12. M. Blasone, A. Capolupo, F. Terranova and G. Vitiello, Phys. Rev. D **72** 013003 (2005).
13. S.M. Bilenky and B. Pontecorvo, Phys. Rep. **41**, 225 (1978).
    S.M. Bilenky and S.T. Petcov, Rev. Mod. Phys. **59**, 671 (1987).
14. M. Blasone, A. Capolupo, S. Capozziello, S. Carloni and G. Vitiello, Phys. Lett. A **323**, 182 (2004); Braz. J. Phys. 35: 455-461, (2005), AIP Conf. Proc. **751**, 208 (2005) [hep-th/0410196].
15. G. Barenboim and N. E. Mavromatos, Phys. Rev. D **70** (2004) 093015

On The Rindler Horizon Energy

Hristu Culetu

Ovidius University, Department of Physics, B-dul Mamaia 124, 8700 Constanta, Romania

Abstract. A nonvanishing value for the Rindler horizon energy is obtained, by an analogy with the "near horizon" Schwarzschild spacetime, contrary to Padmanabhan's result based on the formalism of Euclideanization of the Einstein action. We show that the Rindler horizon energy is given by the same formula E = α/2 obtained by Padmanabhan for the Schwarzschild horizon, where α is the gravitational radius.

Keywords: observer dependent horizon, the Holographic Principle, one way membrane, uniformly accelerated observer..
PACS: 04.70.Dy, 04.50.+h, 04.60.-m.

1. INTRODUCTION

One of the basic features of classical gravity is the generation of surfaces acting as one – way membranes. The typical example is given by the Schwarzschild black hole which has a closed observer – independent surface (event horizon). The de Sitter spacetime has also a compact surface which is however dependent upon the observer [1]. Surprisingly, a one way membrane may be induced even in flat spacetime, with the appearance of an observer – dependent horizon (the non – compact Rindler horizon).

The problem is whether one can associate thermodynamic parameters (temperature, entropy and energy) to that horizon, as for the black hole.

2. THE ELECTROMAGNETIC ANALOGY

One of the authors who developed a model on the subject was T. Padmanabhan [1,3]. He has used the class of metrics

$$ds^2 = -f(r)dt^2 + f^{-1}(r)dr^2 + dL_{\perp}^2 \qquad (1)$$

as a canonical ensemble at constant temperature T = 1/β, with $dL_{\perp}^2$ - the transverse two – dimensional metric. He calculated the partition function

$$Z(\beta) \propto \exp(S - \beta E) \qquad (2)$$

where S and E are the entropy and energy of the system, respectively. The function f(r) is vanishing on some surface $r = \alpha$ (event horizon) and f '(α) is finite. In this case T = |f '(α)| /4π. Padmanabhan computed S and E using the expression of Z(β) in terms of the (Euclidean) gravitational action and found that

$$S = \frac{1}{4} 4\pi\alpha^2 = \frac{A_{hor}}{4} \; ; \qquad E = \frac{\alpha}{2} \qquad (3)$$

For a spacetime with planar symmetry (for instance, the Rindler geometry), he established that S = (1/4) $A_\perp$ while the energy is vanishing ($A_\perp$ is a finite part of the infinite transverse area). We try in this paper to show that the expression E = α/2 is valid for the Rindler spacetime, too. F. Alexander and U. Gerlach [2] showed that the electric field of a uniformly accelerated charge e induces on the event horizon a surface charge density such that the total charge on the surface is –e. In other words, the event horizon behaves like a conductive surface. Moreover, an attractive force between the point charge and the horizon appears, its expression proving to be equal to the radiation reaction force from the r.h.s. of the Lorentz – Dirac equation. By analogy with the electromagnetic case, we suggest to consider that the Rindler horizon contains an (observer dependent) surface density σ given by

$$g = 4\pi G \sigma \qquad (4)$$

where g is the proper acceleration of the hyperbolic observer From now on we take the fundamental constants to be unity ($G = c = k_B = h = 1$).
Among others, Padmanabhan [5] (see also [4]) has observed that the relation (4) "makes physical sense because the accelerated observer will attribute such a surface energy density as the source of the apparent gravitational acceleration". It is, however, in contradiction with his E = 0 result obtained in the paper [1] for the Rindler spacetime.

3. THE "NEAR HORIZON" BLACK HOLE GEOMETRY

To show that $E = \alpha/2$ is true also for the Rindler horizon, we look for an analogy with Schwarzschild's geometry

$$ds^2 = -\left(1 - \frac{2m}{r}\right)dt^2 + \frac{dr^2}{1 - \frac{2m}{r}} + dL_\perp^2 \tag{5}$$

It is well known [5] that near the horizon $r = 2m$ of the black hole the metric (5) may be written in the form

$$ds^2 \approx -\frac{r-2m}{2m}dt^2 + \frac{2m}{r-2m}dr^2 + dL_\perp^2 \tag{6}$$

By means of the coordinate transformation

$$\sqrt{r-2m} = \frac{\rho}{2\sqrt{2m}} \tag{7}$$

eq. (6) becomes

$$ds^2 = -\left(\frac{1}{4m}\right)^2 \rho^2 dt^2 + d\rho^2 + dL_\perp^2 \tag{8}$$

which is exactly the Rindler spacetime. The proper acceleration is $g = 1/4m$, i.e. the surface gravity of the black hole.

Let us consider now a slightly different transformation. Instead of (7) we could introduce

$$\sqrt{r-2m} = \frac{\rho - 4m}{2\sqrt{2m}} \tag{9}$$

In other words, we take $\rho = 1/g$ for $r = 2m$. In this case eq. (8) looks like

$$ds^2 = -(g\rho - 1)^2 dt^2 + d\rho^2 + dL_\perp^2 \tag{10}$$

Hence, the horizon is translated from $\rho = 0$ to $\rho = 1/g$, with $\rho \geq 1/g$.

4. THE RINDLER HORIZON ENERGY

It is clear that the surface gravity $\kappa = 1/4m$ of the black hole plays a similar role with the proper acceleration g of the uniformly accelerated observer. Keeping in mind that the energy of the black hole is given by $E = m = \alpha/2 = 1/4\kappa$, we have therefore for the Rindler horizon

$$E(g) = \frac{1}{4g} = \frac{\alpha}{2} \tag{11}$$

(we have, of course, tacitly assumed that all the energy of the black hole is located at the horizon r = α. A similar situation is faced when the Vilenkin – Ipser – Sikivie (VIS) [6,7] domain wall is studied. As Chamblin and Eardley [8] have noticed, "we think of a VIS spacetime as an inflating universe where all of the vacuum energy has been concentrated on the sheet of the domain wall").
With all fundamental constants, eq. (11) appears as $E = (c^4/4G)(c^2/g)$.
The surface energy density on the horizon looks now as

$$\sigma = \frac{1}{4g}\frac{1}{4\pi(4m^2)} = \frac{g}{4\pi} \tag{12}$$

whence $g = 4\pi\sigma$, as expected (for instance, from Gauss' theorem).
The fact that Rindler's horizon might contain certain energy is also stressed by Padmanabhan in (9) where he proved that the scalar curvature is nonvanishing and is concentrated at the horizon when we treat the Rindler geometry as a limiting case of a family of metrics without horizons (see also [10]).
Having established that the energy of the "horizon membrane" is given by (11), we pass now to the thermodynamic parameters. The temperature is, of course, the Davies – Unruh expression $T_U = g/2\pi$ (it is well known that the Unruh radiation comes from the horizon of the accelerated observer). For the entropy of the Rindler horizon we use the "holographic" [11] expression $S = A_{hor}/4$. One obtains

$$S(g) = 4\pi m^2 = \frac{\pi}{4g^2} \tag{13}$$

It is easy to check that the thermodynamic relation $dE = T\, dS$ is obeyed.
Even though we started with an analogy between the Schwarzschild and Rindler spacetimes near the black hole horizon, we believe the Rindler horizon contains a surface energy given by eq. (12) and generated by the agent who accelerated the test

particle. We also stress that eq. (11) preserves its form even for the planar horizon of an uniformly accelerated observer, in Cartesian coordinates. This might be justified basing on the fact that, for an observer located very near the black hole horizon, it appears as a flat surface (the linear dimension of the observer is considered to be much less than Schwarzschild's radius).

In our coordinates (10), it looks like a "bubble" (domain wall?) located at a distance $\rho = 1/g$. When the expression (12) for σ is written as

$$\sigma = \frac{T}{2l_P^2} \qquad (14)$$

where l_P is the Planck length, we observe that in every "surface element" l_P^2 we find a normal mode average energy of $T/2$.

It is worth to note a relation between the classical expression (11) for the Rindler energy and the quantum one [12] (with all fundamental constants, for clarity)

$$E_{quan} = \frac{hc}{2\frac{c^2}{g}} \qquad (15)$$

Since $E_{class} \propto 1/g$ and $E_{quan} \propto g$, we have

$$E_{quan} E_{class} \approx \varepsilon_P^2 \qquad (16)$$

In other words, the Planck energy ε_P is of the order of the geometrical mean of the two, playing the role of a "boundary" between the two domains of energies.

5. CONCLUSIONS

We showed in this paper that Padmanabhan's expression $E = |\alpha/2|$ for the Schwarzschild and de Sitter horizon energies is valid for the Rindler horizon, too, even in Cartesian coordinates. We may consider that E is the origin of Unruh's radiation which comes from the Rindler horizon. In addition, the corresponding entropy $S(g)$ is finite and the thermodynamic relation $dE = TdS$ is fulfilled.

ACKNOWLEDGMENTS

I would like to thank the Organizers of the ERE05 Conference for inviting me to this very stimulating meeting and for their warm hospitality during my stay in Oviedo. I am also grateful to T. Padmanabhan for some useful comments and suggestions.

REFERENCES

1. T. Padmanabhan, Mod. Phys Letters A17, 923-942 (2002).

2. F. J. Alexander and U. Gerlach, ArXiv : gr-qc/9910086 (1999).
3. T. Padmanabhan, ArXiv : gr-qc/0202080 (2002).
4. J. Makela and A. Peltola, ArXiv : gr-qc/0205128 (2002).
5. T. Padmanabhan, ArXiv : hep-th/0212290 (2002).
6. J. Ipser and P. Sikivie, Phys. Rev. D30, 712-719 (1984).
7. A. Vilenkin, Phys. Rev. D23, 852-857 (1981).
8. A. Chamblin and D.M. Eardley, ArXiv : hep-th/9912166 (1999).
9. T. Padmanabhan, ArXiv : hep-th/0302068 (2003).
10. H. Culetu, Preprint IC/90/307, Trieste (1990).
11. H. Culetu, ArXiv : hep-th/0410133 (2004).
12. H.Culetu, Int .J. Mod. Phys. A18, 4251-4256 (2003).

Accurate simulations of the barmode instability in General Relativity

R. De Pietri*, L. Baiotti†, G. M. Manca* and L. Rezzolla†,**,‡

*Dipartimento di Fisica, Università di Parma and INFN, Parma, Italy
†Max-Planck-Institut für Gravitationsphysik, Albert-Einstein-Institut, Golm, Germany
**SISSA, International School for Advanced Studies and INFN, Trieste, Italy
‡Department of Physics, Louisiana State University, Baton Rouge, USA

Abstract. We present results of simulations in full General Relativity of the dynamical instability against barmode deformations of rapidly and differentially rotating neutron stars. Because of the high accuracy and long-term stability of our code the instability can develop without the introduction of initial, ad-hoc perturbations. This allows us to estimate accurately the threshold for the development of the instability that we determine to be $\beta_c = T/|W| \simeq 0.254$ for polytropic models with $\Gamma = 2$. We also find that the dynamics of the instability sensitively depends on the quantity $\beta - \beta_c$, with persistent bars being possible only for $\beta \gtrsim \beta_c$ and with strong nonlinear hydrodynamical effects emerging when $\beta \gg \beta_c$.

Keywords: Numerical Relativity, Neutron stars, Gravitational waves
PACS: 04.25.Dm, 04.30.Db, 04.40.Dg, 95.30.Lz,

INTRODUCTION

Rapidly rotating compact stars such as neutron stars may be subject to non-axisymmetric rotational instabilities with respect to toroidal modes. This is simply the result of the existence of equilibrium non-axisymmetric configurations that have lower kinetic energies than those relative to oblate axisymmetric spheroids. The threshold for the onset of the instability is traditionally expressed in terms of the dimensionless parameter $\beta \equiv T/|W|$, where T is the rotational kinetic energy and W the gravitational binding energy. An exact treatment of these instabilities exists only for incompressible equilibrium fluids in Newtonian gravity [1, 2], where global dynamical rotational instabilities are seen to arise from non-radial toroidal modes when β exceeds the critical Newtonian value $\beta_{c,N} \simeq 0.27$. Since these instabilities give rise to very strong quadrupolar deformations, they may represent powerful sources of gravitational radiation, possibly detectable by the Earth-based gravitational-wave detectors now in construction or operation. The determination of the threshold for the instability as well as the study of its evolution once it is fully developed, require a fully three-dimensional solution of the nonlinear hydrodynamical equations coupled to the solution of the Einstein equations. However, despite a long history of studies of this problem in Newtonian gravity, rather little is known in relativistic regimes. The first study of dynamical barmode instability in full General Relativity [3] has shown that the critical value for the onset of the instability is lower, with respect to the Newtonian case. This was later confirmed by post-Newtonian calculations [4, 5], which also showed that the critical value β_c varies with the compactness M/R of the star.

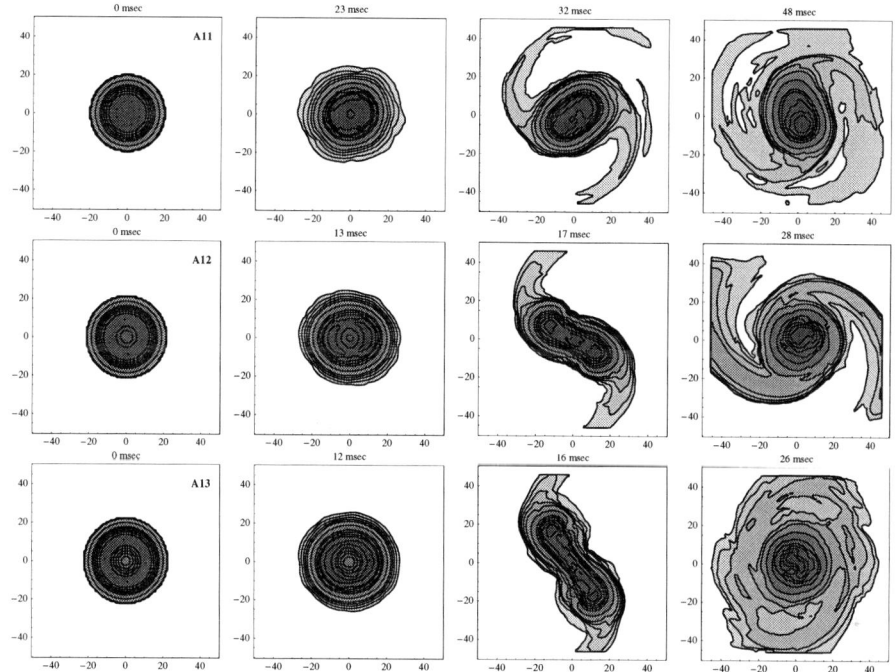

FIGURE 1. Snapshot of the evolution of the isocontours of the rest-mass density for model A11 ($\beta = 0.260$, top row), A12 ($\beta = 0.274$, central row), A13 ($\beta = 0.281$, bottom row). The isocountours refer to $\rho = \rho_{max} 2^{-j}$ ($j = 1, \ldots, 10$), with ρ_{max} changing in each frame.

Whisky, a recently developed numerical code for the accurate solution of the relativistic-hydrodynamics equations in arbitrary background spacetimes has now made it possible to perform accurate simulations of this process in full General Relativity, shedding light on some of its most intricate aspects [6]. The Whisky code [7, 8], is the result of a collaboration among several European Institutes and has been constructed within the Cactus Computational Toolkit [9], developed at the Albert Einstein Institute (Golm) and at the Louisiana State University (Baton Rouge). More specifically, the code solves the general-relativistic hydrodynamics equations in Cartesian coordinates and in a first-order and flux-conservative form [10] using High-Resolution Shock-Capturing methods to ensure minimal numerical dissipation and faithful description of shock dynamics.

SIMULATED MODELS AND RESULTS

The initial stellar models are calculated as stationary and axisymmetric solutions of the Einstein equations with a polytropic equation of state (EOS) $p = K\rho^\Gamma$ with $K = 100$ and $\Gamma = 2$ in the unit system where $c = G = M_\odot = 1$. A degree of differential rotation is introduced by using an angular velocity profile expressed in terms of a dimensionless

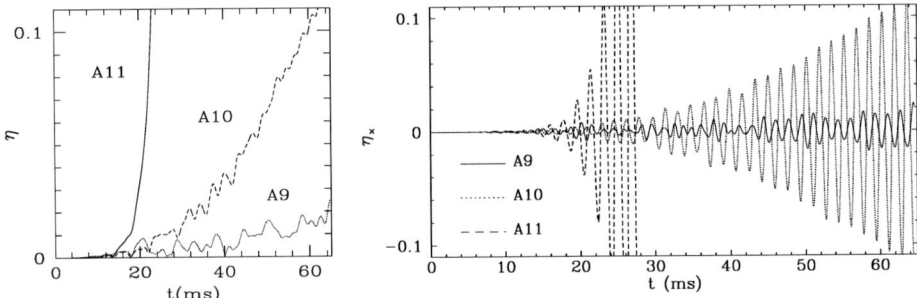

FIGURE 2. Evolution of the distortion parameters η and $\eta_\times$ for models A9 ($\beta = 0.254$), A10 ($\beta = 0.255$) and A11 ($\beta = 0.26$). For clarity, the evolution of models A11 is not shown after 28 msec, when the amplitude would be out of scale and the oscillation frequency would remain unchanged.

parameter $\hat{A}$ which we take to be unity. Also, for an easier comparison with equivalent models discussed in the literature, we use the same sequence of models with constant rest mass $M_* = 1.51 M_\odot$ introduced in ref. [11]. In this way the sequence encompasses models with different M/R ratios, different central densities, different ratios between polar and equatorial radii and values of the β ranging from 0.189 to 0.281.

Once calculated as initial equilibrium models, the rotating stars are evolved using the "ideal fluid" EOS $p = (\Gamma - 1)\rho \varepsilon$, where ε is the specific internal energy. For all of the results reported here the hydrodynamics equations have been solved employing the Marquina flux-formula and a third-order PPM reconstruction [8]. The Einstein field equations, on the other hand, have been evolved using the ICN evolution scheme, the "1+log" slicing condition and the "Gamma-driver" shift conditions [8]. All models have been evolved over several tens of rotational periods, with the stable ones preserving accurate solutions over more than 50 axial rotational periods.

The onset and development of the instability has been monitored using the distortion parameters $\eta_\times$ and η_+ providing a measure of the quadrupolar deformation in ρ^*

$$\eta_+ \equiv \frac{I^{xx} - I^{yy}}{I^{xx} + I^{yy}}, \qquad \eta_\times \equiv \frac{2I^{xy}}{I^{xx} + I^{yy}}, \qquad (1)$$

where I^{ij} is the quadrupole tensor

$$I^{ij} \equiv \int d^3x \, \rho^* \, x^i x^j. \qquad (2)$$

In Figure 1 we report three snapshots of the evolution of the rest-mass density for three of the unstable models A11, A12, A13, characterized by $\beta = 0.260$, 0.274 and 0.281, respectively. Note that in all cases a well-defined bar forms together with spiral arms that help redistribute the excessive angular momentum. Note also that the bars in the three different examples have different extensions (the maximum values of $\eta_\times$ for models A11, A12, A13 are 0.47, 0.78, and 0.85, respectively) and they do not develop all at the same time, since the growth-time for the instability depends inversely on $\beta - \beta_c$. More specifically, during the exponentially-growing phase the evolution of the distortion parameter $\eta_\times(t)$ is very well approximated by a fitting function $\eta_\times(t) = \text{const.} \exp(t/\tau_B) \sin(2\pi f_B t + \text{const.})$, where $\tau_B = 2.69, 1.15, 0.95$ msec and $f_B = 547, 494, 454$ Hz for models A11, A12 and A13, respectively. Finally, note that nonlinear hydrodynamical effects related to coupling among the different modes in the spectral decomposition of the energy density emerge during the simulation, suppressing the instability on a timescale much smaller that the one associated to radiation reaction damping [6]. This process has been verified not to depend on resolution, on the choice of symmetries, on the position of the outer boundaries nor on the use of a polytropic EOS with the same index.

For model A10, characterized by a value of $\beta = 0.255$, the bar grows very slowly and does not appear to reach the saturation value within the 70 msec during which the simulation was carried out. The exponential growth of the bar is nevertheless evident and can be estimated to correspond to a growth-time $\tau_B \simeq 21$ msec with a frequency $f_B \simeq 587$ Hz. A10 is also the model with the smallest value of β which showed a distinct growth of a bar within a dynamical timescale (*i.e.* a timescale that is comparable with the rotational one). Its adjacent model A9, in fact, characterized by $\beta = 0.254$, has a distortion parameter that is growing at most linearly in time. This is shown in Figure 2, whose left panel offers a view of evolution of the global distortion parameter $\eta \equiv \sqrt{\eta_+^2 + \eta_\times^2}$ for models A9, A10 and A11, while the right panel shows the equivalent evolution for η_+. Hence, we conclude that $\beta_c \simeq 0.254$ marks a transition between dynamically unstable and dynamically stable relativistic stars modelled as polytropes with index $\Gamma = 2$ and rotating differentially with $\hat{A} = 1$.

ACKNOWLEDGMENTS

We are grateful to Ian Hawke, Alessandro Nagar and Christian Ott for useful discussions and suggestions. The calculations were performed on the *Albert100* and *Albert2* clusters at the University of Parma and on the *CLX* cluster at CINECA, Bologna.

REFERENCES

1. S. Chandrasekhar, *Ellipsoidal Figures of Equilibrium*, Yale Univ. Press, New Haven, 1969.
2. S. L. Shapiro, and S. A. Teukolsky, *Black Holes, White Dwarfs, and Neutron Stars*, Wiley, New York, 1983.
3. M. Shibata, T. W. Baumgarte, and S. L. Shapiro, *Astrophys. J.* **542**, 453–463 (2000).
4. M. Saijo, M. Shibata, T. W. Baumgarte, and S. L. Shapiro, "Dynamical Bar Instability in Relativistic Rotating Stars," in *20th Texas Symposium on relativistic astrophysics*, edited by J. C. Wheeler, and H. Martel, 2001, vol. 586 of *AIP Conf. Proc.*, p. 766.
5. M. Saijo, M. Shibata, T. W. Baumgarte, and S. L. Shapiro, *Astrophys. J.* **548**, 919–931 (2001).
6. L. Baiotti, R. De Pietri, G. M. Manca, and L. Rezzolla, *in preparation* (2005).
7. L. Baiotti, I. Hawke, P. J. Montero and L. Rezzolla, *Mem. Soc. Astron. It. Suppl.*, **1**, 327 (2003)
8. L. Baiotti, I. Hawke, P. J. Montero, F. Löffler, L. Rezzolla, N. Stergioulas, J. A. Font, and E. Seidel, *Phys. Rev. D* **71**, 024035 (2005).
9. Cactus Computational Toolkit, www.cactuscode.org.
10. F. Banyuls, J. A. Font, J. M. Ibánez, J. M. Martí, and J. A. Miralles, *Astrophys. J.* **476**, 221 (1997).
11. N. Stergioulas, T. A. Apostolatos, and J. A. Font, *Mon. Not. R. Astron. Soc.* **352**, 1089–1101 (2004).

Geodesic Completeness around Sudden Singularities

L. Fernández-Jambrina[*] and R. Lazkoz[†]

[*]E.T.S.I. Navales, Universidad Politécnica de Madrid, Arco de la Victoria s/n, E-28040 Madrid, Spain
[†]Física Teórica, Facultad de Ciencia y Tecnología, Universidad del País Vasco, Apdo. 644, E-48080 Bilbao, Spain

Abstract. In this talk we analyze the effect of recently proposed classes of sudden future singularities on causal geodesics of FLRW spacetimes. Geodesics are shown to be extendible and just the equations for geodesic deviation are singular, although tidal forces are not strong enough to produce a Big Rip.

Keywords: Sudden singularities, Big Rip, Cosmology, Geodesics
PACS: 04.20.Dw, 98.80.Jk

INTRODUCTION

The experimental evidence of accelerated expansion of the universe (supernovae type Ia, redshift of distant objects and fluctuations of background radiation) implemented in cosmological ghost models, $\rho + p < 0$, where ρ is the density of the matter content and p is the pressure, leads to an unexpected kind of singularities, named as "big rip", since they would destroy galactic structures and even atoms due to accelerated expansion [1] of the universe.

In FLRW cosmologies these singularities are characterized by an infinite scale factor at a finite proper time, $a(t_s) \to \infty$.

These FLRW models make use of perfect fluid equations of state that violate all energy conditions (weak, strong and dominant), since $\rho + p < 0$.

SUDDEN SINGULARITIES

Barrow [2] suggests that such behaviour may be attained even if the strong energy condition is fulfilled

$$\rho > 0, \qquad \rho + 3p > 0,$$

by rejecting linear equations of state, $p \neq w\rho$. We may see it by taking a look at Friedmann equations for FRLW cosmologies with Hubble constant $H := a'/a$,

$$3H^2 = \rho - \frac{k}{a^2},$$

$$\rho' + 3H(\rho + p) = 0,$$

$$\frac{a''}{a} = -\frac{\rho + 3p}{6}.$$

If we require $a(t_s)$, $H(t_s) < \infty$ (that is, finite $a(t_s)$, $a'(t_s)$), we obtain finite density $\rho(t_s) < \infty$, but still a singularity may occur if $p(t) \to \infty$ at t_s. Of course, this means $a''/a \to -\infty$.

Barrow achieves this behaviour with the following model,

$$ds^2 = -dt^2 + a(t)\left\{\frac{dr^2}{1-kr^2} + r^2\left(d\theta^2 + \sin^2\theta d\phi^2\right)\right\},$$

$$a(t) = 1 + \left(\frac{t}{t_s}\right)^q (a_s - 1) - \left(1 - \frac{t}{t_s}\right)^n,$$

with constants $a_s = a(t_s)$, $0 < q \leq 1$, $1 < n < 2$, $k = 0, \pm 1$.

Of course, since

$$a''(t) = q(q-1)\frac{a_s - 1}{t_s^q}t^{q-2} - \frac{n(n-1)}{t_s^2(1-t/t_s)^{2-n}} \to -\infty,$$

there is a Big Bang singularity at $t = 0$ and a "sudden" singularity at $t = t_s$.

As we required, since ρ and p are positive, weak and strong energy conditions are fulfilled. But $\rho - p$ is negative at some time. Therefore the dominant energy condition is violated [3]. Further examples of this behaviour have been shown by Barrow [4], Dabrowski, for inhomogenous models [5] and Nojiri and Odintsov, by considering quantum corrections [6]. The sudden singularity is even "milder" if a'' is finite and higher derivatives diverge for larger values of n.

GEODESIC COMPLETENESS

These models therefore show a curvature singularity at t_s, though both a and H are finite there. This behaviour is different from Big Rip singularities found in phantom cosmologies.

However, the usual definition for singularities refers to incomplete causal geodesics [7]. It is reasonable to ask whether causal geodesics are incomplete in Barrow's models.

Since FLRW spacetimes are homogeneous and isotropic, they have six independent isometry generators,

$$\xi_1 = \frac{\sin\theta\cos\phi}{f(r)}\partial_r + \frac{\cos\theta\cos\phi}{rf(r)}\partial_\theta - \frac{\sin\phi}{rf(r)\sin\theta}\partial_\phi,$$

$$\xi_2 = \frac{\sin\theta\sin\phi}{f(r)}\partial_r + \frac{\cos\theta\sin\phi}{rf(r)}\partial_\theta + \frac{\cos\phi}{rf(r)\sin\theta}\partial_\phi,$$

$$\xi_3 = \frac{\cos\theta}{f(r)}\partial_r - \frac{\sin\theta}{rf(r)}\partial_\theta,$$

$$\zeta_1 = \cos\phi\,\partial_\theta - \cot\theta\sin\phi\,\partial_\phi,$$

$$\zeta_2 = \sin\phi\,\partial_\theta + \cot\theta\cos\phi\,\partial_\phi,$$

$$\zeta_3 = \partial_\phi,$$

which produce six constants of geodesic motion for a geodesic parametrized by its proper time τ,

$$P_1 = a(t)\left\{\frac{r(\cos\theta\cos\phi\,\dot\theta - \sin\theta\sin\phi\,\dot\phi)}{f(r)} + f(r)\sin\theta\cos\phi\,\dot r\right\},$$

$$P_2 = a(t)\left\{\frac{r(\cos\theta\sin\phi\,\dot\theta + \sin\theta\cos\phi\,\dot\phi)}{f(r)} + f(r)\sin\theta\sin\phi\,\dot r\right\},$$

$$P_3 = a(t)\left(f(r)\cos\theta\,\dot r - r\sin\theta\,\dot\theta/f(r)\right),$$

$$L_1 = a(t)r^2\left(\cos\phi\,\dot\theta - \sin\theta\cos\theta\sin\phi\,\dot\phi\right),$$

$$L_2 = a(t)r^2\left(\sin\phi\,\dot\theta + \sin\theta\cos\theta\cos\phi\,\dot\phi\right),$$

$$L_3 = a(t)r^2\sin^2\theta\,\dot\phi.$$

Geodesic differential equations for timelike ($\delta = 1$) and lightlike ($\delta = 0$) reduce to first order,

$$\dot r^2 = \delta + \frac{P^2 + kL^2}{a(t)},$$

$$\dot r = \frac{P_1\sin\theta\cos\phi + P_2\sin\theta\sin\phi + P_3\cos\theta}{a(t)f(r)},$$

$$\dot\theta = \frac{L_1\cos\phi + L_2\sin\phi}{a(t)r^2},$$

$$\dot\phi = \frac{L_3}{a(t)r^2\sin^2\theta},$$

$$P^2 = P_1^2 + P_2^2 + P_3^2, \qquad L^2 = L_1^2 + L_2^2 + L_3^2.$$

The system may be further simplified, since due to spherical symmetry every geodesic may be fit in the hypersurface $\theta = \pi/2$, with $L_1 = L_2 = 0 = P_3$, by a suitable choice of the coordinates, then

$$\dot r^2 = \delta + \frac{P^2 + kL^2}{a(t)},$$

$$\dot r = \frac{P_1\cos\phi + P_2\sin\phi}{a(t)f(r)},$$

$$\dot\phi = \frac{L_3}{a(t)r^2},$$

These equations are not singular for finite $a(t)$. Therefore, we realize that geodesics just see the Big Bang singularity at $t = 0$, but not the sudden singularity at $t = t_s$ [8]. Furthermore, since a, a' are finite at t_s, the acceleration vector of the geodesic, $(\ddot t, \ddot r, \ddot\theta, \ddot\phi)$, which comprises the effect of inertial forces, is also regular. Only the third derivative is singular at t_s, but we just require first and second derivatives to define geodesic equations.

TIDAL FORCES

Causal geodesics in such universes do not see the singularities but through geodesic deviation effects, since they are due to the Riemann tensor. Point particles travelling along causal geodesics do not experience any singularity, but extended objects might suffer infinite tidal forces at $t = t_s$.

According to Tipler's definition [9] a strong curvature singularity is encountered at a point p if every volume element defined by three linearly independent, vorticity-free, geodesic deviation vectors along every causal geodesic through p vanishes at this point.

Clarke and Krolak [10] provide necessary and sufficient conditions for the appearance of strong curvature singularities. If a causal geodesic meets a strong singularity at a value τ_s,

$$\int_0^\tau d\tau' \int_0^{\tau'} d\tau'' |R^i{}_{0j0}(\tau'')|,$$

will diverge along the geodesic on approaching τ_s.

In the case of sudden singularities, the components of the Riemann tensor diverge as a'', since a' and a are finite; and in the worst case they diverge as a power $n - 2$, for $1 < n < 2$. Therefore after one integration of the components of the Riemann tensor, the power will be positive and the integral will not diverge. Hence sudden singularities are not strong according to Tipler definition and therefore tidal forces do not crush all finite bodies. The spacetime may be extended across sudden singularities and cannot be considered the final fate of these universes [8].

CONCLUSIONS

We have shown that causal geodesics are not affected by sudden future singularities, since these singularities are not seen by geodesic equations.

Furthermore, considering just curvature singularities, it has been shown that they are weak according to Tipler's definition and therefore finite objects are not necessarily torn on crossing the singularities. This is in contrast with Big Rip singularities and therefore sudden singularities produce no Big Rip.

ACKNOWLEDGMENTS

L. F.-J. is supported by the Spanish Ministry of Education and Science through research grant FIS2005-05198. R.L. is supported by the University of the Basque Country through research grant UPV00172.310-14456/2002 and by the Spanish Ministry of Education and Science through research grant FIS2004-01626.

REFERENCES

1. R. R. Caldwell, *Phys. Lett.* **B545**, 23 (2002).
2. J. D. Barrow, *Class. Quan. Grav.* **21**, L79 (2004).
3. K. Lake, *Class. Quan. Grav.* **21**, L129 (2004).
4. J. D. Barrow, *Class. Quan. Grav.* **21**, 5619 (2004).
5. M. P. Dabrowski, *Phys. Rev.* **D71**, 103505 (2005).
6. S. Nojiri, S.D. Odintsov, *Phys. Lett.* **B595**, 1 (2004).
7. S. W. Hawking, G. F. R. Ellis, *The Large Scale Structure of Space-time*, Cambridge University Press, Cambridge, 1973.
8. L. Fernández-Jambrina, R. Lazkoz, *Phys. Rev.* **D70**, 121503 R1 (2004).
9. F. J. Tipler, *Phys. Lett.* **A64**, 8 (1977).
10. C. J. S. Clarke, A. Królak, *Journ. Geom. Phys.* **2**, 17 (1985).

Coll Positioning systems: a two-dimensional approach

Joan Josep Ferrando

Departament d'Astronomia i Astrofísica, Universitat de València, 46100 Burjassot, València, Spain

Abstract. The basic elements of Coll positioning systems (n clocks broadcasting electromagnetic signals in a n-dimensional space-time) are presented in the two-dimensional case. This simplified approach allows us to explain and to analyze the properties and interest of these relativistic positioning systems. The positioning system defined in flat metric by two geodesic clocks is analyzed. The interest of the Coll systems in gravimetry is pointed out.

Keywords: relativistic positioning systems
PACS: 04.20.-q, 95.10.Jk

INTRODUCTION

The relativistic positioning systems were introduced by B. Coll a few years ago at the Spanish Relativity Meeting celebrated in Valladolid [1]. Remember that these systems are defined by four clocks broadcasting their proper time. Here we will name them for short 'Coll systems'. In a 'long contribution' to these proceedings B. Coll explains the interest, characteristics and good qualities of these relativistic positioning systems in the generic four-dimensional case. In this short communication we present a two-dimensional approach to the Coll systems.

The two-dimensional approach should help us understand better how these relativistic systems work and the richness of the elements of the Coll systems. Indeed, the simplicity of the 2-dimensional case allows us to use precise and explicit diagrams which improve the qualitative comprehension of the positioning systems. Moreover, two-dimensional examples admit simple and explicit analytic results.

Nevertheless, it is worth remarking that the two-dimensional case has particularities and results that cannot be generalized to the generic four-dimensional case. Consequently, the two-dimensional approach is suitable for learning basic concepts about positioning systems, but it does not allow us to study some specific positioning features that necessarily need a three- or a four-dimensional approach.

In the second section we introduce the basic elements of a Coll system: emission coordinates and its other essential physical components. In the third section we explain the analytic method to obtain the emission coordinates from an arbitrary null coordinate system and we use it to develop the positioning system defined in flat space-time by two geodesic clocks. We finish with some comments about other positioning subjects that we are considering at present.

BASIC CONCEPTS AND ESSENTIAL PHYSICAL COMPONENTS

In a two-dimensional space-time, let γ_1 and γ_2 be the world lines of two clocks measuring their proper times τ^1 and τ^2 respectively. Suppose they broadcast them by means of electromagnetic signals, and that these signals reach each other of the world lines. The future light cones (here reduced to pairs of 'light' lines) cut in the region between both emitters and they are tangent outside. Thus, these proper times do not distinguish different events on the emission null geodesics of the exterior region.

The internal region, bounded by the emitter world lines, defines a coordinate domain, the *emission coordinate domain* Ω. Indeed, every event on this domain can be distinguished by the times (τ^1, τ^2) received from the emitter clocks. In other words, the past light cone of every event on the emission domain cuts the emitter world lines at $\gamma_1(\tau^1)$ and $\gamma_2(\tau^2)$ respectively: then $\{\tau^1, \tau^2\}$ are the *emission coordinates* of this event. An important property of the emission coordinates we have defined is that they are null coordinates. The plane $\{\tau^1\} \times \{\tau^2\}$ in which the different data of the positioning system can be transcribed is called the *grid* of the positioning system.

An observer γ travelling throughout the emission coordinate domain and equipped with a receiver which allows to read the proper times (τ^1, τ^2) at each point of his trajectory is a user of this positioning system.

In defining the emission coordinates we have introduced the first essential physical components of a Coll system:

- The principal emitters γ_1, γ_2, which broadcast their proper time τ^1, τ^2.
- The users γ, travelling in the emitter coordinate domain Ω, receive the emitted times $\{\tau^1, \tau^2\}$ (their emitter coordinates).

These elements define a generic, free and immediate location system (it can be defined in a generic space-time; it can be defined without knowing the gravitational field; a user knows his coordinates without delay).

Any user receiving continuously the *user's positioning data* $\{\tau^1, \tau^2\}$ may extract his trajectory, $\tau^2 = F(\tau^1)$, in the grid. Nevertheless, whatever the user be, these data are insufficient to construct both of the two emitter trajectories.

In order to give to any user the capability of knowing the emitter trajectories in the grid, the positioning system must be endowed with a device allowing every emitter to also broadcast the proper time it is receiving from the other emitter:

- The emitters γ_1, γ_2 are also transmitters: they receive the signals (such as a user) and broadcast them.
- The users γ also receive the transmitted times $\{\bar{\tau}^1, \bar{\tau}^2\}$.

In other words, the clocks must be allowed to broadcast *their emission coordinates* and then, any user receiving continuously the *emitter's positioning data* $\{\tau^1, \tau^2; \bar{\tau}^1, \bar{\tau}^2\}$ may extract from them the equations $\bar{\tau}^2 = \varphi_1(\tau^1)$ and $\bar{\tau}^1 = \varphi_2(\tau^2)$ of the emitter trajectories. A positioning system so endowed will be called an *auto-located positioning system*.

Eventually, the positioning system can be endowed with complementary devices. For example, in obtaining the dynamic properties of the system:

- The emitters γ_1, γ_2 can carry accelerometers and broadcast their acceleration.
- The users γ can also receive the emitter acceleration data $\{\alpha_1, \alpha_2\}$.

In some cases, it can be useful that the users generate their own data: they can carry a clock that measures their proper time τ and an accelerometer that measures their acceleration α.

Thus, a Coll positioning system can be performed in such a way that any user can obtain a subset of the user data: $\{\tau^1, \tau^2; \bar{\tau}^1, \bar{\tau}^2; \alpha_1, \alpha_2; \tau, \alpha\}$.

POSITIONING WITH GEODESIC EMITTERS IN FLAT METRIC

Let us assume the *proper time history of two emitters* to be known in a null coordinate system $\{u, v\}$:

$$\gamma_1 \equiv \begin{cases} u = u_1(\tau^1) \\ v = v_1(\tau^1) \end{cases} \qquad \gamma_2 \equiv \begin{cases} u = u_2(\tau^2) \\ v = v_2(\tau^2) \end{cases} \qquad (1)$$

We can introduce the proper times as coordinates $\{\tau^1, \tau^2\}$ as follows:

$$u = u_1(\tau^1), \qquad v = v_2(\tau^2) \qquad (2)$$

This change defines *emission null coordinates* in the *emission coordinate domain* $\Omega \equiv \{(u, v) \; / \; F_2^{-1}(v) \leq u, \; F_1(u) \leq v\}$. In the region outside Ω this change also determines null coordinates which are an extension of the emission coordinates. But in this region the coordinates are not physical, i.e. are not the emitted proper times of the *principal emitters* γ_1, γ_2.

Now we use this procedure for the case of two *geodesic* emitters γ_1, γ_2 in *flat spacetime*. In inertial null coordinates $\{u, v\}$ the proper time parametrization of the emitters are:

$$\gamma_1 \equiv \begin{cases} u = \lambda_1 \tau^1 \\ v = \frac{1}{\lambda_1} \tau^1 + v_0 \end{cases} \qquad \gamma_2 \equiv \begin{cases} u = \lambda_2 \tau^2 + u_0 \\ v = \frac{1}{\lambda_2} \tau^2 \end{cases} \qquad (3)$$

Then, the emitter coordinates $\{\tau^1, \tau^2\}$ are defined by the change:

$$u = u_1(\tau^1) = \lambda_1 \tau^1, \qquad v = v_2(\tau^2) = \frac{1}{\lambda_2} \tau^2 \qquad (4)$$

>From here we can obtain the metric tensor in emitter coordinates $\{\tau^1, \tau^2\}$ and we obtain: $ds^2 = \lambda \, d\tau^1 d\tau^2$, $\lambda \equiv \frac{\lambda_1}{\lambda_2}$. On the other hand, in emission coordinates $\{\tau^1, \tau^2\}$, the equations of the emitter trajectories are:

$$\gamma_1 \equiv \begin{cases} \tau^1 = \tau^1 \\ \tau^2 = \varphi_1(\tau^1) \equiv \frac{1}{\lambda} \tau^1 + \tau_0^2 \end{cases} \qquad \gamma_2 \equiv \begin{cases} \tau^1 = \varphi_2(\tau^2) \equiv \frac{1}{\lambda} \tau^2 + \tau_0^1 \\ \tau^2 = \tau^2 \end{cases} \qquad (5)$$

Let γ be a user of this positioning system. What information can this user obtain from the public data? Evidently (τ^1, τ^2) place the user on the user grid, and $(\bar{\tau}^1, \bar{\tau}^2)$, $\bar{\tau}^i = \varphi_j(\tau^j)$, place the emitters on the user grid. On the other hand, the metric component could be obtained from the emitter's positioning data $\{\tau^1, \tau^2; \bar{\tau}^1, \bar{\tau}^2\}$ at two events. The space-time interval is:

$$ds^2 = \sqrt{\frac{\Delta\tau^1 \Delta\tau^2}{\Delta\bar{\tau}^1 \Delta\bar{\tau}^2}} \, d\tau^1 d\tau^2 \tag{6}$$

DISCUSSION AND WORK IN PROGRESS

We finish this talk with some comments about other positioning subjects we are studying at present. Firstly, the interest of the Coll systems in gravimetry. If we suppose that the user has no previous information on the gravitational field, what metric information can a user obtain from the public and proper user data? Can a user do gravimetry by using our positioning system? We have shown that [2]:

- The public data $\{\tau^1, \tau^2; \bar{\tau}^1, \bar{\tau}^2; \alpha_1, \alpha_2\}$ determine the space-time metric interval and its gradient along the emitter trajectories.
- The public-user data $\{\tau^1, \tau^2; \tau, \alpha\}$ determine the space-time metric interval and its gradient along the user trajectory.

The development of a general method that offer a good estimation of the gravitational field from this information is still an open problem, but some preliminary results show its interest in determining the parameters in a given (parameterized) model [3].

On the other hand, some circumstances can lead to take another point of view: the user knows the space-time in which he is immersed (Minkowski, Schwarzschild,...) and we want to study the information that the data received by the user offer. We have undertaken this problem for the flat case an we have obtained interesting preliminary results. In particular, we have shown that [4]:

- *If a user receives the emitter positioning data $\{\tau^1, \tau^2; \bar{\tau}^1, \bar{\tau}^2\}$ along his trajectory and the acceleration of one of the emitters during a sole echo interval (i.e., travel time of a two-way signal from an emitter to the other), then this user knows: his local unities of time and distance, the metric interval in emission coordinates everywhere, his own acceleration and the acceleration of the principal emitters, the change between emission and inertial coordinates, his trajectory and the emitter trajectories in inertial coordinates.*

ACKNOWLEDGMENTS

This work has been supported by the Spanish Ministerio de Educación y Ciencia, MEC-FEDER project AYA2003-08739-C02-02.

REFERENCES

1. B. Coll, "Elements for a theory of relativistic coordinate systems. Formal and physical aspects" in *Reference Frames and Gravitomagnetism* (World Scientific, Singapore, 2000), p. 279.

2. B. Coll B., J. J. Ferrando, and J. Morales, (submitted Phys. Rev. D).
3. B. Coll B., J. J. Ferrando, and J. Morales, (to be submitted Phys. Rev. D).
4. J. J. Ferrando, and J. Morales,, "Two-dimensional constructions", in *Relativistic Coordinates, Reference and Positioning Systems*, Notes School Salamanca 2005.

Non-abelian Yang-Mills in Kundt spacetimes

Andrea Fuster

NIKHEF, Kruislaan 409, 1098 SJ, Amsterdam

Abstract. We present new exact solutions of the Einstein-Yang-Mills system. The solutions are described by a null Yang-Mills field in a Kundt spacetime. They generalize a previously known solution for a metric of *pp* wave type. The solutions are formally of Petrov type III.

Keywords: Einstein-Yang-Mills, exact solution, non-abelian, Kundt, type III, *pp* waves
PACS: 04.40.Nr, 04.20.Jb

INTRODUCTION

The Einstein-Yang-Mills (EYM) theory in four dimensions has been extensively studied in various contexts. One has to make the distinction between solutions which are effectively abelian and truly non-abelian ones. In the first case they are just embeddings of Einstein-Maxwell (EM) solutions. A procedure to construct abelian-like solutions of EYM from known EM ones was given in [1]. On the other hand, the study of non-abelian EYM has been mainly devoted to static ($\equiv$ time independent) particle-like solutions: solitons and hairy black holes (for a review and references see [2]).
The first non-abelian and non-static Yang-Mills solutions in flat space were given by Coleman in [3]. They were later generalized by Güven in [4] to non-abelian EYM solutions where spacetime was described by the *pp* wave metric. Recently, Coleman's solution was embedded in Petrov type III Kundt spacetimes [5], which can be regarded as a generalization of *pp* waves. We will summarize this result here, while a physical interpretation of these solutions remains an open question[1].

NON-ABELIAN YANG-MILLS IN FLAT AND *PP* WAVE SPACETIMES

We first review the results by Coleman and Güven. We use light-cone coordinates u, v and complex conjugate coordinates z, $\bar{z}$ in the transverse plane. Consider a gauge field of the form:

$$A_u = (\alpha^a \equiv \lambda^a(u)z + \bar{\lambda}^a(u)\bar{z})\,T_a, \quad A_\mu = 0 \text{ for } \mu \neq u \tag{1}$$

---

[1] See, however, [5] for some hints coming from the geodesic equations.

Here a is the gauge index and λ^a are arbitrary bounded functions of u. The Yang-Mills equation is trivially satisfied for such a field:

$$\alpha_{,z\bar{z}}^a = 0 \tag{2}$$

These solutions are known as non-abelian plane waves. When gravity is incorporated one has to solve the coupled EYM equations instead:

$$R_{\mu\nu} - \frac{1}{2}g_{\mu\nu}R = 8\pi T_{\mu\nu}^{\text{YM}} \tag{3}$$

$$\nabla_\lambda F^{\lambda\rho} - [A_\kappa, F^{\kappa\rho}] = 0 \tag{4}$$

The YM energy-momentum tensor reads:

$$T_{\mu\nu}^{\text{YM}} = \frac{\gamma_{ab}}{4\pi}(F_{\mu\eta}^a F_\nu^{b\,\eta} - \frac{1}{4}g_{\mu\nu}F_{\delta\eta}^a F^{b\,\delta\eta}) \tag{5}$$

Here, γ_{ab} is the invariant metric of the Lie group and $F_{\mu\eta}^a = \nabla_{[\mu}A_{\eta]}^a + f_{bc}^a A_\mu^b A_\eta^c$. Güven proposed a more general ansatz for the YM field where the α^a are arbitrary real functions of $(u, z, \bar{z})$. The corresponding curved YM equation reduces to the flat one when spacetime is of the pp wave type:

$$ds^2 = 2\,dz\,d\bar{z} - 2\,du\,dv - H(u,z,\bar{z})\,du^2 \tag{6}$$

There are two reasons why this happens. First, the connection coefficients $\Gamma_{\lambda\sigma}^\lambda$ entering the spacetime covariant derivative in (4) vanish for the pp wave metric. And second, there are no field strength components of the form $F^{u\rho}$ in this geometry. Equation (2) is trivially solved by any field of the form:

$$\alpha^a(u,z,\bar{z}) = \chi^a(u,z) + \bar{\chi}^a(u,\bar{z}) \tag{7}$$

χ^a being an arbitrary complex function. Such a YM field is null [6]. It is not difficult to see that the Einstein equation is satisfied as well. The only non-zero components of the Ricci and YM energy-momentum tensor are R_{uu} and T_{uu}^{YM} while the Ricci scalar vanishes. Eq. (3) reads:

$$H_{,z\bar{z}} = 2\gamma_{ab}\,\alpha_{,z}^a \alpha_{,\bar{z}}^b \tag{8}$$

The solution of the coupled system is given by (7) and a pp wave spacetime specified by the function:

$$H(u,z,\bar{z}) = f(u,z) + \bar{f}(u,\bar{z}) + 2\gamma_{ab}\chi^a \bar{\chi}^b \tag{9}$$

Here f is an arbitrary complex function. Note that the only solution for which the energy-momentum tensor is bounded throughout spacetime occurs when non-abelian plane waves are recovered, $\chi^a = \lambda^a(u)z$. On the other hand, the solution becomes effectively abelian in the limit $\alpha^a(u,z,\bar{z}) = \beta^a \alpha(u,z,\bar{z})$, where $(\alpha(u,z,\bar{z}), ds^2)$ is a solution of EM and the parameters β^a are such that $\gamma_{ab}\beta^a\beta^b = 1$. In any other case the solution has a fully non-abelian character.

NON-ABELIAN YANG-MILLS IN KUNDT SPACETIMES

We will extend the previous result to a more general spacetime in Kundt's class. This class of metrics is described by the line element:

$$ds^2 = -2du\left(dv + W\,dz + \overline{W}\,d\bar{z} + H\,du\right) + 2P^{-2}dzd\bar{z}, \quad P_{,v}=0 \qquad (10)$$

Here, P and H are real functions; W is complex. They are algebraically special, i.e., of Petrov type II, D, III, N or O. We will consider type III metrics. In this case $P \equiv 1$ can be taken without loss of generality. Type III vacuum solutions are well-known. A marked difference with respect to pp waves is the dependence of the function H on the light-cone coordinate v. Further two distinct cases have to be distinguished according to whether the function W entering (10) depends on v as well or not. In this last case pp waves arise in the type N limit. The less known Kundt waves appear as the type N limit in the $W_{,v} \neq 0$ case. Details up to here can be found in [7].

The motivation to study exact solutions of gravity-coupled theories described by Kundt metrics is the following. It is well-known that pp waves play an important role in supergravity and string theory. The main reason being the vanishing of the corresponding scalar curvature invariants of all orders. Kundt spacetimes of type III are generalizations of pp waves and have been shown to be the most general metrics with that remarkable property [8]. They might therefore have interesting applications in supergravity and string theory similar to pp waves.

In what follows we briefly describe the new embedding of the Yang-Mills field (7) in Kundt spacetimes of type III. In this geometry the field is null as well. We make the same distinction in terms of the v-dependence of the function W as for vacuum solutions. It can be shown that the curved YM equation reduces in both cases to (2), the reasons being similar to the pp wave case. Again, there are no $F^{u\rho}$ components of the field strength. Respecting the spacetime derivative, in the first case the only non-zero connection coefficients are precisely $\Gamma_{\lambda u}^{\lambda}$. In the second case there are additional coefficients $\Gamma_{\lambda z}^{\lambda}$, $\Gamma_{\lambda \bar{z}}^{\lambda}$ which however cancel each other. The YM energy-momentum tensor is seen to be the same one as in the pp wave case. We give below the precise form of the equations and the corresponding solutions.

Solutions with $W_{,v} = 0$

Consider the spacetime (10) specified by:

$$P=1, \quad W=W(u,\bar{z}), \quad H = H^0(u,z,\bar{z}) + \frac{1}{2}\left(W_{,\bar{z}}+\overline{W}_{,z}\right)v \qquad (11)$$

Here W is an arbitrary complex function and H^0 is real. Eq. (3) reads:

$$H^0_{,z\bar{z}} - \mathrm{Re}\left(W_{,\bar{z}}^2 + W W_{,\bar{z}\bar{z}} + W_{,\bar{z}u}\right) = 2\gamma_{ab}\,\alpha_{,z}^a\,\alpha_{,\bar{z}}^b \qquad (12)$$

The solution of the coupled system is given by the YM field (7) and:

$$H^0(u,z,\bar{z}) = f(u,z) + \bar{f}(u,\bar{z}) + 2\gamma_{ab}\chi^a\bar{\chi}^b + \mathrm{Re}\left\{\left(W_{,u}+WW_{,\bar{z}}\right)z\right\} \qquad (13)$$

The type N reduction occurs for $\Psi_3 = (1/2)W_{,\bar{z}\bar{z}} = 0$. In this limit the generalized pp waves in [4] are recovered.

Solutions with $W_{,v} \neq 0$

The second class of solutions is characterized by the v-dependence of W. The functions in (10) are now given by:

$$P = 1, \quad W = W^0(u,z) - \frac{2v}{z+\bar{z}}, \quad H = H^0(u,z,\bar{z}) + \frac{W^0 + \overline{W}^0}{z+\bar{z}}v - \frac{v^2}{(z+\bar{z})^2} \quad (14)$$

In principle W^0 is an arbitrary complex function. In order to solve the Einstein equation explicitly we however consider the specific function $W^0 = g(u)z$, $g(u)$ being an arbitrary complex function; this is the simplest function for which the solution is of type III. We further limit the YM field to non-abelian plane waves. Eq. (3) becomes:

$$(z+\bar{z})\left(\frac{H^0 + g\bar{g}z\bar{z}}{z+\bar{z}}\right)_{,z\bar{z}} - g(u)\bar{g}(u) = 2\gamma_{ab}\lambda^a(u)\bar{\lambda}^b(u) \quad (15)$$

The complete solution is described by the YM field $\alpha^a \equiv \lambda^a(u)z + \bar{\lambda}^a(u)\bar{z}$ and:

$$H^0 = (f(u,z) + \bar{f}(u,\bar{z}))(z+\bar{z}) - g\bar{g}z\bar{z} + \sigma(u)(z+\bar{z})^2 \{\ln(z+\bar{z}) - 1\} \quad (16)$$

Here σ is the real function $\sigma(u) = 2\gamma_{ab}\lambda^a(u)\bar{\lambda}^b(u) + g(u)\bar{g}(u)$. The corresponding vacuum solution ($\lambda^a \equiv 0$) has to our knowledge not been considered before. The type N reduction occurs for $\Psi_3 = \bar{g}/(z+\bar{z}) = 0$. In this case generalized Kundt waves arise.

CONCLUSIONS

We present two (classes of) new exact solutions of the four-dimensional EYM system. The solutions are non-static and have a fully non-abelian character. They might be of interest in supergravity and string theory, in the fashion of pp waves (work in progress).

REFERENCES

1. P. B. Yasskin, *Phys. Rev.* **D12**, 2212–2217 (1975).
2. M. S. Volkov, and D. V. Gal'tsov, *Phys. Rept.* **319**, 1–83 (1999), hep-th/9810070.
3. S. R. Coleman, *Phys. Lett.* **B70**, 59 (1977).
4. R. Gueven, *Phys. Rev.* **D19**, 471–472 (1979).
5. A. Fuster, and J.-W. van Holten, *Phys. Rev.* **D72**, 024011 (2005), gr-qc/0505159.
6. J. Tafel, *Lett. Math. Phys.* **12**, 163 (1986).
7. H. Stephani, D. Kramer, M. MacCallum, C. Hoenselaers, and E. Herlt (2003), cambridge, UK: Univ. Pr. 701 P.
8. V. Pravda, A. Pravdova, A. Coley, and R. Milson, *Class. Quant. Grav.* **19**, 6213–6236 (2002), gr-qc/0209024.

Stability of Self-Similar Spherical Accretion

José Gaite

Instituto de Matemáticas y Física Fundamental, CSIC, Serrano 113bis, 28006 Madrid, Spain

Abstract. Spherical accretion flows are simple enough for analytical study, by solution of the corresponding fluid dynamic equations. The solutions of stationary spherical flow are due to Bondi. The questions of the choice of a physical solution and of stability have been widely discussed. The answer to these questions is very dependent on the problem of boundary conditions, which vary according to whether the accretor is a compact object or a black hole. We introduce a particular, simple form of stationary spherical flow, namely, self-similar Bondi flow, as a case with physical interest in which analytic solutions for perturbations can be found. With suitable no matter-flux-perturbation boundary conditions, we will show that acoustic modes are stable in time and have no spatial instability at $r = 0$. Furthermore, their evolution eventually becomes *ergodic-like* and shows no trace of instability or of acquiring any remarkable pattern.

Keywords: Accretion, hydrodynamic stability, black holes
PACS: 97.10.Gz, 47.20.-k, 04.70.-s

RELATIVISTIC POTENTIAL FLUID FLOW

We consider the adiabatic downfall of a perfect fluid onto a compact spherical body or a non-rotating black hole. In particular, we intend to analyse the stability of simple spherical, stationary flows (Bondi flows) [1]. General stability arguments have been given by Garlick [2] and Moncrief [3], but there are no exact solutions for perturbations, except in the WKB approximation. So we further restrict ourselves to self-similar flows, as a case amenable to analytic treatment and with physical interest.

We begin with a summary of the theory of relativistic potential fluid flow and its linear preturbations, introducing the *sound metric*. Then we proceed to the Newtonian limit, sufficient for our purposes. In this limit, we obtain the *self-similar* Bondi flows and the perturbation equations. These equations can be solved in terms of Bessel functions. We study their initial and boundary problems, and so we draw conclusions on their stability.

Let us consider the perfect fluid equations, namely, the energy-momentum tensor $T^{\mu\nu} = (\rho + p)u^\mu u^\nu + pg^{\mu\nu}$, $u^\mu u_\mu = -1$. and thermodynamic equations $h = (p + \rho)/n$, $dp = n(dh - Tds)$, where n is the number density and h is the enthalpy per particle (we have $u^\mu s_{;\mu} = 0$). The equations of motion are the conservation equations $T^{\mu\nu}{}_{;\nu} = 0$, $(nu^\mu)_{;\mu} = 0$. Let us further consider isentropic solutions and *potential flow* [3], such that $\omega_{\mu\nu} = (hu_\mu)_{;\nu} - (hu_\nu)_{;\mu}$ fulfills $P^\mu_\alpha P^\nu_\beta \omega_{\mu\nu} = 0$, where $P_{\mu\nu} = g_{\mu\nu} + u_\mu u_\nu$. Then we have $\omega_{\mu\nu} = 0 \Rightarrow hu_\mu = \psi_{,\mu}$ for some function ψ. The equations of motion become

$$\left(\frac{n}{h}\psi^{;\mu}\right)_{;\mu} = 0,$$

where n is expressed in terms of h by the equation of state: $n = \left.\frac{\partial p}{\partial h}\right|_s$, and $h^2 = \psi_{,\mu}\psi^{,\mu}$.

Linear Perturbations and Sound Metric

The linear perturbations of the equation for the scalar potential give the following scalar wave equation:
$$\nabla^\mu \delta\psi_{,\mu} = 0,$$
where ∇_μ is the covariant derivative with respect to the *sound metric*

$$\mathscr{G}_{\mu\nu} = \frac{n}{h}\frac{c}{c_s}\left[g_{\mu\nu} - \left(1 - \frac{c_s^2}{c^2}\right)u_\mu u_\nu\right], \quad \frac{c_s^2}{c^2} = \left.\frac{\partial p}{\partial \rho}\right|_s.$$

Therefore, causality in sound propagation is determined by $\mathscr{G}_{\mu\nu}$ (characteristics, etc), and the symmetries of $\mathscr{G}_{\mu\nu}$ are the ones common to $g_{\mu\nu}$ and u^μ.

From the linear perturbation equation:

$$\nabla_\nu \mathscr{T}^\nu_\mu = 0, \quad \mathscr{T}^\nu_\mu = \frac{1}{2}\left[\delta\psi_{,\mu}\delta\psi_{,\kappa}\mathscr{G}^{\nu\kappa} - \frac{1}{2}\delta^\nu_\mu \mathscr{G}^{\kappa\sigma}\delta\psi_{,\kappa}\delta\psi_{,\sigma}\right],$$

where $\mathscr{T}^\nu_\mu$ is the energy-momentum tensor of scalar waves. We consider *stationary* flow $\Rightarrow$ conserved energy:

$$E = -2\int d^3x(-\det\mathscr{G})^{1/2}\mathscr{T}^t_t = \frac{1}{2}\int d^3x(-\det\mathscr{G})^{1/2}\left[-\mathscr{G}^{tt}(\delta\psi_{,t})^2 + \mathscr{G}^{ij}\delta\psi_{,i}\delta\psi_{,j}\right].$$

In addition, we are interested in *spherical* flow $\Rightarrow$ conserved angular momentum.

NEWTONIAN SELF-SIMILAR SPHERICAL FLOW

When $r \gg R \geq GM/c^2$ and $|u^i| \ll 1$, potential flow boils down to $\mathbf{v} = \nabla\psi$ and [4]

$$\left[\frac{\partial}{\partial t} + \mathbf{v}\cdot\nabla + \frac{c_s^2}{\rho}\nabla\cdot\left(\frac{\rho\mathbf{v}}{c_s^2}\right)\right]\left(\frac{\partial}{\partial t} + \mathbf{v}\cdot\nabla\right)\delta\psi = \frac{c_s^2}{\rho}\nabla\cdot(\rho\nabla\delta\psi).$$

This is an equation of non-homogeneous wave propagation. The law of conservation of acoustic energy is simply $\frac{\partial E}{\partial t} = \nabla\cdot\mathbf{W}$, where

$$E = \frac{\rho}{2c_s^2}\left[(\partial_t\delta\psi)^2 + c_s^2(\nabla\delta\psi)^2 - (\mathbf{v}\cdot\nabla\delta\psi)^2\right],$$

and the acoustic energy flux current is

$$\mathbf{W} = \frac{\rho}{c_s^2}\partial_t\delta\psi\left[\mathbf{v}\partial_t\delta\psi - c_s^2\nabla\delta\psi + \mathbf{v}\cdot(\mathbf{v}\cdot\nabla\delta\psi)\right] = -\partial_t\delta\psi\,\delta\mathbf{j},$$

with $\mathbf{j} = \rho\mathbf{v}$ the matter flux current.

Basic flow: Bondi solutions

For stationary spherical flow we have the radial mass conservation and Euler equations

$$\frac{d}{dr}(r^2\rho v) = 0, \quad v\frac{dv}{dr} = -\frac{1}{\rho}\frac{dp}{dr} - \frac{GM}{r^2}.$$

Integrating the first one, $4\pi r^2\rho v = \mathcal{K}_1$ (constant). We further use the polytropic equation of state $p = \mathcal{K}_2\rho^\gamma$. We define the accretion radius $\mathcal{R} = GM/(c_s)_\infty^2$, and hence the non-dimensional variables

$$x = \frac{r}{\mathcal{R}}, \quad y = \frac{\rho}{\rho_\infty}, \quad u = \frac{v}{(c_s)_\infty}, \quad \lambda = \frac{\mathcal{K}_1}{4\pi\mathcal{R}^2\rho_\infty(c_s)_\infty}.$$

Writing the equations of motion with these variables and solving for u:

$$\frac{\lambda^2}{2}x^{-4}y^{-2} + n(y^{1/n} - 1) = x^{-1},$$

where the adiabatic index $n = 1/(\gamma - 1)$.

Self-similar solutions

Assume $x \ll 1 \Rightarrow \frac{\lambda^2}{2}x^{-3}y^{-2} + ny^{1/n}x = \frac{\lambda^2}{2}x^{-3+2n}z^{-2n} + nz = 1$, where $z = y^{1/n}x$. If $n = 3/2$, we can solve for $z(\lambda) \Rightarrow y = z(\lambda)^{3/2}x^{-3/2}$, $u = \lambda z(\lambda)^{-3/2}x^{-1/2}$ (power laws). Then, for $n = 3/2 \Leftrightarrow \gamma = 5/3$,

$$\begin{aligned}\rho(r) &= \alpha r^{-3/2}, \\ v(r) &= \beta r^{-1/2}, \\ c_s^2(r) &= \mathcal{K}_2\frac{5}{3}\alpha^{2/3}r^{-1} = \sigma^2 r^{-1},\end{aligned}$$

so the Mach number is given by $\mathcal{M} = v(r)/c(r) = \beta/\sigma$ (constant).

Linear Perturbations

We try the separation of variables in spherical coordinates: $\delta\psi(r,\theta,\varphi,t) = R(r)Y_{lm}(\theta,\varphi)e^{-i\omega t}$. So we obtain the radial differential equation

$$(\sigma^2 - \beta^2)r^2 R'' + (\frac{\sigma^2 - \beta^2}{2} + 2i\beta\omega r^{3/2})rR' + \left[-l(l+1)\sigma^2 + \omega^2 r^3 + i\beta\omega r^{3/2}\right]R = 0.$$

Note that it is not of Sturm-Liouville type. Its general solution is

$$R(r) = r^{1/4}e^{-i\mu r^{3/2}}[C_1 J_\nu(\kappa r^{3/2}) + C_2 J_{-\nu}(\kappa r^{3/2})],$$

where $\mu = \frac{2\beta\omega}{3(\sigma^2-\beta^2)}$, $\kappa = \frac{2\sigma\omega}{3(\sigma^2-\beta^2)}$, $\nu = \frac{\sqrt{[1+16l(l+1)]\sigma^2-\beta^2}}{6\sqrt{\sigma^2-\beta^2}}$.

We need appropriate boundary conditions at two radii. In the self-similar case, we must use $r_1 = 0$ and $r_2 \approx \mathscr{R} \to \infty$, but we keep the latter finite to have a discrete spectrum. Since $\mathbf{W} = -\partial_t \delta\psi \, \delta\mathbf{j}$, holding the mass flow, as done in Ref. [5], is equivalent to holding the energy flow. So, using the variable $z = \kappa r^{3/2}$, we impose

$$z^{-1/6}\left(C_1[J_\nu(z) + 6zJ'_\nu(z)] + C_2[J_{-\nu}(z) + 6zJ'_{-\nu}(z)]\right) = 0$$

at r_1, r_2, and so we obtain the κ-spectrum. Remarkably, we have regularity at $r_1 = 0$, namely, the physical quantities $\delta v_r/v$, $\delta v_{\theta,\varphi}/v$, $\delta\rho/\rho$ stay finite.

Evolution of radial eigen-functions

Since our boundary problem is not of Sturm-Liouville type, we have no eigenfunction orthogonality. Let us turn to the first order equations for radial perturbations:

$$\frac{\partial \delta\rho}{\partial t} + \frac{1}{r^2}\partial_r\left[r^2(\delta\rho\, v + \rho\, \delta v)\right] = 0,$$

$$\frac{\partial \delta v}{\partial t} + \partial_r\left(\frac{c^2}{\rho}\delta\rho + v\delta v\right) = 0$$

$$\Leftrightarrow \quad \frac{d(A \cdot x)}{dr} = i\omega x, \quad A = \begin{pmatrix} v(r) & r^2\rho(r) \\ \frac{c(r)^2}{r^2\rho(r)} & v(r) \end{pmatrix},$$

with $x = (r^2\delta\rho, \delta v)$. If $y = A \cdot x$, $U = \begin{pmatrix} 0 & 1 \\ 1 & 0 \end{pmatrix}$, then we get the orthogonality relation $\int y_n^* \cdot U \cdot A^{-1} \cdot y_m \, dr \propto \delta_{nm}$.

Then we express the initial condition $y(r)$ as a sum of orthogonal modes:

$$y(r) = \sum_{n=-\infty}^{\infty} c_n y_n(r) = c_0 y_0 + 2\,\mathrm{Re}\left[\sum_{n=1}^{\infty} c_n y_n(r)\right].$$

Due to orthogonality, $c_n = \langle y_n, y\rangle/\langle y_n, y_n\rangle$. The time evolution is given by $c_n(t) = c_n e^{-i\omega_n t}$. The norm $\langle y, y\rangle$ is invariant and, in fact, is proportional to the energy. The energy spectrum $\{|c_n|^2\}_{n=0}^{\infty}$ is invariant but the generic evolution of the correlation between the phases is to decrease with time, leading to *quasi-ergodicity* and, therefore, (marginal) stability.

REFERENCES

1. H. Bondi, *Mon. Not. R. Astron. Soc.*, **112**, 195 (1952)
2. A. R. Garlick, *Astron. & Astrophys.*, **73**, 171 (1979)
3. V. Moncrief, *Astrophys. J.*, **235** 1038–1046 (1980).
4. I. G. Kovalenko and M. A. Eremin, *Mon. Not. R. Astron. Soc.*, **298**, 861–870 (1998)
5. J. A. Petterson, J. Silk and J. P. Ostriker, *Mon. Not. R. Astron. Soc.*, **191**, 571 (1980)

Coupling Einstein-Rosen Waves to Matter: The Massless Scalar Field Case

J. Fernando Barbero G.*, Iñaki Garay* and Eduardo J. S. Villaseñor[†,*]

*Instituto de Estructura de la Materia, C.S.I.C., Serrano 123, 28006 Madrid, Spain
[†]Grupo de Modelización y Simulación Numérica, Escuela Politécnica Superior, Universidad Carlos III de Madrid, Avda. de la Universidad 30, 28911 Leganés, Spain

Abstract. We describe the quantization of Einstein-Rosen waves coupled to a cylindrical massless scalar. We obtain a close form for the Hamiltonian and for the unitary quantum evolution operator of the system. This result allows us to discuss several issues relevant to quantum general relativity, such as the causal structure of the space-time at quantum scales or the two point function of the fields.

Keywords: Cylindrical symmetry, quantum Einstein-Rosen waves.
PACS: 04.60.Ds, 04.60.Kz, 04.62.+v.

General relativity is the best theory of gravity that we have, so it would be very interesting to construct a quantum theory of gravity approaching it at the classical limit. As is well known this is not an easy task because of important technical and conceptual issues that must be overcome. Some of them may even be at the heart of the non perturbative non-renormalizability of the theory, in particular the combination of diffeomorphism invariance with the presence of an infinite number of degrees of freedom. These difficulties are indeed important but do not preclude us from learning about quantum gravity by other indirect approaches. One particularly fruitful way to advance is to consider systems with certain symmetries that, in some sense, keep many of the important features of the full theory and allow us to study specific problems without the difficulties present there. One well known example of these systems is provided by the Bianchi models used in quantum cosmology. Although they are very interesting they have a finite number of degrees of freedom, so it is not possible to associate a field theory to them. For our purposes it is more useful to consider the Einstein-Rosen waves [1, 2]. This system is a genuine field theory (it has an infinite number of degrees of freedom), it is diffeomorphism invariant in the radial direction, can be solved exactly both classically and quantum mechanically, and it is rich enough to display interesting behavior. It has been studied since the seventies and has received a lot of attention in the nineties [3, 4]. For the sake of the present paper the results about the causal structure of the space-time shown in [5, 6] are of special relevance.

The main point of this article is to show that it is possible to enrich the model – and make it more useful to study quantum gravitational effects– by coupling a scalar matter field [7]. This gives us a concrete example where the effects of backreaction can be taken into account exactly. Also, the presence of a matter field provides us with an external probe that can be useful to study the quantum geometry and the emergence of its classical counterpart. Here we will focus on the obtention of the Hamiltonian and the obtention of the unitary quantum evolution operator for the system. We will briefly discuss some applications of these results to the study of microcausality and two-point functions of the fields.

The starting point to get the Hamiltonian of our system (Einstein-Rosen waves coupled to a cylindrically symmetric massless scalar field Φ_s) is the four dimensional Einstein-Hilbert action coupled to the matter field with cylindrical symmetry:

$$^4S = \frac{1}{16\pi G_N} \int_{\mathcal{M} \times I} d^4x \sqrt{|^4g|} \left[R - \frac{1}{2} {}^4g^{ab} \nabla_a \Phi_s \nabla_b \Phi_s \right]$$
$$+ \frac{1}{8\pi G_N} \int_{\partial(\mathcal{M} \times I)} d^3x (\sqrt{|^3g|} K - \sqrt{|^3g^0|} K^0).$$

In the last expression we have included the appropriate boundary terms needed to have a well defined variational principle. $I \equiv [z_1, z_2]$ is a closed interval in the direction of the translational Killing vector $\xi^a \equiv \partial_z$, and $\mathcal{M}$ is a 3-dimensional manifold. K and K^0 are the extrinsic curvatures defined at the boundary by the dynamical metric ${}^4g_{ab}$ and the fiducial metric ${}^4g^0_{ab}$ that we use to define an origin for the energy of the system, ensure that the action is finite, and fix the asymptotic behavior of the fields in such a way that the Minkowski metric has zero energy. Finally ${}^3g_{ab}$ and ${}^3g^0_{ab}$ are the induced metrics on the boundary.

We start by applying the Geroch formalism to perform a symmetry reduction along the translational Killing vector. Then, after a conformal transformation of the metric, we get the following equivalent action for the model in three dimensions:

$$^3S = \frac{1}{16\pi G_3} \int_{\mathcal{M}} d^3x \sqrt{|g|} \left[{}^3R - \frac{1}{2} g^{ab} \nabla_a \phi_g \nabla_b \phi_g \right. \tag{1}$$
$$\left. - \frac{1}{2} g^{ab} \nabla_a \phi_s \nabla_b \phi_s \right] + \frac{1}{8\pi G_3} \int_{\partial \mathcal{M}} d^2x [\sqrt{|h|} K - \sqrt{|h^0|} K^0].$$

Here, g_{ab} is the 3-dimensional metric of the spatial manifold $\mathcal{M}$ and ϕ_g is a scalar field that carries the gravitational degrees of freedom of the model [3]; it is defined in terms of the translational Killing vector ξ^a as $\phi_g \equiv \log(g_{ab} \xi^a \xi^b)$. The coupling constant G_3 that appears is the gravitational constant per unit length along the symmetry axis. In the following, we choose units such that $\hbar = c = 8G_3 = 1$. The most striking feature of (1) is the fact that the scalar field term and the gravity field term have *exactly the same form*, although the nature of these fields in 4-dimensions is very different. It is important to note that the two fields are coupled through the metric despite the fact that the expression for the action has no cross terms involving both the gravitational and the matter scalars.

The Hamiltonian of the system can be obtained by following the procedure developed in [3] for the case without matter. This is possible because the terms in the action involving the gravitational and scalar fields have the same form. The final description of the reduced Hamiltonian system is the following. The reduced phase space is coordinatized by the canonical pairs of variables $\phi_s(r), p_s(r), \phi_g(r)$ and $p_g(r)$ and the reduced Hamiltonian is $H = 2(1 - e^{-\gamma_\infty/2})$, with

$$\gamma_\infty = \frac{1}{2} \int_0^\infty dr \, r \left[\phi_g'^2 + \frac{p_g^2}{r^2} + \phi_s'^2 + \frac{p_s^2}{r^2} \right].$$

Notice that γ_∞ is the Hamiltonian for two massless cylindrically symmetric fields evolving in a fictitious Minkowskian background. As we see the full Hamiltonian is a non linear function of the sum of two free Hamiltonians so we are indeed dealing with an interacting system, albeit one with a non-local interaction.

We proceed now to quantize by introducing the following creation and annihilation operators $A_g^\dagger(k) \equiv a_g^\dagger(k) \otimes 1_s$ and $A_s^\dagger(k) \equiv 1_g \otimes a_s^\dagger(k)$. They are defined in a Hilbert space built as a tensor product of two Fock spaces $\mathscr{F}_g$ and $\mathscr{F}_s$; $\mathscr{H} = \mathscr{F}_g \otimes \mathscr{F}_s$, with a vacuum state $|\Omega\rangle = |0\rangle^g \otimes |0\rangle^s$. The operators $a_{g,s}^\dagger(k)$ and $a_{g,s}(k)$ are defined as the creation and annihilation operators acting on the corresponding Fock spaces $\mathscr{F}_g$ and $\mathscr{F}_g$; they satisfy the usual commutation relations. The states $|0\rangle^{g,s}$ are the vacuum states annihilated by $a_{g,s}(k)$. The expressions of the fields and momenta in terms of the creation and annihilation operators are

$$\phi_{g,s}(R) = \frac{1}{\sqrt{2}} \int_0^\infty dk\, J_0(Rk)[a_{g,s}(k) + a_{g,s}^\dagger(k)],$$

$$p_{g,s}(R) = \frac{iR}{\sqrt{2}} \int_0^\infty dk\, k J_0(Rk)[a_{g,s}^\dagger(k) - a_{g,s}(k)].$$

They satisfy the commutation relations $[\hat{\phi}_{g,s}(R), \hat{p}_{g,s}(R')] = i\delta(R, R')$. With this construction, we conclude that the quantum Hamiltonian is

$$\hat{H} = 2\left\{1 - \exp\left[-\frac{1}{2}\int_0^\infty dk\, k[A_g^\dagger(k)A_g(k) + A_s^\dagger(k)A_s(k)]\right]\right\},$$

where the normal ordering of the exponent is needed to avoid having a zero Hamiltonian. As in the classical case, the Hamiltonian is a non trivial function of the sum of the Hamiltonians for two massless, cylindrically symmetric scalar fields in 2+1 dimensions, H_0^g and H_0^s. Once we have the exact form of the Hamiltonian, that is the energy and the generator of the time evolution, we construct the evolution operator (from $t = t_1$ to $t = t_2$) in a straightforward way

$$U(t_2 - t_1) = \exp\left[-2i(t_2 - t_1)(1 - e^{-\frac{1}{2}[H_0^g + H_0^s]})\right]. \tag{2}$$

We can deduce from this expression that there is no conversion of quanta of one type into another, and of course, we have the evolution for arbitrary values of t_1 and t_2. In particular, when $t_1 \to -\infty$ and $t_2 \to +\infty$, it defines the S-matrix of the system. It is interesting to remark that although we have an interacting theory, this solution is exact and non perturbative.

We want to study now the causal structure of the space-time through the study of field commutators. The main result concerning this point is that the vacuum expectation value of the commutator has the same value for the gravitational and for the scalar field:

$$\langle \Omega | [\hat{\phi}_{g,s}(R', t'), \hat{\phi}_{g,s}(R, t)] | \Omega \rangle = -\frac{i}{2} \int_0^\infty dk\, J_0(R'k) J_0(Rk) \sin[(t' - t)E(k)],$$

where $E(k) = 2(1 - e^{-k/2})$.

It is important to remark here that the kind of information that we are obtaining through the consideration of the microcausality of the system is certainly related to the one encoded in the metric but is not directly derived from the metric operator and its matrix elements. The fact that the information about the smearing of the light cone is the same for ϕ_s and for ϕ_g is a natural consistency condition on microcausality. The emergence of the cylindrical light cone structure corresponding to the quantization of a cylindrical massless scalar field in a 2+1-dimensional Minkowskian background can be observed both in the gravitational and matter sectors by using the techniques described in [6].

The availability of a matter field allows us to explore the quantum geometry using it as a kind of external probe. A possible approach to this problem is to interpret the position space two-point function $\langle\Omega|\hat{\phi}_s(R_2,t_2)\hat{\phi}_s(R_1,t_1)|\Omega\rangle$ as an approximate probability amplitude for a particle created at the 1+1 space-time point (R,t) to be found at (R',t'). As in the case of the field commutator a closed form for it can be written. The most interesting feature of this two point function, with the suggested interpretation, is the fact that there is a large probability to find the scalar particle near the axis. This is a gravitational effect and not a consequence of the cylindrical symmetry, because if we consider instead a 2+1 dimensional system, with a cylindrically symmetric massless scalar field but without gravity, this effect does not appear.

There are other interesting questions that will be looked at in the future, such as critical phenomena in 2+1 cylindrical gravitational collapse and problems in black hole physics. For example, we can try to relax the radial asymptotic flatness condition in order to allow the presence of the self-similar solutions needed to discuss critical collapse [8]. It could be also possible to use the results for the massless case as a starting point to introduce other types of fields, such as massive scalars or electromagnetic fields.

ACKNOWLEDGMENTS

We want to thank A. Ashtekar, L. Garay, J. M. Martín García, G. A. Mena Marugán and M. Varadarajan for fruitful discussions. I. G. is supported by a Spanish Ministry of Science and Education FPU research assistantship. This work is also supported by the Spanish MEC under the research grants BFM2002-04031-C02-02 and FIS2005-05736-C03-02.

REFERENCES

1. A. Einstein, and N. Rosen, *J. Franklin Inst.*, **223**, 43 (1937).
2. K. Kuchař, *Phys. Rev. D*, **4**, 955 (1971).
3. A. Ashtekar, and M. Pierri, *J. Math. Phys. (N.Y.)*, **37**, 6250 (1996).
4. A. Ashtekar, and M. Varadarajan, *Phys. Rev. D*, **50**, 4944 (1994).
5. J. F. Barbero G., G. A. Mena Marugán, and E. J. S. Villaseñor *Phys. Rev. D*, **67**, 124006 (2003).
6. J. F. Barbero G., G. A. Mena Marugán, and E. J. S. Villaseñor *J. Math. Phys. (N.Y.)*, **45**, 3498 (2004).
7. J. F. Barbero G., I. Garay, and E. J. S. Villaseñor *Phys. Rev. Lett.*, **95**, 51301 (2005).
8. A. Wang, *Phys. Rev. D*, **68**, 064006 (2003).

Non Spherical Collapse of Scalar Field Dark Matter

Argelia Bernal* and F. Siddhartha Guzmán[†]

*Departamento de Física, Centro de Investigación y de Estudios Avanzados del IPN, AP 14-740, 07000 México D.F., México.
[†]Instituto de Física y Matemáticas, Universidad Michoacana de San Nicolás de Hidalgo. Edificio C-3, Cd. Universitaria, C.P. 58040 Morelia, Michoacán, México.

Abstract. We evolve the Schrödinger-Poisson system of equations for axisymmetric non-rotating scalar field initial configurations. It is shown that for that kind of initial data, spherically symmetric equilibrium configurations are late-time attractors which are reached through the emission of scalar field bursts. These results are relevant within the structure formation mechanism of the scalar field dark matter model or any of the Bose condensate approaches to the dark matter problem, because we show the natural tendency of scalar field configurations to become spherical.

Keywords: Cosmology:dark matter -Galaxies:halos -Galaxies:formation
PACS: 04.25.Dm, 95.30.Sf, 95.35.+d, 98.62.Ai, 98.80.-k

Introduction. Numerous observations have provided accumulative evidence of the existence of dark matter, however at the moment nothing is said about its nature. The model that has shown to be consistent with cosmological observations is the so called Cold Dark Matter (CDM) model, which shows a suitable scenario of structure formation in the Universe, even though the nature of dark matter particles is not specified, it just demands that the dark matter must be a cold dust. This situation opens a wide range of possibilities.

The favorite candidates to play the role of dark matter particles come from the supersymmetric models of particles. With this kind of dark matter particles, the CDM model is very successful at cosmic scales. However, the model fails at galactic level because it predicts cuspy density profiles for galactic halos, among other issues (see [1] and references therein). This is in contradiction with several new high resolution observations which show almost flat core density profiles [2, 3, 4, 5, 6, 7]. Another discrepancy between CDM predictions and observations seems to be the age of the oldest galaxies in the universe (see [8]).

Different alternative dark matter models have appeared in order to sort those problems out, and some of them propose a hypothesis about the nature of dark matter. The scalar field dark matter (SFDM) is one of such models, which assumes the dark matter is made of ultralight spinless particles that condensate and collapse gravitationally in order to form structures [9, 10, 15, 13, 16, 17]. One of the advantages of the scalar field dark matter approach is that at cosmic scales, when combined with a cosmological constant, it reproduces the successes of the ΛCDM model in the free field case, that is, when the potential of the scalar field is $V(\Phi) = m^2\Phi^2$, where Φ is the scalar field. The cosmological behavior of the scalar field dark matter is detailed in [9, 10]. An extra bonus is obtained at galactic scales, for instance, in [11] it was shown that the equations of motion for the SFDM Lagrangian, the Einstein Klein-Gordon (EKG) equations, have regular, asymptotically flat and time dependent solutions for the space-time. The resulting field configurations are called Oscillatons. In [12, 13, 14] numerical simulations were made in order to show that arbitrary initial configurations of scalar field collapses and forms a gravitationally bound object with a regular density profile everywhere, fair to say, in the strong gravity regime.

Because at galactic scales gravitational fields are weak, it is appropriate to work within the Newtonian Limit of the EKG equations, the so called Schrödinger-Poisson (SP) system. Even though the relativistic scalar field is real, the classical limit of the Klein-Gordon equation is the Schrödinger one with a complex wave function oscillating with the fundamental frequency of the relativistic field [19]. This approximation should work for the evolution of a initial density profile after the time of *turnaround*, that is, in the epoch when the overdensity fluctuation starts to evolve independently from the cosmological expansion [15].

An important feature of the SP system is that is has equilibrium stationary solutions that have been shown to be late time attractors for spherically symmetric initial configurations with otherwise arbitrary density profiles [19]. What we show in this manuscript is that such equilibrium configurations are late time attractors for initially non-rotating axisymmetric density profiles as well, thus generalizing the attractor nature of the spherically symmetric equilibrium configurations. This result indicates that the gravitational collapse of self-gravitating Bose condensates made of ultralight spinless particles, tolerates no spherically symmetric initial density profiles.

In this work we want to address the collapse of a non-spherical initial overdensity because this could provide important information about the galactic formation in the frame of the SFDM model. In the next section we summarize the important quantities for equilibrium configurations, later on we describe the code used and show the evolution of axially symmetric initial configurations. Finally in the last section we draw some comments and conclusions.

Equilibrium configurations for the SP system. The SP system is obtained through a post-Newtonian expansion. The background metric is considered to be Minkowski and we assume that the scalar field behavior is Newtonian, then its total energy is comparable to its rest energy and assume $\Phi = e^{-i\gamma t}\phi(\vec{x})$ with $\gamma \sim m$, where m is the scalar field mass. With this considerations in mind, the SP equations in spherical symmetry read

$$\begin{aligned}(x\phi)_{,xx} &= 2x(U-\gamma)\phi \\ (xU)_{,xx} &= x\phi^2\end{aligned} \quad (1)$$

where U is the Newtonian potential, x is the radial coordinate, and where we are using Planck units. The first of the equations is the Schrödinger equation and the second is the Poisson equation in spherical coordinates. This system turns into an eigenvalue problem for γ provided the boundary conditions of $\phi = 0$ at infinity and a smooth gravitational potential at the origin (for details see [19]). Solutions of this eigenvalue problem are stationary solutions with a time independent $\rho = \Phi\Phi^*$ and therefore a time independent gravitational potential U.

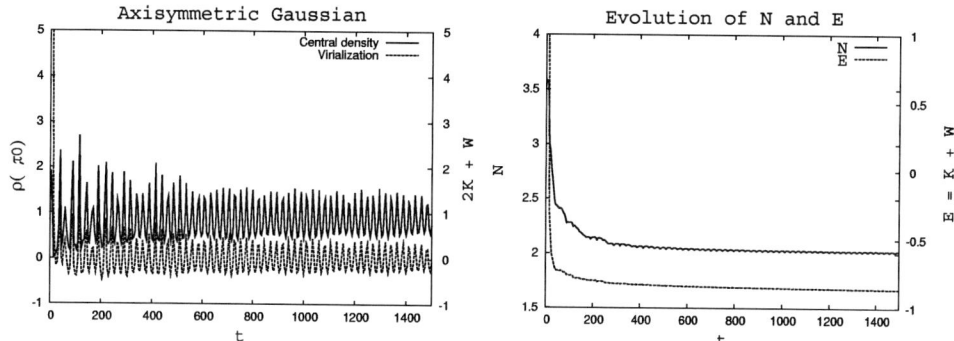

FIGURE 1. On the left panel we show the central density of the configuration and how it converges to an assymptotic value and the evolution of the quantity $2K+W$ which after a certain finite time starts converging to zero. On the right panel we show the evolution of the number of particles, where the emission of bursts of particles can be observed, and the value of the total energy, which becomes a negative constant as expected for any bounded object.

Important physical quantities of configurations that evolve according to the SP system are: the number of particles $N = m\int 4\pi r^2 dr \phi^2$, the expectation value of the kinetic $K = -(1/2)\int \Phi^* \partial_y^2 (y\Phi) y dy$, potential $W = (1/2)\int \rho U y^2 dy$ and total $E = K + W$ energies. The number of particles allows us to calculate the material content in a region of space, and the quantity $2K+W$ determines whether the system is dynamically virialized or not.

Numerical evolution of Non-Spherically symmetric initial profiles. In order to evolve such configurations we developed a code that uses cylindrical coordinates (r,z) and solve the following SP system where now $\Phi = \Phi(r,z)$

$$i\frac{\partial \Phi}{\partial t} = -\frac{1}{2}\left(\frac{\partial^2 \Phi}{\partial r^2} + \frac{1}{r}\frac{\partial \Phi}{\partial r} + \frac{\partial^2 \Phi}{\partial z^2}\right) + U\Phi \qquad (2)$$

$$\frac{\partial^2 \Phi}{\partial r^2} + \frac{1}{r}\frac{\partial \Phi}{\partial r} + \frac{\partial^2 \Phi}{\partial z^2} = \Phi^* \Phi \qquad (3)$$

We have tested this code by evolving equilibrium configurations and observed consistent results with the predictions of perturbation theory in spherical symmetry. Moreover we have obtained second order convergence in all our runs. In the outermost points we have used a sponge similar to those used in spherical symmetry [19], which absorbs the modes reflected by the boundaries of our finite domain.

Using this new code, in Figure 1 we show the result of the evolution for the initial profile given by $\Phi(0,r) = \cos^2(r)e^-$. It can be observed how the central density stabilizes around a constant value and how the system approaches a dynamically virialized state. The number of particles also approaches a constant value and the total energy achieves its constant value before any other quantity.

Conclusions. Using a new code that solves the SP system with axial symmetry, we showed that non-spherically symmetric initial density profiles evolve towards an equilibrium configuration, which is virialized. This conclusion is achieved by looking at the quantity $2K+W$, which after a while starts oscillating around zero and the number of particles N and total energy E stabilize around fixed values.

The process of dynamical virialization is pretty much the gravitational cooling [12, 15, 19] where the system relaxes through the emission of scalar field particles. We expect that for initial profiles involving quadrupolar contributions there will be emission of gravitational radiation as well [20], although this is a study under research.

ACKNOWLEDGMENTS

A.B. acknowledges partial support from CONACyT. F.S.G.was supported by projects CIC-UMSNH-4.9 and PROMEP-UMICH-PTC-121. Runs were carried out in the Ek-bek cluster of the "Laboratorio de Supercómputo Astrofísico (LASUMA)" at CINVESTAV-IPN.

REFERENCES

1. Jeremiah P. Ostriker, and Paul J. Steinhardt, *Science*, **300**, 1909-1913 (2003).
2. W. J. G. de Blok, and S. S. McGaugh, *Mon. not. R. Astron. Soc.*, **209**, 533 (1997).
3. W. J. G. de Blok, S. S. McGaugh, A. Bosma, and V. C. Rubin, *Astrophys. J.* **552**, L23 (2001).
4. S. S. McGaugh, V. C. Rubin, and E. de Block, *Astron. J.* **122**, 2831 (2001).
5. W. J. G. de Blok, S. S. McGaugh, and V. C. Rubin, *Astron. J.* **122**, 2396 (2001).
6. P. A. S. Blais-Ouellette, and C. Carignan, *Astron. J.* **121**, 1952 (2001).
7. A. D. Bolato, J. D. Simon, A. Leroy, and L. Blotz, *Astrophys. J.* **565**, 238 (2002).
8. k. Glazebrook et al., *The GEMINI Deep Survey: III. The evolution of galaxy stellar masses*, astro-ph/0310193.
9. T. Matos, and L. A. Ureña-López, *Class. Quantum Grav.* **17**, L75 (2000).
10. T. Matos, and L. A. Ureña-López, *Phys, Rev. D* **63**, 063506 (2000).
11. L. A. Ureña-López, T. Matos, and R. Becerril, *Class. Quantum Grav.* **19**, 1-19 (2002).
12. E.Seidel, and W-M.Suen, *Phys, Rev. Lett* **66**, 1659 (1991).
13. M. Alcubierre, F. S. Guzmán, T. Matos, D. Núñez, L. A. Ureña, and P. Wiederhold. *Class. Quantum Grav.* **19**, 5017 (2002).
14. M. Alcubierre, R. Becerril, F. S. Guzmán, T. Matos, D. Núñez, and L. A. Ureña, *Class. Quantum Grav.* **20**, 2883 (2003).
15. F. S. Guzmán, and L. A. Ureña, *Phys, Rev. D* **68**, 024023 (2003).
16. A. Arbey, J. Lesgourges and P. Salatti. *Phys. Rev. D* **64**, 123528 (2001). *Ibid* **64**, 083514 (2002). *Ibid* **68**, 023511 (2003).
17. J. P. Mbelek, *Atron. Astrophys.* **424**, 761 (2004).
18. E. Seidel, and W-M. Suen, *Phys, Rev. D* **42**, 384 (1990); L Balakrishna, E. Seidel, and W-M. Suen, *ibid.* **58**, 104004 (1998);
19. F. S. Guzmán, and L. A. Ureña, *Phys, Rev. D* **69**, 124033 (2004).
20. R. Ferrell and M. Gleiser, *Phys, Rev D* **40**, 2524 (1989).

May we use the LLR as a redshift indicator for the gamma-ray bursts?

Mimoza Hafizi* and Robert Mochkovitch[†]

*Dept. of Physics, University of Tirana, Albania
[†]IAP-CNRS, Paris, France

Abstract. We use a simple pulse model to investigate the origin of the time lag-luminosity relation (LLR) discovered by Norris (2000). We investigate the use of the LLR as a distance indicator and find not accurate enough to yield the redshift of a specific burst

INTRODUCTION

Gamma-ray bursts (GRB) are the most powerful explosions in the universe. The flux of radiated energy is of the order of $10^{51} erg/s$. They are at cosmological distances. The problem of their distance remained unsolved until the discoveries of the afterglows by Beppo-Sax, in 1997. Redshifts are now obtained from optical spectra of the afterglow itself or of the host galaxy when the afterglow has faded away. It is expected that recently launched SWIFT will provide up to 100 redshifts/yr. Using the few redshifts known until now, it became possible to calibrate relations linking absolute burst outputs (luminosity or total radiated energy) to quantities directly available from the observations in gamma-rays. One of these quantities is the time lag, the delay between the soft emission part of the GRB and the high energy emission part. A time lag-luminosity relation was discovered by Norris *et al.* [1], where time lags were computed by cross-correlating burst profiles in BATSE band 1 (20-50 keV) and 3 (100-300 keV). They found that high luminosity GRBs exhibit small lags Δt and proposed a power law relation:

$$L = 1.3 \times 10^{53} (\frac{\Delta t}{0.01s})^{-1.14} erg.s^{-1}. \tag{1}$$

Both of quantities, luminosity and the time lag, are redshift corrected. Another relation found by Amati *et al.* [2].

$$\varepsilon_{iso} = 10^{52} \left[\frac{<E_p>}{200 keV}\right]^{2.17} erg \tag{2}$$

shows the correlation between the isotropic radiated energy ε_{iso} and the global peak energy $<E_p>$.

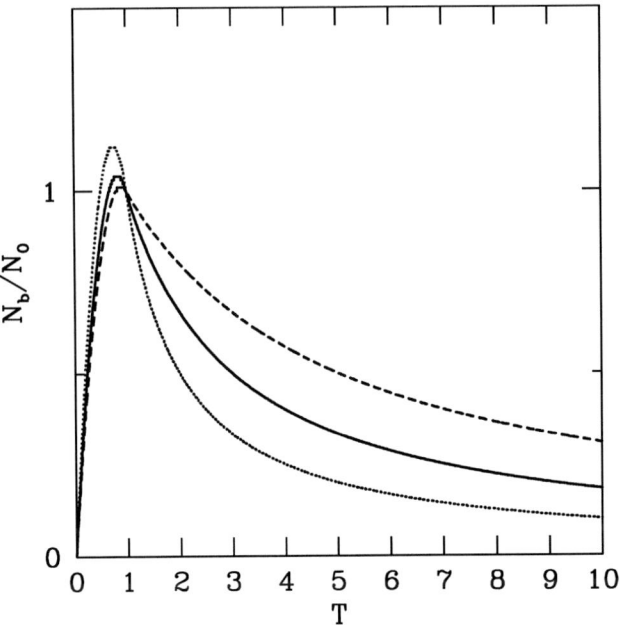

FIGURE 1. Bolometric pulse profiles as functions of time $T = t/\tau$ for different values of $Q = t_0/\tau$.

THE PULSE MODEL

Our model is a simple pulse one. The number of photons produced per unit time and unit energy is

$$N(E,t) = A(t)B\left[\frac{E}{E_p(t)}\right], \quad (3)$$

where $B(x)$ is the Band spectrum [3] with peak energy $E_p(t)$ and two slopes, low energy one α and high energy one β, which we keep fixed in time. The bolometric photon rate

$$N_b(t) = \int_0^\infty N(E,t)dE, \quad (4)$$

is a decreasing function of time. As shown by Ryde and Svensson [4], based on the works of Golenetskii et al. [5] and Liang and Kargatis [6], $E_p(t)$ and $N_b(t)$ for $t > t_0$ are described by the following decreasing functions:

$$N_b(t) = \frac{N_0}{1 + \frac{t-t_0}{\tau}} \quad (5)$$

and

$$E_p(t) = \frac{E_0}{(1 + \frac{t-t_0}{\tau})^\delta}. \quad (6)$$

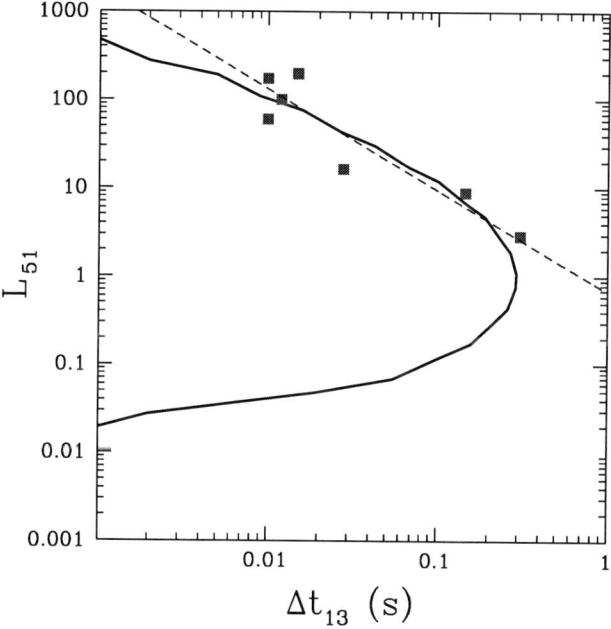

FIGURE 2. LLR obtained with our model compared to Norris *et al.* (2000) data points. The dashed line represents the fit of the LLR proposed by Norris *et al.* [1].

For $t < t_0$ we adopt parabolic form for $N_b(t)$ and linear extension for $E_p(t)$. We obtain the bolometric pulse profiles shown in fig. 1 for different values of parameters. With $N_b(t)$ and E_p being known, we obtain the pulse shape in any band $[E_i, E_j]$. So, we compute time lags Δt_{13} by cross correlating pulse profiles in two energy bands 1 and 3.

For obtaining the L_{max} of the pulse, we make use of the Amati relation [2] for obtaining the proportionality between $L_b(t)$ and $N_b(t)E_p(t)$. Fig. 2 shows the LLR obtained with our model compared to Norris *et al.* [1].

THE LLR AS A REDSHIFT INDICATOR

Instead of the lags, the observations give $\Delta t_{13}^{obs} = (1+z)\Delta t_{13}$, not sufficient to yield an estimate of the luminosity. We can make use of an additional constraint, the observed photon peak flux P_{max}. Theoretically, with these two contraints we can estimate the luminosity of a specific burst.

As we see from the Fig.3, the constant P_{max} lines are rather steep, which make the redshift determination quite uncertain. The spread of pulse temporal parameters Q, δ, of spectral parameters α, β and the Amati relation are likely to add a large uncertainty in the redshift determination. We believe that the use of the LLR to obtain the distance of a single event remains a dangerous exercise.

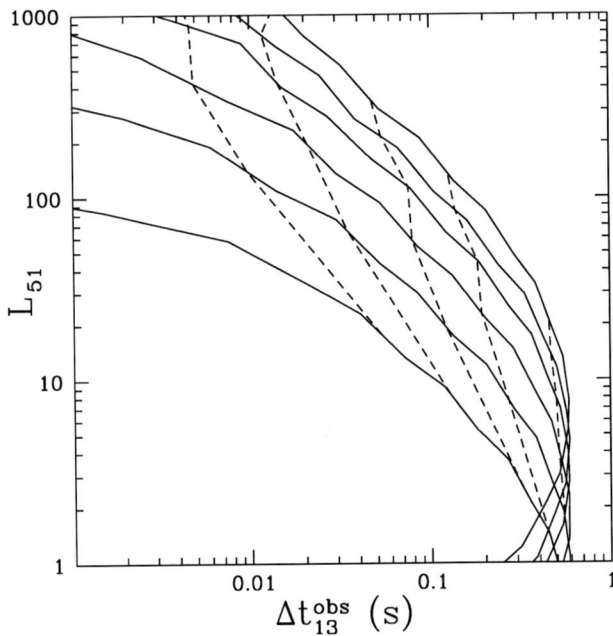

FIGURE 3. Lag-luminosity relation for pulses with $t_0(1+z) = 4s$, $\delta = 0.75$ and $Q = 0.5$. The six full lines correspond to z=0.5 (lower curve), 1, 1.5, 2, 2.5 and 3; the dashed lines correspond to a given observed photon peak flux: 15, 10, 5, 3 and 1.

REFERENCES

1. J. Norris, G. Marani, J. Bonnell, Ap. J. **534**, 248 (2000)
2. L. Amati, F. Frontera, M. Tavani et al. A&A **390**, 81 (2002)
3. D.L. Band et al., Ap. J. **413**, 281 (1993)
4. F. Ryde, R. Svensson, Ap. J. **529**, L13 (2000)
5. S.V. Golenetskii, E.P. Mazets, R.L. Aptekar et al., Nature **306**, 451 (1983)
6. E. Liang, V. Kargatis, Nature **381**, 49 (1996)

Explicit multipole moments of axisymmetric stationary spacetimes

Magnus Herberthson

Matematiska institutionen, Linköpings Universitet, SE-581 83 Linköping, Sweden.
e-mail: maher@mai.liu.se

Abstract. We consider multipole moments of axisymmetric stationary asymptotically flat spacetimes. We show how the tensorial recursion of Geroch and Hansen can be reduced to a recursion of scalar functions. We also demonstrate how a careful choice of the conformal factor collects all moments into one complex valued function on **R**, where the moments appear as the derivatives at 0. As an application, we calculate the moments of the Kerr solution.

Keywords: multipole moments, stationary, axisymmetric
PACS: 04.20.Ha, 04.20.Jb

THE PROBLEM

A static or stationary (asymptotically flat) spacetime can be characterised by its *multipole moments*. Even in the axisymmetric case, these moments can be complicated to calculate beyond the first few orders. The problem is to simplify these calculations and collect all moments into one function.

THE DEFINITION OF GEROCH AND HANSEN

Here we quote the definition given by Geroch and Hansen, [1], [2]. Thus let (M, g_{ab}) be a stationary spacetime with time-like Killing vector field ξ^a. We define the norm $\lambda = -\xi^a \xi_a$, and the twist ω through $\nabla_a \omega = \varepsilon_{abcd} \xi^b \nabla^c \xi^d$. If V is the 3-manifold of trajectories, the metric g_{ab} (with signature $(-,+,+,+)$) induces the positive definite metric $h_{ab} = \lambda g_{ab} + \xi_a \xi_b$ on V. It is required that V is asymptotically flat, i.e., there exists a 3-manifold $\tilde{V}$ and a conformal factor Ω satisfying

(i) $\tilde{V} = V \cup \Lambda$, where Λ is a single point
(ii) $\tilde{h}_{ab} = \Omega^2 h_{ab}$ is a smooth metric on $\tilde{V}$
(iii) At Λ, $\Omega = 0, \tilde{D}_a \Omega = 0, \tilde{D}_a \tilde{D}_b \Omega = 2\tilde{h}_{ab}$,

where $\tilde{D}_a$ is the derivative operator associated with $\tilde{h}_{ab}$. On M, and/or V we define the scalar potential

$$\phi = \phi_M + i\phi_J, \quad \phi_M = \frac{\lambda^2 + \omega^2 - 1}{4\lambda}, \quad \phi_J = \frac{\omega}{2\lambda}$$

The multipole moments of M are defined on $\tilde{V}$ as certain derivatives of the scalar potential $\tilde{\phi} = \phi/\sqrt{\Omega}$ at Λ. Following [2], we put $P = \tilde{\phi}$ and define the sequence $P, P_{a_1}, P_{a_1 a_2}, \ldots$ of tensors recursively:

$$P_{a_1 \ldots a_n} = C[\tilde{D}_{a_1} P_{a_2 \ldots a_n} - \frac{(n-1)(2n-3)}{2} \tilde{R}_{a_1 a_2} P_{a_3 \ldots a_n}]. \tag{1}$$

Here $C[\,\cdot\,]$ stands for taking the totally symmetric and trace-free part, and $\tilde{R}_{ab}$ denotes the Ricci tensor of $\tilde{V}$.

The multipole moments of M are then defined as the tensors $P_{a_1 \ldots a_n}$ at Λ. The requirement that all $P_{a_1 \ldots a_n}$ be totally symmetric and trace-free makes the actual calculations non-trivial. In the axisymmetric case, however, we will see that the tensorial recursion can be replaced by a scalar recursion.

THE AXISYMMETRIC CASE

In the axisymmetric case, the metric can be written in the standard form:

$$ds^2 = -\lambda(dt - W d\varphi)^2 + \lambda^{-1}(R^2 d\varphi^2 + e^{2\beta}(dR^2 + dZ^2)).$$

With $\tilde{\rho}, \tilde{z}$ and r, θ given by $\tilde{\rho} = \frac{R}{R^2 + Z^2} = r \sin\theta$, $\tilde{z} = \frac{Z}{R^2 + Z^2} = r \cos\theta$, we put as comformal factor $\Omega = r^2 e^{-\beta} e^{\kappa}$, where κ is smooth with $\kappa(\Lambda) = 0$. The rescaled metric $\tilde{h}_{ab}$ on $\tilde{V}$ is seen to be $\tilde{h}_{ab} \sim ds^2 = e^{2\kappa}(\tilde{\rho}^2 e^{-2\beta} d\varphi^2 + d\tilde{\rho}^2 + d\tilde{z}^2) = e^{2\kappa}(r^2 \sin^2\theta e^{-2\beta} d\varphi^2 + dr^2 + r^2 d\theta^2)$. Here e^{κ} expresses the conformal freedom. Different values of $\kappa'(\Lambda)$ correspond to 'expansions around different points', i.e., translates the moments as explained in [1]. Higher order terms of κ changes the tensor fields $P_{a_1 \ldots a_n}$, but do not affect the moments, i.e. $P_{a_1 \ldots a_n}$ at Λ.

A scalar recursion on $\mathbf{R}^2$

In the axisymmetric case, where the symmetry axis $\sim z^a$, each moment $P_{a_1 a_2 \ldots a_n}(\Lambda) \propto C[z_{a_1} z_{a_2} \ldots z_{a_n}]$, i.e. the moments are given by the sequence $\{m_n\}_n$ where m_n is defined by $P_{a_1 a_2 \ldots a_n}(\Lambda) = m_n C[z_{a_1} z_{a_2} \ldots z_{a_n}]$.

Furthermore, we introduce a vector field η^a on $\tilde{V}$ with the properites that
a) For all tensors $T_{a_1 \ldots a_n}$, $\eta^{a_1} \ldots \eta^{a_n} T_{a_1 \ldots a_n} = \eta^{a_1} \ldots \eta^{a_n} C[T_{a_1 \ldots a_n}]$
b) At Λ, $P_{a_1 \ldots a_n}$ is determined by $\eta^{a_1} \ldots \eta^{a_n} P_{a_1 \ldots a_n}$
c) $\eta^a \tilde{D}_a \eta^b$ is parallel to η^b.

In particular, a) will be accomplished if we chose η^a to be a complex null vector, i.e., $\eta^a \eta_a = 0$, since terms involving the metric in $C[\cdot]$ will vanish. As one can check explicitly, one may chose $\eta^a = (\frac{\partial}{\partial \tilde{z}})^a - i(\frac{\partial}{\partial \tilde{\rho}})^a$. which gives $\eta^a \tilde{D}_a \eta^b = 2\eta^b \eta^c \tilde{D}_c \kappa$.

Defining $f_n = \eta^{a_1} \eta^{a_2} \ldots \eta^{a_n} P_{a_1 a_2 \ldots a_n}$ and $f_0 = P = \tilde{\phi} = \Omega^{-\frac{1}{2}} \phi$, the recursion (1) becomes [3]

$$f_n = \eta^a \tilde{D}_a f_{n-1} - 2(n-1) f_{n-1} \eta^a \tilde{D}_a \kappa - \frac{(n-1)(2n-3)}{2} \eta^a \eta^b \tilde{R}_{ab} f_{n-2} \tag{2}$$

The multipole moments are given by the values $f_0(0), f_1(0), f_2(0), \ldots$.

A scalar recursion on R

The scalar recursion above can be further simplified as follows. For a function $g : \mathbf{R}^2 \to \mathbf{C}$ which is real analytic near 0, we define the *leading order function* g_L on $\mathbf{R}$ via

$$g_L(r) = g(r, -ir).$$

It then follows that

$$(\eta^a \tilde{D}_a g)_L(r) = g'_L(r),$$

and that the recursion (2) becomes

$$y_n = y'_{n-1} - 2(n-1)\kappa'_L y_{n-1} - \frac{(n-1)(2n-3)}{2} M y_{n-2},$$

where $M(r) = \beta''_L - (\beta'_L)^2 + \frac{2}{r}\beta'_L - \kappa''_L + (\kappa'_L)^2$, $y_n = (f_n)_L$, and $m_n = y_n(0)$. (β comes from the metric.) The recursion simplifies if $M \equiv 0$, and one finds that this is the case if κ is chosen such that

$$\kappa_L(r) = -\ln(1 - r\int_0^r \frac{e^{2\beta_L(r)} - 1}{r^2} dr - rC) + \beta_L(r).$$

With this choice of κ, the recursion becomes $y_n = y'_{n-1} - 2(n-1)\kappa'_L y_{n-1}$. With a new radial coordinate ρ implicitly defined through $\rho(r) = re^{\kappa_L - \beta_L}$, and with $y(\rho) = \tilde{\phi}_L(r(\rho))$, one finally obtains

$$m_n = \frac{d^n y}{d\rho^n}(0).$$

This means that all multipole moments of the spacetime are encoded in the derivatives of y at 0.

Example, the Kerr solution

The Boyer-Lindquist coordinates gives the metric on V as

$$ds^2 = \frac{\tilde{r}^2 - 2m\tilde{r} + a^2 \cos^2\theta}{\tilde{r}^2 - 2m\tilde{r} + a^2} d\tilde{r}^2 + (\tilde{r}^2 - 2m\tilde{r} + a^2\cos^2\theta)d\theta^2 + (\tilde{r}^2 - 2m\tilde{r} + a^2)\sin^2\theta d\varphi^2.$$

We then define r through $\tilde{r} = r^{-1}(1 + mr + \frac{1}{4}(m^2 - a^2)r^2)$, put $\tilde{z} = r\cos\theta$, $\tilde{\rho} = r\sin\theta$ and use as conformal factor $\tilde{\Omega} = \frac{r^2}{\sqrt{(1 - \frac{1}{4}(m^2 - a^2)r^2)^2 - a^2 r^2 \sin^2\theta}}$.

One then verifies the following expressions:

$$\beta = \frac{1}{2}\ln\left(1 - \frac{(4a\tilde{\rho})^2}{(4 - (m^2 - a^2)(\tilde{z}^2 + \tilde{\rho}^2))^2}\right) \Rightarrow \beta_L = \frac{1}{2}\ln(1 + a^2 r^2)$$

451

$$\tilde{\phi} = \frac{m(1+\frac{1}{4}(m^2-a^2)(\tilde{z}^2+\tilde{\rho}^2)) - ia\tilde{z}}{((1-\frac{1}{4}(m^2-a^2)(\tilde{z}^2+\tilde{\rho}^2))^2 - a^2\tilde{\rho}^2)^{\frac{3}{4}}} \Rightarrow \tilde{\phi}_L = \frac{m(1-iar)}{(1+a^2r^2)^{3/4}}$$

Moreover, κ_l is found to be $\kappa_L = -\frac{1}{2}\ln\left(\frac{(r^2a^2+r\kappa'(0)-1)^2}{r^2a^2+1}\right)$, which gives $\rho(r) = \frac{r}{1-r\kappa'(0)-r^2a^2}$, $r(\rho) = \frac{\sqrt{(\rho\kappa'(0)+1)^2+4a^2\rho^2}-\rho\kappa'(0)-1}{2\rho a^2}$ and finally

$$y(\rho) = \frac{m}{\sqrt{1+(2ia+\kappa'(0))\rho}}$$

Again, the derivatives of y at 0 gives the multipole moments. For instance, when $\kappa'(0) = 0$, we have

$$\sum_{n=0}^{\infty} \frac{m_n \rho^n}{n!} = y(\rho) = \sum_{n=0}^{\infty} \frac{m(2ia\rho)^n\sqrt{\pi}}{n!\Gamma(\frac{1}{2}-n)}. \tag{3}$$

With also $a = 0$, we have the Schwarzschild solution expanded around 'the center of mass'. Indeed, in that case $y(\rho) = m$ so that we have a non-zero monopole (if $m > 0$) and all higher moments 0.

The Hansen potential in the static case

In the static case ($\omega = 0$), one can compare the potential suggested by Geroch with the different potential suggested by Hansen (for the stationary case). Expressed in the same framework thay are $\phi_H = \frac{1-\lambda^2}{4\lambda}$ and $\phi_G = \lambda^{-1/4}(1-\sqrt{\lambda})$ respectively. It is known that these potentionals produce the same moments, but in th axisymmetric case it also follows easily with our notation. Namely, one readily finds that

$$\tilde{\phi}_H = \frac{\tilde{\phi}_G}{4}(\Omega\tilde{\phi}_G^2 + 2)\sqrt{\Omega\tilde{\phi}_G^2 + 4},$$

Since $\Omega_L = 0$, $(\tilde{\phi}_H)_L = (\tilde{\phi}_G)_L$, and consequently ϕ_H and ϕ_G produce the same moments.

REFERENCES

1. Geroch, R., *Multipole Moments. II. Curved Space*, J. Math. Phys., **11**, 2580 (1970).
2. Hansen, R.O., *Multipole moments of stationary space-times*, J. Math. Phys., **15**, 46 (1974).
3. Bäckdahl, T., Herberthson, M., *Explicit multipole moments of stationary axisymmetric space-times*, Class. Quantum Grav. **22**, 3585 (2005).

ISA - An Accelerometer to Detect the Disturbing Accelerations Acting on the Mercury Planetary Orbiter of the BepiColombo ESA Cornerstone Mission to Mercury: on Ground Calibration.

V. Iafolla[a], D.M., Lucchesi[a,b] S. Nozzoli[a], F. Santoli[a], M. Fois[a], M. Persichini[a].

[a] *Istituto di Fisica dello Spazio Interplanetario (IFSI/INAF), Via Fosso del Cavaliere, 100, 00133 Roma, Italy.*
[b] *Istituto di Scienza e Tecnologie della Informazione (ISTI/CNR), Via Moruzzi, 1, 56124 Pisa, Italy*

Abstract. To reach the ambitious goals of the Radio Science Experiment of the BepiColombo space mission to Mercury, among which the planet structure and rotation and test Einstein's theory of General Relativity (GR) to an unprecedented accuracy, an accelerometer has been selected to fly on-board the MPO (Mercury Planetary Orbiter), the main spacecraft of the two to be placed around the innermost planet of our solar system around 2017. The key rôle of the on-board accelerometer is to remove from the list of unknowns the non-gravitational accelerations that disturbs the pure gravitational orbit of the MPO spacecraft in the strong radiation environment of Mercury. In this way the "corrected" orbit of the MPO may be regarded as a geodesic in the field of Mercury. Then, thanks to the very precise tracking from Earth, the possibility to study Mercury's center-of-mass around the Sun and estimate several parameters related to the planet structure and verify the theory of GR. The selected accelerometer named ISA (Italian Spring Accelerometer) is an high sensitive instrument with an intrinsic noise of $10^{-10} g_\oplus / \sqrt{Hz}$ (with $g_\oplus \cong 9.8 \, m/s^2$) in the frequency band $3 \cdot 10^{-5} - 10^{-1} \, Hz$. ISA is a three axis accelerometer with a characteristic configuration, in order to minimize the disturbing accelerations due to the gravity-gradients and the apparent forces on the Nadir pointing MPO spacecraft. Because of the complex and strong radiation environment of Mercury, the modelling of the non-gravitational acceleration is quite difficult, while, with the use of ISA accelerometer we are able to gain a factor 100 in accuracy. In this brief paper we will focus on the characteristics of the ISA accelerometer, on its positioning on-board the MPO and in particular to the techniques for on ground calibration, avoiding the effects of the Earth gravity.

Keywords: Experimental Tests of Gravitational Theories, Mechanical Instruments, Computer Modelling and Simulation, Mercury
PACS: 04.80.Cc, 07.10.-h, 07.05.Tp, 96.30.Dz

INTRODUCTION

The main objectives that can be achieved with the Radio Science Experiments (RSE) [4] planned for the ESA cornerstone mission to Mercury are the following one [5,6]: to determine the global gravity field of Mercury and its temporal variations due to solar tides, in order to constrain the internal structure of the planet; to determine the local gravity anomalies, in order to constrain the mantle structure of the planet and the interface between mantle and crust; to determine the rotation state of Mercury, in order to constrain the size and the physical state of the core of the planet; to evaluate the heliocentric orbit of the Mercury's center–of–mass with better accuracy than today, in order to improve the determination of the PPN (Parameterized Post–Newtonian) parameters of GR. The spherical harmonic coefficients of the gravity field of the planet will be determinate (at least) up to degree and order 25; the obliquity of the planet with an accuracy of 4 arcsec and the C_m/C (ratio between mantle and planet moment of inertia) to 0.05 or better; geoid surface to 10 cm over spatial scales of 300 km. More specific indication of the results possible to be achieved for the relativity and gravitational physics, are; PN parameter γ, controlling the deflection of light and the time delay of ranging signals, to $2.5 \cdot 10^{-6}$; PN parameter β, controlling the relativistic advance of Mercury's perihelion, to $5 \cdot 10^{-6}$; PN parameter η, controlling the gravitational self-energy contribution to the gravitational mass, to $2 \cdot 10^{-5}$; the gravitational oblateness of the Sun ($J2$) to $2 \cdot 10^{-9}$; the time variation of the gravitational constant ($d(\ln G)/dt$) to $3 \cdot 10^{-13}$ $years^{-1}$. The measurements to be performed to obtain the indicated objectives are: range and range–rate derivations of the MPO position with respect to Earth–bound radar station(s) (and then of Mercury center–of–mass around the Sun); measurements of the non–gravitational signals acting on the MPO by means of an on–board accelerometer; measurement of the MPO absolute attitude, by means of a Star–Tracker; measure of the angular displacements of reference points on the solid surface of the planet, by means of an high resolution camera.

THE ACCELEROMETER

ISA (Italian Spring Accelerometer) is a three axis accelerometer [1,2] with an intrinsic noise of $10^{-10} g_\oplus / \sqrt{Hz}$ (with $g_\oplus \cong 9.8 m/s^2$) in the frequency band $3 \cdot 10^{-5} - 10^{-1}$ Hz, required for the RSE. In figure-1 it is shown a schematic draw of the mechanical part of the ISA accelerometer. It is constituted of three orthogonal sensing elements, each for every components of the acceleration acting on the MPO. The three centers of mass (com) of each element proof mass are aligned along the MPO rotation axis, in order to minimize the effects of the inertial acceleration and gravity-gradients due to the rotation of the satellite (nadir pointed) [3].

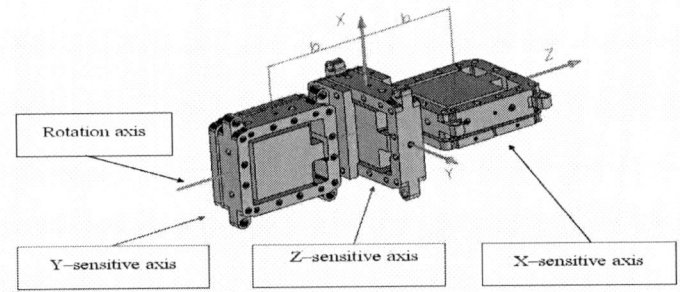

FIGURE 1. Schematic draw of the ISA accelerometer, showing its mechanical part.

The instrument is placed on the MPO center of mass (COM) making it coincident with the com of the z element. Because of the complex and strong radiation environment of Mercury, for instance the solar irradiance varies between $14.448\,W/m^2$ to $6272\,W/m^2$, during a Mercury's year of about $88\,days$ and the direct solar radiation pressure acceleration reaches a value of about $10^{-6}\,m/s^2$, no modeling of the non-gravitational perturbations will guarantee the goals of the Radio Science Experiment. Indeed, with the ISA accelerometer we are able to gain more than a factor 10 in accuracy with respect to the best modeling of the non-gravitational perturbations. In figure-2 are shown the radial component of the acceleration acting on the MPO (due to the direct solar radiation pressure effect) and its spectral analysis, numerically evaluated with a simulation. At the orbital period ($2.3h$) it is about 2 orders–of–magnitude larger than the requested mission accuracy of $\cong 10^{-8}\,m/s^2$.

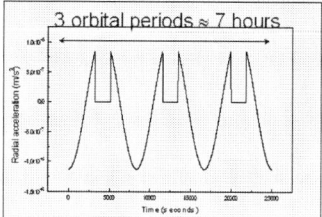 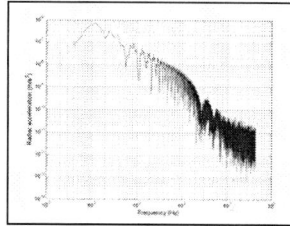

FIGURE 2. The radial Component of the direct solar radiation pressure acting on the MPO, displayed in time and frequency domain, numerically evaluated with a simulation.

GROUND CALIBRATION

One of the main activities concerning the development of the accelerometer is related to its testing and calibration on ground. The analysis of the source noise and the theoretical evaluation of the performances are not satisfactory. Hence, direct experimental tests are necessary, relating big efforts to overcome the problems due to the presence of the seismic noise and to the Earth's gravity acceleration. To measure the accelerometer transducer factor it is necessary to apply a known acceleration to each sensing elements and controlling its output. For ground calibrations, the accelerometer is installed on a base equipped with three support points, two of them electronically controlled with micrometry screws. Adjusting these screws, the gravity acceleration can be set parallel to a sensitive axis (e.g. z) and perpendicular to the other two axes (x, y). In these conditions the proof mass with sensitive axis along the vertical is subjected to acceleration equal to $1g_\oplus$; the other two proof masses are sensitive to a component of $g_\oplus$ depending on the angles ϑ_x, ϑ_y, for the horizontal axes. For small variations of ϑ_x, ϑ_y, their values in radians corresponds to the acceleration, expressed in $g_\oplus$ unit, to which the proof masses have been subjected; controlling the unit output, it is possible to calibrate the apparatus. A problem connected to this kind of calibration, when the torsional frequency of the oscillator is quite low ($3.5Hz$ in our case), is again connected to the presence of the Earth gravity acceleration that change the frequency of the harmonic oscillator, as indicated in the following formula:

$$\omega_0 = \sqrt{\left[\left(\frac{mgR}{I}\right)+\left(\frac{K_{mecc}}{I}\right)\right]}$$

where m is the value of the proof mass, $g_\oplus$ the gravity acceleration on Earth, R the length of the physical pendulum, K_{mecc} the torsional elastic constant and I the moment of inertia of the proof mass. To overcome this problem it is necessary to operate in a condition in which the proof mass is in a "flag" position whit $R = 0$.

The possibility to perform a measure of the intrinsic noise of the apparatus is through the measurement of the seismic noise level in a place where this level is very low (underground laboratory) and using two elements with their sensitive axis aligned, so that the seismic noise can be regarded as a true signal and the difference of them, representing the differential acceleration acting on the system, can be regarded as an upper limit in the intrinsic noise of each single axis. In Figure 3 it is shown the difference of signals recorded by two ISA elements disposed with their sensitive axes parallel. As is it possible to see, the level of residues is well below the sensitivity required by the BepiColombo mission more or less in the entire frequency band.

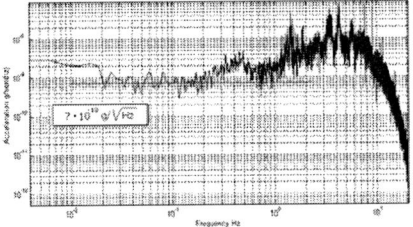

FIGURE 3. Difference of seismic signals recorded by two ISA elements disposed with their sensitive axes parallel, this difference can also be take as an upper limit of each accelerometer elements intrinsic noise.

CONCLUSIONS

The use of ISA accelerometer to detect the not inertial acceleration on the MPO permits a-posteriori to reconstruct the orbit of the satellite with a very high accuracy and so to perform very accurate determination of planetary parameters and to test General Relativity at an unprecedented level of accuracy. The experimental measurements show that the ISA accelerometer has the right sensitivity to fit the requirements imposed by the RSE. The procedures for its calibration on ground are not so difficult, due to the fact that ISA is composed of separate elements for each component of acceleration, permitting to dispose each sensitive axis perpendicular to the Earth gravity acceleration.

ACKNOWLEDGEMENTS

The authors would like to acknowledge the financial support from the Italian Space Agency to the IFSI (Istituto di Fisica dello Spazio Interplanetario).

REFERENCES

1. Fuligni, F., Iafolla, V., Milyukov, V., Nozzoli, S., Experimental gravitation and geophysics. *Il Nuovo Cimento* **20 C (5),** 637–642, 1997.
2. Iafolla, V., Nozzoli, S., Italian spring accelerometer (ISA) a high sensitive accelerometer for ''BepiColombo'' ESA CORNERSTONE. *Plan. Space Science*, 49, 1609–1617, 2001.
3. Iafolla,V., Lucchesi, D.M., Nozzoli, S., On the ISA Accelerometer Positioning inside the Mercury Planetary Orbiter, *Plan. Space Science*, 2004.
4. Iess, L., Boscagli, G., Advanced radio science instrumentation for the mission BepiColombo to Mercury. *Plan. Space Science*, 49, 1597–1608, 2001.
5. Milani, A., Rossi, A., Vokrouhlický, D. Villani, D., Bonanno, C., Gravity field and rotation state of Mercury from the BepiColombo Radio Science Experiments. *Plan. Space Science*, 49, 1579–1596, 2001.
6. Milani, A, Vokrouhlicky, D., Villani, D., Bonanno, C., Rossi, A., Testing general relativity with the Bepicolombo radio science experiment, *Phys. Rev.* D 66, 2002.

Relic gravitational waves and cosmic accelerated expansion

Germán Izquierdo

Departamento de Física, Universidad Autónoma de Barcelona,
08193 Bellaterra (Barcelona) Spain

Abstract. The possibility of reconstructing the whole history of the scale factor of the Universe from the power spectrum of relic gravitational waves (RGWs) makes the study of these waves quite interesting. First, we explore the impact of a hypothetical era -right after reheating- dominated by mini black holes and radiation that may lower the spectrum several orders of magnitude. Next, we calculate the power spectrum of the RGWs taking into account the present stage of accelerated expansion and an hypothetical second dust era. Finally, we study the generalized second law of gravitational thermodynamics applied to the present era of accelerated expansion of the Universe.

Keywords: Gravitational waves, dark energy, accelerated expansion
PACS: 04.30.-w, 98.80.-k

INTRODUCTION

The future detection of relic gravitational waves is expected to provide us with invaluable information about the instant of their decoupling from other fields, i.e., about 10^{-43} seconds after the Big Bang. The relic gravitational waves (RGWs) are generated by parametric amplification of the quantum vacuum during the expansion of the Universe.

The equation governing the evolution of the RGWs is the so-called Lifshitz equation $\mu''(\eta) + \left(k^2 - \frac{a''(\eta)}{a(\eta)}\right)\mu(\eta) = 0$, which can be interpreted as the equation of an harmonic oscillator parametrically excited by the term a''/a [1]. When $k^2 \gg a''/a$, Liftshitz equation becomes the equation of the simple pendulum and consequently the amplitude of the wave decreases adiabatically as a^{-1} in an expanding universe. In the opposite regime, $k^2 \ll a''/a$, the dominant solution is proportional to the scale factor and the amplitude remains constant with the expansion of the universe. This phenomenon is called "superadiabatic" or "parametric" amplification of gravitational waves [2]. Another approach to the RGWs amplification relies on the method of Bogoliubov coefficients and uses the adiabatic vacuum approximation (for details see, e.g., [3]). Here we evaluate the number of RGWs created in the adiabatic vacuum approximation in a universe which experiences three successive stages of evolution: an initial de Sitter stage followed by a stage dominated by the radiation and, finally, a stage dominated by the non-relativistic matter up to the present day. The power spectrum, defined from the energy density of the RGWs as $d\rho_g(\omega) = P(\omega)d\omega$, reads [3, 4]

$$P(\omega) \sim \begin{cases} 0 & (\omega(\eta_0) > 2\pi(a_1/a_0)H_1), \\ \omega^{-1}(\eta_0) & (2\pi(a_2/a_0)H_2 < \omega(\eta_0) < 2\pi(a_1/a_0)H_1), \\ \omega^{-3}(\eta_0) & (2\pi H_0 < \omega(\eta_0) < 2\pi(a_2/a_0)H_2). \end{cases} \quad (1)$$

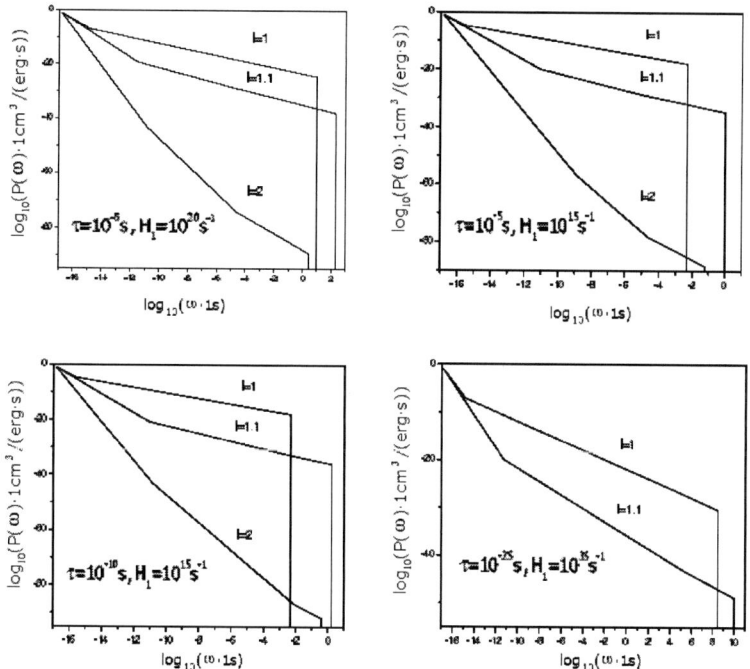

FIGURE 1. GWs spectrum for an expanding universe with a "MBHs+rad" era for certain values of l, τ and H_1.

RELIC GRAVITATIONAL WAVES AND MINI BLACK HOLES

As is well known, mini black holes (MBHs) can be created by quantum tunnelling from the hot radiation and coexist with it until their evaporation [5]. It is reasonable to expect that, at this point, a steady state in the very early Universe would be achieved where the total energy density is shared between the black holes and radiation whence $\rho = \rho_{BH} + \rho_R$ and, consequently, the total pressure is $p = p_R = (\gamma - 1)\rho$, where the constant γ lies in the interval $1 \leq \gamma < 4/3$. If the density of MBHs is large enough to dominate the expansion of the Universe, then $\gamma \simeq 1$. In the opposite case, the Universe expansion is dominated by the radiation, $\gamma \simeq 4/3$. During the "MBHs+rad" era, one finds from the Einstein equations that $a(\eta) \propto \eta^l$, where $l = 2/(3\gamma - 2)$ and $1 < l \leq 2$ [6].

According to this, it seems reasonable to assume a four-stage model of universe, initially de Sitter, then dominated by a mixture of MBHs and radiation, then dominated by the radiation after the evaporation of the MBHs and finally dominated by dust until today. The only free parameters considered here are l, the duration of the "MBHs+rad" era, τ, and the Hubble factor in the de Sitter era, H_1.

The power spectrum for this four-stage scenario is plotted in figure 1. The spectrum predicted for the three-stage model of the previous section is shown for comparison (labelled as $l = 1$). Parameters τ and H_1 are chosen assuming that each spectrum has the maximum value allowed by the CMB anisotropy data at the frequency $\omega = 2\pi H_0 = 2.24 \times 10^{-18} s^{-1}$. The predicted power spectrum of the four-stage model is lower than the three-stage one by several orders of magnitude. In the plot of the bottom-right panel the power spectrum for $l = 2$ is excluded as it predicts a RGWs energy density at the matter-radiation decoupling that generates via the Sach-Wolfe effect a CMB anisotropy larger than observed.

PRESENT ACCELERATED EXPANSION

Some models of dark energy predict that the present accelerated phase of cosmic expansion governed by $a(\eta) \propto \eta^l$ with $l \leq -1$ is transitory and that the expansion, sooner or later, will be dominated by "dust"[1] again [7]. This five-stage model of universe (de Sitter-radiation era-dust era-dark energy era-"second dust" era) predicts a current RGWs power spectrum that is not at variance with the one of the three-stage scenario but evolves differently. As the universe expands in the three-stage model the Hubble volume is continuously increasing and new RGWs are reentering it and contributing to de power spectrum. Meanwhile, in the dark energy scenario during the accelerated era the Hubble volume decreases, the RGWs begin to leave it and cease to contribute to the power spectrum [8]. Once the universe reaches the second dust era the Hubble radius begins to grow again and the RGWs reenter it in a l dependent way. Figure 2 shows the evolution of the RGWs density parameter, defined as $\Omega_g = \rho_g/\rho_c$.

RELIC GRAVITATIONAL WAVES ENTROPY AND GSL

During the present accelerated era of expansion the number density of RGWs is decreasing with time and, eventually, all the RGWs will leave the Hubble radius.

It seems reasonable to assume that the entropy of the RGWs depends on their number present in the considered volume [10]. Thus, the entropy of the RGWs decreases with time. In the simplest assumption the entropy density of RGWs is proportional to their number density, i.e. $s_g = An_g$. The generalized second law of gravitational thermodynamics[2] will be satisfied if the constant A is lower than certain bound, which is model dependent [9]. This is the only information at our disposal on this proportionality constant.

---

[1] These models predict that the dark energy will evolve in a way that mimics the expansion of a universe dominated by non relativistic matter.
[2] According to he generalized second law (GSL) of gravitational thermodynamics, the entropy of the event horizon plus its surroundings (in our case, the entropy in the volume enclosed by the horizon) cannot decrease.

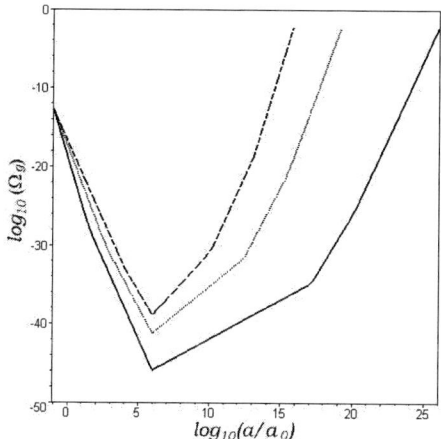

FIGURE 2. Evolution of the density parameter Ω_g with the scale factor in the five-stage scenario (De Sitter inflation-radiation-dust-dark energy-second dust era) from the beginning of the dark energy era. The solid, dotted and dashed lines correspond to $l = -1$, $l = -1.5$ and $l = -2$, respectively.

CONCLUSIONS

The MBH four-stage scenario predicts a much lower power spectrum than the conventional three-stage scenario for the same H_1. The free parameters of this scenario are constrained by the CMB anisotropy data. Although the current power spectrum of the RGWs in the dark energy four-accelerated scenario is not at variance with that of the three-stage scenario, the RGWs density parameter evolves differently. Its future evolution may also help discern between different dark energy decaying models. Assuming that the entropy density of the RGWs is proportional to their number density, the GSL is fulfilled provided a condition over the proportionality constant is met.

REFERENCES

1. E. Lifshitz, J. Phys. USSR **10**, 116 (1946).
2. L.P. Grishchuk, Zh. Eksp. Teor. Fiz. **67**, 825 (1975) [Sov. Phys. JETP 40, 409 (1975)].
3. B. Allen, Phys. Rev. D **37**, 2078 (1987).
4. L.P. Grishchuk, Class. Quantum Grav. **10**, 2449 (1993).
5. D. Gross, M.J. Perry and G.L. Yaffe, Phys. Rev. D **25**, 330 (1982); J.I. Kapusta, Phys. Rev. D **30**, 831 (1984).
6. G. Izquierdo and D. Pavón, Phys. Rev D **68**, 124005 (2003).
7. U. Alam, V. Sahni, A. A. Starobinsky, JCAP04(2003)002; M. Sami and T. Padmanabhan, Phys. Rev. D **67**, 083509 (2003); R. Kallosh and A. Linde, Phys. Rev. D **67**, 023510 (2003).
8. G. Izquierdo and D. Pavón, Phys. Rev. D **70**, 084034 (2004).
9. G. Izquierdo and D. Pavón, Phys Rev. D **70**, 127505 (2004).
10. M. Gasperini and M. Giovannini, Class. Quantum Grav. **10**, L133, (1933); R. Brandenberger, V. Mukhanov and T. Prokopec, Phys. Rev. D **48**, 2443 (1993); A. Nesteruk and A. Ottewill, Class. Quantum Grav. **12**, 51 (1995).

DOUBLY SPECIAL RELATIVITY: A NEW RELATIVITY OR NOT?

Nosratollah Jafari* and Ahmad Shariati[†]

*Institute for Advanced Studies in Basic Sciences, P.O. Box 1159, Zanjan 45195, Iran
[†]Department of Physics, Alzahra University, Tehran 19938-91167, Iran

Abstract. Double Special Relativity theories are the relativistic theories in which the transformations between inertial observers are characterized by two observer-independent scales of the light speed and the Planck length. We study two main examples of these theories and want to show that these theories are not the new theories of relativity, but only are re-descriptions of Einstein's special relativity in the non-conventional coordinates.

Keywords: Doubly special relativity; Planck length; Lorentz Invariance
PACS: 03.30.+p, 11.30.Cp

INTRODUCTION

It seems that the Planck length l_p has a crucial role in the Quantum Gravity. In some scenarios of quantum gravity like loop quantum gravity the Planck length or Planck scales act as a threshold for quantum effects in the spacetime, beyond which the usual description of spacetime breaks down. Thus, it seems that the value of l_p must have the same value in all inertial frames and this is in conflict with Einstein's Special Relativity[1, 5]. Doubly Special Relativity(DSR) has proposed for solving this puzzle [2, 4, 5].

The Magueijo -Smolin(Ms) [2, 3] and Amelino-Camelia [4, 5] DSRs are two main examples of these theories that take much attraction recently. Here, we want to investigate these theories further.

MAGUEIJO-SMOLIN DSR

Let's now explain briefly the Magueijo - Smolin (Ms) transformations: Magueijo and Smolin have looked for a non-linear representation of the Lorentz group that remains the Planck length as an invariant. If we denote the ordinary Lorentz boosts by

$$L_{ab} = p_a \frac{\partial}{\partial p^b} - p_b \frac{\partial}{\partial p^a}, \tag{1}$$

then this representation can be obtained by using the modified generators of boosts as

$$K^i \equiv L_0^i + l_p p^i D, \tag{2}$$

here D is the dilatation generator

$$D = p_a \frac{\partial}{\partial p_a}.$$

But, the rotation generators will be the unmodified $J^i \equiv \varepsilon^{ijk} L_{jk}$. Also, the Lorentz Algebra remains intact:

$$[J^i, K^j] = \varepsilon^{ijk} K_k, \quad [K^i, K^j] = \varepsilon^{ijk} J_k, \quad [J^i, J^j] = \varepsilon^{ijk} J_k.$$

By acting the modified generators of boots on the momentum space, we obtain the Magueijo-Smolin (MS)transformations between two inertial systems which are in relative motion with constant speed along the common x-axis as:

$$p'_0 = \frac{\gamma(p_0 - \frac{v}{c} p_x)}{1 + l_p(\gamma - 1) p_0 - l_p \gamma \frac{v}{c} p_x}, \tag{3}$$

$$p'_x = \frac{\gamma(p_x - \frac{v}{c} p_0)}{1 + l_p(\gamma - 1) p_0 - l_p \gamma \frac{v}{c} p_x}, \tag{4}$$

$$p'_y = \frac{p_y}{1 + l_p(\gamma - 1) p_0 - l_p \gamma \frac{v}{c} p_x}, \tag{5}$$

$$p'_z = \frac{p_z}{1 + l_p(\gamma - 1) p_0 - l_p \gamma \frac{v}{c} p_x}, \tag{6}$$

These transformations have many new features [2, 3]. For example, they do not preserve the usual quadratic invariant on momentum space. But, there is a modified invariant:

$$\|p\|^2 = \frac{\eta_{ab} p_a p_b}{(1 - l_p p_0)^2} \tag{7}$$

Also, these transformations remain invariant the light speed "c" and the Planck length "l_p" as desired. This property can be seen from MS transformations and equation (7). But, looking closer at these transformations we can see by changing 4-momentum p_μ to

$$\pi_\mu = \frac{p_\mu}{1 - l_p p_0}$$

the MS transformations become

$$\begin{cases} \pi'_0 = \gamma(\pi_0 - \frac{v}{c} \pi_x) \\ \pi'_x = \gamma(\pi_x - \frac{v}{c} \pi_0) \\ \pi'_y = \pi_y \\ \pi'_z = \pi_z \end{cases} \tag{8}$$

These are the same ordinary Lorentz transformations for momentum space. Therefore, *the MS transformations are probably only re-description of the usual Lorentz transformations in the non-conventional coordinates* [6, 7]. This fact can also be seen from the MS momentum space diagram in the next page. On that figure the MS momentum p_0 and p_1 are drawn as horizontal and radial lines with respect to the π_0 and π_1 (ordinary) momentum.

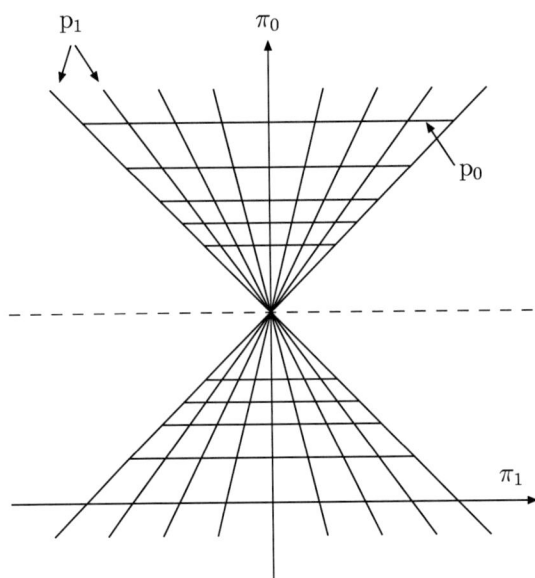

FIGURE 1. Magueijo - Smolin(MS) spacetime diagram in terms of π coordinates. MS coordinates p_0 and p_1 is drawn as horizontal and radial lines.

AMELINO-CAMELIA DSR

The other main example of DRS theories is the Amelino-Camelia DSR. Amelino-Camelia was the first physicist that wanted to solve the mentioned puzzle: *How could l_p play a role in the structure of spacetime without violating Special Relativity?* For this he modified the basic postulates of the Einstein's Relativity as:

1. The laws of physics involve a fundamental velocity scale "c" and a fundamental length "l_p".

2. Each inertial obsrever can establish the value of l_p(same value for all inertial observer) by determining the dispersion relation for photons, which takes the form $E^2 - c^2 p^2 + f(E, p; l_p) = 0$, where f is the same for all inertial observers and in particular all inertial observers agree on the leading l_p dependence of f: $f(E, p; l_p) \simeq l_p c p^2 E$.

If we find that the dispersion relation takes the form of

$$2E^2[\cosh(\frac{E}{E_p}) - \cosh(\frac{m}{E_p})] = \vec{p}^{\,2} e^{E/E_p} \qquad (9)$$

by some reasoning or by the experimental analysis.
The next step will be finding the deformed boost transformations which leave the above dispersion relation as an invariant. In ordinary special relativity the boosts can be described by

$$B_a = i p_a \frac{\partial}{\partial E} + iE \frac{\partial}{\partial p_a} \qquad (10)$$

By assuming that the modified generators also obey the same Lorentz algebra and they preserve the dispersion relation (10) as an invariant, we find the following modified generators of boosts

$$B_a = ip_a \frac{\partial}{\partial E} + i \left(\frac{1}{2E_p} \vec{p}^{\,2} + E_p \frac{1 - e^{-2E/E_p}}{2} \right) \frac{\partial}{\partial p_a} - i \frac{p_a}{E_p} \left(p_b \frac{\partial}{\partial p_b} \right). \tag{11}$$

Please, note that the rotation generators also remain intact. One can easily obtain the finite boost transformations that relate the observations of two observers by integrating the familiar differential equations

$$\frac{dE}{d\xi} = i[B_a, E], \quad \frac{dp_b}{d\xi} = i[B_a, p_b].$$

But, here as in the Magueijo-Smolin transformations case we can see that by defining the new variables ε an π through the relations

$$\frac{\varepsilon}{\mu} = \frac{e^{E/E_p} - \cosh(m/E_p)}{\sinh(m/E_p)}, \quad \frac{\pi}{\mu} = \frac{p \, e^{E/E_p}}{E_p \sinh(m/E_p)} \tag{12}$$

all relations come back to the ordinary special relativity forms. For example, the modified boost will take the form of usual Lorentz boost:

$$B_a = i\pi_a \frac{\partial}{\partial \varepsilon} + i\varepsilon \frac{\partial}{\partial \pi_a}.$$

Note that the transformations (12) are non-singular and we can define the variables ε and π.

So, it seems that the Amelino-Camelia DSR like the MS ones is only a re-description of the special relativity in the non-convenutal coordinates.

CONLCUSION

From the above discussion it seems that Doubly Special theories are only re-descriptions of the special relativity in the non-Cartesian and non-conventional coordinates.

REFERENCES

1. J. Kowalski-Glikman, " Doubly Special Relativity: A Kinematics of Quantum Gravity? ", `arXiv: hep-th/0209264`.
2. J. Magueijo and L. Smolin, " Lorentz invariance with an invariant energy scale", *Phys. Rev. Lett.* **88** (2002) 190403.
3. J. Magueijo and L. Smolin, " Generalized Lorentz invariance with an invariant energy scale", *Phys. Rev.* **D 67** (2003) 044017.
4. G. Amelino-Camelia," Relativity in space-time with stucture governerd by an observer-independent(Planckian) Lenght scale", *Int. J. Mod. Phys.* **D**11(2002) 35-60.
5. G. Amelino-Camelia," Doubly-Special Relativity: First Results and Key Problems", *Int. J. Mod. Phys.* **D**11(2002) 1643.
6. N. Jafari and A. Shariati," Operational Indistinguishably of Varying Speed of Light Theories", `Int. J. Mod. Phys.` **D**13 (2004) 709-716.
7. D. V. Ahluwalia-Khalilova, "Fermions, bosons, and locality in special relativity with two invariant scales", `arXiv: gr-qc/0207004`.

3+1 decomposition of quasi-equilibrium black hole boundary conditions

José Luis Jaramillo

Laboratoire de l'Univers et de ses Théories, UMR 8102 du C.N.R.S., Observatoire de Paris, F-92195 Meudon Cedex, France

Abstract. We present a 3+1 decomposition of the boundary conditions to be satisfied by a black hole whose horizon is in quasi-equilibrium, according to the characterization provided by the Isolated Horizon formalism. Such a decomposition can be useful in the study of black hole dynamics as a Cauchy problem. We illustrate these boundary conditions through the numerical construction of initial data for spacetimes containing a black hole.

Keywords: isolated horizons, trapped surfaces, initial data, quasi-equilibrium

Given a 3+1 decomposition of spacetime by spatial surfaces Σ_t, we aim at prescribing boundary conditions on inner excised spheres $\mathscr{S}_t$ in such a way that: *i)* $\mathscr{S}_t$ represent the horizon of a black hole which is in quasi-equilibrium but contained in an otherwise dynamical space-time, and *ii)* these boundary conditions are geometrically justified.

The underlying main motivation is the construction of initial data (γ_{ij}, K^{ij}) for binary black holes in quasi-circular orbits on a initial slice Σ_0. However, as a preliminary step, the specific goal here is to construct initial data for a single black hole whose horizon is in quasi-equilibrium.

We firstly present the geometrical boundary conditions derived from the Isolated Horizon formalism (see [2] for a review), and then consider their numerical implementation performed by using the pseudo-spectral methods in the Lorene C++ library [3].

GEOMETRICAL QUASI-EQUILIBRIUM BOUNDARY CONDITIONS

We will denote by n^μ the timelike unit vector normal to the initial surface Σ_0, the lapse function by N and the shift vector by β^μ, in such a way that the evolution vector t^μ is written as: $t^\mu = Nn^\mu + \beta^\mu$.

Conformal thin sandwich approach

In order to construct the initial data we make use of the Conformal Thin Sandwich (CTS) approach [4]. The 3-metric and the extrinsic curvature are decomposed as follows

$$\gamma_{ij} = \Psi^4 \tilde{\gamma}_{ij} \; , \quad K^{ij} = \Psi^{-4}\left(\tilde{A}^{ij} + \frac{1}{3}K\gamma^{ij}\right) \; , \tag{1}$$

where $K = \gamma^{ij}K_{ij}$ and $\tilde{A}^{ij} = \frac{1}{2N}\left[\left(\tilde{D}^i\beta^j + \tilde{D}^j\beta^i - \frac{2}{3}\tilde{D}^k\beta^k\tilde{\gamma}^{ij}\right) + \dot{\tilde{\gamma}}^{ij}\right]$ (with $\dot{\tilde{\gamma}}^{ij} \equiv \mathscr{L}_t\tilde{\gamma}^{ij}$ and $\tilde{D}_i$ the Levi-Civita connection associated with $\tilde{\gamma}_{ij}$). Inserting these decompositions into the constraint equations, the Hamiltonian constraint becomes an elliptic equation on Ψ, whereas the momentum constraint gives rise to an elliptic equation for the shift β^i. Finally, prescribing the value of $\mathscr{L}_t K = \dot{K}$ we obtain an elliptic equation for N (see Ref. [4]). At the end of the day, we need to prescribe inner boundary conditions for (Ψ, β^i, N), in such a way that $\mathscr{S}_t$ represents a horizon in quasi-equilibrium. For doing so, we make use of the Isolated Horizon framework.

Isolated Horizon formalism

This geometrical construction is indeed devised for characterizing a black hole in quasi-equilibrium. It provides a quasi-local description of the black hole, something very convenient from the numerical point of view, since no access to global spacetime notions is generally available. The Isolated Horizon framework is structured in hierarchical levels corresponding to different "degrees" of quasi-equilibrium.

Non-Expanding Horizons. The minimal notion of a horizon in quasi-equilibrium is provided by the first level in the Isolated Horizon hierarchy, namely a Non Expanding Horizon (NEH). A NEH tries to capture geometrically the idea of an apparent horizon evolving into apparent horizons of the same area. Denoting by s^μ the unit vector normal to $\mathscr{S}_t$ and lying in Σ_0 (and pointing to spatial infinity), the "outgoing" and "ingoing" null vectors at $\mathscr{S}_t$ are given respectively by $\ell^\mu \propto n^\mu + s^\mu$ and $k^\mu \propto n^\mu - s^\mu$. The NEH structure is infinitesimally characterised by the vanishing of the expansion $\theta_{(\ell)}$ and the shear $\sigma_{(\ell)}$ associated with the "outgoing" null normal ℓ^μ. Denoting the metric induced on $\mathscr{S}_t$ by $q_{\mu\nu}$ ($=\gamma_{\mu\nu} - s_\mu s_\nu$), this means

$$\theta_{(\ell)} = q^{\mu\nu}\nabla_\mu \ell_\nu = 0 \quad \text{and} \quad \sigma_{(\ell)} = q^\rho{}_\mu q^\sigma{}_\nu \nabla_\rho \ell_\sigma - \frac{1}{2}\theta_{(\ell)}q_{\mu\nu} = 0 \ . \tag{2}$$

Inserting the conformal decomposition (1) in $\theta_{(\ell)}$ and $\sigma_{(\ell)}$, together with the decomposition of the shift into its radial and tangent parts to $(\mathscr{S}_0, \tilde{\gamma})$, i.e. $\beta^i = \tilde{b}\tilde{s}^i - V^i$, leads to the following conditions on the constrained parameters of the CTS formulation

$$0 = 4\tilde{s}^i \tilde{D}_i \ln\Psi + \tilde{D}_i \tilde{s}^i + \Psi^{-2}K_{ij}\tilde{s}^i\tilde{s}^j - \Psi^2 K \ , \tag{3}$$

$$0 = \left[\mathscr{L}_t \tilde{q}^{\mu\nu} - \frac{1}{2}(\mathscr{L}_t \ln\tilde{q})\tilde{q}^{\mu\nu}\right] + \left[{}^2\tilde{D}^\mu V^\nu + {}^2\tilde{D}^\nu V^\mu - ({}^2\tilde{D}_\rho V^\rho)\tilde{q}^{\mu\nu}\right] \tag{4}$$
$$+ \left[(N\Psi^{-2} - \tilde{b})\left(\tilde{H}^{\mu\nu} - \frac{1}{2}\tilde{q}^{\mu\nu}\tilde{H}\right)\right] ,$$

where ${}^2\tilde{D}_\mu$ is the connection associated with the metric $\tilde{q}_{\mu\nu} = \Psi^{-4}q_{\mu\nu}$ on $\mathscr{S}_0$ and $\tilde{H}_{\mu\nu} = q^\rho{}_\mu \tilde{D}_\rho \tilde{s}_\nu$ is the (conformal) extrinsic curvature of $\mathscr{S}_0$ as a hypersurface in Σ_0. The first line (3) provides a boundary condition for the conformal factor Ψ. Regarding the vanishing of the shear, we proceed by canceling independently the three terms in

TABLE 1. Quasi-equilibrium horizon boundary conditions (b. c.) on $\mathscr{S}_0$.

NEH b.c.	$\theta_{(\ell)} = 0$	$4\tilde{s}^i \tilde{D}_i \ln \Psi + \tilde{D}_i \tilde{s}^i + \Psi^{-2} K_{ij} \tilde{s}^i \tilde{s}^j - \Psi^2 K = 0$	Ψ
		$^2\tilde{D}^\mu V^\nu + {}^2\tilde{D}^\nu V^\mu - ({}^2\tilde{D}_\rho V^\rho) \tilde{q}^{\mu\nu} = 0$	V^μ
	$\sigma_{(\ell)} = 0$	$\tilde{b} = N\Psi^{-2}$	$\tilde{b}_{(a)}$
		$2\tilde{s}^k \tilde{D}_k \tilde{b} - \tilde{b}\tilde{H} = 3Nh_1 - {}^2\tilde{D}_k V^k - 2V^k \tilde{D}_{\tilde{s}} \tilde{s}_k - NK$	$\tilde{b}_{(b)}$
WIH b.c.	$\kappa_{(\ell)} = $ constant, $\mathscr{L}_\ell N = h_2$	constant $= s^i D_i N - N K_{ij} s^i s^j + h_2$	$N_{(a)}$
	$^2 D^\mu \Omega_\mu = h_3$	$^2\Delta \ln N = {}^2 D^\rho (q^\mu{}_\rho K_{\mu\nu} s^\nu) + h_4$	$N_{(b)}$

brackets. The first term can be set to zero by an appropriate choice of the free parameters in the CTS approach, a choice that is indeed motivated from quasi-equilibrium considerations. The vanishing of the second term characterizes the vector V^i as a conformal symmetry on $\mathscr{S}_0$. Once this conformal symmetry is chosen, it can be used as a Dirichlet condition for V^i. Finally, in order to cancel the last term in (4), we can either impose $\tilde{b} = N\Psi^{-2}$ or $\tilde{H}^{\mu\nu} - \frac{1}{2}\tilde{q}^{\mu\nu}\tilde{H} = 0$. The first option provides a boundary condition for the normal part of the shift. The second possibility, known as *umbilical* condition, can be enforced by an appropriate choice of the free data in the initial data problem, but in that case we must look for a convenient boundary condition for $\tilde{b}$. Following [6] (see also [1]) this can be accomplished by prescribing the value of $K_{ij} s^i s^j$ on $\mathscr{S}_0$: $K_{ij} s^i s^j = h_1$. This translates into the boundary condition

$$2\tilde{s}^k \tilde{D}_k \tilde{b} - \tilde{b}\tilde{H} = 3Nh_1 - {}^2\tilde{D}_k V^k - 2V^k \tilde{D}_{\tilde{s}} \tilde{s}_k - NK \ . \tag{5}$$

Weakly Isolated Horizons. Once a NEH structure is imposed on the world-tube $\mathscr{H}$ of the apparent horizon $\mathscr{S}_0$, an intrinsic 1-form ω_μ can be defined on $\mathscr{H}$. The next level in the Isolated Horizon hierarchy follows from imposing ω_μ to be "time" independent: $\mathscr{L}_\ell \omega_\mu = 0$. This defines a Weakly Isolated Horizon (WIH). Defining the "surface gravity" $\kappa_{(\ell)} = l^\mu \omega_\mu$, the WIH condition is equivalent to $\kappa_{(\ell)} = $ constant. Translating this condition in terms of 3+1 fields results in

$$s^i D_i N - N K_{ij} s^i s^j + h_2 = \text{constant} \ , \tag{6}$$

where $h_2 = \mathscr{L}_\ell N$ must be prescribed on $\mathscr{S}_0$.

The WIH structure permits to fix the slicing of $\mathscr{H}$. Consequently, inner boundary conditions for N follow from a WIH. We simply point out here that, once $\mathscr{S}_0$ is given, this can be accomplished either by determining the divergence of the projection of ω_μ on $\mathscr{S}_0$ (i.e. $^2 D^\mu \Omega_\mu = h_3$, with $\Omega_\mu \equiv q^\rho{}_\mu \omega_\rho$), or by imposing $\mathscr{L}_\ell \theta_{(k)} = h_4$. We refer the reader to Ref. [1] for details. We shall not consider here condition $\mathscr{L}_\ell \theta_{(k)} = 0$ (see [5] for an extended analysis including this condition).

We summarize the NEH and WIH boundaries for the constrained parameters in Table 1.

TABLE 2. Well-posedness of different combinations of boundary on $\mathscr{S}_0$.

	$N_{(a)}$ with $h_2 = 0$	$N_{(b)}$ with $h_3 = 0$	$N_{(c)}$	$N_{(d)}$
$b_{(a)}$	degenerated	well-posed	well-posed	well-posed
$b_{(b)}$	well-posed	well-posed	well-posed	well-posed

NUMERICAL IMPLEMENTATION

The mathematical analysis of the well-posedness of the combinations of inner boundary conditions following from Table 1, is rather difficult due to their coupled, non-linear nature. We are forced to carry out such an analysis in a numerical way. Besides, we note that boundary conditions for the lapse demand the prescription of a free function h_i. It is therefore sensible to consider the value of the lapse as an "effective" boundary condition itself. This is the strategy in Ref. [7] (see [5] for a discussion on the limitations of this strategy), and according to it we include two more additional *effective* boundary conditions for the lapse: $N =$ constant (denoted by $N_{(c)}$) and $N = \frac{1}{2\Psi}$ (denoted by $N_{(d)}$).

Keeping fixed the boundary conditions for Ψ and V^μ in Table 1, we have numerically constructed initial data for the other combinations of boundary conditions, making use of pseudo-spectral methods implemented in the Lorene library. In particular, we have considered flat *and also* non-flat conformal metrics $\tilde{\gamma}_{ij}$ and, on the ther hand, maximal ($K = 0$) and non-maximal slicings. As a result of these tests, we can conclude that all considered combinations of boundary conditions provide a unique solution to the initial data problem, except the combination $b_{(a)} - N_{(a)}$, which is degenerated (as it was anticipated in [7]). The rest of the combinations seem to provide a well-posed elliptic problem. We summarize these results in Table 2.

Together with the presentation of the Isolated Horizon boundary conditions in an explicit 3+1 form, the most important results in this work are: 1) concluding the degeneracy of the combination $b_{(a)} - N_{(a)}$ (we insist in the fact that no particular boundary condition is degenerated, but the full combination itself), and 2) the successful numerical implementation of the boundary condition $K_{ij}s^i s^j = h_1$.

ACKNOWLEDGMENTS

The author acknowledges the support of the Marie Curie Intra-European contract MEIF-CT-2003-500885 within the 6th European Community Framework Programme.

REFERENCES

1. E. Gourgoulhon, J.L. Jaramillo, *A 3+1 perspective on null hypersurfaces and isolated horizons*, to be published in Physics Reports.
2. A. Ashtekar and B. Krishnan, Living Rev. Relativity **7**, 10 (2004) [Online article]: cited on 3 January 2005, http://www.livingreviews.org/lrr-2004-10.
3. http://www.lorene.obspm.fr.

4. J.W. York : *Conformal "thin-sandwich" data for the initial-value problem of general relativity*, Phys. Rev. Lett. **82**, 1350 (1999).
5. J.L. Jaramillo and F. Limousin : *Numerical implementation of isolated horizon boundary conditions*, in preparation.
6. S. Dain, J.L. Jaramillo, and B. Krishnan, Phys. Rev. D **71**, 064003 (2005).
7. G.B. Cook, H. Pfeiffer, Phys. Rev. D **70**, 104016 (2004)

From General Gravity to Einstein's

Kurt Just[*] and William Stoeger[†]

[*] *Department of Physics, University of Arizona, Tucson AZ 85721 <just@physics.arizona.edu>*
[†] *Vatican Observatory Research Group, Steward Observatory, University of Arizona, Tucson AZ 85721 <wstoeger@as.arizona.edu>*

Abstract. Not only Einstein's general ideas about gravity, but also his specific field equation (E) have been excellently confirmed in astronomy. Hence one also applies it to other macroscopic problems (hydrodynamic stellar models) and to quantum gravity. There, however, (E) can only be tested in a tiny region just off the mass shell $K^2 = 0$ of free gravitons. One nevertheless believes (E) to hold even at 'high energies' (the popular expression for large, Lorentz-invariant $\sqrt{|K^2|}$), perhaps for all $\sqrt{|K^2|} < M_{Pl}$ with Planck's $M_{Pl} = (8\pi\, G_{Newton})^{-1/2}$. We rather accept (E) only as the infrared limit of a more basic equation (W), which involves Weyl's conformal curvature tensor. We cannot justify (W) or specify its behavior at low and high K by observations or experiments; only connections with particle theories can help. We thus have examined the effect of gravity on the infrared structure of the electron. As a most striking and firm conclusion, we find the huge M_{Pl} totally irrelevant at high 'energies'. It merely arises in the infrared transition from the general (W) to the well observed (E).

Keywords: <Gravity, Quantum>
PACS: 04.60-m

HOW FAR IS EINSTEIN'S EQUATION CONFIRMED ?

We do not doubt Einstein's metric description (general gravity). How far, though, can we trust his familiar field equation

$$G_{\alpha\beta} + \kappa_E T_{\alpha\beta} = 0 \quad \text{with} \quad \kappa_E := 8\pi\, G_N = 1/M_{Pl}^2 \quad ? \qquad 1(1)$$

Even its nonlinearities have been well confirmed, but mainly by macroscopic solutions of the 'non-quantized' equation $G_{\alpha\beta} = 0$. Their integration constants are determined by matter, which only enters through *integrals* of a classical $T_{\alpha\beta}$, such as stellar masses and angular momenta. That energy tensor itself is used in cosmology and in relativistic hydrodynamics (for numerical stellar models), but never for precise tests. For these, the most significant effect is caused by the sun on visible starlight. At a distance R, it gets deflected by

$$\theta = M_\odot \kappa_E / 2\pi R < 10^{-5} \quad \text{because} \quad R > R_\odot \approx 7 \cdot 10^{10}\,\text{cm}. \qquad 1(2)$$

This also presents the only problem which is understood classically (light bent by 1(2) when moving on a geodesic) as well as by quantum theory. There it can be modelled [1] by a photon (of energy $\omega \approx 2\,\text{eV}$), interacting with a single graviton in Born's approximation. This gives it the spacelike momentum transfer

$$\mu_K := \sqrt{-K^2} = \omega\theta < 4 \cdot 10^{-8} m_e \approx 5 \cdot 10^5 \, \text{eV} \,. \qquad 1(3)$$

In Section 2 we explain how precise observations and experiments test Einstein's equation 1(1) only in the tiny neighborhood 1(3) around the mass shell $K^2 = 0$ of the free graviton. Away from this, many modifications of 1(1) appear possible; but we examine just one proposal. Relating it in Section 3 to Quantum Induction (QI), we compare it in Section 4 with the path from QCD to Nuclear Physics. We also mention a successful application to the *infrared* structure of the electron.

MODIFICATIONS PERMITTED BY OBSERVATIONS

For discussing 1(3) in quantum theory, it suffices to linearize the equations of gravity in

$$h^{\alpha\beta}(x) := g^{\alpha\beta}(x) - \eta^{\alpha\beta} \,, \qquad \text{restricted by} \qquad h^{\alpha\beta}_{,\beta} = 0 \,. \qquad 2(1)$$

Its propagator $\int e^{iKx} dx \langle | T h^{\alpha\beta}(x) h_{\sigma\tau}(0) | \rangle$ contains the analytic functions

$$\widetilde{F}_J(K^2) = \int_0^\infty \rho_J(\zeta) d\zeta / (K_+^2 - \zeta) \qquad (K_+^2 := K^2 + i\varepsilon \ \text{with} \ \varepsilon \to +0) \,. \qquad 2(2)$$

While spin $J = 1$ has been excluded in 2(1), the static force of gravity requires not only $J = 2$, but also $J = 0$ (as established [2] long ago). This does not deny that gravitational *waves* far from their sources are transverse and traceless (having only the helicities $\pm J$ of spin $J = 2$).

In a suitable notation, 1(1) with $T_{\alpha\beta} = 0$ and linearized in 2(1) implies

$$\widetilde{F}_J^E(K^2) \approx 2\kappa_E / K_+^2 \,, \qquad \text{hence} \qquad \rho_J^E(\zeta) = 2\kappa_E \, \delta(\zeta) \,. \qquad 2(3)$$

Einstein's non-linearities (graviton loops) are of great interest [3]. They contribute to 2(3) some 'radiative' corrections $\sigma_J(\zeta)$, but here are *negligible* because they affect 2(2) only at $|K^2| < 10^{-30} m_e^2$. Recent experiments [4] test Newton's (hence Einstein's) force at distances around 10^{-3} cm, which in 1(3) correspond to $-K^2 \approx (10^{-9} m_e)^2$. Until 10^{-5} cm can be reached, the light deflection 1(2) by the sun provides the test of 2(2) which leads farthest off $K^2 = 0$. In view of 1(3), neither observations nor experiments can prevent us from modifying Einstein's equation 1(1) so that 2(3) changes at $\sqrt{|K^2|} > 10^{-7} m_e$. We want to do so only for $\sqrt{|K^2|} > m_e$, hence in the region of and above particle masses. Such changes could be admitted in many ways; thus our rather definite proposal must be motivated by consistency and simplicity.

GRAVITY AS INDUCED BY DIRAC'S FIELD

In QI [5], the basic Bose fields of Higgs, Yang-Mills and Einstein are 'induced' by the Ψ for leptons and quarks. Then gravity basically obeys

$$W_{\alpha\beta} + T_{\alpha\beta} = 0 , \quad \text{where} \quad 2\int \sqrt{g}\, W_{\alpha\beta} \delta g^{\alpha\beta} = \delta \int \sqrt{g}\, C_{\alpha\beta\sigma\tau} C^{\alpha\beta\sigma\tau} \qquad 3(1)$$

(with $g := -\det\{g_{\mu\nu}\}$). While gravity is related to Weyl's conformal curvature tensor, $T_{\alpha\beta}$ is the same energy tensor as in 1(1), but now expressed by 'quantized' matter and light. Actions involving $C_{\alpha\beta\sigma\tau}$ (and other parts of $R_{\alpha\beta\sigma\tau}$) quadratically have been proposed long ago for making gravity renormalizable [6], but never as in 3(1) with *definite* coefficients (even their signs had been unknown). Whereas 3(1) is here not as fundamental as Dirac's equation for Ψ, its conformal invariance is broken by the particle masses in the $T_{\alpha\beta}$ of Quantum Theory.

There is no region of K^2 where 3(1) makes sense with $T_{\alpha\beta} = 0$ (unlike 1(1) which near $K^2 = 0$ is most useful in 'vacuum'). Therefore it appears irrelevant that 2(3) would be replaced by

$$\widetilde{F}_2^W(K^2) = K_+^{-4} \quad \text{and} \quad \widetilde{F}_0^W(K^2) = 0 , \qquad 3(2)$$

if 3(1) could be used with $T_{\alpha\beta} = 0$. These fictitious results simply show, however, that 3(1) is useless not only at low or extremely high $|K^2|$, but already at the $K^2 > \Lambda^2$ one commonly considers for GUTs. In this region, QI makes every Bose field *non-canonical* in the sense that its propagator decreases faster than one commonly expects. In contrast to 3(2), gravity thus makes

$$|K^6 \widetilde{F}_j(K^2)| \ll \Lambda^2 \quad \text{for} \quad \sqrt{|K^2|} \gg \Lambda = 3 \cdot 10^{14}\,\text{GeV} \approx M_{Pl}/6700 ,$$
$$\text{where} \quad M_{Pl} := \kappa_E^{-1/2} = (8\pi G_N)^{-1/2} \approx 2 \cdot 10^{18}\,\text{GeV} \approx 4 \cdot 10^{-9}\,\text{kg} . \qquad 3(3)$$

If 3(2) would persist for $K^2 \to 0$, gravity would become strongly infrared divergent. This mathematical disaster is avoided by an 'infrared transition', details of which have not yet been considered. Due to all known observations and experiments, however, the consequence 2(3) of Einstein's equation 1(1) must be reached. Thus it becomes important to determine that *transition*; but no indication of how this occurs has been found. We are sure only about the goal 2(3) and the starting point 3(2) suggested by QI.

RELATIONS WITH PARTICLE THEORY

The required transition resembles that to Nuclear Physics from the QCD established by 'asymptotic freedom' and jets (from quarks and gluons). Only recently, however, 'Chiral Perturbation Theory' has led from quarks to hadrons. Therefore, details linking 3(2) and 2(3) must not be expected soon; but some ideas can already be excluded: The *propagators* 2(2) must lead from 3(2) to 2(3) smoothly; but we have searched in vain for a field equation which interpolates between 3(1) and 1(1). Likewise, Chiral Perturbation Theory does not provide field equations from which QCD and Nuclear Physics result as limits.

As another connection with particle theory, we recognize the effect of gravity on the infrared structure of the *electron*. A tiny deviation from the practically sufficient, sharp spectral distribution $\delta(\eta - m_e^2)$ is expected [7] from the coupling between electron and photon; but no details have been elaborated. Recently, solving Dyson's equation approximately, we have seen that a stronger influence comes from gravity. For the electron mass, we obtain a relative width $\Delta m_e/m_e \approx (m_e/M_{Pl})^2 \approx 10^{-30}$, whereas Maxwell's field yields less than 10^{-100} times this. Neither of the regions with 2(3) or 3(2) contributes as much as the transition between them (in which spin 0 arises).

Our (practically sharp) mass distribution may have the expected abrupt threshold [8]; but a mathematically *smooth* Breit-Wigner function of the same tiny width is equally possible (in spite of the electron's stability). The decision between these physically equivalent, but mathematically very distinct, alternatives must come from other theories. The lengthy calculations for the effects of gravity remain to be published.

Our most surprising and firm conclusion, however, does not depend on any infrared problem, but exclusively on QI. It says that Planck's Mass is *irrelevant* already at $|K^2|$ above the observed particle masses, hence for all 'ultraviolet' problems. This follows from the faster than canonical decrease of all Bose propagators with $\sqrt{|K^2|} > \Lambda$ (making them utterly negligible at $\sqrt{K^2} > M_{Pl} \approx 6700\Lambda$). Thus M_{Pl} only constrains the infrared approach from general gravity to Einstein's 1(1).

REFERENCES

1. M. D. Scadron, *Advanced Quantum Theory*, Springer, New York, 1979, p. 292.
2. M. V. H. Van Dam, *Nucl. Phys.* **B22**, 397–411 (1970).
3. B. H. N.E.J. Bjerrum-Bohr, J.F. Donoghue, *Phys. Rev.* **D68**, 084005 (2003).
4. C. H. E. Adelberger, B. Heckel, *Phys. World* **18**, 41–45 (2005).
5. K. Just and E. Sucipto, *Basic versus Practical Quantum Induction*, in Doebner, Scherer, Schulte, Group 21, **vol. 2**, World Sci., Singapore, 1997, p. 739.
6. K. Stelle, *Gen. Rel. Grav.* **9**, 353–371 (1978).
7. B. Schroer, *Fortschr. Physik* **11**, 1–32 (1963).
8. R. Haag, *Local Quantum Physics*, 2 edn, Springer, Berlin, 1996, p. 280.

Primordial scalar field : a way out of the Lithium over-production.

Julien Larena*, Jean-Michel Alimi* and Arturo Serna[†]

*LUTh, Observatoire de Paris-Meudon, 92195 Paris, France.
[†]Departamento de física, Universidad Miguel Hernández, Elche-03202, Spain.

Abstract. The baryon density of our Universe, $\Omega_b h^2 = 0.0224 \pm 0.0009$, as inferred from the WMAP first year of observations is used to predict the primordial abundances of light elements produced during Big Bang Nucleosynthesis (BBN). Such a baryon density, and a gravitation described by General Relativity (GR), lead to predictions for the abundances of ^4He and D in very good agreement with the observed ones; but they lead to a significant discrepancy between the calculated and the observed ^7Li abundances. Supposing that the standard non gravitational sector is not modified, we consider scalar-tensor theories of gravity, and study their impact on the ^7Li abundance. It is shown that solving the lithium problem requires a non trivial behaviour of the expansion rate of the Universe at the epoch of BBN, in order not to violate the constraints from ^4He and D.

Keywords: Cosmology, Scalar-tensor gravity, Big Bang Nucleosynthesis.
PACS: 98.80.-k, 98.80.Es

INTRODUCTION

The measurement of the Cosmic Microwave Background (CMB) anisotropies by WMAP [1, 2] now provides an accurate estimate of the baryon to photon ratio : $\eta \times 10^{10} = 6.14 \pm 0.25$. This value of η when used to compute the primordial abundances of light elements leads to [3]: $Y_p = 0.2484^{+0.0004}_{-0.0005}$, $\frac{D}{H} = 2.75^{+0.24}_{-0.19} \times 10^{-5}$, $\frac{^7Li}{H} = 3.82^{+0.73}_{-0.60} \times 10^{-10}$, while the observed abundances are : $Y_p = 0.2391 \pm 0.0020$ in [4] and $Y_p = 0.2452 \pm 0.0015$ in [5], $\frac{D}{H} = 2.42^{+0.35}_{-0.25} \times 10^{-5}$ in [6], $\frac{^7Li}{H} = 1.23^{+0.68}_{-0.32} \times 10^{-10}$ in [7] and $\frac{^7Li}{H} = 2.19^{+0.46}_{-0.38} \times 10^{-10}$ in [8]. There is a good agreement for ^4He and D but a large discrepancy for ^7Li. Assuming that $\eta \times 10^{10} = 6.14$, we will face the ^7Li over-production in the framework of scalar-tensor theories of gravity (STT). In STT, the standard nuclear physics applies, and the gravitational sector is modified by the introduction of a scalar field, changing the expansion history of the Universe while preserving the weak equivalence principle. We will show that such theories provide a mechanism that could be responsible for a low ^7Li abundance.

SCALAR-TENSOR COSMOLOGICAL MODEL

Homogeneous and isotropic Universe

For a spatially flat, homogeneous and isotropic Universe, the metric reduces to the Friedman-Robertson-Walker form : $ds_*^2 = g_{\mu\nu}^* dx^\mu dx^\nu = -dt_*^2 + a_*^2(t*)dl^2$. Then, defining $H_* = \frac{1}{a_*(t_*)} \frac{da_*(t_*)}{dt_*}$, the fields equations in STT write:

$$H_*^2 = \frac{8\pi G_* \rho_*}{3} + \frac{1}{3}\dot{\varphi}^2 + \frac{2}{3}V(\varphi) \quad (1)$$

$$\ddot{\varphi} + 3H_*\dot{\varphi} + \frac{dV(\varphi)}{d\varphi} = -4\pi G_* \alpha(\varphi)(\rho_* - 3p_*) \quad (2)$$

where $\alpha(\varphi) \equiv \frac{d\ln(A(\varphi))}{d\varphi}$ is the coupling, $\dot{u} \equiv \frac{du}{dt_*}$, and ρ_* and p_* are respectivelly the energy density and pressure of the matter perfect fluid plus the vacuum contribution (or equivalently the contribution of a cosmological constant). Observable quantities must be computed in a frame in which standard rods and clocks can be used to make measurements, that is conformally related to the one used here: $g_{\mu\nu} = A^2(\varphi)g_{\mu\nu}^*$. In particular, the observable Hubble parameter H writes:

$$H = \frac{1}{A(\varphi)}(H_*(t_*) + \alpha(\varphi(t_*))\dot{\varphi}(t_*)) \quad (3)$$

Dynamics of the scalar field

Following [9] and generalizing their result to a self-interacting scalar field, we introduce $\lambda = \ln a_* + const$, so that $d\lambda = H_* dt_*$. Then, with $u' = \frac{du}{d\lambda}$ and $w_b = \frac{p_*}{\rho_*}$, the evolution equation for φ writes:

$$m_{eff}(\lambda, \varphi, \varphi')\varphi'' + v_{eff}(w_b, \lambda)\varphi' = F_{eff}(w_b, \lambda, \varphi) \quad (4)$$

It is similar to the equation governing the motion of a mechanical oscillator with a varying "mass" m_{eff}, a varying "friction" v_{eff} and a "force" term $F_{eff} = -\frac{\frac{dV(\varphi)}{d\varphi}}{4\pi G_* \rho_*(\lambda)} - (1 - 3w_b)\alpha(\varphi)$. Then, one can define an effective potential $V_{eff}(\varphi, \lambda)$ that verifies the relation :

$$\frac{\partial V_{eff}(\varphi, \lambda)}{\partial \varphi} = \Theta(\varphi, \lambda) + (1 - 3w_b)\alpha(\varphi) \quad (5)$$

The "friction" term is always positive for a flat Universe, and the dynamics of φ is analogous to a damped oscillating motion in $V_{eff}(\varphi, \lambda)$. So, it is important to identify the minima of this effective potential since they are attractors for φ and determine the behavior of the theory at late times. The convergence of the theory towards GR, required by Solar System experiments [13] is possible only if the value of φ that ensures this convergence is a minimum of $V_{eff}(\varphi,\lambda)$ that has two distinct parts. The term due to the coupling function plays a role if $w_b \neq \frac{1}{3}$, that is if we are not in a radiation dominated period: in a theory without self-interaction, φ cannot significantly evolve during BBN. On the contrary, the other term is at play at every time, and its relative importance, when compared with the first term, becomes greater and greater while the Universe expands since $\rho_*(\lambda)$ is a decreasing function of λ.

BBN IN SCALAR-TENSOR COSMOLOGY

Our analysis relies on a numerical computation of the primordial abundances using the reaction rates of [10, 11] and of the NACRE collaboration [12]. The late time Universe is consistent with a standard ΛCDM scenario. The only parameter that differs from the standard BBN in GR is the expansion rate of the Universe H. It impacts on the various nuclear abundances because they are determined by many reactions whose efficiency depends on the ratio of their reaction rates and H. In order to characterize the deviation from GR, it is then convenient to introduce the speed-up factor, i.e. the ratio between the expansion rate and the corresponding expansion rate in GR H_{GR}: $\xi(T) = \frac{H(T)}{H_{GR}(T)}$ where $H(T)$ is defined by (3); when $\xi(T) > 1$ (resp. $\xi(T) < 1$), the Universe expands faster (resp. slower) than in GR at the temperature T. When η is greater than 3×10^{-10}, the reaction that creates ^7Li is $^3He(\alpha,\gamma)^7Be$ followed by a β-decay of ^7Be. So, the final ^7Li abundance is strongly dependent on the ^4He production, and on the efficiency of that reaction. If the expansion starts slower than in GR, and finishes faster, one can benefit from two effects at different epochs : less ^4He is synthezised at the beginning of BBN (because H is smaller than H_{GR} leading to a smaller temperature of freezing-out for the weak interaction processes that interconvert neutrons and protons), and this ^4He is less burned (because H is greater than H_{GR} after the creation of ^4He and until the end of BBN, making $^3He(\alpha,\gamma)^7Be$ less efficient). Besides, the speed-up factor must be non-monotonic, for it is greater than 1 at the end of BBN, but must converge towards 1 at later times because the theory must have converged towards GR at late time. As emphasized in the previous section, a self-interaction term will play the role of a dragging force on φ even during the radiation dominated epoch, then making it evolve during BBN if the order of magnitude of the energy scale for the self-interaction is properly chosen. We now study this possibility by restricting the analysis to a specific STT defined by $\alpha(\varphi) = a\varphi^2$ and $V(\varphi) = \Lambda^2 \varphi^4$. By varying the parameters (a, Λ, φ_i) one sees that the value of φ_i can be chosen in a large range, provided one modifies the other parameters : φ_i cannot be greater than -0.9 and smaller than -1.5. One can also conclude that there is no acceptable theory for $\varphi_i > 0$, because φ_i has to be negative if we want the initial speed-up factor to be less than 1 : $\xi_i = A(\varphi_i) = exp(\frac{a}{3}\varphi^3)$. Moreover, Λ shouldn't be too small (less than $0.2s^{-1}$) if we want the attraction mechanism described above to occur during BBN, at temperatures between $10MeV$ and $1MeV$. Indeed, from equation (4), we know that the attraction mechanism occurs approximately when $\Lambda^2 \sim G_* \rho_*(\lambda)$; for $1MeV < T < 10MeV$, assuming $\varphi_i \sim -1$, since $\rho_*(\lambda) \propto \left(\frac{T}{T_0}\right)^4$, we have with $0.2s^{-1} < \Lambda < 6s^{-1}$. To be more precise, we present a special case with $\dot{\varphi}_i = 0$, $\varphi_i = -1.3$.

Then, we represent, in the (a, Λ) plane the regions of the acceptable theories on the left part of figure (1). The constraints only come from ^4He and D abundances, and the two regions correspond to the possible constraints imposed on ^4He.

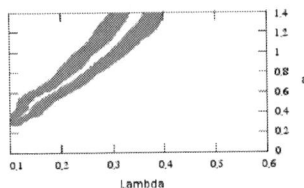 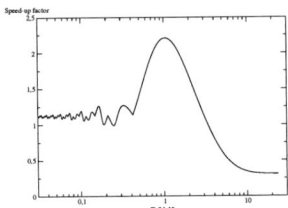

FIGURE 1. Left: Space of the parameters (a, Λ). The acceptable theories are in the shaded regions. Right: Speed-up factor as a function of the temperature for $a = 1$, $\Lambda = 0.3s^{-1}$.

The speed-up factor, shown on the right of figure (1) for the particular choice $(a, \Lambda) = (1, 0.3s^{-1})$ has the behaviour described above.

CONCLUSION

We addressed the ^7Li abundance problem by renewing the standard BBN scenario without modifying the nuclear physics: we only considered a gravitation described by STT rather than GR. We showed that it was possible both to obtain a small ^7Li abundance and to preserve the ^4He and D abundances thanks a non monotonic speed-up factor. The expansion must begin slower than in GR at the beginning of BBN, producing less ^4He ; and it must finish faster at the end of BBN, making the reaction $^3He(\alpha, \gamma)^7Be$ less efficient. Consequently, the ^7Li abundance obtained for the WMAP baryon density could be a trace of a scalar field explicitly coupled to matter at the beginning of our physical Universe, challenging our understanding of the nature of gravity at early epochs: it could then be valuable to study with much care signatures of this scalar field in inflationnary scenarios and CMB physics.

REFERENCES

1. C. Bennett et al, Astrophys. J. S. **148**, 1 (2003)
2. D. Spergel et al., Astrophys. J. S. **148**, 175 (2003)
3. R. Cyburt, B. Fields, and K. Olive, Phys. Lett. B **567**, 227 (2003)
4. V. Luridiana et al., Astrophys. J. **592**, 846 (2003)
5. Y. Izotov et al., Astrophys. J. **527**, 757 (1999)
6. D. Kirkman, D. Tyler, N. Suzuki, J. O'Meara, and D. Lubin, Astrophys. J. S. **149**, 1 (2003)
7. S. Ryan et al, Astrophys. J. **530**, L57 (2000)
8. P. Bonifacio et al, Astron. Astrophys. **390**, 91 (2002)
9. T. Damour, and K. Nordtvedt, Phys. Rev. D **48**, 3436 (1993)
10. G. Caughlan, and W. Fowler, AT. DATA Nucl. Data Tables **40**, 291 (1988)
11. M. Smith, L. Kawano, and R. Malaney, Astrophys. J. S. **85**, 219 (1993)
12. C. Angulo al, Nucl. Phys. A **656**, 3 (1999)
13. B. Bertotti, L. Less, and P. Tortora, Nature **425**, 374 (2003)

Relativistic (covariant) kinetic theory of linear plasma waves and instabilities

M. Lazar[*,†] and R. Schlickeiser[*,**]

[*]*Institut für Theoretische Physik, Lehrstuhl IV: Weltraum- und Astrophysik, Ruhr-Universität Bochum, D-44780 Bochum, Germany*
[†]*"Alexandru Ioan Cuza" University, Faculty of Physics, 6600 Iasi, Romania*
[**]*Centre for Plasma Science and Astrophysics, Ruhr-University, D-44780 Bochum, Germany*

Abstract. The fundamental kinetic description is of vital importance in high-energy astrophysics and fusion plasmas where wave phenomena evolve on scales small comparing with binary collision scales. A rigorous relativistic analysis is required even for nonrelativistic plasma temperatures for which the classical theory yielded unphysical results: e.g. collisonless damping of superluminal waves (phase velocity exceeds the speed of light). The existing nonrelativistic approaches are now improved by covariantly correct dispersion theory. As an important application, the Weibel instability has been recently investigated and confirmed as the source of primordial magnetic field in the intergalactic medium.

Keywords: plasma kinetic equations, magnetized relativistic plasmas, waves, oscilations, micro- and macroinstabilities, waves interaction, nonlinear
PACS: 52.25.Dg; 52.25.Xz; 52.27.Ny; 52.35.Fp; 52.35.Hr; 52.35.Mw; 52.35.Py., 52.35.Qz

INTRODUCTION

As the most interesting phenomena in cosmic or laboratory plasmas the electrostatic and electromagnetic waves (and their instabilities or nonlinear mixing) have received an increasing interest in the last decades. We have to remember here the role of cyclotron waves for the heating process in magnetically confined fusion plasma, and the role of interplanetary plasma fluctuations in cosmic rays transport, or the nonthermal radiation arising in astrophysical plasmas from the interaction of energetic particles with strong cosmic fields.

The fundamental description of plasma waves and instabilities is given by a kinetic approach which is imperious required in collisionless plasmas where the temperature notion has no support. Many fusion or astrophysical plasmas and shocks are collisionless because the wave phenomena evolve on scales small comparing with those of binary collisions (see the Table 8.1 p. 186 in the textbook [1] of Schlickeiser, 2002) and the energy dissipation is mainly dominated by wave-particle interactions rather than particle-particle collisions [2]. The microscopic physics is much more diverse than a macroscopic fluid approach, so that the collisionless damping of both electrostatic and electromagnetic waves in plasma, are exclusively revealed by kinetic plasma physics.

The classical nonrelativistic kinetic theory (Landau, Jackson [3]) is based on the non-relativistic Vlasov-Maxwell equations where particle speeds in excess of the speed of light are permitted. For these reasons some unphysical features arise, and the most noteworthy is the possibility of collisionless (Landau or cyclotron) damping of superluminal waves (with phase velocity greater than speed of light in vacuum, $v_{\text{phase}} = \omega/k \geq c = 3 \times 10^8$ m/s) by resonant interaction with plasma charges. It is clear however that such a situation violates the special theory of relativity, which limits individual plasma particle speeds to subluminal values.

A new relativistic kinetic theory is based first on the correct relativistic (covariant) Vlasov equation where the particle momentum p is not a simple product of the rest mass m_0 and particle velocity. Omitting the Lorentz factor dependence corresponds to the formal limit of an infinitely large speed of light $c \to \infty$. A correct relativistic approach not neglects the Lorentz factor, and it has to keep as the second condition, only the relativistic distribution functions which are vanishing for particle velocities greater than speed of light.

The start point of reformulating the classical theory should be considered the relativistic generalization of equilibrium distribution function (Jütner - 1911) [4]

$$f_J(p) = C \exp[-\mu \sqrt{1 + \gamma^2 v^2/c^2}]; \quad \gamma = \sqrt{1 - \frac{v^2}{c^2}}, \tag{1}$$

which is populated only for particle velocities $|v| < c$ and equal to zero otherwise (the nonrelativistic Maxwell-Boltzmann distribution, $f_{MB}(p)$, is nonvanishing for all particle velocities and $\lim_{c \to \infty} f_J = f_{MB}$ [5]). The existence of superluminal waves has been accepted fifty years ago [6, 7, 8], and also, the first attempts of reformulating the kinetic dispersion theory without violating the special theory of relativity have been done by Trubnikov (1958) [9] in terms of the orbit integral, and for equilibrium Maxwellian plasmas. This formalism is particularly useful in the weak damping limit and the evaluation of the integral equations is quite difficult (almost for damped or growing waves with complex frequency, $\omega = \omega_r + i\omega_i$), leading only to some approximative solutions in the limits of ultrarelativistic or weakly relativistic temperatures [10].

A more general formalism [11] has been chosen by Lerche [8] who did not specify the distribution function for most of his calculations and showed that the superluminal (electrostatic) solutions undergo no Landau damping (having purely real frequency $\omega = \omega_r$). In the last decade Schlickeiser and co-workers continued to develop the relativistic kinetic approach of plasma waves, generalizing first the Trubnikov's approach to a multicomponent plasma and for the whole frequency and wave-number ranges [5, 12, 13, 14], or extending the valences of so-called standard representation [11, 8] to the study of electrostatic and electromagnetic dispersion in hot plasma with isotropic distributions [5, 14, 15, 16, 17] or with anisotropic (bi-axis) distributions [18].

RELATIVISTIC CORRECTIONS

Using mainly the standard formalism, rigorous and noticeable results were obtained first in unmagnetized plasma [15] for both longitudinal and transverse waves and which come to confirm the previous works of Lerche [8]:
- all *superluminal modes* ($\omega/k \geq c$) undergo no growing and no Landau damping.
- *subluminal modes* ($\omega/k < c$): any type of isotropic equilibrium distribution is stable

against any linear excitation (relativistic generalization of Newcomb-Gardner theorem [19])

In magnetized plasma the parallel (with respect to the static magnetic field) propagating waves has been already analyzed [5, 14, 16, 17, 20]. In this case the longitudinal waves are described by the same relations as without magnetic field. But the transverse waves become circularly polarized and the both dispersion relation and cyclotron resonance condition depend on gyrofrequency of plasma charges leading to more complex results:
- *superluminal modes* with $\omega/k < c\sqrt{1+\Omega_{a,0}^2/k^2c^2} = v_{phase}^c$) and all *subluminal modes* ($\omega/kc < 1$) are damped stable waves ($\omega_i < 0$)
- *superluminal modes* with $\omega/kc \geq v_{phase}^c$ undergo no growing or cyclotron damping.

CONFIRMATION - COLLISIONLESS DAMPING

Kinetic theory thus leads not only to a larger number of wave modes, but to a purely kinetic effect which is the collisionless damping of electrostatic waves named Landau damping and given by resonant fitting condition of wave-phase velocity and particle velocity: $v_{phase} = \omega/k \simeq v_\parallel = \mathbf{v} \cdot \mathbf{k}/k$, or cyclotron damping given by the resonant fitting condition of cyclotron wave frequeny and particle gyrofrequency

$$\omega_r' = \Omega_a'; \quad \omega_r' = \frac{\omega_r - k_\parallel v_\parallel}{\sqrt{1-\frac{v_\parallel^2}{c^2}}}; \quad \Omega_a' = \frac{q_a B_0}{m_a' c}. \tag{2}$$

The "prime" quantities are considered in the referential (R') moving with the parallel speed $\mathbf{v}_\parallel \parallel \mathbf{B_0}$ of particles, and the others are considered in the laboratory coordinate system (R). Then, ω_r' is the relativistic Doppler shifted frequency of circularly polarized electric field in the referential (R') where it has to be very closed to the gyrofrequency Ω_a' so that the plasma charges see a nearly constant electric field and a considerable energy transfer (resonance) occur from waves to particles. Applying the Lorentz transforms to the particle mass (of sort a): $m_a' = \gamma m_a \sqrt{1-v_\parallel^2/c^2}$, the resonance condition (2) in the laboratory frame then reads

$$\omega_r - k_\parallel v_\parallel = \frac{\Omega_{a,0}}{\gamma} = \frac{q_a B_0}{\gamma m_a}, \tag{3}$$

which is the well-known condition of cyclotron damping of plasma waves. Now it can be easily found (see in [20]) that circularly polarized waves which are capable to interact with the plasma particles have a maximum value of the parallel component of the phase velocity given by

$$v_{phase}^{max} = c\sqrt{1+\left(\frac{\Omega_{a,0}}{kc}\right)^2} = c\sqrt{1+x_a^2}. \tag{4}$$

This condition is obtained assuming all particle velocities less than speed of light (full agreement with SRT) and therefore it comes to confirm that not all superluminal waves undergo cyclotron damping (cyclotron resonance is relevant for frequencies and not for velocities).

OTHER RESULTS AND PERSPECTIVES

The covariant analysis has been extended to linear unstable waves in thermally anisotropic plasma [18] or to nonlinear wave mixing [21] providing generally relativistic dispersion relations valid in any frame that is not necessary inertial. Comparing with the previous approximatively results the new covariant equations hold for any plasma components, for any value of complex frequency and for the whole range of plasma temperature, from nonrelativistic to ultrarelativistic energies.

The first qualitative (macroscopic and nonrelativistic) investigations have confirmed that Weibel [22] instability could be at the origin of cosmological magnetic field generated in counter-streaming (inter)galactic structures [23]. But a rigorous calculation is required on the basis of the correct relativistic kinetic theory recently elaborated [18] for hot unmagnetized plasma with thermal anisotropy.

ACKNOWLEDGMENTS

This work was partially supported by the Deutsche Forschungsgemeinschaft through the Sonderforschungsbereich 591. M.L. is grateful to the Alexander von Humboldt Foundation for supporting in part this work

REFERENCES

1. R. Schlickeiser, *Cosmic Ray Astrophysics*, Springer-Verlag Berlin Heidelberg, 2002.
2. R.Z. Sagdeev, *Rev. Plasma Phys.*, **4**, 23 (1966).
3. L. Landau, *J. Phys. USSR* **10**, 25 (1946); J.D. Jackson, *J. Nucl. Energy* C, **1**, 171 (1960).
4. F. Jütner, *Ann.Phys.*, **34**, 856 (1911).
5. H. Fichtner and R. Schlickeiser, *Phys. Plasmas*, **2**, 1063 (1995).
6. O. Buneman, *Phys. Rev.*, **112**, 1504 (1958).
7. B.U. Felderhof, *Physica*, **29**, 293 (1963).
8. I. Lerche, *J. Math. Phys.* **8**, 1838 (1967); *Astrophys. J.*, **147**, 689 (1967); *Plasma Phys.*, **11**, 849 (1969)
9. B. A. Trubnikov, in *Plasma Physics and the Problem of Controlled Thermonuclear Reactions*, Vol. 3, ed. M. A. Leontovich, Pergamon, New York, 1958, pp. 122.
10. K. Imre, *Phys. Fluids*, **5**, 459 (1962); I. P. Shkarofsky, *Phys. Fluids*, **9**, 561, 570, (1966); P. L. Pritchett, *J. Geophys. Res.*, **89**, 8957, (1984); P. H. Yoon, *Phys. Fluids B*, **1**, 1336, (1989).
11. G. Bekefi, *Radiation Processes in Plasmas*, Wiley, New York, pp. 227-229, 1966.
12. R. Schlickeiser, *Phys. Plasmas*, **1**, 2119 (1994).
13. R. Schlickeiser, *Astron. Astrophys.*, **294**, 615 (1995).
14. R. Schlickeiser, H. Fichtner, M. Kneller, *J. Geophys. Res.*, **102**, 4725 (1997).
15. R. Schlickeiser, M. Kneller, *J. Plasma Phys.*, **57**, 709 (1997).
16. R. Schlickeiser, *Phys. Scripta*, **T75**, 33 (1998).
17. M. Lazar and R. Schlickeiser, *Can. J. Phys.*, **81**, 1377, (2003).

18. R. Schlickeiser, *Phys. Plasmas*, **11**, 5532, (2004); M. Lazar, R. Schlickeiser, *Phys. Plasmas*, 2005, submitted.
19. C.S. Gardner, *Phys. Fluids*, **6**, 839, 1963.
20. M. Lazar and R. Schlickeiser, *Phys. Scripta*, **T113**, 130, (2004).
21. M. Lazar, I. Merches, *Phys. Lett. A*, **313**, 418, (2003);
22. E. S. Weibel, Phys. Rev. Lett. 2, 83, (1959).
23. R. Schlickeiser *Plasma Phys. Controll. Fusion*, **47**, 205, (2005).

Zero-norm states and stringy symmetries

Chuan-Tsung Chan*, Pei-Ming Ho[†], Jen-Chi Lee**, Shunsuke Teraguchi[‡] and Yi-Yang**

*Physics Division, National Center for Theoretical Sciences, Hsinchu, Taiwan, R.O.C.
[†]Department of Physics, National Taiwan University, Taipei, Taiwan, R.O.C.
**Department of Electrophysics, National Chiao-Tung University, Hsinchu, Taiwan, R.O.C.
[‡]Department of Physics, National Taiwan University and National Center for Theoretical Sciences, Hsinchu, Taiwan, R.O.C.

Abstract. We identify spacetime symmetry charges of string theory from an infinite number of zero-norm states (ZNS) with arbitrary high spin in the old covariant first quantized string spectrum. We give various evidences to support this identification. These include massive sigma-model calculation, Witten string field theory calculation, 2D string theory calculation and, most importantly, three methods of high-energy stringy scattering amplitude calculation. The last calculations explicitly prove Gross's conjectures in 1988 on high energy symmetry of string theory.

INTRODUCTION AND OVERVIEW

One of the fundamental issues of string theory is the spacetime symmetry of the theory. It has long been believed that string theory consists of a huge hidden symmetry or Ward identities. This is strongly suggested by the ultraviolet finiteness of quantum string theory, which contains no free parameter and an infinite number of states. In a local quantum field theory, a symmetry principle was postulated, which can be used to determine the interaction of the theory. In string theory, on the contrary, it is the interaction, prescribed by the very tight quantum consistency conditions due to the extendedness of string, which determines the form of the symmetry.

Historically, the first key progress to understand symmetry of string theory is to study the high energy, fixed angle behavior of string scattering amplitude [1, 2, 3]. This is strongly motivated by the spontaneously broken symmetries in gauge field theories which are hidden at low energy, but become evident in the high-energy behavior of the theory. There are two main conjectures of Gross's [2] pioneer work on this subject. The first one is the existence of an infinite number of linear relations among the scattering amplitudes of different string states that are valid order by order in perturbation theory at high energies. The second is that this symmetry is so powerful as to determine the scattering amplitudes of all the infinite number of string states in terms of the dilaton (tachyon for the case of open string) scattering amplitudes. However, the symmetry charges of his proposed stringy symmetries were not understood and the proportionality constants between scattering amplitudes of different string states were not calculated.

The second key to uncover the fundamental symmetry of string theory was zero-norm states (ZNS) in the old covariant first quantized (OCFQ) string spectrum. It was proposed that [4] spacetime symmetry charges of string theory orginate from an infinite number of ZNS with arbitrary high spin in the spectrum. In the context of σ-model approach of string theory, massive inter-particle symmetries were calculated by using two types of ZNS. Some implications of the corresponding stringy Ward identities on the scattering amplitudes were discussed in [5, 6]. It was recently realized that [7, 8] these symmetries can be reproduced from gauge transformation of Witten string field theory (WSFT) [9] after imposing the no ghost conditions. It is important to note that these symmetries exist only for D=26 thanks to type two ZNS, which is zero-norm only when D=26. On the other hand, ZNS were also shown [10] to carry the spacetime ω_∞ symmetry [11] charges of 2D string theory [12]. This is in parallel with the work of Ref [13] where the ground ring structure of ghost number zero operators was identified in the BRST quantization.

Recently high-energy Ward identities derived from the decoupling of ZNS, which combines the previous two key ideas of probing stringy symmetry, were used to explicitly prove Gross's two conjectures [14, 15, 16, 17, 18]. An infinite number of linear relations among high energy scattering amplitudes of different string states were derived. Moreover, these linear relations can be used to fix the proportionality constants among high energy scattering amplitudes of different string states algebraically at each fixed mass level. Exactly the same results can also be obtained by two other calculations, the Virasoro constraint calculation and the saddle-point calculation. Thus there is only one independent component of high energy scattering amplitude at each fixed mass level. Based on this independent component of high energy scattering amplitude, one can then derive the general formula of high energy scattering amplitude for four arbitrary string states, and express them in terms of that of tachyons. This completes the general proof of Gross's two conjectures on high-energy symmetry of string theory stated above. Incidentally, it was important to discover that the result of saddle-point calculation in [1, 2, 3] was inconsistent with high energy stringy Ward identities of ZNS calculation in [14, 15, 16]. A corrected saddle-point calculation was given in [16], where the missing terms of the calculation in [1, 2, 3] were identified to recover the stringy Ward identities.

ZERO-NORM STATE CALCULATIONS

In this section, we review the calculations of string symmetries from ZNS without taking the high-energy limit. In the OCFQ spectrum of open bosonic string theory, the solutions of physical states conditions include positive-norm propagating states and two types of ZNS. The latter are

$$\text{Type I}: L_{-1}|x\rangle, \text{ where } L_1|x\rangle = L_2|x\rangle = 0, L_0|x\rangle = 0; \tag{1}$$

$$\text{Type II}: (L_{-2} + \frac{3}{2}L_{-1}^2)|\widetilde{x}\rangle, \text{ where } L_1|\widetilde{x}\rangle = L_2|\widetilde{x}\rangle = 0, (L_0+1)|\widetilde{x}\rangle = 0. \tag{2}$$

While type I states have zero-norm at any space-time dimension, type II states have zero-norm *only* at D=26. Some explicit solutions of ZNS can be found in [19]. In the σ-model approach of string theory, a spacetime symmetry transformation $\delta\Phi$ for a background field Φ can be generated by [20]

$$T_\Phi + \delta T = T_{\Phi+\delta\Phi}, \tag{3}$$

where T_Φ is the worldsheet energy momentum tensor with background fields Φ and $T_{\Phi+\delta\Phi}$ is the new energy momentum tensor with new background fields $\Phi + \delta\Phi$. It was shown that [4] for each ZNS, one can construct a δT such that Eq.(3) is satisfied to some order of weak field approximation in the β function calculation. In constrast to the usual σ-model loop expansion (or α' expansion), which is nonrenormalizable for the massive background field, it turns out that weak field approximation is the more convenient expansion to deal with massive background field. An inter-particle symmetry transformation for two high spin states at mass level $M^2 = 4$, for example, can be generated [4]

$$\delta C_{(\mu\nu\lambda)} = (\frac{1}{2}\partial_{(\mu}\partial_\nu\theta_{\lambda)} - 2\eta_{(\mu\nu}\theta_{\lambda)}), \delta C_{[\mu\nu]} = 9\partial_{[\mu}\theta_{\nu]}, \qquad (4)$$

where $\partial_\nu\theta^\nu = 0, (\partial^2 - 4)\theta^\nu = 0$ are the on-shell conditions of the mixed type I and type II D_2 vector ZNS

$$|D_2\rangle = [(\frac{1}{2}k_\mu k_\nu \theta_\lambda + 2\eta_{\mu\nu}\theta_\lambda)\alpha^\mu_{-1}\alpha^\nu_{-1}\alpha^\lambda_{-1} + 9k_\mu\theta_\nu\alpha^{[\mu}_{-2}\alpha^{\nu]}_{-1} - 6\theta_\mu\alpha^\mu_{-3}]|0,k\rangle, \; k\cdot\theta = 0, \qquad (5)$$

and $C_{(\mu\nu\lambda)}$ and $C_{[\mu\nu]}$ are the background fields of the symmetric spin-three and anti-symmetric spin-two states respectively at the mass level $M^2 = 4$. It is important to note that the decoupling of D_2 vector zero-norm state implies simultaneous change of both $C_{(\mu\nu\lambda)}$ and $C_{[\mu\nu]}$, thus they form a gauge multiplet. In WSFT, one can rederive [7, 8] Eq.(4) from the linearized off-shell gauge transformation

$$\delta\Phi = Q_B\Lambda \qquad (6)$$

after imposing the no ghost conditions.

Another evidence to support ZNS as the origin of symmetry charge was demonstrated for the 2D string theory. The spacetime symmetry of 2D string was known to be the w_∞ algebra [11]

$$\int \frac{dz}{2\pi i} \psi^+_{J_1 M_1}(z)\psi^+_{J_2 M_2}(0) = (J_2 M_1 - J_1 M_2)\psi^+_{(J_1+J_2-1)(M_1+M_2)}(0) \qquad (7)$$

generated by the discrete Polyakov states ψ^+_{JM}. An equivalent algebraic structure was the ground ring [13]

$$Q_{J_1,M_1}Q_{J_2,M_2} = Q_{J_1+J_2,M_1+M_2} \qquad (8)$$

proposed by Witten. Alternatively, one can explicitly construct a set of discrete ZNS G^+_{JM} and show that they form a w_∞ algebra [10]

$$\int \frac{dz}{2\pi i} G^+_{J_1 M_1}(z)G^+_{J_2 M_2}(0) = (J_2 M_1 - J_1 M_2)G^+_{(J_1+J_2-1)(M_1+M_2)}(0). \qquad (9)$$

This seems to strongly suggest that ZNS are closely related to the spacetime symmetry of string theory.

HIGH ENERGY ZNS CALCULATIONS

Recently a further evidence to support ZNS as the spacetime symmetry charge of string theory was obtained by taking the high-energy, fixed angle limit of stringy Ward identities derived from the decoupling of ZNS on the scattering amplitudes. The two conjectures of Gross stated above were then explicitly proved. At mass level $M^2 = 4$, in contrast to Eq.(4) which is valid to all energies, one can show the linear relations among the high-energy scattering amplitudes [14, 15]

$$\mathcal{T}_{TTT} : \mathcal{T}_{LLT} : \mathcal{T}_{(LT)} : \mathcal{T}_{[LT]} = 8 : 1 : -1 : -1. \tag{10}$$

In the above equations, we have defined the normalized polarization vectors of the second string vertex to be $e_P = \frac{1}{m_2}(E_2, k_2, 0) = \frac{k_2}{m_2}$, $e_L = \frac{1}{m_2}(k_2, E_2, 0)$ and $e_T = (0, 0, 1)$ in the CM frame contained in the plane of scattering. It turns out that e_P approaches e_L in the high-energy limit. The tensor index of the other three string vertex were suppressed in Eq.(10). For the case of string-tree level $\chi = 1$ with one tensor V_2 and three tachyons $V_{1,3,4}$, all four scattering amplitudes in Eq.(10) were calculated to be $\mathcal{T}_{TTT} = -8E^9 \sin^3 \phi_{CM} \mathcal{T}(3) = 8\mathcal{T}_{LLT} = -8\mathcal{T}_{(LT)} = -8\mathcal{T}_{[LT]}$, where

$$\mathcal{T}(n) = \sqrt{\pi}(-1)^{n-1} 2^{-n} E^{-1-2n} (\sin\frac{\phi_{CM}}{2})^{-3} (\cos\frac{\phi_{CM}}{2})^{5-2n}$$
$$\times \exp(-\frac{s\ln s + t\ln t - (s+t)\ln(s+t)}{2}). \tag{11}$$

In Eq.(11), ϕ_{CM} is the center of momentum scattering angle, s, t are the Mandelstam variables and $M^2 = 2(n-1)$. For the scattering amplitude of general mass level, one first notes that the only states that will survive in the high energy limit at level $M^2 = 2(n-1)$ are of the form

$$|n, 2m, q\rangle \equiv (\alpha_{-1}^T)^{n-2m-2q} (\alpha_{-1}^L)^{2m} (\alpha_{-2}^L)^q |0, k\rangle. \tag{12}$$

For the case of general four tensor scattering amplitude, one has, in the high energy limit, [16, 17, 18]

$$<V_1 V_2 V_3 V_4> = \prod_{i=1}^{4} (-\frac{1}{M_i})^{2m_i+q_i} (\frac{1}{2})^{m_i+q_i} (2m_i - 1)!! \mathcal{T}_{n_1 n_2 n_3 n_4}^{T^1..T^2..T^3..T^4..}, \tag{13}$$

which is calculated algebraically by the decoupling of high-energy ZNS and is thus valid to all string-loop order. $\mathcal{T}_{n_1 n_2 n_3 n_4}^{T^1..T^2..T^3..T^4..}$ in Eq.(13), the generalization of $\mathcal{T}_{TTT}$ at mass level $M^2 = 4$, is the only independent high-energy scattering amplitudes at level $(n_1 n_2 n_3 n_4)$ and was calculated at tree level to be [14, 16]

$$\mathcal{T}_{n_1 n_2 n_3 n_4}^{T^1..T^2..T^3..T^4..} = [-2E^3 \sin\phi_{CM}]^{\Sigma n_i} \mathcal{T}(\Sigma n_i), \tag{14}$$

where n_i is the number of T^i of the $i-th$ vertex operators and T^i is the transverse direction of the $i-th$ particle. One can use a "dual calculation", the Virasoro constraints [17, 18], to reproduce Eq.(13). Finally, a saddle-point calculation [16, 17, 18] can be developed for the string-tree level $\chi = 1$ scattering amplitudes with one tensor V_2 and three tachyons $V_{1,3,4}$ to explicitly justify Eq.(13).

REFERENCES

1. D.J. Gross and P. Mende, Phys. Lett. B197,129 (1987); Nucl.Phys.B303, 407(1988).
2. D.J. Gross, "High energy symmetry of string theory", Phys. Rev. Lett. 60,1229 (1988); Phil.Trans. R. Soc. Lond. A329,401(1989).
3. D.J. Gross and J.L. Manes, "The high energy behavior of open string theory", Nucl.Phys.B326, 73(1989). See section 6 for details.
4. J.C. Lee, Phys. Lett. B241, 336(1990); J.C. Lee, Phys. Rev. Lett. 64, 1636(1990); J.C. Lee and B. Ovrut, Nucl. Phys. B336, 222(1990).
5. J.C. Lee, Prog. of Theor. Phys. Vol. 91, 353(1994).
6. J.C. Lee, Phys. Lett. B337,69(1994).
7. H.C. Kao and J.C. Lee, Phys. Rev. D67, 086003(2003).
8. C.T. Chan, J.C. Lee and Y. Yang, "Anatomy of zero-norm states in string theory", Phys. Rev.D71, 086005 (2005), hep-th/0501020.
9. E. Witten, Nucl. Phys. B268, 253 (1986).
10. T. D. Chung and J. C. Lee, "Discrete gauge states and w_∞ charges in $c = 1$ $2D$ gravity" Phys. Lett. B350, 22 (1995) [arXiv:hep-th/9412095]; "Superfield form of discrete gauge states in c = 1 2-d supergravity" Z. Phys. C **75**, 555 (1997) [arXiv:hep-th/9505107]. J.C. Lee, Euro.Phys. JC1,739(1998).
11. J. Avan and A.Jevicki, Phys. Lett. B266, 35 (1991); B272, 17 (1991). I.R. Klebanov and A.M. Polyakov, Mod. Phys. Lett. A6, 3273 (1991).
12. For a review see I.R. Klebanov and A. Pasquinucci, hep-th/9210105 and references therein.
13. E. Witten, Nucl. Phys. B373, 187 (1992); E.Witten and B. Zwiebach, Nucl. Phys. B377, 55 (1992).
14. C.T.Chan and J.C. Lee, "Stringy symmetries and their high-energy limits", Phys. Lett. B611,193 (2005), hep-th/0312226. J.C. Lee, hep-th/0303012.
15. C.T. Chan and J.C. Lee, Nucl.Phys. B690 (2004)3, hep-th/0401133.
16. C. T. Chan, P. M. Ho and J. C. Lee, "Ward identities and high-energy scattering amplitudes in string theory," Nucl. Phys. B **708**, 99 (2005) [arXiv:hep-th/0410194].
17. C.T. Chan, P.M. Ho, J.C. Lee, S. Teraguchi and Y. Yang, " Solving all 4-point correlation functions for bosonic open string theory in the high energy limit", Nucl.Phys.B725, 352(2005). hep-th/0504138.
18. C.T. Chan, P.M. Ho, J.C. Lee, S. Teraguchi and Y. Yang, "High-energy zero-norm states and symmetries of string theory". hep-th/0505035.
19. J.C. Lee, "Calculations of zero-norm states and reduction of stringy scattering amplitudes", Prog. Theo. Phys. 114, 259(2005), hep-th/0302123.
20. M. Evans and B.A. Ovrut, Phys. Lett. B231,80 (1989).

Data and Diagnostics in *LISA PathFinder*

A. Lobo[*,†], M. Nofrarias[†], J. Ramos[**], J. Sanjuan[†], A. Conchillo[†], J.A. Ortega[†], X. Xirgu[†], H. Araujo[‡], C. Boatella[†], M. Chmeissani[§], C. Grimani[¶], C. Puigdengoles[§], P. Wass[‡], S. Anza[†], M. Díaz Michelena[||], E. García-Berro[**] and R. Pérez del Real[||]

[*]*Instituto de Ciencias del Espacio (ICE), Barcelona, Spain*
[†]*Institut d'Estudis Espacials de Catalunya (IEEC), Barcelona, Spain*
[**]*Universitat Politècnica de Catalunya (UPC), Barcelona, Spain*
[‡]*Imperial College of London (UK)*
[§]*Institut de Física d'Altes Energies (IFAE), Barcelona, Spain*
[¶]*Università di Urbino and INFN (Italy)*
[||]*Instituto Nacional de Técnica Aoeroespacial, Madrid, Spain*

Abstract. The Diagnostics measurement set is a subsystem of the *LTP* which is intended to monitor a number of spurious disturbances, even if the satellite complies with all cleanliness requirements. It monitors thermal and magnetic perturbations, and incorporates a charged particle counter. The purpose of this communication is to discuss the conceptual aspects of the mentioned subsystem, and to summarise its current status of development.

Keywords: Gravity waves, LISA, diagnostics
PACS: 04.80.Nn, 95.55.Ym, 04.30.Nk

INTRODUCTION

LISA (Laser Interferometer Space Antenna) is a space mission which will fly a Gravitational Wave (GW) telescope within the next decade. The main idea of *LISA* is to gain access to a frequency band around 1 mHz, where many interesting sources are expected but earth based detectors are (by far) not sensitive to: many galactic binaries, massive black holes in distant galaxies and (perhaps) primeval GWs are amongst the signals *LISA* is expected to sight, at least.

LISA will measure GW induced phase shifts in beams of laser light bouncing back and forth between freely falling test masses. According to basic theoretical principles (see e.g. [1]), this requires the nominal distance between the test masses to be of the order of $c/2\nu$, where ν is the frequency of the incoming GW and c is the speed of light. For $\nu \sim 1$ mHz this gives an arm length of a few million km. For *LISA*, 5×10^6 km has been selected, and the mission is defined as a formation of three spacecraft in a triangular configuration, 5×10^6 km to the side [2]. For this, a heliocentric orbit, 1 AU from the Sun, is foreseen. The goal sensitivity of the instrument, in *rms* spectral density of noise, is given by

$$S_h(\nu) = 4 \times 10^{-21} \left[1 + \left(\frac{\nu}{3 \text{ mHz}}\right)^2\right] \text{Hz}^{-1/2}, \qquad 10^{-4} \text{Hz} \leq \nu \leq 10^{-1} \text{Hz} \qquad (1)$$

To achieve such a formidable requirement is not easy, indeed: picometre interferometry will be needed, and an extremely quiet environment for the test masses has to be ensured. The latter is provided by a so called *drag-free* system, which consists in a high precision test mass position sensing device, called *Inertial Sensor (IS)*, working in combination with a set of micro-thrusters which maintain the spacecraft in orbit *following* the test masses.

The European Space Agency (*ESA*) has decided to launch a previous technology demonstrator to check that the required technology for *LISA* can be met. The mission is called *LISA PathFinder (LPF)*, and will fly in 2009. Various European countries participate in the mission, amongst them Spain. The Spanish collaboration is funded by Ministerio de Educación y Ciencia through a Project granted to IEEC, Barcelona. The scientific team is formed by the signatories of this report.

Our commitment to *LPF* is the design and delivery for flight of the so called *Data and Diagnostics Subsystem (DDS)*, which consists of a series of items intended to monitor various factors of disturbance inside the payload, the *LISA Test-flight Package (LTP)*. The purpose of these instruments is to provide information to split up the total system readout noise into different components, with the ultimate goal of properly assessing *LISA*'s design. In addition, the *DDS* also provides the *Data Management Unit (DMU)*, with many control and feedback functions, and on-board data analysis duties.

DIAGNOSTICS ELEMENTS

Three types of disturbances have been identified which need to be diagnosed in the *LTP*:

- Thermal fluctuations
- Magnetic fluctuations
- Charged particle incident fluxes

We give below a brief summary of the various diagnostics concepts. An update on their current status of development is deferred to the last section.

Thermal fluctuations

Thermal gradients are a major source of concern since they affect almost every component of the *LTP*, and this impacts the system readout. It is required that thermal fluctuations be below $10^{-4}\,\mathrm{K\,Hz^{-1/2}}$ [3], and the satellite design should be able to comply with this limit. However temperature measurements need to be taken at various strategic spots to gauge the actual temperature environment conditions. Resolution of such measurements must be more exigent, and $10^{-5}\,\mathrm{K\,Hz^{-1/2}}$ is required for them within the measuring bandwidth (*MBW*), which in the case of the *LTP* is slightly shifted relative to *LISA*:

$$\text{MBW for LTP}: \quad 1\,\text{mHz} \leq \nu \leq 30\,\text{mHz} \qquad (2)$$

A total of 22 thermometers will be distributed across the *LTP*, close to the test masses inside the *IS*, in the optical bench and in a few other sensitive places.

But, once the temperature readings are given, next question is: how do we extract useful information from them? We clearly need to know the relationship between temperature and system readout. Mathematical modeling turns out to be insufficiently accurate, so a *calibration* method is resorted to. This one consists in applying thermal control signals to the *LTP*, then measuring its response. Signals of selected spectra are generated by a set of *heaters*, whereby thermal *transfer functions* are determined. A total of 14 heaters in suitably chosen locations will be used, which will produce the required information. The processing of the latter is a non-trivial problem, and is currently under study.

Magnetic fluctuations

Each of the *LTP* test masses is a cube, 46 mm to the side, 1.96 kg of mass, and made of an alloy of 70% gold and 30% platinum. This has a very low magnetic susceptibility, $|\chi| \leq 10^{-5}$, and a very low remanant magnetic moment, too: $|\mathbf{m}_0| \leq 10^{-8}\,\mathrm{A\,m^2}$. Nevertheless, fluctuations of these magnitudes, as well as of the environmental magnetic fields and gradients, will contribute to the system's overall noise. Just like thermal fluctuations, in-flight measurements of magnetic disturbances are necessary for a thorough assessment of mission design, i.e., how much magnetic noise is present in the readout.

Magnetic fields have a special feature which complicates design. This is the fact that, so long as the susceptibility is non-zero, they create magnetisation on the test masses, hence a force on them. If the magnetic field is **B** then force is given by

$$\mathbf{F} = \left\langle \left[\left(\mathbf{m}_0 + \frac{\chi V}{\mu_0} \mathbf{B} \right) \cdot \nabla \right] \mathbf{B} \right\rangle \tag{3}$$

where V is the test mass volume, and $\langle - \rangle$ stands for volume average within the test mass. Equation (3) shows that magnetic field intensity relates *non-linearly* to the associated force, and this has two important consequences: first, magnetic field and gradient *DC* components do have to be properly monitored, as they couple to the each other's fluctuating components. And second, high frequency magnetic field fluctuations result in *DC* —or low frequency— forces because of the quadratic dependence of **F** on **B**. The second should be actually weaker than the first, and this is why *DC* magnetic field and gradient requirements are set at top level [3].

Magnetic diagnostics also include the concept of controlled generation of magnetic fields for in-flight calibration purposes. This is provided by a pair of magnetic induction coils which produce non-homogeneous magnetic fields in the test mass regions.

Radiation monitor

Cosmic rays and certain solar events contain ionising particles which will hit the *LTP* in flight, thus causing spurious signals in the *IS*. These particles are mostly protons, with 10% or less of He nuclei, and a minor component of heavier galactic nuclei and solar ions. Charging rates and the properties of noise caused by charging vary depending on whether the particle flux comes from Galactic Cosmic Rays (*GCR*) or is augmented by

Solar Energetic Particles (*SEP*). The reason is that the two types of radiation present different energy spectra. Although average charging rates are detected by a dedicated measurement provided by the *IS*, temporal *fluctuations* of the *GCR* flux and *SEP* can contaminate the data A particle counter is thus necessary to provide correlations between the flux of energetic particles and the instantaneous charging rates observed in the test masses. In addition, the device must have the ability to distinguish *SEP* events from *GCR*s, and this consequently means it needs to determine the *energy spectra* of the detected particles. Finally, not all charged particles hitting the satellite structure will make it to the test masses, as that structure itself has a certain *stopping power*. The particle counter must only be triggered by those particles having enough energy to reach the TMs, hence it must be properly *shielded*. Simulation work indicates that only ions with energies larger than ~ 100 MeV should be counted. The particle counter together with the above added capabilities is known as Radiation Monitor (RM).

Contrary to the previous diagnostics, the RM does not require in-flight calibration. This will be done on ground by submitting it to laboratory proton beam irradiation.

CONCLUSION

As mentioned in the first section, the *DMU* is also part of the *LTP DDS*. In fact it is *the LTP* computer, and is mission critical, i.e., the *LTP* will not work without it. Its design and construction, including *software*, is however an engineering affair, and as such is under the responsibility of an industrial company (*NTE* in this case). The IEEC Project is contracting this company to the effect. Software developpers are however IEEC personnel, therefore part of the science team.

In October 2005 the *DDS* has undergone a Preliminary Design Review, which has been rather successful, pending a small number of issues that need further analysis. At the time of writing, this is essentially complete, and the next step will be the beginning of hardware procurement and assembly. This will based on prototypes designed, built and tested by the IEEC team. Temperature sensors are thermistors, and are driven and controlled by electronics adapted to the requirements' needs. The Radiation Monitor makes use of two silicon PIN diodes in telescopic configuration, also with associated electronics. A successful proton beam test was passed in November 2005 for the whole device. Magnetic diagnostics is ready for test, as instruments are mostly commercially available. Here, studies of magnetic field mapping are underway on the basis of magnetometer readings.

Acknowledgement: We thank Ministerio de Educación y Ciencia for support under contract ESP2004-01647.

REFERENCES

1. J.A. Lobo, *Classical and Quantum Gravity* **9**, 1385 (1992).
2. A. Hammesfahr et al., *LISA, a Cornerstone mission for the observation of Gravitational Waves*, ESA document ESA-SCI(2000)11 (2000).
3. S. Vitale et al., *Science Requirements and Top-level Architecture Definition for the Lisa Technology Package (LTP) on Board LISA PathFinder*, LPF document LTPA-UTN-ScRD-Iss003-Rev1.

The gravitational wave radiation of pulsating white dwarfs

P. Lorén–Aguilar[*,†], J. Isern[*,†], L.G. Althaus[**], A.H. Córsico[**], J.A. Lobo[*,†] and E. García–Berro[*,‡]

[*]*Institut d'Estudis Espacials de Catalunya, Ed. Nexus, C/ Gran Capità 2, 08034 Barcelona (Spain)*
[†]*Institut de Ciències de l'Espai (CSIC), Campus UAB, Facultat de Ciències, Torre C-5, 08193 Bellaterra (Spain)*
[**]*Facultad de Ciencias Astronómicas y Geofísicas, Universidad Nacional de La Plata, Paseo del Bosque s/n, (B1900FWA) La Plata, Argentina*
[‡]*Departament de Física Aplicada (UPC), Av Canal Olímpic sn, 08860 Castelldefels (Spain)*

Abstract. We compute the emission of gravitational radiation from pulsating white dwarfs. This is done for a standard $0.6 M_\odot$ white dwarf with a liquid CO core and a H-rich envelope, for a massive DA white dwarf with a partially crystallized core for which various $\ell = 2$ modes have been observed (BPM 37093) and for PG 1159-035 for which several quadrupole modes have been observed as well. We find that these stars do not radiate sizeable amounts of gravitational waves through their observed g-modes. We also explore the possibility of detecting gravitational waves radiated by the f-mode and the p-modes. We find that in this case the gravitational wave signal is very large. We also discuss the possible implications for the detection of gravitational waves from pulsating white dwarfs within the framework of future space-borne interferometers like LISA.

Keywords: White dwarf stars, gravitational radiation
PACS: 95.30.Lz, 95.55.Ym

INTRODUCTION

Despite its potential interest, the emission of gravitational waves by pulsating white dwarfs has been little explored up to now, see, however, Ref. [1]. White dwarfs are the most common end-point of the evolution of low- and intermediate-mass stars and, hence, they constitute the most numerous stellar remnants in our Galaxy, outnumbering neutron stars. Among white dwarfs there are three families of variable stars, known as ZZ Ceti (or DAV, with hydrogen-rich envelopes and $T_{\rm eff} \sim 12\,000$ K), V777 Her (or DBV, with helium-rich envelopes and $T_{\rm eff} \sim 25\,000$ K) and GW Vir stars (or variable PG 1159 objects, with envelopes which are rich in C, O and He, and $T_{\rm eff}$ ranging from $\sim 80\,000$ to $150\,000$ K). The typical periods lay in the region of frequencies to which LISA will be sensitive and their luminosity changes have been successfully explained as due to nonradial g-mode pulsations.

BPM 37093 is the most massive pulsating white dwarf ever found [2]. It is a massive ZZ Ceti star with a stellar mass of $\sim 1.05 M_\odot$, and an effective temperature $T_{\rm eff} \simeq 11\,800$ K. One of the most apparent modes of BPM 37093 has a period $P = 531.1$ s, very close to the frequency of maximum sensitivity of LISA and pulsates with $\ell = 2$. The distance to BPM 37093 is known ($d = 16.8$ pc). On the other hand, PG 1159-035, the prototype of the GW Vir class of objects, has a complex spectrum with several

$\ell = 2$ modes [3]. Unfortunately there is no reliable parallax determination for PG 1159-035. In Ref. [4] $d \sim 800^{+600}_{-400}$ pc was obtained, whereas in Ref. [5] $d \simeq 400 \pm 40$ pc was derived. However, a spectroscopic determination of its mass ($M_\star \simeq 0.54 M_\odot$) is available. These are, to the best of our knowledge, the only two known white dwarf pulsators with confirmed quadrupole g-modes. Here we compute the gravitational waves radiated by a typical $0.6 M_\odot$ white dwarf with a CO core and a $10^{-4} M_\star$ H envelope, which we regard as our fiducial model. We also compute the gravitational waves emitted by BPM 37093 and PG 1159-035, the only two known white dwarfs with quadrupole g-modes.

INPUT PHYSICS AND METHOD OF CALCULATION

We compute the nonradial pulsation modes of the white dwarf models with a pulsational code [6, 7] which provides very accurate oscillation frequencies and nonradial eigenfunctions by solving the fourth-order set of equations governing Newtonian, linear, nonradial stellar pulsations in the adiabatic approximation. The white dwarf models needed for our pulsational code were derived using the LPCODE evolutionary code [8, 9]. The perturbed density profile is given by $\rho(\vec{r},t) = \rho_0(r) + \rho'(r) \mathrm{Re}\left(Y_\ell^m(\theta,\phi) e^{i\sigma t}\right)$, where ρ_0 is the unperturbed density profile, $\rho'(r)$ stands for the radial perturbation, $Y_\ell^m(\theta,\phi)$ are the spherical harmonics, and σ is the pulsation. We adopt $\ell = 2$ and $m = 0$. Additionally, it must be taken into account that BPM 37093 has a sizeable crystallized core. When the core of the white dwarf undergoes crystallization the nonradial eigenfunctions are inhibited from propagating in the crystallized region. Consequently, and keeping in mind that for the axisymmetric case $Q_{11} = Q_{22} = -\frac{1}{2} Q_{33}$ and $Q_{ij} = 0$ if $i \neq j$ it can be easily obtained that the dimensionless strain, h_{33}^{TT}, is given by

$$h_{33}^{\mathrm{TT}} \approx 5 \times 10^{-18} \left(\frac{M_\star}{M_\odot}\right) \left(\frac{R_\star}{R_\odot}\right)^2 \left(\frac{\nu}{1\,\mathrm{mHz}}\right)^2 \left(\frac{1\,\mathrm{pc}}{d}\right) \left[A(R_\star) - F_\mu F_R^2 A(R_0)\right] \cos(\sigma t) \quad (1)$$

being $\nu = 2\pi\sigma$, $A(r) \equiv y_4 - 2y_3$, $y_3 \equiv \Phi'/gr$, $y_4 \equiv (1/g) d\Phi'/dr$, $F_\mu \equiv M_0/M_\star$, $F_R \equiv R_0/R_\star$, Φ' the perturbed gravitational potential, g the gravitational acceleration at a given radius, $M_\star$ and $R_\star$ the mass and radius of the star, and M_0 and R_0 the mass and radius of the crystallized core. The luminosity radiated in the form of gravitational waves is:

$$L_{\mathrm{GW}} \approx 10^{36} \left(\frac{M_\star}{M_\odot}\right)^2 \left(\frac{R_\star}{R_\odot}\right)^4 \left(\frac{\nu}{1\,\mathrm{mHz}}\right)^6 \left[A(R_\star) - F_\mu F_R^2 A(R_0)\right]^2 \quad (2)$$

RESULTS

The linear theory of nonradial pulsations does not provide any indication of the value of the fractional change in radius. In addition, $\delta R_\star/R_\star$ is poorly constrained by the observations. Thus, we have adopted $\delta R_\star/R_\star = 10^{-4}$, which is a typical value for pulsating white dwarfs, and reasonably reproduces the amplitude of the observed light curve [10]. In Table 1 we summarize the most important results for several g-modes of the computed models. The maximum dimensionless strain, $h_{\max}$ for BPM 37093 has

TABLE 1. Summary of the gravitational wave emission of the g-modes of BPM 37093 and PG 1159-035. Our fiducial model is also shown for comparison.

Model	$M/M_\odot$	k	P_o (s)	P_c (s)	h_{max}	L_{GW} (erg/s)	$\log(E_K)$ (erg)
BPM 37093	1.10	26	511.7	516.7	4.8×10^{-28}	9.2×10^{18}	46.4
		27	531.1	536.4	5.4×10^{-28}	1.1×10^{19}	46.5
		28	548.4	555.8	5.3×10^{-28}	9.9×10^{18}	46.6
		29	582.0	574.9	6.4×10^{-28}	1.3×10^{19}	46.7
		30	600.7	593.0	6.4×10^{-28}	1.2×10^{19}	46.8
		32	633.5	630.4	5.5×10^{-28}	8.3×10^{18}	46.9
PG 1159-035	0.54	25	352.7	358.9	1.0×10^{-30}	5.0×10^{16}	44.0
		30	423.8	423.2	2.3×10^{-30}	1.7×10^{17}	43.8
		50	694.9	684.5	3.5×10^{-31}	1.6×10^{15}	43.4
		55	734.2	752.9	2.0×10^{-31}	4.1×10^{14}	43.3
		60	812.6	818.1	1.8×10^{-31}	2.8×10^{14}	43.2
		70	968.7	950.1	6.0×10^{-32}	2.5×10^{13}	42.8
$0.6\,M_\odot$	0.6	1	—	66.6	6.9×10^{-25}	1.0×10^{28}	47.0
		10	—	310.3	1.8×10^{-27}	3.1×10^{21}	44.8
		20	—	555.2	1.5×10^{-27}	6.6×10^{20}	45.1

been computed adopting the measured distance to the source ($d = 16.8$ pc), whereas for our fiducial model we have adopted a distance $d = 50$ pc, which is representative of a typical white dwarf. For the case of PG 1159-035 we have adopted a distance of 400 pc. The agreement between the computed and the observed periods is excellent. However, the amplitudes of the dimensionless strains are small and so are the luminosities radiated away in the form of gravitational waves. For comparison we also show the kinetic energy of each mode. This is the reason why we also have computed the gravitational wave radiation of the f-mode and the p-modes.

In order to check whether LISA will be able to detect pulsating white dwarfs we have assumed that the integration time of LISA will be one year. The signal-to-noise ratio, η, is given by:

$$\eta^2 = \int_{-\infty}^{+\infty} \frac{\tilde{h}^2(\sigma)}{S(\sigma)} \frac{d\sigma}{2\pi} \qquad (3)$$

where $S(\sigma) = S_h(\sigma)\tau$ is the sensitivity of LISA, τ is the integration period and $\tilde{h}(\sigma)$ is the Fourier Transform of the dimensionless strain. For a monochromatic gravitational wave $\eta = h(\sigma)/S_h^{1/2}(\sigma)$. We have adopted $\eta = 5$. We have furthermore used the integrated sensitivity of LISA as obtained from http://www.srl.caltech.edu/~shane/sensitivity. The results for our fiducial model are shown in Fig. 1. The g-modes are shown as circles, the f-mode is shown as a square and the p-modes are displayed as triangles. As can be seen LISA will not be able to measure the dimensionless strains of the g-modes, even at a reduced signal-to-noise ratio. Most of the p-modes will not be observed as well, either because their frequencies are too high to be observed by LISA or because they are too weak, but we expect to detect the f-mode, if excited.

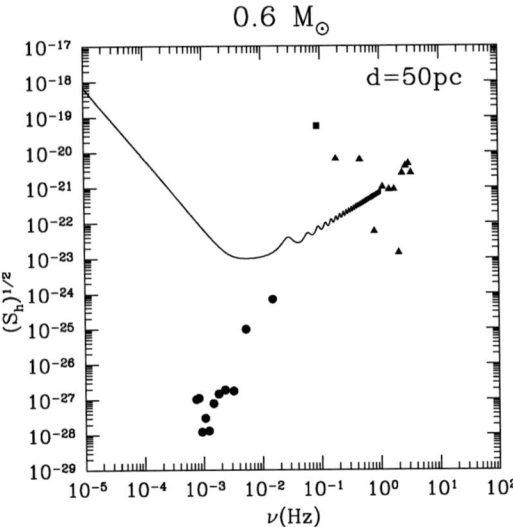

FIGURE 1. A comparison of the signal produced by the quadrupole g-modes of our fiducial model (circles) by the f-mode (square) and by the p-modes (triangles) with the spectral distribution of noise of LISA for a one-year integration period.

CONCLUSIONS

We have shown that the gravitational wave signal of pulsating white dwarfs undergoing nonradial g-mode pulsations is too weak to be observed by future space-borne interferometers, like LISA. Hence, they contribute to the Galactic noise and no individual detections are expected. We have also computed the gravitational wave emission of white dwarfs undergoing nonradial f- and p-mode oscillations, even if these modes have not been observationally detected. We have found that for white dwarfs undergoing this kind of pulsations some detections could be possible.

REFERENCES

1. Y. Osaki, & C.J. Hansen, *ApJ*, **185**, 277 (1973)
2. A. Kanaan, S.O. Kepler, S. O., O. Giovannini, & M. Diaz, M., *ApJ*, **390**, L89 (1992)
3. D.E. Winget, et al., *ApJ*, **378**, 326 (1991)
4. K. Werner, U. Heber, & K. Hunger, *A&A*, **244**, 437 (1991)
5. S.D. Kawaler, & P.A. Bradley, *ApJ*, **427**, 415 (1994)
6. A.H. Córsico, L.G. Althaus, O.G. Benvenuto, & A.M. Serenelli, *A&A*, **380**, L17 (2001)
7. A.H. Córsico, L.G. Althaus, O.G. Benvenuto, & A.M. Serenelli, *A&A*, **387**, 531 (2002)
8. L.G. Althaus, A.M. Serenelli, A.H. Córsico, & M.H. Montgomery, *A&A*, **404**, 593 (2003)
9. L.G. Althaus, A.M. Serenelli, J.A. Panei, A.H. Córsico, E. García-Berro, & C.G. Scćoccola, *A&A*, **435**, 631 (2005)
10. E.L. Robinson, S.O. Kepler, & R.E. Nather, *ApJ*, **259**, 219 (1982)

A combinatorial approach to discrete geometry

L. Bombelli* and Miguel Lorente[†]

University of Mississipi. USA
[†]*Universidad de Oviedo, Spain*

Abstract. We present a paralell approach to discrete geometry: the first one introduces Voronoi cell complexes from statistical tessellations in order to know the mean scalar curvature in term of the mean number of edges of a cell [1]. The second one gives the restriction of a graph from a regular tessellation in order to calculate the curvature from pure combinatorial properties of the graph [2].

Our proposal is based in some epistemological pressupositions: the macroscopic continuous geometry is only a fiction, very usefull for describing phenomena at certain sacales, but it is only an approximation to the true geometry. In the discrete geometry one starts from a set of elements and the relation among them without presuposing space and time as a background.

Keywords: spin networks, combinatorics, space time
PACS: 04.20.Gz, 04.60.Nc, 0.5.50+q

1. Introduction. In recent years some approaches to quantum gravity have suggested the hypothesis of a discrete space time [3] as a consequence of the combinatorial properties of spin networks underlying the structure of space [4] and implemented with the hypothesis of causal sets [5][6].

In our model we have to choose the discrete quantities in such a way that in the continuous limit they become the classical ones. The direct way consists on calculating some continuous quantity in a 2d-manifold. By the inverse way, we can start directly from the graph and find some embedding where the corresponding quantities become analog, like the genus of some graph [7], or the curvature in a triangulated manifold.

According to Bombelli one can scatter points in a Lorentzian manifold, and then keep the statistical distribution of points from which some discrete quantities can be defined, such as curvature, from combinatorial properties of the set of relations [8].

2. Reflection groups and tessellations. Let P a finite sided n-dimensional convex polyhedron in a metric space X, all of whose dihedral angles are submultiple of π. Then the group generated by the reflection of X in the sides of P, $\{S_i\}$ is a discrete reflection group Γ with respect to the polyhedron P.

Let Δ be an n-simplex in X all of whose dihedral angles are submultiple of π. The group Γ generated by the reflections of X in the sides of Δ is an n-simplex discrete reflection group. Notice that X can be S^n, E^n or H^n. The classification of all the irreducible n-simplex (spherical, euclidean and hyperbolic) reflection groups is complete [9].

Assume that $n = 2$. Then Δ is a triangle in X, whose angles $\frac{\pi}{l}, \frac{\pi}{m}, \frac{\pi}{n}$ are submultiple of π. If we call $T(l,m,n)$ the group Γ generated by the reflections in the sides of $\Delta, T(l,m,n)$ is call a triangle reflection group.

If $X = S^2$ the only spherical triangle reflection groups are:

$$T(2,2,2), \quad T(2,2,n)\, n > 2, \quad T(2,3,3), \quad T(2,3,4), \quad T(2,3,5)$$

If $X = E^2$ we have the euclidean triangle reflection groups:

$$T(3,3,3), \quad T(2,4,4), \quad T(2,3,6)$$

If $X = H^2$ we have the hyperbolic triangle reflection groups:

$$T(2,m,n)\ m \geq n \geq 3, \quad T(l,m,n)\ l \geq m \geq n \geq 3$$

Geometrically a reflection can be represented by a linear transformation which fixes an hyperplane pointwise and sends some non zero vector to its negative. In the metric space X we construct vectors $\{\alpha_i\}$ in one to one correspondence to the sides $\{S_i\}$ defined before, in such a way that the angle between α_i and α_j will be compatible with the values of k_{ij}, namely, $\vartheta\left(\alpha_i, \alpha_j\right) = \frac{\pi}{k_{ij}}$, ($k_{ij}$, positive integer o infinite, $k_{ii} = 1$).

In order to construct a reflection with respect to these vectors $\{\alpha_i\}$ we define a non-degenerate symmetric bilinear form on X by the formulas

$$\langle \alpha_i, \alpha_j \rangle = -\cos\frac{\pi}{k_{ij}}$$

This expresion is interpreted to be -1 for $k_{ij} = \infty$. Obviously $\langle \alpha_i, \alpha_i \rangle = 1$, and $\langle \alpha_i, \alpha_j \rangle \leq 0$ for $i \neq j$. For each vector α_i we can define a reflection S_i on X:

$$S_i \beta = \beta - 2\langle \alpha_i, \beta \rangle \alpha_i, \quad \beta \in X$$

clearly $S_i \alpha_i = -\alpha_i$ and all γ satisfying $\langle \alpha_i, \gamma \rangle = 0$ belong to a plane invariant under S_i.

It can be proved that the collection of the polyhedra obtained by the reflections on the side of Δ is a tessellation of X all the n-simplex (compact or non-compact) reflection groups lead to regular tessellation of X.

In Figure 1 we give now some examples of tessellations in H^2 generating by reflecting in the sides of the hyperbolic triangle $T(2,3,8)$.

3. Gauss curvature of continuous tessellations.

Two dimensional tessellations in $X \left(= S^2, E^2 \text{ or } H^2\right)$ are generated by 2-simplex (triangle) reflection group.

In S^2 the geodesic triangle are spherical.

The excess of the interior angles of a spherical triangle is:

$$\varepsilon = \alpha + \beta + \gamma - \pi = \frac{\pi}{l} + \frac{\pi}{m} + \frac{\pi}{n} - \pi$$

It can be proved that this excess is always positive [9]

The area of the triangle $T(x,y,z)$ is

$$\text{Area}\{T(x,y,z)\} = \alpha + \beta + \gamma - \pi = \varepsilon$$

The excess of the interior angles of an euclidean triangle is

$$\varepsilon = \alpha + \beta + \gamma - \pi = 0$$

In H^2 the geodesic triangles are hyperbolic.

The excess of the interior angles of an hyperbolic triangle is

$$\varepsilon = \alpha + \beta + \gamma - \pi = \frac{\pi}{l} + \frac{\pi}{m} + \frac{\pi}{n} - \pi$$

It can be proved that this excess is always negative [9]

The area of an hyperbolic triangle $T(x,y,z)$ is

$$\text{Area}(T) = \pi - (\alpha + \beta + \gamma)$$

We can now apply these results to the curvature of the surfaces corresponding to the 2-dimensional regular tessellations (spherical, euclidean or hyperbolic). According to Gauss-Bonet theorem the excess angle of some geodesic triangle T is equal to the integral of the gaussian curvature K over T

$$\varepsilon = \alpha + \beta + \gamma - \pi = \iint_T K d\sigma$$

where $d\sigma$ is the area element. If K=const.

$$K = \frac{\varepsilon}{A}$$

Applying this formula to the above results, we have:
$K = 1$ for spherical geodesic triangles
$K = 0$ for euclidean triangles
$K = -1$ for hyperbolic geodesic triangle

4. Curvature on planar graphs. A graph is a par $G = \{V,E\}$ where V is a non-empty set of vertices and E an unordered 2-set of vertices, called edges, in such a way that two vertices are incident to an edge.

A graph can be defined in an abstract way using only combinatorial properties of vertices and edges, or can be obtained from geometrical objects. For instance, given a particulr tessellation described in section 2, we keep the edges and vertices of all the triangles and eliminate the embedding manifold in such a way that we are left with the points (vertices) and relations among those (edges). In Figure 2, we have drawn the graphs that we have derived by this method from the tessellations given in Figure 1, where the vertices are represented by points and the edges by arrows. In a graph one can define such elements as path, circuit, length, distance. For instance, in a given graph one may travel from one vertex to another using several edges. the set of the vertices visited in that journey is called a path. These definitions coincide with the standard ones when the graph is embedded in some continuous manifold.

One can define the excess of this triad of vertices as the quantity, in analogy with (1),

$$\delta = \frac{1}{l} + \frac{1}{m} + \frac{1}{n} - 1$$

where $2l, 2m, 2n$ are the number of edges incident in each of the three vertices, which correspond to $2l$–valued, $2m$–valued or $2n$–valued vertices, respectively.

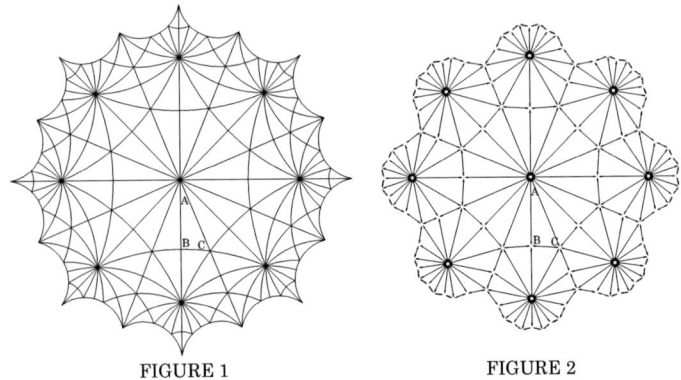

FIGURE 1   FIGURE 2

If we define the spherical, euclidean or hyperbolic graph, that is obtained from a spherical, euclidean or hyperbolic tessellation respectively, we can check

$\delta > 0$, for a spherical graph
$\delta = 0$, for an euclidean graph
$\delta < 0$, for an hyperbolic graph

We define the area and the area of the triad $T(l,m,n)$ in an hyperbolic graph

$$\sigma(T) = 1 - \left(\frac{1}{l} + \frac{1}{m} + \frac{1}{n}\right)$$

Similarly, we define the curvature of a triad $T(l,m,n)$

$$K(T) = \frac{\delta}{\sigma} = \begin{cases} 1, & \text{for a spherical graph} \\ 0, & \text{for an euclidean graph} \\ -1, & \text{for an hyperbolic graph} \end{cases}$$

an expression that can be considered the discrete version of the Gauss-Bonet theorem.

5. Statistical approach to Gaussian curvature. In a random distribution of points we can apply the combinatorial approach to calculate the curvature as in the regular case. We can consider a cell decomposition of a manifold by the embedding of a cell complex Ω, formed by N_0 vertices, N_1 edges, N_2 faces, N_3 tetrahedra and so on, satisfying

$$\sum_{k=0}^{D} (-1)^K N_k(\Omega) = (-1)^D \chi(\Omega)$$

One of the most useful cell decomposition is a triangulation in which all the cell complex are simplicial, satisfying $N_1(\Omega) = \frac{1}{2}(D+1)N_0(\Omega)$. Given a random distribution of points p_i we construct the Voronoi complex as the set of points p_j that are closer to p_i than other point p_j. The Delaunay complex is the dual to Voronoi, and it is formed by all the points p_i and the edges joining them (see Figure 3). A Voronoi complex gives rise to a non-regular tessellation. If we substract the embedding manifold we are left with

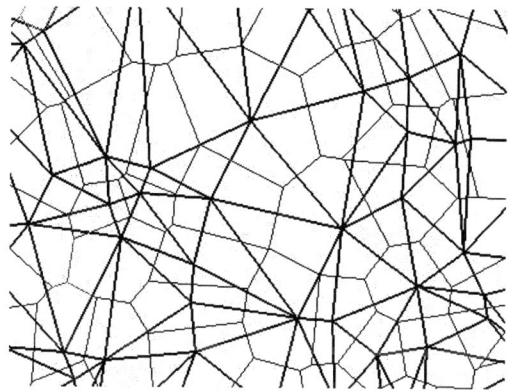

FIGURE 3. Example of 2D Voronoi (thin lines) and Delaunay (thick lines) complexes; the points they are based on are the Delaunay vertices.

a graph, where we can calculate curvature from combinatorial properties of Voronoi complex. Suppose we have a 2-dimensional Voronoi graph Ω with elements N_0, N_1, N_2 satisfying $N_1 = \frac{1}{2}(2+1)N_0$, $N_0 - N_1 + N_2 = \chi(\Omega)$, χ Euler number.

In order to calculate the mean curvature, we use the Gauss-Bonet formula:

$$\bar{N}_1 = 2 \cdot \frac{N_1(\Omega)}{N_2(\Omega)} = G\left(1 - \frac{\chi(\Omega)}{N_2}\right) = G\left(1 - \frac{\int K d\sigma}{4\pi\rho A}\right) = G\left(1 - \frac{K}{4\pi\rho}\right)$$

whence $K = 4\pi\rho\left(1 - \frac{1}{G}\bar{N}_1\right)$, where ρ is the density of the cell complex in Ω

This work was partially supported by Ministerio Educación y Ciencia, grant BFM 2003-00313/FIS

REFERENCES

1. L. Bombelli et al., "Semiclassical quantum gravity: Statistics of Combinatorial Riemannian Geometries", arXiv: gr-qc/0409006.
2. M. Lorente, "A discrete curvature on a planar graph", arXiv: gr-qc/0412094.
3. T. Regge, R.M. Williams, "Discrete Structure in Gravity", J. Math Phys., 44 (2000) 3964-3984.
4. R. Penrose, "Angular momentum: an approach to combinatorial space time" in Quantum Theory and Beyond (ed. T. Bastin), Cambridge U. Press, Cambridge, 1971, pp. 151-181.
5. L. Bombelli, J. Lee, D. Meyer, R.D. Sorkin, "Space time as a causal set", Phys. Rev. Lett. 59 (1987) 521-524. F. Markopoulou, L. Smolin, "Causal evolution of spin networks", Nucl. Phys. B 508 (1997) 409-430.
6. M. Lorente, "Causal spin networks and the Structure of Space and Time", 12th Int. Congress on Logic, Methodology and Philosophy of Science, Universidad de Oviedo, 2003.
7. P. Kramer, M. Lorente "Surface embedding, topology and dualization for spin networks", J. Phys. A: Math. Gen. 35 (2002) 8563-8574.
8. L. Bombelli, "Statistical Lorentzian geometry", arXiv: gr-qc/0002053.
9. J.G. Ratcliffe, Foundation of Hyperbolic Manifolds, Springer, New York 1994, pp. 293-296.

Detector Configurations for Equivalence Principle Tests with Strong Separation of Signal from Noise

E.C. Lorenzini[a], I.I. Shapiro[a], J. Ashenberg[a], C. Bombardelli[a],
P.N. Cheimets[a], V. Iafolla[b], D.M. Lucchesi[b], S. Nozzoli[b], F. Santoli[b],
and S. Glashow[c]

[a]*Harvard-Smithsonian Center for Astrophysics, 60 Garden Street, Cambridge, MA 02138, USA*
[b]*Istituto di Fisica dello Spazio Interplanetario, Via Fosso del Cavaliere 100, 00133 Roma, Italy*
[c]*Boston University, 590 Commonwealth Avenue, Boston, MA 02215, USA*

Abstract. Testing the Equivalence Principle (EP) at a level of accuracy substantially higher than the present state of the art requires resolving a very small signal out of the instrument's intrinsic noise and also the noise associated with the instrument's motion and gravity gradients. In the test of the Equivalence Principle in an Einstein Elevator under development by our team, the acceleration detector spins about a horizontal axis while free falling for about 25 s inside a co-moving capsule released from a stratospheric balloon. The characteristics of the instrument package and the configuration of the detector play a key role in the ability to extract an EP violation signal at the desired threshold level out of dynamics-related noise. Numerical simulations of the detector's dynamics in the presence of relevant perturbations, having assumed realistic errors and construction imperfections, show the merits of the detector configuration selected. The results illustrate that the effects of dynamics and gravity gradients, near or at the signal frequency, can be limited to levels much smaller than the expected threshold sensitivity of the detector.

Keywords: Experimental Tests of Gravitational Theories, Mechanical Instruments, Design of Experiments, Computer Modeling and Simulation, Equivalence Principle
PACS: 04.80.Cc, 07.05.Fb, 07.05.Tp, 07.10.Pz

INTRODUCTION

Our experiment[1] to test the (weak) Equivalence Principle (EP) consists in releasing a differential acceleration detector from the top of an evacuated capsule immediately after release of the capsule itself from a balloon at an altitude of 40 km. The detector is in almost ideal free fall conditions inside the capsule while the latter is slightly decelerated by the rarefied atmosphere at that altitude. We have computed that the detector will take about 25 s to reach the bottom of the capsule after having traveled about 2 m from the top to the bottom of the capsule.

Testing the EP at an accuracy substantially higher than the present state of the art requires resolving a very small signal out of not only white Gaussian noise but also the colored noise associated with the instrument motion. Several papers have dealt in detail with white noise sources in gravitational tests[2,3,4] and only a few have touched upon the latter type of noise[5,6]. In EP tests, the need for signal modulation requires spinning the platform hosting the instrument at a frequency that is well distinct from other frequencies that affect the experiment. However, the rotational motion produces a comparatively high level of dynamics-related noise that needs to be accounted for. The issue of extracting an EP violation signal from the dynamics-related noise is intimately related to the configuration of the acceleration detector, its location inside the hosting platform (i.e., the instrument package), and the construction and centering errors of the proof masses.

A key issue is associated with the off-diagonal terms of the gravity gradient tensor. These terms may be non zero if the spin axis of the detector is not perfectly orthogonal to the local gravity vector. If the elevation angle (relative to the plane normal to the gravity vector) of the spin axis is different from zero, then the off-diagonal terms may generate acceleration contributions at the spin frequency. Diagonal terms of the tensor are not an issue because they are modulated at twice the spin frequency.

Detectors with different designs have different sensitivities to gravity gradient components and the orientation requirement can be strongly relaxed for certain configurations. One of the goals of this study is to develop a design whereby the detector has low sensitivity to both the effects of the damaging gravity gradient forces and the dynamics of the instrument package.

The approach to solving this problem involves the analysis of the effects of the platform dynamics and gravity gradients on the detector's output with good-fidelity dynamics models. Simulation results for different detector configurations were then used to select a design that provides a high degree of rejection of the perturbing effects, through either amplitude attenuation or frequency separation from the signal or both.

DETECTOR CONFIGURATIONS

There are some basic requirements that a differential accelerometer must satisfy to reach the measurement precision necessary for improving the state of the art in testing the Equivalence Principle. Typically, the differential accelerometer needs to have: (a) a low resonance frequency to increase sensitivity; (b) a very high Q-factor and low temperature to reduce the Brownian noise; (c) proof masses with second-order spherical inertia ellipsoids and, for very high test accuracies, higher-order sphericity; and (d) an accurate construction in terms of shape, material homogeneity, and centering of the proof masses. Specifically, gravity-gradient forces acting on non-centered proof masses contain harmonics at the signal frequency and twice the signal frequency. Such forces can easily overtake the violation signal at the threshold levels considered here.

Other key points to be considered in connection with the detector's motion are the geometrical relationship between sensitive axis and spin axis, the restraining of the proof masses, and the placement of the putative violation material within the proof masses. Some of these points have not been addressed systematically in the literature.

Accelerometers can be designed to exploit a purely translational motion as in Fig. 1(a) or a combination of translational and rotational motions of the proof masses as in Fig. 1(b) to measure the differential acceleration. Differential accelerometers under development[7] and prototype instruments built[8] thus far have followed these options.

The location of the two materials within the proof masses is also important. A new approach is to design a detector that responds only to torques (i.e., with a purely rotational motion) as shown in Fig. 1(c). In this configuration, a proof mass made of one homogeneous material would not be able to sense linear accelerations (except for errors between the pivot axis and the center of mass (CM)) but only rotational accelerations. In order to sense an EP violation, we will make the proof mass of two different materials (with the two materials on opposite sides of the pivot axis) so that an EP violation will generate a torque about the pivot axis. Such a proof mass will be highly insensitive to any linear acceleration, inclusive of those associated with gravity-gradient forces, but will react to torques.

Insensitivity to the detector's rotational motion is then provided by the second proof mass (all made of one material and insensitive to EP violations) and with the resonant frequency of its rotational motion ideally equal to that of the first mass. In other words, the first mass is sensitive to EP violations while the second mass acts as the 'dynamics reference' that removes the effects of common-mode rotational motion of the detector.

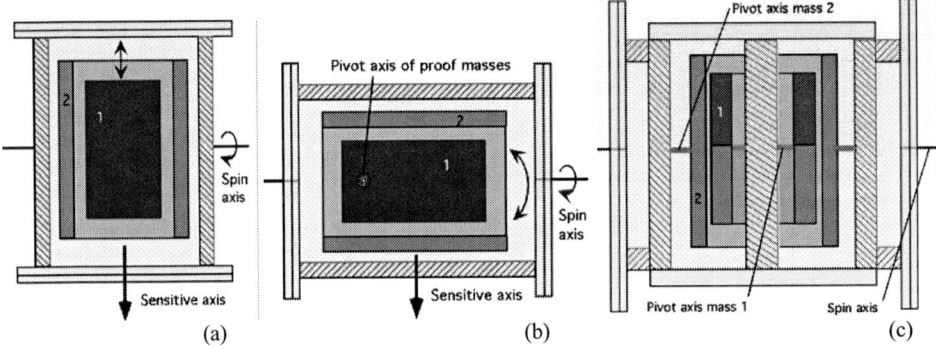

FIGURE 1. Detector configurations grouped by relative motion of the proof masses: (a) translational; (b) translational and rotational; and (c) purely rotational.

A key perturbation acting on each proof mass is the gravity-gradient torque about the pivot axis due to imperfect cylindrical symmetry of the proof masses (i.e., associated with the difference between the 2nd-order moments of inertia). This 2nd-order gravity gradient torque about the pivot axis is modulated at *twice the spin frequency* for a configuration with the spin axis parallel to the pivot axes (see Fig. 1(c)).

NUMERICAL RESULTS

Figure 2 shows numerical results for the purely-rotational configuration with the pivot axes parallel to the spin axis.

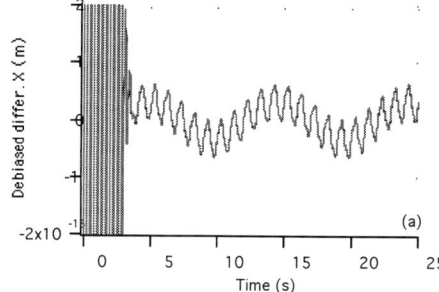 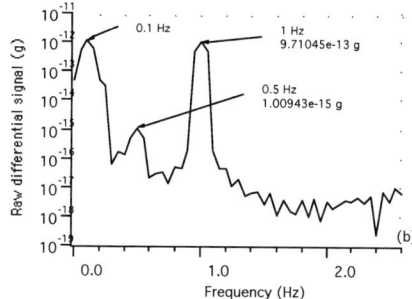

FIGURE 2. Simulated response of rotational detector with pivot and spin axes parallel: (a) differential displacement of proof masses; and (b) spectrum of differential acceleration. Free fall time = 25 s (< 5 s of which is used for damping), spin rate = 0.5 Hz, resonant frequency = 3 Hz, and precession frequency = 0.1 Hz. Detector's construction and release errors are: centering errors: inner proof mass = 10 µm and outer proof mass = 12 µm; velocity error at release (orthogonal to spin) = 0.1 deg/s; spin axis elevation angle = 1 deg; fractional errors in moments of inertia: proof masses = 10^{-4} and instrument package = 10^{-2}; initial oscillation amplitude of proof masses = 50 µm. Simulated EP violation acceleration = 1×10^{-15} g.

A signal of 1×10^{-15} g was recovered by simple frequency analysis out of the dynamics noise and gravity gradient torques. The other configurations do not have this level of noise rejection. Note that the detector's intrinsic noise is not present in this simulation and that the EP violation signal recovered is fivefold smaller than the accuracy goal of our experiment of 5 parts in 10^{15}. The numerical results show that the purely rotational configuration of the detector has low sensitivity to rotational dynamics, gravity gradient forces, and orientation of the spin axis with respect to the local gravity vector. In other words, this detector can be designed with moderate requirements of CM centering and axial symmetries of its proof masses. Likewise, the release mechanism has to satisfy a moderate requirement (of order one degree) on the orientation of the spin axis with respect to the local gravity vector. Thanks to the purely rotational detector configuration, the accuracy of the experiment is no longer limited by the detector's rotational dynamics and gravity gradient effects but rather by other noise sources such as Brownian noise.

CONCLUSIONS

A configuration that makes use of purely rotational motion of the proof masses has the advantage of mitigating strongly the error contributions of gravity gradients and attitude motion of the instrument package, and separating their frequency from the EP violation signal. We conclude that this detector's design leads to a substantial simplification of the requirements for an Equivalence Principle experiment via vertical free fall and, consequently, increases significantly its likelihood of success.

ACKNOWLEDGEMENTS

The authors would like to acknowledge the financial support from NASA Glenn through Grant NAG3-2881 to the Smithsonian Astrophysical Observatory and from the Italian Space Agency through contract I/R/098/02 to the Institute of Interplanetary Space Physics.

REFERENCES

1. Lorenzini, E.C., I.I. Shapiro, F. Fuligni, V. Iafolla, M.L. Cosmo, M.D. Grossi, P.N. Cheimets and J.B. Zielinski, "Test of the Weak-Equivalence Principle in an Einstein Elevator" *Il Nuovo Cimento*, **109B**(1), 1195-1209 (1994).
2. Giffard, R.P., *Physical Review D,* **14**(10), 2478-2486 (1976).
3. Paik, H.J. "Superconducting tensor gravity gradiometer for satellite geodesy and inertial navigation." *J. of the Astronautical Sciences,* **29**(1), 1-18 (1981).
4. Braginsky, V.B. and A.B. Manukin, "Measurements of Weak Forces in Physics Experiments." The University of Chicago Press, 1974.
5. Worden, P., J. Mester and R. Torii, "STEP error model development." *Classical and Quantum Gravity,* **18**, 2543-2550 (2001).
6. Touboul, P., B. Foulon, L. Lafargue, and G. Metris, "The Microscope Mission." *Acta Astronautica,* **50**(7), 433 (2002).
7. Mester, J., R. Torii, P. Worden, N. Lockerbie, S. Vitale and C.W.F. Everitt, "The STEP mission: principle and baseline design." Classical and Quantum Gravity, **18**, 2475-2486 (2001).
8. Iafolla, V., F. Fiorenza, S. Nozzoli, D. Lucchesi, E.C. Lorenzini, I.I. Shapiro, S. Glashow and P. Cheimets, "Development of a high-sensitivity differential accelerometer to be used in the experiment to test the Equivalence Principle in an Einstein Elevator." *Procs. of XXVIII Recontres de Moriond, Gravitational Waves and experimental Gravity,* 22-29 March 2003, Les Arcs, France, 2003.

Obtaining a class of conformally flat pure radiation metrics with cosmological constant using invariant operators.

S. Brian Edgar* and M. P. Machado Ramos[†]

*Matematiska Institutionen, University of Linköping, Sweden.
[†]Departamento de Matemática para a Ciência e Tecnologia, Universidade do Minho, Portugal.

Abstract.
In a recent paper by B. Edgar and M. P. Ramos the method using invariant operators to obtain the type O pure radiation class of metrics was generalized to type N pure radiation solutions. In this work we further generalize the method in determining type O pure radiation metrics with cosmological constant.

Keywords: tetrad formalism, Karlhede classification
PACS: 0420, 1127

THE METHOD

The method of integration using GHP operators [1] pioneered by Held [2], [3] and developed by Edgar and Ludwig [4], [5], [6], [7] has been shown to be particularly useful and efficient in spacetimes where two null directions are picked out by the geometry. Recently an analogous integration method [8], [9] using operators of the GIF formalism of Ramos and Vickers [10], [11], [12] has also been developed and shown to be especially useful and efficient in spaces where only one null direction has been picked out by the geometry. In particular, conformally flat pure radiation spacetimes [8] and a class of Petrov type N pure radiation spacetimes [9] have been investigated successfully in the GIF formalism, and we would expect closely related spaces to these classes also to be suitable for investigation by this method.

In this work we further explore the potential of the GIF operator method by generalizing the earlier investigation of conformally flat pure radiation spaces to include the case of non-zero cosmological constant.

As well as increasing our experience and expertise by further developing the GIF operator integration method to investigate another class of spaces, this particular class is interesting in its own right. The conformally flat pure radiation spaces manifest "extreme" properties [13] regarding the Karlhede classification of spacetimes [14] and so it would be suspected that this closely related class might have similar or even more extreme properties. On the other hand, though spaces may initially appear to be closely related because of their definitions, their structure and properties may be very different. For example, what might appear to be a comparatively insignificant addition of a non-zero cosmological constant has made a significant difference to vacuum Petrov type D spaces [15], [16], [17]. Griffiths et al [18] have pointed out that the introduction of a non-zero cosmological constant for the non-expanding subclass of Kundt's class of spacetimes is not so easily achieved as for the expanding (Robinson-Trautman) subclass, but nevertheless have succeeded in obtaining the complete subclass of Petrov type III of Kundt's class with a non-zero cosmological constant; the introduction of the cosmological constant gives a much richer class. Although these Petrov type III spaces contain the conformally flat spaces implicitly, it will also be instructive to determine whether there are also significant differences when a non-zero cosmological constant is introduced into the explicit class of conformally flat pure radiation spaces.

One of the intriguing aspects of these operator methods in both the GHP and GIF formalisms is that they are especially suited for investigating spaces lacking, or with limited, Killing vectors [7]; this of course complements other tetrad formalism methods, such as the NP formalism [19] which have been particularly useful in spacetimes with multiple Killing vectors. Furthermore, when spacetimes are calculated by means of the operator methods in the GHP and GIF formalisms, it is easy to deduce the existence and properties of any Killing and homothetic Killing vectors, and also straightforward to deduce the Karlhede classification of the spacetime under investigation [7], [8]; such results cannot be deduced so obviously when working in other tetrad formalisms, such as the NP formalism. (However, motivated by the GHP and GIF applications, there are now being developed some applications along these lines in the NP formalism [20].)

The philosophy and techniques of the GIF operator integration procedure have been described in [8], [9].

THE METRIC

The four tables for the zero-weighted functionally independent scalars m, n, a, b obtained from the procedure [21] enable us to write down the tetrad vectors in these coordinates,

$$
\begin{aligned}
l^i &= \frac{1}{Q}\left(0, -\frac{4A_1^{\frac{3}{4}}}{A_2^{\frac{1}{2}}}, 0, 0\right) \\
n^i &= \frac{Q}{a}\left(FA_1^{\frac{1}{4}}, \tilde{L}, \frac{A_1^{\frac{5}{4}}}{A_2^{\frac{1}{2}}}n, HA_1^{\frac{1}{4}}\right) \\
m^i &= P\left(0, 0, -2A_1, -4iA_1^{\frac{1}{2}}(b^2 + \frac{\Lambda}{4})\right) \\
\bar{m}^i &= \bar{P}(0, 0, -2A_1, 4iA_1^{\frac{1}{2}}(b^2 + \frac{\Lambda}{4}))
\end{aligned} \quad (1)
$$

F will be a function of only the one coordinate m, so we will write $F = f_2(m)$. The scalar functions S and H are given by

$$S(m,b) = \frac{f_1(m)}{(b^2 + \frac{\Lambda}{4})^{\frac{1}{2}}} \quad (2)$$

$$H(m,b) = (b^2 + \frac{\Lambda}{4})f_3(m) \quad (3)$$

where $f_1(m), f_2(m)$ and $f_3(m)$ are all arbitrary functions of m — subject to restrictions made in the calculations. Also $F \neq 0$ implies that $f_2(m) \neq 0$ while $S \neq 0$ implies that $f_1(m) \neq 0$; on the otherhand $f_3(m)$ is completely arbitrary and may be zero.

$\tilde{L}$ is given in these coordinates by

$$\tilde{L} = \frac{A_1^{\frac{1}{4}}}{A_2^{\frac{1}{2}}(b^2+\Lambda/4)^{\frac{1}{2}}} f_1(m) + \frac{2\sqrt{2}A_1^{\frac{1}{4}}}{A_2^{\frac{1}{2}}} m - \frac{\Lambda A A_1^{\frac{1}{4}}(\Lambda a^2 - 4)}{4A_2^{\frac{3}{2}}} n^2 - \frac{2A_1^{\frac{3}{4}}}{\Lambda A_2^{\frac{1}{2}}} \tag{4}$$

with $A_1 = \Lambda a^2 + 1/2$ and $A_2 = 1 - \Lambda a^2$.

The metric follows immediately from the equation

$$g^{ij} = 2l^{(i}n^{j)} - 2m^{(i}\bar{m}^{j)} \tag{5}$$

$$g^{ij} = \begin{pmatrix} 0 & -\frac{4A_1 f_2(m)}{aA_2^{1/2}} & 0 & 0 \\ -\frac{4A_1 f_2(m)}{aA_2^{1/2}} & -\frac{8A_1^2 L^*}{aA_2} & -\frac{4A_1^2 n}{aA_2} & -\frac{4A_1(b^2+\Lambda/4)f_3(m)}{aA_2^{1/2}} \\ 0 & -\frac{4A_1^2 n}{aA_2} & -4A_1^2 & 0 \\ 0 & -\frac{4A_1(b^2+\Lambda/4)f_3(m)}{aA_2^{1/2}} & 0 & -16A_1(b^2+\Lambda/4)^2 \end{pmatrix} \tag{6}$$

where

$$L^* = \frac{f_1(m)}{A_1(b^2+\Lambda/4)^{1/2}} + \frac{2\sqrt{2}}{A_1} m - \frac{\Lambda(\Lambda a^2-4)}{4A_1 A_2} an^2 - \frac{2}{A_1^{1/2}\Lambda} \tag{7}$$

We must remember that we have assumed that neither $f_1(m)$, nor $f_2(m)$ are zero, and in addition we have assumed at certain stages in our calculations that $S \neq 0 \neq A_1$. So this metric is not necessarily the most general form for this class of spacetimes.

ACKNOWLEDGMENTS

This work was supported by Officina Mathematica of the University of Minho and Matematiska Institutionen of the University of Linköping. M.P.M.R gratefully acknowledges the financial support and hospitality of Matematiska Institutionen, University of Linköping, where part of this research was carried out. S.B.E would like to thank the Department of Mathematics of the University of Minho for their hospitality and financial support during the elaboration of this work. SBE also acknowledges travel support from the G S Magnuson fond, The Rotal Swedish Academy of Sciences.

REFERENCES

1. R. Geroch, A. Held, R. Penrose. *J. Math. Phys.*, **14**, 874 (1973).
2. A. Held *Gen Rel Grav*, **7**, 177, (1974)

3. A. Held *Comm Math Phys*, **44**, 211, (1975)
4. S B Edgar *Gen Rel Grav*, **24**, 1267, (1992)
5. S.B. Edgar and G. Ludwig. *Gen Rel Grav*, **29**, 1309 (1997).
6. S.B. Edgar and G. Ludwig. *Class. Quantum Grav.*, **17**, 1683 (2000).
7. S.B. Edgar and G. Ludwig. *Gen. Rel. Grav*, **32**, 637 (2000).
8. S.B. Edgar and J.A. Vickers. *Class. Quantum Grav.*, **16**, 589 (1999).
9. S.B. Edgar and M. P. Machado Ramos. *Class. Quantum Grav.*, **22**, 791 (2005).
10. M.P. Machado Ramos and J.A.G. Vickers. *Proc.Roy.Soc.London A*, **1940**, 693 (1995).
11. M.P. Machado Ramos and J.A.G. Vickers. *Class. Quantum Grav.*, **13**, 1579 (1996).
12. M.P. Machado Ramos and J.A.G. Vickers. *Class. Quantum Grav.*, **6**, 1589 (1996).
13. Jim E F Skea. *Class. Quantum Grav.*, **14**, 2393 (1997).
14. A. Karlhede. *Gen. Rel. Grav.*, **12**, 693 (1980).
15. S.R. Czapor and R.G. McLenaghan *Gen Rel Grav*, **19**, 623, (1987)
16. J. Carminati and K. T. Vu *Gen Rel Grav*, **33**, 295, (2001)
17. J. Carminati and K. T. Vu *Gen Rel Grav*, **35**, 263, (2003)
18. J.B. Griffiths, P. Doherty and J. Podolsky. *Class. Quantum Grav.*, **21**, 207 (2004).
19. E. Newman and R. Penrose. *J. Math. Phys.*, **3**, 566 (1962).
20. S.B. Edgar and G. Ludwig. *Class. Quantum Grav.*, **34**, 807 (2002).
21. S.B. Edgar and M. P. Machado Ramos. *preprint* (2005).
22. H. Stephani, D. Kramer, M. MacCallum, C. Hoenselaers and E. Herlt (2003). *Exact Solutions of Einstein's Equations.* (Cambridge: Cambridge University Press).
23. M.P. Machado Ramos. *Class. Quantum Grav.*, **15**, 435 (1998).
24. M.P. Machado Ramos and E. Vaz. *Class. Quantum Grav.*, **17**, 5089 (2000).

Gravitational Lensing of stars orbiting Sgr A*

Valerio Bozza[*,**] and Luigi Mancini[†,**]

*Centro Studi e Ricerche "Enrico Fermi", Roma, Italy
†Dipartimento di Fisica "E.R. Caianiello", Universitá di Salerno, Baronissi (SA), Italy.
**Istituto Nazionale di Fisica Nucleare, Sezione di Napoli, Italy.

Abstract. There are many indications that the center of our Galaxy hosts a supermassive black hole, corresponding to the radio source Sgr A*. Thanks to the observations in the near infrared band, it has been possible to determine the orbits of several stars moving in the neighborhood of the Galactic center. General Relativity tells us that the central black hole, acting as a gravitational lens, bends the light rays of these source stars. As a consequence of this fact, a secondary image and two infinite series of relativistic images will be generated. In the framework of Schwarzschild black hole, we have calculated the light curves for the secondary and the first two relativistic images for each star examined. In this way, we have been able to estimate the best times to observe the secondary images, which will happen when the stars approach the minimum distance from the black hole. The detection of such images by future astronomical instruments will provide very useful information about the physical nature of the Milky Way central black hole.

Keywords: black hole physics - gravitational lensing
PACS: Replace this text with PACS numbers

INTRODUCTION

In the last years, the studies of stellar dynamics in the neighborhood of Sgr A* have played a major role in the establishment of the existence of a supermassive black hole in the Galactic center. At the same time, the measurement of the proper motions of stars very close to the central black hole has led to a full three-dimensional reconstruction of the orbits of six stars, named S2, S12, S14, S1, S8, S13 [1]. The system formed by these stars and the black hole at the Galactic center provides a unique lab for gravitational lensing research, in that we have an evident advantage: we precisely know the source location in space as a function of time and thus the geometry of the lensing configuration. Since we already observe the direct image, at any time we can predict where to look for a secondary lensing image and what its apparent magnitude should be. Thanks to our knowledge of their orbits and light curves, we can easily predict the best time to observe these secondary images. Finally, since the distances between the sources and the lens are known, the information extraction from gravitational lensing observations is much easier and unambiguous.

GRAVITATIONAL LENSING AROUND SGR A*

The Schwarzschild radius of the central black hole is $R_{\text{Sch}} = 2GM_{\text{BH}}/c^2 = 0.071$ AU, taking for its mass the value $M_{\text{BH}} = 3.61 \times 10^6$ M$_\odot$ [1]. The distance of the Sun from the Galactic center is currently estimated to $D_{\text{OL}} \simeq 8$ kpc. Among the stars in the neighborhood of Sgr A*, the minimum distance from the central black hole is reached by S14, which touches 110 AU, still 1545 times greater than R_{Sch}. Hence, we can safely consider, from the gravitational lensing point of view, both the observer and the source as being infinitely distant from the lens.

Consider a generic three-point configuration between source, lens and observer. We indicate by D_{OL}, D_{LS} and D_{OS} the distance between observer and lens, lens and source, observer and source respectively. All the distances are measured in terms of the Schwarzschild radius of the black hole. We denote by γ the angle between the line connecting the source to the lens and the optical axis, and by β the angle formed by the line connecting source and observer with the optical axis. The direct image of the source suffers (weak) gravitational lensing only when γ is very close to zero. The secondary image is formed by photons passing behind the black hole. The impact angles for the photons as seen from the observer and the source are θ and $\bar{\theta}$, respectively. The deflection angle is defined as the angle between the asymptotic directions of motion of the photons, before and after the encounter with the black hole. The lens equation is

$$\gamma = \alpha(\theta) - \theta - \bar{\theta}. \tag{1}$$

The impact parameter u is related to the impact angles θ and $\bar{\theta}$ by $u = D_{\text{OL}} \sin\theta = D_{\text{LS}} \sin\bar{\theta}$. This relation can be used to eliminate $\bar{\theta}$ in favor of θ. The position angle of the source γ runs from 0 (perfect alignment) to π (perfect anti-alignment).

Besides the direct and the secondary image, we also have higher order relativistic images, which form closer to the central black hole. The third order image is formed by photons turning around the black hole and reaching the observer from the same side of the direct image. The fourth order image is formed by photons performing a complete loop around the black hole and reaching the observer from the side of the secondary image. They are described by a lens equation of the type

$$2\pi \pm \gamma = \alpha(\theta) - \theta - \bar{\theta}, \tag{2}$$

Notice that the deflection angle $\alpha(\theta)$ is in the range $[\pi, 2\pi]$ for the third order image and $[2\pi, 3\pi]$ for the fourth order image.

The exact deflection angle for a Schwarzschild black hole as a function of the closest approach distance x_0 is given by the integral

$$\alpha(x_0) = I(x_0) - \pi \tag{3}$$

$$I(x_0) = \int_{x_0}^{\infty} \frac{2R_{\text{Sch}}}{x} \frac{dx}{\sqrt{\frac{x^2}{x_0^2}\left(1 - \frac{R_{\text{Sch}}}{x_0}\right) - \left(1 - \frac{R_{\text{Sch}}}{x}\right)}}, \tag{4}$$

solved by Darwin [2] in terms of elliptic integrals. When $x_0 \gg R_{\text{Sch}}$ we recover the weak field formula $\alpha \sim 2R_{\text{Sch}}/x_0$.

The magnification of an image at angle θ is given by the general formula

$$\mu = \frac{D_{\text{OS}}^2}{D_{\text{LS}}^2} \frac{\sin\theta}{\frac{d\gamma}{d\theta}\sin\gamma}. \tag{5}$$

Armed with these tools, once we know the absolute magnitude and the position of a star around Sgr A*, we can give the position and the apparent magnitude of its four most relevant images. Following this path, we have developed a complete analysis of the aforementioned stars as potential sources for gravitational lensing by the central black hole. We found very interesting results, especially for the star S14, as we will show in the incoming section.

Star S14

This star represents the most interesting lensing candidate, since its orbit is almost edge-on ($i = 97.3$ deg), and its eccentricity brings it very close to the black hole. For the remaining five stars, we refer the reader to [5, 6]. Fig. 1a shows the apparent magnitudes in the K-band of its most important gravitational lensing images as functions of time, throughout the next orbital period. From top to bottom, we have the secondary image, the third order and the fourth order one. Their brightness changes very slowly during the whole orbital period, save for a sharp peak corresponding to the periapse epoch, as it is evident from Fig. 1e, which shows D_{LS} as a function of time. This is easily understood as a consequence of the D_{LS}^{-2} dependence in the magnification formula, Eq. (5).

For S14, the best alignment and anti-alignment times are very close to the periapse epoch and are responsible for two subpeaks, which are unresolved in Fig. 1a, but clearly visible in Fig. 1b, which zooms on the peak. First we have the anti-alignment peak, mostly enhancing the third order image, which reaches the brightness of the secondary image. Then we have the alignment peak, which mostly enhances the secondary and fourth order images.

What makes S14 particularly interesting is the high brightness attained by the secondary image. At the best alignment time the secondary image has K=23. This is because we are close to a weak field gravitational lensing situation. The duration of the main peak is about one month and is determined by the velocity of S14 at the periapse epoch, Fig. 1f. The next brightness peak of the secondary image of S14 will occur in 2038, according to the current estimates of the orbital parameters, and the maximal angular distance from the apparent event horizon is about 0.125 mas. The rapid motion of this image would help to distinguish it from background sources of similar brightness, while its distortion should also be an unambiguous sign of its gravitational lensing nature with respect to fast moving stars around Sgr A*.

As regards the angular resolution in the K-band, the VLT units can be combined to perform interferometry observations with an equivalent baseline of 200 m and a maximal angular resolution of 2.2 mas. Some space missions performing nulling interferometry (TPF, DARWIN) should be launched in the near future. According to the mission designs, some spacecraft should fly in formation at distances of the order of tens of meters. A futuristic development of such idea might lead to much higher resolutions. The baseline needed for 0.1 mas resolution is of the order of several kilometers. High precision formation flying may be achieved by laser ranging and microthrusters in the wake of what is being studied for LISA, where the distance between the spacecraft is 5 million km. Given the present situation and all the technical advances that will soon be exploited in scientific researches, 0.1 mas resolution in the K-band and the detection of secondary gravitational lensing images around Sgr A* seem just the next step after the generation of telescopes currently under study.

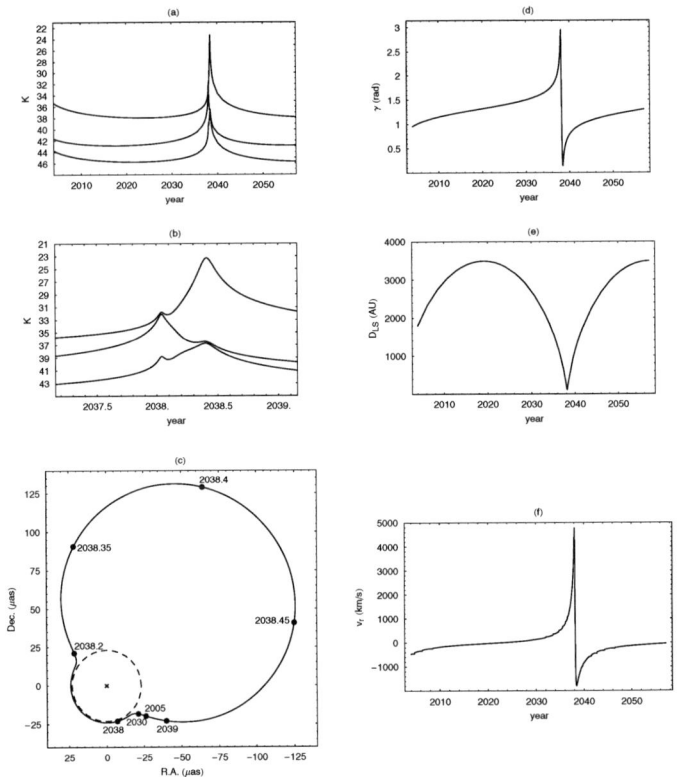

FIGURE 1. (a) Light curves for the gravitational lensing images of S14. From top to bottom, we have the secondary, the third order and fourth order images. (b) Details of the same light curves at the peak epoch. (c) Trajectory of the secondary image around the central black hole (marked by the cross). The dashed circle is its apparent horizon. (d) The angle γ, formed by the line connecting S14 with the central black hole and the optical axis; (e) distance between S14 and Sgr A*; (f) radial velocity of S14.

REFERENCES

1. Eisenhauer, F., Genzel, R., Alexander, T., et al., Astroph. J. **628**, 246 (2005)
2. Darwin, C., Proc. of the Royal Soc. of London **249**, 180 (1959)
3. Bozza, V., Phys. Rev. D **66**, 103001 (2002)
4. Virbhadra, K.S., & Ellis, G.F.R., Phys. Rev. D **62**, 084003 (2000)
5. Bozza, V. & Mancini, L., Astroph. J. **611**, 1045 (2004)
6. Bozza, V. & Mancini, L., Astroph. J. **627**, 790 (2005)

THE PSEUDOSPIN SYMMETRY IN ATOMIC NUCLEI. ANALYSIS OF SOME OF THE EXPLANATIONS PROPOSED

S Marcos[*], M López-Quelle[†], R Niembro[*] and L N Savushkin[**]

[*]*Departamento de Física Moderna, Universidad de Cantabria, E-39005 Santander, Spain*
[†]*Departamento de Física Aplicada, Universidad de Cantabria, E-39005 Santander, Spain*
[**]*Department of Physics, St. Petersburg University for Telecommunications, 191186 St. Petersburg, Russia*

Abstract. We analyse the main explanations of the pseudospin symmetry (PSS) given within the relativistic mean field framework. We make a comparative analysis of the mechanisms responsible for the breaking of the spin and pseudospin symmetries that help to clarify the different nature of these symmetries. We also propose a non-relativistic explanation of the PSS, in which the form of the single-particle central potential and the effect of the spin-orbit interaction play an important role.

Keywords: Pseudospin symmetry
PACS: 24.10.Jv, 21.60.Cs, 21.10.Pc, 24.80.+y.

INTRODUCTION

The PSS was originally associated to the frequently observed quasi-degeneracy of pseudospin doublets in both spherical and deformed nuclei. Recently, it has been considered as a relativistic symmetry, claiming that it is not possible to explain it properly in the non-relativistic framework [1]-[5].

Two single-particle states labelled by "a" and "b" make a pseudospin doublet (PSD) if their radial, orbital, and total angular momentum quantum numbers are related by the equations $n_b = n_a - 1$, $l_b = l_a + 2$, and $j_b = j_a + 1 = l_a + 3/2$, respectively. In the pseudospin formalism, the same pseudo-orbital angular momentum $\tilde{l} = (2j - l)$ is assigned to both states of a PSD. In terms of $\tilde{l}$, the total angular momentum of these states is given by $j = \tilde{l} \mp 1/2$. We shall say a PSD exhibits PSS if its two pseudospin (PS) partners have the same energy. In the same way, we shall say there is spin symmetry (SS) if the two levels of a spin doublet (SD) have the same energy.

The single-particle states in the Dirac-Hartree approximation are obtained from a Dirac equation that can be written as

$$[-i\alpha \cdot \nabla + \beta(M + \Sigma_S) + \Sigma_0]\psi(\vec{r}) = E\psi(\vec{r}), \quad (1)$$

where, $E = M + e$ is the relativistic energy, Σ_S is the scalar self-energy coming from the scalar σ meson and Σ_0 is the vector self-energy coming from the vector ω and ρ mesons and the Coulomb field. For spherical nuclei, the nucleon Dirac spinor $\psi(\vec{r})$ can be written as

$$\psi(\vec{r}) = \frac{1}{r}\begin{pmatrix} iG_a(r)y_{jl}^m(\theta,\phi) \\ F_a(r)y_{j\tilde{l}}^m(\theta,\phi) \end{pmatrix}, \qquad (2)$$

where $\frac{G_a(r)}{r}$ and $\frac{F_a(r)}{r}$ represent the radial parts of its big and small components, respectively. Notice that $\tilde{l} = l \pm 1$ appears in the small component of the spinor.

>From Eqs. (1) and (2), one can get the two following equivalent second order differential equations for the G and F components:

$$G'' - \left[\frac{W'}{W}\left(\frac{G'}{G} + \frac{\kappa}{r}\right) + \frac{l(l+1)}{r^2} + VW\right]G = 0, \qquad (3)$$

$$F'' - \left[\frac{V'}{V}\left(\frac{F'}{F} - \frac{\kappa}{r}\right) + \frac{\tilde{l}(\tilde{l}+1)}{r^2} + VW\right]F = 0, \qquad (4)$$

where, $W = 2M + \Sigma_S - \Sigma_0 + e$, $V = \Sigma_S + \Sigma_0 - e$ and $\kappa = (2j+1)(l-j) = j(j+1) - \tilde{l}(\tilde{l}+1) + 1/4$. The solutions of Eq. (3) with the same number of nodes (n_r) of G and the same value of l form a SD, whereas the solutions of Eq. (4) with the same number of nodes ($\tilde{n}_r$) of F and the same value of $\tilde{l}$ form a PSD. As κ depends on j, the κ terms entering Eqs. (3) and (4) are responsible for the splitting of the spin and pseudospin doublets, i.e., they break the SS and PSS, respectively. However, if W is constant, the factor $W'/W = 0$ (i.e., there is no spin-orbit (SO) interaction) and the SS is restored. Similarly, if V is constant, the factor $V'/V = 0$ and the PSS is restored. Although Eqs. (3) and (4) look very similar each other, there is an essential difference between them, since V becomes zero at some point (r_0) in the nuclear surface. Thus, $V'/V \sim (r-r_0)^{-1}$ for $r \to r_0$ and the $F - \kappa$ term ($\frac{V'}{V}\frac{\kappa}{r}$) in Eq. (4) is singular at $r = r_0$.

At present, the PSS is considered *slightly* broken in nuclei due to some of the following hypotheses: 1) The magnitude of $\Sigma_S + \Sigma_0$ is small [3]-[5], 2) The $F - \kappa$ term is small ($|\frac{V'}{V}\frac{\kappa}{r}| \ll \frac{\tilde{l}(\tilde{l}+1)}{r^2}$) [6]-[9], 3) The compensation of different contributions to the energy in Eq. (4) takes place [10]-[15].

The aim of this contribution is to show that hypotheses 1) and 2) fail to describe the PSS in finite nuclei and to compare the spin and pseudospin symmetries to shed new light for explaining the PSS.

ROLE OF THE QUANTITIES $\Sigma_S + \Sigma_0$ AND $\Sigma_S - \Sigma_0$ IN THE PSEUDOSPIN AND SPIN SYMMETRIES

In the limit $\Sigma_S + \Sigma_0 = 0$, two PS partners, a and b, have the same energy and also F_a and F_b are identical up to a phase [4]. We shall denote this particular type of pseudospin symmetry by PSS*. But the condition $\Sigma_S + \Sigma_0 = 0$ does not allow bound states (except for models with unrealistic values of $\Sigma_S - \Sigma_0$). In real nuclei, $\Sigma_S + \Sigma_0$ is small ($\simeq -50$ MeV) but $\neq 0$. Then, in refs. [3]-[5], it is argued that one can expect $e_a \simeq e_b$ and $F_a \simeq F_b$, i.e., approximate PSS*. It happens that, as the value of $\tilde{n}_r$ of a PSD increases, the PSS* is better satisfied (though F_a and F_b always differ from each other considerably near the singularity point of the $F-\kappa$ term), but PSD's satisfying $e_a \simeq e_b$ with a small value of $\tilde{n}_r$ can be found as well [10]-[15]. However, we have verified that, as $|\Sigma_S + \Sigma_0|$ decreases, the PSS or PSS* are not necessarily improved [14, 15]. On one hand, all PSD's that become degenerate ($e_a = e_b$) for a given quantity $\Sigma_S + \Sigma_0$ split when $|\Sigma_S + \Sigma_0|$ changes, in particular, when it decreases. On the other hand, it is quite obvious that F_a becomes very different from F_b as $\Sigma_S + \Sigma_0$ decreases when e_a or e_b is close to the continuum. Then, we can conclude that neither the PSS nor the PSS* can be based on the smallness of $|\Sigma_S + \Sigma_0|$.

In the limit $\Sigma_S - \Sigma_0 = 0$, two states of a SD have the same energy (i.e., there is SS) and, moreover, their G functions are identical (we denote this special kind of spin symmetry by SS*). In real nuclei, $\Sigma_S - \Sigma_0$ is large and no SS or SS* is expected. In fact, we have large SO splittings, but the G functions of a SD remain very similar (actually, much more similar than the F small components of the PS partners [14]). Moreover, if $|\Sigma_S - \Sigma_0|$ decreases, the SO splittings decrease and the G functions of a SD become more similar, i.e., the SS and SS* are improved. All that means that the relations between the SS or SS* and the magnitude of $\Sigma_S - \Sigma_0$ is quite different to the relation between the PSS or PSS* and the magnitude of $\Sigma_S + \Sigma_0$, due to the different behaviour of the $G - \kappa$ and $F - \kappa$ terms in Eqs. (3) and (4), respectively.

ROLE OF THE κ TERMS IN THE EQUATIONS FOR THE SMALL F AND LARGE G COMPONENTS OF THE DIRAC SPINOR

As we said above, the $F - \kappa$ term in Eq. (4) breaks the PSS. As $V(r_0) = 0$, the $F - \kappa$ term is large around r_0 and, consequently, F_a and F_b differ mostly around r_0. In fact, the $F - \kappa$ term together with the pseudo-centrifugal barrier (PCB) $\frac{\tilde{l}(\tilde{l}+1)}{r^2}$ control the F functions in the nuclear surface, both terms being essential. Then, the $F - \kappa$ term cannot be considered much smaller than the PCB, in contrast with the conclusions of Refs. [6]-[9].

However, the $F - \kappa$ term is roughly an odd function around r_0. This allows a small PS splitting for F_a and F_b quite different. In fact, exact degeneracy needs $F_a \neq F_b$. Then, we have quasi degenerate PSD's, not because the $F - \kappa$ term is small, but due to the compensation of different contributions coming from terms in Eq. (4) different to the $F - \kappa$ term, the details depending on the Σ_S and Σ_0 self-energies [11]-[14]. We say the PSS is a dynamical symmetry for these reasons.

The $G - \kappa$ term in Eq. (3) breaks the SS. As $W'/W \simeq (\Sigma_S' - \Sigma_0')/2M$, the $G - \kappa$ term can be considered as a "small" relativistic term. In fact, $G_a \simeq G_b$ and, though the SO splittings are large, the $G - \kappa$ term behaves as a perturbative one. We can conclude that the $G - \kappa$ term, which breaks the SS and has been considered always large, is smaller than the $F - \kappa$ term, which breaks the PSS and has been considered as a small term by some groups [6]-[9]. Thus, the PSS in nuclei is not due to the smallness of the $F - \kappa$ term neither. As it is shown in ref. [11] it is due to the partial compensation of the contributions of different terms entering Eq. (4). However, the PSS is favoured by the fact that the $F - \kappa$ term is roughly an odd function around its singularity point r_0 and by the flatness of $\Sigma_S + \Sigma_0$ inside the nucleus.

In relation with Eqs. (3) and (4), we can make the following general remark: as the number of nodes n_r ($\tilde{n}_r$) increases, the contribution of the G'' (F'')-term increases, the effect of the $G - \kappa$ ($F - \kappa$) term, relatively, decreases and the SS* (PSS*) is improved.

A SIMPLE EXPLANATION OF THE PSEUDOSPIN SYMMETRY VALID IN THE RELATIVISTIC AND NON-RELATIVISTIC APPROXIMATIONS

The $F - \kappa$ term that splits the PSD's can be considered as a non-relativistic term because it does not become negligible in Eq. (4) as $M \to \infty$. Then, it should be possible to explain the PSS in the non-relativistic framework. In fact, there exist non-relativistic models giving a good shell structure. To understand how the approximate PSS is generated, one can start from a model with $|\Sigma_S - \Sigma_0| = 0$, making a relativistic description unnecessary, and $\Sigma_S + \Sigma_0$ being a harmonic oscillator potential. This model gives degenerate SD's with $G_a = G_b$ (i.e., SS*) and degenerate PSD with $F_a \neq F_b$ (i.e., PSS). Then, although a more realistic quantity $\Sigma_S + \Sigma_0$ breaks the degeneracy of the PSD's, the inclusion of a realistic SO interaction in the non-relativistic case, or choosing appropriately the quantity $|\Sigma_S - \Sigma_0| \neq 0$ in the relativistic one, partially restores the PSS. Thus, we can say there is a continuous way (*perturbative*) that connects the model satisfying exact PSS (but not PSS*) with realistic models of nuclei.

ACKNOWLEDGMENTS

This work has been supported by the DGESIC grant FIS2005-04033.

REFERENCES

1. Arima A, Harvey M and Shimizu K 1969 *Phys. Lett.* **30B** 517
2. Hecht K T, Adler A 1969 *Nucl. Phys.* **A137** 129
3. Ginocchio J N 1997 *Phys. Rev. Lett.* **78** 436; 1999 *Phys. Rep.* **315** 231; 1999 *J. Phys. G: Nucl. Part. Phys.* **25** 617
4. Ginocchio J M, Madland D G 1998 *Phys. Rev.* C **57** 1167
5. Ginocchio J N and Leviatan A 2001 *Phys. Rev. Lett.* **87** 072502.
6. Sugawara-Tanabe K, Arima A 1998 *Phys. Rev.* C **58** R3065
7. Meng J, Sugawara-Tanabe K, Yamaji S, Ring P and Arima A 1998 *Phys. Rev.* C **58** R628
8. Meng J, Sugawara-Tanabe K, Yamaji S, and Arima A 1999 *Phys. Rev.* C **59** 154
9. Sugawara-Tanabe K, Meng J, Yamaji S and Arima A 1999 *J. Phys. G: Nucl. Part. Phys.* **25** 811
10. Marcos S, Savushkin L N, López-Quelle M and Ring P 2000 *Phys. Rev.* C **62** 054309
11. Marcos S, López-Quelle M, Niembro R, Savushkin L N and Bernardos P 2001 *Phys. Lett.* B **513** 30; 2003 *Eur. Phys. J. A* **17** 173
12. Alberto P, Fiolhais M, Malheiro M, Delfino A and Chiapparini M 2001 *Phys. Rev. Lett.* **86** 5015; 2002 *Phys. Rev.* C **65** 034307; Lisboa R, Malheiro M and Alberto P 2003 *Phys. Rev.* C **67** 054305; Lisboa R, Malheiro M, de Castro A S, Alberto P and Fiolhais M 2004 *Phys. Rev.* C **69** 024319
13. López-Quelle M, Savushkin L N, Marcos S, Bernardos P and Niembro R 2003 *Nucl. Phys. A* **727** 269
14. Marcos S, López-Quelle M, Niembro R and Savushkin L N 2004 *Eur. Phys. J. A* **20** 443
15. López-Quelle M, Marcos S, Savushkin L N and Niembro R 2005 *Recent Res. Devel. Physics* **6** 29

First order perturbations of the Einstein-Straus model

Marc Mars*, Filipe C. Mena† and Raul Vera**

*Facultad de Ciencias, Universidad de Salamanca, Plaza de la Merced, 37008 Salamanca, Spain
†Departamento de Matemática, Universidade do Minho, Gualtar, 4710-078 Braga, Portugal
**School of Mathematical Sciences, Dublin City University, Glasnevin, Dublin 9, Ireland

Abstract. We consider the problem of matching a Schwarzschild spacetime with stationary axially symmetric linear perturbations to a FLRW spacetime with arbitrary linear perturbations. We derive the most general matching conditions using the Hodge decomposition on the sphere and give an example of how our formalism can be used to derive general results.

Keywords: Cosmology, matching, perturbations
PACS: 04.20.-q,04.20.Cv,98.80.-k,04.40.-b

INTRODUCTION

The Einstein-Straus model [3] consists of a vacuum spherical static cavity, described by the Schwarzschild metric, embedded in an expanding dust Friedmann-Lemaître-Robertson-Walker (FLRW) model. The matching between both spacetimes is performed across a timelike hypersurface using the standard matching theory. This model is, however, unstable. This has been shown by Mars [5], who proved the following: Given the matching of an arbitrary static region with an arbitrary FLRW region, then the surface of the static part must be a 2-sphere with its center moving along a geodesic in the 3-space of the FLRW background. Furthermore, for standard matter fields, the interior geometry is exactly spherically symmetric [5].

Now, there are two possibilities to generalise such models and study their robustness: the first would be to take non-spherical exact solutions generalising the FLRW exterior and the second to consider non-spherical perturbations of FLRW. The first possibility has already been considered in [9], who found severe restrictions to the models. Therefore it would be important to construct perturbed models.

Perturbed matching conditions have been applied many times in the past [4, 2, 8] however, a matching perturbation theory in general relativity has only recently been developed in its full generality for first order [1, 6] and second order perturbations [6]. An important aspect of general perturbed matchings is that the matching hypersurface can, a priori, be perturbed with vector fields which are independent from the perturbations on the two background spacetimes. Furthermore, these vector fields do not have to be the same at both sides of the matching surface [6].

In this paper, we summarise recent results concerning first order perturbations of the Einstein-Straus model.

MATCHING THEORY IN BRIEF

Let (M^+, g^+) and (M^-, g^-) be the interior and exterior space-times to be matched with oriented boundaries Ω^+ and Ω^-, respectively. Following the standard theory of spacetime matching, given two embeddings $\Phi^\pm : \Omega \to M^\pm$ we can compute the *matching conditions*, which are the equality of the first and second fundamental forms

$$q_{ij}^+ = q_{ij}^- \quad \text{and} \quad k_{ij}^+ = k_{ij}^- \tag{1}$$

on $\Omega^\pm \equiv \Phi^\pm(\Omega)$. Suppose both spacetimes are perturbed such that $g^\pm = g^{(0)\pm} + g^{(1)\pm}$, and that the hypersurfaces $\Omega^\pm$ are perturbed with the vector fields $\vec{Z}^\pm = Q^\pm \vec{n}^{(0)\pm} + \vec{T}^\pm$, where $\vec{n}^{(0)\pm}$ are unit normals to the unperturbed matching hypersurfaces (with the appropriate orientation) and $\vec{T}^\pm$ are tangent to $\Omega^\pm$. In this case, once the background is matched i.e. $q_{ij}^{(0)+} = q_{ij}^{(0)-}$ and $k_{ij}^{(0)+} = k_{ij}^{(0)-}$, the matching conditions are the equality of the perturbed quantities:

$$q^{(1)\pm} = \Phi^{*\pm}(\mathcal{L}_{\vec{Z}^\pm} g^{(0)\pm}) + \Phi^{*\pm}(g^{(1)\pm}) \tag{2}$$

$$k^{(1)\pm} = \frac{1}{2}\Phi^{*\pm}(\mathcal{L}_{\vec{Z}^\pm}\mathcal{L}_{\vec{n}^{(0)\pm}} g^{(0)\pm} + \mathcal{L}_{\vec{n}^{(1)\pm}} g^{(0)\pm} + \mathcal{L}_{\vec{n}^{(0)\pm}} g^{(1)\pm}), \tag{3}$$

where $\vec{n}^{(1)\pm}$ is defined by $\vec{n}^\pm = \vec{n}^{(0)\pm} + \vec{n}^{(1)\pm}$ and this vector is the unit normal of the perturbed matching hypersurface. These expressions can be rewritten in terms of quantities intrinsic to $\Omega^\pm$ [1, 6]. Under spacetime perturbation gauge transformations the tensors $q^{(1)\pm}$ and $k^{(1)\pm}$ are gauge invariant since they are defined intrinsically on $\Omega^\pm$. This is an important aspect as one is free to choose a perturbation gauge on M^+ different from the one on M^-. Furthermore, the two perturbation vectors $\vec{Z}^\pm$ are also gauge dependent and their gauges may be chosen to be different.

PERTURBED SPACETIMES AND MATCHING SURFACES

For the interior, we start by taking a general stationary and axially symmetry metric given by

$$g^{(0)-} = -e^{2U}(dt + A d\phi)^2 + e^{-2U}\left[e^{2k}(d\rho^2 + dz^2) + \rho^2 d\phi^2\right], \tag{4}$$

and the perturbation tensors up to first order, in the weyl gauge, in the following form

$$K^{(1)-} = -2e^{2U}U^{(1)}dt^2 - 4U^{(1)}Ae^{2U}d\phi dt - 2U^{(1)}e^{2U}A^2 d\phi^2 - 2e^{2U}A^{(1)}dtd\phi \tag{5}$$
$$- 2AA^{(1)}e^{2U}d\phi^2 + 2e^{-2U}e^{2k}\left(-U^{(1)} + k^{(1)}\right)(d\rho^2 + dz^2) - 2e^{-2U}U^{(1)}\rho^2 d\phi^2$$

with U, k, $U^{(1)}$, $A^{(1)}$, $k^{(1)}$, being functions of ρ and z, which satisfy the EFEs. The Schwarzschild background is obtained from (4) by setting

$$U = \frac{1}{2}\log\left(1 - \frac{2m}{r}\right), \quad e^{2k} = \frac{r(r-2m)}{(r-m)^2 - m^2\cos^2\theta}, \quad A = 0,$$

$$\rho = r\sin\theta\sqrt{1-\frac{2m}{r}}, \quad z = (r-m)\cos\theta.$$

For the exterior consider a FLRW dust metric as $g^{(0)+} = a^2(\tau)\left(-d\tau^2 + \gamma_{ij}^{(0)}dx^i dx^j\right)$, where $\gamma_{ij}^{(0)}dx^i dx^j = dR^2 + \Sigma^2(R,\varepsilon)(d\theta^2 + \sin^2\theta d\phi^2)$. We perturb the background metric using scalar, vector and tensor perturbations such that $g^+ = g^{(0)+} + g^{(1)+}$ and

$$g_{00}^{(1)+} = -2a^2\Psi, \quad g_{0i}^{(1)+} = a^2(\partial_i W_0 + \tilde{W}_i), \quad g_{ij}^{(1)+} = a^2(-2\Phi\gamma_{ij} + D_{ij}\chi + \nabla_{(i}Y_{j)} + \Pi_{ij}),$$

where ∇ is the covariant derivative of $\gamma_{ij}^{(0)}$, $D_{ij} := \nabla_i\nabla_j - \frac{1}{3}\gamma_{ij}^{(0)}\nabla^2$ and $\nabla^i Y_i = \nabla^i \tilde{W}_i = \nabla^i \Pi_{ij} = \Pi_i^{\ i} = 0$. The evolution and constraint equations for each mode are well known, see e.g. [10].

We take $\{\lambda, \vartheta, \varphi\}$ as variables intrinsic to $\Omega^\pm$. The matching surface is the perturbed with the vector fields $\vec{Z}^\pm = Z^{0\pm}(\lambda,\vartheta,\varphi)\partial_{t^\pm} + Z^{1\pm}(\lambda,\vartheta,\varphi)\partial_{r^\pm} + Z^{2\pm}(\lambda,\vartheta,\varphi)\partial_\theta + Z^{3\pm}(\lambda,\vartheta,\varphi)\partial_\phi$, where $t^+ = \tau$, $t^- = t$, $r^+ = R$ and $r^- = r$. Using (2) and (3) we can readily compute the perturbed first and second fundamental forms on $\Omega^\pm$. However, they would not take into account the underlying geometrical structure of the problem and would therefore be unsuitable for most applications. It is then convenient to decompose the vectors $Z^{2\pm}\partial_\vartheta + Z^{3\pm}\partial_\varphi$ (which are tangent to the S^2 isometry group orbits) according to its Hodge decomposition. There exist functions $S^\pm(\lambda,\vartheta,\varphi)$ and $V^\pm(\lambda,\vartheta,\varphi)$ so that

$$Z^{2\pm}\partial_\vartheta + Z^{3\pm}\partial_\varphi = \left(\partial_\vartheta S^\pm + \sin\vartheta^{-1}\partial_\varphi V^\pm\right)\partial_\vartheta + \left(\sin\vartheta^{-2}\partial_\varphi S^\pm - \sin\vartheta^{-1}\partial_\vartheta V^\pm\right)\partial_\varphi.$$

We can also write each component of $q_{ij}^\pm$ and $k_{ij}^\pm$ using the Hodge decomposition. For example, we can write $q_{\lambda A}^{(1)\pm} = D_A F^\pm + (\star dG^\pm)_A$, where D_A is the covariant derivative on S^2 (with the metric of unit radius) and $\star$ the corresponding Hodge dual. A straightforward calculation gives

$$F^- \stackrel{\Omega}{=} -\left(1 - \frac{2m}{r}\right)\dot{t}Z^{0-} + r^2\frac{\partial S^-}{\partial\lambda} + \frac{\dot{r}Z^{1-}}{1-\frac{2m}{r}}, \quad G^- \stackrel{\Omega}{=} r^2\frac{\partial V^-}{\partial\lambda} + iB^-,$$

$$F^+ \stackrel{\Omega}{=} a^2\left(\frac{\partial S^+}{\partial\lambda}\Sigma^2 - Z^{0+} + W_1 + W_0\right), \quad G^+ \stackrel{\Omega}{=} a^2\left(\frac{\partial V^+}{\partial\lambda}\Sigma^2 + W_2\right),$$

where $t(\lambda), r(\lambda)$ define the unperturbed matching hypersurface from the Schwarzschild side, the dot denotes differentiation with respect to λ, the function B^- is defined by $A^{(1)}(1-2m/r) \equiv \sin\theta\partial_\theta B^-$ and W_1, W_2 by $\tilde{W}_\theta d\theta + \tilde{W}_\phi d\phi = dW_1 + \star dW_2$.

MATCHING CONDITIONS AND RESULTS

After deriving the perturbed first and second fundamental forms in terms of scalars, we obtain a set of twelve equations, among which

$$F^- \stackrel{\Omega}{=} F^+ - a^2 F_0(\lambda), \quad G^- \stackrel{\Omega}{=} G^+ - a^2 G_0(\lambda), \tag{6}$$

where the subindex 0 denotes arbitrary functions arising from the kernel of the Hodge decomposition. These equations give, when combined with other equations that we do not display here,

$$V^+ \stackrel{\Omega}{=} \frac{1}{\Sigma^2}\left[R_0 + R_m Y_1^m - U_2 - Y_2\right] + V^-, \qquad (7)$$

$$W_2 = G_0 - \left(\frac{R_0}{d\lambda} + \frac{dR_m}{d\lambda} Y_1^m\right) + \frac{\partial U_2}{\partial \lambda} + \frac{\partial Y_2}{\partial \lambda} + \frac{B^-\Sigma'\Sigma}{(a\Sigma - 2m)}, \qquad (8)$$

where expressions with subindex m are arbitrary functions of λ and Y_1^m are the $l=1$ spherical harmonics (thus $m = 1, 2, 3$). The functions Y_2 and U_2 arise in the Hodge decomposition of Y_i and Π_{ij}, respectively. Since these matching conditions are fully general for the present spacetimes, this formalism can be used to prove strong results. As an example, we were able to show [7]

Proposition *The most general stationary and axially symmetric first order perturbation of a Schwarzschild metric, matching to an exact FLRW metric across any linearly perturbed non-null surface, must be static.*

This result generalises the recent work of [11]. Our formalism has also an important potential for modeling relevant physical phenomena of compact objects on cosmological backgrounds such as gravitational wave emission, oscillations and collapse.

ACKNOWLEDGMENTS

We thank MEC (Spain) and CRUP (Portugal) for grant E-113/04. FCM thanks FCT (Portugal) for grant SFRH/BPD/12137/2003 and CMAT (Univ. Minho) for support. MM acknowleges finantial support from MEC under the project BFME2003-02121. RV acknowledges the IRCSET (Ireland) fellowship PD/2002/108.

REFERENCES

1. Battye R A & Carter B, *Phys. Lett. A* **357** (1995) 29
2. Chamorro A, *Gen. Rel. Grav.* **20** (1988) 1309
3. Einstein A & Straus E G, *Rev. Mod. Phys.* **17** (1945) 120; *ibid* **18** 148
4. Hartle J, *Ap. J.* **150** (1967) 1005
5. Mars M, *Class. Quant. Grav.* **18** (2001) 3645
6. Mars M, *Class. Quant. Grav.* **22** (2005) 3325
7. Mars M, Mena F C & Vera R, *First order rotating perturbations in equilibrium of the Einstein-Straus model*, in preparation
8. Martin-Garcia J M & Gundlach C, *Phys. Rev.* **D64** (2001) 024012
9. Mena F C, Tavakol R & Vera R, *Phys. Rev.* **D66** (2002) 044004
10. Noh H & Hwang J, *Phys. Rev. D* **69** (2004) 104011
11. Nolan B & Vera R, *Class. Quant. Grav.* (2005)

Spin Foam Models from the Tetrad Integration

A. Miković

Departamento de Matemática, Universidade Lusófona de Humanidades e Tecnologias,
Av. do Campo Grande 376, 1749-024 Lisbon, Portugal

Abstract. We describe a class of spin foam models of four-dimensional quantum gravity which is based on the integration of the tetrad one-forms in the path integral for the Palatini action of General Relativity. In the Euclidian gravity case this class of models can be understood as a modification of the Barrett-Crane spin foam model. Fermionic matter can be coupled by using the path integral with sources for the tetrads and the spin connection, and the corresponding state sum is based on a spin foam where both the edges and the faces are colored independently with the irreducible representations of the spacetime rotations group.

The approach of defining a quantum theory of gravity by using a path integral quantization has been revitalized by the appearance of the idea of spin foams [1]. A spin foam model can be described as a lattice gauge theory for a BF theory, and although a BF theory is a topological theory, the Palatini action of General Relativity (GR) can be represented as a constrained BF theory, where the two-form B is a wedge product of the spacetime tetrade one-forms. This then leads to the idea that the GR path integral could be defined as a modification of the path integral for a topological theory. This was the approach used for the construction of the Barret-Crane (BC) models [2, 3], which culminated when a finite partition function for GR was constructed for any non-degenerate triangulation of the spacetime manifold [4]. However, it was soon realized that one can obtain several finite BC models with different convergence properties [5]. This ambiguity is a problem because it is still not clear which one of these has GR as the classical limit. The source of the ambiguity is the fact that the edge amplitudes of the dual two-complex cannot be fixed in the BC quantization procedure, which then leads to many possible models.

Another problem with the BC type models is that it is difficult to couple matter, especially fermions, because matter fields couple to the tetrades and it is often impossible to rewrite the matter actions coupled to gravity as functionals of the B and matter fields only. One can couple matter to BC models algebraically [7], but the algebraic constraints are not strong enoguh to determine the exact matter amplitudes.

These problems of the BC approach suggest that one should try to find a spin foam model which is based on the integration of the tetrade fields in the GR path integral. Such an approach should be feasible because the Palatini action is quadratic in the tetrads, so that the path integral over the tetrads is Gaussian.

Let us consider the Palatini action

$$S = \int_M \varepsilon_{abcd} e^a \wedge e^b \wedge R^{cd} = \int_M \langle e^2 R \rangle d^4x \quad , \tag{1}$$

where M is the spacetime manifold, e^a are the tetrad one-forms, $R^{ab} = d\omega^{ab} + \omega^a_c \wedge \omega^{cb}$

is the curvature two-form, ω^{ab} is the spin connection one-form and ε_{abcd} is the totally antisymmetric symbol ($\varepsilon_{0123} = 1$). The corresponding path integral can be rewritten formally as

$$Z = \int \mathcal{D}\omega \, \mathcal{D}e \, e^{i\int_M \langle e^2 R \rangle d^4 x} = \int \mathcal{D}\omega \, (\det R)^{-1/2} \quad , \tag{2}$$

where $(\det R)^{-1/2}$ denotes the result of the integration of the tetrads.

The formal expression (2) suggests that one may try to define Z on a triangulation of M as

$$Z = \int \prod_l dA_l \prod_f (\det F_f)^{-1/2} = \int \prod_l dg_l \prod_f \Delta(g_f) \quad , \tag{3}$$

where $A_l = \int_l \omega$, $g_l = e^{A_l}$, $R_f = F_f = \int_f R$, $\det F = (\varepsilon^{abcd} F_{ab} F_{cd})^2$,

$$g_f = e^{F_f} = \prod_{l \in \partial f} g_l \quad , \quad \Delta(g_f) = (\det F_f)^{-1/2} \quad , \tag{4}$$

and the indices l and f stand for the edges and the faces of the dual two-complex of the triangulation. The group function $\Delta(g)$ should be gauge invariant, so that we take

$$\Delta(g) = \sum_\Lambda \Delta(\Lambda) \chi_\Lambda(g) \quad , \tag{5}$$

where $\chi_\Lambda(g)$ is the character for an irreducible representation (irrep) Λ, and the sum is over all irreps of a given category (finite-dimensional or unitary). This then implies that

$$\Delta(\Lambda) = \int_G dg \, \bar{\chi}_\Lambda(g) \Delta(g) \quad . \tag{6}$$

By using the formula

$$\int_G dg \, D^{(\Lambda_1)\beta_1}_{\alpha_1}(g) \cdots D^{(\Lambda_4)\beta_4}_{\alpha_4}(g) = \sum_\iota C^{\Lambda_1 \cdots \Lambda_4(\iota)}_{\alpha_1 \cdots \alpha_4} \left(C^{\Lambda_1 \cdots \Lambda_4(\iota)}_{\beta_1 \cdots \beta_4} \right)^* \quad , \tag{7}$$

where $C^{\Lambda_1 \cdots \Lambda_4(\iota)}_{\alpha_1 \cdots \alpha_4}$ are the components of the intertwiners ι for the tensor product of four irreps and $D^{(\Lambda)}(g)$ are the corresponding representation matrices, we will obtain a state sum of the form

$$Z = \sum_{\Lambda_f, \iota_l} \prod_f \Delta(\Lambda_f) \prod_v A_v(\Lambda_f, \iota_l) \quad , \tag{8}$$

where the vertex amplitude A_v is given by the evaluation of the pentagon spin network, which in the $SU(2)$ case is known as the $15j$ symbol. The state sum (8) is of the same form as in the case of the topological theory given by the BF action; however, the weights we put on the faces are not $\dim \Lambda_f$ but the functions $\Delta(\Lambda_f)$.

These new weights are given by the integrals which are generically divergent, due to $\det F_f = 0$ configurations, so that some kind of regularization must be used. In order to do this, let us write the Lie algebra element F_f as

$$F_f = \vec{E}_f \cdot \vec{K} + \vec{B}_f \cdot \vec{J} \quad , \tag{9}$$

where $\vec{K}$ are the boost generators, while $\vec{J}$ are the spatial rotations generators. The $so(4)$ Lie algebra is a direct sum of two $so(3)$ algebras, and a basis of this decomposition is given by

$$\vec{\mathcal{J}}_{\pm} = \frac{1}{2}(\vec{J} \pm \xi \vec{K}) \quad , \tag{10}$$

where $\xi = 1$ in the Euclidian case and $\xi = i$ in the Minkowski case. From (10) it follows that

$$\det F = (\vec{E} \cdot \vec{B})^2 = \frac{\xi^2}{16}((\vec{E}_+)^2 - (\vec{E}_-)^2) \quad , \tag{11}$$

where $\vec{E}_{\pm} = \vec{B} \pm \frac{1}{\xi}\vec{E}$. Note that $\vec{E}_{\pm}$ are real in the Euclidian case, while in the Minkowski case are complex conjugates ($\vec{E}_+ = \vec{E}_-$). The group function $\Delta(g_f)$ is then given by the expression

$$\Delta(g) = \frac{4}{\xi}((\vec{E}_+)^2 - (\vec{E}_-)^2)^{-1} \quad , \quad g = e^{\vec{E}_+ \cdot \vec{\mathcal{J}}_+ + \vec{E}_- \cdot \vec{\mathcal{J}}_-} = g_+ g_- \quad . \tag{12}$$

Since (12) has a structure of the relativistic momentum square, and the integral (6) is an essentially a Fourier transform, one can use the $i\varepsilon$ regularisation from QFT [6], so that one obtains

$$\Delta(j,l) = -\frac{1}{2\xi}[2\theta(l-j) - \theta(l-j+1) - \theta(l-j-1)] \quad , \tag{13}$$

where j and l are the $SU(2)$ spins. The formula (13) implies that the non-zero coeficients are the ones with $l - j = 0$ or $l - j = \pm 1$, so that one obtains a weight which is concentrated around the simple irreps (j, j). Hence the model can be considered as a generalization of the Barrett-Crane model.

Including matter and the cosmological constant term requires the evaluation of the path integral

$$Z = \int \mathcal{D}e \mathcal{D}\omega \mathcal{D}\psi \exp\left(i \int_M \left(\langle e^2 R\rangle + \lambda \langle e^4\rangle\right) d^4x + iS_m[\psi, e, \omega]\right) \quad , \tag{14}$$

where λ is the cosmological constant, $S_m = \int_M d^4x \mathcal{L}_m$ and $\mathcal{L}_m$ is a function of the tetrads, spin connection, matter fields ψ and their derivatives. Note that in the case of spin-half fermions $\mathcal{L}_m$ is a polynomial in e and ω given by

$$S_m = \int_M \varepsilon_{abcd} e^a \wedge e^b \wedge e^c \wedge \bar{\psi}\left(\gamma^d\left(d + \frac{1}{2}\omega_{rs}\gamma^r\gamma^s\right) + me^d\right)\psi \quad , \tag{15}$$

where γ^a are the Dirac gamma matrices and m is the fermion mass. Hence the path integral (14) can be evaluated at least perturbetively.

When $\mathcal{L}_m$ is a polynomial of the fields and their derivatives, the path integral (14) can be evaluated perturbatively via

$$Z = \lim_{j,J,\chi \to 0} e^{i\lambda \int_M \langle (\delta/\delta j)^4\rangle d^4x + iS_m[-i\delta/\delta\chi, -i\delta/\delta j, -i\delta/\delta J]} Z_0[j, J, \chi] \quad , \tag{16}$$

525

where
$$Z_0[J,j,\chi] = \int \mathcal{D}e\,\mathcal{D}\omega\,\mathcal{D}\psi\, e^{i\int_M(\langle e^2 R\rangle + J_{ab}\omega^{ab} + j_a e^a + \chi^\alpha \psi_\alpha)d^4x} \tag{17}$$

is the generating functional. Z_0 is essentially the gravitational path integral with the sources since the matter integration in (17) gives a delta function $\delta(\chi)$. The tetrade path integral is Gaussian, so that we need to define

$$Z_0[J,j] = \int \mathcal{D}\omega\, e^{i\int_M d^4x\, J_{ab}\omega^{ab}} (\det R)^{-1/2} e^{-i\int_M d^4x \int_M d^4y \langle j(x) R^{-1}(x,y) j(y)\rangle/4} \tag{18}$$

This expression can be defined on a triangulation of M along the lines of the $J = j = 0$ case. However, when the sources are present a more intricate state sum will appear. Guided by the expression (18) we will define Z_0 as

$$Z_0(J,j) = \int \prod_l dg_l\, \mu(g_l, J_l) \prod_f \Delta(g_f, j_\varepsilon) \quad, \tag{19}$$

where
$$\mu(g_l, J_l) = e^{iTr(\omega_l J_l)} \quad, \quad \Delta(g_f, j_\varepsilon) = \Delta(g_f) e^{-i\langle j_\varepsilon F_f^{-1} j_{\tilde\varepsilon}\rangle/4} \quad, \tag{20}$$

and the subscripts ε and $\tilde\varepsilon$ denote two edges of the triangle dual to the face f.

The functions μ and Δ can be expanded as

$$\mu(g_l, J_l) = \sum_{\Lambda_l} \mu(\Lambda_l, J_l) D^{(\Lambda_l)}(g_l) \quad, \quad \Delta(g_f, j_\varepsilon) = \sum_{\Lambda_f} \Delta(\Lambda_f, j_\varepsilon) \chi_{\Lambda_f}(g_f) \quad. \tag{21}$$

The group integrations in (19) can be performed by using the analog of the formula (7) for the tensor product of five irreps. One then obtains a state sum

$$Z_0(J,j) = \sum_{\Lambda_f, \Lambda_l, l_l} \prod_f \Delta(\Lambda_f, j_\varepsilon) \prod_l \mu(\Lambda_l, J_l) \prod_v A_v(\Lambda_f, \Lambda_l, l_l) \quad. \tag{22}$$

This is a novel spin foam state sum, because it involves a dual 2-complex whose edges and faces are independently colored with the irreps of the group. $Z_0(J,j)$ can be understood as an amplitude for a Faynman diagram given by a five-valent graph whose edges carry the irreps Λ_l and the loops carry the irreps Λ_f. The edges have propagators $\mu(\Lambda_l, J_l)$ and each loop carries a weight $\Delta(\Lambda_f, j_\varepsilon)$, while the vertex amplitudes A_v are given by the evaluation of the pentagon spin network with five external edges, where the internal edges carry Λ_f irreps, while the external edges carry Λ_l irreps.

In order to have a physical model, the simplex weights should be defined for the Minkowski case. One way to do this would be to perform an analytic continuation of the Euclidian weights (13) such that $\xi \to i$. In this approach one would work with the same category of representations as in the Euclidian case, i.e. finite-dimensional $SU(2,\mathbf{C}) \times SU(2,\mathbf{C})$ representations, so that the Minkowski weights will be the Euclidian weights times the appropriate factors of i. An alternative approach would be to use the category of unitary $SL(2,\mathbf{C})$ representations. In this case the irreps are infinite-dimensional and can be labeled as (j,ρ) where $2j \in \mathbf{Z}_+$ and $\rho \in \mathbf{R}_+$.

An important next step is to study the convergence of the state sum in the Euclidian and the Minkowski case. Note that if the Euclidian state sum turns out to be divergent, it can be regularized by passing to the quantum group at a root of unity, which is usually done in the case of topological spin foam models. However, since our model is non-topological, using a quantum group regularization is not necessary, so that one can use alternative regularizations, for example a gauge fixing procedure for spin foams [8].

As far as the semiclassical limit is concerned, this is still an unsolved problem for quantum gravity spin foam models. The difficulty is that in the case of non-topological models the partition function state sum is triangulation dependent. This is an obstruction for finding the smooth-manifold limit. One then needs to study triangulations with increasing number of simplexes. Hopefully one could then extract an effective diffeomorphism invariant action. However, a technique must be developed in order to do this.

Another problem is that the formula

$$(\det R)^{-1/2} = \prod_f (\det R_f)^{-1/2} \tag{23}$$

is an approximation. By replacing R with R^* in the Palatini action one obtains a topological gravity theory, while the corresponding state sum has the same weights as in the non-topological case. One can better understand the model by analyzing the $(\det R)^{-1/2}$ operator and the corresponding state sum for the simplest regular triangulation of the four-sphere by six four-simplices[1].

REFERENCES

1. J.C. Baez, *Lect.Notes Phys.* **543**, 25 (2000)
2. J.W. Barrett and L. Crane, *J.Math.Phys.* **39**, 3296 (1998)
3. J.W. Barrett and L. Crane, *Class.Quant.Grav.* **17**, 3101 (2000)
4. L. Crane, A. Perez and C. Rovelli, *Phys.Rev.Lett.* **87**, 181301 (2001).
5. J.C. Baez, J.D. Christensen, T.R. Halford and D.C. Tsang, *Class.Quant.Grav.* **19**, 4627 (2002)
6. A. Miković, Tetrad Spin Foam Model, gr-qc/0504131
7. A. Miković, *Class.Quant.Grav.* **19**, 2335 (2002)
8. L. Freidel and D. Louapre, *Nucl.Phys.* B **662**, 279 (2003); *Class.Quant.Grav.* **21**, 5685 (2004)
9. A.G. Riess et al., *Astrophys. J.*, **607**, 665 (2004).
10. D.N. Spergel et al., *Astrophys. J. Suppl. Ser.*, **148**, 175 (2003).
11. W. Zimdahl, D. Pavón, and L.P. Chimento, *Phys. Lett. B*, **521**, 133 (2001).
12. L.P. Chimento, A.S. Jakubi, D. Pavón, and W. Zimdahl, *Phys. Rev. D*, **67**, 083513 (2003).
13. G. Olivares, F. Atrio-Barandela, and D. Pavón, *Phys. Rev. D*, **71**, 063523 (2005).
14. U. Seljak, and M. Zaldarriaga, *Astrophys. J.*, **469**, 437 (1996). See http://www.cmbfast.org

---

[1] The corresponding dual one-complex is the five-valent hexagon graph consisting of six verticies and 15 edges.

Brownian motion on the relativistic velocity space

E. Minguzzi,†

Department of Applied Mathematics, Florence University, Via S. Marta 3, I-50139 Florence, Italy
INFN, Piazza dei Caprettari 70, I-00186 Roma, Italy

Abstract. The Brownian motion on the relativistic velocity space was introduced in a fully relativistic invariant formalism in a seminal paper by R.M. Dudley. The idea that particles could follow a Brownian motion on the velocity space could prove interesting as the hypothesis is compatible with the equivalence principle. Here we review the results available and in particular show that the Brownian motion can be completely reduced from the frame bundle to the velocity space if a prescription is adopted that the driving Brownian motion Wigner rotates under Lorentz transformation. The asymptotic radial process on the velocity space is also considered showing that it gives rise to a constant outward average acceleration. The possibility of applying this result to the motion of distant galaxies so as to account for the observed acceleration of the universe is commented.

Keywords: Relativistic diffusion, Brownian motion, relativistic velocity space
PACS: 02.50.Ey; 05.40.-a; 05.40.Jc

INTRODUCTION

General relativity is built on the principle of universality of free fall - small test bodies independently of their masses fall along the timelike geodesics of a Lorentzian manifold M. In this work we consider the possibility of a small stochastic perturbation on the motion of the test particle. As we shall see this perturbation may be independent or not of the mass of the test particle. In the former case, although the trajectory of the particle is not determined by the initial conditions, we can say that the probability law of its motion is independent of the mass and hence that there is still universality of free fall in a probabilistic sense.

In order to deal with stochastic perturbations we need the techniques of stochastic calculus. As a first step we have to fix the kind of stochastic perturbation. The first naive idea would be to consider the motion $x(\tau)$ of the particle as a Wiener process (Brownian motion) on the spacetime M i.e. on configuration space. Unfortunately, if $x^\mu(\tau) = B^\mu(\tau)$ then, in Minkowski spacetime, this would imply because of relativistic invariance, $E[x^\mu x^\nu] = D\eta^{\mu\nu}$, where E denotes expectation (we use the spacelike convention $\eta_{00} = -1$). This last equation is meaningless as the covariance matrix is always positive defined. A different approach was conceived by Schay and Dundley [1, 2] who considered a Brownian motion on the relativistic velocity space $\mathbb{H}^3$. The relativistic velocity space $\mathbb{H}^3$ is defined by

$$-(u^0)^2 + (u^1)^2 + (u^2)^2 + (u^3)^2 = -1, \quad u^0 > 0 \tag{1}$$

and represents the space of possible velocities of a particle on spacetime. For simplicity we shall consider only the flat spacetime case where M represents the Minkowski spacetime. In this way the different velocity spaces for each point of spacetime can be identified in a canonical way, $TM = M \times \mathbb{H}^3$. The relativistic velocity space is an hyperboloid on the 4-dimensional tangent space $\mathbb{R}^4$ and hinerits from its metric $\eta_{\mu\nu}du^\mu du^\nu$ a Riemannian metric

$$dl^2 = dr^2 + \sinh^2 r(d\theta^2 + \sin^2\theta d\phi^2) \tag{2}$$

where r, is the rapidity, $\tanh r = v$, and (θ,ϕ) determine the direction of the velocity $\mathbf{v} = \mathbf{u}/u^0$. Since $\mathbb{H}^3$ is a Riemannian space, a Brownian motion (Wiener process) $u(\tau)$ can be defined on it. The idea is to interpret the time parametrization as the proper time (the process is invariant under time translations thus this hypothesis is reasonable) and to define the worldline of the particle as the proper time integral of the Brownian motion on the curved velocity space.

Since the velocity space is a curved Riemannian manifold, a rigorous definition of Brownian motion requires techniques from stochastic analysis on curved manifolds which, unfortunately, are not familiar to physicists. In particular the Brownian motion on the manifold can be expressed as a stochastic process driven by *real* Brownian motions, that is by Brownian motions on $\mathbb{R}^3$. In this work we necessarily focus on some results leaving the details of the various constructions for another work.

ASYMPTOTIC BEHAVIOR OF THE BROWNIAN MOTION

From the physical point of view, the most satisfactory way of constructing the Brownian motion on $\mathbb{H}^3$ is through reduction from the bundle of frames. Each point on the bundle of frames represents an observer (i.e. a orthonormal tetrad associated to it). The stochastic motion on the frame bundle is obtained from successive random boosts, which of course change not only the orientation but also the velocity of the frame. The Lorentz transformation from the frame in stochastic motion to the fixed inertial frame is

$$\Lambda(\tau) = \lim_{N\to+\infty} \prod_{n=1}^{N} e^{\sqrt{\alpha}\Sigma_i K_i \Delta_n B^i} = \lim_{N\to+\infty} [e^{\sqrt{\alpha}\Sigma_i K_i \Delta_1 B^i} \cdots e^{\sqrt{\alpha}\Sigma_i K_i \Delta_N B^i}] \tag{3}$$

with

$$\Delta_n B^i = B^i(\frac{n}{N}\tau) - B^i(\frac{n-1}{N}\tau), \quad i=1,2,3 \tag{4}$$

where K_i are the matrix boost generators, α is the diffusion constant and B^i are three independent Brownian motions satisfying the Itô rule $dB^i dB^j = \delta^{ij} d\tau$.

The previous expression is equivalent to a stochastic differential equations for the tetrad $e_a^\mu(\tau)$ ($e_0^\mu = u^\mu$)

$$du^\mu = \frac{3\alpha}{2} u^\mu d\tau + e_i^\mu \sqrt{\alpha} dB^i, \tag{5}$$

$$de_i^\mu = \frac{\alpha}{2} e_i^\mu d\tau + u^\mu \sqrt{\alpha} dB^i. \tag{6}$$

The worldline $x(\tau)$ is obtained from the integration of the 4-velocity

$$dx^\mu = u^\mu d\tau. \tag{7}$$

We anticipate that its reduction to the velocity space gives a Brownian motion on $\mathbb{H}^3$. This reduction can be obtained from the SDEs above using local coordinates on $\mathbb{H}^3$ and the Itô lemma. However, in local coordinates the SDEs loose their Lorentz invariance at sight. We want here to give a method for reducing the Brownian motion to the velocity space while keeping unaltered this invariance. This is accomplished with the following equation which can be derived from the previous system

$$du^\mu = \frac{3\alpha}{2} u^\mu d\tau + L^\mu_i(u)\sqrt{\alpha}\, d\tilde{B}^i, \tag{8}$$

where $L^\mu_\nu(u)$ is a canonical Lorentz transformation [3, 4] between a frame of 4-velocity u and a frame at rest

$$L^i_k(u) = \delta_{ik} + (\gamma-1)\hat{v}_i\hat{v}_k = \delta_{ik} + \frac{u^i u^k}{1+u^0},$$
$$L^i_0(u) = L^0_i(u) = \gamma v_i = u^i,$$
$$L^0_0(u) = \gamma = u^0,$$

and $\tilde{B}^i$, such that $d\tilde{B}^i = (L^{-1}(u))^i{}_\mu e^\mu_j dB^j$, are suitable real Brownian motions since $(L^{-1}(u))^i{}_\mu e^\mu_j$ is an orthogonal matrix [Theorem 8.4.3][5]. The previous equation depends only on the stochastic process $u^\mu(\tau)$ and not on $e^\mu_i(\tau)$. Moreover it is invariant under Lorentz transformations. It is easy to show, multiplying on the left by a Lorentz matrix Λ, that $u'^\alpha(\tau) = \Lambda^\alpha{}_\mu u^\mu(\tau)$ satisfies the same SDE driven by the Wigner rotated Brownian motions $W(\Lambda,u)^j{}_i \tilde{B}^i$ where $W(\Lambda,u) = L^{-1}(\Lambda u)\Lambda L(u)$ is a Wigner rotation matrix and where we have used the fact that the stochastic rotation of a vectorial Brownian motion is itself a vectorial Brownian motion.

A theorem by Dudley [6] shows that the Brownian motion tends to a asymptotic direction. If the Brownian motion starts ar rest, $u^i(0) = 0$, this asymptotic direction is isotropically distributed over the sphere S^2. Figure 1 summarizes this result. The asymptotic radial process on the velocity space satisfies a SDE of the form [7, 8] ($s = \alpha\tau$)

$$r(s) = r(0) + B(s) + \int_0^s \coth r(s')\, ds', \tag{9}$$

where $r = \tanh^{-1} v$ is the rapidity. It is not difficult to prove the asymptotic behavior

$$\lim_{s\to+\infty} \frac{r(s)}{s} = 1. \tag{10}$$

Now, in a unidirectional motion the proper time differentiation of the rapidity is the proper acceleration, thus the previous equation states that the particle goes towards infinity following a (random) asymptotic direction and with an outward average proper acceleration α.

We suggest that galaxies could perform a Brownian motion on the velocity space, although with a small diffusion constant. This hypothesis would be quite natural if $\alpha = \beta m^\gamma$, $\gamma > 0$, and β is a universal constant, as in this case the diffusion constant

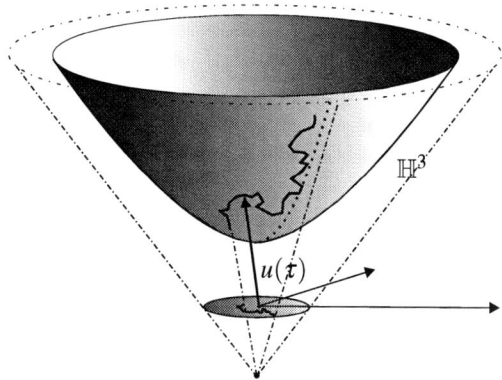

FIGURE 1. The Brownian motion on the Lorentz group reduces to a Brownian motion on the relativistic velocity space $\mathbb{H}^3$. According to Dudley's theorem the velocity tends to a well defined direction represented here by a dotted line. The figure displays also the stereographic projection of the Brownian motion on the sphere S^3.

increases with the mass of the object. The case that preserves the equivalence principle would correspond to $\gamma = 0$. In any case, according to the hypothesis, the galaxies would naturally tend to separate isotropically leading to an expanding universe. Moreover, the asymptotic outward average acceleration would be related with the observed acceleration of the universe that according to this model would be a consequence of the stochastic motion on the velocity space. However, in order to construct a more realistic model, these findings must be first generalized to a curved background.

ACKNOWLEDGMENTS

This work has been partially supported by INFN grant No. 9503/02.

REFERENCES

1. R. M. Dudley, *Ark. Mat.* **6**, 241–268 (1966).
2. R. M. Dudley, *Ark. Mat.* **6**, 575–581 (1967).
3. C. Møller, *The Theory of Relativity*, Clarendon-Press, Oxford, 1962.
4. S. Weinberg, *The Quantum Theory of Fields*, vol. I, Cambridge University Press, Cambridge, 1995.
5. B. Øksendal, *Stochastic differential equations: an introduction with applications*, Springer, Berlin, 2003.
6. R. M. Dudley, *Proc. Nat. Acad. Sci. USA* **70**, 3551–3555 (1973).
7. E. P. Hsu, *Stochastic analysis on manifolds*, AMS, Providence, 2002.
8. E. P. Hsu, A brief introduction to Brownian motion on a Riemannian manifold (2004), lecture notes, Department of Mathematics, Northwestern University.

Quantization of Minimal Strings: a Mechanical Analog

César Gómez*,†, Sergio Montañez*,** and Pedro Resco*,‡

*Instituto de Física Teórica CSIC/UAM, C-XVI Universidad Autónoma, E-28049 Madrid SPAIN
†cesar.gomez@uam.es
**sergio.montannez@uam.es
‡juanpedro.resco@uam.es

Abstract. Recent progress in the study of Liouville field theory opens the possibility to address some problems of quantum gravity using minimal strings as a theoretical laboratory. We present a procedure to embed the minimal string target space into the phase space of an associated mechanical system. By this map quantum effects on the target space correspond to quantum corrections on the mechanical model. This talk is based on [1].

Keywords: Non-critical string theory, Matrix models
PACS: 11.25.Pm, 11.25.-w

INTRODUCTION

One of the goals of string theory is to understand the quantum dynamics and the geometry of target space, and what extent the classical concepts of space and time are valid. We know that the knowledge of of non-perturbative aspects of string theory is crucial for this enterprise and, in fact, this has become a vast topic since the discovery of branes and dualities.

In this talk we deal with non-critical string theories with central charge less than 1, in particular, the so-called minimal string theories. These are important examples of tractable and exactly solvable models of quantum gravity. The fact that they contain many of the features of critical string theory such as D-branes and open-closed string duality makes these theories be interesting laboratories for the study of string theory. Minimal string theories, which have a worldsheet description in terms of minimal conformal field theories coupled to Liouville field theory, were first solved in the early nineties by using a matrix model description [2]. Nevertheless, in the last few years these theories have again received much attention due to later progress in the study of Lioville field theory [3], resulting in a fruitful interplay between the two dual descriptions [4, 5, 6].

Since the central charge of the matter sector of the worldsheet description is less than 1, we do not have a direct target space interpretation. Nevertheless, the information about the target space has been proposed to be encoded into the dynamics of FZZT branes [6]. In this talk we present a new approach to study quantum corrections to the minimal string target geometry, by mapping the FZZT brane data to a quantum mechanical system. By this procedure we associate the classical target space with a curve in the phase space of the mechanical model, in such a way that we can derive the quantum effects of the target from the semiclassical mechanics on phase space as described by the Wigner function. We learn from this analysis the deep interplay between the double scaling limit of matrix models and the resolution of the fold catastrophe.

FZZT BRANES AND THE CLASSICAL TARGET

Let us first recall what is the worldsheet description of $(2,q)$ minimal string theories. The matter sector is given by the corresponding minimal conformal field theory, labeled by its central charge $c_m = 1 - \frac{3(2-q)^2}{q} < 1$, whereas the Liouville sector, the theory for the Liouville field ϕ, is given by

$$S_{LFT} = \frac{1}{4\pi} \int \sqrt{g} \left[\partial \phi \partial \phi + Q\phi R + 4\pi\mu e^{2b\phi} \right] \quad (1)$$

where b is the Louville coupling constant, $Q = \frac{1}{b} + b$ is the background charge and μ is the worldsheet cosmological constant. Since $c_m < 1$, the only thing we can say naively about the target space is that it is given by ϕ. In addition to the fact that there is no time, this space is a strange world because it is not homogeneous. The fact that $Q \neq 0$ implies that the effective string coupling depends exponentially on ϕ, in such a way that at $\phi = -\infty$ strings are free, whereas at $\phi = +\infty$ strings are infinitely strongly coupled. In addition, the μ-term strongly suppresses contributions to the path integral from large positive values of ϕ, giving effectively a wall, which is called Liouville wall.

The point of view we take in this talk is that a better description of target space comes from the moduli space of FZZT branes. This branes are introduced by adding macroscopic loops in the worldsheet with Neumann boundary conditions for ϕ, with a boundary interaction $\oint \mu_B e^{b\phi}$ labeled by a boundary cosmological constant μ_B. This interaction term suppresses contributions to the path integral beyond $\phi = -\frac{1}{b} \log \mu_B$, and it is in this sense for which we interpret μ_B as the target space coordinate. From recent progress in the understanding of LFT with boundaries we can compute the FZZT disk amplitude [4] $\Phi(\mu_B)$ where it can be seen that the parameter μ_B has to be complexified in order to include all types of FZZT branes. If we define now $x = \frac{\mu_B}{\mu}$ and $y = \frac{1}{\mu^{1/2+1/(2b^2)}} \partial_x \Phi$, we find that Φ gives a relation between them

$$2y^2 - 1 - T_q(x) = 0 \quad (2)$$

which defines a Riemann surface. Here T_q are the Chebyshev polynomials of the first kind. This surface has been proposed by [4] to be the classical target space of the theory and has the structure of a double cover of the x-plane. Φ can be interpreted geometrically as the integral $\Phi(x) = \int^x y(x')dx'$ of a one form over a contorn with final point given by μ_B.

FROM THE MATRIX MODEL TO THE MECHANICAL ANALOG

The question at this point is if higher genus corrections modify significantly this target space, and the answer is given by the matrix model description. This description is given by the double scaling limit of a $N \times N$ hermitian one matrix model with partition function

$$Z_m = \frac{1}{vol\,(U(N))} \int dM e^{-\frac{1}{g_m}V(M)} \qquad (3)$$

where the potential $V(M)$ is chosen in such a way that the system has the same critical exponent behavior as the (2,q) minimal string. From the orthonormal polynomials P_n associated with this matrix models we can define a complete set of states $|\psi_n\rangle$ of the Hilbert space of a one particle quantum mechanical system, where the matrix model coupling constant g_m plays the role of the Planck constant $\hbar$. Since it is possible to show that we can compute every matrix model correlator as a mean value over the mixed state $\hat{\rho} = \sum_{n=0}^{N-1} \frac{1}{N}|\psi_n\rangle\langle\psi_n|$, we can say that the mechanical model contains the same information as the matrix model. In particular, the resolvent of the matrix model is equal to

$$R(x) = \frac{1}{N} < \frac{1}{x-M} > = < \frac{1}{x-\hat{q}} > \qquad (4)$$

Since we are considering the mixed state $\hat{\rho}$ to correspond to an equilibrium state, it is natural to impose the states $|\psi_n\rangle$ to be stationary states, that is, eigenstates of a hamiltonian. We can learn something about this hamiltonian from the fact that

$$\psi_N(q) \propto e^{-\frac{V(q)}{2g_m}} < det(q-M) > \qquad (5)$$

On the one hand, at small g_m

$$\psi_N(q) \propto e^{-\frac{V(q)}{2g_m}} e^{<tr \log(q-M)>_o} = e^{-\frac{V(q)}{2g_m}} e^{N \int^q R_o(q')dq'} \qquad (6)$$

On the other hand, at first order in WKB approximation

$$\psi_N(q) \propto e^{\frac{i}{\hbar} \int^q p(q', E_N)dq'} \qquad (7)$$

This leads to a map between the matrix model and a bounded mechanical system at which the resolvent corresponds to a closed curve in phase space

$$-2g_m N R_o(q) + V'(q) = ip(q, E_N) \qquad (8)$$

As an example we can consider the (2,1) minimal string, which is known to correspond to the gaussian matrix model. The associated bounded mechanical system is in this case the harmonic oscillator, where $p(q, E_N) = \pm\sqrt{E_N - q^2}$.

The effect of the double scaling limit is to perform a zoom into the edge of the eigenvalue distribution. By using (8), this corresponds inside the mechanical system to a zoom at the classical turning point, in such a way that the system becomes unbounded, and the parameter that plays the role of $\hbar$ is now the string coupling constant κ. Taking into account that the integral of the resolvent is the matrix model mean value of the macroscopic loop operator, we find that we can map the classical phase space curve of energy $E = 0$ of the mechanical model to the Riemann surface of the minimal string, that is

$$y(q) = ip(q, E = 0) \qquad (9)$$

The usefulness of the map is that it goes beyond the simple classical identification in the sense that we can obtain the non perturbative (all genus) minimal string effects by quantizing the mechanical system. In addition, the all genus FZZT partition function Z_{FZZT}, which corresponds to the matrix model mean value of the exponential of the macroscopic loop operator (the Baker-Akhiezer function of the KP hierarchy), is mapped to the eigenfunction $\psi_{E=0}$ of the mechanical model. This Baker-Akhiezer function is an entire function of x, and therefore we can say that, due to quantum effects, the Riemann surface dissapears. A remarkable fact about the map we have derived is that it lets us to obtain a quantization rule for the number of ZZ branes, which are another type of branes we can introduce in the theory, that we can have in a minimal string background.

THE FOLD CATASTROPHE AND DOUBLE SCALING

The way we can understand the quantum fate of the Riemann surface, that is, the classical target of minimal strings, is by studying the quantum corrections to the Wigner function $f(p,q)$ associated with the classical curve. Since at $\hbar = 0$ $f(p,q)$ becomes a delta function on the curve, we define the target space as the support of f. If we try to compute f by using WKB approximation for the corresponding eigenfunction, and saddle point approximation for the integral, the result is an expression which diverges on the classical curve [7]. This fact is called the fold catastrophe. If we want to obtain the FZZT partition function for example, by integrating the Wigner function

$$|\psi_E|^2 = \int dp f_E(p,q) \tag{10}$$

it is necessary, first of all and in order to perform the zoom at the classical turning point, to resolve the fold catastrophe. For instance, in the (2,1) model the uniform approximation for the Wigner function of the bounded mechanical system gives an expression which, after performing the zoom and the integration, gives

$$\psi_{E=0} = Ai\left(\frac{q+1}{2^{1/3}\hbar^{2/3}}\right) \tag{11}$$

This is the same FZZT partition function we obtain from the double scaling of the matrix model. In the case of $(2, 2k-1)$ models with $k > 1$ the resolution of the WKB singularity at the classical turning point is more complicated due to the fact that the classical turning point is of higher order, but the conclusion is the same: the meaning of the double scaling is the quantum resolution, whatever it is, of the fold catastrophe on the target curve at the classical turning point. In addition, the resolved Wigner function of the unbounded system has generically an oscillatory behavior in the convex part of the phase space plane, and exponential decaying behavior outside. Therefore, we also conclude that the effect of the quantum corrections on the target is not to change the curve, which corresponds to the classical target, into another one, but to spread it into a non commutative phase space.

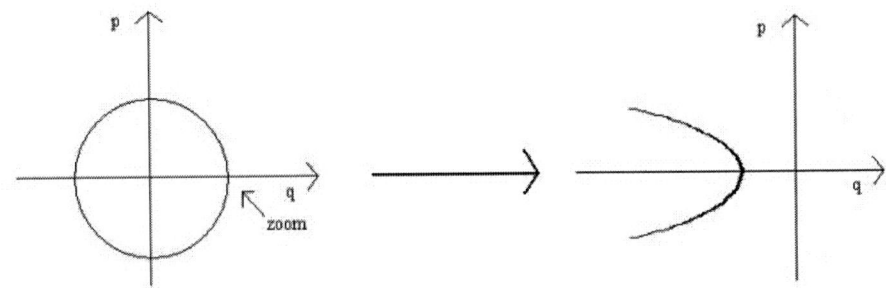

REFERENCES

1. C. Gómez, S. Montañez and P. Resco, "Semi-classical mechanics in phase space: The quantum target of minimal strings," arXiv:hep-th/0506159.
2. P. H. Ginsparg and G. W. Moore, "Lectures on 2-D gravity and 2-D string theory," arXiv:hep-th/9304011. P. Di Francesco, P. H. Ginsparg and J. Zinn-Justin, "2-D Gravity and random matrices," Phys. Rept. **254**, 1 (1995) [arXiv:hep-th/9306153].
3. H. Dorn and H. J. Otto, "Some conclusions for noncritical string theory drawn from two and three point functions in the Liouville sector," arXiv:hep-th/9501019. J. Teschner, "On the Liouville three point function," Phys. Lett. B **363** (1995) 65 [arXiv:hep-th/9507109]. A. B. Zamolodchikov and A. B. Zamolodchikov, "Structure constants and conformal bootstrap in Liouville field theory," Nucl. Phys. B **477** (1996) 577 [arXiv:hep-th/9506136]. V. Fateev, A. B. Zamolodchikov and A. B. Zamolodchikov, "Boundary Liouville field theory. I: Boundary state and boundary two-point function," arXiv:hep-th/0001012. J. Teschner, "Liouville theory revisited," Class. Quant. Grav. **18** (2001) R153 [arXiv:hep-th/0104158]. A. B. Zamolodchikov and A. B. Zamolodchikov, "Liouville field theory on a pseudosphere," arXiv:hep-th/0101152. B. Ponsot and J. Teschner, "Boundary Liouville field theory: Boundary three point function," Nucl. Phys. B **622** (2002) 309 [arXiv:hep-th/0110244].
[4]
4. N. Seiberg and D. Shih, "Branes, rings and matrix models in minimal (super)string theory," JHEP **0402** (2004) 021 [arXiv:hep-th/0312170].
5. D. Gaiotto and L. Rastelli, "A paradigm of open/closed duality: Liouville D-branes and the Kontsevich model," arXiv:hep-th/0312196. D. Kutasov, K. Okuyama, J. w. Park, N. Seiberg and D. Shih, "Annulus amplitudes and ZZ branes in minimal string theory," JHEP **0408** (2004) 026 [arXiv:hep-th/0406030]. J. Ambjorn, S. Arianos, J. A. Gesser and S. Kawamoto, "The geometry of ZZ-branes," Phys. Lett. B **599** (2004) 306 [arXiv:hep-th/0406108]. A. Sato and A. Tsuchiya, "ZZ brane amplitudes from matrix models," JHEP **0502**, 032 (2005) [arXiv:hep-th/0412201].
6. J. Maldacena, G. W. Moore, N. Seiberg and D. Shih, "Exact vs. semiclassical target space of the minimal string," JHEP **0410** (2004) 020 [arXiv:hep-th/0408039].
7. M. V. Berry, "Semi-classical mechanics in phase space: A study of Wigner's function," Phil. Trans. Roy. Soc. Lond. A **287** (1977) 237.

Coordinates and frames from the causal point of view

Juan Antonio Morales Lladosa

Departament d'Astronomia i Astrofísica, Universitat de València, 46100 Burjassot, València, Spain

Abstract. Lorentzian frames may belong to one of the 199 causal classes. Of these numerous causal classes, people are essentially aware only of two of them. Nevertheless, other causal classes are present in some well-known solutions, or present a strong interest in the physical construction of coordinate systems. Here we show the unusual causal classes to which belong so familiar coordinate systems as those of Lemaître, those of Eddington-Finkelstein, or those of Bondi-Sachs. Also the causal classes associated to the Coll light coordinates (four congruences of real geodetic null lines) and to the Coll positioning systems (light signals broadcasted by four clocks) are analyzed. The role that these results play in the comprehension and classification of relativistic coordinate systems is emphasized.

Keywords: space-time frames, causal structure
PACS: 04.20.-q

The main purpose of this short communication is to gain stimulus in the investigation of relativistic coordinate systems from the causal point of view. This issue might be relevant in several situations. For example, to investigate the coordinates which are appropriated to deal with evolution problems throughout horizons. Current 3+1 numerical codes in relativistic hydrodynamics are being implemented using such a coordinates. Also, to consider admissible *cuts* of the space-time others than the very usual "space $\oplus$ time" decomposition. The causal classification of frames may help us to better understand other aspects of the space-time, for instance the "light $\oplus$ light $\oplus$ light $\oplus$ light" decomposition.

Firstly, we remember the causal classification of Lorentzian frames (also presented at the ERE-88 celebrated in Salamanca). Then, we give several examples of causal classes of relativistic coordinates: those associated with the coordinates introduced by Lemaître, by Bondi and Sachs, and by Coll.

In dimension $n = 4$, the causal class of a frame $\{v_1, v_2, v_3, v_4\}$ is defined by a set of 14 characters:

$$\{c_1 c_2 c_3 c_4, C_{12} C_{13} C_{14} C_{23} C_{24} C_{34}, c_1 c_2 c_3 c_4\}$$

c_i being the causal character of the vector v_i, C_{ij} ($i \neq j$) being the causal character of the adjoint 2-plane $\{v_i v_j\}$, and c_i being the causal character of the covectors of the dual coframe $\{\theta^i\}$, $\theta^i(v_j) = \delta^i_j$. The covector θ^i is time-like (resp. space-like) iff the 3-plane generated by $\{v_j\}_{j \neq i}$ is space-like (resp. space-like). This applies for both the Newtonian and the Lorentzian causal structures. In addition, for the later, the covector θ^i is light-like iff the 3-plane generated by $\{v_j\}_{j \neq i}$ is light-like. Elsewhere (see [1] and [2]) we have presented the following result:

__Theorem:__ *i) In the 4-dimensional Newtonian space-time there exist 4, and only 4, causal classes of frames, and ii) In the 4-dimensional relativistic space-time there exist 199, and only 199, causal classes of Lorentzian frames.*

This result provides the causal classification of coordinate systems in Newtonian and in Relativistic physics. A coordinate system $\{x^1, x^2, x^3, x^4\}$ belongs to the causal class $\{c_1c_2c_3c_4, C_{12}C_{13}C_{14}C_{23}C_{24}C_{34}, c_1c_2c_3c_4\}$ if the cobasis $\{dx^1, dx^2, dx^3, dx^4\}$ has causal type $(c_1c_2c_3c_4)$ and has associated four families of coordinate 3-surfaces whose mutual intersections give six families of coordinate 2-surfaces of causal characters $(C_{12}C_{13}C_{14}C_{23}C_{24}C_{34})$ and four congruences of coordinate lines of causal characters $(c_1c_2c_3c_4)$.

As a matter of notation, romanic letters (e, t, l) represent the causal character of vectors (space-like, time-like, light-like, respectively), and capital (E, T, L) and italic letters (*e, t, l*) denote the causal character of 2-planes and covectors, respectively. Accordingly, the four Newtonian causal classes are denoted as:

$\{teee, TTTEEE, teee\}, \{ttee, TTTTTE, eeee\}, \{ttte, TTTTTT, eeee\}, \{tttt, TTTTTT, eeee\}$.

Among the 199 Lorentzian classes, four of them have the same set of causal characters as the Newtonian ones [1]. Next, we consider another examples of relativistic classes.

1.- The three causal classes of Lemaître coordinates. The familiar metric form of the Schwarzschild solution in coordinates $\{t, r, \theta, \phi\}$ is written as

$$ds^2 = -\left(1 - \frac{2m}{r}\right)dt^2 + \frac{1}{1 - \frac{2m}{r}}dr^2 + r^2 d\Omega^2$$

where $r > 2m$ and $d\Omega^2 = d\theta^2 + \sin^2\theta\, d\phi^2$. The coordinate basis $\{\partial_t, \partial_r, \partial_\theta, \partial_\phi\}$ belong to the causal class $\{teee, TTTEEE, teee\}$. Lemaître [3] extended the Schwarzschild solution at the region $0 < r \le 2m$, obtaining a metric form

$$ds^2 = -\left(1 - \frac{2m}{r}\right)dT^2 + 2\varepsilon\sqrt{\frac{2m}{r}}\,dT\,dr + dr^2 + r^2 d\Omega^2 \qquad (\varepsilon = \pm 1)$$

that is regular at $r = 2m$. The causal character of the coordinate lines are given by the sign of the four diagonal elements $g_{\alpha\alpha}$ of the metric $g_{\alpha\beta}$ in this basis. The causal character of the coordinate 2-surfaces is given by the sign of the principal second order minors $g_{\alpha\alpha}g_{\beta\beta} - (g_{\alpha\beta})^2$. And the causal character of the coordinate 3-surfaces is related to the sign of the diagonal elements $g^{\alpha\alpha}$ of the contravariant metric expression $g^{\alpha\beta}$. Consequently, the Lemaître coordinate basis $\{\partial_T, \partial_r, \partial_\theta, \partial_\phi\}$ belong to the causal class $\{teee, TTTEEE, teee\}$ if $r > 2m$, $\{leee, TLLEEE, tlee\}$ if $r = 2m$, or $\{eeee, TEEEEE, ttee\}$ if $r < 2m$.

2.- The thirteen causal classes of Bondi-Sachs coordinates. Any space-time metric may be expressed in the form [4]:

$$g_{\alpha\beta} \equiv g\left(\frac{\partial}{\partial x^\alpha}, \frac{\partial}{\partial x^\beta}\right) = \begin{pmatrix} g_{00} & g_{01} & g_{02} & g_{03} \\ g_{01} & 0 & 0 & 0 \\ g_{02} & 0 & g_{22} & g_{23} \\ g_{03} & 0 & g_{23} & g_{33} \end{pmatrix}$$

For the associated contravariant metric one has $g^{00} = g^{02} = g^{03} = 0$. Therefore, $\{x^0 = \lambda\}$ (with λ a real parameter) are null hypersurfaces. When $g_{22}g_{33} - g_{23}^2 > 0$, the coordinates are called (generalized) Bondi-Sachs coordinates [4] [5], and are usually denoted by $x^0 \equiv u, x^1 \equiv r, x^2 \equiv \theta, x^3 \equiv \phi$. We have that the *the Bondi-Sachs coordinate systems are classified in 13 causal classes:*

$$\{tlee, TTTLLE, leee\}$$

$$\{llee, TTTLLE, leee\} \quad \{elee, TTTLLE, leee\}$$
$$\{llee, TTLLLE, leee\} \quad \{elee, TTLLLE, leee\}$$
$$\{llee, TLLLLE, llee\} \quad \{elee, TLLLLE, leee\}$$

$$\{leee, TLLLEE, llee\} \quad \{elee, TTELLE, leee\}$$
$$\{leee, TLLEEE, llee\} \quad \{elee, TLELLE, leee\}$$
$$\{leee, TLLEEE, tlee\} \quad \{elee, TEELLE, leee\}$$

For example, a coordinate system of the the causal class $\{elee, TEELLE, leee\}$ has associated a family of null coordinate 3-surfaces and three families of time-like 3-surfaces. Their mutual cuts give one family of time-like surfaces, two families of null 2-surfaces and three families of space-like 2-surfaces. The intersections of these surfaces give a congruence of null lines and three congruences of space-like lines.

Note that the Bondi-Sachs coordinates are a generalization of the familiar Eddington-Finkelstein coordinates used in the Schwarzschild space-time. These coordinates belong to the class $\{tlee, TTTLLE, leee\}$ outside the horizon, to the class $\{llee, TLLLLE, llee\}$ at the horizon $r = 2m$, and to the class $\{elee, TEELLE, ltee\}$ inside the horizon. The later is the class $\{leee, TLLEEE, tlee\}$ (as it results when the the first and the second vectors are changed).

3.- The two causal classes of Coll coordinates. Let us consider the classes

$$\{eeee, EEEEE, llll\} \quad \text{and} \quad \{llll, TTTTT, eeee\}$$

A coordinate systems of the first class has associated four families of null 3-surfaces whose mutual cuts give six families of space-like 2-surfaces and four congruences of space-like lines. This class includes the *emission coordinates* of the *Coll positioning systems* [6] [7]. A coordinate system of the second class has associated four families of time-like 3-surfaces whose mutual cuts give six families of time-like 2-surfaces and four congruences of null lines. This class includes the *Coll light coordinates* built from the intersection of four beams of light [8]. We call them the *Coll causal classes*.

	eeee	leee	elee	teee	llee	tlee	ttee	llle	tlle	ttle	lll	tll	ttll	tttl	tttt
eeee	EEEEEE LEEEEE TEEEEE LLEEEE TLEEEE TTEEEE LLLEEE TLLEEE TTLEEE TTTEEE LLLLEE TLLLEE TTLLEE TTTLEE TTTTEE LLLLLE TLLLLE TTLLLE TTTLLE TTTTLE TTTTTE LLLLLL TLLLLL TTLLLL TTTLLL TTTTLL TTTTTL TTTTTT	TTLEEE TTTEEE TTLLEE TTTLEE TTLLLE TTTLLE TTTTLE TTLLLL TTTLLL TTTTLL TTTTTL TTTTTT	TTEELE TEELLE TELLLE TLLLLE TTTLLE TTTTLE	TTTEEE TTTLEE TTTLLE TTTTLE TTTTTE TTTTTT	TTLLLE TTTLLE TTTTLE TTLLLL TTTLLL TTTTLL TTTTTL TTTTTT	TTTTTL TTTTTT	TTTTE TTTTL TTTTTE TTTTTT	TTLLLL TTTLLL TTTTLL TTTTTT	TTTTTL TTTTTL TTTTTT	TTTTTL TTTTTT	TTTTTT	TTTTTT	TTTTTT	TTTTTT	TTTTTT
leee	EEELEE LEELEE TEELEE LELEEE LELLEE TELLEE LLELEE TLELEE TETLEE LLELLE TLELLE TETLLE TTLEEE LLLEEE TLLLEE	TELEE TTEEE TTILLE TTTLEE TTILLE TILLLE TTLLLL TTLLLL TTTLLL TTTTLL	TEELLE TLELLE TLLLLE TILLLE TILLLE	TTTEEE TTLEE	TLLLE										
teee	EEEEEE TEEEEE TLEEEE LLEEEE TLEEEE TTEEEE	TLLEE TTTEE		TTTEE											
llee	EEEEEE LLEEEE LLELEE LLLLEE	ELEEE ELLEE ELLLE LILE			TLLLE										
tlee	EEEEEE LEEEEE ELEEE TLEEEE	TEEE													
ttee	EEEEEE LEEEEE TEEEEE														
llle	EEEEEE LEEEEE LLEEEE														
tlle	EEEEEE LEEEEE														
ttle	EEEEEE														
ttte	EEEEEE														
llll	EEEEEE														
tlll	EEEEEE														
ttll	EEEEEE														
tttl	EEEEEE														
tttt	EEEEEE														

The Coll causal classes {llll, TTTTTT eeee} and {eeee, EEEEEE, llll}. They are dual each other (see references [1] and [2]).

Coll causal classes

The 199 causal classes of relativistic coordinates may be wholly visualized in a table that we call the 199-Table (for details, see [1] and [2]). Here, we reproduce the 199-Table in miniature, but showing the location of the Coll causal classes.

To gain more comprehension of the role that coordinates play in Relativity, the 199 causal classification would be investigated through and through. The causal classes may be associated to admissible *cuts* of the space-time others than the well-known three-space $\oplus$ one-time usual in the at present evolution conception of physics. Other cuts (among the other 198 possible ones) may help us to better understand other aspects of the space-time, and even to wake up our interest for other variations of physical fields than the time-like ones associated to the evolution formalism. A lot of work still remains to be done in this direction.

ACKNOWLEDGMENTS

This work has been supported by the Spanish Ministerio de Educación y Ciencia, MEC-FEDER project AYA2003-08739-C02-02.

REFERENCES

1. J. J. Ferrando and J. A. Morales, "Newtonian and Lorentzian frames", *in Relativistic Coordinates, Reference and Positioning Systems*, Notes School Salamanca 2005.
2. B. Coll and J. A. Morales, Int. Jour. Theor. Phys. **31**, 1045–1062 (1992). See also "Las 199 clases causales de referenciales de espacio-tiempo" in *Actas de los E. R. E-88*, edited by J. Martín and E. Ruíz (Universidad de Salamanca, 1989), 171–180.
3. G. Lemaître, Gen. Rel. and Grav. **29**, 641–680 (1997).
4. R. K. Sachs, Proc. Roy. Soc. **A 270**, 103–126 (1962).
5. S. J. Fletcher and A. W. Lun, Class. Quantum Grav. **20**, 4153–4167 (2003).
6. B. Coll, "Relativistic Positioning Systems" (in these proceedings).
7. J. J. Ferrando, "Coll Positioning Systems: a two-dimensional approach" (in these proceedings).
8. B. Coll, "Coordenadas Luz en Relatividad" in *Trobades Científiques de la Mediterrània: Actes E.R.E-85*, edited by A. Molina (Servei de Publicacions de l'ETSEIB, Barcelona, 1985), 29–39. An English translation of this article entitled *Light coordinates in relativity* is available in http//www.coll.cc/.

Perturbation theory and stability analysis for string-corrected black holes in arbitrary dimensions

F. Moura

Service de Physique Théorique, Orme des Merisiers, CEA/Saclay, F91191 Gif-sur-Yvette Cedex and Centre de Physique Théorique, École Polytéchnique, F91128 Palaiseau Cedex, France

Abstract. We develop the perturbation theory for R^2 string-corrected black hole solutions in d dimensions. After having obtained the master equation and the α'-corrected potential under tensorial perturbations of the metric, we study the stability of the Callan, Myers and Perry solution under these perturbations.

Keywords: Black Holes in String Theory
PACS: 04.70.Dy, 11.25.-w

String theory low energy effective actions have three different types of contributions, with different origins. The classical terms come from the expansion in α' (world-sheet loops). The quantum terms depend on the string coupling constant $g_s = e^\phi$; they can be perturbative (coming from space-time loops) and non-perturbative. In this work we consider only the classical α' corrections, neglecting any kind of string quantum correction. Both the bosonic and the heterotic string theories have corrections already at the first order in α', which are at most quadratic in the Riemann tensor. In these corrections we neglect the Ricci terms, which would only contribute in a higher order in α'; we are only considering an effective action which is perturbative in α'. All these theories also have antisymmetric tensors in their massless spectra, which can always be consistently set to zero. That will be the case in the α'-corrected black hole solution we use in this work. Although these theories lie respectively in 26 or 10 space-time dimensions, we will consider in this article black hole space-times in generic d-dimensions. This way we take, as our effective action in the Einstein frame [1],

$$\frac{1}{2\kappa^2} \int \sqrt{-g} \left[R - \frac{4}{d-2} (\partial^\mu \phi) \partial_\mu \phi + e^{\frac{4}{2-d}\phi} \frac{\lambda}{2} R^{\mu\nu\rho\sigma} R_{\mu\nu\rho\sigma} \right] d^d x + \text{fermion terms}, \quad (1)$$

with $\lambda = \frac{\alpha'}{2}, \frac{\alpha'}{4}, 0$ for bosonic, heterotic and superstrings, respectively.

The corrected bosonic equations of motion for the dilaton and the graviton are, to this order,

$$\nabla^2 \phi - \frac{\lambda}{4} e^{\frac{4}{2-d}\phi} \left(R_{\rho\sigma\lambda\tau} R^{\rho\sigma\lambda\tau} \right) = 0, \quad (2)$$

$$R_{\mu\nu} + \lambda e^{\frac{4}{2-d}\phi} \left(R_{\mu\rho\sigma\tau} R_\nu{}^{\rho\sigma\tau} - \frac{1}{2(d-2)} g_{\mu\nu} R_{\rho\sigma\lambda\tau} R^{\rho\sigma\lambda\tau} \right) = 0. \quad (3)$$

We are interested in studying the behaviour of a string-corrected black hole solution under perturbations. We will be studying these perturbations in generic d spacetime dimensions [2], taking as background metric

$$ds^2 = -f(r)dt^2 + f^{-1}(r)dr^2 + r^2 d\Omega_{d-2}^2 \tag{4}$$

with $d\Omega_{d-2}^2 = \gamma_{ij}(\theta) d\theta^i d\theta^j$, $\gamma_{ij} = g_{ij}/r^2$ being the metric of a $(d-2)$-sphere S^{d-2}.

One can in general consider perturbations to the metric and any other physical field of the system under consideration. General tensors of rank at most 2 on the $(d-2)$-sphere can be uniquely decomposed in their scalar, vectorial and (for $d > 4$) tensorial components. In this work we only consider tensorial (in S^{d-2}) perturbations to the metric, given by $h_{\mu\nu} = \delta g_{\mu\nu}$ (as we will show, we can consistently set the tensorial perturbation to the dilaton to 0). These perturbations are worked out in [3], where it is shown that they can be written as

$$h_{ij} = 2r^2(y^a) H_T(y^a) \mathcal{T}_{ij}(\theta^i), \quad h_{ir} = h_{it} = 0, h_{rr} = h_{tr} = h_{tt} = 0 \tag{5}$$

with $\mathcal{T}_{ij}$ satisfying

$$\left(\gamma^{kl} D_k D_l + k_T\right) \mathcal{T}_{ij} = 0, \quad D^i \mathcal{T}_{ij} = 0, \quad g^{ij} \mathcal{T}_{ij} = 0. \tag{6}$$

D_i is the S^{d-2} covariant derivative; $\mathcal{T}_{ij}$ are the eigentensors of the S^{d-2} laplacian; on the same reference [3], it is also shown that the eigenvalues are given by $-k_T = 2 - l(l+d-3), l = 2,3,4\ldots$

We actually need the variation of the components of the Riemann tensor. Using the components of $h_{\mu\nu}$ given in (5) and the Palatini equation, one gets

$$\begin{aligned}\delta R_{ijkl} &= [(2f-1)H_T + f\partial_r H_T]\left(g_{il}\mathcal{T}_{jk} - g_{ik}\mathcal{T}_{jl} - g_{jl}\mathcal{T}_{ik} + g_{jk}\mathcal{T}_{il}\right) \\ &\quad + r^2 H_T \left(D_i D_l \mathcal{T}_{jk} - D_i D_k \mathcal{T}_{jl} - D_j D_l \mathcal{T}_{ik} + D_j D_k \mathcal{T}_{il}\right)\end{aligned} \tag{7}$$

$$\delta R_{itjt} = \left[-r^2 \partial_t^2 H_T + \frac{1}{2} ff' r^2 \partial_r H_T + ff' r H_T\right] \mathcal{T}_{ij} \tag{8}$$

$$\delta R_{irjr} = \left[-r\frac{f'}{f}H_T - \frac{1}{2}r^2\frac{f'}{f}\partial_r H_T - 2r\partial_r H_T - r^2 \partial_r^2 H_T\right] \mathcal{T}_{ij} \tag{9}$$

$$\delta R_{abcd} = 0. \tag{10}$$

Using the explicit form of the Riemann tensor and the variations (5) and (7-10), one can perturb the field equations (2) and (3). From (2), we are able to show that we can consistently set the dilaton perturbation $\delta\phi = 0$. By perturbing (3) we are able to determine the equation for H_T, which is given by

$$\left(1 - 2\lambda\frac{f'}{r}\right)(\partial_t^2 H_T - f^2 \partial_r^2 H_T)$$
$$+ \left[(2-d)\frac{f^2}{r} - ff' + \lambda\left(4(4-d)\frac{f^2}{r^3}(1-f) + 4\frac{f^2}{r^2}f' + 2\frac{f}{r}f'^2\right)\right]\partial_r H_T$$

$$+ \left[k_T\frac{f}{r^2}\left(1+\frac{4\lambda}{r^2}(1-f)\right) - 2\frac{ff'}{r} + 2(d-2)\frac{f}{r^2} - 2(d-3)\frac{f^2}{r^2}\right.$$
$$+ \left.\lambda\frac{f}{r^2}\left(\frac{2+2d}{r^2} - 4(d-1)\frac{f}{r^2} + 2(d-3)\frac{f^2}{r^2} - \frac{(f'')^2}{d-2}r^2\right)\right]H_T = 0. \quad (11)$$

We now write the equation above in the form of a master equation

$$\frac{\partial^2 \Phi}{\partial r_*^2} - \frac{\partial^2 \Phi}{\partial t^2} =: V_T \Phi. \quad (12)$$

For that, first we write the perturbation equation in terms of the tortoise coordinate r_*, defined by $dr^*/dr = 1/f$. As carefully explained in [5], following the procedure introduced in [4] we derive our master function and potential:

$$\Phi = \frac{H_T}{\sqrt{f}}\exp\left(\int \frac{\frac{f'}{f} + \frac{d-2}{r} + \frac{4}{r^3}(d-4)\lambda(1-f) - \frac{4}{r}\lambda f' - \frac{2}{rf}\lambda f'^2}{2 - \frac{4}{r}\lambda f'}dr\right)$$

$$V_T(f) = \left(\frac{1}{1-2\lambda\frac{f'}{r}}\right)^2\left(1+\frac{4\lambda}{r^2}(1-f)\right)\left[\frac{d-4}{4r^2}\left(1+\frac{4\lambda}{r^2}(1-f)\right) + \frac{2\lambda f'' - 1}{2r^2}\right]$$
$$+ \frac{1}{1-2\lambda\frac{f'}{r}}\left[(k_T+2)\frac{f}{r^2} + 2(d-3)\frac{f(1-f)}{r^2} + \frac{d-8}{2}\frac{ff'}{r} - \frac{\lambda}{d-2}f(f'')^2\right.$$
$$+ \left. 4\lambda(k_T+2)\frac{f(1-f)}{r^4} + 2(d-3)\lambda\frac{f(1-f)^2}{r^4} + 2(d-4)\lambda\frac{f(1-f)f'}{r^3}\right]$$
$$+ \frac{ff'}{r} + (d-4)\frac{f^2}{r^2}. \quad (13)$$

This is the potential for tensor perturbations of any kind of R^2-corrected black holes in d dimensions, in terms of which the perturbation equation (11) is written as a "master equation" like (12).

To study the stability of a solution, we use the "S-deformation approach" first introduced in [3] and developed in [4]. After having obtained the potential $V_T(f)$, we assume that its solutions are of the form $\Phi(r_*,t) = e^{i\omega t}\phi(r_*)$, such that $\partial \Phi/\partial t = i\omega\Phi$. The master equation is then written in the Schrödinger form $A\Phi = \omega^2\Phi$, and a solution to the field equation is then stable if the operator A is positive definite with respect to the following inner product:

$$\langle\phi,A\phi\rangle = \int_{-\infty}^{+\infty}\bar{\phi}(r_*)\left[-\frac{d^2}{dr_*^2}+V\right]\phi(r_*)\,dr_* = \int_{-\infty}^{+\infty}\left[\left|\frac{d\phi}{dr_*}\right|^2 + V|\phi|^2\right]dr_*.$$

Defining $D = \frac{d}{dr_*} - \frac{fH_T}{\Phi}\frac{d}{dr}\left(\frac{\Phi}{H_T}\right)$ and after some algebraic tricks [5], we are left with

$$\langle\phi,A\phi\rangle = \int_{-\infty}^{+\infty}|D\phi|^2\,dr_* + \int_{-\infty}^{+\infty}\frac{Q}{f}|\phi|^2\,dr_*,$$

with

$$\frac{Q}{f} = \frac{1}{1-2\lambda\frac{f'}{r}}\frac{1}{r^2}\left[(k_T+2)\left(1+\frac{4\lambda}{r^2}(1-f)\right)+(2d-6)(1-f)\left(1+\frac{\lambda}{r^2}(1-f)\right)\right.$$
$$\left. - 2rf' - \frac{\lambda}{d-2}(f'')^2 r^2\right] \qquad (14)$$

All that is necessary to guarantee the stability is to check the positivity of $\frac{Q}{f}$.

We considered the R^2-corrected black hole solution of the type of (4) studied in [1]. Its only free parameter is μ, which is related to the classical ADM black hole mass through $m_{cl} = \frac{(d-2)A_{d-2}}{\kappa^2}\mu$, A_n being the area of the unit n−sphere.

For the classical Schwarzschild-Myers-Perry solution, we have $f(r) = 1 - \frac{2\mu}{r^{d-3}}$. In order to introduce the α'-corrections to this solution, we choose a coordinate system in which the position of the horizon, given by $r = (2\mu)^{\frac{1}{d-3}} =: r_H$, is not changed. According to [1] $f(r)$ is given, in this coordinate system, by

$$f(r) = \left(1-\left(\frac{r_H}{r}\right)^{d-3}\right)\left[1-\lambda\frac{(d-3)(d-4)}{2}\frac{r_H^{d-5}}{r^{d-1}}\frac{r^{d-1}-r_H^{d-1}}{r^{d-3}-r_H^{d-3}}\right]. \qquad (15)$$

This solution has as free parameters the inverse string slope λ, the black hole mass parameter μ (or, equivalently, the horizon radius r_H) and the spacetime dimension d. Since λ is a perturbative parameter, we should take it small (say $\lambda \ll 1$), for the potential to make sense. For small values of λ, for each value of d between 5 and 10, and for a wide range of values of μ, we have studied numerically and made plots of $\frac{Q}{f}$ as it is given by (14), and we always found positive values. From this numerical study we conclude that this solution is stable under tensor perturbations for every relevant spacetime dimension, for every value of the black hole mass.

Having obtained the potential for the tensorial perturbations, one can study the black hole emission spectrum, namely the quasi-normal modes, the scattering (reflection and transmission coefficients) and the greybody factors. This work is in progress [5].

ACKNOWLEDGMENTS

This work has been supported by a Chateaubriand scholarship from EGIDE and by fellowship BPD/14064/2003 from Fundação para a Ciência e a Tecnologia, and is part of a joint project with Ricardo Schiappa [5].

REFERENCES

1. C. G. Callan, R. C. Myers and M. J. Perry, *Nucl. Phys.* **B311**, 673 (1989).
2. H. Kodama, A. Ishibashi and O. Seto, *Phys. Rev.* **D62**, 064022 (2000).
3. A. Ishibashi and H. Kodama, *Prog. Theor. Phys.* **110**, 901 (2003).
4. G. Dotti and R. J. Gleiser, *Phys. Rev.* **D72**, 044018 (2005).
5. F. Moura and R. Schiappa, SPHT-T05/51, to appear.

Gravitational Collapse in Higher Curvature Theory

Masato Nozawa* and Hideki Maeda[†]

*Department of Physics, Waseda University, 3-4-1 Okubo, Shinjuku-ku, Tokyo 169-8555, Japan
[†]Advanced Research Institute for Science and Engineering, Waseda University, Shinjuku, Tokyo 169-8555, Japan

Abstract. We find an exact solution in dimensionally continued gravity in arbitrary dimensions which describes the gravitational collapse of a null dust fluid. Considering the situation that a null dust fluid injects into the initially anti-de Sitter spacetime, we show that a naked singularity can be formed. In even dimensions, a massless ingoing null naked singularity emerges. In odd dimensions, meanwhile, a massive timelike naked singularity forms. These naked singularities can be globally naked if the ingoing null dust fluid is switched off at a finite time; the resulting spacetime is static and asymptotically anti-de Sitter spacetime. The curvature strength of the massive timelike naked singularity in odd dimensions is independent of the spacetime dimensions or the power of the mass function. This is a characteristic feature in Lovelock gravity.

Keywords: Gravitational Collapse, Cosmic Censorship, Lovelock Gravity
PACS: 04.20.Dw, 04.20.Jb, 04.50.+h

INTRODUCTION

It has been shown that spacetimes necessarily have a singularity under physically reasonable conditions [1]. Gravitational collapse is one of the presumable scenarios that singularities are formed. In order for spacetimes not to be pathological, Penrose made a celebrated proposal, the *cosmic censorship hypothesis* (CCH) [2, 3], that is, singularities which are formed in a physically reasonable gravitational collapse should not be seen by distant observers. Although there is a long history of research on the final fate of the gravitational collapse, we are far from having achieved consensus on the validity of the CCH, which is one of the most important open problems in general relativity.

Lately it has been of great importance to consider higher-dimensional spacetimes. There exists a natural extension of general relativity in higher dimensions, Lovelock gravity [4]. Higher-order curvature terms come into effect where gravity becomes very strong. One of the present authors has discussed the gravitational collapse of a null dust fluid in Gauss-Bonnet gravity [5]. The purpose of the present work is to analyze how higher-order Lovelock terms modify the final fate of gravitational collapse in comparison to the Gauss-Bonnet or general relativistic cases.

Bañados, Teitelboim and Zanelli have proposed a method which simplifies the analysis in Lovelock gravity [6]. An exact solution in the theory of gravity, called dimensionally continued gravity (DC gravity), representing a static vacuum solution, has been found [6, 7], which is a higher-dimensional generalization of BTZ solution [9]. We extend our investigations on the gravitational collapse of a null dust fluid in Gauss-Bonnet gravity [5] into DC gravity in order to see the effects of higher-order Lovelock terms on the final fate of gravitational collapse. A detailed analysis is reported in [8].

NULL DUST SOLUTION IN DC GRAVITY

Lovelock gravity is a natural extension of general relativity in $D(\geq 3)$-dimensional spacetimes, whose action is given by $I = \int \mathscr{L}_D + I_m$ with

$$\int \mathscr{L}_D \equiv \kappa \sum_{p=0}^{[(D-1)/2]} \alpha_p I_p, \quad I_p \equiv \int_{\mathscr{M}} \varepsilon_{a_1 \cdots a_D} \mathscr{R}^{a_1 a_2} \wedge \cdots \wedge \mathscr{R}^{a_{2p-1} a_{2p}} \wedge e^{a_{2p+1}} \wedge \cdots \wedge e^{a_D}, \tag{1}$$

The $[(D+1)/2]$ real constants α_p in the action have dimensions $[\text{length}]^{-(D-2p)}$. The special combinations of Lovelock coefficients in DC gravity are given by [6]

$$\alpha_p = \begin{cases} \dfrac{1}{D-2p} \binom{n-1}{p} l^{-D+2p}, & \text{for } D = 2n-1, \\[2mm] \binom{n}{p} l^{-D+2p}, & \text{for } D = 2n, \end{cases} \tag{2}$$

where $n \equiv [(D+1)/2]$. We find an exact solution of DC gravity in D-dimensional spacetimes representing a radially ingoing null dust fluid:

$$ds^2 = -f(v,r)dv^2 + 2dvdr + r^2 d\Sigma_{k,D-2}^2 \tag{3}$$

with

$$f(v,r) = \begin{cases} k - (2M(v)/r)^{1/(n-1)} + l^{-2}r^2, & \text{for } D = 2n, \\ k - M(v)^{1/(n-1)} + l^{-2}r^2, & \text{for } D = 2n-1. \end{cases} \tag{4}$$

From the field equations, the energy density of the null dust fluid is given by

$$\rho(v,r) = \frac{1}{\Omega_{D-2} r^{D-2}} \dot{M} \tag{5}$$

both in odd and even dimensions. $\dot{M} \geq 0$ is required due to the weak energy condition. We shall call the solution (3) DC-Vaidya solution. When $M = \text{const.}$, Eq. (3) describes a static solution, which we shall call DC-BTZ solution.

NAKED SINGULARITY FORMATIONS

We consider the situation in which a null dust fluid radially falls into the initial AdS spacetime ($M(v) = 0$) at $v = 0$ in even dimensions at first. We set $M(v) = M_0 v^q$ for simplicity, where $M_0(> 0)$ and $q(\geq 1)$. Then a central singularity appears at $r = 0$ for $v > 0$ both in odd and even dimensions.

We first consider the even dimensional case. We find the radial null geodesics emanating from the singularity with asymptotic form $v \simeq K_1 r$. Along these null geodesics, the energy density for the null dust fluid (5) and the Kretschmann scalar $I_1 = R_{\mu\nu\rho\sigma}R^{\mu\nu\rho\sigma}$ diverge as $r \to 0$ for $1 \leq q < D-1$. Thus, the spacetime represents the formation of a naked singularity.

In order to see whether the singularity is globally naked, we consider the situation in which the null dust fluid is switched off at $v = v_f > 0$. If $v_f \simeq 0$, the singularity can be globally naked. The Penrose diagram of the gravitational collapse is drawn in Fig. 1 (1) for the globally naked singularity formation.

Next we examine the odd dimensional case. We choose the mass function as the power-law form. The singularity $r = 0$ might be naked for $0 \leq v \leq v_{AH}$, where $v_{AH} \equiv M_0^{-q}$. We find null geodesics from the singularity with the asymptotic form $v \simeq v_0 + K_2 r$ for $0 \leq v_0 < v_{AH}$. Along the radial null geodesics from the singularity $v = r = 0$, the Kretschmann scalar diverges for $D \geq 5$. In the meanwhile, it is finite for $D = 3$. However, the energy density of the null dust fluid (5) diverges along the null geodesics for $D \geq 3$. On the other hand, the radial null geodesics from the singularity $r = 0$, $0 < v < v_{AH}$ are singular null geodesics for any $q(\geq 1)$. Along them, the Kretschmann scalar and the energy density of the null dust fluid diverge for $D \geq 5$ and $D \geq 3$, respectively.

In the case of $0 < v_f < v_{AH}$, the singularity is timelike and a Penrose diagram of the gravitational collapse is shown in Fig. 1 (2). In the case of $v_f \geq v_{AH}$, the singularity is timelike for $0 < v < v_{AH}$. For $0 < v_f \leq v_{AH}$, the singularity is always globally naked. In the case of $v_f > v_{AH}$, the singularity can be globally naked if $v_f \simeq v_{AH}$. The possible Penrose diagrams are depicted in Fig. 1 (3) or (4) for $v_f > v_{AH}$ and in Fig. 1 (5) or (6) for $v_f = v_{AH}$.

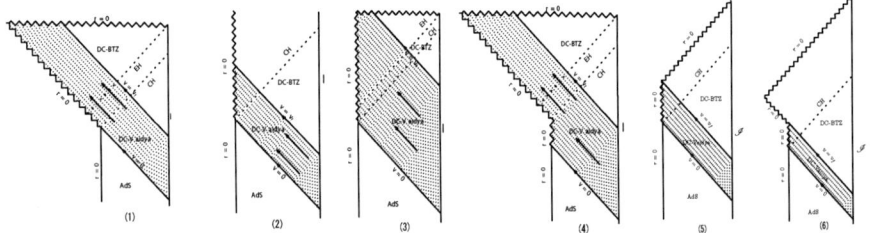

FIGURE 1. Possible Penrose diagrams

STRENGTH OF NAKED SINGULARITIES

In this section, we investigate the strength of naked singularities along the radial null geodesics. We calculate

$$\psi \equiv R_{\mu\nu} k^\mu k^\nu = -\frac{2(D-2)\dot{f}}{rf^2}(k^r)^2, \qquad (6)$$

where $k^\mu = dx^\mu/d\lambda$. In four dimensions, the strong curvature condition and the limiting focusing condition are satisfied along an affinely parametrized geodesic if $\lim_{\lambda \to 0} \lambda^2 \psi > 0$ and $\lim_{\lambda \to 0} \lambda \psi > 0$, respectively [10, 11]. W obtain

$$\lim_{\lambda \to 0} \lambda^2 \psi = \frac{s_1(D-2)}{(1+s_1)^2} > 0 \qquad \text{for } q = 1, D = 2n \qquad (7)$$

$$\lim_{\lambda \to 0} \lambda^{2(1-(q-1)/(D-2))} \psi = s_2(D-2) > 0, \qquad \text{for } q > 1, D = 2n \qquad (8)$$

We next consider the odd dimensional case. For the singularity $r = 0$ and $0 < v < v_{AH}$, we find

$$\lim_{\lambda \to 0} \lambda \psi = s_3(D-2). \tag{9}$$

For the singularity $v = r = 0$ in odd dimensions, we obtain

$$\lim_{\lambda \to 0} \lambda^{2(1-q/(D-1))} \psi = s_4(D-2). \tag{10}$$

It is emphasized that the strength depends both on the spacetime dimensions and the power of the mass function, except for the singularity $r = 0$ and $0 < v < v_{AH}$ in odd dimensions.

CONCLUSIONS AND DISCUSSIONS

We analyzed the $D(\geq 3)$-dimensional gravitational collapse of a null dust fluid in DC gravity. We supposed that (i) the power-law mass function $M(v) = M_0 v^q$, where $M_0 > 0$ and $q \geq 1$, and (ii) the null geodesics obey a power law near the singularity.

We found that globally naked singularities can be formed in DC gravity. Furthermore, the final states of the gravitational collapse differ substantially depending on whether the spacetime dimensions are odd or even. A massless ingoing null naked singularity can appear in the even-dimensional case for $1 \leq q < D-1$. In the odd-dimensional case, on the other hand, a massive timelike naked singularity is formed for any $q(\geq 1)$. These naked singularities can be globally naked. As a result, the formation of a naked singularity cannot be avoided in DC gravity, nor in general relativity. In DC gravity, massive timelike singularities appear in odd dimensions. The formation of a massive timelike singularity in odd dimensions is considered to be a characteristic feature in Lovelock gravity.

We also investigated the strength of naked singularities. The strength of the naked singularity at $v = r = 0$ depends on D and q both in odd and even dimensions. In contrast, around the massive timelike singularity at $r = 0$ and $0 < v < v_{AH}$ in odd dimensions, ψ diverges as λ^{-1}, independent of D or q. Massive timelike singularities diverging as λ^{-1} might be salient features in Lovelock gravity.

REFERENCES

1. S. W. Hawking and G. F. R. Ellis, *The large scale structure of space-time* (Cambridge University Press, 1973)
2. R. Penrose, *Riv. Nuovo Cim.* **1**, 252 (1969)
3. R. Penrose, in *General Relativity, an Einstein Centenary Survey*, edited by S.W. Hawking and W. Israel (Cambridge University Press, Cambridge, England, 1979), p. 581.
4. A. Lovelock, J. Math. Phys. **12** 498 (1971)
5. H. Maeda, gr-qc/0504028
6. M. Bañados, C. Teitelboim and J. Zanelli, Phys. Rev. D. **49** 975 (1994)
7. R-G. Cai and K. Soh, Phys Rev. D. **59** 044013 (1999)
8. M. Nozawa and H. Maeda, gr-qc/0510070
9. M. Bañados, C. Teitelboim and J. Zanelli, Phys. Rev. Lett. **69** 1849 (1992)
10. C. J. S. Clarke and K. Królak, J. Geom. Phys. **2** 127 (1985)
11. K. Królak, J. Math. Phys. **28** 138 (1987)

Constraining dark energy interacting models with WMAP

Germán Olivares*, Fernando Atrio-Barandela† and Diego Pavón*

*Departamento de Física, Universidad Autónoma de Barcelona, Spain
†Departamento de Física Teórica, Universidad de Salamanca, Spain

Abstract. We determine the range of parameter space of an interacting quintessence (IQ) model that best fits the luminosity distance of type Ia supernovae data and the recent WMAP measurements of Cosmic Microwave Background temperature anisotropies. Models in which quintessence decays into dark matter provide a clean explanation for the coincidence problem. We focus on cosmological models of zero spatial curvature. We show that if the dark energy (DE) decays into cold dark matter (CDM) at a rate that brings the ratio of matter to dark energy constant at late times, the supernovae data are not sufficient to constrain the interaction parameter. On the contrary, WMAP data constrain it to be smaller than $c^2 < 10^{-2}$ at the 3σ level. Accurate measurements of the Hubble constant and the dark energy density, independent of the CMB data, would support/disprove this set of models.

INTRODUCTION

Recent observational data suggest the Universe has entered a period of accelerated expansion [1]. The ΛCDM model is the simplest model that best fits the supernovae luminosity distance and CMB anisotropy spectrum data [1, 2]. Though it looks a very serious candidate it raises two main theoretical problems. First, it assumes the energy density of the vacuum but according to quantum field theory it should be larger than observed by 120 orders of magnitude. Secondly, there is the coincidence problem, namely: *why the energy densities of matter and dark energy -scaling so differently with expansion- happen to be of the same order precisely today?* To solve the latter problem models featuring an interaction between quintessence and cold dark matter has been proposed [3, 4].

THE INTERACTING QUINTESSENCE MODEL

To account for the coincidence problem Zimdahl et al. [3] postulated an interaction between cold dark matter and dark energy (a quintessence scalar field).

$$\dot{\rho}_x + 3H(1+w_x)\rho_x = -\delta, \qquad \dot{\rho}_{cdm} + 3H\rho_{cdm} = \delta. \qquad (1)$$

This brings the ratio between their energy densities, $r \equiv \rho_{cdm}/\rho_x$, to a stable constant value at late times. The interaction term, δ, may depend on H, ρ_x and ρ_{cdm}. A simple ansatz is $\delta = 3Hc^2(\rho_x + \rho_{cdm})$, where the constant c^2 is the interaction parameter. From Eqs. (1) the aforesaid ratio is constrained to evolve between two stationary values,

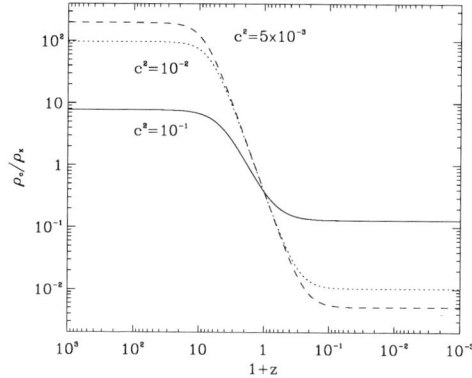

FIGURE 1. Evolution of the energy density ratio $r = \rho_{cdm}/\rho_x$ from an unstable maximum towards a stable minimum (at late times) for different values of c^2. As current value we take $r_0 = 0.42$.

$r_\pm = -\frac{1+w_x}{c^2} \pm \sqrt{\frac{w_x^2}{4c^4} + \frac{w_x}{c^2}}$, where r_+ denotes the unstable value and r_- the stable one -see Ref. [4] for details. As seen from Fig. 1 the stronger the interaction, the smaller ρ_{cdm} in the past. Obviously, such a dissimilar evolution of ρ_{cdm} as compared to the ΛCDM model, gives rise to a different expansion rate and a different evolution of the metric perturbations. The former modifies the luminosity distance of SNIa. The latter should be observed in the CMB anisotropy spectrum. Below we constrain the IQ model with the SNIa data of Riess et al. [1] and the CMB anisotropy data collected by the WMAP satellite [2].

OBSERVATIONAL CONSTRAINTS

To constrain the IQ model with the SNIa data we assume the following priors: a spatially flat model ($\Omega_k = 0$) and $-1 < w_x < -0.6$. As seen from Fig. 2 the lack of data for redshifts $z > 1.8$ does not constrain c^2 [5].

To constrain the model with the WMAP data we adopt a phenomenological point of view and impose that the interaction exists along the streamlines of the fluids (dark energy and CDM), i.e.,

$$u^\nu \nabla_\mu T^\mu_{\nu\,(cdm)} = -u^\mu \nabla_\mu \rho_{cdm} - 3H(\rho_{cdm} + P_{cdm}) = -\delta. \tag{2}$$

The net momentum transfer between CDM and dark energy alongside the accelerated expansion, limits the growth of the CDM perturbations. We implement the perturbations equations (2) together with $\nabla_\mu T^\mu_{\nu\,(x)} = -\nabla_\mu T^\mu_{\nu\,(cdm)}$ in the CMBfast code [6] and assume a flat model with ($\Omega_k = 0$), universes older than 12 Gyrs, BBN, and $-1 < w_x < 0.6$.

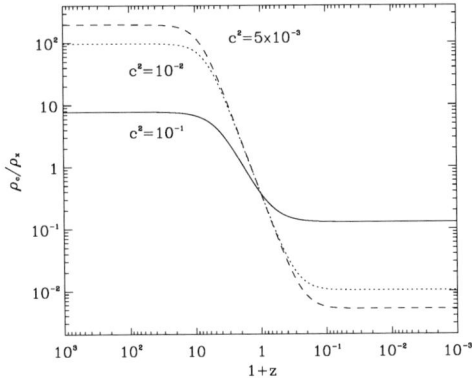

FIGURE 2. Joint confidence intervals at 68%, 95% and 99.9% C.L. of the IQ model fitted to the "gold" sample of SNIa data of Riess et al. [1]. For convenience, the c^2 axis is represented using a logarithmic scale and it has been cut to $c^2 \leq 10^{-4}$, though models with $c^2 = 0$ have been included in the analysis.

TABLE 1. Mean values and 1σ C.L. of the parameters of the IQ and Λ CDM models.

Parameter	IQ model	ΛCDM model *
Interaction parameter, c^2	$< 5 \times 10^{-3\dagger}$	0**
CDM density/critical density, Ω_c	0.4 ± 0.1	0.29 ± 0.07
Baryon density/critical density, Ω_b	0.063 ± 0.012	0.047 ± 0.006
Hubble constant, h_0	0.61 ± 0.06	0.72 ± 0.05
Spectral tilt, n_s	0.96 ± 0.03	0.99 ± 0.04
Amplitude, A	0.9 ± 0.1	0.9 ± 0.1

* WMAP team results [2].
† Upper limit at 1σ C.L.
** This is a prior of the ΛCDM model.

By means of a Monte Carlo Markov chain we reconstruct the surface of the likelihood in the 7D-space parameter defined by the cosmological parameters $c^2, \Omega_x, w_x, \Omega_b, H_0, n_s, A$. Table 1 summarizes our results. Due to the interaction the CDM redshift is smaller than in the ΛCDM model, therefore a higher value of Ω_{cdm} is needed to have enough gravitational potential at matter-radiation decoupling. Moreover, in order not to alter the height of the acoustic peaks the Ω_b value increases. Otherwise, the small mean value of H_0 would suppress them. The likelihood shows clearly that the interaction parameter c^2 is compatible with the WMAP data (see Fig. 3). Nevertheless, non-interacting models are also allowed, unless the amplitude, A, is not a parameter (see Ref. [5]). Thus, c^2 is bounded from above. At 1σ confidence level $c^2 < 5 \times 10^{-3}$. The spectral tilt, n_s, and the amplitude of the perturbations at the horizon size, A, are in perfect agreement with those of the WMAP team [2].

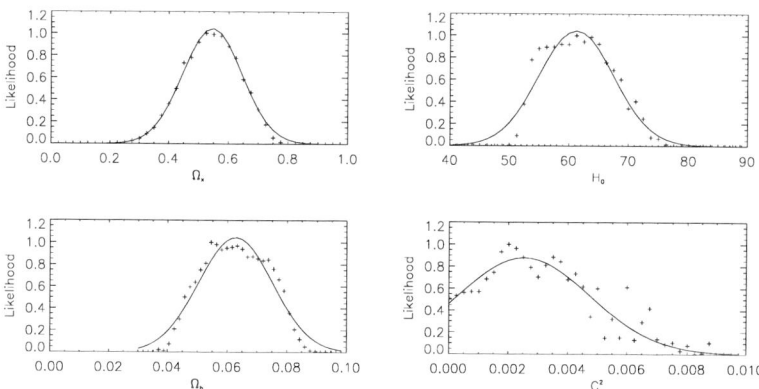

FIGURE 3. Crosses represent the likelihood computed from 50,000 models, marginalized over all parameters but one. The solid line represents the best Gaussian fit.

CONCLUSIONS

We have presented an interacting cosmological model that solves the coincidence problem. Despite the seemingly odd results of Tab. 1 it is in perfect agreement with SNIa data [1] and CMB data [2]. Moreover, the lack of power at the lower multipoles becomes a less serious problem. Indeed, because of the smaller redshift evolution of CDM energy density the gravitational potential evolves slower than in the ΛCDM model. Thus, a smaller Integrated Sachs-Wolfe effect produces less power in the CMB anisotropy spectrum. A next generation of accurate measurements of the Hubble parameter and the dark energy density, independent of the CMB data, are needed to support/disprove this kind of models.

ACKNOWLEDGMENTS

We thank the organizers of the XXVIIIth edition of the "Encuentros Relativistas Españoles" for this opportunity. This research was partly supported by the Spanish Ministry of Science and Technology under Grants BFM2003-06033, BFM2000-1322 and the Junta de Castilla y León (projects SA002/03, SA010C05).

REFERENCES

1. A.G. Riess et al., *Astrophys. J.*, **607**, 665 (2004).
2. D.N. Spergel et al., *Astrophys. J. Suppl. Ser.*, **148**, 175 (2003).
3. W. Zimdahl, D. Pavón, and L.P. Chimento, *Phys. Lett. B*, **521**, 133 (2001).
4. L.P. Chimento, A.S. Jakubi, D. Pavón, and W. Zimdahl, *Phys. Rev. D*, **67**, 083513 (2003).
5. G. Olivares, F. Atrio-Barandela, and D. Pavón, *Phys. Rev. D*, **71**, 063523 (2005).
6. U. Seljak, and M. Zaldarriaga, *Astrophys. J.*, **469**, 437 (1996). See http://www.cmbfast.org

Non-Adiabatic Radiating Collapse in de Sitter Spacetime

N. Özdemir

ITU Faculty of Science and Letters Department of Physics, 34469, Istanbul, Turkey and Feza Gürsey Institute Emek Mah. Rasatane Yolu No: 68, 34684 Çengelköy, Istanbul, Turkey

Abstract. Gravitational collapse of a radiating spherical symmetric object in the presence of the cosmological constant Λ is studied.

Keywords: Non-adiabatic radiating spacetime, Vaidya de Sitter spacetime.
PACS: 04.20.Jb, 04.20.Cv

INTRODUCTION

It is interesting to find realistic solutions to the radiating stars in astrophysics. A spherically symmetric non-rotating stellar object accepts Vaidya spacetime [1] as its exterior during the radiating collapse. The stellar object divides the spacetime into two distinct regions: interior and exterior regions which admit stellar surface as their boundaries. In general, to consider the spacetime as a whole these two distinct regions should match on the boundary surface and they must satisfy the junction conditions. In [2], the general spherically symmetric interior of the Vaidya's radiating metric is studied and junction conditions are given.

In this work, we consider non-adiabatic radiating collapse of a spherically symmetric object in the presence of the cosmological constant Λ. We also give an example of Friedmann-like radiating collapse in de Sitter spacetime and calculate physical parameters such as energy density, pressure, heat flux and total luminosity.

The spacetime exterior to the radiating sphere is described by the outgoing Vaidya metric

$$ds_+^2 = -(1 - \frac{2m(v)}{\hat{r}})dv^2 - 2dvd\hat{r} + \hat{r}^2(d\theta^2 + \sin^2\theta d\phi^2) \tag{1}$$

where $m(v)$ represents the mass of the sphere measured by distant observer and is function of retarded time coordinate v and $_+$ ($_-$) refers to exterior (interior) region entities. The energy momentum tensor related to metric (1) is that of pure radiation. The shear free motion of the fluid in the sphere's interior allows the use of isotropic coordinates in which the line element of the interior is given by

$$ds_-^2 = -A(t,r)^2 dt^2 + B(t,r)^2(dr^2 + r^2 d\theta^2 + r^2 \sin^2\theta d\phi^2). \tag{2}$$

Here A and B are assumed to be positive functions of the radial coordinate r and time coordinate t. We consider the energy-momentum tensor of the stellar object's interior as

$$T^-_{\mu\nu} = (\mu + p)u_\mu u_\nu + pg_{\mu\nu} + q_\mu u_\nu + q_\nu u_\mu \tag{3}$$

where μ is the mass energy density, p is the isotropic pressure, $u^\mu = 1/A\,\delta^\mu_t$ is the fluid's four velocity and q^μ the energy flux vector which can be interpreted as heat flux since there is no particle flux relative to the fluid. Due to the orthogonality condition $u^\mu q_\mu = 0$ and spherical symmetry of the system the heat flux vector has only radial component. The exterior and interior spacetimes match on a spherical hypersurface Σ at $r = const.$ and junction conditions state the equality of the first fundamental and discontinuity of the second fundamental forms

$$[ds^2]_\Sigma = [ds^2_-]_\Sigma = [ds^2_+]_\Sigma, \quad [K_{ij}] - g_{ij}[K] = \tau_{ij}. \tag{4}$$

respectively. Here, $[K_{ij}] = K^+_{ij} - K^-_{ij}$ and $[K] = g^{ij}[K_{ij}]$. In case the surface energy momentum tensor $\tau_{ij} = 0$, in our problem, it reduces to the equality of the extrinsic curvatures
$[K_{ij}] \equiv 0$ or $K^+_{ij} = K^-_{ij}$.

RADIATING VAIDYA DE SITTER SPACETIME

We consider radiating collapsing stellar object in Vaidya de Sitter spacetime and by using junction conditions on the boundary surface the effects of the cosmological constant on the physical parameters are shown. The metric of the Vaidya de Sitter spacetime is

$$ds^2_+ = -\left(1 - \frac{2m(v)}{\hat{r}} - \frac{\Lambda}{3}\hat{r}^2\right) dv^2 - 2\,dv\,d\hat{r} + \hat{r}^2(d\theta^2 + \sin^2\theta\,d\phi^2). \tag{5}$$

Matching of the first fundamental forms and normal vector to the boundary hypersurface can be given as

$$\left(1 - \frac{2m(v)}{\hat{r}^2} - \frac{\Lambda\hat{r}^2}{3} - \frac{2\,d\hat{r}}{dv}\right)_\Sigma = \left(\frac{1}{\dot{v}^2}\right)_\Sigma, \quad n^+_\alpha = \frac{1}{A}(-\dot{\hat{r}},\dot{v},0,0), \tag{6}$$

and the extrinsic curvature K^+_{ij} to Σ can be calculated as

$$K^+_{\tau\tau} = \left(\frac{\ddot{v}}{A\dot{v}} - \frac{\dot{v}\,m}{A\,\hat{r}^2} + \Lambda\left(-\frac{\dot{v}}{3A\hat{r}^2} + \frac{A\hat{r}\dot{v}}{3A + mA\dot{v}^3}\right)\right)_\Sigma, \tag{7}$$

$$K^+_{\theta\theta} = \frac{1}{\sin^2\theta}K^+_{\phi\phi} = \frac{1}{A}\left(\dot{v}\hat{r}\left(1 - \frac{2m}{\hat{r}} - \frac{\Lambda}{3}\hat{r}^2\right) + \hat{r}\dot{\hat{r}}\right)_\Sigma. \tag{8}$$

The line element of the interior spacetime can be chosen in the most general spherically symmetric form which is given by (2) and the energy-momentum tensor for the interior region becomes

$$T_{tt} = \mu + \Lambda, \quad T_{tr} = T_{rt} = -AB^2 q,$$
$$T_{rr} = T_{\theta\theta}/r^2 = T_{\phi\phi}/r^2\sin^2\theta = p - \Lambda \tag{9}$$

where μ is the energy density, p is the isotropic pressure, $u^t = 1/A$ is the fluid's four velocity and $q^r = q$ is the heat flux. The Einstein field equations for the interior spacetime are

$$\underbrace{(\mu + \Lambda)}_{\mu_\Lambda} = 3\frac{\dot{B}^2}{B^2} - \frac{A^2}{B^2}\left(2\frac{B''}{B} - \left(\frac{B'}{B}\right)^2 + \frac{4}{\rho}\frac{B'}{B}\right), \quad q = \frac{2}{B^2}\left(\frac{\dot{B}}{AB}\right)'$$

$$\underbrace{(p - \Lambda)}_{p_\Lambda} = \frac{1}{B^2}\left[\frac{B'^2}{B^2} + 2\frac{A'B'}{AB} + \frac{2}{\rho}\left(\frac{B'}{B} + \frac{A'}{A}\right)\right] - \frac{1}{A^2}\left(2\frac{\ddot{B}}{B} + \frac{\dot{B}^2}{B^2} - 2\frac{\dot{A}\dot{B}}{AB}\right).$$

Now, let us consider junction conditions. In the interior region the surface Σ is the boundary to the material distribution and has the equation $f^-(r,t) = r - r_\Sigma$, where $r_\Sigma = const$. The extrinsic curvature K_{ij}^- are obtained as

$$K_{\tau\tau}^- = -\frac{1}{AB}\frac{\partial A}{\partial r}|_\Sigma, \quad K_{\theta\theta}^- = \frac{1}{\sin^2\theta}K_{\phi\phi}^- = r\frac{\partial(rB)}{\partial r}|_\Sigma \quad (10)$$

and the surface equation for the exterior region (1) is $f^+(\hat{r},v) = \hat{r} - \hat{r}_\Sigma$, with $\hat{r}_\Sigma = const$. The junction conditions lead to relations $\hat{r}_\Sigma(v) = \mathcal{R}$ and

$$\left(2\frac{d\hat{r}}{dv} + 1 - \frac{2m}{\hat{r}}\right)_\Sigma = \left(\frac{A^2}{\dot{v}^2}\right)_\Sigma, \quad n_\alpha^+ = 1/A(\hat{r},\dot{v},0,0). \quad (11)$$

The extrinsic curvature K_{ij}^+ to Σ can be calculated as

$$K_{\tau\tau}^+ = \frac{1}{A}\left(\frac{\ddot{v}}{\dot{v}} - \dot{v}\frac{m}{\hat{r}^2}\right)_\Sigma, \quad K_{\theta\theta}^+ = \frac{1}{\sin^2\theta}K_{\phi\phi}^+ = \frac{1}{A}\left(\dot{v}\hat{r}(1 - \frac{2m}{\hat{r}}) + \hat{r}\dot{\hat{r}}\right)_\Sigma. \quad (12)$$

By using the matching conditions and arranging equations we get the physical quantities as follows:

$$Br|_\Sigma = \hat{r}|_\Sigma, \quad \left(-\frac{1}{AB}\frac{\partial A}{\partial r}\right)_\Sigma = \frac{1}{A}\left(\frac{\ddot{v}}{\dot{v}} - \frac{\dot{v}m}{\hat{r}^2} + \Lambda(\frac{r\dot{v}}{3} + \frac{2\dot{v}^3m}{3A^2})\right)_\Sigma,$$

$$m(v) = -\frac{Br}{2}\left(\frac{r^2 B'}{B^2} - \frac{r^2 \dot{B}}{A^2} + \frac{2rB'}{B} - \frac{\Lambda}{3}\frac{r^2 B^2}{3}\right),$$

$$q|_\Sigma = p_\Lambda B + \frac{2\Lambda}{B} + \frac{2\Lambda}{rB^2}\left(1 + \frac{r}{B}\frac{\partial B}{\partial r} + \frac{r}{A}\frac{\partial B}{\partial t}\right)^{-2}\mathcal{F}|_\Sigma$$

where

$$\mathcal{F} = \left(-\frac{1}{3r^2 B^2} + \frac{r^3 B \dot{B}^2}{2A^2} - \frac{\Lambda}{6}r^3 B^3 - r^2 B' - \frac{r^3 B'}{2B}\right).$$

As $\Lambda \to 0$, we recover [3]'s result.

Furthermore, physically reasonable solutions should satisfy the conditions: Energy density and pressure must decrease outward, heat flux must increase outward, and rate of change of the physical radius must decrease for collapsing configuration.

EXAMPLE

A Friedman-like non-adiabatic radiating collapse in de Sitter spacetime is examined as a realistic example of these type of spacetimes. We generalize the example of [4] to de Sitter spacetime and calculate the physical quantities in this example. By taking $A = 1$ in (2) we get Friedman-like spacetime

$$B = \frac{M}{2b}\left(\frac{1-b^2 h(t)}{1-r^2 h(t)}\right) u(t)^2 \qquad (13)$$

where $u(t) = (6t/m)^{1/3}$, $h(t) = a e^{u(t)}$, a, b and M are constants. The physical quantities are

$$\mu = -\Lambda + \frac{12}{u^4 M^2}\left(\frac{2}{u} - \frac{(b^2-r^2)h}{(1-b^2 h)(1-r^2 h)}\right)^2 - \frac{4b^2 h}{(1-b^2 h)^2},$$

$$p = \Lambda + \frac{4(b^2-r^2)h}{u^4 M^2 (1-b^2 h)(1-r^2 h)}\left(\frac{8}{u} + \frac{5}{1-r^2 h} - \frac{1}{(1-b^2 h)} - 2\right)$$

$$+ \frac{16}{u^4 M^2} \frac{b^2 h}{(1-b^2 h)}$$

$$q = \frac{1}{u^4 M^2} \frac{16 b r h}{(1-b^2 h)(1-r^2 h)}, \quad L = -\frac{dm}{dt}\frac{1}{\dot{v}}, \quad \dot{v} = \frac{A^2}{r\dot{B}+A[1+rB'/B]}.$$

The horizon of the interior region is the solution of the equation which makes denominator of the luminosity "0".

We see that as $r \to 0$, pressure and heat flow goes to zero as expected and the physical radius decreases with time. In this work, it is shown that the cosmological constant has the imprints on the physical parameters such as density, pressure and the total luminosity. It will be interesting to find interior solution to the generalized and higher dimensional Vaidya Sitter spacetimes given in [5].

REFERENCES

1. Vaidya, P.C. 1953, *Nature*, **171**, 260.
2. Fayos, F., Jaen, X., Llanta, E., Senovilla, J.M.M., 1992 *Phys Rev* **D 45**, 2732
3. Santos, N.O., 1985, *Mont. Not. Astr. Soc.* **216**, 403-410.
4. Govender, M., Maharaj, S., Maartens, R., 1998, *Class. Quant. Grav.* **15**, 323.
5. Wang, A., Wu, Y., 1999, *General Relativity And Gravitation* **31**, 107-114.

Solving the Nosé-Hoover thermostat for Nuclear Pasta

M. Ángeles Pérez García[1]

Departamento de Física Fundamental, Universidad de Salamanca, Plaza de la Merced s/n
E-37008 Salamanca Spain

Abstract. At densities just below nuclear saturation density, there may be possible non-uniform spatial configurations of neutron rich matter. In this work we present a calculation using molecular dynamics techniques for a nuclear system interacting via a semiclassical potential depending on both positions and momenta and kept at fixed temperature by using the Nosé-Hoover Thermostat.

Keywords: nuclear astrophysics, dense matter, nuclear pasta, crust, supernovae
PACS: 97.60.Bw, 97.60.Jd

MODEL HAMILTONIAN

At nuclear densities ranging from 10^{12} to 10^{14} g/cm^3 a rich variety of nuclear species are favoured over the fluid configuration. This type of matter is known as *nuclear pasta* and it is believed to exist in Supernovae matter and neutron star crusts [1]. Attractive short-range strong interactions correlate nucleons into nuclei. However, nuclear sizes are limited by long-range repulsive Coulomb interactions and thermal excitations.

In this work we are interested in evaluating a nuclear system under the constraint of the Nosé-Hoover thermostat. This thermostat has been widely studied for position-dependent potentials but not as much for potentials dependent on both positions and momenta.

We model an electrically charge-neutral system with a fixed number of nucleons (protons and neutrons), A, and electrons providing a neutralizing background.

The Hamiltonian under the Nosé-Hoover method for the extended system in this case can be written as [2]

$$H_{\text{NH}} = \sum_{i=1}^{A} \frac{\mathbf{P}_i^2}{2m_i} + V(R_{ij}, P_{ij}) + \frac{s^2 p_s^2}{2Q} + g\frac{\ln s}{\beta} \qquad (1)$$

where $V(R_{ij}, P_{ij})$ is the potential which depends on both positions and momenta and it is described below, s is the extended position variable, p_s is the momentum conjugate to s, Q is the thermal inertial parameter corresponding to a coupling constant between the system and the thermostat and takes a value $Q \sim 10^6$-10^8 MeV $(fm/c)^2$, g is a parameter to be determined as $3A$ by a condition for generating the canonical ensemble in the classical molecular dynamics simulations, ξ is the thermodynamic friction coefficient and β is defined as $\beta = 1/k_B T$.

The total potential energy of the system, V, consists of a sum of two-body interactions

$$V = V_{had} + V_{Coulomb} + V_{Pauli} \qquad (2)$$

where

$$V_{had} = \sum_{i<j} a e^{-R_{ij}^2/\Lambda} + \left[b + c\tau_i \tau_j\right] e^{-R_{ij}^2/2\Lambda} \qquad (3)$$

$$V_{Coulomb} = \sum_{i<j} \frac{e^2}{R_{ij}} e^{-R_{ij}/\lambda} \frac{(1+\tau_i)}{2} \frac{(1+\tau_j)}{2}, \qquad (4)$$

$$V_{Pauli} = d \left(\frac{\hbar}{q_0 p_0}\right)^3 \sum_{i,j(\neq i)} \exp\left[-\frac{(R_{ij})^2}{2q_0^2} - \frac{(P_{ij})^2}{2p_0^2}\right] \delta_{\tau_i \tau_j} \delta_{\sigma_i \sigma_j}, \qquad (5)$$

Here the distance between the particles in phase space is denoted by $R_{ij} = |\mathbf{R}_i - \mathbf{R}_j|$, $P_{ij} = |\mathbf{P}_i - \mathbf{P}_j|$ and τ_i represents the ith-nucleon isospin projection on z-axis ($\tau = +1$ for protons and $\tau = -1$ for neutrons). $V_{Coulomb}$ corresponds to the screened Coulomb interaction. The screening length, λ, that results from the slight polarization of the electron gas is arbitrarily set to $\lambda = 10 \ fm$ as in previous works [5]. V_{Pauli} is the Pauli potential that incorporates phase space repulsion for fermions by means of the Kronecker deltas in spin and isospin [3].

The model parameters (a, b, c, d, q_0, p_0 and Λ) have been adjusted to reproduce the saturation density and binding energy per nucleon of symmetric nuclear matter and neutron matter, and the binding energy of finite nuclei. All these properties were computed at temperature set at $T = 1$ MeV as described below. The parameter set employed is displayed in Table 1.

TABLE 1. Model parameters.

a (MeV)	b (MeV)	c (MeV)	d (MeV)	q_0 (fm)	p_0 (MeV/c)	Λ (fm^2)
133	-47	11	29	3	120	1.5

According to the hamiltonian in eq.(1) the equations of motion for each nucleon yield,

$$\frac{d\mathbf{R}_i}{dt} = \frac{\partial H_{NH}}{\partial \mathbf{P}_i} = \frac{\mathbf{P}_i}{m_i} + \frac{\partial V}{\partial \mathbf{P}_i}, \qquad (6)$$

$$\frac{d\mathbf{P}_i}{dt} = -\frac{\partial H_{NH}}{\partial \mathbf{R}_i} = -\frac{\partial V}{\partial \mathbf{R}_i} - \xi \mathbf{P}_i, \qquad (7)$$

$$\frac{1}{s}\frac{ds}{dt} = \frac{1}{s}\frac{\partial H_{NH}}{\partial p_s} = \frac{1}{Q}\frac{\partial H_{NH}}{\partial \xi} = \xi, \qquad (8)$$

$$\frac{d\xi}{dt} = \frac{1}{Q}\left\{\sum_{i=1}^{A}\left(\frac{\mathbf{P}_i^2}{m_i} + \mathbf{P}_i \cdot \frac{\partial V}{\partial \mathbf{P}_i}\right) - \frac{g}{\beta}\right\}, \xi \equiv \frac{sp_s}{Q}, \qquad (9)$$

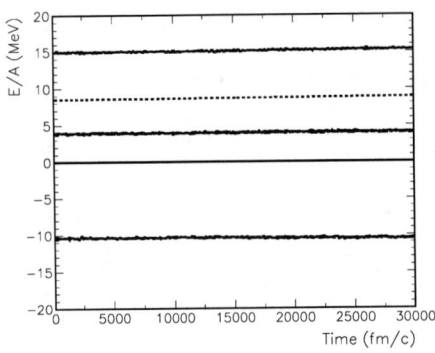
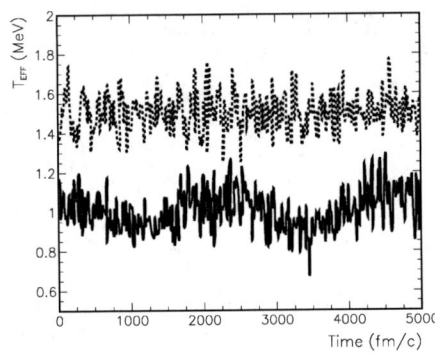

FIGURE 1. (Left panel) Energy per nucleon versus simulation time at $T = 1\,MeV$, $n_b = 0.016 fm^{-3}$, $Y_e = 0.2$ and A=1,000 particles. (Right panel) Effective temperature for a configuration of A=200 particles at $T = 1\,MeV$, $n_b = 0.016 fm^{-3}$ and $Y_e = 0.2$ for $Q = 10^6\,MeV(fm/c)^2$ (upper) and $Q = 10^8\,MeV(fm/c)^2$ (lower). See text for details.

To minimize finite size effects we use periodic boundary conditions as usual. After solving the equations of motion, using an integrator algorithm [4] we can compute the conserved energy, E_{NH}

$$E_{NH} = \sum_{i=1}^{A} \frac{P_i^2}{2m} + V + \frac{1}{2}Q\xi^2 + gkT\ln s = K + V + E_\xi + E_T \quad (10)$$

Then an effective temperature can be defined as

$$T_{\text{eff}} = \frac{2}{3Ak_B} \sum_{i=1}^{A} \frac{1}{2} \mathbf{P}_i \cdot \frac{d\mathbf{R}_i}{dt} \quad (11)$$

that fluctuates around the desired initially set temperature T. The simulations presented in this work were carried out with fixed baryonic number density, n_b, lepton fraction, $Y_e = \frac{n_e}{n_b}$ and a fixed number of nucleons, A, initially placed in a cubic box of side $L = (A/n_b)^{1/3}$ at random. Typical thermalization times are of order $10^5 fm/c$.

SIMULATION RESULTS

As an example of typical low density conditions we consider $n_b = 0.016 fm^{-3}$ which is about a tenth of normal nuclear density, a temperature $T = 1\,MeV$ and a typical electron fraction $Y_e = 0.2$. The energy of the system is conserved, according to eq. (10), as can be seen in the plot of energy per particle versus time on left panel in Fig. 1. From top to bottom we plot E_T, E_{NH}, K, E_ξ, V per nucleon. The initial time for the plot is set once the system is thermalized. The system exhibits characteristic oscillations in the thermostat variables due to the value of Q, that it is associated with the heat capacity of the system.

On the right panel in Fig. 1 we plot the effective temperature for the same case as before and comparing two values of Q. The upper curve (dashed line) corresponds to

$Q = 10^6$ MeV $(fm/c)^2$ and the lower curve (solid line) to $Q = 10^8$ MeV $(fm/c)^2$. The temperature is set to T=1 MeV for both curves. The upper curve has been biased by adding an offset of 0.5 MeV for the sake of clarity. By decreasing Q the temperature control is better but leads to rapid oscillations in the energy that must be carefully considered when studying the dynamical response in energy modes of the system [6].

In the numerical simulation the energy of the system may suffer a small drift with time due to the accuracy of the algorithm used. We have checked that the relative extended energy error at time t, defined as $\frac{\Delta E}{E} = \left|\frac{E(t)-E(0)}{E(0)}\right|$ is $\frac{\Delta E}{E} = 10^{-4}$ for time length $\Delta t = 5.10^3$ fm/c using timesteps $dt = 0.025$ fm/c and $Q = 10^6$ $MeV(fm/c)^2$. This error increases somewhat with timestep size. We use timesteps in the range $dt = 0.01 - 0.1$ fm/c for our simulations in this work.

CONCLUSIONS

In this work we have employed Molecular Dynamics techniques to solve a hamiltonian model for neutron rich matter with an interaction potential depending on positions and momenta. We have simulated this phase at a given density and keeping fixed the temperature by using the Nosé-Hoover thermostat. We find that a clustered phase is formed and that by changing the heat capacity of the system in the range $Q = 10^6 - 10^8$ $MeV(fm/c)^2$ a good temperature control of the temperature is achieved. Induced oscillations in the thermostat variables must be considered with further detail when studying excitation energy modes of this system.

ACKNOWLEDGMENTS

The author is grateful to Charles J. Horowitz, J. Vigo-Aguiar and B. Wade for helpful comments. This work has been partially supported by the Spanish Ministry of education under project BFM2003-021121.

REFERENCES

1. D. G. Ravenhall, C. J. Pethick, and J. R. Wilson, Phys. Rev. Lett. **50**, 2066 (1983). M. Hashimoto, H. Seki, and M. Yamada, Prog. Theor. Phys. **71**, 320 (1984).
2. S. Bond, B. Leimkulher, B. Laird, Journal of Computational Physics **151** 114 (1999) and references therein.
3. G. Peilert, J. Randrup, H. Stocker, W. Greiner Phys. Let. B **260**, 271 (1991).
4. J. Vigo-Aguiar, Int.J.Appl.Math. Vol1, 8, 911, (1999).
5. C. J. Horowitz, M. A. Perez-Garcia, D. K. Berry, and J. Piekarewicz, Phys. Rev. C **72** 035801 (2005).
6. C. J. Horowitz, M. A. Perez-Garcia, in preparation.

Quantum Collapse in Quark Stars?

A. Pérez Martínez*, H. Pérez Rojas* and H. J. Mosquera Cuesta[†]

ICIMAF, Calle E esq 15 No. 309 Vedado, CUBA
[†]*Centro Brasileiro de Pesquisas Físicas, Laboratório de Cosmologia e Física Experimental de Altas Energias, Rua Dr. Xavier Sigaud 150, Urca, CEP 22290-180, Rio de Janeiro, Brazil*

Abstract.
Quark matter is expected to exist in the interior of compact stellar objects as neutron stars or even the more exotic strange stars. Bare strange quark stars and (normal) strange quark-matter stars, those possessing a baryon (electron-supported) crust, are hypothesized as good candidates to explain the properties of a set of peculiar stellar sources. In this presentation, we modify the MIT Bag Model by including the electromagnetic interaction. We also show that this version of the MIT model implies the anisotropy of the Bag pressure due to the presence of the magnetic field. The equations of state of degenerate quarks gases are studied in the presence of ultra strong magnetic fields. The behavior of a system made-up of quarks having (or not) anomalous magnetic moment is reviewed. A structural instability is found, which is related to the anisotropic nature of the pressures in this highly magnetized matter.

Keywords: magnetic field, stars-quarks, anisotropic pressures
PACS: 97.60Bw,98.80Ft,26.30+k,12.20-m

INTRODUCTION

The relation between strong magnetic fields and dense matter is a subject that have attracted so much attention recently, especially after the observations of peculiar X-ray emission from anomalous X-pulsars (AXPs) and low energy γ-ray radiation from soft gamma-ray repeaters (SGRs). The central engine of these radiations is believed to be a neutron star or 'quark star' endowed with a magnetic field larger than $10^{13.5}$ G. The x-ray binaries dubbed as Galactic Black Hole Candidates have also been recently suggested as possessing a strange star as its primary Ref. [1].

Witten Ref. [2], and later Farhi and Jaffe Ref. [3], showed that for strange matter, the binding energy could be lower than for Fe over a rather wide range of QCD parameters. The matter in this condition, extremely dense, exhibits novel properties which are worthwhile to study. Strong magnetic fields are known to exist in the interior of compact stars. Notwithstanding, only a few attempts to study the behavior of quark matter permeated by a magnetic field have been performed so far Ref. [4].

The scope of this paper is to show that for 'quark matter' acted upon by a strong magnetic field the same anisotropic behavior obtained earlier in Ref. [5]- [6] holds. For the sake of simplicity, we focus here on the case of non-strange quark matter, as a degenerate Fermi gas of quarks. Nonetheless, our treatment and conclusions prove to be independent of the specific model being used. To study the degenerate quark gas in presence of a magnetic field we considers the quark field interacting with magnetic field via its charge and also considered that quarks have anomalous magnetic moment.

BAG MODEL IN PRESENCE OF A MAGNETIC FIELD

The Lagrangian density including the MIT B_{bag} in presence of a magnetic field should be written as[1] $\mathcal{L}_{Bag} = [\overline{\psi}(i\gamma_\mu(\partial^\mu - ie_q A^\mu) - m_q)\psi - 1/4 \mathcal{F}_{\mu\nu}\mathcal{F}^{\mu\nu} - B_{Bag}]\theta_v(x) - \frac{1}{2}\overline{\psi}\psi\Delta_s$

To obtain the equation of state of magnetized quark matter the starting point is the statistical average of energy-momentum tensor.[2]

Let us note that in the context of the modified MIT Bag model to describe magnetized quark matter, only what happens inside the bag is interesting, so we will put $\theta_v = 1$ in the Lagrangian and thermodynamical potential.

$\mathcal{T}_{\mu\nu}$ has the form $\mathcal{T}_{\mu\nu} = \left(T\frac{\partial\Omega}{\partial T} + \sum \mu_i \frac{\partial\Omega}{\partial \mu_i}\right)\delta_{4\mu}\delta_{4\nu} + 4F_\mu^\lambda F_{\nu\lambda}\frac{\partial\Omega}{\partial F^2} - \delta_{\mu\nu}\Omega$, where for the sake of simplicity we assume $\Omega = \Omega_q - B_{bag}$, and Ω_q is the thermodynamical potential of the magnetized degenerate quark gas, q denotes the species of quarks. The aim of this note is to analyze magnetic effects, so in what follow we consider the magnetic field pointing in the direction x_3.

The diagonal components of the tensor $\mathcal{T}_{\mu\nu}$ corresponds to the energy density and the pressures, albeit the last ones are anisotropic due to the magnetic field $P_\perp = P_3 - \mathcal{M}B$ and $P_3 = -\Omega$. $P_\perp \leq P_3$ if the magnetization is a positive quantity, and thus the behavior of the gas is paramagnetic.

The condition of quantum-magnetically-induced transverse collapse: $P_\perp = 0$, discussed in Ref. [5]-[6], implies for the magnetically modified MIT Bag model, the relation $B_{bag}^\perp = -\Omega_q - \mathcal{M}_q B$ and $B_{bag}^\parallel = -\Omega_q$.

This would mean an anisotropic bag pressure (since the nucleon is deformed by the magnetic field), and in place of Ω we would have $\Omega\delta_{\mu\nu} \to \delta_{\mu\nu}\Omega_q - B_{\mu\nu}^{Bag}$. Notice, nonetheless, that in the absence of the magnetic field the pressure becomes isotropic and we recover the condition of stability (or instability) of the Bag model given by $B_{bag} = -\Omega_q$. Thus, we conclude that a magnetic field brings in an instability to the system, but it is compatible with the Bag Model if we consider that the Bag is not isotropic and the nucleons are deformed in a prolate-shape.

QUARK MATTER EOS IN PRESENCE OF MAGNETIC FIELD

Our aim in this section is to discuss the collapse described in the previous section deriving the thermodynamical quantities of the degenerate quark gas in presence of an ultra strong magnetic field.

The quark spectrum in presence of B has the form $E_q = \sqrt{p_3^2 + \left(\sqrt{2e_q B n + m_q^2} + \eta Q_q B\right)^2}$ where e_q is the quark's charge, m_q represents the quark masses, $\eta = \pm$ are the eigenvalues corresponding to two orientations of magnetic moment, and Q_q anomalous magnetic

---

[1] ψ-wave function of quarks m_q- quark masses and e_q- quark charges. B_{bag}-Bag constant. The parameter $\theta_v = 1, 0$ inside (outside) the Bag. $\partial\theta_v/\partial x^\nu = n_\nu \Delta_s$, with Δ_s being the surface δ-function and n_ν is a space-like unit vector normal to the surface, and $\mathcal{F}_{\mu\nu}$ defines the electromagnetic (Maxwell) tensor.

[2] To do that we use standard methods of finite temperature quantum field theory Ref. [7]

moment of quarks. We also define $y_q = Q_q/m_q$, $b_q = 2e_q/m_q^2$ as relative quantities and $x_q = \mu_q/m_q$, $g_q(x_q, B, n) = \sqrt{x_q^2 - h_q(B,n)^2}$ and $h_q(B,n) = \sqrt{b_q B n + 1} + \eta y_q B$ as dimensionless ones.[3]

For the sake of simplicity we study here nonstrange matter but the conclusions for strange matter are esentially the same. In this case: $Q_u = 1.82\mu_N$, $Q_d = -0.9\mu_N$, being μ_N the nuclear magneton $\mu_N = hc/m_n$.

Thus, for the quarks thermodynamical potential we get [4]

$$\Omega = \sum \Omega_q, \quad \Omega_q = -\Omega_q^0 B \sum_n \sum_{\pm\eta}^{n_{max}} \left[x_q g_q - h_q^2 \ln \frac{x_q + g_q}{h_q} \right], \quad (1)$$

The magnetization is then given as $\mathcal{M} = \sum \mathcal{M}_q$

$$\mathcal{M}_q = \mathcal{M}_q^0 \sum_n \sum_{\pm\eta} \left\{ g_q x_q - \left(h_q^2 + 2h_q B \left[\frac{b_q n}{2\sqrt{b_q B n + 1}} + \eta y_q \right] \right) \ln \frac{x_q + g_q}{h_q} \right\}, \quad (2)$$

The density of particles has the form $N = \sum_q N_q$ and $N_q = N_q^0 \left(\frac{B}{B_q^c} \right) \sum_n \sum_{\pm\eta} g_q(x_q, B, n)$
The charge neutrality in this case is given by the expression $N_d = 2N_u$.

By bringing the relations for the pressures given before together with the expressions for the Bag pressures we obtain for the anisotropic pressures the following expressions

$$B_{bag}^{\parallel} = P_{\parallel} = \frac{e_q m_q^2 B}{4\pi^2 \hbar c} \sum_n \sum_{\pm\eta} \left[x_q g_q - h_q^2 \ln \frac{x_q + g_q}{h_q} \right], \quad (3)$$

$$B_{bag}^{\perp} = P_{\perp} = \frac{2 e_q m_q^2 B^2}{\pi^2 \hbar c} \sum_n \sum_{\pm\eta} \left(2h_q \left[\frac{b_q n}{2\sqrt{b_q B n + 1}} + \eta y_q \right] \right) \ln \frac{x_q + g_q}{h_q}. \quad (4)$$

Let us remark that we can also study a model of a degenerate quark gas without taking into account the quark magnetic moment. This means to make $y_q = 0$ in all the equations above. We recover in that case the expressions for all the thermodynamical quantities. In Fig.1 is shown the phenomenon of anisotropy of the pressures. In this case, we observe that the limiting case $P_{\perp} = 0$ is possible to achieve for typical values of the density $N = 10^{39}$ cm^{-3} and B-field typical of the interior of millisecond spinning just-born neutron stars $B \sim 10^{17}$ G

The relation between the particle density and magnetic field strength that fulfills the condition $P_{\perp} = 0$ is the following $N_q(B) = 2N_q^0 y_q^{1/2} \left(\frac{B^{3/2}}{B_q^c} \right)$.

---

[3] Thermodynamical quantities have been written here in the cgs system and m_q has energy dimension.
[4] $\Omega_q^0 = \frac{e_q m_q^2}{4\pi^2(\hbar c)^2}$, $n_{max} = I\left(\frac{(x_q - \eta y_q B)^2 - 1}{b_q B} \right)$ where I is an integer part function $B_q^c = \frac{m_{u,d}^2}{(e_q \hbar c)}$, $\mathcal{M}_q^0 = \frac{e_q m_q^2}{4\pi^2(\hbar c)^2}$, $N_q^0 = \frac{m_q^3}{(4\pi^2(\hbar c)^3)}$

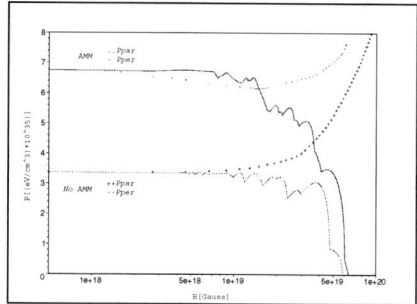

FIGURE 1. Anisotropy of pressures for the two cases studied above: a gas of quarks (u,d) having anomalous magnetic moment, and without it. The first being more stable as the pressure $P_\perp$ goes to zero for a large magnetic field.

CONCLUSIONS

We have explored the behavior of a quark gas in the presence of extremely large magnetic fields $B \sim m_q^2/e_q$. We used a version of the MIT Bag Model which includes the electromagnetic interaction between quarks. From it we verify that the B_{bag} can be replaced by an anisotropic tensor. We confirm that the instability due to the strong magnetic field discussed in Ref. [5]-[6] is present also in this case. Finally, we show that the degenerate quark gas with anomalous magnetic moment is more stable than the quark gas without it. A more extensive version of this note can be found in Ref. [8]

ACKNOWLEDGMENTS

A.P.M would like to thank the organizers of ERE05 for financial support and specially Prof J. Diaz Alonso who made a valuable effort to be able my presence in the Conference. I also thanks the hospitality of the University of Oviedo in particular the local organizer of the ERE05. I offer a big clap to Prof Lysiane Mornas and others members of Physics Department (professors and students).

REFERENCES

1. J. E. Horvath and G. Lugones, A & A 422, L1-L4 (2004)
2. E. Witten, Phys. Rev. **D 30** 272 (1984) (1971)
3. E. Farhi and R. L. Jaffe, Phys. Rev. **D 30** 2379 (1984)
4. S. Mandal, S. Chakrabarty, IJMP **D**, Vol 13, No. 6, 1157 (2004)
5. M. Chaichian, S. Masood, C. Montonen, A. Pérez Martínez, H. Pérez Rojas, Phys. Rev. Lett. **84**, 5261 (2000).
6. A. Perez Martínez, H. Pérez Rojas and H. J. Mosquera Cuesta, Eur. Phys. J. C 29, 111 (2003).
7. E. S. Fradkin, Quantum Field Theory and Hydrodynamics Proceedings of Lebedev Institute (edited by Consultants Bureau New York Vol. 29) (1967).
   astro-ph/0011148 (2000).
8. A. Pérez Martínez, H. Pérez Rojas, H. J. Mosquera Cuesta Int J. Mod Phys **D** (2005) Vol. 14, No 10 (2005).

Quantum degrees of freedom of a region of spacetime

Federico Piazza

Institute of Cosmology and Gravitation, University of Portsmouth, UK

Abstract. The entropy of black holes, the holographic principle and the thermodynamics of de Sitter space suggest that the total number of fundamental degrees of freedom associated with any finite-volume region of space may be finite. The naive picture of a short distance cut-off, however, is hardly compatible with the dynamical properties of spacetime, let alone with Lorentz invariance. Considering the regions of space just as general "subsystems" may help clarifying this problem. In usual QFT the regions of space are, in fact, associated with a tensor product decomposition of the total Hilbert space into "subsystems", but such a decomposition is given a priori and the fundamental degrees of freedom are labelled, already from the beginning, by the spacetime points. We suggest a new strategy to identify "localized regions" as "subsystems" in a way which is intrinsic to the total Hilbert-space dynamics of the quantum state of the fields.

Keywords: Quantum Gravity
PACS: 04.60.-m, 11.10.-z, 03.67.-a

In quantum field theory (QFT), independent degrees of freedom are associated with each localized and space-like separated region of spacetime. In the presence of a UV cut-off, the number of degrees of freedom is finite and generally proportional to the volume of the region considered. As soon as the effects of gravity are taken into account, such a picture is challenged by a number of semi-classical interrelated arguments:

1) According to the *holograpic principle* [1], the maximal entropy within a region is finite and proportional to the area – rather than the volume – of the region. Although entanglement entropy very generally exhibits area scaling [2], a localized maximally entropic state can be explicitely constructed [3] with entropy proportional to the volume; by applying the holographic bound one is forced to restrict the effective Fock space inside the region in a non-local way (see, however, [4] for a different view).

2) Locality is questioned, at various levels, in several (e.g. [5]) proposed solutions of the *black hole information-loss paradox*. There are also solid arguments [6] against local QFT whenever describing field configurations that have non-negligeable effects on the background metric (e.g. when they tend to form closed trapped surfaces).

3) It has been argued that the Hilbert space describing *quantum gravity in asymptotically de Sitter* (dS) space is of finite dimension [7]. Since the volume of the dS spatial sections changes with time, how can the fundamental degrees of freedom be local and separated by a fixed, short distance cut-off?

4) More generally, there is an evident friction between the naive picture of a fixed short distance cut-off and the dynamical nature of spacetime. One may simply resolve to associate, to each finite volume of space, an infinite number of degrees of freedom, an infinite entropy, infinite information etc... But this seems hardly compatible with the *finiteness of the black hole entropy*, which is a well established result (see e.g. [8] for a

review), explicitly worked out within string theory for certain extremal cases.

In the hope of possibly gaining some insight into these issues, in this note I give a concise account of a preliminary attempt [9] to consider locality and spacetime itself under a novel perspective. As a matter of fact, any spacetime measurement is made by the mutual relations between objects, fields, particles etc... Any operationally meaningful assertion about spacetime is therefore intrinsic to the degrees of freedom of the matter (i.e. non-gravitational) fields and concepts such as locality, "proximity", "contiguity", should, at least in principle, be operationally definable entirely within the dynamics of the matter fields. In this respect, the usual approach of QFT follows quite an opposite route: the fundamental degrees of freedom are labelled already from the beginning by the spacetime points and locality is given *a priori* as an attribute of the class of sub-systems that are to be considered.

In quantum mechanics the Hilbert space of a composite system is the direct product of the Hilbert spaces of the components. Therefore, each possible way we can intend a given system (say "the Universe") as made of subsystems [10], mathematically corresponds to a *tensor product structure* (TPS) of its Hilbert space:

$$\mathcal{H}_{\text{Universe}} = \mathcal{H}_A \otimes \mathcal{H}_B \otimes \mathcal{H}_C \otimes \dots . \tag{1}$$

A decomposition into subsystems such as (1) can be assigned without any reference to the locality or to the geometric properties of the components. Of course, an arbitrarily assigned TPS on a Hilbert space does not have much significance on its own: without any observable/operator of some definite meaning it looks impossible to extract any physical information about the system. However, by knowing the dynamics inside $\mathcal{H}_{\text{Universe}}$, i.e. the unitary operator $U(t_2, t_1)$ that, in the Schroedinger representation, evolves the state vectors according to $|\Psi(t_2)\rangle = U(t_2, t_1)|\Psi(t_1)\rangle$, we can, at least, follow the evolution of the *correlations* between the subsystems. Correlations play a central role in Everett's view of quantum mechanics. In his seminal dissertation [12], the relation between quantum correlations and mutual information is deeply exploited and measurements are consistently described as appropriate unitary evolutions that increase the degree of correlation between two subsystems: the "measured" and the "measuring". By taking Everett's view one can try to re-interpret the evolution of the system $|\Psi(t)\rangle$ as measurements actually going on between the different parties A, B, C etc... It is very compelling that locality itself and the usual local observables of direct physical interpretation may be eventually picked out within such an abstract scheme.

The central issue is to characterize the class of TPSs that single out "localized subsystems"; the ones, in other words, naturally associated with space-like separated regions of spacetime[1]. As a source of correlation/information we only consider *quantum entanglement*, which, for a bipartite system AB in a pure state $|\Psi\rangle$, is measured by the *von Neumann entropy* $S(A) = -\text{Tr}_A(\rho_A \log_2 \rho_A)$, where $\rho_A = \text{Tr}_B|\Psi\rangle\langle\Psi|$. The space of the

---

[1] Here t plays the role of a global time parameter, and therefore the class of "local subsystems" that we aim to define belongs to some given spacelike slicing. One should be able to recover *a posteriori* the complete general covariance, as in the Hamiltonian formulation of field theories where a spatial slicing is initially required. The difference between the "external" and "inaccessible" time parameter t and the time as perceived by the observers is, on the other hand, discussed elsewhere [9] (but see also [11]).

TPSs is spanned by the elements of a group. By taking, for instance, $\mathcal{H}_{\text{Universe}}$ of dimensions d^N and each of the N subsystems (1) of dimensions d, the group is $U(d^N)/U(d)^N$. The smaller the dimensions of the subsystems, the finer-grained the description that is given.

The most elementary type of spacetime relation that one can try to define intrinsically from the dynamics of general subsystems is that of *mutual spacetime coincidence*, what may be intuitively viewed as "being in the same place at the same time" or just "having been in touch". Inspired by the known local character of physical laws, we attempt to define coincidence by means of physical interactions, i.e. to define two parties as "having been coincident" if they "have physically interacted" with each other. By choosing the production of entanglement as a "measure" of interaction a sufficient condition for spacetime coincidence can be given:

> *spacetime coincidence (sufficient condition)*: If before the instant t_1 the subsystems A and B are in a pure state *i.e.* $S(A; t < t_1) = S(B; t < t_1) = 0$ and, at a later time t_2 they are entangled, $S(A; t_2) = S(B; t_2) > 0$, without, during all the process, having mixed with anything else, $S(AB; t < t_2) = 0$, then A and B *have been coincident* with each other between t_1 and t_2.

As a sufficient condition, the one above stated is rather strict: there are a number of physical situations that one would legitimately consider as "spacetime coincidence relations" but do not fit into the above definition. Most notably, two systems may interact with each other while being already entangled with something else; i.e. not initially in a pure state. In this case, however, it is hard to give a quantitative definition of contiguity because a reliable measure of entanglement for multipartite systems is still a matter of debate.

Note that, if A and B have been coincident, coincidence generally applies also to many other "larger" systems containing A and B as subsystems. The opposite can also be true. Say that A is itself a composite system *i.e.* $\mathcal{H}_A = \mathcal{H}_{A1} \otimes \mathcal{H}_{A2} \otimes \ldots$; we may discover that the sub-subsystem $A2$ is in fact "responsible" at a deeper level for the coincidence between A and B. Again, the smaller the dimension of the systems which coincidence is recognized to apply, the more finely grained the spacetime description we are able to give.

The notion of coincidence can be applied to subsystems belonging to arbitrary TPSs and therefore possibly maximally unlocalized, such as, in ordinary QFT in flat space, those associated with the modes of given momentum. In the TPS of the localized subsystems, however, during an infinitesimal lapse of time, each subsystem creates new correlations with the smallest possible number of other subsystems: its "neighbors". We argue therefore the following generic property of the TPSs that single out localized subsystems:

> *Locality Conjecture:* "Localized subsystems" have the minimum tendency to create *coincidence* relations with each other: the tensor product structure that singles out *localized systems* is the one in which the entanglement of initially completely factorized states *minimally* grows during time evolution.

In [9] generic interacting second quantized models with a finite number of fermionic degrees of freedom have been considered. The symmetries of the Hamiltonian (in this case the conservation of number of particles) dramatically restrict the possible TPS choices. By applying the above conjecture to a one-dimensional Heisenberg spin chain the tensor product structure usually associated with "position" was recovered. While referring to that paper for more details, in the following we finally briefly comment on how, according to this general approach, one should re-consider the relation between the quantum degrees of freedom and the regions of spacetime.

The Hilbert space of a QFT with a short distance cut-off and formulated within a finite-size Universe can be written as $\mathcal{H} = \otimes_{i=1}^{N} \mathcal{H}_i$, where $\mathcal{H}_i$ is the Hilbert space of each of the N bosonic or fermionic modes associated with the points of space. A region of space is thus associated to a subset R of the N points and therefore to the Hilbert space $\mathcal{H}_R = \otimes_{i \in R} \mathcal{H}_i$. Note that, due to the presence of the cut-off, there are only a finite number of possible "regions" because there are a total finite number of points. For generic states, the instantaneous tendency of the subsystem R to create new correlations with the "outside" is roughly proportional to its boundary, since only the points at the boundary contribute in this process. A sphere is, therefore, the ideal "localized subsystem" according to the principle of minimal tendency to entanglement, since, for a given dimensionality/volume, it is the shape with minimum area. The locality conjecture, therefore, applies already in this framework by excluding regions of space which are disconnected or very spread in only one or two directions. We are arguing, on the other hand, that the tendency to entanglement should be minimized not simply among the finite number of possible partitions of the N points: restricting only to the set of partitions of the type $\mathcal{H}_R = \otimes_{i \in R} \mathcal{H}_i$ just reflects our prejudicial – and operationally meaningless – idea of a spacetime pre-existing and independent of the state of the fields. Therefore, the minimization problem should be applied, for a given state, to the entire infinite number of partitions (tensor product structures) of the total Hilbert space.

REFERENCES

1. G. 't Hooft, arXiv:gr-qc/9310026; L. Susskind, J. Math. Phys. **36**, 6377 (1995)
2. See, among many others, L. Bombelli, R. K. Koul, J. H. Lee and R. D. Sorkin, Phys. Rev. D **34**, 373 (1986); M. Srednicki, Phys. Rev. Lett. **71**, 666 (1993); H. Casini, Class. Quant. Grav. **21**, 2351 (2004); S. Das and S. Shankaranarayanan, arXiv:gr-qc/0511066.
3. U. Yurtsever, Phys. Rev. Lett. **91**, 041302 (2003).
4. R. V. Buniy and S. D. H. Hsu, arXiv:hep-th/0510021.
5. L. Susskind, L. Thorlacius and J. Uglum, Phys. Rev. D **48**, 3743 (1993); C. R. Stephens, G. 't Hooft and B. F. Whiting, Class. Quant. Grav. **11**, 621 (1994); G. T. Horowitz and J. Maldacena, JHEP **0402**, 008 (2004).
6. S. B. Giddings and M. Lippert, Phys. Rev. D **69**, 124019 (2004).
7. T. Banks, arXiv:hep-th/0007146; E. Witten, arXiv:hep-th/0106109; M. K. Parikh and E. P. Verlinde, JHEP **0501**, 054 (2005).
8. See e.g., for a review, T. Damour, arXiv:hep-th/0401160.
9. F. Piazza, arXiv:hep-th/0506124.
10. P. Zanardi, Phys. Rev. Lett. **87** 077901 (2001).
11. D. N. Page and W. K. Wootters, Phys. Rev. D **27**, 2885 (1983); T. Banks, Nucl. Phys. B **249**, 332 (1985); D. N. Page, NSF-ITP-89-18.
12. H. Everett III Princeton PhD thesis, in B. DeWitt, R. Graham eds. "The many-worlds Interpretation of Quantum Mechanics" Princeton Series in Physics, Princeton University Press (1973); see also H. Everett, Rev. Mod. Phys. **29**, 454 (1957);

Would Closed Timelike Curves Help to Do Quantum Cloning?

A.R. Plastino and C. Zander

Department of Physics, University of Pretoria, Pretoria 0002, South Africa

Abstract.
The possibility of performing quantum cloning in the presence of closed time like curves is considered. We analyze a model of the quantum cloning process involving a source qubit, a target qubit, and a quantum copy machine which is itself a composite quantum system. A subsystem of the copy machine is allowed to traverse a closed time like curve. Under these circumstances the evolution of the chronology respecting parts of the system is nonlinear. This suggests (Deutsch, 1991) that quantum cloning might be possible. We prove that, if only one subsystem of the copy machine is allowed to traverse a closed time like curve and the copy machine's Hilbert space is of finite dimension, universal quantum cloning is impossible.

Keywords: Closed Timelike Curves, Quantum Cloning, Quantum Information
PACS: 01.55.+b, 03.67.-a, 89.70.+c, 05.90.+m

INTRODUCTION

The ideas and methods of quantum information theory [1] provide an interesting framework for the study of certain aspects of the physics of closed timelike curves [2]. Quantum computation processes with part of the quantum data traversing closed timelike curves lead to a new physical model of computation [3], and also to various physical effects with profound implications for the foundations of quantum theory [2]. It has been conjectured [2] that the nonlinear evolution of chronology respecting qubits interacting with closed timelike curve qubits may be used to overcome the celebrated quantum no cloning theorem [4, 5]. This theorem plays a fundamental role in quantum information theory [1] and has been the focus of an intensive research activity [6, 7, 8, 9, 10]. The aim of the present contribution is to investigate the alluded to conjecture, by recourse to the analysis of specific models of the cloning process.

The Quantum No-cloning Theorem constitutes a hallmark feature of quantum information. It states that quantum information cannot be cloned: an unknown quantum state of a given (source) system cannot be perfectly duplicated while leaving the state of the source system unperturbed [4, 5]. No unitary (quantum mechanical) transformation exists that can perform the process

$$|\psi\rangle \otimes |0\rangle \otimes |\Sigma\rangle \longrightarrow |\psi\rangle \otimes |\psi\rangle \otimes |\Sigma_\psi\rangle, \qquad (1)$$

for arbitrary source states $|\psi\rangle$ (in the above equation $|0\rangle$ and $|\Sigma\rangle$ denote, respectively, the initial standard states of the target qubit and of the copy machine. $|\Sigma_\psi\rangle$ is the final state of the copy machine). In other words, universal quantum cloning is not permitted by the basic laws of quantum mechanics. The impossibility of universal quantum cloning can be proved in two different ways. One can show that it is not compatible with the linearity of quantum evolution or that it is not compatible with the unitarity of quantum evolution.

PROPERTIES OF THE FIDELITY DISTANCE

In our discussion of the cloning process in the presence of closed timelike curves we are going to use the fidelity distance between two quantum states (represented by two density matrices) of a given quantum system [1]. Here we are going to review briefly the main properties of this measure that we are going to use afterwards (for a detailed discussion of the fidelity distance see [1]).

The fidelity distance between two quantum states is given by [1]

$$F[\rho,\sigma] = \text{tr}\sqrt{\rho^{1/2}\sigma\rho^{1/2}}. \tag{2}$$

In the particular case that one of the states is pure we have,

$$F[|\psi\rangle,\rho] = \sqrt{\langle \psi | \rho | \psi \rangle}. \tag{3}$$

A fundamental property of the fidelity measure is that it remains constant under unitary transformations,

$$F[U\rho U^\dagger, U\sigma U^\dagger] = F[\rho,\sigma]. \tag{4}$$

If we have a composite system AB, the distance between two density matrices describing two states of the composite system is lower or equal than the distance between the marginal density matrices associated with one of the subsystems,

$$F[\rho_{AB}, \sigma_{AB}] \leq F[\rho_A, \sigma_A]. \tag{5}$$

Finally, the distance between two factorizable density matrices complies with,

$$F[\rho_0 \otimes \sigma_0, \rho_1 \otimes \sigma_1] = F[\rho_0, \rho_1] F[\sigma_0, \sigma_1]. \tag{6}$$

CTC AND QUANTUM CLONING

We consider a quantum cloning process where the copy machine is a composite system constituted by two subsystems A and B. The subsystem B is allowed to traverse a closed time like curve. The joint density matrix describing the state of the source qubit, target qubit, and the subsystem A of the copy machine (all three assumed to be chronology respecting) evolves, due to their interaction with subsystem B, nonlinearly. A successful cloning process would be of the form,

$$|\psi_i\rangle \otimes |0\rangle \otimes \rho_A \otimes \rho_B^{(i)} \longrightarrow |\psi_i\rangle \otimes |\psi_i\rangle \otimes \sigma_{AB}^{(i)}, \tag{7}$$

where $|0\rangle$ and ρ_A are, respectively, standard initial states of the target qubit and the subsystem A of the copy machine. The process (7) is assumed to be described by an unitary transformation.

The subsystem B (traversing a closed timelike curve) verifies the consistency condition [2],

$$\rho_B^{(i)} = \text{tr}_A\left(\sigma_{AB}^{(i)}\right). \tag{8}$$

Notice that this consistency condition implies that the initial state of the subsystem B (traversing a closed timelike curve is not independent of the initial state of the source qubit). Furthermore, due to the aforementioned consistency condition, the evolution of the source and target qubits (as well as the subsystem A of the copy machine) are nonlinear. Consequently, the usual argument for the quantum no-cloning theorem based on the linearity of quantum evolution can not be applied here. Instead, we consider the behavior of the fidelity distance between two realizations of the cloning process. The fidelity distance between density matrices is given by equation (2).

We assume that two states $\langle\psi_1|$ and $\langle\psi_2|$ can be successfully cloned. Comparing the fidelity distances between the corresponding initial and final states of the cloning process, and using the basic properties of the Fidelity distance already mentioned, one gets,

$$\begin{aligned}|\langle\psi_1|\psi_2\rangle| F\left[\rho_B^{(1)},\rho_B^{(2)}\right] &= |\langle\psi_1|\psi_2\rangle|^2 F\left[\sigma_{AB}^{(1)},\sigma_{AB}^{(2)}\right] \\ &\leq |\langle\psi_1|\psi_2\rangle|^2 F\left[\rho_B^{(1)},\rho_B^{(2)}\right]\end{aligned} \tag{9}$$

In order to satisfy this inequality, at least one of the following conditions must be fulfilled,

- $\langle\psi_1|\psi_2\rangle = 0$.
- $F\left[\rho_B^{(1)},\rho_B^{(2)}\right] = 0$.

In the first case we have orthogonal source states which, as happens with standard, linear quantum evolution, can be cloned. The second case can be realized for an appropriate choice of the unitary transformation (7) [11]. Consequently, it is possible to clone two non-orthogonal states in the presence of a closed timelike curve. However, if the subsystem B of the copy machine has a Hilbert space of finite dimension N, the maximum number of non- orthogonal states that can be cloned by the present scheme is N [11].

CONCLUSIONS

We have explored the possibility of quantum cloning in the presence of closed timelike curves. We con- sidered a cloning process where a subsystem of the copy machine is allowed to travel in a closed timelike curve. For this kind of cloning process we proved that

- Contrary to what occurs in standard, linear quantum evolution, it is possible to clone non orthogonal states.
- The maximum number of non orthogonal states that can be cloned is limited by the dimension of the Hilbert space associated with the part of the copy machine traveling on a closed timelike curve.

As a final remark, it is important to stress that the cloning process based upon closed timelike curves can not be used to implement faster that light signalling: due to the non-linear character of the process we have just discussed, a mixture of two pure states of the target qubit does not evolve into the corresponding mixture of the final states generated by those initial states separately [11].

ACKNOWLEDGMENTS

The financial assistance of the National Research Foundation (NRF; South African Agency) towards this research is hereby acknowledged. Opinions expressed and conclusions arrived at, are those of the authors and are not necessarily to be attributed to the NRF.

REFERENCES

1. M. Nielsen and I. Chuang, *Quantum Computation and Information*, Cambridge University Press, Cambridge, 2000.
2. D. Deutsch, *Phys. Rev. D* **44**, 3197 (1991).
3. D. Bacon, *Phys. Rev. A* **70**, 032309 (2004).
4. W.K. Wootters and W. H. Zurek, *Nature* **299**, 802 (1982).
5. D. Dieks, *Phys. Lett. A* **92**, 271 (1982).
6. H. Barnum, C.M. Caves, C.A. Fuchs, R. Jozsa, and B. Schumacher, *Phys. Rev. Let.* **76**, 2818 (1996).
7. V. Buzek and M. Hillery, *Phys. Rev. A* **54**, 1844 (1996).
8. N. Gisin, *Phys. Lett. A* **242**, 1 (1998).
9. A. Daffertshofer, A. R. Plastino, and A. Plastino, *Phys. Rev. Lett.* **88**, 210601 (2002).
10. A.R. Plastno and A. Daffertshfer, *Phys. Rev. Lett.* **93**, 13871 (2004).
11. C. Zander and A.R. Plastino, in preparation, 2006.

Gravity from Lorentz Symmetry Violation

Robertus Potting

CENTRA and Physics Department, FCT, University of the Algarve, 8005-139 Faro, Portugal

Abstract. In general relativity, the masslessness of gravitons can be traced to symmetry under diffeomorphisms. In this talk, we consider another possibility, whereby the masslessness arises from spontaneous violation of Lorentz symmetry.

Keywords: Lorentz symmetry, spontaneous symmetry breaking, gravitation
PACS: 04.20.Cv, 11.30.Cp, 11.15.Ex, 95.30.Sf

In the standard approach to general relativity, the Einstein equations are derived using geometrical notions of Riemannian space-time such as curvature. Another approach exists [1], in which the starting point is the linearized free equation of motion for a spin-2 particle (the graviton)

$$R^L_{\mu\nu} \equiv K_{\mu\nu\alpha\beta} h^{\alpha\beta} = 0 \qquad (1)$$

which can be obtained from a free lagrangian $\mathcal{L}^0$. From Newton's law we know gravity is coupled to matter. The way to take this into account consistently is to add a term proportional to the matter energy-momentum tensor. In the case of free gravitons described by (1) we have no matter, and the equation is consistent as it stands. Nevertheless, gravitons represent energy-momentum, and thus should be expected to contribute as well to the right-hand-side of (1). The resulting, modified equation of motion can be obtained from a cubic lagrangian $\mathcal{L}^1$. In turn, the cubic term in this lagrangian gives another contribution to the energy-momentum tensor, which should again be added to the right-hand-side of (1). This leads to a quartic lagrangian $\mathcal{L}^2$. This process continues indefinitely and in the limit one recovers an equation that can be summarized by the full Einstein equation in free space

$$R_{\mu\nu} = 0. \qquad (2)$$

A convenient one-step algorithm that leads from (1) to (2) has been derived by Deser [2].

In this derivation, the reason for starting with a symmetric field $h^{\mu\nu}$ is easily understood: it is needed in the action to couple to the symmetric energy-momentum tensor $T^{\mu\nu}$. What is the reason for the masslessness of gravitons?

The usual argument that is given is one of symmetry. For instance, in the electroweak model, the photon is massless because of an unbroken $U(1)$ gauge symmetry. In QCD, the gluons are massless because of the $SU(3)$ gauge symmetry. Similarly, in general relativity the masslessness of the graviton is taken to be a consequence of diffeomorphism symmetry.

However, an alternative reason exists as to why a particle might be massless, which has to do with the breaking of a symmetry, rather than its presence. The Nambu-

Goldstone theorem states that, with some mild assumptions, there must be a massless particle whenever a continuous global symmetry of an action isn't a symmetry of the vacuum [3]. This result is readily understood by considering an action with a scalar potential V which has its minimum for nonzero field values. Considering a vacuum in which the field assumes such a constant value, a symmetry of the theory shifts the vacuum to another, equivalent, minimum. Thus V has at least one flat direction; excitations around the vacuum in that direction correspond to massless particles, the Nambu-Goldstone modes.

A well-known example of the realization of this mechanism is the (approximate) masslessness of the pion due to the spontaneous breaking of chiral symmetry in the sigma model. A more recent example involves the so-called "bumblebee" model [4, 5] with Lagrangian

$$\mathscr{L}_B = -\frac{1}{4}B_{\mu\nu}B^{\mu\nu} - \lambda(B_\mu B^\mu \pm b^2) - B_\mu J^\mu. \tag{3}$$

Here $B_{\mu\nu} = \partial_\mu B_\nu - \partial_\nu B_\mu$, λ is a Lagrange multiplier field and J^μ is an external current. The equation of motion of the latter forces the vector B_μ to assume a nonzero vacuum value b_μ, breaking spontaneously the Lorentz symmetry. Thus we expect massless Goldstone modes in the flat direction of the potential, which can be obtained by Lorentz transformations of the vacuum value:

$$\delta B^\mu \equiv A^\mu = B^\mu - b^\mu \approx \varepsilon^{\mu\nu} b_\nu. \tag{4}$$

Here the $\varepsilon^{\mu\nu}$ generate a Lorentz transformation. Expressing (3) in terms of A_μ yields

$$\mathscr{L}_B \to \mathscr{L}_{NG} \approx -\frac{1}{4}F_{\mu\nu}F^{\mu\nu} - A_\mu J^\mu - b_\mu J^\mu \tag{5}$$

subject to the constraint $b_\mu A^\mu = 0$. Thus we obtain electrodynamics in the axial gauge! Note that, in this model, the masslessness of the photon arises, not by the presence of a $U(1)$ gauge symmetry (the Lagrangian (3) is not gauge invariant), but by the spontaneous breaking of Lorentz symmetry.

Instead of the Lagrange multiplier term in (3) one can take a smooth scalar potential with a nonzero minimum. In that case we have, apart from the Nambu-Goldstone fluctuations in the flat direction, radial fluctuations corresponding to a massive particle.

As it turns out, spontaneous breaking of Lorentz symmetry can be employed as well to obtain a massless graviton [6]. To this effect, we consider the Lagrangian density

$$\mathscr{L}_C = \frac{1}{2}C^{\mu\nu}K_{\mu\nu\alpha\beta}C^{\alpha\beta} - V(C^{\mu\nu}). \tag{6}$$

Here, $C^{\mu\nu}$ is a symmetric two-tensor, $K_{\mu\nu\alpha\beta}$ is the usual quadratic kinetic operator for a massless spin-2 field, and $V(C^{\mu\nu})$ is a scalar potential constructed from $C^{\mu\nu}$ and $\eta_{\mu\nu}$. This theory is invariant under local Lorentz transformations and under diffeomorphisms. The potential V ensures $C^{\mu\nu}$ takes some vacuum value $c^{\mu\nu}$, which we will assume to be nonzero. This spontaneously breaks local Lorentz and diffeomorphism invariances. The massless Nambu-Goldstone fields are the excitations $\delta C_{\mu\nu} = C_{\mu\nu} - c_{\mu\nu}$ about this solution, generated by the broken symmetries and maintaining the potential minimum:

$$\delta C_{\mu\nu} = C_{\mu\nu} - c_{\mu\nu} \approx \varepsilon_\mu{}^\alpha c_{\alpha\nu} + \varepsilon_\nu{}^\alpha c_{\alpha\mu} \equiv M_{\mu\nu}{}^{\alpha\beta}\varepsilon_{\alpha\beta} \tag{7}$$

with $M_{\mu\nu\alpha\beta} = \frac{1}{2}(\eta_{\mu\alpha}c_{\nu\beta} + \eta_{\nu\alpha}c_{\mu\beta} - \eta_{\mu\beta}c_{\nu\alpha} - \eta_{\nu\beta}c_{\mu\alpha})$. For generic $c_{\mu\nu}$, these fluctuations satisfy the four (linearized) constraints

$$\delta C^\mu{}_\nu (c^m)^\nu{}_\mu = 0 \tag{8}$$

with $m = 0, 1, 2, 3$.

The equations of motion that follow from varying the lagrangian density (6) with respect to the independent degrees of freedom $\varepsilon_{\mu\nu}$ are $M^{\mu\nu\rho\sigma}K_{\mu\nu\alpha\beta}\delta C^{\alpha\beta} = 0$, that can be solved using Fourier decomposition. The solutions obey the massless wave equation

$$\partial^\lambda \partial_\lambda \delta C_{\mu\nu} = 0, \tag{9}$$

subject to the Lorenz condition $\partial^\mu \delta C_{\mu\nu} = 0$. The Lorenz condition fixes four of the initial six independent degrees of freedom carried by the Lorentz generators, leaving two massless propagating degrees of freedom.

It is not difficult to show that this theory matches the usual description of a massless graviton field $h^{\mu\nu}$ in Minkowski spacetime. Consider the Lagrange density for a free massless graviton $h^{\mu\nu}$:

$$\mathscr{L}_h = \frac{1}{2} h^{\mu\nu} K_{\mu\nu\alpha\beta} h^{\alpha\beta}. \tag{10}$$

Initially $h^{\mu\nu}$ has ten degrees of freedom. The Lagrangian (10) is invariant under local diffeomorphisms

$$\delta h_{\mu\nu} = \partial_\mu \xi_\nu + \partial_\nu \xi_\mu. \tag{11}$$

Consequently, we can choose four gauge fixing conditions. Rather than adopting the usual transverse-traceless gauge, we pick a different gauge that yields a direct match with the new theory. For generic $c_{\mu\nu}$, we choose the conditions

$$h^\mu{}_\nu (c^m)^\nu{}_\mu = 0 \tag{12}$$

with $m = 0, 1, 2, 3$. From the equations of motion $K_{\mu\nu\alpha\beta} h^{\alpha\beta} = 0$ and the gauge conditions one finds that the solutions satisfy the usual wave equation for a massless field $\partial^\lambda \partial_\lambda h_{\mu\nu} = 0$, as well as the Lorenz condition $\partial^\mu h_{\mu\nu} = 0$. This leaves $10 - 4 - 4 = 2$ propagating degrees of freedom, and an explicit match with the theory of the cardinal field. Once more, we see that while the symmetry structure of the two theories is radically different, the equations are in direct correspondence at low energy.

The full nonlinear Einstein equations can be obtained by insisting on a consistent coupling to the energy-momentum tensor, by using a version of Deser's one-step procedure referred to above.

While reproducing the Einstein equations at lowest order, the new theory differs from general relativity in various ways.

In the pure-gravity sector, the new theory has subleading corrections to the Einstein equations in vacuum. They are of higher order in the Riemann tensor, and thus very small for nearly Minkowski spacetimes (and thus in laboratory and solar-system tests). However, they may lead to significant deviations from general relativity in extreme environments such as black holes, or the early Universe.

There are effects in the matter sector because of couplings of the type $c_{\mu\nu}T^{\mu\nu}$ involving the matter energy-momentum tensor. A framework for the comprehensive treatment of such effects exists that maintains standards of consistency such as stability and microcausality [7]. Numerous experimental searches looking for Lorentz-violating signals within this framework are currently under way.

Most interestingly, maybe, are structural differences related to excitations of the cardinal field in the non-flat directions. These will play a role at very high energies (presumably close to the Planck scale) and temperatures. Also, at high temperatures the potential V acquires corrections that restore local Lorentz symmetry by shifting the minimum of the effective potential to zero cardinal field value. These effects will have profound implications for the very early Universe.

It is clear that the quantum properties of the new model will differ significantly from those of general relativity. It the case of the bumblebee model it has been shown that nonpolynomial and superficially unrenormalizable potentials V can become renormalizable and stable when quantum corrections are included [8], a result that can be extended to the current gravity theory.

ACKNOWLEDGMENTS

This talk is based on work done in collaboration with V. Alan Kostelecký. R.P. acknowledges financial support from the Fundação para a Ciência e a Tecnologia (Portugal), and wishes to thank the Physics Department of Indiana University for hospitality.

REFERENCES

1. R. Kraichnan, MIT thesis, 1947; *Phys. Rev.*, **98**, 1118 (1955); A. Papapetrou, *Proc. Roy. Irish Acad.*, **52A**, 11 (1948); S. N. Gupta, *Proc. Phys. Soc. London*, **A65**, 608 (1952); R. P. Feynman, *Chapel Hill Conference*, 1956.
2. S. Deser, *Gen. Rel. Grav.*, **1**, 9 (1970).
3. Y. Nambu, *Phys. Rev. Lett.*, **4**, 380 (1960); J. Goldstone, *Nuov. Cim.*, **19**, 154 (1961); J. Goldstone, A. Salam, and S. Weinberg, *Phys. Rev.*, **127**, 965 (1962).
4. V. A. Kostelecky and S. Samuel, *Phys. Rev. Lett.*, **63**, 224 (1989); *Phys. Rev. D*, **40**, 1886 (1989).
5. R. Bluhm and V. A. Kostelecký, *Phys. Rev. D*, **71**, 065008 (2004).
6. V. A. Kostelecký and R. Potting, *Gen. Rel. Grav.*, **37**, 1675 (2005).
7. D. Colladay and V. A. Kostelecký, *Phys. Rev. D*, **58**, 116002 (1998); V. A. Kostelecký, *Phys. Rev. D*, **69**, 105009 (2004).
8. B. Altschul and V. A. Kostelecký, *Phys. Lett. B*, **628**, 106 (2005).

Numerical simulations of the gravitational wave background produced by binaries

Neus Puchades and Diego Sáez

Departamento de Astronomía y Astrofísica, Universidad de Valencia, Burjassot, Valencia, Spain

Abstract. The incoherent gravitational waves produced by the elements of some toy distributions of binaries are numerically superimposed and, then, the effect of the resulting signal on a detector formed by three particles (which move along geodesics in the solar system space-time) is calculated. The numerical code created in this preliminary work must be improved with the essential aim of performing appropriate simulations to be used –for data analysis– in LISA mission for the detection of gravitational waves (NASA/ESA).

Keywords: Background radiations:gravitational waves—Stars:binaries— Methods:numerical
PACS: 98.70.Vc, 04.30.-w, 04.80.Nn, 97.80.-d

INTRODUCTION

The Laser Interferometer Space Antena (LISA) will be sensitive to gravitational pulses lasting between ~ 10 seconds and ~ 3 hours and, moreover, binaries radiate gravitational waves with a period identical to that of their orbital motion; hence, only compact binaries with small enough periods are significant for LISA mission. There are various kinds of these binary systems [1]: X-ray bursters, polars, intermediate polar, cataclysmic variables, and so on. Although some of these binaries will finish in a violent collision, they are almost stable during very long periods of time. We are interested in the background of gravitational waves produced by compact binaries during their long quasi-stable evolutions. The number of sources creating this background is almost constant as a result of the low rate of binary coalescence.

The free parameters of a binary system are: mass of the primary component M_1, mass of the secondary component M_2, eccentricity ε, orbital period P_{orb}, and distance D between Earth and the binary system. Along this paper, binaries similar to 4U1820-30 and AM Herculis are considered. The first of these binaries belong to the X-ray bursters type and it is formed by a neutron star ($M_1 = 2.3 M_\odot$) and a white dwarf ($M_2 = 0.067 M_\odot$). Its orbital period is ~ 11.4 minutes and it is located at 6.13 Kpc from Earth. The second binary: AM Herculis (polars type) involves a white dwarf ($M_1 = 0.6 M_\odot$) and a low-mass star with Roche lobe overflow ($M_2 = 0.2 M_\odot$). Its period is ~ 3.09 hours and its distance to us is $\sim 79\ pc$. The eccentricity is considered as a free parameter in both cases. Other parameters are varied when appropriate.

In spite of the short orbital periods of the above compact binaries, the velocity of the stars with respect to the center of mass is $V \leq 0.01c$, where c is the speed of light; hence, these binaries can be described by using the slow motion approximation; then, in the transverse traceless (TT) gauge (see e.g. [2]), metric perturbations have the form:

$$h_{ij}^{TT} = \frac{G}{c^4}\frac{2}{D}\ddot{I}_{ij}\left[t - \frac{D}{c}\right], \qquad (1)$$

where G is the gravitational constant, and $\ddot{I}_{ij}$ are the second time derivatives of the traceless inertial tensor components. These derivatives are to be calculated at retarded positions; namely, at time $t' = t - \frac{D}{c}$, where t is the observation time. The components I_{ij} of the inertial tensor are:

$$I_{ij} = \sum_{A=1,2} M_A \left[X_{Ai}X_{Aj} - \frac{1}{3}\delta_{ij}X^2\right]. \qquad (2)$$

Our detector is assumed to be formed by three particles A, B, and C which move along time geodesics forming and equilateral triangle (LISA structure). The TT gauge can be realized in a system of cartesian coordinates (x',y',z') attached to the detector, in which the origin coincides with particle A and the z' axis is orthogonal to the detector plane. In this situation, coordinates of particle B at times t and 0 are related as follows:

$$X_B'^i(t) - X_B'^i(0) = \frac{1}{2}X_B'^j(0)\left[h_{ij}^{TT}(t)\right]_A, \qquad (3)$$

where the h_{ij}^{TT} quantities are calculated at point A. From this formula, it follows that oscillations in the h_{ij}^{TT} quantities leads to oscillations in the relative position of particles A and B (with related frequencies and amplitudes). Relative distance variations $\Delta l/l$ are calculated for pairs AB and AC.

THE CODE AND ITS APPLICATIONS

Our code works as follows: (1) each binary is described in a cartesian coordinate system (x,y,z) attached to it, in which the origin is the primary star (M_1), the z axis is perpendicular to the orbital plane, and the x axis is parallel to periastron direction, (2) tensor quantities are written in a new system of coordinates with its origin in the center of mass and with axis parallel to the x',y',z' axis attached to the detector, (3) the operator $P_{ij} = \delta_{ij} - n_i n_j$ – where n_i are the components of the unit direction vector of z' – is used to project tensor quantities on the detector plane, (4) the trace of the resulting h_{ij} is then taken off to get h_{ij}^{TT}, and finally, (5) the signals of all the binaries are added and Eq. (3) is used to calculate the detector response.

Hereafter, θ and ϕ are the spherical coordinates defining the direction orthogonal to the orbital plane in the cartesian reference associated to the detector; hence, $\theta = 0$ ($\theta = \pi/2$) means that the detector and orbital planes are parallel (orthogonal).

In order to test the code, it is first applied to an unique binary. Four cases (A, B, C, and D) are considered. In all these cases, it is assumed that mass M_2 is initially located in the positive direction of the x-axis which points towards the periastron. For an unique binary, this assumption does not imply any lost of generality. The condition $\theta = \phi = 0$ is satisfied in cases A and B. In case A, the binary has the parameters and location assigned to AM Herculis in the introduction, but the eccentricity is varied (three ε values are considered). The detector response $\Delta l/l$, for the pair AB, is presented in the top left panel of Fig. (1), where there are peaks whose amplitude increases as the eccentricity does. For $\varepsilon = 0.9$, the duration of the pulses associated to the peaks is $\sim T/10$, these pulses are produced while the secondary is either close to the periastron or close to the apastron; namely, while the accelerations take on their maximum values; of course, the greater the eccentricity, the greater (smaller) the accelerations (the interval of time elapsed near the periastron or apastron) and, consequently, the greater (smaller) the amplitude (duration) of the associated pulse. This means that large eccentricities lead to pulses whose duration is much smaller than the orbital period; hence, this type of pulses can be detected by LISA for some binary systems having orbital periods of a few days. The distribution of eccentricities in the set of binary systems is then crucial to look for the properties of the gravitational wave background produced by these systems. In the second case (hereafter called case B), the binary has the parameters assigned to 4U1820-30, but it is located at the same distance as AM Herculis (79 pc) for comparisons. The eccentricity takes on the same values as in first case. Since 4U1820-30 is more compact than AM Herculis, its accelerations (characteristic times) are greater (smaller) than those of AM Herculis. Results are presented in the top right panel of Fig. (1), where the amplitudes (pulse durations) are greater (smaller) than those of the top left panel. Cases C and D are identical to case A but the eccentricity $\varepsilon = 0.5$ has been fixed, whereas angle θ has been varied in case C and angle ϕ in case D. Results of case C (D) are displayed in the left (right) middle panel of Fig. (1). From these panels it follows that: (1) peak amplitudes decrease by a factor close to $8/3$ as angle theta varies from $\theta = 0$ to $\theta = \pi/2$ (see above for interpretation), and (2) amplitudes weekly depend on angle ϕ. Finally, the code is applied to superimpose the signals of two sets of five AM Herculis type binaries. These binaries have different eccentricities. In case E (F) the orbital periods are identical (different). In both cases, the orientation of the orbital plane and the initial position of the stars in their orbits are random. Results of case E (F) are presented in the left (right) panel of Fig. (1) for the pairs AB and AC. A complicated signal with the common period of all the superimposed binaries is obtained in case E (bottom left panel); however, the signal appearing in the bottom right panel (case F) is not periodic because the periods of the binaries are different.

The code gives reasonable results consistent with expectations. Parameters ε and θ are important. Realistic distributions of each type of binaries must be now considered and the associated backgrounds of gravitational waves must be statistically analyzed.

ACKNOWLEDGMENTS

This work has been supported by the Spanish MEC (project AYA2003-08739-C02-02 partially funded with FEDER) and also by the Generalitat Valenciana (grupos03/170). Calculations were carried out at the Centro de Informática de la Universidad de Valencia (CERCA and CESAR). One of us N.P. thanks the Spanish MEC for a fellowship.

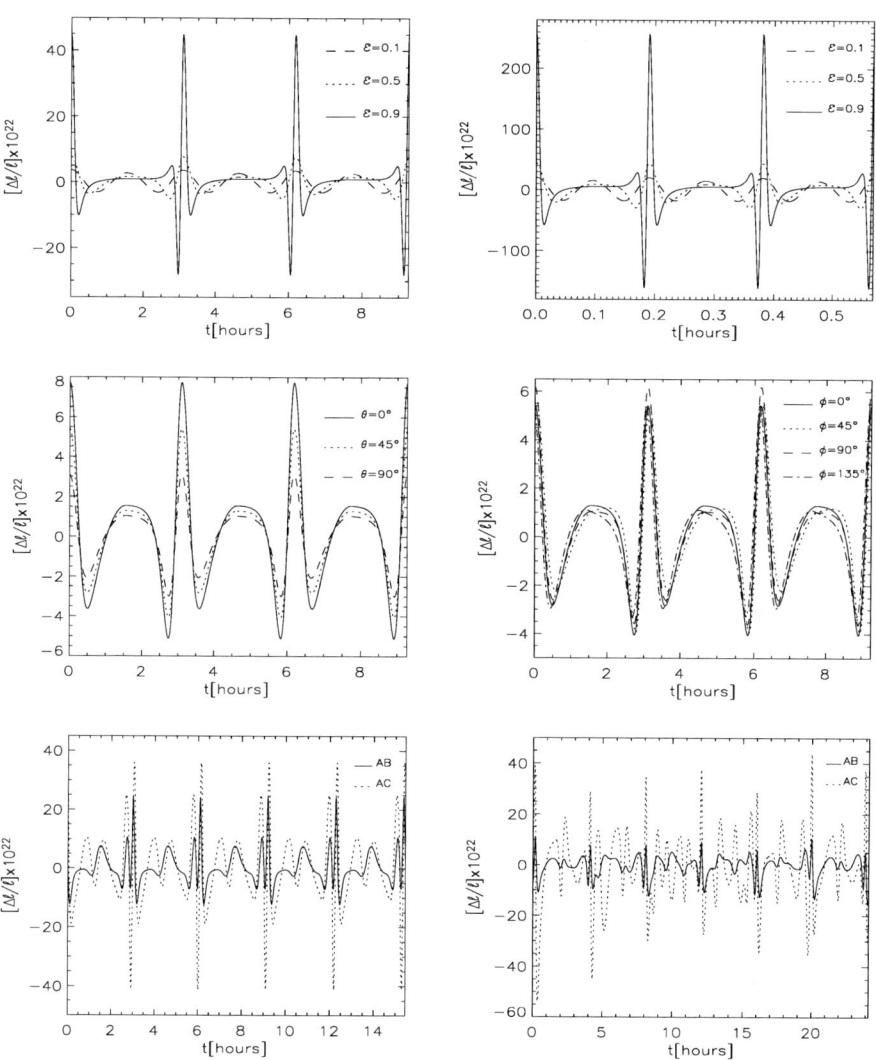

FIGURE 1. Quantity $\Delta l/l \times 10^{22}$ v.s. time in hours: top left, top right, middle left, middle right, bottom left, and bottom right panels corresponds to cases A, B, C, D, E, and F defined in the text, respectively

REFERENCES

1. V. M. Lipunov, K. A. Postnov and M. E. Prokhorov, *The scenario machine: Binary star population synthesis*, Harwood Academic Publishers, Amsterdam, 1996.
2. C. W. Misner, K. S. Thorne and J. A. Wheeler, *Gravitation*, W.H. Freeman and Company, New York, 1995.

Determination of Stellar Shape via Microlensing

Nicholas J. Rattenbury*, Phil Yock[†] and Ian Bond**

Jodrell Bank Observatory, Manchester, England
[†]*Department of Physics, The University of Auckland, New Zealand*
**Institute of Information and Mathematical Sciences, Massey University, New Zealand*

Abstract. Einstein predicted that light from a distant "source" star would be deflected by the gravitational field of an intervening "lens" star: the phenomenon known as gravitational microlensing. The lens star produces magnified and distorted images of the source, and as the lens passes between the observer and the source, the magnification changes. For lens systems in our Galaxy, events occur on timescales of weeks to months.

Lens systems comprised of more than one object can produce complex light curves. Such light curves can be analysed to obtain information about both the lens and the source systems. These analyses include the detection of low-mass extra-solar planets and the limb-darkening characteristics of distant stars. In this paper, we present the results from one extreme microlensing event for which the limb-darkening of the 5 kpc distant source star was determined, as well as limits on the shape of the projected source star profile. The effective resolution of these measurements is approximately 0.04 microarcsec.

Keywords: microlensing, stars, photometry
PACS: 95.85.Kr, 97.10.Kc, 98.62.Sb, 95.75.-z

MICROLENSING EVENT MOA 2002-BLG-33

Abe et al (2003) reported a microlensing event designated MOA 2002-BLG-33 (hereafter MOA-33), where a distant solar-like star was lensed by a binary star [1]. The light curve was measured in two passbands (infrared and visual, see Figure 1) and used to determine the characteristics of the binary lens system, and also the limb-darkening on the source star. In that analysis it was assumed that the source star was spherical. However, if a star is rotating sufficiently rapidly, it may present an elliptical aspect. Here we present a more general analysis in which possible ellipticity of the source star profile is allowed for.

The lens system in event MOA-33 was a close binary star. At one stage during the event the lens system centre-of-mass was almost co-linear with the source star, as seen from Earth, resulting in a very high magnification of the source star. In addition, during the time peak magnification for event MOA-33 the source star passed through the small central "caustic" of the binary lens[1].

Figure 2 shows the central caustic due to the binary lens and the circular source star used to model event MOA-33 by Abe et al (2003). At event time $HJD - 2450000 \simeq$

---

[1] In gravitational lensing with binary or more complex lenses, "caustic" curves (closed curves) form on which the magnification is extremely high. As the source passes through caustic lines, abrupt but continuous changes in source flux are observed.

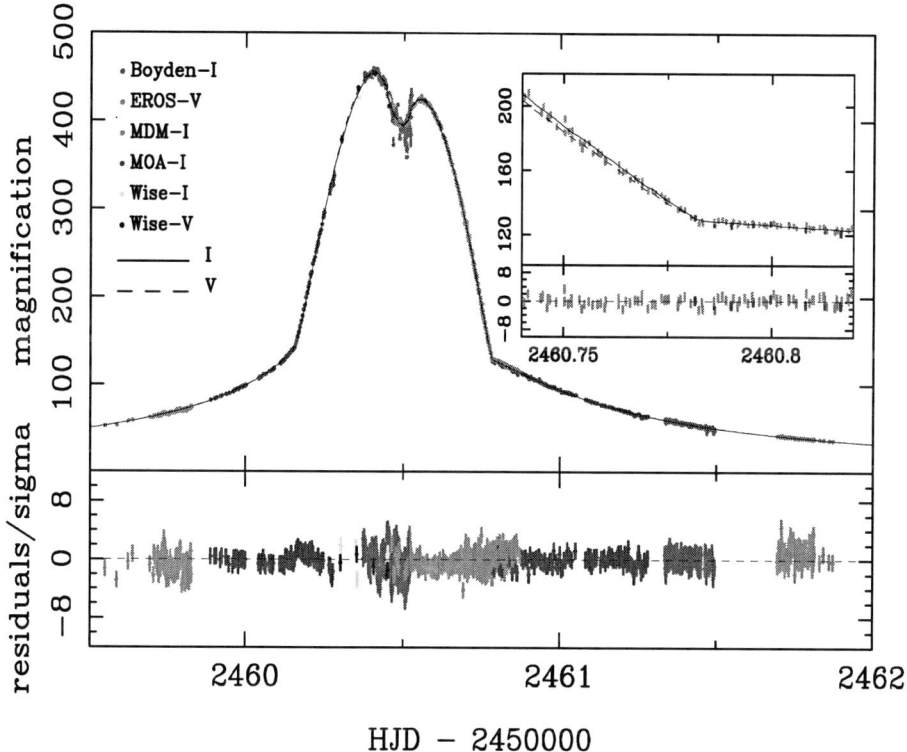

FIGURE 1. Observed light curve, best-fitting model and residuals from the model for event MOA-33 [1]. Data were obtained from five telescopes: the Boyden 1.5m (South Africa, 473 points), the EROS 1m (La Silla, 186 points), the MDM 2.4m (Kitt Peak, 260 points), the MOA 0.6m (New Zealand, 480 points) and the Wise 1.0m (Israel, 10 points). Figure reproduced from [1].

2460.5, the source star was closely bounded by the caustic curve. Any significant excursion from a circular source star profile will be apparent due to this unique situation.

ELLIPTICAL SOURCE PROFILE MODELLING

In order to model the source star profile as an ellipse instead of a circle, we introduce two further parameters in the light curve model algorithm: the profile eccentricity, e, and position angle β, measured with respect to the source star track (see inset Figure 2).

Event MOA-33 was modelled using light curves generated using the inverse ray-shooting technique [2], implemented on a cluster computer. The best fitting light curves were found by minimising χ^2 using a Monte Carlo Markov Chain technique [3]. To set limits on the source shape the values of e and β were fixed on a 2D grid and all parameters were allowed to vary in order to minimise χ^2.

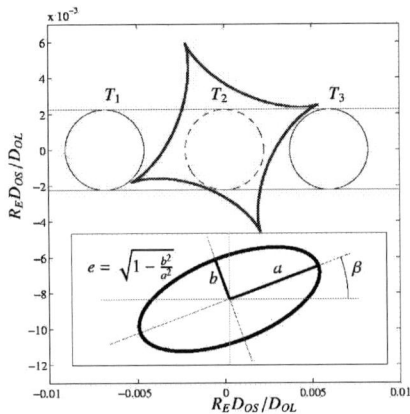

FIGURE 2. The caustic curves (asteroid) of the binary lens of event MOA-33. The source star (circles), shown to scale, has a relative motion horizontally from left to right. The axes are in units of the Einstein ring projected to the source plane. The circular source star profile is shown at the moments of the caustic entry and exit, corresponding to rapid changes in magnification. During a finite time interval the caustic curve completely enclosed the source star (dashed line). Any significant deviation from a circular source profile will produce significant features in the light curve. Inset: the current work uses an elliptical source profile, with apparent eccentricity, e, and position angle β, measured with respect to the source star track.

RESULTS

The χ^2 minimisation process was initially carried out with all 11 physical parameters introduced by Abe et al (2003) allowed to float over a coarse grid of values for the source shape parameters e and β. It was found that the limb-darkening coefficients did not vary significantly from the values obtained by Abe et al (2003) for a circular source star. Subsequent χ^2 minimisation was carried out for 40 values of e ranging from 0 to 0.485 and 41 values of β ranging from $-90°$ to $90°$ with the limb-darkening coefficients held fixed at values found by Abe et al (2003), allowing all other parameters to vary. The process was repeated six times to eliminate statistical artifacts. Figure 3 shows the values of χ^2 map as a function of the source parameters e and β.

Using the 95% contour line in Figure 3, and accounting for the possible rotation of the binary lens during the event, we set the following limits on the source profile eccentricity: $e = 0.17 \pm 0.17$.

DISCUSSION AND CONCLUSION

While the above limits on source profile eccentricity do not exclude a circular source, the method outlined here indicates that a certain fraction of microlensing events can be used to constrain the profile of the source star. Simulations of binary lens microlensing events show that around 0.1% of all microlensing events may have the geometry necessary for constraining the source profile in the manner described in this work. We note that this

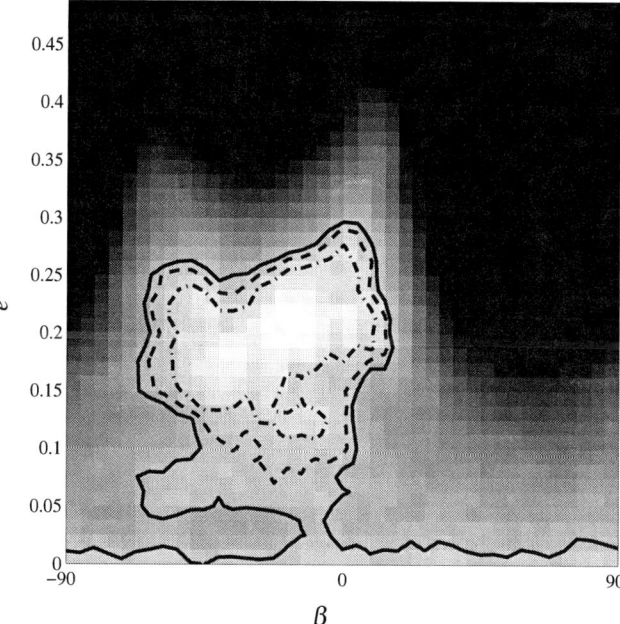

FIGURE 3. $\Delta\chi^2$ values from replacing the circular source profile used by Abe et al (2003) with an elliptical source star profile. Limb darkening was held fixed at the values found by Abe et al (2003), but all other fitting parameters are allowed to float, using a Markov Chain Monte Carlo model fitting algorithm. The colour code varies from white (~ -20) to light grey (~ 30), dark grey (~ 50), and black ($\gtrsim 90$). The contours are the 99% (solid), 95% (dashed) and 68% (dot-dash) confidence limits.

would require time sampling of the light curve of a similar order to that acheived for event MOA-33.

ACKNOWLEDGMENTS

The authors thank the EROS collaboration for the use of their data. The MOA project is supported by the Marsden Fund of New Zealand, the Ministry of Education, Culture, Sports, Science and Technology (MEXT) of Japan, and the Japan Society for the Promotion of Science (JSPS).

REFERENCES

1. F. Abe, and 32 co-authors, A&A **411**, L493–L496 (2003).
2. J. Wambsganss, MNRAS **284**, 172–188 (1997).
3. N. J. Rattenbury, I. A. Bond, J. Skuljan, and P. C. M. Yock, MNRAS **335**, 159–169 (2002).

Vacuum self-magnetization?

H. Pérez Rojas and E. Rodríguez Querts

*Instituto de Cibernética, Matemática y Física,
Calle E No. 309, esq. a 15 Vedado, C. Habana, Cuba*

Abstract. We study vacuum properties in a strong magnetic field as the zero temperature and zero density limit of quantum statistics. For charged vector bosons (W bosons) the vacuum energy density diverges for $B > B_c = m_w^2/e$, leading to vacuum instability. A logarithmic divergence of vacuum magnetization is found for $B = B_c$, which suggests that if the magnetic field is large enough, it is self-consistently maintained, and this mechanism actually prevents B from reaching the critical value B_c. For virtual neutral vector bosons bearing an anomalous magnetic moment, the instability of the ground state for $B > B'_c = m_n^2/q$ also leads to the vacuum energy density divergence for fields $B > B'_c$ and to the magnetization divergence for $B = B'_c$. The possibility of virtual electron-positron pairs bosonization in strong magnetic field and the applicability of the neutral bosons model to describe the virtual positronium behavior in a magnetic field are discussed. We conjecture that this could lead to vacuum self-magnetization in QED.

Keywords: magnetic field, QED vacuum
PACS: 95.30.Cq, 98.80.Cq, 12.20.Ds

INTRODUCTION

Macroscopic bodies become unstable when its rest energy is approximately equal to the internal one $Mc^2 \simeq U$. Classical instability is produced when the gravitational and rest energies of a body of mass M and radius R are of the same order $Mc^2 \sim GM^2/R$, leading to gravitational collapse. Quantum magnetic collapse for macroscopic magnetized objects has been also claimed to occur for high magnetic fields [1], [2], [3], when the magnetic energy density is of order of the internal energy density.

Quantum instability is produced e.g. in atoms when the energy gap for $e^- - e^+$ pair creation in the electric field created by the atomic nucleus vanishes for the atomic number $Z = 1/\alpha$: the vacuum boils [4]. On the other hand, the usual electroweak vacuum in an external magnetic field B is not stable for fields greater than some critical value $B_{wc} = m_w^2/e \sim 10^{24} G$ since the $W^\pm$ also vanishes [5], [6].

Here we want to address the problem of quantum vacuum stability in a magnetic field and the possibility of QED vacuum self magnetization, starting from the quantum statistical point of view and methods. We obtain vacuum energy, magnetization and pressure densities as the zero temperature and zero chemical potential limit of the quantum relativistic energy-momentum tensor of matter in the temperature formalism.

QUANTUM VACUUM IN AN EXTERNAL MAGNETIC FIELD

Relativistic quantum statistical averages in the limit $T \to 0$, $\mu \to 0$ (see e.g. [7]) leads to the quantum field averages in vacuum, $<<..>> \to <..>$. The contribution of observable particles, given by the statistical term $\Omega_s(T,\mu)$ in the expression for the total thermodynamic potential $\Omega = \Omega_s + \Omega_0$, vanishes in that limit, and the remaining term leads to the zero point energy of vacuum Ω_0. The non zero components of the total energy-momentum tensor in an external field $B_j = B\delta_{j3}$ are $\mathscr{T}_1^1 = \mathscr{T}_2^2 = -\Omega - B\mathscr{M}$, $\mathscr{T}_3^3 = -\Omega$, and $\mathscr{T}_4^4 = -(TS + \mu N + \Omega) = -U$, where $\mathscr{T}_j^i$ $(i,j = 1,2,3)$ are the anisotropic pressures[1], $\mathscr{M} = -\partial\Omega/\partial B$ is the magnetization, $S = -\partial\Omega/\partial T$ is the entropy density, $N = -\partial\Omega/\partial\mu$ is the density of particles minus antiparticles and U the internal energy density. We consider vacuum in a magnetic field as the limit of a medium for which $S = N = 0$, but other quantities, like $\mathscr{M}$ and $\mathscr{T}_j^i$ are non zero [3], [8], and their previous expressions are valid for $\Omega = \Omega_0$.

Charged vector bosons

For W bosons in a magnetic field the energy levels are $E_{W0} = \sqrt{p_3^2 + m_W^2 - 2eB}$, $E_W(n) = \sqrt{p_3^2 + m_W^2 + 2eB(n + \frac{1}{2})}$, $n = 0, 1, 2...$ [9]. The regularized zero point energy density (an Euler-Heisenberg like term) and the vacuum magnetization in the tree level approximation are[2] $\Omega_{0W} = -\frac{e^2 B^2}{16\pi^2}\int_0^\infty e^{-B_c x/B} f(x)\frac{dx}{x^2} < 0$ and $\mathscr{M}_{0W} = -2\frac{\Omega_{0W}}{B} + \frac{em_W^2}{16\pi^2}\int_0^\infty e^{-B_c x/B} f(x)\frac{dx}{x} > 0$, respectively, where $f(x) = \left(\frac{1+2\cosh 2x}{\sinh x} - \frac{3}{x} - \frac{7x}{2}\right)$. The system becomes unstable, that is Ω_{0W} diverges, for $B > B_c$. Vacuum shows a paramagnetic behavior, described by $\mathscr{M}_{0W}$, which diverges logarithmically for $B \to B_c$ (Fig. 1); by equating $B = 4\pi\mathscr{M}_{0W}$, one can obtain a vacuum self-magnetization for some $B \approx B_c$ (ferromagnetism). This suggests that the instability may be avoided if vacuum self-magnetize [11], [12].

Neutral vector bosons

It is believed that neutron stars magnetic fields could be produced due to ferromagnetic spin coupling of neutrons. The boson state resulting from such pairing is more favorable energetically, since its Gibbs free energy is smaller than that of the original neutron system. For neutral vector bosons with an anomalous magnetic moment we assume that $E_{nb}(\eta) = \sqrt{p_3^2 + p_\perp^2 + m_{nb}^2 + \eta qB(p_\perp^2 + m_{nb}^2)^{1/2}}$, $\eta = -1, 0, 1$ (see [13]). The zero point regularized energy density looks like[3] $\Omega_{0nb} = -\frac{(qm_{nb}B)^2}{8\pi^2}(I_{02} + $

---

[1] The anisotropy is due to the arising of a negative transverse pressure, generated by an axial "force". This force is the quantum analog of the Lorentz force, arising when the magnetic field acts on charged particles having non-zero spin, leading thus to a magnetization parallel to **B**.
[2] Note that the ground state $E_{W0}(p_3 = 0)$ vanishes for $B = B_c = m_W^2/e = 1.09 \times 10^{24} G$, and becomes imaginary for $B > B_{wc}$ [10].
[3] The ground state $E_{nb}(\eta = -1, \mathbf{p} = 0)$ vanishes for $B = B_c' = m_{nb}/q$, and becomes imaginary for $B > B_c'$, in analogy to the charged case, leading to the instability (divergence of Ω_{0nb} for $B > B_c'$). Note also that

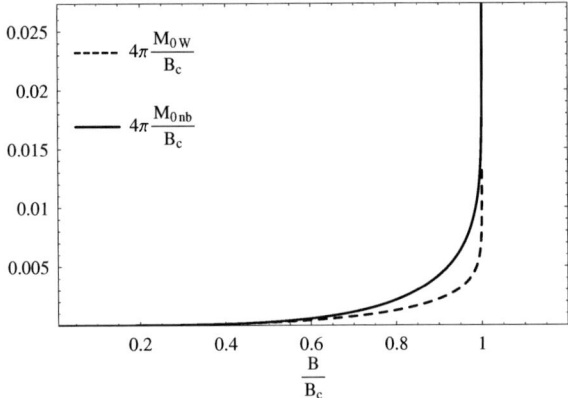

FIGURE 1. Vacuum magnetization for W bosons ($B_c = m_W^2/e$) and for neutral vector bosons ($B_c = B'_c = m_{nb}/q$), assuming that $m_{nb} = 2m_e$, $q = 2\mu_B$.

$I_{13} + I_{20}$), where $I_{0k} = \int_0^\infty e^{-\frac{B'_c}{B}x}(\cosh x - 1)\frac{dx}{x^k}$, $I_{1k} = \int_0^\infty e^{-\frac{B'_c}{B}x}(\cosh x - 1 - \frac{x^2}{2})\frac{dx}{x^k}$ and $I_{2k} = \int_0^\infty \int_0^\infty e^{-\frac{B'_c y^2}{Bx}}\left(\sinh y - y - \frac{y^3}{6}\right)\frac{y^{2k}}{x^{2+k}}dudx$, with $y = x(u+1)$. Vacuum instability arises for $B > B'_c = \frac{m_{nb}}{q}$. Neutron boson vacuum magnetization $\mathcal{M}_{0nb} = -2\frac{\Omega_{0nb}}{B} + \frac{qm_{nb}^3}{8\pi^2}(I_{01} + I_{12} + I_{21}) > 0$ diverges for $B \to B'_c$, due to the behavior of the states $E_{nb}(p_\perp = 0)$ (Fig. 1). This means that the neutral vector boson vacuum can also self-consistently maintain the magnetic field, for some $B \approx B'_c$.

QED vacuum

For $B \sim B_{ec} = m_e^2/e = 4.41 \times 10^{13}G$, the QED vacuum polarization effects, like the creation of electron-positron pairs by a photon, become important [4]. The electron-positron vacuum shows a paramagnetic behavior. Photons coexist with mutually independent virtual $e^\pm$ pairs and with bound $e^\pm$ virtual states (positronium), which is related to the singular behavior of the polarization operator $\Pi_{\mu\nu}$ near the thresholds for $e^\pm$ pair creation [14]. We want to address to the problem of the positronium (bound electron-positron pair) vacuum in a magnetic field, by considering the bosonic state as a (virtual) particle.

The positronium mass is[5] $m_p(B) = 2m_e - \triangle\varepsilon$. For the ground state, $\triangle\varepsilon$ reaches high

---

$-((qm_{nb}B)^2/8\pi^2)I_{02}$ is the contribution of the states $E_{nb}(p_\perp = 0)$.
[4] For an electron-positron vacuum in an external magnetic field, in the tree level approximation, $\mathcal{M}_{0e} = -2\Omega_{0e}/B - (e^2 B_c/8\pi^2)\int_0^\infty e^{-B_c x/B} F(x)_{HE} dx > 0$, but $\mathcal{M}_{0e} \ll B$ (see [8])
[5] $\triangle\varepsilon \ll m_e$ is the binding energy, which is due to the Coulomb interaction between the electron and the positron. In presence of the high magnetic field, the positronium particle moves parallel to B.

values when the distance between the Larmor orbits of the electron and the positron tends to zero [15]. The approximate expression for the energy of this state (orthopositronium) is $E_p = (p_3^2 + m_p^2 - qBm_p)^{1/2}$, with $q = 2\mu_B$. This means the bosonization of the pair resulting from the parallel and antiparallel spin coupling of virtual electrons and positrons, leading to a neutral boson with a magnetic moment $q = 2\mu_B$ and confined to move parallel to the field B, in states $E_{nb}(\eta = -1, \mathbf{p}_\perp = 0)$. These states lead to a logarithmic divergence of the neutral boson vacuum magnetization for $B \to B_{pc} = 2m_e^2/e$. But then there is a real solution for the Eq. $B = -4\pi\mathcal{M}_{0p}$ for certain $B \approx 8.82 \times 10^{13}$: vacuum self-magnetization might be possible in QED.

CONCLUSIONS

Charged vector bosons (W bosons) ground state energy leads to an instability of electroweak vacuum for $B > B_c = qm_W^2/e$. It is saved by equating $B = 4\pi\mathcal{M}$, since the field can be self-consistently maintained, the system becomes a ferromagnet. This mechanism actually prevents B from reaching B_c. For neutral vector bosons with an anomalous magnetic moment q the ground state also shows an instability for $B > B_c' = m_{nb}^2/q$. $\mathcal{M}_{0nb}$ also diverges for $B \to B_c'$ and, as a consequence, can be self-consistently maintained, keeping $B < B_c'$. We conjecture this mechanism might be applied to magnetized QED vacuum by assuming virtual positronium as the neutral vector particle with anomalous magnetic moment. Such phase transition would mean a spontaneus arising of an "order parameter", or symmetry breaking of vacuum.

REFERENCES

1. A. Pérez Martínez, H. Pérez Rojas and H. J. Mosquera Cuesta, *Eur.Phys.J.*, **C29**,111-123 (2003)
2. R.Gonzalez Felipe, H. J. Mosquera Cuesta, A. Pérez Martínez, H. Pérez Rojas, *Chin.J.Astron.Astrophys.*, **5**,399 (2005)
3. M. Chaichian, S. Masood, C. Montonen, A. Pérez Martínez, H. Pérez Rojas, *Phys. Rev. Lett.*, **84**, 5261 (2000).
4. L. I. Schiff, *Quantum mechanics*, New York, McGraw-Hill Book Co.(1955).
5. J. Ambjorn and P. Olesen, *Nucl. Phys.* **B 315** 606-614 (1989).
6. J. Ambjorn and P. Olesen, *Nucl. Phys.* **B 330** 193-204 (1990).
7. E. S. Fradkin, *Quantum Field Theory and Hydrodynamics*, Proc. of the P.N. Lebedev Inst. **No. 29**, Consultants Bureau (1967).
8. H. Perez Rojas, E. Rodriguez Querts, hep-ph/0402213.
9. H. Perez Rojas, *Acta Phys. Pol.*, **B17**, 861 (1986).
10. W. Heisenberg and H. Euler, *Z. Phys.*, **98**, 714 (1936)
11. E. Rodriguez Querts, A. Martin Cruz and H. Perez Rojas, *Int. J. Mod. Phys.*, **A17**, 561-573 (2002)
12. E. Rodriguez Querts, H. Perez Rojas and Perez Martinez, *Int. J. Mod. Phys.* , **D13**, 1261-1265(2004).
13. H. Perez Rojas, A. Perez Martinez and H. Mosquera Cuesta, *Bose-Einstein condensates of neutral particles with non-zero magnetic moment in strong magnetic fields*, to be published.
14. A. E. Shabad, *Ann. Phys.*, **90**, 166-195(1975).
15. A. E. Shabad, V. V. Usov, *Astrophys. and Space Sci.*, **117**, 309-325(1985).

Differential Rotation of *R*-Modes

Paulo M. Sá* and Brigitte Tomé*

*Departamento de Física and Centro Multidisciplinar de Astrofísica – CENTRA,
Universidade do Algarve, Campus de Gambelas, 8005-139 Faro, Portugal
E-mail: pmsa@ualg.pt, btome@ualg.pt

Abstract. Recent work has shown that differential rotation, producing large scale drifts of fluid elements along stellar latitudes, is an unavoidable feature of r-modes. We discuss the role of this differential rotation in the evolution of the r-mode instability of newly born, hot, rapidly rotating neutron stars.

Keywords: r-mode instability, differential rotation, neutron stars
PACS: 04.40.Dg, 95.30.Lz, 97.10.Kc, 97.10.Sj

R-modes are non-radial pulsation modes of rotating stars that are driven unstable by gravitational radiation reaction in all rotating perfect-fluid stars [1]. In newly born, hot, rapidly rotating neutron stars the r-mode instability is active for a wide range of relevant temperatures and angular velocities of the star, with the consequence that most of the angular momentum of the star is carried away by gravitational radiation and the star spins down to just a small fraction of its initial angular velocity [2, 3]. Gravitational radiation emitted during this spin-down phase could be detected by advanced versions of the laser interferometer detectors LIGO and VIRGO [4, 5].

Of crucial importance for a successful detection of gravitational radiation from such sources is the ability to predict the form of the gravitational-wave signals. In order to develop these waveform templates a model of the evolution of the r-mode instability is needed. Such a model was first proposed in Ref. [4]. Within this model it is assumed that the time evolution of the system (star and r-mode perturbation) is characterized by just two parameters: the angular velocity of the star $\Omega(t)$ and the amplitude of the r-mode $\alpha(t)$. Then, it is assumed that the angular momentum of the star, which is the sum of the angular momentum of the unperturbed star and the angular momentum of the r-mode perturbation, decreases through the emission of gravitational radiation. The angular momentum of the r-mode is assumed to be equal to the so-called canonical angular momentum, a quantity whose evaluation only requires the knowledge of the solution of the *linearized* hydrodynamic equations (note that at the time this model was proposed a nonlinear extension of the r-mode solution was not yet available). Within the model of Ref. [4], it was also assumed that the energy of the r-mode (in the co-rotating frame) increases due to the emission of gravitational radiation and decreases due to the dissipative effects of viscosity. The above assumptions lead to a system of differential equations determining the time evolution of the angular velocity of the star and of the amplitude of the mode.

This model works pretty well during the initial (linear) phase of the evolution, which lasts for about 500 sec: the amplitude of the mode, α, grows exponentially on the gravitational radiation timescale and the angular velocity of the star, Ω, decreases very slowly on the viscous dissipation timescale. However, during the nonlinear phase, the evolution of the r-mode instability cannot be described appropriately by this model; it simply does not provide a physical mechanism to saturate the amplitude of the r-mode. This problem is circumvented by saturating α by hand, i.e., before all the angular momentum of the star is radiated away by gravitational radiation, α is fixed at an arbitrary constant value, α_{sat}. With α fixed, the angular velocity of the star Ω decreases rapidly on the gravitational timescale to values which depend critically on the saturation value of the mode's amplitude. Choosing α_{sat} appropriately one can achieve agreement between the final value of the angular velocity of the star predicted by the model and the value inferred from astronomical observations.

The arbitrariness in the choice of the saturation value of the mode's amplitude is an unpleasant feature of the model proposed in Ref. [4]; it would be desirable, of course, that α saturates, due to some nonlinear hydrodynamic effect, in a natural way. This is, as will be explained below, what happens in a modified model of the evolution of the r-mode instability for certain values of the initial amount of differential rotation.

The existence of differential rotation of kinematical nature induced by r-modes was suggested in Ref. [6] and soon afterwards confirmed by numerical studies, carried both in general relativistic [7] and Newtonian hydrodynamics [8]. However, only recently an analytical solution, representing differential rotation of r-modes that produces large scale drifts of fluid elements along stellar latitudes, was found within the nonlinear Newtonian theory up to second order in the mode's amplitude [9].

The knowledge of the nonlinear solution allows us to calculate the physical angular momentum of the r-mode perturbation up to second order in the mode's amplitude. As shown in Ref. [9], since differential rotation *does*, in general, contribute to the physical angular momentum of the r-mode perturbation, this quantity is not equal to the canonical angular momentum. Therefore, the model of Ref. [4], which assumes that the physical angular momentum is simply the canonical angular momentum, should be modified. Such modification and the investigation of the role of differential rotation in the evolution of the r-mode instability was performed in Ref. [10] and will be described here succinctly.

Let us assume that the total angular momentum of the star is the sum of the angular momentum of the unperturbed star and the full physical angular momentum of the r-mode perturbation, the latter being given by [10]

$$\delta^{(2)} J_{r\text{-}mode} = \frac{1}{2}(4K+5)\tilde{J}MR^2\alpha^2(t)\Omega(t). \qquad (1)$$

In deriving the previous expression, only the most unstable $l = 2$ r-mode was considered and it was assumed that the mass density ρ and the pressure p of the fluid are related by a polytropic equation of state $p = k\rho^2$, with k such that the mass and the radius of the star take the values $M = 1.4 M_\odot$ and $R = 12.53$ km, and $\tilde{J} = 0.01635$. The constant K is related to the initial amount of differential rotation associated with the r-mode and its value lies in the interval $-5/4 \leq K \ll 10^{13}$. The upper limit is related to the fact that one wishes to impose the condition that the initial absolute value of the physical angular momentum of the perturbation is much smaller than the initial value of the angular momentum of the unperturbed star. For $K < -5/4$, the total angular momentum of the star decreases and eventually becomes negative, as the amplitude of the mode grows due

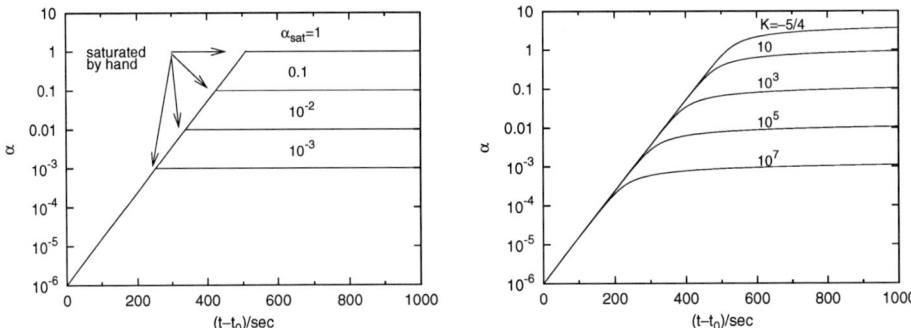

FIGURE 1. Time evolution of the amplitude of the r-mode, α, for different values of α_{sat} in the model of Ref. [4] (left panel) and for different values of K in our model (right panel).

to the gravitational radiation instability. To avoid this unphysical situation, we consider only $K \geq -5/4$, for which the total angular momentum of the star is always positive.

Within this modified model, in the initial stages of the evolution of the r-mode instability, α increases exponentially as [10]

$$\alpha(t) \simeq \alpha_0 \exp\left\{0.027 \left(\frac{\Omega_0}{\Omega_k}\right)^6 \left(\frac{t-t_0}{\sec}\right)\right\}, \quad (2)$$

while for later times α increases very slowly as

$$\alpha(t) \simeq 2.48 \left(\frac{\Omega_0}{\Omega_k}\right)^{3/5} \left(\frac{t-t_0}{\sec}\right)^{1/10} \frac{1}{\sqrt{K+2}}, \quad (3)$$

where α_0 and Ω_0 are the initial values of the mode's amplitude and of the star's angular velocity, respectively, and Ω_k is the Keplerian angular velocity at which the star starts shedding mass at the equator. The smooth transition between the regimes (2) and (3) occurs for $t - t_0 \simeq$ few $\times 10^2$ seconds (see Fig. 1). Equations (2) and (3) reveal that the amplitude of the r-mode saturates in a natural way and that the saturation amplitude depends critically on the parameter K, i.e., on the initial amount of differential rotation present in the model. This contrasts with the results of Ref. [4], where the amplitude of the r-mode has to be saturated by hand at an arbitrary value (see Fig. 1).

Let us now turn our attention to the time evolution of the angular velocity of the star. In the initial stages of the evolution, Ω decreases as [10]

$$\frac{\Omega(t)}{\Omega_0} \simeq 1 - 0.13(K+2)\alpha_0^2 \exp\left\{0.054 \left(\frac{\Omega_0}{\Omega_k}\right)^6 \left(\frac{t-t_0}{\sec}\right)\right\}, \quad (4)$$

while for later times Ω decreases slowly as

$$\frac{\Omega(t)}{\Omega_0} \simeq 1.30 \left(\frac{\Omega_0}{\Omega_k}\right)^{-6/5} \left(\frac{t-t_0}{\sec}\right)^{-1/5}, \quad (5)$$

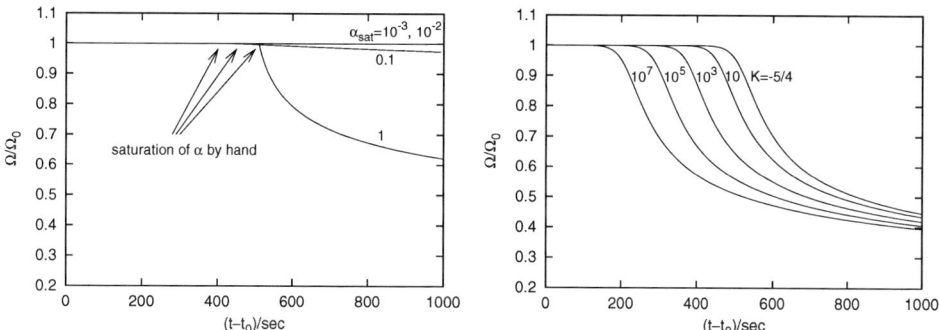

FIGURE 2. Time evolution of the angular velocity of the star, Ω, for different values of α_{sat} in the model of Ref. [4] (left panel) and for different values of K in our model (right panel).

the smooth transition between the regimes (4) and (5) occurring for $t - t_0 \simeq$ few $\times 10^2$ seconds (see Fig. 2). Remarkably, in the later phase of the evolution, the angular velocity Ω does not depend on the value of K, and, consequently, does not depend on the saturation value of α. As can be seen in Fig. 2, already at $t - t_0 = 1000$ s, for values of K ranging from $-5/4$ to 10^7, the angular velocities are not very different. For later times, any difference becomes negligibly small. This contrasts with the results obtained in Ref. [4], where the value of Ω depends critically on the choice of α_{sat} (see Fig. 2).

After about one year of evolution (i.e., at the time viscosity effects become dominant and start damping the r-mode) the angular velocity of the star becomes $0.042\Omega_k$ (for any K and $\Omega_0 = \Omega_k$), in good agreement with the inferred initial angular velocity of the fastest pulsars associated with supernovae remnants [10].

ACKNOWLEDGMENTS

This work was supported in part by the *Fundação para a Ciência e a Tecnologia* (FCT), Portugal. BT acknowledges financial support from FCT through grant PRAXIS XXI/BD/21256/99.

REFERENCES

1. N. Andersson, *Astrophys. J.* **502**, 708 (1998).
2. L. Lindblom, B. J. Owen and S. M. Morsink, *Phys. Rev. Lett.* **80**, 4843 (1998).
3. N. Andersson, K. D. Kokkotas and B. F. Schutz, *Astrophys. J.* **510**, 846 (1999).
4. B. J. Owen, L. Lindblom, C. Cutler, B. F. Schutz, A. Vecchio and N. Andersson, *Phys. Rev. D.* **58**, 084020 (1998).
5. P. Arras, E. E. Flanagan, S. M. Morsink, A. K. Schenk, S. A. Teukolsky and I. Wasserman, *Astrophys. J.* **591**, 1129 (2003).
6. L. Rezzolla, F. K. Lamb and S. L. Shapiro, *Astrophys. J. Lett.* **531**, L139 (2000).
7. N. Stergioulas and J. A. Font, *Phys. Rev. Lett.* **86**, 1148 (2001).
8. L. Lindblom, J. E. Tohline and M. Vallisneri, *Phys. Rev. Lett.* **86**, 1152 (2001).
9. P. M. Sá, *Phys. Rev. D.* **69**, 084001 (2004).
10. P. M. Sá and B. Tomé, *Phys. Rev. D* **71**, 044007 (2005).

Comparing Rees-Sciama and Integrated Sachs Wolfe effects

D. Sáez*, N. Puchades*, M.J. Fullana[†] and J.V. Arnau**

*Departamento de Astronomía y Astrofísica, Universidad de Valencia, Burjassot, Valencia, Spain
[†]Institut de Matemàtica Multidisciplinar, Universitat Politècnica de València, València, Spain
**Departamento de Matemática Aplicada, Universidad de Valencia, Burjassot, Valencia, Spain

Abstract. Both the Rees-Sciama (RS) and the Integrated Sachs-Wolfe (ISW) effects are produced by the peculiar gravitational potential of cosmological inhomogeneities. This potential fully defines a scalar perturbation of the Robertson-Walker metric in the longitudinal gauge. The RS (ISW) effect is produced by nonlinear (linear) structures which grow as a result of gravitational instability. The anisotropies corresponding to these two effects are compared for a wide range of angular scales. N-body simulations and a certain ray-tracing procedure –recently proposed– are used to study the RS effect.

Keywords: cosmic microwave background—cosmology:theory—large-scale structure of the Universe
PACS: 98.70.Vc, 98.80.Cq, 98.80.Es

INTRODUCTION

A realization of the so-called convergence model is assumed along this paper. The Hubble constant is $H_0 = 100h\ km\ s^{-1} Mpc^{-1}$ with $h = 0.71$. A cosmological constant whose density parameter is $\Omega_\Lambda = 0.73$ produces the current acceleration of the universe. The density parameters of cold dark matter and baryonic matter are $\Omega_d = 0.23$ and $\Omega_b = 0.04$, respectively; hence, the density parameter of matter is $\Omega_m = \Omega_b + \Omega_d = 0.27$. In this flat universe, quantum fluctuations of the inflationary field generated a gaussian distribution of adiabatic energy density perturbations with Zel'dovich spectrum. The power spectrum of these perturbations –obtained with CMBFAST [1]– is normalized using the condition $\sigma_8 = 0.93$ [2]. Space comoving coordinates are denoted x^i.

The partial time derivative of the peculiar gravitational potential $\phi(x^i,t)$ produces temperature contrasts $\Delta T/T = (T - T_B)/T_B$, where T_B is the background temperature. Given an observation direction $\vec{n}$, the associated contrast is given by the following formula:

$$\frac{\Delta T}{T}(\vec{n}) = 2 \int_{t_L}^{t_0} \frac{\partial \phi(x^i,t)}{\partial t} dt \ . \qquad (1)$$

This integral is to be calculated along the background null geodesics

$$x^i = \lambda(z) n^i \ , \qquad (2)$$

where z is the redshift and

$$\lambda(z) = H_0^{-1} \int_0^z \frac{db}{[\Omega_m(1+b)^3 + \Omega_\lambda]^{1/2}} . \tag{3}$$

In a flat universe with cosmological constant, linear structures contribute to the integral in Eq. (1). This contribution is the so called Integrated Sachs-Wolfe (ISW) effect, whereas the contribution from nonlinear inhomogeneities is the Rees-Sciama (RS) effect. In next section, the angular power spectra of the ISW [3] and RS [4] effects are calculated and compared for a wide range of ℓ values.

COMPARING ISW AND RS EFFECTS

Since the ISW (RS) effect is produced by linear (nonlinear) scales, some considerations about linearity are worthwhile. Our discussion is based on the root mean square (rms) of the relative mass fluctuations inside an sphere of radius R (randomly placed in the universe). This rms value is given by the following formula:

$$\sigma(R) = \langle (\delta M/M)^2 \rangle^{1/2} = \frac{1}{2\pi^2} \int_0^\infty k^2 P(k) W(k) dk ; \tag{4}$$

function $W(k) = \frac{9}{y^6}(\sin y - y \cos y)^2$, where $y = kR$, is the window function of the R-sphere [6]. The $\sigma(R)$ values required below are numerically calculated by using Eq. (4) and the power spectrum $P(k) = \langle |\delta_k|^2 \rangle$ given by CMBFAST. It is usually assumed that overdensities with a size $2R$ are linear for $\sigma(R) \leq 1$. Following this criterium, regions having sizes greater than $30\, Mpc$ [$\sigma(15) \simeq 0.7$] are assumed to be linear along this paper and, accordingly, in plane wave expansions, linear (nonlinear) calculations must involve spatial scales greater (smaller) than $\sim 60\, Mpc$. This separation scale, $k_s = 2\pi/60$, is appropriate here because the nonlinear scales –producing the RS effect– are studied in boxes of $256\, Mpc$, in which, scales smaller than $60\, Mpc$ are expected to be well described. On the contrary, under the assumption that linearity requires e.g. $\sigma(R) \sim 0.2$, linear perturbations must have sizes larger than $120\, Mpc$ [$\sigma(60) \simeq 0.2$], and the new separation scale, $k_s = 2\pi/240$, is too large to be described inside our boxes. In such a case, scales between $60\, Mpc$ and $240\, Mpc$ would require another treatment (mildly nonlinear regime).

Since the ISW effect [3] is produced by linear energy density fluctuations, analytical calculations are possible. The growing mode of these fluctuations is

$$D_1(z) = \frac{1}{x} \left[\frac{2}{x} + x^2\right]^{1/2} \int_0^x \left[\frac{2}{y} + y^2\right]^{-3/2} dy , \tag{5}$$

where

$$x = \left[\frac{2\Omega_\lambda}{\Omega_m}\right]^{1/3} (1+z)^{-1} . \tag{6}$$

595

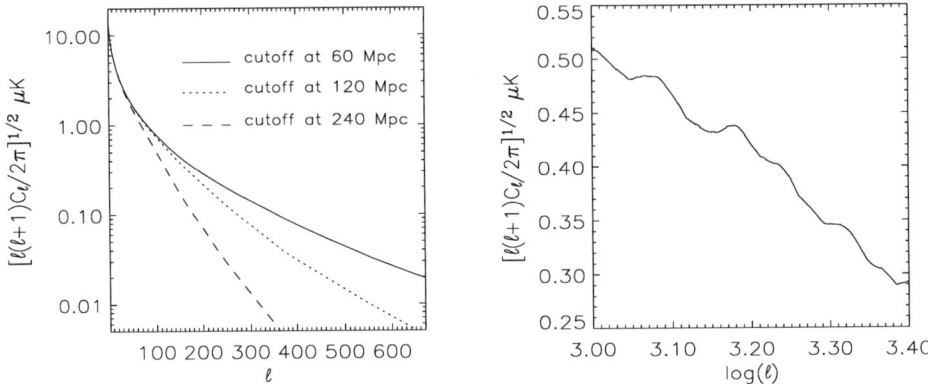

FIGURE 1. Left (right) panel displays the ISW (RS) angular power spectra in μK, v.s. ℓ ($log(\ell)$)

The angular power spectrum of the ISW effect is given by the following formula [5]:

$$C_\ell = N \int_0^\infty \frac{P(k)}{k^2} \xi_\ell^2(k) dk , \qquad (7)$$

where:

$$\xi_\ell(k) = \int_{\lambda_0}^{\lambda_L} j_\ell[\lambda k] \frac{d}{d\lambda}\left[(1+z)D_1(z)\right] d\lambda ; \qquad (8)$$

symbol j_ℓ stands for the spherical Bessel function of order ℓ, and

$$N = \frac{18 H_0^4}{\pi}\left[\frac{\Omega_m}{D_1(0)}\right]^2 . \qquad (9)$$

The integrals in Eqs. (7) and (8) are numerically calculated to get the angular power spectrum of the ISW effect. The integral of the r.h.s. of Eq. (7) is first performed in the k-interval $[k_{min}, k_{max}]$, where $k_{max} = 2\pi/60$ and $k_{min} = 2\pi/60000$; thus, the effect of nonlinear scales with sizes smaller than 60 Mpc is not considered at all (cutoff at 60 Mpc). The Solid line of the left panel of Fig. (1) shows the resulting spectrum. Quantity $\Delta_T = [\ell(\ell+1)C_\ell/2\pi]^{1/2}$ appears to be absolutely negligible (smaller than $\sim 0.02~\mu K$) for $\ell > 670$. The cutoff is also performed for 120 and 240 Mpc in order to study the contributions of different spatial scales to the ISW effect. One easily see that: (a) only scales greater than $\sim 240~Mpc$ contribute to the ISW effect for $\ell \leq 80$ and, (b) for these ℓ values, quantity $\Delta_T(\ell)$ is greater than $\sim 1~\mu K$.

Now, let us consider the RS effect. In this case, numerical nbody simulations are used to evolve strongly nonlinear structures inside a box of 256 Mpc size. A grid with 512 points per edge is used in the simulations; hence, the cell size is 0.5 Mpc, and the spatial resolution of the PM code is $\sim 1~Mpc$. A set of 512^3 particles is evolved in our box. Resolution is not good enough to describe the cluster cores in detail, but it suffices to get a very good potential for RS calculations. The Universe is then covered by simulation boxes and, consequently, it is an artificial periodic universe where CMB photons move.

Periodicity magnifies the RS effect, but it has been proved that this magnification becomes negligible if photons move along special directions. Following these directions, photons cross successive boxes through distinct statistically independent zones. The preferred directions defined in references [7] and [8] are used here. We have verified that the main part of the RS effect is produced while photons travel from the initial redshift 5 to present time. During this time interval, the CMB photons cross around 30 boxes ($\sim 8000\ Mpc$). Relative temperature variations are calculated for many directions which regularly pixelise a map whose angular size is $\sim 2° \times 2°$ (the image of a box face located at redshift 5). The resulting maps (20 of them) are analyzed to get an averaged angular power spectrum in the ℓ interval [1000, 2500], this spectrum is presented in the right panel of Fig. (1). Quantity Δ_T ranges from 0.29 to 0.51; hence, it takes on values much greater than those corresponding to the ISW effect in the same ℓ interval [extrapolation in left panel of Fig. (1)]. These values are too small to be detected in current and scheduled experiments (WMAP, PLANCK and so on).

In reference [4], a certain approach was used to get the RS angular power spectrum in some models of structure formation. Afterwards, other methods ([9], [10], [7]) based on ray-tracing through numerical simulations were described. These methods allow us to describe nonlinear evolution with high resolution and, consequently, they could be useful to improve on previous estimates [4] of the RS angular power spectrum, at least, for very large ℓ values; namely, for very small angular scales which are affected by strongly nonlinear structures; after using the method of reference [7] with the highest resolution allowed by computational cost, it must be recognized that the resulting spectrum [right panel of Fig. (1)] has not any relevant feature at large ℓ values (could we find some of these features with other n-body codes allowing higher resolutions in our box?). For $\ell > 1000$, our spectrum is very similar to those of [4], which decrease as ℓ increases. For $\ell < 1000$ the approach of reference [4] should work; hence, after extrapolation to these ℓ-values, our spectrum should be qualitatively similar to those presented in [4]; namely, it should have a maximum for a ℓ value of a few hundreds and, in this maximum, Δ_T would be only a little greater that 0.5 (its value at $\ell = 1000$).

In conclusion, taking into account the above discussion and the spectra of Fig. (1) and paper [4], one esaily concludes that: (i) the ISW effect clearly dominates for $\ell < 40$ ($\Delta_T > 2\ \mu K$), (ii) ISW and RS effects are comparable for $40 < \ell < 200$ and, (iii) RS clearly dominates for $\ell > 200$. Detection only seems to be possible in case (i), in which the contribution of other dominant effects is $\Delta_T < 35\ \mu K$.

ACKNOWLEDGMENTS

This work has been supported by the Spanish MEC (project AYA2003-08739-C02-02 partially funded with FEDER) and also by the Generalitat Valenciana (grupos03/170). Calculations were carried out at the Centro de Informática de la Universidad de Valencia (CERCA and CESAR). One of us N.P. thanks the Spanish MEC for a fellowship.

REFERENCES

1. U. Seljak, and M. Zaldarriaga, *ApJ*, **469**, 437 (1996)
2. V. Eke, S. Cole, and C.S. Frenk, *MNRAS*, **282**, 263 (1996)
3. W. Hu, and S. Dodelson, *Annu. Rev. Astron. Astrophys.*, **40**, 171 (2002)
4. U. Seljak, *ApJ*, **463**, 1 (1996)
5. M.J. Fullana, and D. Sáez, *New Astronomy*, **5**, 10 (2000)
6. P.J.E. Peebles, *The Large Scale Structure of the Universe*, Princeton University press, Princeton, 1980
7. P. Cerdá-Durán, V. Quilis, and D. Sáez, *Phys. Rev.*, **69D**, 043002 (2004)
8. L. Antón, P. Cerdá-Durán, V. Quilis, and D. Sáez, *ApJ*, **628**, 1 (2005)
9. B. Jain, U. Seljak U, and S. White, *ApJ*, **530**, 547 (2000)
10. M. White, and W. Hu, *ApJ*, **537**, 1 (2001)

Supersymmetric quantization in midisuperspace

Alfredo Macías[*], Hernando Quevedo[†] and Alberto Sánchez[*]

[*]*Departamento de Física, Universidad Autónoma Metropolitana–Iztapalapa*
A.P. 55–534, México D.F. 09340, México
[†]*Instituto de Ciencias Nucleares, Universidad Nacional Autónoma de México*
A.P. 70-543, México D.F. 04510, México

Abstract. We use the canonical quantization of $N = 1$ supergravity for the case of Gowdy T^3 cosmological models. We show the existence of physical states for these models, and obtain the wave function of the Universe. We also analyze the behavior of the wave function of the universe near the classical curvature singularity of these models.

Keywords: Supersymmetry, $N = 1$ Supergravity
PACS: 04.60.Kz, 04.65.+e, 12.60.Jv, 98.80.Hw

INTRODUCTION

Since the pioneering works of DeWitt and Misner [1], the amount of effort spent in studying the quantization of cosmological model has been enormous. DeWitt and Misner began the **mini-superspace quantization/quantum cosmology** program in order to describe the evolution of cosmological spacetimes as trajectories in the finite dimensional sector of the superspace. Since then the minisuperspace models and their quantum version have been extensively studied. In 1987 Macias, Obregon and Ryan [2] introduced the supersymmetric quantum cosmology approach by applying $(N = 1)$ supergravity to quantum minisuperspaces. In 1994 Carrol, Freedman, Ortiz, and Page [3] showed that there are no physical states in $N = 1$ supergravity, unless there exists an infinite set of gravitino modes. In 1998 Macias, Mielke and Socorro [4] showed that there are no-physical states in supersymmetric quantum cosmology. On the other hand, as it is well known the canonical quantization procedure does not provide time evolution. In this work we consider the problem of finding physical states and the time evolution considering a specific midisuperspace described by Gowdy T^3 cosmological models in the context of $N = 1$ supergravity, and allowing that the time evolution be driven by a classical solution. We will find an explicit expression for the wave function of the universe and we will show that these considerations leads to the avoidance of the classical singularity.

$N = 1$ SUPERGRAVITY AND POLARIZED GOWDY T^3 MODELS

The Hamiltonian form of the $N = 1$ Supergravity [5] can be written as follows: $H = N\mathscr{H}_\perp + N^i \mathscr{H}_i + \frac{1}{2}\omega_{0AB}J^{AB} + \overline{\Psi}_0 S$ where $\mathscr{H}_\perp$, $\mathscr{H}_i$, J^{AB} and S are the Hamiltonian, diffeomorphism, rotational Lorentz and supersymmetric constraints, respectively. They are constructed from the canonical variables only and do not depend on the corresponding multipliers: $N = e_0^{\ 0}$ (the lapse function), $N_i = e_i^{\ 0}$ (the shift vector), ω_{0AB} (normal component of the connection) and $\overline{\Psi}_0$ (normal component of the gravitino field). The supergravity constraints satisfy the usual algebra discovered by Teitelboim [6] and, according to Dirac's canonical quantization procedure [7], the physical states $|\Psi\rangle$ must be annihilated by the corresponding constraint operators, i.e. : $S|\Psi\rangle = 0$, $\mathcal{H}_A|\Psi\rangle = 0$, $J_{AB}|\Psi\rangle = 0$, where the constraint $S|\Psi\rangle = 0$, on account of the Teitelboim's algebra ($\{S(x), \overline{S}(x')\} = \gamma^A \mathcal{H}_A \delta(x, x')$), and $S|\Psi\rangle = 0$ implies that $\mathcal{H}_A|\Psi\rangle = 0$, therefore, we have to focus only on the Lorentz J_{AB} and on the supersymmetric S constraints. Using the densitized local components $\phi_a = e e_a^{\ \alpha} \Psi_\alpha$, as the basic gravitino field, the Lorentz constraint can be written in the form: $J_{AB} = \frac{1}{2} \phi_{[A}^T \phi_{B]}$. The supersymmetric constraint has the form: $S = \varepsilon^{0\alpha\beta\delta} \gamma_5 \gamma_\alpha D_\beta \Psi_\delta$, where a factor ordering is usually chosen and for the γ matrices we use the real Majorana representation.

The metric for the polarized Gowdy T^3 models can be written as [9]: $ds^2 = e^{-\frac{\lambda}{2}+\frac{\tau}{2}}(e^{-2\tau}d\tau^2 - d\chi^2) - e^{-\tau}[e^P d\sigma^2 + e^{-P}d\delta^2]$, where P and λ depend on the non-ignorable coordinates τ and χ. The vacuum field equations for this line element read: $P_{\tau\tau} - e^{-2\tau}P_{\chi\chi} = 0$, $\lambda_\tau = P_\tau^2 + e^{-2\tau}P_\chi^2$ and $\lambda_\chi = 2P_\chi P_\tau$. Using separation of variables we can solve the equation for P to obtain the solution: $P(\tau, \chi) = \sum_{n=0}^{\infty} [A_n \cos(n\chi) + B_n \sin(n\chi)][C_n J_0(ke^{-\tau}) + D_n N_0(ke^{-\tau})]$, where A_n, B_n, C_n and D_n are arbitrary constants and J_0, N_0 are Bessel functions. The function λ can be calculated by quadratures once P is known. In order to analyze the behavior of the polarized Gowdy models we use the general "asymptotically velocity term dominated" (AVTD) solution [8] which has been shown to represent the general behavior near the singularity. The AVTD behavior implies that at the singularity all spatial derivatives of the field equations can be neglected and only the temporal behavior is relevant. This leads to a "truncated" set of differential equations which in the case of polarized Gowdy T^3 models can be obtained by neglecting all the derivatives with respect to the spatial coordinate χ in the vacuum field equations. The general solution to this "truncated" system is given by [8]: $P_{AVTD} = \ln[a(e^{-c\tau} + b^2 e^{c\tau})]$, $\lambda_{AVTD} = \lambda_0 + c^2\tau$, where a, b, c, and λ_0 are arbitrary real constants. The singularity situated at $\tau \to \infty$ is characterized by a blow up of the curvature which is determined by the behavior of the AVTD solution.

PHYSICAL STATES AND AVTD BEHAVIOR

Assuming the wave function of the universe $|\Psi\rangle = \begin{pmatrix} \Psi_I \\ \Psi_{II} \\ \Psi_{III} \\ \Psi_{IV} \end{pmatrix}$, the Lorentz constraint leads to the following system of equations [9]: $J_{12}\Psi_{III} = -J_{13}\Psi_{IV}$, $J_{12}\Psi_{II} = J_{23}\Psi_{IV}$ and $J_{13}\Psi_{II} = -J_{23}\Psi_{III}$, where J_{ab} are the Lorentz generators. Therefore, the components Ψ_I, Ψ_{II}, Ψ_{III}, and Ψ_{IV} of the wave function must be considered as 4×1 matrices. The relation between the components of the Lorentz generators can be solved as a product of γ-matrices, under the condition that the corresponding Lorentz algebra is preserved [10]. Using the standard generators of the standard rotation group $O(3)$ and solving explicitly the Lorentz constraint we obtain for the wave function of the universe the expression [10]

$$|\Psi\rangle = \begin{pmatrix} \Psi_I \\ \Psi_{II} \\ \Psi_{III} \\ \Psi_{IV} \end{pmatrix} = E \begin{pmatrix} a_0 \\ b_0 \\ c_0 \\ d_0 \end{pmatrix}, \qquad (1)$$

where E is a function to be determined, $\mathbf{a}_0$ is an arbitrary 4-vector and b_0, c_0 and d_0 are arbitrary constants. The densitized form of the gravitino field [9] reads: $\psi_1 = e^{-\frac{1}{2}(\lambda+\tau)}\phi_1$, $\psi_2 = e^{-\frac{1}{4}(\lambda+5\tau)}\phi_2$, $\psi_3 = e^{-\frac{1}{4}(\lambda+5\tau+2P)}(\phi_3 + e^P Q\phi_2)$. The explicit form of the supersymmetric constraint S is

$$S = e^{-\frac{1}{4}(\lambda+7\tau)}S_1 + e^{-\frac{1}{2}(\lambda+2\tau-P)}S_2, \qquad (2)$$

where $S_1 = (\partial_\chi - \frac{1}{4}\lambda_\chi + \frac{1}{4}P_\chi)\Gamma^1 + (\partial_\chi - \frac{1}{4}\lambda_\chi - \frac{1}{4}P_\chi)\Gamma^2 + \frac{1}{2}e^P Q_\chi \Gamma^3$, $S_2 = (e^{-P}\Gamma^4 + Q\Gamma^5)\partial_\sigma - \Gamma^5\partial_\delta$ and Γ^i are 4×4 matrices. The physical states are given as the non-trivial solutions of the equation $S|\Psi\rangle = 0$, or equivalently $S_1|\Psi\rangle = 0$ and $S_2|\Psi\rangle = 0$. If we assume that the wave function depends only on the spatial coordinate χ, the constraint $S_2|\Psi\rangle = 0$ is identically satisfied and we only need to solve $S_1|\Psi\rangle = 0$. For T^3 polarized Gowdy cosmological models the supersymmetric constraint $S|\Psi\rangle = 0$ leads to a set of first order partial differential equations for the components of the wave function of the universe [9]. When we consider the special case where the components of the wave function are independent of the spatial coordinates σ and δ it is straightforward to show that the general solution is given by

$$|\Psi\rangle = \begin{pmatrix} \Psi_I \\ \Psi_{II} \\ \Psi_{III} \\ \Psi_{IV} \end{pmatrix} = e^{\frac{1}{4}(\lambda \mp iP)} \begin{pmatrix} a_0 \\ b_0 \\ c_0 \\ d_0 \end{pmatrix}, \qquad (3)$$

where a_0, b_0, c_0, and d_0 are arbitrary constants satisfying the relationships $a_0^2 + b_0^2 = 0$ and $c_0^2 + d_0^2 = 0$, whose solution reads $a_0 = \pm ib_0$ and $c_0 = \mp id_0$. Now we can analyze the behavior of the wave function at the classical singularity. The wave function of the universe in the AVTD sector reads: $|\Psi\rangle_{AVTD} \sim \exp(\lambda_{AVTD}/2) = \exp[(\lambda_0 + c^2\tau)/2]$. Since the metric function λ_{AVTD} diverges linearly as $\tau \to \infty$, we see that the corresponding wave function diverges near the classical singularity. However, if we consider the special solution: $P = A_1 \cos\chi J_0(e^{-\tau}) + A_2 \cos 2\chi J_0(2e^{-\tau})$, we can integrate the field equations for λ and obtain:

$$\begin{aligned}\lambda &= -A_1^2 e^{-\tau} J_0(e^{-\tau}) J_1(e^{-\tau}) \sin^2\chi + \frac{8}{3} A_1 A_2 e^{-\tau} J_0(2e^{-\tau}) J_1(e^{-\tau}) \cos^3\chi \\ &+ \frac{4}{3} A_1 A_2 e^{-\tau} J_0(e^{-\tau}) J_1(2e^{-\tau}) \cos\chi (\cos 2\chi - 2) - 2A_2^2 e^{-\tau} J_0(2e^{-\tau}) J_1(2e^{-\tau}) \sin^2 2\chi \\ &- \frac{1}{2} A_1^2 e^{-2\tau} [J_1(e^{-\tau})^2 - J_0(e^{-\tau}) J_2(e^{-\tau})] - 2A_2^2 e^{-2\tau} [J_1(2e^{-\tau})^2 - J_0(2e^{-\tau}) J_2(2e^{-\tau})].\end{aligned}$$

Using this result, it is easy to show that the wave function of the universe remains constant and finite for all values of χ and τ. Fig. 1 shows this behavior. Consequently,

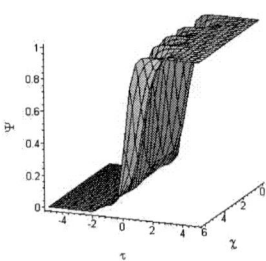

FIGURE 1. Behavior of the wave function of the universe. The AVTD singular behavior at $\tau \to \infty$ has disappeared and instead it arises a well–behaved function with a constant value.

we observe that the classical cosmological singularity of this particular Gowdy model has been removed. We can conclude that the presence of the fermionic gravitino field solves the singularity problem in this particular midisuperspace model.

CONCLUSIONS

We used the canonical formalism of the $N = 1$ supergravity to study the quantum cosmological behavior of polarized Gowdy T^3 models. We obtained the wave function of the universe for these models and we show that the presence of the gravitino field removes the classical initial singularity.

ACKNOWLEDGMENTS

This work was supported by CONACyT grants 42191–F and 36581–E.

REFERENCES

1. B. S. DeWitt, Phys. Rev. **160**, 1113 (1967); C. W. Misner, Phys. Rev. **186**, 1319 (1969).
2. A. Macías, O. Obregón and M. P. Ryan, Class. Quantum Grav. **4** 1477 (1987).
3. S.M. Carroll, D. Z. Freedman, M.E. Ortiz, and D. Page, Nucl. Phys. **B423** 661 (1994).
4. A. Macías, E.W. Mielke, and J. Socorro, Phys. Rev. **D57** 1027 (1998).
5. M. Pilati, Nucl. Phys. **B132** 138 (1978).
6. C. Teitelboim, Phys. Lett. **B69** 240 (1977); Phys. Rev. Lett. **38** 1106 (1977); R. Tabensky and C. Teitelboim, Phys. Lett. **B69** 453 (1977).
7. P.A.M. Dirac: *Lectures on Quantum Mechanics* (Academic Press, New York, 1965).
8. A. Sánchez, A. Macías and H. Quevedo, J. Math. Phys. **45**, 1849 (2004).
9. A. Macías, H. Quevedo, and A. Sánchez, *Gowdy T^3 Cosmological Models in $N = 1$ Supergravity*, gr-qc/0505013.
10. A. Macías, H. Quevedo, and A. Sánchez, *On the local Lorentz invariance in $N = 1$ supergravity*, gr-qc/0506038.

Light scattering test regarding the relativistic nature of heat

A. Sandoval-Villalbazo* and L.S. García-Colín[†]

*Departamento de Física y Matemáticas, Universidad Iberoamericana Lomas de Santa Fe 01210 México D.F., México
[†]Departamento de Física, Universidad Autónoma Metropolitana México D.F., 09340 México and El Colegio Nacional, Centro Histórico 06020 México D.F., México

Abstract. The dynamic structure factor of a simple relativistic fluid is calculated. The coupling of acceleration with the heat flux present in Eckart's version of irreversible relativistic thermodynamics is examined using the Rayleigh Brillouin spectrum of the fluid. A modification of the width of the Rayleigh peak associated to Eckart's picture of the relativistic nature of heat is predicted and estimated.

Keywords: Rayleigh scattering, relativistic thermodynamics
PACS: 0.11.44.,

INTRODUCTION

The interpretation of heat in the theory of relativity is a question that has often been ignored in the literature. Besides Tolman's exhaustive study of the problem dating from seventy years ago [1] it is hardly met in a book on the subject. In 1940, C. Eckart proposed a formalism to deal with the relativistic properties of fluids constructing an energy-matter tensor including heat [2]. Since this inclusion allows an identification of heat as mechanical energy, there is a conflict with the standard tenets of irreversible thermodynamics.

On the other hand, the present authors introduce heat into the formalism [3] using the same methodology as it appears in Meixner's version of irreversible thermodynamics [4]. This idea has not yet met universal acceptance. In this paper we propose an experiment using the scattering of light by a relativistic fluid which would throw light on the two proposals. It is shown that the Rayleigh peak ought to be broadened by a measurable correction if Eckart is correct. It is perhaps likely that its features may be detected in some features of the cosmic background radiation.

BASIC FORMALISM IN ECKART'S THERMODYNAMICS

The starting point in Eckart's irreversible thermodynamics is the stress-energy tensor:

$$T^\alpha_\beta = \frac{n\varepsilon}{c^2} u^\alpha u_\beta + p h^\alpha_\beta + \Pi^\alpha_\beta + \frac{1}{c^2} q^\alpha u_\beta + \frac{1}{c^2} u^\alpha q_\beta. \qquad (1)$$

Here, ε is the internal energy per particle, u is the hydrodynamic four-velocity, p is the local pressure, Π^α_β is the Navier tensor and q^α is the heat four-flux. $h^\alpha_\beta = \delta^\alpha_\beta +$

$\frac{1}{c^2}u^\alpha u_\beta$ is the spatial projector, where $u^\alpha u_\alpha = -c^2$ is the speed of light. The dissipative contributions to the stress-energy tensor in Eq. (1), according to Eckart's formalism, are assumed to satisfy the orthogonality relations:

$$u_\alpha \Pi^\alpha_\beta = u^\beta \Pi^\alpha_\beta = 0, \qquad u_\alpha q^\alpha = u^\beta q_\beta = 0. \qquad (2)$$

Eqs. (2) prevent the existence of dissipation in the time axis. The number of particles per unit of volume n satisfies a conservation equation, so that

$$\dot{n} + nu^\alpha_{;\alpha} = \dot{n} + n\theta = 0. \qquad (3)$$

The total energy balance equation is obtained from the expression $u^\beta T^\alpha_{\beta;\alpha} = 0$, which as shown in Ref. [3] reads as

$$n\dot{\varepsilon} + p\theta + q^\alpha_{;\alpha} + \frac{1}{c^2}\dot{u}_\alpha q^\alpha + u^\beta_{;\alpha}\Pi^\alpha_\beta = 0. \qquad (4)$$

The entropy balance equation is obtained by means of the local equilibrium hypothesis, $s = s(n,\varepsilon)$, so that

$$n\dot{s} = n\left(\frac{\partial s}{\partial n}\right)_\varepsilon \dot{\varepsilon} + n\left(\frac{\partial s}{\partial \varepsilon}\right)_n \dot{n}, \qquad (5)$$

where s is the entropy per particle. Using well known thermodynamical relations and Eqs. (3-4), Eq.(5) becomes after some rearrangements,

$$(nsu^\alpha + \frac{q^\alpha}{T})_{;\alpha} = -\frac{u^\beta_{;\alpha}\Pi^\alpha_\beta}{T} - \frac{q^\alpha T_{,\alpha}}{T^2} - \frac{1}{c^2}\frac{\dot{u}_\alpha q^\alpha}{T}, \qquad (6)$$

which leads to identify the total entropy flux as $J^\alpha_{[S]} = nsu^\alpha + \frac{q^\alpha}{T}$, and the local entropy production Σ as

$$\Sigma = -\frac{\theta\Pi}{T} - \frac{\sigma^\beta_\alpha \overset{o}{\Pi}^\alpha_\beta}{T} - \frac{q^\alpha}{T^2}(T_{,\alpha} + \frac{T\dot{u}_\alpha}{c^2}). \qquad (7)$$

In Eq. (7) Π^α_β is assumed symmetric and has been separated into its scalar and traceless components Πh^α_β and $\overset{o}{\Pi}^\alpha_\beta$. The traceless symmetric part of the velocity gradient is denoted as σ^β_α. As mentioned in Ref. [3], linear constitutive equations consistent with Eqs. (2) read:

$$\Pi = -\eta_B \theta, \qquad (8)$$

$$\overset{o}{\Pi}^{\alpha\beta} = -\eta_s h^{\mu\alpha} h^{\nu\beta} \sigma_{\mu\nu}, \qquad (9)$$

$$q^\alpha = -k_{th} h^{\alpha\beta}(T_{,\beta} + \frac{T\dot{u}_\beta}{c^2}). \qquad (10)$$

In Eqs. (8-10) the transport coefficients are η_B, the bulk viscosity, η_s the shear viscosity and k_{th} is the thermal conductivity. The entropy production (7) and the constitutive equation (10) differ from their counterparts in Meixner's scheme [3] by the presence of the acceleration term $\frac{T \dot{u}_\beta}{c^2}$ whose appearance can be traced back to the inclusion of heat in the energy momentum tensor, a feature not present in Meixner's formalism. The effect of this term in the RB spectrum will be the central subject of the next section.

LINEARIZED EQUATIONS AND THE RB SEPCTRUM.

As it has been repeatedly shown in the literature [5], the transport equations derived from Meixner's scheme arise from the conservation laws supplemented by the constitutive equations (8-10). If a local variable X is assumed to posses an equilibrium value X_o and a fluctuation around this equilibrium value δX, then we write $X = X_o + \delta X$. Assuming a fluid with vanishing hydrodynamic velocity $u^\alpha = (0,0,0,c)$, the linearized particle conservation equation (3) reads

$$\frac{\partial}{\partial t}(\delta \rho) + \rho_o \delta \theta = 0. \tag{11}$$

The linearized internal energy equation (4) can be written in terms of the temperature, leading to the expression

$$\rho_o \frac{\partial}{\partial t}(\delta T) + \frac{T_o(\gamma - 1)}{\rho_o} \frac{\partial}{\partial t}(\delta \rho) - \gamma D_{th}[\nabla^2(\delta T) + \frac{T_o}{c^2} \frac{\partial \theta}{\partial t}] = 0, \tag{12}$$

where $\gamma = \frac{c_p}{c_\rho}$ is the heat capacities ratio and $D_{th} = \frac{k_{th}}{\rho_o C_\rho}$ is the thermal diffusivity. The linearized divergence of the equation of motion reads:

$$\frac{\partial}{\partial t}(\delta \theta) + \frac{C_T^2}{\rho_o} \nabla^2(\delta \rho) - \alpha C_T^2 \nabla^2(\delta T) - D_v \nabla^2(\delta \theta) = 0. \tag{13}$$

$C_T^2 = \left(\frac{\partial p}{\partial \rho}\right)_T$ is the square of the isothermal speed of sound, $\alpha = -\frac{1}{\rho_o}\left(\frac{\partial \rho}{\partial T}\right)_p$ is the thermal expansion coefficient and $D_v = \frac{\eta}{\rho_o}$ is the kinematic viscosity; for an ideal gas $\alpha = \frac{1}{T_o}$. Volumetric viscosity has been neglected as well as second order contributions to the fluctuations. On the other hand, the linearized equations that arise from Meixner's formalism resemble equations (11-13) except for the term $\frac{C_T^2}{c^2} D_{th} \gamma \omega^2$ which appears in the energy equation. The two sets of equations may be solved, as shown in Ref. [5] by going into the $(\vec{q}, \omega)$ space. The procedure is the standard one, a system of three equations for the three unknowns $\hat{T}(\vec{q}, \omega)$, $\hat{\rho}(\vec{q}, \omega)$ and $\hat{\theta}(\vec{q}, \omega)$ is obtained. The corresponding determinant reads as follows:

$$A = \begin{bmatrix} \omega & \rho_o & 0 \\ \frac{C_T^2}{\rho_o} q^2 & \omega + D_v q^2 & -\frac{C_T^2}{T_o} q^2 \\ \frac{T_o}{\rho_o}(\gamma - 1)\omega & -\gamma \frac{D_{th} T_o}{c^2} \omega & \omega + \gamma D_{th} q^2 \end{bmatrix}. \tag{14}$$

The matrix (14), has been written for Eckart's equations, which differ only from its Meixner's counterpart in the 3 − 2 component of the matrix A. The full dispersion relation, $\det(A) = 0$, reads:

$$\omega^3 + (D_v q^2 + \gamma D_{th} q^2 - \frac{C_T^2}{c^2} D_{th} q^2 \gamma)\omega^2 + \\ (-C_T^2 q^2 \gamma + D_{th} D_v q^4 \gamma)\omega - C_T^2 D_{th} q^4 \gamma = 0. \tag{15}$$

The only addition in the dispersion relation arising from the inclusion of heat in the stress tensor regarding the Rayleigh peak is $\frac{C_T^2}{c^2} D_{th} q^2 \gamma$ in the ω^2 term. The ordinary width of the Rayleigh peak corresponds to the real solution of Eq. (15) when $\gamma \to 1$, namely, $\omega \simeq -D_{th} q^2$.

The modification to this root is proportional to $\frac{C_T^2}{c^2}$ and the new root can be fairly approximated as:

$$\omega \sim -D_{th} q^2 \left[1 + \frac{C_T^2}{c^2}\right]. \tag{16}$$

Thus, for an ideal gas, the relative change of the width in the Rayleigh peak at temperature T is proportional to the relativistic parameter $z = \frac{kT}{mc^2}$. This parameter is a natural measure of relativistic effects in systems such as the hot electron gas present in galaxy clusters.

FINAL REMARKS

In a typical light scattering experiment [5], the width of the central peak is of the order of $10^7 Hz$. The absolute associated change in this width, according to Eckart's version of relativistic irreversible thermodynamics would be about $10^3 Hz$ for a an ideal electron gas with temperature $T \sim 10^8 K$, with a non-vanishing thermal conductivity. This change is, in principle, measurable and a way to verify or discard experimentally Eckart's picture of the relativistic nature of heat.

ACKNOWLEDGMENTS

The authors wish to thank A.L. García Perciante for valuable comments and discussions. This work has been supported by CONACyT project 41081-F and FICSAC, México (PFSA).

REFERENCES

1. R.C. Tolman, "Relativity, Thermodynamics and Cosmology", Dover Publications Inc., N.Y. (1987).
2. C. Eckart, Phys. Rev. 58, 919 (1940).
3. L.S. García Colín and A. Sandoval-Villalbazo, J. Non. Eq. Thermodyn. (2005) (in press) [gr-qc/0503047]
4. S.R. de Groot and P. Mazur, "Non-equilibrium thermodynamics", Dover Publications Inc., Mineola, N.Y. (1984).
5. B.J. Berne and R. Pecora, "Dynamic light scattering", Dover Publication Inc, Mineola., N.Y. (2000).

The initial conditions of the Universe and holography

Pablo Díaz*, Miguel Angel Per* and Antonio Seguí*

*Departamento de Física Teórica
Universidad de Zaragoza. 50009-Zaragoza. Spain*

Abstract. We address the initial conditions for an expanding cosmology using the holographic principle. For the case of a closed model, the old prescription of Fishler and Susskind, that uses the particle horizon to encode the bulk degrees of freedom, can be implemented for accelerated models with enough acceleration. As a bonus we have singularity free bouncing models. The bound is saturated for co-dimension one branes dominated universes.

Keywords: Holography, Cosmology, Singularities, Quintessence, Closed Models.
PACS: 11.25.-w, 98.80.-k, 95.36.+x, 98.80.Qc

INTRODUCTION

Holography is one of the underlying principles of a full theory of quantum gravity [1]. This principle states that if the gravitational field is quantized, the degrees of freedom of a physical system can be mapped to other degrees of freedom of a dual theory now living on the boundary of the original one. The paradigmatic realization of holography is the AdS/CFT correspondence [2]. In the semiclassical regime, when the curvature scale is larger than the Planck length, the residual principle adopts the form of different sorts of entropy bounds. A characteristic of all of them is that the number of degrees of freedom of a physical system does not scale any more with the volume of the system but with the area that covers the system. This drastic reduction in the number of operative degrees of freedom is due to the formation of black holes. If we try to pack a given amount of information in smaller regions, because information is carried by energy, we pack an amount of energy in regions that in the limit can collapse to form a black hole; the information is lost at all effects.

The first time that the holographic principle was used to understand the FRW cosmological models was in [3]; the prescription was to map the different degrees of freedom traversing the past light cone of the particle horizon (PH), on the PH area. Concretely, the prescription was to impose that the total amount of the entropy of the cosmic fluid traversing the past light cone of the PH was smaller than one quarter of the area of the PH at any cosmic time. For flat and open FRW universes, this covariant bound was effective; however, for closed spatial models, the bound is violated. The reason is that for closed models the PH, that at the beginning of the expansion grows, because the spatial sections are three spheres, at a given time begins to shrink diminishing its area, being less and less effective to store the increasing amount of entropy crossing its past light cone. In the limit, the area goes to zero as it reaches the antipodal point that constitutes a focusing point for the light cone. To solve this and other problems, Bousso introduced its covariant entropy bound prescription using the light cone as the recipient of the entropy and comparing this amount with the area of some *preferred screens* that were appropriately defined [4]. A crucial characteristic of the preferred screens was to define what constitutes its *interior*. This characteristic was able to solve the problems previously commented for the Fischler Susskind entropy bounds on closed FRW models.

Although the Bousso prescription is operationally effective for dealing with closed FRW models, its physical interpretation is not clear. When the PH traverses the apparent horizon, its interior changes side from the north hemisphere (where the big bang takes places) to the south one; this is because the interior is defined as the region where the null geodesic congruence orthogonal to the boundary has negative expansion. So, using Bousso's prescription we are relating the entropy that traverses the *future* light cone to the area of the PH using an inversion of the arrow of time that seems not clear. We study in this communication the possibility of continuing the use of the PH as the depositary of the degrees of freedom carried by the cosmic fluid that traverses its *past* light cone. This will restrict the nature of the cosmic fluid.

CLOSED QUINTESSENCE MODELS ARE SINGULARITY FREE.

In order to map the information traversing the past light cone of the PH onto itself, we need that the size of the PH never shrinks. To accomplish this, the expansion must be fast enough to prevent the horizon recolapse; in this way the area of the PH never decreases. So we need to deal with accelerated expansions. The expansion is accelerated if $\ddot{R} > 0$, where $R(t)$ is the scale factor of the universe. This occurs if $-1 \leq \omega < -1+2/n$, where n is the number of spatial dimensions and ω appears in the equation of state of the cosmic fluid relating pressure and energy density $p = \omega \rho$.

It is interesting to note that the accelerated closed models do not present initial nor final singularities; in fact the relation between the scale factor R and the conformal cosmic time $d\eta = dt/R(t)$ as derived from the Friedman equations is

$$R(\eta) = R_m (\sin \frac{n(1+\omega) - 2}{2} \eta)^{\frac{2}{n(1+\omega)-2}}, \tag{1}$$

and presents two different regimes: if the exponent $\frac{2}{n(1+\omega)-2}$ is positive, which corresponds to a decelerated expansion, the scale factor (1) expands from an initial singularity at the big bang, attains a maximum value R_m and then shrinks symmetrically until a final singularity is reached at the big crunch. If the exponent in (1) is negative, i.e. with values that corresponds to accelerated expansion, there is not initial, no final singularity and the arrow of time extends along the complete real line, describing a bouncing model, where the scale factor attains its minimum (bounce) at $R(t=0) = R_m$ given by

$$R_m = R_0 (\frac{1}{1 - \Omega_0^{-1}})^{\frac{1}{n(1+\omega)-2}}; \tag{2}$$

R_0 and Ω_0 is the size and density of the universe at a given reference time t_0. This bouncing models are also known as oscillating models of second kind [5].

We see that these accelerated models, needed to realize the Fischler Susskind entropy bound, has as a bonus that they are singularity free. However we will see that not all the accelerated models are able to give enough room in the PH to encode the amount of entropy traversing its past light cone; we need enough accelerated models that we study in the next section.

THE BOUND IS SATURATED FOR CO-DIMENSION 1 BRANE FLUIDS

We need to compare the area of the PH (measured in four times the Planck area) and the total flux of entropy that traverses the past light cone of the horizon. We suppose that the expansion is adiabatic and the evolution of the density of entropy is given by

$$s(t) = s_0 \frac{R_0^n}{R(t)^n}. \tag{3}$$

We also use the comoving coordinate χ given by

$$\chi = \int_0^r \frac{dr'}{\sqrt{1-r'^2}}, \tag{4}$$

so that $r(\chi) = \sin(\chi)$. We rewrite (1) using $\alpha(n, \omega)$ defined by

$$\alpha(n, \omega) = \frac{n(1+\omega)}{n(1+\omega) - 2}, \tag{5}$$

so that, because $\alpha < 0$ (acceleration),

$$R(\eta) = \frac{R_m}{(\sin \frac{\eta}{1+|\alpha|})^{1+|\alpha|}}, \tag{6}$$

where now

$$R_m = R_0(1-\Omega_0)^{\frac{2}{1+|\alpha|}}. \tag{7}$$

The *big bang* takes place for the smaller value for scale factor R_m, which can be arbitrarily small and that corresponds to a value of the conformal time given by $\eta_{BB} = \pi(1+|\alpha|)/2$. The comoving coordinate of the PH will be

$$\chi_{PH}(\eta) = \eta - \eta_{BB} = \eta - \frac{\pi}{2}(1+|\alpha|). \tag{8}$$

The scale factor (6) diverges for a value of the conformal time $\eta_\infty = (1+|\alpha|)\pi$. Using (8), the value of the angle χ that asymptotically localizes the horizon is

$$\chi_{PH}(\eta_\infty) = \eta_\infty - \eta_{BB} = \frac{\pi}{2}(1+|\alpha|). \tag{9}$$

If we want to avoid the re-convergence of the light cone in the antipodal point, the value of $\chi_{PH}(\eta_\infty)$ must be limited by π. As a consequence, the value of α must be greater than

-1; note that this result is independent on the number of spatial dimensions n. Using (5) we obtain

$$\omega < \frac{1}{n} - 1. \tag{10}$$

The limiting case of $\alpha = -1$ has been also studied [6] obtaining, for the quotient between the entropy that traverses the past light cone of the horizon S_{PH} and the area of this horizon A_{PH},

$$\frac{S_{PH}}{A_{PH}}(\eta) = \frac{s_m}{4} \tan^2 \frac{\eta}{2} (\eta - \pi - \sin \eta \cos \eta), \tag{11}$$

where s_m, of order one in Planck units, is the reference spatial density of entropy in the bounce; the future infinity corresponds to $\eta_\infty = 2\pi$. The value (11) remains finite on this range. Now we allow the PH to reach the antipodal point $\chi_{PH}(\eta_\infty) = \pi$, but the divergence of the scale factor is stronger, giving rise a still infinite area of the PH.

We have also [6] studied the behavior of $\frac{S_{PH}}{A_{PH}}(\eta)$ near the bouncing time. Taking the limit of the general expression

$$\frac{S}{A}(\chi_{PH}) = s_m \frac{\chi_{PH} - \sin \chi_{PH} \cos \chi_{PH}}{\sin^2 \chi_{PH}} \left(\cos \frac{\chi_{PH}}{1-\alpha}\right)^{2(1-\alpha)} \tag{12}$$

for χ near the *big bang*, (12) remains finite approaching $2s_m \chi_{PH}/3$.

It is interesting to note that the value that saturates the bound corresponds to the limiting value $\alpha = -1$ which, in terms of the parameter of the equation of state is $\omega = 1/n - 1$; such value corresponds to the relation between energy density (ρ) and pressure (minus tension) of a gas of co-dimension one branes [7].

ACKNOWLEDGMENTS

This work has been partially supported by MCYT (Spain) under grant FPA2003-02948.

REFERENCES

1. G. 'tHooft, gr-qc/9310006. L. Susskind, hep-th/9409089, *J. Math. Phys.* **36** (1995) 6377.
2. J. M. Maldacena, hep-th/9711200, *Adv. Theor. Math. Phys.* **2** (1998) 231 and E. Witten, hep-th/9802150, *Adv. Theor. Math. Phys.* **2** (1998) 253.
3. W. Fischler and L. Susskind, hep-th/9806039.
4. R. Bousso, hep-th/9905177, *JHEP* **9907** (1999) 004.
5. See for example J. V. Narlikar: *Introduction to Cosmology*; Cambridge University Press, Cambridge (1993).
6. P. Díaz, M. A. Per, and A. Seguí. To appear.
7. A. Karch and L. Randall, hep-th/0506053, *Phys. Rev. Lett.* **95** (2005) 161601.

Rigid Motions in Relativity: Applications

Daniel Soler

Oinarrizko Zientziak Saila,Goi-Eskola Politeknikoa, Mondragon Unibertsitatea, Spain

Abstract. The concept of rigid motion, given the previous work [*Clas. Quantum Grav.* **21**, 3067,(2004)], is here applied to some specific spacetimes: In particular the rigid rotating disc with constant angular velocity in Minkowski space-time is analyzed, a new approach to the Ehrenfest paradox is given as well as a new explanation of the Sagnac effect. Finally the anisotropy of the speed of light and its measurable consequences in a reference frame co-moving with the Earth are discussed.

Keywords: Rigid Motion, Rotating Frames, Sagnac Effect, Ehrenfest Paradox
PACS: 04.20.Cv, 04.80.Cc, 02.40.Hw , 02.40.Ky

The notion of rigid motion is often related to the existence of a class of time-like 3-congruences admitting a metric ($\bar{g}$) that is projectable onto the reference space [1]. However, in our view $\bar{g}$ must also satisfy the axiom of free movability [2]. This condition requires $\bar{g}$ to be maximally symmetric, i.e. to have constant curvature.

In ref. [1] we proved the existence of a wide enough class of congruences admitting maximally symmetric projectable metric which can be obtained by a constriction transformation from the quotient metric ($\hat{g}_{\alpha\beta} = g_{\alpha\beta} + u_\alpha u_\beta$):

$$g_{\alpha\beta} = \varphi \left(\hat{g}_{\alpha\beta} + \varepsilon \mu_\alpha \mu_\beta \right) \tag{1}$$

where φ and $\|\mu\|^2 := g^{\alpha\beta} \mu_\alpha \mu_\beta$ fulfil a previously chosen arbitrary constraint.

Any of these congruences can be used to model the kind of motion followed by a real bodies rigid "enough", and the corresponding spatial metric $\bar{g}$ will be interpreted as the metric that is embodied by standard rods in the lab frame.

This election provides an instrument for measuring distances directly, that is, independent of speed of light and, therefore, of the Einstein convention of simultaneity. In other words, we do not interpret the quotient metric as the space metric. This allows us to study the uniformly rotating rigid disc from a new point of view. We explain our approach to the Ehrenfest paradox, and we compare it with some other explanations available in recently published papers.

Even though we have eliminated the interpretation of the quotient metric as the "physical spatial metric", this object has not lost its importance. On the contrary, it can be interpreted as the "optical metric" and, as a consequence,the possibility of measuring distances and times with two independent metrics, respectively $\bar{g}$ and $\hat{g}$, makes sense to question the isotropy of the speed of light. This analysis allows us to interpret physically the 1-form μ appearing in the constriction transformation (1).

Finally we study the anisotropy of the speed of light in a uniformly rotating reference frame, and in a reference frame co-moving with the Earth[1], giving our interpretation of the well known Sagnac effect [4] for the rotating disc case, and a theoretical explanation of the Brillet and Hall [5] experimental results. This is done in a very similar way to Bel et all [6, 7], but with different experimental predictions.

ROTATING DISC

Let ds^2 be the line element of Minkowski spacetime in cylindric coordinates. The motion of a reference frame with constant angular velocity ω is a timelike Killing congruence with parametric equations: $\{T = t, R = r, \Theta = \theta + \omega t, Z = z\}$

Using a system of coordinates $\{t, r, \theta, z\}$ adapted to the rotating frame of reference, the line element becomes

$$ds^2 = -(1 - r^2\omega^2)dt^2 + 2r^2\omega d\theta dt + dr^2 + r^2 d\theta^2 + dz^2 \qquad (2)$$

Therefore, introducing the 1-form

$$\mu = \frac{\omega r^2}{\sqrt{1 - \omega^2 r^2}} d\theta \qquad (3)$$

it is obvious that the $\bar{g} := \hat{g} - \mu\mu = dr^2 + r^2 d\theta^2 + dz^2$ is a flat metric.

Ehrenfest Paradox

Since 1909, there has been a vast literature devoted to this paradox, but according to recent published papers, it seems that the physics involved in a rigid rotating disc is not clear at all.

According to (2), a rotating platform defines a non-time orthogonal physical frame, being impossible using Einstein protocol to synchronize clocks in a finite closed path. This forces to split spacetime into a space and a time relatively to a reference frame and the space points are identified with the timelike worldlines of the congruence defining de reference frame. Therefore, by definition, the disc does not suffer any deformation.

In this sense we follow the same approach as Rizzi and Ruggiero [8, 9] and Rodrigues and Sharif [10], but reaching to a very different conclusion, because they use the quotient metric ($\hat{g}$) as the metric of the space of reference instead of the Euclidean metric ($\bar{g}$).

When the quotient metric is used as the space metric, although the disc does not suffer any deformation, when the rim of the disc is measured in the space of reference, one obtains $L' = \frac{2\pi R}{\sqrt{1 - \omega^2 R^2}}$ and $R' = R$. The physical interpretation of that fact [9] is that the rods along the rim contract, while the circumference does not. Moreover, neither the rods along the radius, nor the radius itself contract

In our opinion, if neither the circumference itself nor the radius contract, all the troubles are caused by the rods used, precisely because this metric does not satisfies the free mobility axiom. Moreover, it does not make sense to state that rods along the rim

---

[1] See [3] for details

and along the radius have the same length if the free mobility condition is not assumed, as far as this assertion is based on the free mobility axiom.

We claim that instead of $\hat{g}$, the space metric must be $\bar{g}$, the constant curvature projectable metric. In this case, the space of the disc is flat ($L' = 2\pi R$).

ANISOTROPY OF THE SPEED OF LIGHT

The election of the quotient metric as the space metric is physically based on the Einstein convention of simultaneity, and the associated measuring protocol, that is based on the assumption of the isotropy and constancy of the speed of light, and the well known fact that the physical time spent by light in an infinitesimal round trips is, in adapted coordinates, $d\tau = \sqrt{\hat{g}_{ij}dx^i dx^j}$

However, the constancy of c is not so well-established fact in global measurements. Think for instance about the Sagnac effect, that states that the transit time for a light ray to go around a closed path enclosing a non-null area depends on the sense of the curve followed by the light ray.

Leaving aside some interpretations [2], in our opinion, this behaviour shows that in global measurements the speed of light is anisotropic. This anisotropy is predicted even when the quotient metric is used to compute space distance, as it can be seen in the above mentioned references [8, 9, 10], this apparent contradiction arising from the impossibility of synchronizing clocks around a finite closed trip in non-time orthogonal frames.

Given a rigid reference frame, there are two distinguished spatial metrics: first, the one belonging to the reference frame g, which is materialized by the reference body and is rigid, and the second one, the quotient metric $\hat{g}$, which is embodied in radar signals and is not generally preserved. Thus, experiments can be devised to compare these two metrics: those used for measuring the anisotropy of the speed of light in vacuum. According to our interpretation, the *one-way speed* of a light signal is

$$v = \sqrt{\frac{\bar{g}_{ij}dx^i dx^j}{\hat{g}_{ij}dx^i dx^j}} \qquad (4)$$

including the relation $\bar{g}_{ij} = \exp(\theta)(\hat{g}_{ij} + \varepsilon\mu_i\mu_j)$ and working in an suitable basis of the space we obtain[3]

$$v = \frac{1}{\sqrt{\exp(-\theta) - \varepsilon(\vec{M}\cdot\vec{\gamma})^2}} \approx 1 + \frac{1}{2}\theta - \frac{1}{2}M^a M^b \gamma_a \gamma_b \qquad (5)$$

---

[2] See for instance [11], where the isotropy of the speed of light, locally and globally, is claimed. To reach this conclusion an ad hoc correction in time measurements must be introduced. In [12] a good criticism to this approach can be found.

[3] Where the last approximation is justified if θ and M are small, and where $\vec{\gamma}$ defines the direction of propagation.

Hence, we conclude that *in any point of the space, the speed of light is isotropic in the plane orthogonal to* μ.

Sagnac effect

In the rotating disc case, since for light signals $ds^2 = 0$, it is easy to obtain the physical times spent by light in their round trips: $\tau_\pm = \sqrt{1 - \omega^2 R^2} \frac{2\pi R}{1 \mp \omega R}$, where the signs $\pm$ refer to the two possible paths around the periphery. Therefore, using $(\bar{g})$ for the space distance we obtain

$$c_\pm = \frac{-\omega R \pm 1}{\sqrt{1 - \omega^2 R^2}}. \tag{6}$$

while using $\hat{g}$, as in [10], the obtained values are $c_\pm = \frac{1}{\omega R \pm 1}$ and an accuracy of second order in (ωR) is needed to decide between both approaches.

The speed of light in a frame co-moving with the Earth

Finally, following Bel and Molina [7, 6] we consider the linear approximation of the Earth's gravitational field:

$$ds^2 = -(1 - 2U)dt^2 + 2A_i dx^i dt + (1 + 2U_G)\delta_{ij} dx^i dx^j$$

and the Killing congruence defined by $u = \frac{1}{\sqrt{1-2U}} \partial_t \approx (1 + U)\partial_t$.

It is easy to find the 1-form μ and the scalar θ such that $\bar{g}$ is an euclidean metric.

Since the interferometer is usually kept horizontal, $\gamma^1 = \cos A$, $\gamma^2 = \sin A$ and $\gamma^3 = 0$, thus, the speed of light is given by $v = a_0 + a_2 \sin 2A + b_2 \cos 2A$ where $a_2 = \frac{1}{2} M_1 M_2$ and $b_2 = \frac{1}{4}(M_1^2 - M_2^2)$. Therefore, the parameter characterizing the anisotropy is $\sqrt{a_2^2 + b_2^2} \approx 3.6 \times 10^{-13}$.

In [5] a value of 2.1×10^{-13} was obtained experimentally for this quantity, while in the above mentioned works of Bel et al., the values 2.5×10^{-12} and 6.2×10^{-13} are obtained in their two approaches.

REFERENCES

1. J. Llosa, and D. Soler, *Class. Quantum Grav.* **21**, 3067–3094 (2004), gr-qc/0305085.
2. E. Cartan, *Leçons sur la géometrie des espaces de Riemann*, Paris: Gauthier-Villars, 1951.
3. D. Soler (2005). To be published, gr-qc/0511041
4. M. G. Sagnac, *C. R. Acad. Sci. Paris* **157**, 708 (1913).
5. A. Brillet, and J. L. Hall, *Phys. Rev. Lett.* **42**, 549–552 (1979).
6. L. Bel, *Proc. Encuentros Relativistas Españoles-1998: Relativity and Gravitation in General*, edited by A. Molina, J. Martin, E. Ruiz, and F. Atrio., World Scientific, Singapore,1999, gr-qc/9812062.
7. L. Bel, and A. Molina, *Nuovo Cim.* **B115**, 577–586 (2000), gr-qc/9806099.
8. G. Rizzi, and M. Ruggiero, *Found.Phys.* **32**, 1525–1556 (2002), gr-qc/0207104.
9. M. Ruggiero, *Eur. Jour.Phys.* **24**, 563–573 (2003).
10. W. Rodrigues, and M. Sharif, *Found. of Physics* **31**, 1767–1783 (2001), math-ph/0302008.
11. G. Rizzi, and A. Tartaglia, *Found. Phys.* **28**, 1663 (1998), gr-qc/9805089.
12. R. Klauber, *Found.Phys.Lett.* **11**, 405–443 (1998), gr-qc/0103076.

Asymptotic Flatness and Algebraically Special Metrics

W. Natorf* and J. Tafel*

Institute of Theoretical Physics, Warsaw University, ul. Hoża 69, 00-681, Warsaw, Poland

Abstract. We examine conditions assuring that an algebraically special metric is asymptotically flat at future null infinity. Using approximate Bondi–Sachs coordinates we give explicit formulae for the Bondi mass aspect and we interpret one of the Einstein equations as the energy loss formula.

INTRODUCTION

Shortly after Trautman's work on gravitational radiation [1] Bondi [2, 3] introduced the notion of quasilocal gravitational energy density (so-called Bondi mass aspect). Total energy of an isolated system was defined as an integral of this density over a sphere at infinity. The Einstein vacuum equations imply that the energy decreases due to gravitational radiation. This approach to gravitational radiation had been further generalized by Sachs [4].

In this short communication we summarize our results [5] on properties of algebraically special metrics which fall into the class of Bondi–Sachs metrics. Such metrics can play an important role in a search for explicit solutions describing gravitational radiation from bounded sources, as well as in general approach to such systems (see [6] and references therein.)

ASYMPTOTIC FLATNESS

We follow a weakened version of Penrose's definition of asymptotic flatness. Let $\mathscr{\tilde M}$ denote the physical spacetime with metric $\tilde g$ of signature $+---$. We call it asymptotically flat at future null infinity $\mathscr{I}^+$ iff

(i) there exists an unphysical spacetime $\mathscr{M}$ with metric g such that $\mathscr{\tilde M} \subset \mathscr{M}$ and $g = \Omega^2 \tilde g$, $\Omega > 0$ on $\mathscr{\tilde M}$.
(ii) the boundary of $\mathscr{\tilde M}$ in $\mathscr{M}$ contains the three–dimensional null submanifold $\mathscr{I}^+$ such that $\Omega \hat= 0$ on $\mathscr{I}^+$, $\Omega^{|\mu} \Omega_{|\mu} \hat= 0$, $d\Omega \hat{\neq} 0$ (a hat over "=" means an equality on $\mathscr{I}^+$). g is regular and nondegenerate up to $\mathscr{I}^+$.
(iii) $\mathscr{I}^+$ has the structure of $\mathbb{R} \times \mathbb{S}_2$, where the factor $\mathbb{R}$ is generated by a future directed null vector field $v \hat= -\Omega^{|\mu} \partial_\mu$.
(iv) $\varphi^* g = -g_S$ where $\varphi: \mathscr{I}^+ \to M$ is the natural embedding and g_S is the standard metric on $\mathbb{S}_2$.
(v) the Ricci tensor of $\tilde g$ obeys $\tilde R_{\alpha\beta} \hat= 2q \Omega_{|\alpha} \Omega_{|\beta}$, where q is a function (possibly $q = 0$).

For spacetime satisfying above definition one can (in principle) construct the Bondi coordinates [7] $x^0 = u'$, $x^1 = r$, x^A such that the physical metric components have the following expansions in the powers of $r = \Omega^{-1}$:

$$\tilde{g}_{00} = 1 - 2M'r^{-1} + O(r^{-2}), \quad \tilde{g}_{01} = 1 + O(r^{-1}),$$
$$\tilde{g}_{0A} = \psi_A + \kappa_A r^{-1} + O(r^{-2}), \quad \tilde{g}_{AB} = -s_{AB} + n_{AB}r^{-1} + O(r^{-2}). \tag{1}$$

If r satisfies the so-called luminosity condition, i.e. if $\det \tilde{g}_{AB} = \det s_{AB}$, the coefficient M' is the Bondi mass aspect and the gravitational energy at the retarded time u' is given by

$$E(u') = \frac{1}{4\pi} \int_{u'=\text{const}} M' d\sigma. \tag{2}$$

The vacuum Einstein equations imply the energy loss formula

$$E_{,u'} = -\frac{1}{32\pi} \int_{u'=\text{const}} n_{AB,u'} n^{AB}{}_{,u'} d\sigma \tag{3}$$

(usually written in terms of the Bondi news function [3, 4]).

ALGEBRAICALLY SPECIAL METRICS

By virtue of the Goldberg–Sachs theorem the degenerate principal null vector of the Weyl tensor of an algebraically special metric is tangent to a shear-free congruence of null geodesics. In suitably chosen coordinates $u, r, \xi, \bar{\xi}$ the metric reads [8]

$$\tilde{g} = 2\kappa(H\kappa + dr + W d\xi + \bar{W} d\bar{\xi}) - 2\frac{r^2 + \Sigma^2}{P^2} d\xi d\bar{\xi}, \tag{4}$$

where

$$\kappa = du + L d\xi + \bar{L} d\bar{\xi}, \quad W = -(r + i\Sigma)L_{,u} + i\partial\Sigma$$
$$H = -r(\log P)_{,u} - \frac{mr + M\Sigma}{r^2 + \Sigma^2} + P^2 \text{Re}[\partial(\bar{\partial}\log P - \bar{L}_{,u})] \tag{5}$$
$$\Sigma = -\frac{1}{2}P^2\sigma = \frac{i}{2}P^2(\partial\bar{L} - \bar{\partial}L), \quad \partial = \partial_\xi - L\partial_u$$

and P, m and L are two real and one complex function independent of r. The Einstein vaccum equations reduce to

$$P^2 \text{Re}\left[\partial\bar{\partial}\Sigma - 2(\partial\Sigma)\bar{L}_{,u} - \Sigma(\partial\bar{L})_{,u} + 2\Sigma(\partial\bar{\partial}\log P - \partial\bar{L}_{,u})\right] = M, \tag{6}$$

$$\partial(m + iM) = 3(m + iM)L_{,u} \tag{7}$$

and

$$(mP^{-3})_{,u} = P\text{Re}\{[\partial + 2(\partial\log P - L_{,u})]\partial I\}, \tag{8}$$

where

$$I = \bar{\partial}(\bar{\partial}\log P - \bar{L}_{,u}) + (\bar{\partial}\log P - \bar{L}_{,u})^2. \tag{9}$$

Let ξ and $\bar{\xi}$ be interpreted as the complex stereographic coordinates on $\mathbb{S}_2$ such that $g_S = 2(1+\xi\bar{\xi}/2)^{-2}d\xi d\bar{\xi}$. Define

$$\hat{P} = (1+\xi\bar{\xi}/2)^{-1}P. \tag{10}$$

The function $\hat{P}/r$ can be chosen as the conformal factor Ω. Regularity of the metric $\tilde{g}$ can be achieved by assuming that the form κ and the functions $\hat{P}$, m and M are regular on $\mathbb{R} \times \mathbb{S}_2$. These conditions are sufficient for (4) to be asymptotically flat [5].

BONDI PARAMETERS

Tafel [9] proved that the so-called modified Bondi mass aspect $\hat{M}$ (differing from M by a divergence on $\mathbb{S}_2$ and thus defining the same total energy) can be expressed solely in terms of the conformal factor Ω which appears in the definition of asymptotic flatness. Moreover, one does not have to assume that $r = \Omega^{-1}$ satisfies the Bondi's luminosity condition. $\hat{M}$ can be obtained using the "corrected" conformal factor $\Omega' = \Omega + \eta\Omega^2$. The coefficient η depends on the choice of an approximate Bondi time $\hat{u}$. The formula for $\hat{M}$ is

$$\hat{M} = \lim_{\Omega' \to 0} \Omega'^{-1}(1 - \tfrac{1}{2}\Omega'^{-1}\tilde{\Box}\Omega'^{-1}), \tag{11}$$

where $\tilde{\Box}$ is the covariant wave operator corresponding to $\tilde{g}$. In this approach it is also easy to obtain the tensor $n_{AB,u'}$ which corresponds to the Bondi news function. In terms of $\hat{M}$ the energy loss formula (3) reads

$$\hat{M}_{,u'} = -\frac{1}{8}n_{AB,u'}n^{AB}{}_{,u'}. \tag{12}$$

In [5] we applied this approach to algebraically special metrics (4). Let $\hat{P}$, m and M be regular on $\mathbb{R} \times \mathbb{S}_2$. The null generator of $\mathscr{I}^+$ is $v = \hat{P}^{-1}\partial_u$ and the approximation of the Bondi retarded time u' can be chosen as

$$\hat{u} = \int \hat{P} du. \tag{13}$$

Then

$$\eta = -\tfrac{1}{2}\tilde{\Delta}\hat{u}, \tag{14}$$

where $\tilde{\Delta} = \tfrac{1}{2}(1+\tfrac{1}{2}\xi\bar{\xi})^2\{\partial, \bar{\partial}\}$. The modified mass aspect reads

$$\hat{M} = \frac{m}{\hat{P}^3} + \frac{1}{2}(\tilde{\Delta}+2)\eta + \frac{3}{2\hat{P}}\left(\frac{\Sigma^2}{\hat{P}^2}\right)_{,u} + \\ + 2(1+\tfrac{1}{2}\xi\bar{\xi})^2 \operatorname{Im}\left[(\bar{\partial}\log\hat{P} - \bar{L}_{,u})\partial(\hat{P}^{-1}\Sigma)\right] \tag{15}$$

while the news tensor is

$$n_{AB,u'} = \hat{P}^{-1}\left[-\{D_A - L_{A,u}, D_B - L_{B,u}\}\hat{P} + \\ + s_{AB}(1+\tfrac{1}{2}\xi\bar{\xi})^2\{\partial - L_{,u}, \bar{\partial} - \bar{L}_{,u}\}\hat{P}\right], \tag{16}$$

where $D_A = \nabla_A - L_A$ and ∇ is the Levi–Civita connection corresponding to $-g_S$ and L_A are coefficients of the decomposition $\omega = \mathrm{d}u + L_A \mathrm{d}x^A$. One can further continue this calculus [5] to obtain the angular momentum aspect κ_A which occurs in (1).

Let $V = (1 + \xi\bar{\xi}/2)\hat{u}$ be the potential introduced by Robinson and Robinson [10]. In terms of V (11) reads

$$\hat{M} = m\hat{P}^{-3} - (1+\xi\bar{\xi}/2)^3 \mathrm{Re}(\partial\partial\bar{\partial}\bar{\partial}V) \tag{17}$$

while (16) takes the form

$$n_{AB,u'}\mathrm{d}x^A\mathrm{d}x^B = -2\bar{I}\mathrm{d}\xi^2 - 2I\mathrm{d}\bar{\xi}^2, \quad I = -P^{-1}\bar{\partial}^2 V. \tag{18}$$

Substituting (18) and (17) to (8) shows that the latter equation is equivalent to

$$\hat{P}^{-1}\hat{M}_{,u} = -\hat{P}^{-1}(\partial^2 V)_{,u}\hat{P}^{-1}(\bar{\partial}^2 V)_{,u} \tag{19}$$

(note that here $\hat{P}^{-1}\partial_u = \partial_{u'}$). Thus, one can interpret (8) as the mass loss formula for algebraically special metrics. This interpretation seems to be overlooked in the literature.

REFERENCES

1. A. Trautman, "Lectures on General Relativity. 1. Boundary Conditions in Gravitational Radiation Theory" *GRG* **34**, 721 (2002).
2. H. Bondi, "Gravitational Waves in General Relativity", *Nature* (London),**186**, 535 (1960).
3. H. Bondi, M.G.J. van der Burg and A.W.K. Metzner *Proc. R. Soc. A* **269**, 21 (1962).
4. R.K. Sachs, "Gravitational waves in General Relativity: VIII. Waves in asymptotically flat space-times", *Proc. R. Soc. A* **270**, 103 (1962).
5. W. Natorf, J. Tafel, "Asymptotic flatness and algebraically special metrics", *Class. Quantum Grav.* **21**, 5397 (2004).
6. C. Kozameh, E.T. Newman and G. Silva-Ortigoza, "Twisting null geodesic congruences, scri, H-space and spin-angular momentum", *Class. Quantum Grav.* **22**, 4679 (2005).
7. L.A. Tamburino and J.H. Winicour, "Gravitational fields in finite and conformal Bondi frames", *Phys. Rev.* **150** 1039, (1966).
8. H. Stephani, D. Kramer, M. MacCallum, C. Hoenselaers and E. Herlt, *Exact Solutions to Einstein's Field Equations*, Second Edition, Cambridge University Press, 2003.
9. J. Tafel, "Bondi mass in terms of the Penrose conformal factor", *Class. Quantum Grav.* **17**, 4397 (2000).
10. Robinson I. and Robinson J. R., *Int. J. Theor. Phys.* **7** (1969), 231.
11. A.G. Riess et al., *Astrophys. J.*, **607**, 665 (2004).
12. D.N. Spergel et al., *Astrophys. J. Suppl. Ser.*, **148**, 175 (2003).
13. W. Zimdahl, D. Pavón, and L.P. Chimento, *Phys. Lett. B*, **521**, 133 (2001).
14. L.P. Chimento, A.S. Jakubi, D. Pavón, and W. Zimdahl, *Phys. Rev. D*, **67**, 083513 (2003).
15. G. Olivares, F. Atrio-Barandela, and D. Pavón, *Phys. Rev. D*, **71**, 063523 (2005).
16. U. Seljak, and M. Zaldarriaga, *Astrophys. J.*, **469**, 437 (1996). See http://www.cmbfast.org

On nonanticommutative sigma models

Luis Álvarez-Gaumé* and Miguel A. Vázquez-Mozo[†,**]

Theory Unit, Physics Department CERN, CH-1211 Geneva 23, Switzerland
[†]*Depto. de Física Fundamental, Universidad de Salamanca, Plaza de la Merced s/n, E-37008 Salamanca, Spain*
[**]*Instituto Universitario de Física Fundamental y Matemáticas (IUFFyM), Universidad de Salamanca, Salamanca, Spain*

Abstract. We discuss classical aspects of nonanticommutative Kähler sigma models in two dimensions.

Keywords: Noanticommutative field theories, superspace, two-dimensional sigma models.
PACS: 11.30.Pb, 11.10.Nx.

The study of deformations of quantum field theories has been the subject of renewed attention in later years due to their role in compactifications of string/M-theory. In particular noncommutative quantum field theories are known to emerge as low-energy effective descriptions of type-II string theories on D-brane backgrounds in the presence of a constant antisymmetric tensor field along the directions of the D-brane [1].

In the case of supersymmetric theories it is possible in principle to consider more general deformations than the "ordinary" noncommutative ones, $[x^\mu, x^\nu] = i\theta^{\mu\nu}$. In particular, when described in superspace formalism, one can consider also deforming the anticommutator of the fermionic coordinates [2]

$$\{\theta^a, \theta^b\} = \mathscr{C}^{ab}, \qquad \mathscr{C}^{ab} \in \mathbb{C}. \tag{1}$$

As with noncommutative field theories, here there is also a stringy connection. This class of deformations describe the low-energy dynamics of type-II strings in the background of a self-dual constant graviphoton background. Consistency requirements force the deformation (1) to leave the anticommutator of the antiholomorphic fermionic coordinates untouched, $\{\bar\theta^{\dot a}, \bar\theta^{\dot b}\} = 0$. One is therefore obliged to work in Euclidean space where $\bar\theta^{\dot\alpha} \neq (\theta^\alpha)^\dagger$. In the following we will refer to barred quantities as "antiholomorphic", although one should keep always in mind that these are not the complex conjugate of the corresponding unbarred ones.

Here we report on our work on nonanticommutative deformations of two-dimensional sigma models with $\mathcal{N} = 2$ supersymmetry [4]. We consider then Euclidean superspace with coordinates $(y^\pm, \theta^\pm, \bar\theta^\pm)$, where $y^\pm$ are the so-called chiral coordinates, defined in terms of the "ordinary" ones $x^\pm$ by $y^\pm = x^\pm - i\theta^\pm \bar\theta^\pm$. Next, we deform superspace anticommutators by

$$\{\theta^+, \theta^-\} = \frac{1}{M}, \tag{2}$$

where M sets the energy scale at which the deformation effects are sizeable. At the same time we leave all other (anti)commutators undeformed, including the commutator of the bosonic coordinates $[y^+, y^-] = 0$. It is important to stress at this point that, unlike the general deformation (1) that in four dimensions breaks the Euclidean group SO(4), in two dimensions the deformation (2) preserve the two-dimensional Euclidean group SO(2), acting on the the fermionic coordinates as $\theta^\pm \longrightarrow e^{\pm\frac{i}{2}\vartheta}\theta^\pm$, $(0 \leq \vartheta \leq 2\pi)$.

The anticommutator (2) induces the following deformation of the algebra of supercharges

$$Q_\pm^2 = \overline{Q}_\pm^2 = \{Q_+, Q_-\} = 0, \quad \{Q_\pm, \overline{Q}_\pm\} = 2P_\pm, \quad \{\overline{Q}_+, \overline{Q}_-\} = \frac{4}{M}P_+P_-. \tag{3}$$

The deformed supersymmetry algebra preserves SO(2) and, moreover, it only amounts to a central extension by a term proportional to the quadratic Casimir of the Euclidean group P_+P_-. If we forget for a moment that we are working in Euclidean signature and consider the centrally extended algebra (3) in two-dimensional Minkowski space-time, we find after a simple algebra that the central extension imposes a cutoff in the spectrum of the theory set by the scale of the deformation M

$$P_+P_- \leq \frac{M^2}{4}. \tag{4}$$

It would be very interesting to know if there is physical realization of the algebra (3) in two-dimensional Lorentzian signature, since such a theory would be automatically equipped with a built-in cutoff.

The algebra of supercovariant derivatives $D_\pm, \overline{D}_\pm$ is left undeformed by (2), as well as the anticommutation relations of these with the supercharges. This implies that one can still define the notion of chiral ($\overline{D}_\pm\Phi = 0$) and antichiral ($D_\pm\overline{\Phi} = 0$) superfields. In order to deal with the theory in the deformed superspace it is extremely convenient to define the Weyl map between functions on superspace and operators in some Hilbert space. This is done by introducing the operators $\hat{Q}^\pm$, satisfying the deformed algebra $\{\hat{Q}^+, \hat{Q}^-\} = \frac{1}{M}$, and defining the map

$$f(\theta^\pm) \longrightarrow \hat{f} \equiv -\int d^2\eta\, e^{-(\eta_+\hat{Q}^+ + \eta_-\hat{Q}^-)}\tilde{f}(\eta_\pm), \tag{5}$$

where $\tilde{f}(\eta_\pm)$ is the Fourier transform of $f(\theta^\pm)$ (see [4] for details and conventions). The noncommutative algebra of functions on the deformed superspace is then mapped into the algebra of their symbols

$$f(\theta) \star g(\theta) \longrightarrow \widehat{fg}, \tag{6}$$

where the $\star$-product is defined by

$$f(\theta) \star g(\theta) \equiv f(\theta)\exp\left[-\frac{\mathscr{C}^{ab}}{2}\frac{\overleftarrow{\partial}}{\partial\theta^a}\frac{\overrightarrow{\partial}}{\partial\theta^b}\right]g(\theta). \tag{7}$$

In order to construct the deformed Lagrangian of a (2,2) nonanticommutative sigma model the starting point is the Kähler potential of the undeformed model $\mathcal{K}(\Phi,\overline{\Phi})$, defined through its series expansion around $\Phi = 0$, $\overline{\Phi} = 0$. Replacing the chiral and antichiral superfields by their corresponding Weyl symbols $\widehat{\Phi}$, $\widehat{\overline{\Phi}}$ one obtains the function $\mathcal{K}(\widehat{\Phi},\widehat{\overline{\Phi}})$. The deformed Kähler potential is then obtained by applying the inverse Weyl map

$$\mathcal{K}(\widehat{\Phi},\widehat{\overline{\Phi}}) \longrightarrow \mathcal{K}(\Phi,\overline{\Phi})_\star. \tag{8}$$

The same technique can be used to compute the holomorphic and antiholomorphic superpotentials. This procedure, apart from simplifying the calculations, allows a resummation of the infinite series found in Refs. [3] for the deformed Kähler sigma model. When expressed in terms of component fields, the resulting Lagrangian density can be split into two terms $\mathcal{L} = \mathcal{L}_0 + \mathcal{L}_1$ where the first part $\mathcal{L}_0$ is given by the usual (2,2) Lagrangian with the following replacement of the Kähler potential (see [4])

$$\mathcal{K}(\varphi^i,\overline{\varphi}^{\bar{\imath}}) \longrightarrow \mathcal{K}_0(\varphi^i,F^i,\overline{\varphi}^{\bar{\imath}}) \equiv \int_{-\frac{1}{2}}^{\frac{1}{2}} d\xi\, \mathcal{K}\left(\varphi^i + \frac{\xi}{M}F^i, \overline{\varphi}^{\bar{\imath}}\right) \tag{9}$$

On the other hand, the second piece of the Lagrangian $\mathcal{L}_1$ is expressed in terms of the following deformation of the Kähler potential

$$\mathcal{K}_1(\varphi^i,F^i,\overline{\varphi}^{\bar{\imath}}) \equiv \int_{-\frac{1}{2}}^{\frac{1}{2}} \xi d\xi\, \mathcal{K}\left(\varphi^i + \frac{\xi}{M}F^i, \overline{\varphi}^{\bar{\imath}}\right) \tag{10}$$

and contain only operators of canonical dimension 3, suppressed by an overall power of M.

For the antiholomorphic superpotential $\overline{\mathcal{W}}(\overline{\varphi}^{\bar{\imath}})$ one finds that the deformation induces no correction whatsoever, in accordance with the corresponding situation in the four-dimensional nonanticommutative Wess-Zumino model [2]. On the other hand, the holomorphic superpotential, when expressed in components, has the usual expression with the only replacement

$$\mathcal{W}(\varphi^i) \longrightarrow \mathcal{W}_0(\varphi^i,F^i) \equiv \int_{-\frac{1}{2}}^{\frac{1}{2}} d\xi\, \mathcal{W}\left(\varphi^i + \frac{\xi}{M}F^i\right) \tag{11}$$

From the previous results we learn that the deformation of the superspace anticommutator (2) induces a kind of holomorphic fuzziness in target space, since in the deformed Lagrangian both the Kähler potential and the holomorphic superpotential are not evaluated at the point φ^i but their values are given by the average of the function around the point φ^i with an amplitude set by F^i/M. Since the vacuum expectation value of the auxiliary field F^i can be seen as the order parameter setting the scale of supersymmetry breaking, the size of the fuzziness is given by the ratio between the two supersymmetry breaking scales, the one set by the value of the auxiliary field and the one given by the energy scale of the deformation of the ordinary (2,2) theory.

In [4] we confined ourselves to a classical analysis of the nonanticommutative sigma model. The next question to be answered is of course how the deformation interferes with the quantum properties of the model, specially the proportionality of the one-loop beta-function with the Ricci tensor of the target space manifold

$$\mu \frac{dg_{i\bar{k}}}{d\mu} \sim R_{i\bar{k}} + \dots \tag{12}$$

In spite of the apparent complication of the dependence of the deformed Lagrangian on the auxiliary field F^i, it is actually possible to solve the equations of motion of F^i and $\overline{F}^{\bar{i}}$. The resulting expressions are specially simple in the cases when the target space has dimension two and four [5]. In the two dimensional case, surprisingly, the result is that the on-shell Lagrangian does not contain any new terms depending on the deformation scale M. Hence the on-shell theory is identical to the undeformed model and the beta function is the same one at all orders.

In the case of a four-dimensional target space, the on-shell Lagrangian contains two extra terms of canonical dimension 4 with four holomorphic fermionic fields each. On general grounds it can be seen that these M-dependent terms give rise to vertices that cannot contribute to the one-loop beta functions, although contributions can be expected in principle at higher orders. In general it can be seen that, because of the presence of fermion bilinears in the expression of the auxiliary field F^i, the infinite series found in [3] actually truncate to a finite number of terms. A systematic analysis of the effect of the nonanticommutative deformation on the quantum properties of the Kähler sigma models will be presented elsewhere [5].

ACKNOWLEDGMENTS

We thank the organizers of the 2005 Spanish Relativity Meeting the opportunity of presenting our work. M.A.V.-M. acknowledges the support of Spanish Science Ministry grants FPA 2002-02037, BFM 2003-02121 and FPA2005-04823.

REFERENCES

1. N. Seiberg and E. Witten, *String Theory and Noncommutative Geometry*, J. High Energy Phys. **09** (1999) 032.
2. N. Seiberg, *Noncommutative Superspace, $\mathcal{N} = \frac{1}{2}$ Supersymmetry, Field Theory and String Theory*, J. High Energy Phys. **06** (2003) 010.
3. B. Chandrasekhar and A. Kumar, *$D = 2$, $\mathcal{N} = 2$, Supersymmetric Theories on Nonanticommutative Superspace*, J. High Energy Phys. **03** (2004) 013.
   B. Chandrasekhar, *$D = 2$, $\mathcal{N} = 2$, Supersymmetric Sigma Models on Non(anti)commutative Superspace*, Phys. Rev. **D70** (2004) 125003.
4. L. Alvarez-Gaumé and M. A. Vázquez-Mozo, *On Nonanticommutative $\mathcal{N} = 2$ Sigma-Models in Two Dimensions*, J. High Energy Phys. **04** (2005) 007.
5. L. Alvarez-Gaumé, F. Meyer and M. A. Vázquez-Mozo, work in progress.

Axially symmetric equilibrium regions in FLRW universes

B. C. Nolan* and R. Vera*

School of Mathematical Sciences, Dublin City University, Dublin 9, Ireland

Abstract. The study of the matching of stationary and axisymmetric spacetimes with Friedmann-Lemaître-Robertson-Walker spacetimes preserving the axial symmetry is presented. We show, in particular, that any orthogonally transitive stationary and axisymmetric region in FLRW must be static, irrespective of the matter content. Therefore, previous results on static regions in FLRW cosmologies apply. As a result, the only stationary and axisymmetric vacuum region that can be matched to a FLRW (non-static) spacetime is a spherically symmetric region of Schwarzschild. This constitutes another uniqueness result for the Einstein-Straus model (as well as its Oppenheimer-Snyder counterpart), and hence another indication of its unsuitability as an answer to the influence of the cosmic expansion on local physics.

Keywords: Cosmology, voids, matching, stationary, axial symmetry
PACS: 04.20.Jb, 04.20.Cv, 98.80.-k, 04.40.Nr

INTRODUCTION AND SUMMARY

In dealing with isolated systems in general relativity, one usually assumes that the system is resident in an asymptotically flat space-time. This is an idealisation, as the universe at large is clearly not asymptotically flat. Recognising this fact, one is then confronted with the issue of the impact of cosmology on local physics, see for example Ellis' recent review [1].

Asymptotically flat systems in general relativity can be studied via the introduction of idealised structures (surfaces) representing space-like and null infinity. A central idea of Ellis' programme for the study of the influence of cosmology on local physics is that these surfaces should be replaced by surfaces at a finite distance from the local system, representing what he refers to as 'finite infinity', and beyond which a cosmological model provides the appropriate description of space-time. This surface should have certain characteristics. For example, the gravitational field encountered must be sufficiently small in some quantifiable way and there should be limits on the radiation and matter content of the surface guaranteeing that the system interior to the surface is indeed isolated to a sufficiently high degree of approximation. The point of view of the present work is that a natural candidate for the construction of finite infinity is a matching hypersurface conjoining portions of two different space-times, one (the interior) corresponding to the local system and the other (the exterior) corresponding to a cosmological model.

This is not a new idea, nor is the idea that the cosmological background may influence local systems. For example, soon after the discovery of the expansion of the universe, McVittie questioned the influence of the expansion on planetary orbits [2]. His approach involved determining an exact solution of Einstein's field equations that represents the Schwarzschild solution embedded in a Friedmann-Lemaître-Robertson-Walker (FLRW) background. However, it is not clear that the interpretation of this solution as representing a point mass embedded in a FLRW background is entirely accurate, as the resulting space-time has troubling global pathologies [3], and it relies on a flawed interpretation of the radius of the planetary orbits (see Section 3.3 of [4]). The first self-consistent and formally correct study of this problem was given by Einstein and Straus [5]. Here, they showed that the Schwarzschild solution can be matched across a co-moving time-like boundary of a dust FLRW universe. The resulting structure is referred to as the Einstein-Straus (ES) vacuole, and the conclusion is that there is no influence from the cosmic background onto the static interior.

However, there are many drawbacks associated with this model. First, exact spherical symmetry of the vacuole is required. Second, the mass parameter of the Schwarzschild region is naturally and directly related to the radius of the vacuole and to the density of the FLRW background. This means that the model is highly inflexible [6]. Thirdly, the model is unstable against radial perturbations [4]. Therefore, it is highly desirable to have at hand generalisations of the ES model. Regarding the mass relation, a direct generalisation exists by using a (radiative) Vaidya interior [7]. Nevertheless, spherical symmetry still plays a fundamental role.

Unfortunately, the ES model has shown itself to be remarkably reluctant to admit non-spherical generalisations. Attempts so far have emphasised the equilibrium nature of the interior region, while moving away from the constraint of spherical symmetry. The matching of a static cylindrically symmetric region with a FLRW universe was found to be impossible [8]. Similarly, Mars considered *any* static configuration [9, 10] for a possible interior. The results of these studies play a central role in the present work, and can be summarised as follows: at each instant of cosmic time, the static region has a spherically symmetric boundary. However, the centre of these spheres moves, and their radius change, and so the overall configuration is not spherically symmetric in general. In fact, the whole configuration is axially symmetric. Furthermore, imposing a structure on the matter distribution of the interior region (including the cases of vacuum and perfect fluid) implies that full spherical symmetry is obtained. In particular, the only static vacuole that may be embedded in a FLRW universe is the spherically symmetric ES vacuole.

Given that, for reasons outlined above, one would like to be able to embed into a FLRW background the vacuum gravitational field of a non-spherical isolated system, and in particular, a non-spherical isolated system in equilibrium, one must look for a get-out clause to release us from the no-go results quoted above. Our attempt to do this involves studying *stationary* rather than static configurations. Thus rotation is allowed, but no gravitational radiation. We study the standard model of a rotating system in equilibrium, the class of stationary axially symmetric gravitational fields. We can express our results briefly and prosaically as follows: it doesn't work. More precisely, we find that *the stationary region must in fact be static*, and so Mars' results apply. In particular we can conclude that *the only stationary axially symmetric vacuum region that can be matched with a FLRW universe must be spherically symmetric and is therefore an ES vacuole*. The restrictions and instabilities mentioned above then also apply. This prevents the embedding of a rotating Kerr cavity in FLRW. In fact, by the dual local nature of the matching of spacetimes, interchanging the interior with the exterior [7], this result also states that FLRW cannot be a source of Kerr. It must be stressed that the only assumption used is the natural preservation of an axial symmetry across the matching.

In the contribution to the last EREs, [11] we already started with the study of the general problem of matching stationary and axisymmetric spacetimes to FLRW spacetimes while preserving an axial symmetry, and found some partial results. In the following we present a brief sketch of the proof of the final result. We refer to [12] for a full and precise account of the results and the detailed proofs.

SKETCH OF THE LAST STEPS OF THE PROOF

Let us recall that for the sake of generality no specific matter content is in the stationary region. The only assumption on the stationary region is that the orbits of the group admit orthogonal surfaces, the so-called circularity condition, which is ensured in the most interesting cases: vacuum, electrovacuum, perfect fluid without convective motions, and others. The group is then said to act orthogonally transitively (OT). The OT stationary and axisymmetric region $(\mathscr{W}^{SX}, g^{SX})$ is characterised by the existence of a coordinate system $\{T, \Phi, x^M\}$ ($M, N, \ldots = 2, 3$) in which the line-element for the metric g^{SX} outside the axis reads [13]

$$ds^2_{SX} = -e^{2U}(dT + Ad\Phi)^2 + e^{-2U}W^2 d\Phi^2 + g_{MN} dx^M dx^N,$$

where U, A, W and g_{MN} are functions of x^M, the axial Killing vector field is given by $\vec{\eta} = \partial_\Phi$, and a timelike (future-pointing) Killing vector field is given by $\vec{\xi} = \partial_T$. Special attention is given to the vector field $\vec{\zeta}$,

$$\vec{\zeta} = \vec{\xi} - \frac{g^{SX}(\vec{\eta}, \vec{\xi})}{g^{SX}(\vec{\eta}, \vec{\eta})} \vec{\eta},$$

which is orthogonal to the hypersurfaces of constant T as well as orthogonal to $\vec{\eta}$ by construction. $\vec{\zeta}$ is a Killing vector only when $\Omega \equiv g^{SX}(\vec{\eta}, \vec{\xi})/g^{SX}(\vec{\eta}, \vec{\eta})$ is constant, in which case the region $(\mathscr{W}^{SX}, g^{SX})$ is *static*. Regarding the FLRW region, there is a coordinate system in which the line element reads

$$ds^2_{RW} = -dt^2 + a^2(t) ds^2_{\mathscr{M}},$$

where $(\mathscr{M}, g_{\mathscr{M}})$ is a complete, simply connected, three-dimensional Riemannian manifold of constant curvature.

Recalling [9, 10] (see also [11]), the proofs are based on the decomposition of the matching conditions into constraint and evolution equations. Using only the constraint matching conditions, we showed in [11] that given a spacelike slicing of the matching hypersurface defined by $T = const.$ on the FLRW side, it has to coincide with the $t = const.$ slices on the stationary side. Furthermore, each of these slices must be two-spheres. In other words, at any value of the cosmological time (and at a corresponding value of T) we either have a stationary region inside a ball in FLRW or a ball of FLRW embedded in a stationary spacetime.

The complete resolution of the problem is based on showing first that $\vec{\zeta}$ is nowhere tangent to Σ except, possibly, at points in the axis. This is therefore also true for the stationary Killing vector $\vec{\xi}$ by construction. The vector $\vec{\zeta}$ is then used as a rigging vector [14] to compute the rest of the matching conditions, and together with the properties at the axis of symmetry, one finds $g^{SX}(\vec{\eta},\vec{\eta})\vec{\lambda}(\Omega) = 0, g^{SX}(\vec{\eta},\vec{\eta})\vec{m}(\Omega) = 0$, where $\vec{\lambda}$ and $\vec{m}$ form with $\vec{\eta}$ a tangent basis of Σ except for the points at the axis. Recalling that $\vec{\eta}(\Omega) = 0$ we have that Ω is constant all over Σ by continuity. This is, in turn, used to show that $\vec{\zeta}$, and hence $\vec{\xi}$ is transverse to Σ everywhere. Lie transport along $\vec{\xi}$ is finally used to extend the fact that Ω is continuous to the stationary region around Σ. Therefore, $\vec{\zeta}$ is a Killing vector there and thus $(\mathscr{W}^{SX}, g^{SX})$ must be *static*.

As a consequence, Mars' results in [10] follow, and thus: firstly, Σ in FLRW –given $a(t)$– is determined in terms of an arbitrary function $f(t)$. Secondly, the stationary region is in fact static, and, together with Σ there, is determined in terms of $f(t)$ and $a(t)$. And finally, imposing an electrovacuum content in the stationary (static) region determines $f(t)$ in such a way so that the whole region must be spherically symmetric. We refer to [10, 9, 12] for the explicit expressions.

These results show, in particular, that *the only axially symmetric* vacuum *region in equilibrium in FLRW is Schwarzschild, giving rise to the Einstein-Straus model*. And by the interior/exterior duality [7], that *FLRW cannot be the source of a rotating black hole*.

ACKNOWLEDGMENTS

RV acknowledges the Irish Research Council for Science, Engineering and Technology postdoctoral fellowship PD/2002/108.

REFERENCES

1. G. F. R. Ellis *New Astron. Rev.*, **46**, 645-657 (2002).
2. G. C. McVittie *Mon. Not. Roy. Astr. Soc.*, **93**, 325-339 (1933).
3. B. C. Nolan *Phys. Rev. D*, **58**, 064006 (1998).; *Class. Quantum Grav.*, **16**, 1227-1254 (1999).; *Class. Quantum Grav.*, **16**, 3183-3191 (1999).
4. A. Krasiński *Inhomogeneous Cosmological Models* Cambridge University Press, Cambridge, 1997
5. A. Einstein and E. G. Straus *Rev. Mod. Phys.*, **17**, 120-124 (1945).
6. W. B. Bonnor *Mon. Not. Roy. Astr. Soc.*, **282**, 1467-1469 (1996).
7. F. Fayos, J. M. M. Senovilla and R. Torres *Phys. Rev. D*, **54**, 4862-4872 (1996).
8. J. M. M. Senovilla and R. Vera *Phys. Rev. Lett.*, **78**, 2284-2287 (1997).
9. M. Mars *Phys. Rev. D*, **57**, 3389-3400 (1998).
10. M. Mars *Class. Quantum Grav.*, **18**, 3645-3663 (2001).
11. B. C. Nolan, R. Vera "On global models for finite rotating objects in equilibrium in cosmological backgrounds" in *Proceedings of the Spanish Relativity Meetings '04* to appear.
12. B. C. Nolan, R. Vera *Class. Quantum Grav.*, **22**, 4031-4050 (2005).
13. H. Stephani, D. Kramer, M. A. H. MacCallum, C. Hoenselaers and E. Herlt *Exact solutions of Einstein's field equations. Second Edition*, Cambridge University Press, Cambridge, 2003
14. M. Mars and J. M. M. Senovilla *Class. Quantum Grav.*, **10**, 1865-1897 (1993).

Separation of variables and exact solution of the Dirac equation in some cosmological space-times

Víctor M. Villalba[1]

Centro de Física, IVIC, Apdo 21827 Caracas 1020A, Venezuela

Abstract. We apply the algebraic method of separation of variables in order to reduce the Dirac equation to a set of coupled first-order ordinary differential equations. We obtain the sufficient conditions for partial or complete separability corresponding to homogeneous cosmological backgrounds.

Keywords: Dirac equation, Separation of variables
PACS: 03.65.Pm

The Dirac equation is a system of coupled partial differential equations which is separable in a very restricted set of metrics. Among the spacetimes where the separability of the Klein-Gordon and Dirac equations has been studied one can mention the Stäckel spaces, which are those metrics where the Hamilton-Jacobi equations is separable. A systematic classification of the gravitational backgrounds where the Dirac equation is separable with the help of the algebraic method is presented in ref. [1, 2].

The covariant generalization of the Dirac equation in curved space-time is [3]

$$\{H\}\tilde{\Psi} = \{\tilde{\gamma}^\mu(\partial_\mu - \Gamma_\mu) + m\}\tilde{\Psi} = 0 \qquad (1)$$

where Γ_μ are the spin connections.

$$\Gamma_\lambda = \frac{1}{4}g_{\mu\alpha}[(\partial b_\nu^\beta/\partial x^\lambda)a_\beta^\alpha - \Gamma^\alpha_{\nu\lambda}]s^{\mu\nu} \qquad (2)$$

with $\tilde{\gamma}_\mu = b_\mu^\alpha \gamma_\alpha$, $\tilde{\gamma}^\mu = a_\beta^\mu \gamma^\beta$ and $s^{\mu\nu} = \frac{1}{2}(\tilde{\gamma}^\mu \tilde{\gamma}^\nu - \tilde{\gamma}^\nu \tilde{\gamma}^\mu)$.

The algebraic method of separation of variables [1] consists in reducing the Dirac equation to a sum of commuting differential operators in the form:

$$\{H\}\tilde{\Psi} = \Im\{H\}SS^{-1}\tilde{\Psi} = \left(\hat{K}_1 + \hat{K}_2\right)\Phi = 0, \left[\hat{K}_1, \hat{K}_2\right]_- = 0 \qquad (3)$$

The line element

$$ds^2 = -dt^2 + a^2(t)\left(dx^2 + b^2(x)\left(dy^2 + c^2(y)dz^2\right)\right) \qquad (4)$$

corresponds to a non-factorizable metric. It is not possible to find four first-order differential operators commuting with the Dirac Hamiltonian. In order to separate variables we choose the following curved gamma matrices:

---

[1] email:villalba@ivic.ve

$$\tilde{\gamma}^0 = \gamma^0, \tilde{\gamma}^1 = a^{-1}\gamma^1, \tilde{\gamma}^2 = a^{-1}b^{-1}\gamma^2, \tilde{\gamma}^3 = a^{-1}b^{-1}c^{-1}\gamma^3 \tag{5}$$

where γ^μ are the standard Dirac flat matrices, The Dirac equation (1) takes the form

$$\left\{ \gamma^0 \partial_t + \frac{1}{a}\gamma^1 \partial_x + \frac{1}{ab}\gamma^2 \partial_y + \frac{1}{abc}\gamma^3 \partial_z + m \right\} \tilde{\Psi} = 0 \tag{6}$$

where we have introduced the spinor $\tilde{\Psi} = a^{-3/2}b^{-1}c^{-1/2}\Psi$. We proceed to separate variables using a pairwise scheme. Eq. (6) can be rewritten as sum of two commuting operators $\hat{K}_1, \hat{K}_2$

$$\hat{K}_1 = -iab\left[\gamma^0 \partial_t + \frac{\gamma^1}{a}\partial_x + m\right]\gamma^1\gamma^0, \quad \hat{K}_2 = -i\left[\gamma^2 \partial_y + \frac{\gamma^3}{c}\partial_z\right]\gamma^1\gamma^0, \quad \Phi = \gamma^1\gamma^0\tilde{\Psi} \tag{7}$$

The expression $\hat{K}_2 \Phi + k\Phi = 0$ can be separated as follows

$$\hat{K}_3 = c\left[\gamma^2\gamma^3 \partial_y + ik\gamma^3\gamma^1\gamma^0\right], \quad \hat{K}_4 = \partial_z \tag{8}$$

where

$$[\hat{K}_3, \hat{K}_4] = 0, \quad (\hat{K}_3 + \hat{K}_4)\Omega = 0, \quad \Omega = \gamma^3\gamma^1\gamma^0\Phi \tag{9}$$

The separation of variables in the equation $\hat{K}_1\Phi = k\Phi$ cannot be achieved in terms of first order commuting differential operators. We rewrite this equations as follows:

$$\left\{ \left[\gamma^0 \partial_x + i\frac{k}{b}\right] + a\left[\gamma^0 \partial_t + m\right]\gamma^1\gamma^0 \right\} \Phi = \{\hat{L}_1 + \hat{L}_2\gamma^1\gamma^0\}\Phi = 0 \tag{10}$$

The operators $\hat{L}_1$ and $\hat{L}_2$ commute. Let us introduce the auxiliary function η defined by,

$$\Phi = \left(a\left[\gamma^0 \partial_t + m\right]\gamma^1\gamma^0 + \left[\gamma^0 \partial_x - i\frac{k}{b}\right] \right)\eta \tag{11}$$

$$\{\hat{N}_1 + \hat{N}_2\}\eta = 0 \quad [\hat{N}_1, \hat{N}_2] = 0 \tag{12}$$

the operators $\hat{N}_1$ and $\hat{N}_2$ are:

$$\hat{N}_1 = [a(\gamma^0 \partial_t + m)][a(-\gamma^0 \partial_t + m)], \quad \hat{N}_2 = \left[\gamma^0 \partial_x + i\frac{k}{b}\right]\left[\gamma^0 \partial_x - i\frac{k}{b}\right] \tag{13}$$

using the representation

$$\gamma^0 = \begin{pmatrix} -i & 0 \\ 0 & i \end{pmatrix}, \gamma^1 = \begin{pmatrix} 0 & \sigma^3 \\ \sigma^3 & 0 \end{pmatrix}, \gamma^2 = \begin{pmatrix} 0 & \sigma^2 \\ \sigma^2 & 0 \end{pmatrix}, \gamma^3 = \begin{pmatrix} 0 & \sigma^1 \\ \sigma^1 & 0 \end{pmatrix}, \tag{14}$$

we can write the spinor η as follows

$$\eta = \begin{pmatrix} u(t)U(x) \\ v(t)V(x) \\ w(t)W(x) \\ z(t)Z(x) \end{pmatrix} \tag{15}$$

applying (11) to (15), we have that Φ takes the form:

$$\Phi = \begin{pmatrix} (-\delta^{-1}+\sigma)\,u(t)W(x)f(y) \\ (\delta^{-1}+\sigma)\,u(t)W(x)g(y) \\ (\delta-\sigma^{-1})\,w(t)U(x)f(y) \\ (-\delta+\sigma^{-1})\,w(t)U(x)g(y) \end{pmatrix} exp(ik_z z) \quad (16)$$

$$(d_x - \frac{k}{b})W = i\lambda/\sigma U, (d_x + \frac{k}{b})U = i\lambda \sigma W, a(d_t - im)u = \lambda \delta w, a(d_t + im)w = -\lambda/\delta u \quad (17)$$

The functions $f(y)$ and $g(y)$ satisfy the system of equations

$$(-\delta^{-1}+\sigma)(\partial_y - \frac{k_z}{c})f(y) = ik(\delta^{-1}+\sigma)g(y), \quad (18)$$

$$(\delta^{-1}+\sigma)(\partial_y + \frac{k_z}{c})g(y) = ik(-\delta^{-1}+\sigma)f(y) \quad (19)$$

The Robertson-Walker line element can be written as $ds^2 = e^{\alpha(\eta)}\{dr^2 + \xi^2(r)d\Omega^2 - d\eta^2\}$, where the radial factor ξ takes the values $\xi(r) = r, \sinh(r)$ and $\sin(r)$ for spatially flat, open and closed universes respectively. Choosing to work in the diagonal tetrad a^α_β we obtain that the Dirac equation in the Robertson-Walker background field reads [4]

$$(\gamma^0 \partial_\eta + \gamma^1 \partial_r + \frac{\gamma^2 \partial_\vartheta}{\xi(r)} + \frac{\gamma^3 \partial_\varphi}{\xi(r)\sin\theta} + me^{\alpha/2})\Psi = 0 \quad (20)$$

where Ψ is related to $\tilde{\Psi}$ by $\tilde{\Psi} = \xi(r)^{-1}(\sin\theta)^{-1/2}e^{-3\alpha/4}\Psi$. Eq. (20) can be written in the form (3) with [4]

$$\hat{K}_1 = -i\{\gamma^2\partial_\theta + \frac{\gamma^3}{\sin\theta}\partial_\varphi\}\gamma^1\gamma^0, \quad \hat{K}_2 = -i\xi(r)\{(\gamma^0\partial_0 + \gamma^1\partial_1) + me^{\alpha/2}\}\gamma^1\gamma^0 \quad (21)$$

$$\Phi = \gamma^1\gamma^0\Psi, \quad S\gamma^\nu S^{-1} = h^\nu_\beta \tilde{\gamma}^\beta = \gamma^\nu_c, \quad S\Psi = \Psi_c \quad (22)$$

where γ^ν_c are Dirac matrices in the Cartesian tetrad gauge. The matrix transformation S reads:

$$S = \exp(-\frac{\varphi}{2}\gamma^1\gamma^2)\exp(-\frac{\theta}{2}\gamma^3\gamma^1)\mathcal{N}, \quad \mathcal{N} = \frac{1}{2}(\gamma^1\gamma^2 + \gamma^2\gamma^3 + \gamma^3\gamma^1 + 1) \quad (23)$$

where

$$\left(\frac{d}{d\eta} - ime^{\alpha/2}\right)u = -i\lambda v, \quad \left(\frac{d}{d\eta} + ime^{\alpha/2}\right)v = -i\lambda u \quad (24)$$

Making the coordinate transformation

$$e^{-z} = \cosh r - \sinh r \cos\theta, \quad e^{-z}x = \sin\theta\cos\phi\sinh r, \quad e^{-z}y = \sin\theta\sin\phi\sinh r \quad (25)$$

the spatially open Robertson-Walker metric takes the form [5]

$$a^{-2}(\eta)ds^2 = -d\eta^2 + dz^2 + e^{-2z}(dx^2 + dy^2) \quad (26)$$

The Dirac equation in the metric (26) takes the form

$$(\gamma^0 \frac{\partial}{\partial \eta} + \gamma^1 e^z \left(\frac{\partial}{\partial x} - A_1(y) \right) + \gamma^2 e^z \frac{\partial}{\partial y} + \gamma^3 \frac{\partial}{\partial z} + M\alpha(\eta))\Psi = 0, \quad (27)$$

where we have introduced the spinor $\tilde{\Psi} = a(\eta)^{-3/2} e^z \Psi$. $\gamma^3 \gamma^0 \Psi = \Phi$ and

$$\hat{K}_1 \Phi = \left\{ \gamma^2 \frac{\partial}{\partial y} + \gamma^1 \left(\frac{\partial}{\partial x} - iA_1(y) \right) \right\} \gamma^3 \gamma^0 \Phi = ik\Phi, \quad (28)$$

$$\hat{K}_2 \Phi = e^z \left\{ \gamma^0 \frac{\partial}{\partial \eta} + \gamma^3 \frac{\partial}{\partial z} + M\alpha(\eta) \right\} \gamma^3 \gamma^0 \Phi = -ik\Phi. \quad (29)$$

In order to separate variables in Eq. (29) we rewrite it in the following form $(\hat{L}_1 \gamma^3 \gamma^0 + \hat{L}_2) \Phi = 0$ where $\hat{L}_1$ and $\hat{L}_2$ are two commuting differential operators given by the expressions

$$\hat{L}_1 = \gamma^0 \frac{\partial}{\partial \eta} + M\alpha(\eta), \ \hat{L}_2 = \gamma^0 \frac{\partial}{\partial z} + ike^z, \quad (30)$$

we introduce the auxiliary spinor $\mathscr{Y}$

$$(\hat{L}_1 \gamma^3 \gamma^0 + \tilde{L}_2) \mathscr{Y} = \Phi, \ \tilde{L}_2 = \gamma^0 \frac{\partial}{\partial z} - ike^z \quad (31)$$

We obtain that $\mathscr{Y}$ satisfies the equation

$$\{\hat{M}_1 + \hat{M}_2\} \mathscr{Y} = 0, \text{with } [\hat{M}_1, \hat{M}_2] = 0, \quad (32)$$

and

$$\left(-\frac{\partial^2}{\partial z^2} - i\gamma^0 k e^z + k^2 e^{2z} + \tilde{\lambda} \right) \mathscr{Y} = 0, \ \left(\frac{\partial^2}{\partial \eta^2} + \gamma^0 M \frac{d\alpha(\eta)}{d\eta} + M^2 \alpha^2(\eta) - \tilde{\lambda} \right) \mathscr{Y} = 0, \quad (33)$$

where $\tilde{\lambda}$ is a separation constant.

ACKNOWLEDGMENTS

The author wishes to express his gratitude to the Alexander von Humboldt Stiftung for financial support.

REFERENCES

1. I. E. Andrushkevich, and G. V. Shishkin, Theor. Math. Phys. **70**, 204 (1987), G. V. Shishkin, and V.M. Villalba, J. Math. Phys. **30**, 2132 (1989)., V. M. Villalba, Mod. Phys. Lett A. **8**, 2351 (1993).
2. M. N. Hounkonnou, and J. E. B. Mendy, J. Math. Phys **40**, 3827 (1999)., J. Math. Phys **40**, 4240 (1999).
3. D. Brill, and J. A. Wheeler, Rev. Mod. Phys. **29**, 465 (1957).
4. V. M. Villalba, and U. Percoco, J. Math. Phys. **31**, 715 (1990).
5. V. M. Villalba, and E. Isasi Catalá. J. Math. Phys. **43**, 4909 (2002).

PART IV: POSTERS

On the concept of relative velocity

V. J. Bolós

Dep. Matemáticas, Facultad de Ciencias, Universidad de Extremadura.
Avda. de Elvas s/n. 06071–Badajoz, Spain,
e-mail: vjbolos@unex.es

Abstract. Four different concepts of "relative velocity" are given intrinsically in general relativity. The *kinematic* and *kinematic2* relative velocities are defined in the framework of spacelike simultaneity, obtaining some general properties and interpretations. The kinematic relative velocity generalizes the usual concept of relative velocity when the two observers β, β' are at the same event. On the other hand, the kinematic2 relative velocity does not generalize this concept, but it is physically interpreted as the variation of the *relative position* of β' with respect to β along the world line of β. Analogously, the *spectroscopic* and *astrometric* relative velocities are defined and studied in the framework of observed (lightlike) simultaneity. Next, we give some relations between these concepts in special relativity. Finally, we show some fundamental examples in Schwarzschild and Robertson-Walker space-times.

Keywords: Relative velocity, Radial velocity, Simultaneity, Relative position
PACS: 04.20.Cv, 02.40.Hw

An *observer* in the space-time M is determined by a timelike world line β, and the events of β are the *positions* of the observer. It is usual to identify an observer with its world line, and so β is an observer. The *4-velocity* of the observer is a future-pointing timelike unit vector field U defined in β and tangent to β. Given an event p, the 4-velocity of an observer at p is given by a future-pointing timelike unit vector u. It is also usual to identify an observer with its 4-velocity, since they are defined reciprocally.

The relative velocity of an observer with respect to another observer is well defined (i.e. intrinsically defined) only when these observers are at the same event: given two observers u and u' at the same event p, there exists a unique vector $v \in u^\perp$ and a unique positive real number γ such that $u' = \gamma(u+v)$. As consequences, we have $0 \leq \|v\| < 1$ and $\gamma := -g(u',u) = \frac{1}{\sqrt{1-\|v\|^2}}$. We will say that v is the *relative velocity of u' observed by u*, and γ is the *gamma factor* corresponding to the velocity $\|v\|$. So, we have

$$v = \frac{1}{-g(u',u)} u' - u.$$

Let u, u' be two observers at p, q respectively such that $q \in L_{p,u} := \exp_p u^\perp$. The *kinematic relative velocity of u' with respect to u* is the unique vector $v_{\text{kin}} \in u^\perp$ such that $\tau_{qp} u' = \gamma(u + v_{\text{kin}})$, where γ is the gamma factor corresponding to the velocity $\|v_{\text{kin}}\|$. It is given by

$$v_{\text{kin}} := \frac{1}{-g(\tau_{qp}u',u)} \tau_{qp} u' - u.$$

Given u an observer at p, and a simultaneous event $q \in L_{p,u}$, the *relative position of q with respect to u* is $s := \exp_p^{-1} q$. We can generalize this definition for two observers β and β': let β, β' be two observers and let U be the 4-velocity of β. The *relative position of β' with respect to β* is the vector field S defined on β such that S_p is the relative position of q with respect to U_p (in the sense of the previous definition), where $p \in \beta$ and q is the unique event of $\beta' \cap L_{p,U_p}$. Let β, β' be two observers, let U be the 4-velocity of β, and let S be the relative position of β' with respect to β. The *kinematic2 relative velocity of β' with respect to β* is the projection of $\nabla_U S$ onto $U^\perp$, i.e. it is the vector field

$$V_{\text{kin2}} := \nabla_U S + g\left(\nabla_U S, U\right) U$$

defined on β.

We are going to define the spectroscopic relative velocity analogously to the definition of the kinematic relative velocity: let u, u' be two observers at p, q respectively such that $q \in E_p^- := \exp_p C_p^-$, (where C_p^- is the past-pointing light cone in $T_p M$) and let λ be a light ray from q to p. The *spectroscopic relative velocity of u' observed by u* is the unique vector $v_{\text{spec}} \in u^\perp$ such that $\tau_{qp} u' = \gamma (u + v_{\text{spec}})$, where γ is the gamma factor corresponding to the velocity $\|v_{\text{spec}}\|$. It is given by

$$v_{\text{spec}} := \frac{1}{-g\left(\tau_{qp} u', u\right)} \tau_{qp} u' - u.$$

Analogously to the definition of the kinematic2 relative velocity (using E_p^- instead of $L_{p,u}$), we can define the *relative position of β' observed by β*, denoted by S_{obs}. In the same way, the *astrometric relative velocity of β' observed by β* is the vector field

$$V_{\text{ast}} := \nabla_U S_{\text{obs}} + g\left(\nabla_U S_{\text{obs}}, U\right) U$$

defined on β.

In the Minkowski space-time, using the previous notation, it is satisfied $V_{\text{kin2}} = \left(1 + g\left(S, \nabla_U U\right)\right) V_{\text{kin}}$.

Moreover, if $S_{\text{obs}} \neq 0$ then $V_{\text{ast}} = \|S_{\text{obs}}\| \nabla_U U + \dfrac{1}{1 + g\left(V_{\text{spec}}, \frac{S_{\text{obs}}}{\|S_{\text{obs}}\|}\right)} V_{\text{spec}}$.

For a more detailed information, and some fundamental examples in Schwarzschild and Robertson-Walker space-times, see [1] and [2].

ACKNOWLEDGMENTS

This work has been partially supported by the the Spanish Comisión Interministerial de Ciencia y Tecnología (CICYT, MTM2004-06226).

REFERENCES

1. V. J. Bolós, *J. Geom. Phys.*, (In press). Preprint: `gr-qc/0501085`.
2. V. J. Bolós, Preprint: `gr-qc/0506032`.

Mathematical Properties of the Elasticity Difference Tensor

E.G.L.R. Vaz* and Irene Brito*

Departamento de Matemática para a Ciência e Tecnologia, Universidade do Minho, Campus de Azurém, 4800-058 Guimarães, Portugal - evaz@mct.uminho.pt

Abstract. A tetrad, adapted to the principal directions of the unstrained reference tensor, is chosen and the elasticity difference tensor, as introduced in [1], is decomposed along those directions. The second order tensors obtained are studied and an example is presented.

Keywords: elasticity; classification
PACS: 04.20.-q

INTRODUCTION

Here we will consider a continuous medium possessing elastic properties, the collection of all its idealized particles being the 3-dimensional space X - the material space. (M, g) represents the space-time manifold, i.e. M is a four-dimensional, connected, Hausdorff manifold and g is a Lorentzian metric with signature $(-+++)$ such that $g = -\mathbf{u} \otimes \mathbf{u} + h$, where: (i) $h = \pi^* k$, π^* being the usual canonical projection onto X and k being a metric in X; (ii) $\pi^{-1}(p \in X)$ defines a timelike curve in M having $\mathbf{u}$ as unit tangent vector field and represents the flowline of p; (iii) $\pi : U \subset M \longrightarrow X$ describes a state of matter.

Following [1], for an unrelaxed state of matter the unstrained reference tensor [2] can be written as $k_{ab} = n_1^2 x_a x_b + n_2^2 y_a y_b + n_3^2 z_a z_b$, the scalar fields n_1, n_2, n_3 being related to the eigenvalues along the principal directions of k_b^a. An orthonormal tetrad $\{\mathbf{u},\mathbf{x},\mathbf{y},\mathbf{z}\}$, with $\mathbf{u}$ timelike and $\mathbf{x},\mathbf{y},\mathbf{z}$ unit spacelike vector fields aligned with the eigenvectors of k_a^b, will be used. On a local coordinate system, the metric g can be written as

$$g_{ab} = -u_a u_b + x_a x_b + y_a y_b + z_a z_b. \tag{1}$$

In order to study elasticity properties of the space-time, the authors in [1] define the elasticity difference tensor:

$$S^a_{bc} = \frac{1}{2} k^{am}(D_b k_{mc} + D_c k_{mb} - D_m k_{bc}),$$

where D denotes projected covariant derivative associated to g. A classification of S will certainly be interesting for the characterization of the elasticity properties of the space-time. In order to do so, we decompose S^a_{bc} along the principal directions of k^a_b:

$$S^a_{bc} = M_{bc} x^a + N_{bc} y^a + P_{bc} z^a.$$

The second order symmetric tensors M, N, P are now investigated.

MAIN RESULTS. AN EXAMPLE

The following results for M_{bc} were obtained, the proofs being in [4].

Theorem 1 *The general form of M_{bc} is given by*

$$M_{bc} = u^m(x_{m;(b}u_{c)} + u_{(b}x_{c);m}) + x_{(b;c)} - x^m x_{(c}x_{b);m} + \gamma_{011}\, u_{(b}x_{c)} - \gamma_{010}\, u_b u_c$$
$$+ \frac{1}{n_1}[2n_{1,(b}x_{c)} + 2n_{1,m}u^m u_{(b}x_{c)} + n_{1,m}x^m x_b x_c]$$
$$+ \frac{1}{n_1^2}\{-x^m(z_b z_c n_3 n_{3,m} + y_b y_c n_2 n_{2,m}) + n_2^2[(\gamma_{021} - \gamma_{120})u_{(b}y_{c)} + x^m(y_{m;(b}y_{c)} - y_{(b}y_{c);m})]$$
$$+ n_3^2[(\gamma_{031} - \gamma_{130})u_{(b}z_{c)} + x^m(z_{m;(b}z_{c)} - z_{(b}z_{c);m})]\},$$

where γ_{abc} are the rotation coefficients and a comma represents a partial derivative.

Theorem 2 *$\mathbf{x}$ is an eigenvector of M_{bc} iff n_1 remains invariant along the directions of $\mathbf{y}$ and $\mathbf{z}$, i.e. $\Delta_y(\log n_1) = \Delta_z(\log n_1) = 0$, where Δ_y represents the intrinsic derivative along $\mathbf{y}$. The corresponding eigenvalue is $\lambda = \Delta_x(\log n_1)$.*

Theorem 3 *$\mathbf{y}$ is an eigenvector of M_{bc} iff n_1 remains invariant along the direction of $\mathbf{y}$, i.e. $\Delta_y(\log n_1) = 0$, and $\frac{1}{2}\gamma_{132}[-(n_3^2/n_1^2) + 1] + \frac{1}{2}\gamma_{123}[1 - (n_2^2/n_1^2)] + \frac{1}{2}\gamma_{231}[(n_3^2/n_1^2) - (n_2^2/n_1^2)] = 0$. The corresponding eigenvalue is $\lambda = -(n_2/n_1^2)\Delta_x n_2 + \gamma_{122}[-(n_2^2/n_1^2) + 1]$.*

Theorem 4 *$\mathbf{z}$ is an eigenvector of M_{bc} iff n_1 remains invariant along the direction of $\mathbf{z}$, i.e. $\Delta_z(\log n_1) = 0$, and $\frac{1}{2}\gamma_{123}[1 - (n_2^2/n_1^2)] + \frac{1}{2}\gamma_{132}[1 - (n_3^2/n_1^2)] + \frac{1}{2}\gamma_{231}[(n_3^2/n_1^2) - (n_2^2/n_1^2)] = 0$. The corresponding eigenvalue is $\lambda = -(n_3/n_1^2)\Delta_x n_3 - \gamma_{133}[(n_3^2/n_1^2) - 1]$.*

Similar results have been obtained by the authors for N and P [4].

The following example illustrates the results above. We consider a spherically symmetric metric g written in local coordinates t, r, θ, ϕ as $ds^2 = -dt^2 + dr^2 + r^2 d\theta^2 + r^2 \sin^2\theta d\phi^2$ (see [3], p.186). If a radial deformation is considered such that $d\tilde{s}^2 = -dt^2 + n^2(r)[dr^2 + r^2 d\theta^2 + r^2 \sin^2\theta d\phi^2]$, the only non-zero components of the elasticity difference tensor are $S^r_{rr} = S^\theta_{\theta r} = S^\phi_{\phi r} = \frac{1}{n(r)}\frac{dn(r)}{dr}$ and $S^r_{\phi\phi} = -\frac{r^2\sin^2(\theta)}{n(r)}\frac{dn(r)}{dr} = \sin^2(\theta)S^r_{\theta\theta}$. Then $M_{bc} = \lambda_1(x_b x_c - y_b y_c - z_b z_c)$, $N_{bc} = 2\lambda_2(x_b y_c + x_c y_b)$ and $P_{bc} = 2\lambda_3(x_b z_c + x_c z_b)$, where $\lambda_1 = \lambda_2 = \lambda_3 = \frac{1}{n(r)}\frac{dn(r)}{dr}$. Therefore, the eigenvalue associated with the eigenvector **u** vanishes identically. The remaining eigenvectors are: (i) {**x,y,z**} for M_{bc}, $\frac{1}{n(r)}\frac{dn(r)}{dr}$, $-\frac{1}{n(r)}\frac{dn(r)}{dr}$, $-\frac{1}{n(r)}\frac{dn(r)}{dr}$ being the corresponding eigenvalues, so that the Segre type is $\{1,1(11)\}$; (ii) {**x+y,x-y,z**} for N_{bc} with eigenvalues $\frac{1}{n(r)}\frac{dn(r)}{dr}$, $-\frac{1}{n(r)}\frac{dn(r)}{dr}$ and zero, respectively, the Segre type being $\{1,111\}$; (iii) {**x+z,x-z,y**} for P_{bc} with eigenvalues $\frac{1}{n(r)}\frac{dn(r)}{dr}$, $-\frac{1}{n(r)}\frac{dn(r)}{dr}$ and zero, respectively, the Segre type being then $\{1,111\}$.

REFERENCES

1. M. Karlovini, and L. Samuelsson, *Class.Quantum Grav.*, **20** (2003).
2. B. Carter and H. Quintana, *Proc. R. Soc.*, A 331, **57** (1972).
3. R. Inverno, *Introducing Einstein's Relativity*, Clarendon Press, Oxford (1996).
4. E.G.L.R. Vaz and Irene Brito, preprint.

Scalar Field Coupled to Gravity in Spherically Symmetric Background in $(2+1)$ Dimensions

D. Daghan and A. H. Bilge

Department of Mathematics, Faculty of Science and Letters, Istanbul Technical University, Maslak, 80626 Istanbul, Turkey, daghand@itu.edu.tr, bilge@itu.edu.tr.

Abstract. We obtain an exact solution for the Einstein's equations with cosmological constant coupled to a scalar, static particle in static, "spherically" symmetric background in $2+1$ dimensions.

Keywords: Black holes; Choptuik formation; singularity.
PACS: 83C80

INTRODUCTION

The Einstein's equations with cosmological constant, coupled to a scalar, static field ϕ, in a static spherically symmetric background in $(2+1)$ dimensions described by the metric

$$ds^2 = -e^{2X(r)}dt^2 + e^{2X(r)}dr^2 + e^{-2Y(r)}d\theta^2 \qquad (1)$$

have been previously studied in reference [1] using a perturbative asymptotic analysis of the field equations. In references [2]-[5], similar systems in $2+1$ dimensions were studied numerically and analytically.

In this note we obtain an exact solution for this problem with appropriate coordinate transformations [to be published in GRG]. This exact solution provides an exact model for the cutting and gluing procedure presented in [6].

EINSTEIN'S FIELD EQUATIONS

The original equations are reduced to the following second order system

$$X' = \frac{1}{2}ce^Y - \frac{1}{2}[c^2e^{2Y} + 4(-\Lambda e^{2X} + \lambda^2 e^{2Y})]^{\frac{1}{2}},$$
$$Y' = \frac{1}{2}ce^Y + \frac{1}{2}[c^2e^{2Y} + 4(-\Lambda e^{2X} + \lambda^2 e^{2Y})]^{\frac{1}{2}},$$
$$\phi' = \frac{\lambda}{\sqrt{2\pi}}e^Y \qquad (2)$$

where $'$ denotes differentiation with respect to r, Λ is the cosmological constant, c and λ are integration constants, $c \neq 0$ but can be either positive or negative.

Taking $\Lambda = -1$ and by using the transformation

$$e^X = \frac{1}{2}\rho \, \cos\varphi, \quad e^Y = \frac{1}{\sqrt{c^2 + 4\lambda^2}}\rho \, \sin\varphi, \quad \mu = \frac{c}{\sqrt{c^2 + 4\lambda^2}}, \qquad (3)$$

we obtain the following equivalent system of ODE's

$$\rho' = \frac{1}{2}\rho^2[\mu\sin\varphi - \cos 2\varphi],$$
$$\varphi' = \rho\sin\varphi\cos\varphi. \tag{4}$$

Then, we obtain the exact solution in the new coordinate frame as

$$ds^2 = -\left[\frac{1}{2}\rho\cos\varphi\right]^2 dt^2 + \left[\frac{1}{2}\frac{1}{\sin\varphi}\right]^2 d\varphi^2 + \left[\frac{\sqrt{c^2+4\lambda^2}}{\rho\sin\varphi}\right]^2 d\theta^2 \tag{5}$$

where

$$\rho(\varphi) = \frac{\rho_0}{\sqrt{2}}[1+\sin\varphi]^{\frac{\mu}{2}}[\sin\varphi]^{-\frac{1}{2}}[\cos\varphi]^{\frac{-1-\mu}{2}}, \tag{6}$$

and ρ_0 is an integration constant. By using further coordinate transformations we obtain

$$ds^2 = -\left[\frac{\cosh\chi}{\sinh\chi}\right]^{\mu}\cosh\chi\sinh\chi\, d\tilde{t}^2 + d\chi^2 + (c^2+4\lambda^2)\left[\frac{\cosh\chi}{\sinh\chi}\right]^{-\mu}\cosh\chi\sinh\chi\, d\tilde{\theta}^2. \tag{7}$$

The term $\sqrt{c^2+4\lambda^2} = \frac{c}{\mu}$ induces a scaling of θ which can be interpreted as inducing a topological defect. For $\lambda = 0$, $\mu = \frac{c}{|c|} = \pm 1$. Taking $c = 1$ we have AdS solution

$$ds^2 = -\cosh^2\chi\, d\tilde{t}^2 + d\chi^2 + \sinh^2\chi\, d\tilde{\theta}^2 \tag{8}$$

as given by Eqn. (1.1) in Ref. [6]. The analytic solution to the complete problem investigated here has been previously obtained in reference [7]. With the signature $(+,-,-)$ our metric (7) coincides with equation (3.7) in [7].

ACKNOWLEDGMENTS

The authors would like to thank Professors G. Clement and A. Fabbri for pointing out that our solution as given by Eqn. (7), has been obtained previously in Ref. [7], and also they would like to thank Professor M. Hortacsu for illuminating discussions and for pointing out reference [6].

REFERENCES

1. Birkandan, T and Hortacsu, M. (2003). Three Dimensional Gravity in the Presence of Scalar Fields, *Gen. Rel. and Grav.* **35**, pp.457-466.
2. Pretorius, F. and Choptuik, W.M. (2000). Gravitational collapse in 2+1 dimensional AdS spacetime, *Phys. Rev. D* **62**, 124012.
3. Garfinkle, D. (2001). *Phys. Rev. D* **63**, 044007.
4. Husain, V and Olivier, M. (2001). *Class. Quant. Grav.* **18**.
5. Clement, G., Fabbri, A.: *Class. Quant. Grav.* **18**,3665 (2001)[gr-qc/0101073].
6. Matschull, H-J. (1999). Black hole creation in 2+1 dimensions, *Class. Quant. Grav.* **16**, pp.1069-1095.
7. Clement, G., Fabbri, A.: The cosmological gravitating σ model: solitons and black holes. *Class. Quant. Grav.* **17**,2437 (2000)[gr-qc/9912023].

Non-equivalence of newtonian and relativistic second order perturbations

Filipe C. Mena

Mathematical Institute, University of Oxford, 24-29 St. Giles', Oxford OX1 3LB, U.K.
and
Departamento de Matemática, Universidade do Minho, Gualtar, 4710-078 Braga, Portugal

Abstract. I show how the correspondence between the relativistic and newtonian frameworks, present for FLRW first order perturbation theory, is bound to be broken at the second order.

Keywords: Cosmology, perturbations, newtonian limit
PACS: 98.80.Jk,98.80.Es,95.30.Sf

A fundamental question in General Relativity is how the Newtonian limit can be achieved and a great deal has been done to study this issue, see e.g. [2, 3, 5]. An important aspect is to analyse how accurate are the newtonian approximations in the cosmological context, where they are frequently used to study structure formation [4].

Structure formation is mainly modelled with perturbed Friedmann-Lemaitre-Roberson-Walker (FLRW) spacetimes and although the linear theory is already well studied [8], non-linear perturbations have only recently become a prime focus of attention thanks to the need for higher accuracy to analyse experimental data [6].

It is well known that the linearised Einstein field equations for particular FLRW models can be reduced to the linearised newtonian equations, in the absence of vector and tensor perturbation modes. Recently, Hwang & Noh [5] claimed that there is also such correspondence between the two theories at second order. In particular, they found an equivalence between the density contrast evolution equations of both theories, in the case of a perturbed flat dust FLRW model, using the comoving C-gauge.

Following Hwang & Noh, I took the flat dust FLRW metric

$$ds^2 = a^2(\tau)(d\tau^2 + \delta_{ij}dx^i dx^j)$$

perturbed up to second order:

$$g_{00} = -a^2(1+2\psi+\psi^{(2)})$$
$$g_{0i} = a^2(w_i + \frac{1}{2}w_i^{(2)})$$
$$g_{ij} = a^2\left(\delta_{ij} - (2\phi + \phi^{(2)})\delta_{ij} + \chi_{ij} + \frac{1}{2}\chi_{ij}^{(2)}\right)$$

and, at each order, I used the standard decompositions

$$w_i = \partial_i w + w_i^\perp$$
$$\chi_{ij} = D_{ij}\chi + \chi_{(i;j)}^\perp + \chi_{ij}^T$$

with $D_{ij} = \partial_i \partial_j - \frac{1}{3}\delta_{ij}\nabla^2$ and $\partial^i w_i^\perp = \partial^i \chi_i^\perp = \partial^i \chi_{ij}^T = \chi_i^{Ti} = 0$. At first order, I assumed that the only non-zero perturbation modes were scalars.

I found that, at second order, the correspondence between the perturbation variables in the two theories is strongly gauge dependent but, indeed in the comoving C-gauge, the density contrast evolution equations agree in the two theories. I then computed some relativistic kinematic quantities as well as the weyl tensor. At first order, these quantities have a clear newtonian correspondence, with the magnetic part H_{ij} of the Weyl tensor being zero. This is crucial since there is no newtonian counterpart for H_{ij}. At second order, my results for the shear and weyl tensor in the comoving C-gauge are [7]

$$\sigma_{ij}^{(2)} = a[v_{,ij}^{(2)} - \frac{1}{3}\delta_{ij}\nabla^2 v^{(2)} - 2\phi\sigma_{ij} + \frac{1}{2}\chi_{ij}^{(2)T'}]$$

$$H_{ij}^{(2)} = \eta_{(j}{}^{kl}\left(2(w_{,(i)}\phi_{,k}))_{,l} - 2\phi_{,(i)}w_{,kl)} + w_{,i)kl}^{(2)} + 2w_{(i),k)l}^{(2)\perp} + \frac{1}{2}\chi_{i)k,l}^{(2)T}\right)$$

$$E_{ij}^{(2)} = D_{ij}(\phi^{(2)} + w^{(2)'}) + w_{(i;j)}^{(2)\perp'} - \frac{1}{2}(\chi_{ij}^{(2)T''} - \nabla^2 \chi_{ij}^{(2)T})$$
$$+ \partial_\tau \left(2w_{,(i}\phi_{,j)} - 2w_{,k}\phi^{,k}\delta_{ij}\right) - 2\frac{a'}{a}\left(2w_{,(i}\phi_{,j)} - 2w_{,k}\phi^{,k}\delta_{ij}\right) + \frac{2}{a}\phi' D_{ij}w$$
$$- w^{,k}D_{ij}w_{,k} - \delta^{kl}D_{ik}wD_{lj}w + \frac{1}{2}D_{ij}(w^{,k}w_{,k}) + \delta_{ij}D_l^k wD^l{}_k w,$$

where the prime denotes differentiation with respect to τ and v is the scalar part of the velocity perturbation. Although there are no vector or tensor modes at first order, these are inevitably generated at second order by couplings of first order scalar modes (see e.g. [1]). Mode coupling also has the consequence of introducing non-newtonian-like terms in $\sigma_{ij}^{(2)}$ and $E_{ij}^{(2)}$ and, more importantly, of generating $H_{ij}^{(2)}$. Therefore, the first order correspondence between the two frameworks breaks down at second order. This result also shows the non-linear instability of the silent spacetimes.

ACKNOWLEDGMENTS

Part of this work was done in collaboration with Reza Tavakol and Marco Bruni. I thank British Council/CRUP (Portugal) for grant B-13/03, FCT (Portugal) for grant SFRH/BPD/12137/2003 and CMAT (Universidade do Minho) for support.

REFERENCES

1. Bruni M, Mena F C & Tavakol R, *Class. Quant. Grav.* **19** (2002) L23
2. Ehlers J, *Akad. Wiss. Lit. Mainz, Abhandl. Math.–Nat. Kl.*, **11** (1961)
3. Ellis G F R, in *Cargèse Lectures in Physics Vol. 6*, ed. Schatzman E (Gordon and Breach, 1973)
4. Coles P & Lucchin F, *Cosmology: The origin and evolution of cosmic structure* (Wiley, 1995)
5. Hwang & Noh, *Phys. Rev.* **D72** (2005) 044011
6. Matarrese S, Mollerach S & Bruni M, *Phys. Rev.* **D58**, 043504 (1998).
7. Mena F C, Tavakol R & Bruni M, "Comparison of first and second order perturbations in newtonian and relativistic frameworks", in preparation
8. Mukhanov V F, Feldman H A & Brandenberger R H, *Phys. Rep.* **215** (1992) 203

Emission Coordinates and the Central Observer

José M. Pozo

SYRTE–CNRS, Observatoire de Paris. 61, Avenue de l'Observatoire. F-75014 Paris.

Abstract. 4 emitters broadcasting an increasing electromagnetic signal generate a space-time coordinate system, called *emission coordinates*, which constitutes an immediate relativistic positioning system. In this work we explain two local properties of them: the existence of a natural splitting for the metric, which separates its 6 degrees of freedom into 2 types of parameters, and the existence of a unique observer (the *central observer*) and vierbein field, satisfying a natural geometric condition.

Keywords: Positioning, Emission coordinates, Central observer, Equifacial tetrahedra
PACS: 04.20.Cv

INTRODUCTION

Four emitters broadcasting an increasing electromagnetic signal generate a system of space-time coordinates, the so called *emission coordinates*. The most natural case is the one in which the emitted signal is the proper time of the emitter. The emission coordinates of an event (the 4 signals received) can be immediately known by this event, thus they constitute an immediate relativistic positioning system. Indeed, its physical construction is very similar to the one realized by the GPS or the future GALILEO, where the emitters are satellites. But, although many different *relativistic corrections* (from both restricted and general relativity) are used for the current precisions, these systems are conceived as classical (Newtonian). The study of emission coordinates is aimed to develop a fully relativistic theory of positioning and reference systems. A complete theory should be able to substitute the nowadays classical perturbative approach to the satellite navigation, providing a different framework for experimental tests of general relativity.

Emission coordinates and the associated positioning systems has been extensively studied in 2-dimensional space-times [1, 2], and some global and local properties have been obtained for 3 and 4 dimensions [3, 4]. In this work we briefly explain two interesting local properties of 4-dimensional emission coordinates.

SPLITTING OF THE METRIC IN EMISSION COORDINATES

The first property of emission coordinates is that the natural covectors are all light-like, $g^{AA} = 0$, and future directed, $g^{AB} > 0$ $\forall A \neq B$ (signature -2). The signature of the metric is Lorentzian iff the *triangular* inequalities $A < B+C$, $B < A+C$, $C < A+B$, for $A \equiv \sqrt{g^{23}g^{14}}$, $B \equiv \sqrt{g^{13}g^{24}}$, $C \equiv \sqrt{g^{12}g^{34}}$, are satisfied. These conditions lead to a natural splitting of the metric:

$$g_{AB} = \mu_A \hat{g}_{AB} \mu_B \quad \text{and} \quad g^{AB} = \mu^A \hat{g}^{AB} \mu^B,$$

where the scaling paremeters are defined by the norm of the natural vectors (all space-like): $\mu_A \equiv \sqrt{-g_{AA}}$ and $\mu^A \equiv 1/\mu_A$. The "normalized" metrics have the symmetric form

$$(\hat{g}_{AB}) = -\begin{pmatrix} 1 & Z & Y & X \\ Z & 1 & X & Y \\ Y & X & 1 & Z \\ X & Y & Z & 1 \end{pmatrix} \quad \text{and} \quad (\hat{g}^{AB}) = \frac{1}{2abc}\begin{pmatrix} 0 & c & b & a \\ c & 0 & a & b \\ b & a & 0 & c \\ a & b & c & 0 \end{pmatrix}, \quad \text{with}$$

$X = \cos\theta_{23}$, $Y = \cos\theta_{13}$, $Z = \cos\theta_{12}$ and $a = \sin\theta_{23}$, $b = \sin\theta_{13}$, $c = \sin\theta_{12}$.

The angles are the ones formed by each pair of natural vectors, which satisfy

$$\theta_{AB} = \theta_{CD} \quad \forall A,B,C,D \neq \quad \text{and} \quad \theta_{12} + \theta_{13} + \theta_{23} = 2\pi \quad \text{with} \quad 0 < \theta_{AB} < \pi.$$

This splitting separates the 6 degrees of freedom of the metric into 2 types of parameters (scalar functions): 4 *scale* parameters, μ^A, which are proportional to the rescaling of the signal emitted by the corresponding satellite and independent of the others, and 2 *shape* parameters, 2 independent angles, θ_{12}, θ_{13} for instance, which are completely independent of the signal emitted, they only depend on the world-line of the satellites and on the conformal structure of the space-time.

THE CENTRAL OBSERVER AND THE CENTRAL TETRAD

Let us look for an observer (time-like congruence) who see the emitters in a symmetric configuration. We can map the spacial apparent directions of the signals arriving to each point, into its unit celestial sphere. This defines a tetrahedron (4 non-coplanar points). Asking for this configuration to be a regular tetrahedron is too restrictive, but we can ask for an *equifacial* tetrahedron (all the faces of equal area (and shape, indeed)). An observer seeing the signals with this configuration will be called *central*.

Theorem 1 *The central observer exists and is unique:* $u^A = \mu^A \sqrt{\frac{a+b+c}{8abc}}$.

Besides, an equifacial tetrahedron defines 3 orthogonal axes. Hence, the 4 world-lines of the emitters determine also a unique (up to signs) field of (*central*) orthonormal tetrads.

ACKNOWLEDGMENTS

J.M.Pozo acknowledge the support of the postdoc fellowship EX-2004-0090, from the spanish *Ministerio de Educación y Ciencia*, and the project BFM2003-07076.

REFERENCES

1. B. Coll, *Reference Frames and Gravitomagnetism*, Proc. of the Spanish Relativistic Meeting ERE2000, eds. J. F. Pascual-Sanchez et al., World Scientific, Singapore, 2001, pp. 53–65.
2. B. Coll, *JSR 2002*, eds. N. Capitaine and M. Stavinschi, Pub. Observatoire de Paris, 2002, pp. 34–38.
3. J. M. Pozo, and B. Coll, *Intern. school: Relativistic Coordinates, Reference and Positioning Systems*, held in in Salamanca, February 2005. The results will be collected in a series of papers in preparation.
4. B. Coll, and J. A. Morales, *J. Math. Phys.*, **32**, 2450–2455 (1991).

Perturbations of slowly rotating relativistic stars

Isabel Rica-Méndez* and Adamantios Stavridis[†]

*Institut für Astronomie und Astrophysik, Theoretische Astrophysik, Universität Tübingen,
Auf der Morgenstelle 10, D-72076 Tübingen, Germany
[†]Department of Physics, Aristotle University of Thessaloniki, 54124 Greece

Abstract. In our presentation we focus on oscillations of slowly rotating neutron stars. For that case we show in particular results of the numerically calculated time evolution of the fluid- and the metric-perturbations of the linearized Einstein field equations, thus without the frequently used Cowling approximation. By including also the metric-perturbations we are able to determine the error of the Cowling approximation and increase the accuracy of the resulting frequencies.

Keywords: stellar perturbations; oscillation frequencies
PACS: 04.40.Dg; 97.10.Sj; 97.60.Jd

INTRODUCTION

Compact stellar objects like neutron stars are promising candidates for gravitational wave sources. For measuring purpose it is important to determine first theoretically the frequencies of the axial and polar oscillation modes. Several models of oscillating neutron stars have been considered intensively before [1, 2]. Most of the full relativistic models were based on the Cowling approximation. We present numerical results of the Einstein field equations for an oscillating, slowly rotating neutron star without the Cowling approximation. One of our main aims was to examine the reliability and the shortcomings of that often used approximation for oscillation frequencies in the relativistic regime.

For that purpose we considered the Einstein field equations in the ADM-form in deriving the perturbation equations for the oscillations. The linearized perturbation equations consist of the dynamical equations that govern the metric and the extrinsic curvature perturbations and the constraint equations, which allow us to construct physically valid initial data. Due to spherical symmetry we expand the complete set of perturbation variables into spherical harmonics in order to eliminate the angular dependence. Thus, we obtain a set of partial differential equations for the perturbation variables, which only depend on time and radial coordinate and the order of the spherical harmonics. In contrast to the non-rotating case, rotation induces a coupling between polar and axial equations. After introducing the BCL gauge one obtains in total fourteen evolution equations plus four constraints, which constitute a first-order hyperbolic system of coupled wave equations. Moreover there is a coupling between the fluid and the spacetime perturbation functions.

Introducing the BCL gauge instead of the Regge-Wheeler gauge is necessary because in the Regge-Wheeler gauge the perturbation equations resulting from the linearized Einstein field equations would be in a form that cannot be used to perform numerical time evolutions.

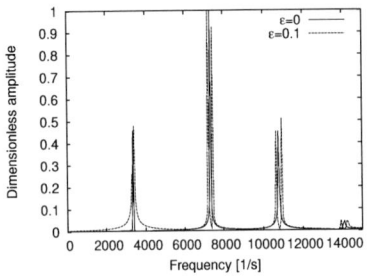 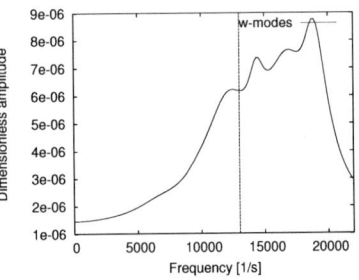

FIGURE 1. Frequency spectra of a perturbed fluid component for a rotating star and the spacetime w-modes for a non-rotating star in the case of non-Cowling approximation.

NUMERICAL RESULTS

The numerical results for polytropic stellar models for the fluid and the spacetime modes are shown in Fig.1. The left graph shows the frequency spectrum of a perturbed fluid component for a rotating star with $\varepsilon = 0.1$ compared to the non-rotating case. The splitting of the frequencies of the non-rotating modes is clearly visible. The f-mode and the first three p-modes are plotted in the graph.

The right graph of Fig.1. shows the frequency spectrum of the spacetime w-modes for a non-rotating star. The broadening of the frequency peaks is due to the high damping time of the oscillations.

Our results for the oscillation frequencies from the full relativistic equations for the fluid modes as well as for the spacetime modes are in consistency with previous results [3, 4].

The accuracy of the Cowling approximation, which we have obtained by comparing the results of the full system of equations with our further numerical calculations including the Cowling approximation, is in accordance with previous results of [5], where the error with respect to the full system of equations was calculated in the frequency domain. From our calculations we obtain a deviation of 18% for the lower frequency domain whereas for the higher frequency domain the resulting deviation is about 1.5%. So, the Cowling approximation is more accurate for the higher frequencies.

ACKNOWLEDGMENTS

This work was supported by Deutsche Forschungsgemeinschaft (SFB/TR 7).

REFERENCES

1. J. Ruoff, A. Stravidis and K.D. Kokkotas, *Mon. Not. R. Astr. Soc.*, **332**, 676 (2002).
2. G. Allen, N. Andersson, K.D. Kokkotas and B.F. Schutz, *Phys. Rev. D*, **58**, 124012 (1997).
3. J. Ruoff, *Phys. Rev. D*, **63**, 064018 (2001).
4. N. Andersson, K.D. Kokkotas and B.F. Schutz, *Mon. Not. R. Astr. Soc.*, **280**, 1230 (1996).
5. S. Yoshida and Y. Kojima, *Mon. Not. R. Astr. Soc.*, **35**, 315–341 (1997).

Tomographic Approach to Quantum Cosmology

Cosimo Stornaiolo

Istituto Nazionale di Fisica Nucleare, Sezione di Napoli–Complesso Universitario di Monte S. Angelo Edificio N' via Cinthia, 45 – 80126 Napoli

Abstract. We present a version of Quantum Cosmology in symplectic tomographic representation.

Keywords: Mathematical and relativistic aspects of cosmology; quantum cosmology. State reconstruction, quantum tomography
PACS: 98.80.Hw, 03.65.Wj.

Recently [1] a novel representation for quantum mechanics has been proposed. In this formulation the properties of a quantum system are described by a fair probability distribution function which is related to the wave function $\psi(x)$ by the following relation

$$\mathscr{W}(X,\mu,\nu) = \frac{1}{2\pi|\nu|}\left|\int \psi(y)e^{\frac{i\mu}{2\nu}y^2-\frac{iX}{\nu}y}dy\right|^2. \qquad (1)$$

The advantage of using the tomographic approach to quantum mechanics is that that tomograms, 1) *given the set of tomograms $\mathscr{W}(X,\mu,\nu)$ for all the values of μ and ν, it describes completely the quantum state of a system i.e. it gives the same information of the corresponding wave function $\psi(x)$;* 2) *tomograms are probability distribution functions;* 3) *they are observables and follow a classical-like evolution equation;* 4) *tomograms are functions defined on a phase space, so they can be compared with classical tomograms.*

This approach to quantization suggested to apply it to Quantum Cosmology [2] in order to have a better understanding of the evolution from a quantum universe to the classical one we observe.

Quantum cosmology was introduced to study some properties of quantum gravity. To this aim it is restricted to a minisuperspace, i.e. the set of metrics of homogeneous spaces, where the gravitational field is described by few degrees of freedom. Therefore quantum gravity is reduced to quantum mechanical problem. In particular, in the 3+1 approach, Schrö dinger equation is substituted by the Wheeler-de Witt equation.

$$\hat{\mathscr{H}}\psi = 0 \qquad (2)$$

Where $\hat{\mathscr{H}}$ is the operator corresponding to the gravitational Hamiltonian constraint. The solution of this equation is the wave function of the universe.

In the tomographic representation one can either substitute equation (2 by an equation for the tomogram [2] or quantize directly the equations of motion, after imposing the constraint(s)[3]. We discuss here this second approach.

Let us consider a homogeneous and isotropic cosmological model

$$ds^2 = a^2(\eta)\{-d\eta^2 + (1-kr^2)^{-1}(dx^2+dy^2+dz^2)\} \qquad (3)$$

with a perfect fluid, whose the equation of state is $P = (\gamma - 1)\rho$ and the cosmological constant Λ is fixed to zero. Given the cosmological equations

$$\frac{a''}{a^3} - \frac{a'^2}{a^4} = -\frac{4\pi G}{3}(\rho + 3P), \qquad \frac{a'^2}{a^4} + \frac{k}{a^2} = \frac{8\pi G}{3}\rho, \qquad (4)$$

where $a(\eta)$ is the expansion factor of the universe, $'$ is derivation with respect to the conformal time η and $k = -1, 0, 1$. It is easy to show [3] [4] that using the equation of state one can reduce equations (4) to

$$w'' + k\chi^2 w = 0, \qquad (5)$$

where $w = a^\chi$ and $(\chi = \frac{3}{2}\gamma - 1)$. According to the value of k this equation describes a harmonic oscillator, a free particle or a "repulsive" oscillator. The corresponding quantum systems can be easily described in terms of tomograms.

Let us consider for simplicity the case k=0. In this case the equation for the tomogram (see [1]) is

$$\frac{\partial \mathscr{W}(X, \mu, \nu, t)}{\partial t} - \mu \frac{\partial \mathscr{W}(X, \mu, \nu, t)}{\partial \nu} = 0 \qquad (6)$$

with initial conditions satisfying the uncertainty principle. The solution of equation (6) has the form

$$\mathscr{W}(X, \mu, \nu, t) = \int \Pi^{free}(X, \mu, \nu, t, X', \mu', \nu', t_0) \mathscr{W}(X', \mu', \nu', t_0) dX' d\mu' d\nu'. \qquad (7)$$

where

$$\Pi^{free}(X, \mu, \nu, t, X', \mu', \nu') = \delta(X - X')\delta(\mu' - \mu)\delta(\nu' - \nu - \mu t). \qquad (8)$$

is the transition probability function which describes the evolution of tomogram of the universe $\mathscr{W}(X, \mu, \nu, t)$ from the initial conditions $\mathscr{W}(X, \mu, \nu, t_0)$. Relation (7) is invertible. So if we "measure", from the present observations, the tomograms of the universe, we can also obtain information of the primordial quantum states of the universe. Work is in progress.

ACKNOWLEDGMENTS

I am very grateful to Prof. V. I. Man'ko and Prof. G. Marmo for helpful discussions.

REFERENCES

1. S. Mancini, V. I. Manko, P. Tombesi, Quant. Semiclass. Opt. **7**, 615 (1995); S. Mancini, V. I. Manko, P. Tombesi, Phys. Lett. **A213**, 1, (1996); for a review see V. I. Manko "Conventional Quantum Mechanics Without Wave Function and Density Matrix" : quant-ph/9902079 (1999)
2. V. I. Manko, G. Marmo and C. Stornaiolo, Gen. Rel. Grav. **37**, 99 (2005) [arXiv:gr-qc/0307084];
3. V. I. Manko, G. Marmo and C. Stornaiolo, "Cosmological dynamics in tomographic probability representation" to appear in Gen. Rel. Grav.
4. N. A. Lemos, J. Math. Phys. **37** (1996) 1449 [arXiv:gr-qc/9511082]; V. Faraoni, Am. J. Phys. **67** (1999). 732 [arXiv:physics/9901006].

Anti-Newtonian universes do not exist

Lode Wylleman[1]

Faculty of Applied Sciences TW16, Ghent University, Galglaan 2, 9000 Gent, Belgium

Abstract. It is proved that irrotational dust solutions with vanishing electric part of the Weyl tensor are necessarily FLRW (conjectured by Maartens et al., 1998). A link with further work is made.

Keywords: Irrotational dust, purely magnetic Weyl curvature, non-existence
PACS: 04.20.Ex

Non-vacuum irrotational dust solutions ($\rho \neq 0, \omega^a = 0, p = 0$) of the Einstein perfect fluid field equations

$$R_{ab} - \frac{1}{2} R g_{ab} + \Lambda g_{ab} = \rho u_a u_b. \tag{1}$$

with vanishing electric part $E_{ab} = C_{acbd} u^c u^d$ of the Weyl tensor C_{abcd} were called *anti-Newtonian* universes in [1] (wherein Λ was set to zero). For non-vacuum dust, the fluid flow lines are geodesics ($\dot{u}^a = 0$). The remaining covariant dynamical quantities are then the matter density ρ, volume-expansion scalar θ, shear tensor σ_{ab} and magnetic part $H_{ab} = \eta_{cdaf} C^{cd}{}_{be} u^e u^f$ of the Weyl tensor. In an initial value formulation of GR, the Bianchi identities and the Ricci identity for u^a (where the field equations (1) are taken as an algebraic definition of the Ricci tensor R_{ab}) are covariantly split into propagation equations and constraint equations [2]. For anti-Newtonian universes and in a streamlined, index-free notation the two respective sets read:

$$\dot{\rho} = -\rho\theta, \quad \dot{\theta} = -\frac{1}{3}\theta^2 - \mathrm{tr}(\sigma^2) - \frac{1}{2}\rho + \Lambda, \quad \dot{\sigma} = -\frac{2}{3}\theta\sigma - \hat{\sigma^2}, \quad \dot{H} = -\theta H + 3\widehat{\sigma H}$$

$$\mathrm{div}\,\sigma = \frac{2}{3} D\theta, \quad \mathrm{curl}\,\sigma = H, \quad \mathrm{div}\,H = 0, \quad [\sigma, H] = -\frac{1}{3} D\rho, \quad \mathrm{curl}\, H = \frac{1}{2}\rho\sigma.$$

Herein tr(S) denotes the trace of a 2-tensor S and $\hat{S}$ its spatial, symmetric and trace-free part, $(ST)_{ab} = S_a{}^c T_{cb}$, $[\sigma, H]$ is the vector dual to the (antisymmetric) covariant commutator of the 2-tensors σ and H, $\dot{A} = \nabla_u A$ the covariant time derivative of an arbitrary tensor A, D the covariant spatial derivative, and curl and div the related 'curl' and divergence operators (see also e.g. [3] for definitions and properties). The first three constraints propagate consistently [4]. Propagation of the last one (coming from div$E = 0$) gives rise to a chain of new equations; from the first time derivative of this constraint, it was concluded in [1] that there are no non-trivial anti-Newtonian

---

[1] research assistant supported by the Fund for Scientific Research Flanders(F.W.O.), e-mail: lwyllema@cage.ugent.be.

linearisations around an FLRW spacetime (for $\Lambda = 0$). The time derivative of the fourth constraint (coming from $\dot{E} = 0$) or, equivalently, the divergence of the last one, gives rise to a new integrability condition

$$\rho D\theta + \left(\frac{1}{2}\sigma - \frac{1}{3}\theta\right) D\rho = 0, \tag{2}$$

with $(\sigma D\rho)_a = \sigma_a{}^b D_b \rho$. Again, propagation of (2) yields an in principle infinite chain of compatibility equations. In [1], the first and second time derivative were calculated and the general structure of the nth derivative was given, but apparently the dramatic consequence of this chain was overlooked.

With (2) as starting point, we were able to prove that there exist no non-FLRW anti-Newtonian universes at all, irrespective as to whether Λ is zero or not. On considering the first 6 time derivatives of (2) (computed in Maple) and going over to a shear eigentetrad at the end, it follows that $D\rho = D\theta = 0$. But then the fourth constraint reads $[\sigma, H] = 0$ (i.e., σ and H commute), and taking the divergence of this equation yields

$$\mathrm{div}\,[\sigma, H] = \mathrm{tr}(H\,\mathrm{curl}\,\sigma) - \mathrm{tr}(\sigma\,\mathrm{curl}\,H) = \mathrm{tr}(H^2) - \frac{1}{2}\rho\,\mathrm{tr}(\sigma^2) = 0, \tag{3}$$

where the first equality is a general identity (see [3]) and the second one follows from the second and fifth constraint. In a second step, on calculating the first five covariant time derivatives of (3) in Maple and turning to shear and Weyl eigenvalues, it follows that $\sigma = H = 0$, and this finishes the proof.

The interested reader can find the full paper (bearing the same title) via http://www.arxiv.org/find/gr-qc. By a combination of the above technique and the explicit use of the spatial Einstein equations in a shear tetrad approach, irrotational geodesic (IG) perfect fluids with magnetic Weyl tensor and obeying a non-constant equation of state were meanwhile classified as well. The result is that non-degenerate shear is inconsistent, whereas for degenerate shear and Petrov type I one explicit new metric is found. The corresponding paper 'Irrotational and geodesic perfect fluids with an equation of state' is accessible via the same link. In it, Petrov type I IG's with purely *electric* Weyl tensor and with non-constant equation of state are also classified, thereby finishing a subcase of the analysis in [5].

REFERENCES

1. R. Maartens, W.M. Lesame, and G.F.R. Ellis, *Class. Quantum Grav.*, **15**, 1005–1017(1998).
2. G.F.R. Ellis, "Relativistic Cosmology", in *General Relativity and Cosmology*, edited by R.K. Sachs, Academic, New York, 1971, 104-149.
3. R. Maartens, and B.A. Bassett *Class. Quantum Grav.*, **15**, 705–717 (1998).
4. R. Maartens *Phys. Rev. D*, **55**, 463–467 (1997).
5. A. Barnes, and R.R. Rowlingson *Class. Quantum Grav.*, **6**, 949–960 (1989).

CONFERENCE PROGRAM

TUESDAY 6th: PLENARY SESSION
Chairman: Brandon CARTER

09:00-09:30   Registration

09:30-10:00   Welcome address

10:00-11:30   **Jürgen EHLERS**
100 Years of Relativity - Foundations, Results, Problems

11:30-12:00   Welcome cocktail

12:00-13:00   **Thibault DAMOUR**
100 Years of Relativity: Was Einstein 100% Right?

13:00-15:00   Lunch

15:00-16:00   **Jean Pierre LUMINET**
The Shape of Space from Einstein to WMAP Data

TUESDAY 6th: PARALLEL SESSION #1
Chairman: Rémi HAKIM

16:00-16:40   **Jean EISENSTAEDT**
Relativity: The Roots

16:40-17:00   Coffee Break

17:00-17:40   **Francois DE GANDT**
Geometry and Experience : Hilbert, Husserl, Einstein

17:40-18:20   **Robert DISALLE**
The Philosophical Origins of Relativity Theory, Reconsidered

TUESDAY 6th: PARALLEL SESSION #2
Chairman: Thibault DAMOUR

16:00-16:40   **Marta BURGAY**
The Double Pulsar J0737-3039A/B

16:40-17:00   Coffee Break

17:00-17:40　**Diego PAVÓN**
Holographic Dark Energy and Present Cosmic Acceleration

17:40-18:00　**Antonio CAPOLUPO**
Vacuum Energy and Cosmological Constant

18:00-18:20　**F. Siddhartha GUZMAN**
Virialization and Interaction of Scalar Field Structures

TUESDAY 6th: PARALLEL SESSION #3
Chairman: Miguel Ángel Ramos OSORIO

17:00-17:40　**Bayram TEKIN**
Spin and Energy of Higher Dimensional Kerr-AdS Black Holes

17:40-18:00　**Filipe MOURA**
Stability and Spectrum of String-Corrected Black Holes

18:00-18:20　**José Robel ARENAS SALAZAR**
Black Hole Entanglement Entropy

TUESDAY 6th: PARALLEL SESSION #4
Chairman: Diego SAÉZ

17:00-17:20　**Miguel LORENTE**
A Combinatorial Approach to Discrete Geometry

17:20-17:40　**Aleksandar MIKOVIĆ**
Tetrade Spin Foam Model

17:40-18:00　**Federico PIAZZA**
Towards a Pre-Geometric Perspective?

TUESDAY 6th: POSTER SESSION (18:20 - 19:30)

Vanessa C. de ANDRADE
Gravitation and Duality Symmetry

Vicente J. BOLÓS
On the Concept of Relative Velocity

Irene BRITO
Mathematical Properties of the Elasticity Difference Tensor

Durmus DAGHAN
Scalar Field Coupled to Gravity in Spherically Symmetric Background in (2+1) Dimensions

Allan E.K. LIM
A Graph-Theoretic Algorithm for the Determination of Polynomial Syzygies Relating Tensor Invariants

Filipe MENA
Newtonian versus Relativistic Second Order Perturbations

Luis NUÑEZ
Variable Eddington Factor and Radiating Slowly Rotating Bodies in General Relativity

Francisco J. PEÑA
Teorías No-Conmutativas Formuladas en el Espacio de Caminos

Viktor PERVUSHIN
Quantum Gravity as Theory of Superfluidity

José María POZO
Emission Coordinates and the Central Observer

Isabel RICA MÉNDEZ
Slowly Rotating Relativistic Star Perturbations

Özgür SARIOGLU
Gödel-Type Metrics in Various Dimensions

Cosimo STORNAIOLO
Tomographic Approach to Quantum Cosmology

Huba Lazlo SZÖCS
The Generalized Linear Transformations of Special Relativity Theory for Inertial Systems and some Problems of Measure Transferring and their Consequencies

Lode WYLLEMAN
Anti-Newtonian Universes Do Not Exist

WEDNESDAY 7th: PLENARY SESSION
Chairman: Thibault DAMOUR

09:30-10:30 **Tomás ORTÍN**
The Supersymmetric Vistas of the Supergravity Landscape

10:30-11:00 Coffee Break

11:00-12:00 **Rémi HAKIM**
Relativistic Statistical Mechanics: Past and Present Status

12:00-13:00 **David WANDS**
Brane-World Cosmology

13:00-15:00 Lunch

WEDNESDAY 7th: PARALLEL SESSION #1
Chairman: Yolanda LOZANO

15:00-15:40 **Eric BERGSHOEFF**
Ten-Dimensional Supergravity Revisited

15:40-16:20 **Laurent HOUART**
Kac-Moody Algebras in Gravity and M-theories

16:20-16:40 **José EDELSTEIN**
Chern-Simons gravity and M theory

16:40-17:00 Coffee Break

17:00-17:20 **Jen-Chi LEE**
High Energy Zero-Norm States and Symmetries of String Theory

17:20-17:40 **Sergio MONTAÑEZ**
Quantization of Minimal Strings: a Mechanical Analog

WEDNESDAY 7th: PARALLEL SESSION #2
Chairman: Kerstin KUNZE

15:00-15:40 **José M.M. SENOVILLA**
Second-Order Symmetric Lorentzian Manifolds

15:40-16:00 **José GAITE**
Stability of Self-Similar Spherical Accretion Flows

16:00-16:20	**Nese OZDEMIR**	

16:00-16:20 **Nese OZDEMIR**
Non-Adiabatic Radiating Collapse in de Sitter Spacetime

16:20-16:40 **Leonardo FERNÁNDEZ-JAMBRINA**
Geodesic Completeness around Sudden Singularities

16:40-17:00 Coffee Break

17:00-17:20 **Raül VERA**
Axially Symmetric Equilibrium Regions in FLRW Universes

17:20-17:40 **Filipe MENA**
First Order Perturbations of the Einstein-Straus Model

17:40-18:00 **Magnus HERBERTHSON**
Explicit Multipole Moments of Axisymmetric Space-Times

18:00-18:20 **Thomas BÄCKDAHL**
Static Axisymmetric Spacetimes with Prescribed Multipole Moments

18:20-18:40 **Jacek TAFEL**
Asymptotic Flatness and Algebraically Special Metrics

WEDNESDAY 7th: PARALLEL SESSION #3
Chairman: Miguel Ángel VÁZQUEZ-MOZO

15:00-15:40 **Bartolomé COLL**
Relativistic Positioning Systems

15:40-16:00 **Joan FERRANDO**
Coll Positioning Systems: a Two-Dimensional Approach

16:00-16:20 **Juan Antonio MORALES**
Coordinates and Frames from the Causal Point of View

16:20-16:40 **Daniel SOLER**
Reference Frames and Rigid Motions in Relativity: Applications

16:40-17:00 Coffee Break

17:00-17:40 **Luca LUSANNA**
The Objectivity of Spacetime: Dirac Observables and Gauge Variables for the Gravitational Field

17:40-18:20 **Alberto CHAMORRO**
On the Physical Meaning of the Principle of General Covariance

18:20-18:40 **Nosratollah JAFARI**
Doubly Special Relativity: a New Relativity or Not?

WEDNESDAY 7th: PARALLEL SESSION #4
Chairman: Diego PAVÓN

15:00-15:40 **Günter NIMTZ**
Do Evanescent Modes (Tunneling) Violate Einstein Causality?

15:40-16:00 **Nicholas RATTENBURY**
Determination of Stellar Shape in Gravitational Microlensing Events

16:00-16:20 **Luigi MANCINI**
Gravitational Lensing of Stars in the Central Arcsecond of our Galaxy

16:20-16:40 **Mimoza HAFIZI**
May we use the LLR as a Redshift Indicator for the Gamma-Ray Bursts?

16:40-17:00 Coffee Break

17:00-17:20 **Alberto LOBO**
Data and Diagnostics Systems for LISA Pathfinder

17:20-17:40 **Enrico LORENZINI**
Detector Configurations for Equivalence Principle Tests with Strong Separation of Signal from Noise

17:40-18:00 **Valerio IAFOLLA**
ISA - An Accelerometer to Detect the Disturbing Accelerations Acting on the Mercury Planetary Orbiter

THURSDAY 8th: PLENARY SESSION
Chairman: Jürgen EHLERS

09:30-10:30 **Malcolm MACCALLUM**
Finding and Using Exact Solutions of the Einstein Equations

10:30-11:00 Coffee Break

11:00-12:00 **Brandon CARTER**
Einstein's "Ugly Duckling": the Theory of Black Holes

12:00-13:00 **Luc BLANCHET**
General Relativity and the Two-Body Problem

13:00-15:00 Lunch

THURSDAY 8th: PARALLEL SESSION #1
Chairman: Alberto CHAMORRO

15:20-16:00   **Frank MEYER**
A Gravity Theory on Noncommutative Spaces

16:00-16:20   **Alexander NESTEROV**
Black Hole Entropy in Nonassociative Geometry

16:20-16:40   **Miguel Ángel VÁZQUEZ-MOZO**
Nonanticommutative Kahler sigma models

16:40-17:00   Coffee Break

17:00-17:20   **Hristu CULETU**
On the Rindler Horizon Energy

17:20-17:40   **Antonio BROTAS**
Fishing in Black Holes

17:40-18:00   **Kurt JUST**
From General Gravity to Einstein's

THURSDAY 8th: PARALLEL SESSION #2
Chairman: José María IBÁÑEZ

15:00-15:40   **Sascha HUSA**
Hyperboloidal Initial Data and Evolution

15:40-16:00   **Carles BONA**
Approximate Killing vectors

16:00-16:20   **Carlos PALENZUELA**
Geometrical Gauge Conditions for Numerical Relativity

16:20-16:40   **Luisa BUCHMAN**
Numerical Relativity using a Frame Based Approach

16:40-17:00   Coffee Break

17:00-17:20   **Diego SÁEZ**
Comparing the Rees-Sciama and Sachs-Wolfe Integrated Effects

17:20-17:40   **José Luis JARAMILLO**
3+1 Decomposition of Quasi-Equilibrium Black Hole Boundary Conditions

17:40-18:20   **Simonetta FRITTELLI**
Boundary Conditions for the Einstein Equations in the BSSN Formulation

THURSDAY 8th: PARALLEL SESSION #3
Chairman: Fernando ATRIO BARANDELA

15:20-16:00 **Alicia SINTES**
All Sky Search for Gravitational Waves Emitted by Unknown Isolated Neutron Stars

16:00-16:20 **Neus PUCHADES**
Numerical Simulations of the Gravitational Wave Background Produced by Binaries

16:20-16:40 **Paulo SÁ**
Differential Rotation of R-Modes

16:40-17:00 Coffee Break

17:00-17:20 **Roberto DE PIETRI**
Simulation of Dynamical Bar-Mode Instability and its Threshold in General Relativity

17:20-17:40 **Pablo LORÉN-AGUILAR**
Gravitational wave radiation of pulsating white dwarfs revisited: the case of BPM 37093 and PG 1159

THURSDAY 8th: PARALLEL SESSION #4
Chairman: Miguel LORENTE

16:00-16:20 **Robertus POTTING**
Gravity from Lorentz Violation

16:20-16:40 **Victor VILLALBA**
Separation of Variables and Exact Solution of the Dirac Equation in Some Cosmological Space-Times

16:40-17:00 Coffee Break

17:00-17:40 **Brian EDGAR**
A Weighted De Rham Operator Acting on Arbitrary Tensor Fields and their Local Potentials

17:40-18:20 **Felix FINSTER**
Decay of Solutions of the Wave Equation in the Kerr Geometry

THURSDAY 8th – 18:30-20:00: Meeting of the SEGRE

FRIDAY 9th: PLENARY SESSION
Chairman: Luc BLANCHET

09:00-10:00 **Luis ÁLVAREZ GAUMÉ**
Black Holes in the AdS/CFT Correspondence. Gravity and Matrix Models

10:00-10:30 Coffee Break

10:30-11:30 **Rashid SUNYAEV**
Accreting neutron stars and black holes: their place and role in the Galaxy and in the Universe

11:30-12:30 **Joseph SILK**
Observing Dark Matter

12:30-13:30 **Enrique MARTÍNEZ- GONZÁLEZ**
Recent Developments in the Study of the Cosmic Microwave Background Anisotropies

13:30-15:00 Lunch

FRIDAY 9th: PARALLEL SESSION #1
Chairman: Enrique ÁLVAREZ

15:00-15:40 **Jorge ALFARO**
Quantum Gravity and Lorentz Invariance Violation in the Standard Model

15:40-16:20 **Daniel LITIM**
Fixed Points of Quantum Gravity

16:20-16:40 **Iñaki GARAY**
Exact Quantization of Einstein-Rosen Waves Coupled to Massless Scalar Matter

16:40-17:00 Coffee Break

17:00-17:20 **Diego BLAS**
Bigravity and Massive Gravity

17:20-17:40 **Alberto SÁNCHEZ**
Gowdy T^3 Cosmological Models in $N=1$ Supergravity

17:40-18:00 **Angel Ricardo PLASTINO**
Closed Timelike Curves and Quantum Cloning

FRIDAY 9th: PARALLEL SESSION #2
Chairman: Vladimir MANKO

15:20-16:00 **Kerstin KUNZE**
Stochastic Inflation with a Coloured Noise Term

16:00-16:20 **Eugeny BABICHEV**
B-Inflation

16:20-16:40 **Antonio SEGUÍ**
The Initial Conditions of the Universe and Holography

16:40-17:00 Coffee Break

17:00-17:20 **Andrea FUSTER**
Type III Einstein-Yang-Mills Solutions

17:20-17:40 **M. Piedade MACHADO RAMOS**
Obtaining Exact Solutions with Cosmological Constant Using Invariant Operators

17:40-18:00 **Masato NOZAWA**
Gravitational Collapse in Higher Curvature Theory

FRIDAY 9th: PARALLEL SESSION #3
Chairman: Joaquín DIAZ ALONSO

15:20-15:20 **Elizabeth RODRIGUEZ QUERTS**
Vacuum Self Magnetization?

15:20-15:40 **Aurora PÉREZ MARTÍNEZ**
Quark Stars and Quantum-Magnetically Induced Collapse

15:40-16:00 **Ángeles PÉREZ GARCÍA**
Dynamical Reponse in Frustrated Nuclear Systems

16:00-16:20 **Jean-Christophe CAILLON**
Nuclear Matter Equation of State in Relativistic Nonlinear Models: Results and Applications

16:20-16:40 **Saturnino MARCOS**
The Pseudospin Symmetry in Atomic Nuclei. Analysis of Some of the Explanations Proposed

16:40-17:00 Coffee Break

17:00-17:20 **Marian LAZAR**
Relativistic (Covariant) Kinetic Theory of Linear Plasma Waves and Instabilities

17:20-17:40 **João Carlos FERNANDES**
Heat Transfer in Theory of Relativity

17:40-18:00 **Alfredo SANDOVAL-VILLALBAZO**
Light Scattering Test Regarding the Relativistic Nature of Heat

18:00-18:20 **Ettore MINGUZZI**
Brownian Motion on the Relativistic Velocity Space

FRIDAY 9th: PARALLEL SESSION #4
Chairman: Luigi TOFFOLATTI

15:00-15:20 **Fernando ATRIO-BARANDELA**
Nature of the Integrated Sachs-Wolf effect: Has it Been Measured in WMAP First Year Data?

15:20-16:00 **Germán OLIVARES**
Constraining Dark Energy Interacting Models with WMAP

15:40-16:00 **Germán IZQUIERDO**
Relic Gravitational Waves and Cosmic Accelerated Expansion

16:00-16:20 **Julien LARENA**
Primordial Scalar Field : a Way Out of the Lithium Over-Production?

16:20-16:40 **María RODRÍGUEZ MARTÍNEZ**
Constraining Lorentz Violation with GRB

SATURDAY 10th: PLENARY SESSION
Chairman: Jean Pierre LUMINET

09:00-10:00 **José Maria IBÁÑEZ**
Current Issues in Numerical Relativitic (Magneto-)Hydrodynamics

10:00-10:30 Coffee Break

10:30-11:30 **Enrique ÁLVAREZ**
Gravitons in the Mist

11:30-12:30 **John STACHEL**
Albert Einstein: A Man for the Next Millenium?

LIST OF PARTICIPANTS

Jorge ALFARO
Dept. de Física Teórica C-XI
Facultad de Ciencias
Univ. Autónoma de Madrid
Cantoblanco, 28049, Madrid
Spain
jorge.alfaro@uam.es
and
Facultad de Física,
Pontificia Universidad Católica de Chile
Casilla 306, Santiago 22
Chile
(permanent address)
jalfaro@puc.cl

Enrique ÁLVAREZ
Departamento de Física Teórica
Modulo C-XI
Facultad de Ciencias
Universidad Autónoma de Madrid
Cantoblanco, 28049 Madrid
Spain
enrique.alvarez@uam.es

Enrique ÁLVAREZ-GAUMÉ
Physics Department
Theory Division
CH-1211 Geneva 23
Switzerland
luis.alvarez-gaume@cern.ch

Vanessa Carvalho de ANDRADE
Instituto de Física
Universidade de Brasília
Campus Darcy Ribeiro
Caixa Postal 04455
70919-970 Brasília DF
Brazil
andrade@fis.unb.br

José Robel ARENAS-SALAZAR
Observatorio Astronómico
Universidad Nacional de Colombia
Ciudad Universitaria (Cra. 30, Calle 45)
Bogotá
Colombia
jrarenass@unal.edu.co

Fernando ATRIO-BARANDELA
Física Teórica
Universidad de Salamanca
Plaza de la Merced s/n
37008 Salamanca
Spain
atrio@usal.es

Evgeny BABICHEV
Max-Planck-Institut für Physik
(Werner-Heisenberg-Institut)
Foehringer Ring 6
80805 München
Germany
babichev@mppmu.mpg.de

Thomas BÄCKDAHL
Matematiska Institutionen
Linköpings Universitet
SE-581 83 Linköping
Sweden
thbac@mai.liu.se

Jesús Fernando BARBERO GONZÁLEZ
Instituto de Estructura de la Materia
CSIC
C/Serrano 123
28006 Madrid
Spain
fbarbero@iem.cfmac.csic.es

Alan BARNES
Computer Science Group
Aston University
Birmingham
B4 7ET
UK
barnesa@aston.ac.uk

Göran BERGQVIST
Matematiska Institutionen
Linköpings Universitet
SE-58183 Linköping
Sweden
gober@mai.liu.se

Eric BERGSHOEFF
Centre for Theoretical Physics
University of Groningen
Nijenborgh 4
9747 AG Groningen
The Netherlands
E.A.Bergshoeff@rug.nl

Luc BLANCHET
Gravitation et Cosmologie (GReCO)
Institut d'Astrophysique de Paris
98 bis boulevard Arago
75014 Paris
France
blanchet@iap.fr

Diego BLAS
Universitat de Barcelona
Bóbila N 4 Atic 2
Spain
dblas@ffn.ub.es

Vicente José BOLÓS LACAVE
Dpto. Matemáticas
Facultad de Ciencias
Universidad de Extremadura
Avda. de Elvas s/n
06071 Badajoz
Spain
vjbolos@unex.es

Carles BONA GARCÍA
Departament de Fisica
Ed. Mateu Orfila
Universitat de les Illes Balears
07122 Palma de Mallorca
Spain

Matthieu BRASSART
Observatoire de Paris
Laboratoire Univers et Théories
92195 Meudon cedex
France
brassart@mesiog.obspm.fr

Irene BRITO
Departamento de Matemática
para a Ciência e Tecnologia
Universidade do Minho
Campus de Azurém
4800 058 Guimarães
Portugal
ireneb@mct.uminho.pt

Antonio BROTAS
Departamento de Física
Instituto Superior Técnico
Av. Rovisco Pais
1000 Lisboa
Portugal
brotas@fisica.ist.utl.pt

Luisa BUCHMAN
Astrophysics and Gravitation Group 3266
Jet Propulsion Laboratory
Mail Stop 169-327
4800 Oak Grove Dr.
Pasadena, CA 91109
USA
Luisa.Buchman@jpl.nasa.gov

Marta BURGAY
INAF
Osservatorio Astronomico di Cagliari
Loc. Poggio dei Pini, Strada 54
09012 Capoterra (CA)
Italy
burgay@ca.astro.it

Jean-Christophe CAILLON
Centre d'Etudes Nucléaires
de Bordeaux-Gradignan
Le Haut Vigneau, BP 120
33175 Gradignan Cedex
France
caillon@cenbg.in2p3.fr

Abel CAMACHO QUINTANA
Physics Department
Universidad Autónoma
Metropolitana-Iztapalapa
Av. San Rafael Atlixco No. 186
Col. Vicentina
Delegacion Iztapalapa, Mexico DF
Mexico
acq@xanum.uam.mx

Antonio CAPOLUPO
Dipartimento di Fisica and INFN
Universitá di Salerno
I-84100 Salerno
Italy
capolupo@sa.infn.it

Brandon CARTER
Laboratoire Univers et Théories (LUTH)
Observatoire de Paris-Meudon
F-92195 Meudon
France
Brandon.Carter@obspm.fr

Alberto CHAMORRO
Departamento de Física Teórica
Facultad de Ciencias
Universidad del País Vasco
Apartado 644
48080 Bilbao
wtpchbea@lg.ehu.es

Manuel Ángel COBAS ALONSO
Departamento de Física
Universidad de Oviedo
Avda Calvo Sotelo 18,
E-33007 Oviedo (Asturias)
Spain

Bartolomé COLL
Systèmes de référence relativistes
Observatoire de Paris
61, Avenue de l'Observatoire
75014 Paris
France
bartolome.coll@obspm.fr

Jorge CONDE
Instituto de Física Teórica
Facultad de Ciencias, Módulo C-XVI, 318
Universidad Autónoma de Madrid,
28049 Cantoblanco, Madrid
Spain
jorge.conde@uam.es

Isabel CORDERO
Depto de Astronomía y Astrofísica
Universidad de Valencia
C/ Dr. Moliner, 50
46100-Burjassot (Valencia)
Spain
icorca@alumni.uv.es

Hristu CULETU
Ovidius University
Dept.of Physics
B-dul Mamaia 124
8700 Constanta
Romania
hculetu@yahoo.com

Durmuş DAGHAN
Department of Mathematics
Faculty of Science and Letters
Istanbul Technical University
Maslak, 80626, Istanbul
Turkey
daghand@itu.edu.tr

Thibault DAMOUR
Institut des Hautes Études
Scientifiques (IHES)
35, Route de Chartres,
F-91440 Bures-sur-Yvette
France

François DE GANDT
Université Lille III
UMR 8163 "Savoirs et Textes"
BP 60149
59653 Villeneuve d'Ascq
France
francois.de-gandt@wanadoo.fr

Roberto DE PIETRI
Dipartimento di Fisica
Universita di Parma
via Parco delle Scienze 7/A
43100 Parma (PR)
Italy
depietri@fis.unipr.it

Joaquín DIAZ ALONSO
Departamento de Física
Universidad de Oviedo
Avda Calvo Sotelo 18,
E-33007 Oviedo (Asturias)
Spain
joaquin@string1.ciencias.uniovi.es

Alberto DIEZ TEJEDOR
Departamento de Física Teórica
Universidad del País Vasco
Apartado 644
E-48080 Bilbao
Spain
wtbditea@lg.ehu.es

Robert DISALLE
Department of Philosophy
University of Western Ontario
London, Ontario N6A 3K7
Canada
rdisalle@uwo.ca

José EDELSTEIN
Department of Particle Physics
Universidade de Santiago de Compostela
E-15782, Santiago de Compostela
Spain
jedels@usc.es

Brian EDGAR
Mathematics Department
Linköping University
Linköping
Sweden
bredg@mai.liu.se

Jürgen EHLERS
Max-Planck-Institut fr̈ Gravitationsphysik
Albert Einstein Institut
Am Mühlenberg 1
14476 Golm
Germany
juergen.ehlers@aei.mpg.de

Jean EISENSTAEDT
Observatoire de Paris
SYRTE/UMR8630-CNRS
61, avenue de l'Observatoire
75014 Paris, France
Jean.Eisenstaedt@obspm.fr

João Carlos FERNANDES
Departamento de Física
Instituto Superior Técnico
Av. Rovisco pais
1096. Lisboa Codex
Portugal
joao.carlos@tagus.ist.utl.pt

Leonardo FERNÁNDEZ-JAMBRINA
Universidad Politécnica de Madrid
ETSI Navales
Arco de la Victoria s/n
28040-Madrid, Spain

Joan FERRANDO
Departament d'Astronomia i Astrofisica
Universitat de València
Avda Vicent Andrés Estellés
46100 Burjassot (Valencia)
Spain
joan.ferrando@uv.es

Felix FINSTER
NWF I - Mathematik
Universität Regensburg
93040 Regensburg
Germany
Felix.Finster@mathematik.uni-regensburg.de

Pierluigi FORTINI
Dipartimento di Fisica
Università di Ferrara
Via Saragat 1
44100 Ferrara
Italy
fortini@fe.infn.it

Simonetta FRITTELLI
Department of Physics
Duquesne University
Pittsburgh, PA 15282
USA
simo@mayu.physics.duq.edu

Carlos A. FUERTES
Universidad de Salamanca
Plaza de la Merced s/n
37008 Salamanca
Spain
cafuertes@yahoo.com

Andrea FUSTER
NIKHEF
Kruislaan 409
1098 SJ Amsterdam
The Netherlands
fuster@nikhef.nl

José GAITE
Instituto de Matemáticas
y Física Fundamental del CSIC
C/ Serrano 113 bis
Madrid 28006
Spain
jgaite@imaff.cfmac.csic.es

Iñaki GARAY ELIZONDO
Instituto de Estructura de la Materia
CSIC
C/Serrano 123
28006 Madrid
Spain
igael@iem.cfmac.csic.es

F. Siddharta GUZMAN
Instituto de Fisica y Matematicas
Edificio C-3 Cd Universitaria
Universidad Michoacana
de San Nicolas de Hidalgo
Morelia Michoacan, CP 58060
Mexico
guzman@ifm.umich.mx

Mimoza HAFIZI
University of Tirana
Fakulteti i Shkencave të Natyrës
Departamenti i Fizikës
Tiranë
Albania
mimozah@excite.com

Rémi HAKIM
Laboratoire Univers et Théories (LUTH)
Observatoire de Paris-Meudon
F-92195 Meudon
France
Remi.Hakim@gmail.com

Magnus HERBERTHSON
Matematiska Institutionen
Linköpings Universitet
SE-581 83 Linköping
Sweden
maher@mai.liu.se

Laurent HOUART
Service de Physique Théorique
et Mathématique and
International Solvay Institutes
Université Libre de Bruxelles
Campus Plaine CP 231
Boulevard du Triomphe
B-1050 Bruxelles
Belgium
lhouart@ulb.ac.be

Sasha HUSA
MPI for Gravitational Physics
Am Muehlenberg 1
D-14476 Potsdam
Germany
shusa@aei.mpg.de

Valerio IAFOLLA
Istituto di Fisica dello
Spazio Interplanetario
IFSI / INAF
Via del Fosso del Cavaliere, 100
00133 Rome
Italy
iafolla@ifsi.rm.cnr.it

Jesús IBÁÑEZ
Depto. Física Teórica
Facultad de Ciencias
Universidad del País Vasco
Apdo. 644
48080 Bilbao
Spain
wtpibmej@lg.ehu.es

José María IBÁÑEZ
Departament d'Astronomía
i Astrofísica (DAA)
Universitat de Valencia
Edifici d'Investigació Jeroni Muñoz
C/ Dr. Moliner, 50
E-46100 Burjassot (Valencia)
Spain
Jose.M.Ibanez@uv.es

Germán IZQUIERDO
Departamento de Física
Facultad de Ciencias, Edificio Cc
Universidad Autónoma de Barcelona
08193 Bellaterra (Barcelona)
Spain
german.izquierdo@uab.es

Nosratollah JAFARI
Institute for Advanced Studies
in Basic Sciences (IASBS)
P.O.Box: 45195-1159
Zanjan, Iran
njafary@iasbs.ac.ir

Bert JANSSEN
Departamento de Física
Teórica y del Cosmos
Universidad de Granada
Campus de Fuente Nueva s/n
E-18071 Granada
Spain
bjanssen@ugr.es

José Luis JARAMILLO
LUTH
Observatoire de Paris-Meudon
92195 Meudon
France
jose-luis.jaramillo@obspm.fr

Kurt JUST
Department of Physics
University of Arizona
Tucson AZ 85721
USA
just@physics.arizona.edu

Georgios KOFINAS
Departament de Fisica Fonamental
Universitat de Barcelona
Avenida Diagonal 647
E-08028 Barcelona
Spain
kofinas@ffn.ub.es

Kerstin KUNZE
Física Teórica
Universidad de Salamanca
Plaza de la Merced s/n
37008 Salamanca
Spain
kkunze@usal.es

Julien LARENA
LUTH
Observatoire de Paris-Meudon
5, Place J. Janssen
92190 Meudon
France
julien.larena@obspm.fr

Joey LATTA
Dalhousie University
Halifax, Nova Scotia
B3H 3J5
Canada

Marian LAZAR
Institut für Theoretische Physik
Lehrstuhl IV
Ruhr-Universität Bochum
D-44780 Bochum
Germany
mlazar@tp4.rub.de, mlazar@uaic.ro

Ruth LAZKOZ
Departamento de Física Teórica
Facultad de Ciencia y Tecnología
Universidad del País Vasco
Apartado 644
48080 Bilbao
Spain
ruth.lazkoz@ehu.es

Serge LECLERCQ
Université de Mons-Hainaut
Service de Mécanique et Gravitation
20, place du Parc
B-7000 MONS
Belgium
serge.leclercq@umh.ac.be

Jen-Chi LEE
Department of Electrophysics
National Chiao-Tung University
Hsin-Chu, Taiwan
jcclee@cc.nctu.edu.tw

Allan E.K. LIM
Mathematics and Computational
Theory Group,
School of Information Technology
Deakin University,
Geelong VIC 3217
Australia
aekl@deakin.edu.au

Daniel LITIM
School of Physics
University Southampton
Highfield
SO17 1BJ, UK
daniel.litim@cern.ch

Alberto LOBO
CSIC
Instituto de Ciencias del Espacio
Edificio Nexus
Gran Capitán 2-4
08034 Barcelona
Spain
Alberto.Lobo@ub.edu

Pablo LORÉN AGUILAR
Institut de Ciències de l'Espai (CSIC)
Campus UAB
Facultat de Ciències
Torre C-5
08193 Bellaterra, Spain
loren@fa.upc.edu

Miguel LORENTE
Departamento de Física
Universidad de Oviedo
Avda Calvo Sotelo 18,
E-33007 Oviedo (Asturias)
Spain
lorentemiguel@uniovi.es

Enrico LORENZINI
Harvard-Smithsonian
Center for Astrophysics
60 Garden Street, MS80
Cambridge, MA 02138
USA
elorenzini@cfa.harvard.edu

Yolanda LOZANO
Departamento de Física
Universidad de Oviedo
Avda Calvo Sotelo 18,
E-33007 Oviedo (Asturias)
Spain
yolanda@string1.ciencias.uniovi.es

Jean Pierre LUMINET
Laboratoire Univers et Théories (LUTH)
Observatoire de Paris-Meudon
F-92195 Meudon
France
Jean-Pierre.Luminet@obspm.fr

Luca LUSANNA
Sezione INFN di Firenze
Polo Scientifico
Via Sansone 1
50019 Sesto Fiorentino
Italy
lusanna@fi.infn.it

Malcolm MACCALLUM
School of Mathematical Sciences
Queen Mary, University of London
Mile End Road
London E1 4NS
UK
m.a.h.maccallum@qmul.ac.uk

Maria da Piedade MACHADO RAMOS
Departamento de Matemática
para a Ciência e Tecnologia
Universidade do Minho
Campus de Azurém
4810 Guimarães
Portugal
mpr@mct.uminho.pt

John MADORE
LPT, Univ. de Paris Sud
F-91400 Orsay
France
madore@zephyr.th.u-psud.fr

Luigi MANCINI
Dipartimento di Fisica
"E.R. Caianiello"
Università degli Studi di Salerno
Via S. Allende
I-84081 Baronissi (SA)
Italy
lmancini@sa.infn.it

Vladimir S. MANKO
Departamento de Física
Centro de Investigación y
de Estudios Avanzados del IPN
A.P. 14-740
07000 México D.F.
Mexico
vsmanko@fis.cinvestav.mx

Saturnino MARCOS
Universidad de Cantabria
Dpto. de Física Moderna
Facultad de Ciencias
Avda. de Los Castros s/n
39005 Santander
Spain
marcoss@unican.es

Jesús MARTÍN MARTÍN
Física Teórica
Universidad de Salamanca
Plaza de la Merced s/n
37008 Salamanca
Spain
chmm@usal.es

Enrique MARTÍNEZ GONZÁLEZ
Instituto de Fisica de Cantabria
Universidad de Cantabria
Av. Los Castros s/n
39005 Santander
martinez@ifca.unican.es

Patrick MEESSEN
CERN
Physics Department, Theory
1211 Geneve 23
Switzerland
patrick.meessen@cern.ch

Filipe MENA
Departamento de Matematica
Universidade do Minho
Campus de Gualtar
4710 Braga
Portugal
f.mena@qmul.ac.uk

Ana MENÉNDEZ MANJÓN
Departamento de Física
Universidad de Oviedo
Avda Calvo Sotelo 18,
E-33007 Oviedo (Asturias)
Spain

Frank MEYER
Max-Planck Institute for Physics
(Werner-Heisenberg Institute)
Foehringer Ring 6
80805 Muenchen
Germany
Frank.Meyer@physik.uni-muenchen.de

Aleksandar MIKOVIĆ
Departamento de Matemática
Universidade Lusófona de
Humanidades e Tecnologias
Av. do Campo Grande, 376
1749-024 Lisbon
Portugal
amikovic@ulusofona.pt

Ettore MINGUZZI
Departamento de Matemáticas
Universidad de Salamanca
Plaza de la Merced 1-4
E-37008 Salamanca
Spain
minguzzi@usal.es

Sergio MONTAÑEZ NAZ
Instituto de Física Teórica UAM-CSIC
Módulo C-XVI, 3ª planta
Facultad de Ciencias
Universidad Autónoma de Madrid
Cantoblanco, 28049 Madrid
Spain
sergio.montannez@uam.es

Juan Antonio MORALES LLADOSA
Departament d'Astronomia i Astrofísica
Facultat de Matemátiques
Universitat de València
46100, Burjassot (València)
Spain
antonio.morales@uv.es

Lysiane MORNAS
Departamento de Física
Universidad de Oviedo
Avda Calvo Sotelo 18,
E-33007 Oviedo (Asturias)
Spain
lysiane@fisi24.ciencias.uniovi.es

Filipe MOURA
Departamento de Matemática
Instituto Superior Técnico
Av. Rovisco Pais
P-1049-001 Lisboa
Portugal
fmoura@math.ist.utl.pt

Alexander I. NESTEROV
Departamento de Física, CUCEI
Universidad de Guadalajara
Av. Revolución 1500
Guadalajara, CP 44420
Jalisco, México
nesterov@cencar.udg.mx

Agustín NIETO
Departamento de Física
Universidad de Oviedo
Avda Calvo Sotelo 18,
E-33007 Oviedo (Asturias)
Spain
Agustin.Nieto@cern.ch

Günter NIMTZ
II. Physikalisches Institut
Universität zu Köln
Zülpicher Straße 77
D-50937 Köln
Germany
G.Nimtz@uni-koeln.de

Masato NOZAWA
Department of Physics
Waseda University
3-4-1 Okubo
Shinjuku-ku
Tokyo 169-8555
Japan
nozawa@gravity.phys.waseda.ac.jp

Germán OLIVARES PULIDO
Departamento de Física
Facultad de Ciencias
Universidad Autónoma de Barcelona
08193 Bellaterra (Barcelona)
Spain
german.olivares@uab.es

José Antonio ORTEGA RUIZ
Institut d'Estudis
Espacials de Catalunya
Edificio Nexus
C/ Gran Capità 2-4
08034 Barcelona
Spain
ortega@ieec.fcr.es

Tomás ORTÍN
Instituto de Física Teórica
Facultad de Ciencias, C-XVI
Universidad Autonoma de Madrid
C.U. Cantoblanco
28049-Madrid
Spain
tomas.ortin@cern.ch

Alejandro OSCOZ
Instituto de Astrofísica de Canarias
C/ Vía Láctea s/n
38200 - La Laguna
Tenerife
aoscoz@iac.es

Miguel Angel Ramos OSORIO
Departamento de Física
Universidad de Oviedo
Avda Calvo Sotelo 18,
E-33007 Oviedo (Asturias)
Spain
osorio@string1.ciencias.uniovi.es

Nese OZDEMIR
Istanbul Technical University
Faculty of Science and Letters
34469 Maslak, Istanbul
Turkey
and
Feza Gursey Institute
34684 Cengelkoy, Istanbul
Turkey
nozdemir@itu.edu.tr

Carlos PALENZUELA LUQUE
Louisiana State University
Department of Physics & Astronomy
202 Nicholson Hall
Baton Rouge, LA 70803
USA
carlos@baton.phys.lsu.edu

J.-Fernando PASCUAL-SÁNCHEZ
Dept. Matemática Aplicada
Facultad de Ciencias
Universidad de Valladolid
Valladolid, 47005
Spain
jfpascua@maf.uva.es

Diego PAVÓN
Departamento de Física,
Facultad de Ciencias, Edificio Cc,
Universidad Autónoma de Barcelona,
08193 Bellaterra (Barcelona)
Spain
Diego.Pavon@uab.es

Francisco José PEÑA BENITEZ
Universidad Simón Bolívar
Edificio Física y Electrónica I
Departamento de Física
Valle de Sartenejas
Estado Miranda, municipio Baruta
Caracas
Venezuela
joeben16@gmail.com

Ángeles PÉREZ GARCÍA
Universidad de Salamanca
Plaza la Merced s/n
37008 Salamanca
Spain
mperezga@usal.es

Aurora María PÉREZ MARTÍNEZ
Instituto de Cibernética
Matemática y Física
Calle E esq a 15 No. 309
CP 10400 C. Habana
Cuba
aurora@icmf.inf.cu

Viktor PERVUSHIN
Joint Institute for Nuclear Research
Joliot-Curie 6
141980 Dubna
Moscow region
Russia
pervush@theor.jinr.ru

Federico PIAZZA
Institute of Cosmology and Gravitation
Mercantile House
Hampshire Terrace
University of Portsmouth
Portsmouth, PO1 2EG
United Kingdom
federico.piazza@port.ac.uk

Angel Ricardo PLASTINO
Department of Physics
University of Pretoria
Pretoria 0002
South Africa
vdfsarp9@uib.es

Robertus POTTING
Universidade do Algarve
Departamento de Física
Faculdade de Ciências e Tecnologia
Campus de Gambelas
8005-139 Faro
Portugal
rpotting@ualg.pt

José María POZO
SYRTE - Bâtiment A
Observatoire de Paris
61, Avenue de l'Observatoire
75014 Paris
France
Jose-Maria.Pozo@obspm.fr

Neus PUCHADES
Depto de Astronomía y Astrofísica
Universidad de Valencia
C/ Dr. Moliner, 50
46100-Burjassot (Valencia)
Spain
neus.puchades@uv.es

Nicholas RATTENBURY
Jodrell Bank Observatory
Dept. of Physics & Astronomy
The University of Manchester
Macclesfield, Cheshire SK11 9DL
UK
njr@jb.man.ac.uk

Isabel RICA MÉNDEZ
Institut für Theoretische Astrophysik
Universität Tübingen
Auf der Morgenstelle 10
72076 Tübingen
Germany
isabel@tat.physik.uni-tuebingen.de

Maria RODRIGUEZ MARTINEZ
Hebrew University
Givat Ram, 91904
Jerusalem, Israel
mrm@phys.huji.ac.il

Elizabeth RODRÍGUEZ QUERTS
Instituto de Cibernética
Matemática y Física
Calle E esq a 15 No. 309
CP 10400 C. Habana
Cuba
elizabeth@icmf.inf.cu

Diego RUBIERA GARCÍA
Departamento de Física
Universidad de Oviedo
Avda Calvo Sotelo 18,
E-33007 Oviedo (Asturias)
Spain

Milton RUIZ MENESES
Instituto de Ciencias Nucleares
Universidad Nacional Autónoma de México
Circuito Exterior C.U.
A. postal: 70-543 04510
México, D.F.
ruizm@nucleares.unam.mx

Eduardo RUIZ CARRERO
Física Teórica
Universidad de Salamanca
Plaza de la Merced s/n
37008 Salamanca
Spain
eruiz@usal.es

Paulo SÁ
Universidade do Algarve
Departamento de Física
Faculdade de Ciências e Tecnologia
Campus de Gambelas
8005-139 Faro
Portugal
pmsa@ualg.pt

Diego SÁEZ
Depto de Astronomía y Astrofísica
Universidad de Valencia
C/ Dr. Moliner, 50
46100-Burjassot (Valencia)
Spain
diego.saez@uv.es

Moises SÁEZ BELTRÁN
Instituto de Física Teórica
Facultad de Ciencias, C-XVI
Universidad Autonoma de Madrid
C.U. Cantoblanco
28049-Madrid
Spain
proportional@gmail.com

Alberto SANCHEZ MORENO
Universidad Autónoma Metropolitana
Av. San Rafael Atlixco No. 186,
Col. Vicentina
Delegación Iztapalapa
C. P. 09340 México, D.F.
asan@xanum.uam.mx

Alfredo SANDOVAL-VILLALBAZO
Universidad Iberoamericana
Depto. de Física y Matemáticas
Prolongación Reforma 880
México DF 01210
México
alfredo.sandoval@uia.mx

Josep SANJUAN
IEEC - Institut d'Estudis
Espacials de Catalunya
Edifici Nexus - Despatx 201
C/ Gran Capità 2-4
08034 Barcelona
Spain
sanjuan@ieec.fcr.es

Bahtiyar Özgür SARIOGLU
Department of Physics
Middle East Technical University
06531, Ankara
Turkey
sarioglu@metu.edu.tr

Antonio SEGUÍ
Departamento de Física Teórica
Facultad de Ciencias
Universidad de Zaragoza
50009 Zaragoza
Spain
segui@unizar.es

José M M SENOVILLA
Departamento de Física Teórica
Facultad de Ciencia y Tecnología
Universidad del País Vasco
Apartado 644
48080 Bilbao
Spain
josemm.senovilla@ehu.es

Joseph SILK
Astrophysics
Denys Wilkinson Building
Keble Road
Oxford, OX1 3RH,
UK
silk@astro.ox.ac.uk

Alicia M. SINTES
Departament de Fisica
Universitat de les Illes Balears
Cra. Valldemossa Km. 7.5
E-07122 Palma de Mallorca
Spain
sintes@aei.mpg.de

Daniel SOLER
Mondragon GOI Eskola Politeknikoa
Loramendi, 4 Apdo. 23
20500 Arrasate-Mondragon
Gipuzkoa
Spain
dsoler@eps.mondragon.edu

John STACHEL
Physics Department
and Center for Einstein Studies
Boston University
Commonwealth Avenue
Boston MA 02215
USA
stachel@bu.edu

Cosimo STORNAIALO
INFN-Sezione di Napoli
Complesso Universitario
di Monte S. Angelo
Via Cinthia 80126 Napoli
Italy
cosmo@na.infn.it

Juan Pablo SUÁREZ CURIESES
Departamento de Física
Universidad de Oviedo
Avda Calvo Sotelo 18,
E-33007 Oviedo (Asturias)
Spain

María SUÁREZ DIEZ
Departamento de Física
Universidad de Oviedo
Avda Calvo Sotelo 18,
E-33007 Oviedo (Asturias)
Spain

Marcos SUÁREZ GONZÁLEZ
Departamento de Física
Universidad de Oviedo
Avda Calvo Sotelo 18,
E-33007 Oviedo (Asturias)
Spain

Rashid A. SUNYAEV
Max Planck Institut für Astrophysik
Postfach 1317
D-85741 Garching
Germany
sunyaev@mpa-garching.mpg.de
and
Space Research Institute
Russian Academy of Sciences
Profsoyuznaya St. 84/32
Moscow, Russia
sunyaev@hea.iki.rssi.ru

Jacek TAFEL
Institute of Theoretical Physics
University of Warsaw
Ul. Hoza 69
00681 Warsaw
Poland
tafel@fuw.edu.pl

Bayram TEKIN
Department of Physics
Middle East Technical University
Ankara 06531
Turkey
btekin@metu.edu.tr

Luigo TOFFOLATTI
Departamento de Física
Universidad de Oviedo
Avda Calvo Sotelo 18,
E-33007 Oviedo (Asturias)
Spain
ltoffolatti@uniovi.es

Manuel A. VALLE
Departamento de Física Teórica
Universidad del País Vasco
Apartado 644
E-48080 Bilbao
Spain
manuel.valle@ehu.es

Ramón VÁZQUEZ LORENZO
Faculty of Mathematics
Department of Geometry and Topology
University of Santiago de Compostela
15782 - Santiago de Compostela
Spain
ravazlor@usc.es

Miguel Angel VÁZQUEZ-MOZO
Universidad de Salamanca
Plaza de la Merced s/n
37008 Salamanca
Spain
Miguel.Vazquez-Mozo@cern.ch

Raül VERA JIMÉNEZ
School of Mathematical Sciences
Dublin City University
Dublin 9
Ireland
raul.vera@dcu.ie

Victor M. VILLALBA
Instituto Venezolano
de Investigaciones Científicas
Centro de Física
Apdo 21827
Caracas 1020A
Venezuela
villalba@ivic.ve

Andrés VIÑA
Departamento de Física
Universidad de Oviedo
Facultad de Ciencias
Avda Calvo Sotelo s/n
33007 Oviedo
vina@uniovi.es

David WANDS
Institute of Cosmology and Gravitation
Mercantile House
Hampshire Terrace
University of Portsmouth
Portsmouth, PO1 2EG
UK
david.wands@port.ac.uk

Lode WYLLEMAN
University of Ghent
Department of Mathematical Analysis
Galglaan 2 9000 Gent
Belgium
lwyllema@cage.ugent.be

Author Index

A

Alfaro, J., 247
Alimi, J.-M., 475
Althaus, L. G., 493
Alvarez, E., 3
Álvarez-Gaumé, L., 9, 619
Anza, S., 489
Araujo, H., 489
Arenas-Salazar, J. R., 385
Arnau, J. V., 594
Ashenberg, J., 502
Atrio-Barandela, F., 389, 550

B

Bäckdahl, T., 393
Baiotti, L., 416
Barbashov, B. M., 362
Barbero G., J. F., 437
Bergshoeff, E., 255
Bernal, A., 441
Bilge, A. H., 637
Blanchet, L., 10
Blas, D., 397
Blasone, M., 406
Boatella, C., 489
Bolós, V. J., 633
Bombardelli, C., 502
Bombelli, L., 497
Bond, I., 582
Bozza, V., 511
Breuer, H.-P., 314
Brito, I., 635
Burgay, M., 263

C

Caillon, J. C., 402
Camilo, F., 263
Capolupo, A., 406
Capozziello, S., 406
Carter, B., 29
Chamorro, A., 271
Chan, C.-T., 484
Cheimets, P. N., 502

Chmeissani, M., 489
Coll, B., 277
Conchillo, A., 489
Córsico, A. H., 493
Culetu, H., 410

D

Daghan, D., 637
D'Amico, N., 263
Damour, T., 51
De Gandt, F., 285
De Pietri, R., 416
de Roo, M., 255
Díaz, P., 607
Díaz Michelena, M., 489

E

Edgar, S. B., 291, 507

F

Fernández-Jambrina, L., 420
Ferrando, J. J., 424
Fois, M., 453
Freire, P. C. C., 263
Fullana, M. J., 594
Fuster, A., 429

G

Gaite, J., 433
Garay, I., 437
García-Berro, E., 489, 493
García-Colín, L. S., 603
Génova-Santos, R., 389
Glashow, S., 502
Gómez, C., 532
Grimani, C., 489
Guzmán, F. S., 441

H

Hafizi, M., 445
Hakim, R., 63
Herberthson, M., 449
Hernández-Monteagudo, C., 389
Ho, P.-M., 484
Houart, L., 298
Husa, S., 306

I

Iafolla, V., 453, 502
Ibáñez, J. M., 100
Isern, J., 493
Izquierdo, G., 458

J

Jafari, N., 462
Jaramillo, J. L., 466
Joshi, B. C., 263
Just, K., 471

K

Kerstan, S., 255
Kramer, M., 263
Kunze, K. E., 314

L

Labarsouque, J., 402
Larena, J., 475
Lazar, M., 479
Lazkoz, R., 420
Lee, J.-C., 484
Litim, D., 322
Lobo, A., 489
Lobo, J. A., 493
López-Quelle, M., 515
López-Villarejo, J. J., 3
Lorén-Aguilar, P., 493
Lorente, M., 497
Lorenzini, E. C., 502
Lorimer, D. R., 263
Lucchesi, D. M., 453, 502
Luminet, J.-P., 115

Lusanna, L., 330
Lyne, A. G., 263

M

MacCallum, M. A. H., 129
Machado Ramos, M. P., 507
Macías, A., 599
Maeda, H., 546
Manca, G. M., 416
Manchester, R. N., 263
Mancini, L., 511
Marcos, S., 515
Mars, M., 519
Martínez-González, E., 144
McLaughlin, M. A., 263
Mena, F. C., 519, 639
Meyer, F., 340
Miković, A., 523
Minguzzi, E., 528
Mochkovitch, R., 445
Montañez, S., 532
Morales Lladosa, J. A., 537
Mosquera Cuesta, H. J., 562
Moura, F., 542

N

Natorf, W., 615
Niembro, R., 515
Nimtz, G., 348
Nofrarias, M., 489
Nolan, B. C., 623
Nozawa, M., 546
Nozzoli, S., 453, 502

O

Olivares, G., 550
Ortega, J. A., 489
Ortín, T., 162
Özdemir, N., 554

P

Pastor, J., 402
Pavón, D., 356, 550

Per, M. A., 607
Pérez del Real, R., 489
Pérez, García, M. A., 558
Pérez Martínez, A., 562
Pérez Rojas, H., 562, 586
Persichini, M., 453
Pervushin, V. N., 362
Piazza, F., 566
Plastino, A. R., 570
Possenti, A., 263
Potting, R., 574
Pozo, J. M., 641
Puchades, N., 578, 594
Puigdengoles, C., 489

Q

Quevedo, H., 599

R

Ramos, J., 489
Rattenbury, N. J., 582
Resco, P., 532
Rezzolla, L., 416
Rica-Méndez, I., 643
Riccioni, F., 255
Rodríguez Querts, E., 586

S

Sá, P. M., 590
Sáez, D., 578, 594
Sánchez, A., 599
Sandoval-Villalbazo, A., 603
Sanjuan, J., 489
Santoli, F., 453, 502
Savushkin, L. N., 515
Schlickeiser, R., 479
Schneemann, C., 306
Seguí, A., 607
Senovilla, J. M. M., 291, 370
Serna, A., 475
Shapiro, I. I., 502
Shariati, A., 462
Silk, J., 182
Sintes , A. M., 378
Sivak, H. D., 63

Soler, D., 611
Stachel, J., 195
Stairs, I. H., 263
Stavridis, A., 643
Stoeger, W., J., 471
Stornaiolo, C., 645

T

Tafel, J., 615
Tejeiro-Sarmiento, J. M., 385
Teraguchi, S., 484
Tomé, B., 590

V

Vaz, E. G. L. R., 635
Vázquez-Mozo, M. A., 619
Vera, R., 519, 623
Villalba, V. M., 627
Villaseñor, E. J. S., 437
Vitiello, G., 406
Vogel, T., 306

W

Wands, D., 228
Wass, P., 489
Wylleman, L., 647

X

Xirgu, X., 489

Y

Yang, Y., 484
Yock, P., 582

Z

Zakharov, A. F., 362
Zander, C., 570
Zenginoğlu, A., 306
Zimdahl, W., 356
Zinchuk, V. A., 362